Lecture Notes in Computer Science 4831

Commenced Publication in 1973
Founding and Former Series Editors:
Gerhard Goos, Juris Hartmanis, and Jan van Leeuwen

Boualem Benatallah   Fabio Casati
Dimitrios Georgakopoulos
Claudio Bartolini   Wasim Sadiq
Claude Godart (Eds.)

Web Information Systems Engineering – WISE 2007

8th International Conference on
Web Information Systems Engineering
Nancy, France, December 3-7, 2007
Proceedings

 Springer

Volume Editors

Boualem Benatallah
The University of New South Wales, Sydney, NSW 2052, Australia
E-mail: boualem@cse.unsw.edu.au

Fabio Casati
University of Trento, 38050 Trento, Italy
E-mail: casati@dit.unitn.it

Dimitrios Georgakopoulos
Telcordia, Austin Research Center, Austin, TX 78701, USA
E-mail: dimitris@research.telcordia.com

Claudio Bartolini

HP Laboratories, Palo Alto, CA 94304, USA
E-mail: claudio.bartolini@hp.com

Wasim Sadiq
SAP Australia, Brisbane, QLD 4000, Australia
E-mail: wasim.sadiq@sap.com

Claude Godart
LORIA-ECOO, 54506 Vandœuvre-lès-Nancy, France
E-mail: claude.godart@loria.fr

Library of Congress Control Number: 2007939800

CR Subject Classification (1998): H.4, H.2, H.3, H.5, K.4.4, C.2.4, I.2

LNCS Sublibrary: SL 3 – Information Systems and Application, incl. Internet/Web
and HCI

ISSN   0302-9743
ISBN-10   3-540-76992-7 Springer Berlin Heidelberg New York
ISBN-13   978-3-540-76992-7 Springer Berlin Heidelberg New York

Springer is a part of Springer Science+Business Media

springer.com

Typesetting: Camera-ready by author, data conversion by Scientific Publishing Services, Chennai, India
Printed on acid-free paper   SPIN: 12197309   06/3180   5 4 3 2 1 0

Preface

WISE 2007 was held in Nancy, France, during December 3–6, hosted by Nancy University and INRIA Grand-Est. The aim of this conference was to provide an international forum for researchers, professionals, and industrial practitioners to share their knowledge in the rapidly growing area of Web technologies, methodologies and applications. Previous WISE conferences were held in Hong Kong, China (2000), Kyoto, Japan (2001), Singapore (2002), Rome, Italy (2003), Brisbane, Australia (2004), New York, USA (2005), and Wuhan, China (2006).

The call for papers created a large interest. Around 200 paper submissions arrived from 41 different countries (Europe: 40%, Asia: 33%, Pacific: 10%, North America: 7%, South America: 7%, Africa: 2%). The international Program Committee selected 40 full-papers (acceptance rate of 20%) and 18 short papers (acceptance rate of 9%). As a result, the technical track of the WISE 2007 program offered 13 sessions of full-paper presentation including one industrial session and 5 sessions of short papers. The selected papers cover a wide and important variety of issues in Web information systems engineering such as querying; trust; caching and distribution; interfaces; events and information filtering; data extraction; transformation and matching; ontologies; rewriting, routing and personalization; agents and mining; quality of services and management and modelling. A few selected papers from WISE 2007 will be published in a special issue of the *World Wide Web Journal*, by Springer. In addition, $1000 value was awarded to the authors of the paper selected for the "Yahiko Kambayashi Best Paper." We thank all authors who submitted their papers and the Program Committee members and external reviewers for their excellent work.

Finally, WISE 2007 included two prestigious keynotes given by Eric Billingsley, eBay research and Lutz Heuser, SAP research, one panel, one preconference tutorial, and six workshops.

We would also like to acknowledge the local organization team, in particular Anne-Lise Charbonnier and François Charoy. We also thank Mohand-Said Hacid and Mathias Weske as Workshop Chairs, Manfred Hauswirth as Panel Chair, Mike Papazoglou as Tutorial Chair, Olivier Perrin, Michael Sheng and Mingjun Xiao as Publicity Chairs, Qing Li, Marek Rusinkiewicz and Yanchun Zhang for the relationship with previous events and the WISE Society, and Ustun Yildiz for his work in the editing proceedings.

We hope that the present proceedings will contain enough food for thought to push the Web towards many exciting innovations for tomorrow's society.

September 2007

Boualem Benatallah
Fabio Casati
Dimitrios Georgakopoulos
Claudio Bartolini
Wasim Sadiq
Claude Godart

Organization

General Chairs   Claude Godart, France
                                  Qing Li, China

Program Chairs   Boualem Benatallah, Australia
                                  Fabio Casati, Italy
                                  Dimitrios Georgakopoulos, USA

Industrial Program Chairs   Claudio Bartolini, USA
                                  Wasim Sadiq, Australia

Workshop Chairs   Mohand-Said Hacid, France
                                  Mathias Weske, Germany

Tutorial Chair   Mike Papazoglou, The Netherlands

Panel Chair   Manfred Hauswirth, Ireland

Publicity Chairs   Olivier Perrin, France
                                  Michael Sheng, Australia
                                  Mingjun Xiao, China

Publication Chair   Claude Godart, France

Wise Society Representatives   Yanchun Zhang, Australia
                                  Marek Rusinkiewicz, USA

Local Organization Chair   François Charoy, France

Local Organization Committee   Anne-Lise Charbonnier, France
                                  Laurence Félicité, France
                                  Nawal Guermouche, France
                                  Olivier Perrin, France
                                  Mohsen Rouached, France
                                  Hala Skaf, France
                                  Ustun Yildiz, France

Program Committee

Karl Aberer, Switzerland   Rafael Alonso, USA
Marco Aiello, The Netherlands   Toshiyuki Amagasa, Japan

Sponsoring Institutions

emisa.org

Table of Contents

Session 1: Querying

On Tree Pattern Query Rewriting Using Views 1
   Junhu Wang, Jeffrey Xu Yu, and Chengfei Liu

Querying Capability Modeling and Construction of Deep Web
Sources 13
   Liangcai Shu, Weiyi Meng, Hai He, and Clement Yu

Optimization of Bounded Continuous Search Queries Based on Ranking
Distributions 26
   Dirk Kukulenz, Nils Hoeller, Sven Groppe, and Volker Linnemann

Session 2: Trust

Evaluating Rater Credibility for Reputation Assessment of Web
Services... 38
   Zaki Malik and Athman Bouguettaya

An Approach to Trust Based on Social Networks.................... 50
   Vincenza Carchiolo, Alessandro Longheu, Michele Malgeri,
   Giuseppe Mangioni, and Vincenzo Nicosia

A New Reputation-Based Trust Management Mechanism Against False
Feedbacks in Peer-to-Peer Systems 62
   Yu Jin, Zhimin Gu, Jinguang Gu, and Hongwu Zhao

Session 3: Caching and Distribution

Freshness-Aware Caching in a Cluster of J2EE Application Servers 74
   Uwe Röhm and Sebastian Schmidt

Collaborative Cache Based on Path Scores 87
   Bernd Amann and Camelia Constantin

Similarity-Based Document Distribution for Efficient Distributed
Information Retrieval.. 99
   Sven Herschel

Session 4: Interfaces

BEIRA: A Geo-semantic Clustering Method for Area Summary 111
   Osamu Masutani and Hirotoshi Iwasaki

Building the Presentation-Tier of Rich Web Applications with
Hierarchical Components 123
    Reda Kadri, Chouki Tibermacine, and Vincent Le Gloahec

WeBrowSearch: Toward Web Browser with Autonomous Search 135
    Taiga Yoshida, Satoshi Nakamura, and Katsumi Tanaka

Session 5: Events and Information Filtering

A Domain-Driven Approach for Detecting Event Patterns in E-Markets:
A Case Study in Financial Market Surveillance 147
    Piyanath Mangkorntong and Fethi A. Rabhi

Adaptive Email Spam Filtering Based on Information Theory 159
    Xin Zhang, Wenyuan Dai, Gui-Rong Xue, and Yong Yu

Time Filtering for Better Recommendations with Small and Sparse
Rating Matrices 171
    Sergiu Gordea and Markus Zanker

Session 6: Data Extraction, Transformation, and Matching

A Survey of UML Models to XML Schemas Transformations 184
    *Eladio Domínguez, Jorge Lloret, Beatriz Pérez, Áurea Rodríguez,
    Ángel L. Rubio, and María A. Zapata*

Structural Similarity Evaluation Between XML Documents and
DTDs 196
    Joe Tekli, Richard Chbeir, and Kokou Yetongnon

Using Clustering and Edit Distance Techniques for Automatic Web
Data Extraction 212
    *Manuel Álvarez, Alberto Pan, Juan Raposo, Fernando Bellas, and
    Fidel Cacheda*

Session 7: Ontologies

A Semantic Approach and a Web Tool for Contextual Annotation of
Photos Using Camera Phones 225
    *Windson Viana, José Bringel Filho, Jérôme Gensel,
    Marlène Villanova-Oliver, and Hervé Martin*

Formal Specification of OWL-S with Object-Z: the Dynamic Aspect 237
    Hai H. Wang, Terry Payne, Nick Gibbins, and Ahmed Saleh

An Approach for Combining Ontology Learning and Semantic Tagging
in the Ontology Development Process: eGovernment Use Case   249
   Ljiljana Stojanovic, Nenad Stojanovic, and Jun Ma

Session 8: Rewriting, Routing, and Personalisation

Term Rewriting for Web Information Systems – Termination and
Church-Rosser Property   261
   Klaus-Dieter Schewe and Bernhard Thalheim

Development of a Collaborative and Constraint-Based Web
Configuration System for Personalized Bundling of Products and
Services   273
   Markus Zanker, Markus Aschinger, and Markus Jessenitschnig

SRI@work: Efficient and Effective Routing Strategies in a PDMS   285
   Federica Mandreoli, Riccardo Martoglia, Wilma Penzo,
   Simona Sassatelli, and Giorgio Villani

Session 9: Agents and Mining

Learning Management System Based on SCORM, Agents and
Mining   298
   Carlos Cobos, Miguel Niño, Martha Mendoza, Ramon Fabregat, and
   Luis Gomez

A Web-Based Learning Resource Service System Based on Mobile
Agent   310
   Wu Di, Cheng Wenqing, and Yan He

Wikipedia Mining for an Association Web Thesaurus Construction   322
   Kotaro Nakayama, Takahiro Hara, and Shojiro Nishio

Session 10: QoS and Management

Economically Enhanced Resource Management for Internet Service
Utilities   335
   Tim Püschel, Nikolay Borissov, Mario Macías, Dirk Neumann,
   Jordi Guitart, and Jordi Torres

Enhancing Web Services Performance Using Adaptive Quality of
Service Management   349
   Abdelkarim Erradi and Piyush Maheshwari

Coverage and Timeliness Analysis of Search Engines with Webpage
Monitoring Results   361
   Yang Sok Kim and Byeong Ho Kang

Session 11: Modeling

Managing Process Customizability and Customization: Model,
Language and Process 373
    Alexander Lazovik and Heiko Ludwig

A WebML-Based Approach for the Development of Web GIS
Applications.. 385
    *Sergio Di Martino, Filomena Ferrucci, Luca Paolino,
    Monica Sebillo, Giuliana Vitiello, and Giuseppe Avagliano*

An Object-Oriented Version Model for Context-Aware Data
Management 398
    Michael Grossniklaus and Moira C. Norrie

Session 12: Topics

Extending XML Triggers with Path-Granularity 410
    Anders H. Landberg, J. Wenny Rahayu, and Eric Pardede

A Replicated Study Comparing Web Effort Estimation Techniques 423
    *Emilia Mendes, Sergio Di Martino, Filomena Ferrucci, and
    Carmine Gravino*

Development Process of the Operational Version of PDQM 436
    Angélica Caro, Coral Calero, and Mario Piattini

A New Reputation Mechanism Against Dishonest Recommendations in
P2P Systems 449
    Junsheng Chang, Huaimin Wang, Gang Yin, and Yangbin Tang

Industrial Session

A Framework for Business Operations Management Systems............ 461
    Tao Lin, Chuan Li, Ming-Chien Shan, and Suresh Babu

A Practical Method and Tool for Systems Engineering of
Service-Oriented Applications 472
    Lisa Bahler, Francesco Caruso, and Josephine Micallef

A Layered Service Process Model for Managing Variation and Change
in Service Provider Operations.................................. 484
    Heiko Ludwig, Kamal Bhattacharya, and Thomas Setzer

Short Paper Session 1

Providing Personalized Mashups Within the Context of Existing Web
Applications... 493
    Oscar Díaz, Sandy Pérez, and Iñaki Paz

Wooki: A P2P Wiki-Based Collaborative Writing Tool   503
Stéphane Weiss, Pascal Urso, and Pascal Molli

Creating and Managing Ontology Data on the Web: A Semantic Wiki
Approach   513
Chao Wang, Jie Lu, Guangquan Zhang, and Xianyi Zeng

Web Service Composition: A Reality Check   523
Jianguo Lu, Yijun Yu, Debashis Roy, and Deepa Saha

Short Paper Session 2

MyQoS: A Profit Oriented Framework for Exploiting Customer
Behavior in Online e-Commerce Environments   533
Ahmed Ataullah

Task Assignment on Parallel QoS Systems   543
*Luis Fernando Orleans, Carlo Emmanoel de Oliveira, and
Pedro Furtado*

Autonomic Admission Control for Congested Request Processing
Systems   553
Pedro Furtado

Towards Performance Efficiency in Safe XML Update.   563
Dung Xuan Thi Le, and Eric Pardede

Short Paper Session 3

From Crosscutting Concerns to Web Systems Models   573
*Pedro Valderas, Vicente Pelechano, Gustavo Rossi, and
Silvia Gordillo*

Generating Extensional Definitions of Concepts from Ostensive
Definitions by Using Web   583
*Shin-ya Sato, Kensuke Fukuda, Satoshi Kurihara,
Toshio Hirotsu, and Toshiharu Sugawara*

Modeling Distributed Events in Data-Intensive Rich Internet
Applications.   593
*Giovanni Toffetti Carughi, Sara Comai,
Alessandro Bozzon, and Piero Fraternali*

Privacy Inspection and Monitoring Framework for Automated Business
Processes   603
Yin Hua Li, Hye-Young Paik, and Jun Chen

Short Paper Session 4

Digging the Wild Web: An Interactive Tool for Web Data
Consolidation   613
Max Goebel, Viktor Zigo, and Michal Ceresna

A Web-Based Learning Information System - AHKME 623
   Hugo Rego, Tiago Moreira, and Francisco José Garcia

A Recommender System with Interest-Drifting...................... 633
   Shanle Ma, Xue Li, Yi Ding, and Maria E. Orlowska

Short Paper Session 5

Improving Revisitation Browsers Capability by Using a Dynamic
Bookmarks Personal Toolbar 643
   José A. Gámez, Juan L. Mateo, and José M. Puerta

Hierarchical Co-clustering for Web Queries and Selected URLs........ 653
   Mehdi Hosseini and Hassan Abolhassani

Navigating Among Search Results: An Information Content
Approach 663
   Ramón Bilbao and M. Andrea Rodríguez

Author Index 673

On Tree Pattern Query Rewriting Using Views*

Junhu Wang[1][**], Jeffrey Xu Yu[2], and Chengfei Liu[3]

[1] Griffith University, Gold Coast, Australia
J.Wang@griffith.edu.au
[2] Chinese University of Hong Kong, Hong Kong, China
yu@se.cuhk.edu.hk
[3] Swinburne University of Technology, Melbourne, Australia
cliu@ict.swin.edu.au

Abstract. We study and present our findings on two closely related problems on XPATH rewriting using views when both the view and the query are tree patterns involving $/, //$ and $[]$. First, given view V and query Q, is it possible for Q to have an equivalent rewriting using V which is the union of two or more tree patterns, but not an equivalent rewriting which is a single pattern? This problem is of both theoretical and practical importance because, if the answer is no, then, to answer a query completely using the views, we should use more efficient methods, such as the PTIME algorithm of [13], to find the equivalent rewriting, rather than try to find the union of all contained rewritings and test its equivalence to Q. Second, given a set $\mathcal{V}$ of views, we want to know under what conditions a subset $\mathcal{V}'$ of the views are redundant in the sense that for *any query* Q, the contained rewritings of Q using the views in $\mathcal{V}'$ are contained in those using the views in $\mathcal{V} - \mathcal{V}'$. Solving this problem can help us to, for example, choose the minimum number of views to be cached, or better design the virtual schema in a mediated data integration system, or avoid repeated calculation in query optimization. We provide necessary and sufficient conditions for the second problem, based on answers to the first problem. When the views produce comparable answers, we extend our findings to include the case where the intersection of views, in addition to the individual views, are used in the rewriting.

1 Introduction

Query rewriting using views has many applications including data integration, query optimization, and query caching [6]. A view is an existing query whose answer may or may not have been materialized. Given a new query, the problem is to find another query using only the views that will produce correct answers to the original query. Usually two types of rewritings are sought: *equivalent rewritings* and *contained rewritings*. An equivalent rewriting produces all answers

---

* The work described in this paper was supported by grant from the Research Grant Council of the Hong Kong Special Administrative Region, China (CUHK418205).
** Work done while visiting the Chinese University of Hong Kong.

B. Benatallah et al. (Eds.): WISE 2007, LNCS 4831, pp. 1–12, 2007.

to the original query, while a contained rewriting may produce only part of the answers. Both types of rewritings have been extensively studied in the relational database context [6,10].

More recently rewriting XML queries using XML views has attracted attention because of the rising importance of XML data [13,8,7,11]. Since XPATH lies in the center of all XML languages, the problem of rewriting XPATH queries using XPATH views is particularly important. Some major classes of XPATH expressions can be represented as tree patterns [1,9]. Among previous work on rewriting XPATH queries using views, Xu et al [13] studied equivalent rewritings for several different classes of tree patterns, while Mandhani and Suciu [8] presented results on equivalent rewritings of tree patterns when the tree patterns are assumed to be minimized. Lakshmanan et al [7] studied maximal contained rewritings of tree patterns where both the view and the query involve $/,//$ and $[]$ only (these XPATH expressions correspond to tree patterns in $P^{\{/,//,[]\}}$ [9]), both in the absence and presence of non-recursive schema graphs - a restricted form of DTDs.

In this paper, we study two closely related problems on XPATH rewritings using views, for queries and views in $P^{\{/,//,[]\}}$. The first problem is about the form of equivalent rewritings: given view V and query Q, is it possible for Q to have an equivalent rewriting using V which is the union of two or more tree patterns, but not an equivalent rewriting which is a single tree pattern? This problem is of both theoretical and practical importance because, if the answer is no, then, to completely answer a query using the view, we can use more efficient methods such as the PTIME algorithm of [13] to find the equivalent rewriting, rather than try to find the union of all contained rewritings and test its equivalence to Q. The second problem is what we call the *redundant views* problem. Given a set $\mathcal{V}$ of views, we want to know under what conditions a subset $\mathcal{V}'$ of the views are redundant in the sense that for *any query* Q, the contained rewritings of Q using the views in $\mathcal{V}'$ are contained in those using the views in $\mathcal{V} - \mathcal{V}'$. Solving this problem can help us to, for example, choose the minimum number of views to be cached, or better design the virtual schema in a mediated data integration system, or avoid useless computation in query optimization. We identify a necessary and sufficient condition for $\mathcal{V}'$ to be redundant, based on our results to the first problem. In the case that all the views in $\mathcal{V}$ produce comparable answers, we show it is possible to use the intersection of views to rewrite a query, when there are no rewritings using any single view. We extend our results on the two problems to include this scenario.

Our main contributions are:

- We show that for queries and views in $P^{\{/,//,[]\}}$, if there is no equivalent rewriting in the form of a single tree pattern, then there is no equivalent rewriting in the form of a union of tree patterns, and this is true even in the presence of non-recursive, non-disjunctive DTDs. But the result no longer holds for larger classes of tree patterns or if there is a recursive DTD.
- When multiple views exist, we provide a necessary and sufficient condition for identifying redundant views.

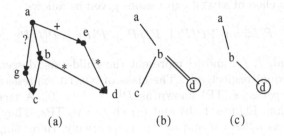

Fig. 1. Example DTD and TPs consistent with the DTD

– We extend tree pattern rewriting using single views to that using intersections of views, and extend the above results to include such rewritings.

The rest of the paper is organized as follows. Section 2 provides the terminology and notations. Section 3 presents our result on the form of equivalent rewritings. Based on this result, Section 4 proves a necessary and sufficient condition under which some views are redundant. Section 5 studies the two problems when rewriting using intersection of views is included. Section 6 discusses related work. Finally Section 7 concludes the paper.

2 Preliminaries

2.1 DTDs, XML Trees, and Tree Patterns

Let Σ be an infinite set of tags. We adopt the similar notations used in [7], and model a DTD as a connected directed graph G (called a *schema graph*) as follows. (1) Each node is labeled with a distinct label in Σ. (2) Each edge is labeled with one of 1, ?, +, and *, which indicate "exactly one", "one or zero", "one or many", and "zero or many", respectively. Here, the default edge label is 1. (3) There is a unique node, called the root, which has an incoming degree of zero. Because a node in a schema graph G has a unique label, we also refer to a node by its label. If the graph is acyclic, then the DTD is said to be *non-recursive*. We will use DTD and schema graph interchangeably. A schema graph (DTD) example is shown in Figure 1 (a).

An XML document is a node-labeled, unordered tree[1]. Let v be a node in an XML tree t, the label of v is denoted by $label(v)$. Let $N(t)$ (resp. $N(G)$) denote the set of all nodes in XML tree t (resp. schema graph G), and $rt(t)$ (resp. $rt(G)$) denote the root of t (resp. G). A tree t is said to conform to schema graph G if (1) for every node $v \in N(t)$, $label(v) \in \Sigma$, (2) $label(rt(t)) = label(rt(G))$, (3) for every edge (u, v) in t, there is a corresponding edge $(label(u), label(v))$ in G, and (4) for every node $v \in N(t)$, the number of children of v labeled with x is constrained by the label of the edge $(label(v), x)$ given in G. We denote the set of all XML trees conforming to G by T_G.

---

[1] Order and attributes in XML documents are disregarded in this paper.

We consider a class of XPATH expressions given as follows.

$$P \ ::= \tau \ | \ P/P \ | \ P//P \ | \ P[P] \ | \ P[//P]$$

Here, $\tau \in \Sigma$, and, $/$, $//$, and $[\ ]$ represent the child-axis, descendant-axis, and branching condition, respectively. The class of XPATH expressions corresponds to a set of *tree patterns* (TP) known as $P^{\{/,//,[]\}}$ in [9]. A tree patten has a tree representation. Figures 1 (b) and (c) show two TPs. They correspond to XPATH expressions $a/b[c]//d$ and $a/b/d$, respectively. Here, single and double lines represent child-axis ($/$), called /-edge, and descendant-axis ($//$), called //-edge, respectively. A circle denotes the output of P (called the distinguished node of P). Below, given a TP P, like [7], we use d_P to denote the distinguished node.

Let $N(P)$ (resp. $rt(P)$) denote the set of all nodes in a TP P (resp. the root of P). A *matching* of P in an XML tree t is a mapping δ from $N(P)$ to $N(t)$ which is (1) **label-preserving**, i.e., $\forall v \in N(P)$, $label(v) = label(\delta(v))$, (2) **root-preserving**, i.e., $\delta(rt(P)) = rt(t)$, and (3) **structure-preserving**, i.e., for every edge (x, y) in P, if it is a /-edge, then $\delta(y)$ is a child of $\delta(x)$; if it is a //-edge, then $\delta(y)$ is a descendant of $\delta(x)$, i.e, there is a path from $\delta(x)$ to $\delta(y)$. Each matching δ produces a subtree of t rooted at $\delta(d_P)$, denoted $sub^t_{\delta(d_P)}$, which is also known as an *answer* to the TP. We use $P(t)$ to denote the set of all answers of P on t:

$$P(t) = \{sub^t_{\delta(d_P)} \mid \delta \text{ is a matching of } P \text{ in } t\} \tag{1}$$

Let T be a set of XML trees. We use $P(T)$ to denote the union of answer sets of Q on the trees in T. That is, $P(T) = \bigcup_{t \in T} P(t)$. In addition, when we discuss TPs in the presence of DTD G, we will implicitly assume every TP P is consistent with G, that is, there is $t \in T_G$ such that $P(t) \neq \emptyset$.

2.2   TP Rewriting Using Views

A TP P is said to be *contained* in another TP Q, denoted $P \subseteq Q$, if for every XML tree t, $P(t) \subseteq Q(t)$ (Refer to Eq. (1)). Given $P, Q \in P^{\{/,//,[]\}}$, $P \subseteq Q$ iff there is a containment mapping from Q to P [1]. Recall: a *containment mapping* (CM) from Q to P is a mapping h from $N(Q)$ to $N(P)$ that is label-preserving, root-preserving, structure-preserving (which now means that for every /-edge (x, y) in Q, $(h(x), h(y))$ is a /-edge in P, and for every //-edge (x, y), there is a path from $h(x)$ to $h(y)$), and output-preserving, which means $h(d_Q) = d_P$.

Contained Rewriting and Maximal Contained Rewriting: A *view* is an existing TP. Let V be a view and Q be a TP. A *contained rewriting* (CR) of Q using V is a TP Q' such that when evaluated on the subtrees returned by V, Q' gives correct answers to Q. More precisely, for any XML tree t, $Q'(V(t)) \subseteq Q(t)$, and (2) there exists some t such that $Q'(V(t)) \neq \emptyset$. Let Q' be a CR of Q. By definition the root of Q' must be labeled with $label(d_V)$. We use $V \oplus Q'$ to represent the *expansion* of Q', which is the TP obtained by merging $rt(Q')$ and

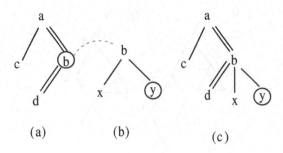

Fig. 2. (a) the view V, (b) the CR Q', (c) the expansion $V \oplus Q'$

d_V (see Figure 2). Clearly $V \oplus Q' \subseteq Q$. If $V \oplus Q' \supseteq Q$ also holds, Q' is called an *equivalent rewriting* (ER) of Q using V [13]. The *maximal contained rewriting* (MCR), denoted MCR(Q,V), is the union of all CRs of Q using V [7]. We use EMCR(Q,V) to denote the union of expansions of the CRs in MCR(Q,V).

In the presence of DTDs: The concepts of TP containment, CR, and MCR extend naturally to the case where a DTD is available. Given a DTD G and two TPs P and Q, P is said to be *contained in* Q *under* G, denoted $P \subseteq_G Q$, if for every XML tree $t \in T_G$, $P(t) \subseteq Q(t)$. A *contained rewriting (resp. equivalent rewriting) of* Q *using view* V *under* G is a TP Q' such that (1) for any XML tree $t \in T_G$, $Q'(V(t)) \subseteq Q(t)$ (resp. $Q'(V(t)) = Q(t)$), (2) for some $t \in T_G$, $Q'(V(t)) \neq \emptyset$. A MCR of Q using V under G is the union of all CRs of Q using V under G.

3 Form of Equivalent Rewritings

Let $V \in P^{\{/,//,[]\}}$ be the view, and $Q \in P^{\{/,//,[]\}}$ be the query. By definition, EMCR$(Q,V) \subseteq Q$. In the best case, EMCR(Q,V) may be equivalent to Q (In this case, we say the MCR of Q using V is equivalent to Q). The question arises now whether it is possible for Q to have no ER (which is a single TP) using V, but it has a MCR (which is a union of TPs) using V which is equivalent to Q. In other words, any single CR of Q using V is not equivalent to Q, but the union of all CRs is.

The answer to the above question is "no", and this is even true in the presence of non-recursive DTDs. We prove this result below.

Theorem 1. *Let* $V, Q \in P^{\{/,//,[]\}}$ *be the view and query respectively. When there are no DTDs, if* Q *has a MCR using* V *which is equivalent to* Q, *then it has a single CR using* V *which is equivalent to* Q.

The proof of Theorem 1 follows directly from an interesting property of TPs in $P^{\{/,//,[]\}}$ stated in the theorem below.

Theorem 2. *For tree patterns* $P, P_1, \ldots, P_n \in P^{\{/,//,[]\}}$, *if* $P \subseteq \bigcup_{i=1}^n P_i$, *then there exists* $i \in [1, n]$ *such that* $P \subseteq P_i$.

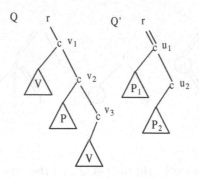

Fig. 3. Q and Q' as in [9]

To prove the above theorem, we need the concept of boolean tree patterns. A *boolean pattern* [9] is a pattern with no output node. Let P be a boolean pattern. For an XML tree t, the result of evaluating P on t, denoted $P(t)$, is either TRUE or FALSE. $P(t)$ is TRUE if and only if there is a matching of P in t. For two boolean patterns P_1 and P_2, $P_1 \subseteq P_2$ means $P_1(t) =$ TRUE implies $P_2(t) =$ TRUE for any XML tree t. If P_1 and P_2 are in $P^{\{/,//,[]\}}$, then $P_1(t) \subseteq P_2(t)$ iff there is a homomorphism from P_2 to P_1 (Recall: a homomorphism is the same as a containment mapping, except it does not need to be output-preserving) [9]. In addition, $P_1 \cup P_2$ returns $P_1(t) \vee P_2(t)$ for any t.

We first prove the following lemma.

Lemma 1. *For boolean patterns* $P, P_1, \ldots, P_n \in P^{\{/,//,[]\}}$, *if* $P \subseteq \bigcup_i P_i$, *then there exists* $i \in [1, n]$ *such that* $P \subseteq P_i$.

Proof. We prove the lemma for the case $k = 2$. The case $k > 2$ is similar.

Using Lemma 1 of [9], we can construct two boolean patterns Q and Q' such that $P \subseteq P_1 \cup P_2$ iff $Q \subseteq Q'$. Q and Q' are as shown in Figure 3, where V is a pattern which is contained in both P_1 and P_2. Furthermore, V does not have *-nodes (see [9] for how to construct V). Since $P \subseteq P_1 \cup P_2$ implies $Q \subseteq Q'$, and there is no *-node in Q', as given in [9], we know there is a homomorphism from Q' to Q. Now examine the structure of Q and Q', any homomorphism from Q' to Q must map the nodes u_1 and u_2 either to v_1 and v_2 respectively, or to v_2 and v_3 respectively. In the former case, there will be a homomorphism from P_2 to P; in the later case there will be a homomorphism from P_1 to P. Therefore, either $P \subseteq P_1$ or $P \subseteq P_2$.

We are now ready to prove Theorem 2.

Proof (proof of Theorem 2).

We denote by P', P_i' ($i \in [1, n]$) the boolean patterns obtained from P, P_i by attaching a child node labelled with a *distinct* label z to the distinguished nodes of P and P_i respectively (Since Σ is an infinite set of tags, z exists, refer to Figure 4). Let us denote the new nodes in P and P_i by z_P and z_{P_i} respectively.

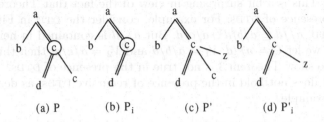

(a) P   (b) P_i (c) P'   (d) P'$_i$

Fig. 4. Tree patterns P, P_i and the boolean patterns P' and P_i'

We show that $P \subseteq \bigcup_{i \in [1,n]} P_i$ implies $P' \subseteq \bigcup_{i \in [1,n]} P_i'$. Let t be any XML tree. For every matching h of P' in t, there is a matching of P in t which is the one obtained by restricting h to all nodes in P except z_P. Since $P \subseteq \bigcup_{i \in [1,n]} P_i$, there exists an $i \in [1, n]$ such that there is a matching f of some P_i in t and $f(d_{P_i}) = h(d_P)$. This matching f can clearly be extended to P_i' - simply let $f(z_{P_i}) = h(z_P)$. Therefore $P' \subseteq \bigcup_{i \in [1,n]} P_i'$.

By Lemma 1, there exists $i \in [1, n]$ such that $P' \subseteq P_i'$. Therefore, there is a homomorphism from P_i' to P'. This homomorphism implies a containment mapping from P_i to P. Hence $P \subseteq P_i$.

Because of Theorem 1, if we cannot find a single CR of Q using V which is equivalent to Q, then it is impossible to find a union of CRs of Q using V which is equivalent to Q. This suggests that, if we want to use the results of a view to **completely** answer query Q (for all XML trees), we should always use a more efficient algorithm, such as Lemma 4.8 of [13], to find an equivalent rewriting of Q using V. If we cannot find such an equivalent rewriting, then it is impossible to answer Q completely using the results of V.

Theorem 1 can be easily extended to the case when there are multiple views.

Theorem 3. *Let* $V_1, \ldots, V_n \in P^{\{/,//,[]\}}$ *be views and* $Q \in P^{\{/,//,[]\}}$ *be a query. If* $Q \subseteq \mathsf{EMCR}(Q, V_1) \cup \cdots \cup \mathsf{EMCR}(Q, V_n)$, *then there exists* $i \in [1, n]$ *such that there is an equivalent rewriting of* Q *using* V_i.

Theorem 1 (but not Theorem 3) is still true in the presence of non-recursive DTDs. This is because an equivalent rewriting is also a MCR, and when both Q and V are in $P^{\{/,//,[]\}}$, the MCR of Q using V under a non-recursive DTD is a

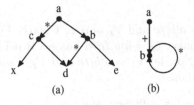

(a)   (b)

Fig. 5. Example schema graphs

single TP [7]. This is a bit surprising in view of the fact that Theorem 2 is not true in the presence of DTDs. For example, consider the DTD in Figure 5 (a). Under the DTD, $a//d \subseteq a/b/d \cup a/c/d$. But $a//d$ is contained in neither $a/b/d$ nor $a/c/d$. If we let $Q = a//d$, $V_1 = a/b/d$ and $V_2 = a/c/d$, then this serves as an example to show Theorem 3 is not true in the presence of DTDs.

Theorem 1 does not hold in the presence of recursive DTDs, as demonstrated by the next example.

Example 1. Consider the query $Q = a//b$ and the view $V = a/b$ under the recursive DTD in Figure 5 (b). Q has two CRs: $Q'_1 = b$ and $Q'_2 = b//b$. The expansions of these CRs are a/b and $a/b//b$ respectively. Under the DTD, Q is equivalent to the union of a/b and $a/b//b$, but it is not equivalent to either one of them. That is, there is no single CR of Q using V which is equivalent to Q, but the union of the two CRs is equivalent to Q.

We point out that Theorem 1 is not true if V and the rewritings have the wildcard $*$ (see details in the full version of this paper).

4   Redundant Views

In this section we assume there are multiple views $V_1, \ldots, V_n$ in $P^{\{/,//,[]\}}$. We now ask the question when a subset of views $V_{i_1}, \ldots, V_{i_k}$ are redundant in the sense that for every query Q, the MCRs of Q using $V_{i_1}, \ldots, V_{i_k}$ are all contained in the union of the MCRs of Q using the other views. Formally we have

Definition 1. *Let* $V_1, \ldots, V_n \in P^{\{/,//,[]\}}$ *be views and* $k < n$. *If for every* TP $Q \in P^{\{/,//,[]\}}$,

$$\bigcup_{i=1}^{k} \mathsf{EMCR}(Q, V_i) \subseteq \bigcup_{j=k+1}^{n} \mathsf{EMCR}(Q, V_j)$$

then we say the views $V_1, \ldots, V_k$ *are* redundant.

Intuitively, when considering CRs the redundant views can be ignored because all answers returned by CRs using the redundant views can be returned by CRs using other views.

One might wonder how the redundant views problem is related to the view containment problem $\cup_{i=1}^{k} V_i \subseteq \cup_{j=k+1}^{n} V_j$. As we show in the next example, the condition $\cup_{i=1}^{k} V_i \subseteq \cup_{j=k+1}^{n} V_j$ is neither sufficient nor necessary for $V_1, \ldots, V_k$ to be redundant.

Example 2. (1) Let $V_1 = a[b]/c$ and $V_2 = a/c$. Clearly $V_1 \subseteq V_2$. But V_1 is not redundant, because the query $Q = a[b]/c/d$ has a CR using V_1, but it does not have a CR using V_2. (2) Now let $V_1 = a[b]/c$ and $V_2 = a[b]$. It is easy to verify $V_1 \nsubseteq V_2$, but V_1 is redundant.

We now provide the following sufficient and necessary condition for redundant views.

Theorem 4. *Given* $V_1, \ldots, V_n \in P^{\{/,//,[]\}}$, $V_1, \ldots, V_k$ *(k < n) are redundant iff for every* $i \in [1, k]$, *there exists* $j \in [k+1, n]$ *such that* V_i *has an equivalent rewriting using* V_j.

Proof. (only if) we only need to consider the query $Q = V_i$ for $i \in [1, k]$. Clearly there is an equivalent rewriting of V_i using itself. Therefore $\bigcup_{j=1}^{k} \text{EMCR}(V_i, V_j) = V_i$ for $i \in [1, k]$. By definition, $V_1, \ldots, V_k$ are redundant implies that

$$\bigcup_{i=1}^{k} \text{EMCR}(V_i, V_j) \subseteq \bigcup_{i=k+1}^{n} \text{EMCR}(V_i, V_j).$$

Thus

$$V_i \subseteq \bigcup_{j=k+1}^{n} \text{EMCR}(V_i, V_j)$$

for all $i \in [1, k]$. By Theorem 3, there is an equivalent rewriting of V_i using some V_j $(j \in [k+1, n])$.

(if) Suppose for every $i \in [1, k]$, there exists $j \in [k+1, n]$ such that V_i has an equivalent rewriting using V_j. Let V_i' be the equivalent rewriting of V_i using V_j. By definition V_i is equivalent to $V_j \oplus V_i'$. For any TP $Q \in P^{\{/,//,[]\}}$, if Q' is a CR of Q using V_i, then $V_i \oplus Q' \subseteq Q$. Hence $V_j \oplus (V_i' \oplus Q') = (V_j \oplus V_i') \oplus Q'$ is contained in Q. Therefore $V_i' \oplus Q'$ is a CR of Q using V_j, and the expansion of this CR is equivalent to the expansion of Q'. Therefore, $\text{EMCR}(Q, V_i) \subseteq \text{EMCR}(Q, V_j)$. This implies

$$\bigcup_{i=1}^{k} \text{EMCR}(Q, V_i) \subseteq \bigcup_{j=k+1}^{n} \text{EMCR}(Q, V_j).$$

When DTDs exist, the condition in Theorem 4 is still sufficient but not necessary. The proof of sufficiency is similar to the case when there are no DTDs. The non-necessity is shown by the DTD in Figure 5 (a), and the views $V_1 = a/b$, $V_2 = a/c$, and $V_3 = a//d$. Clearly V_3 is redundant under the DTD, but it does not have an equivalent rewriting using either V_1 or V_2.

A special case is when all the views $V_1, \ldots, V_n$ are *compatible*, that is, their distinguished nodes have identical labels. In this case, if V_i is redundant, then by Theorem 4, there is $j \neq i$ such that V_i has an equivalent rewriting using V_j. That is, there is a TP Q' such that $V_i = V_j \oplus Q'$. However, since V_i and V_j have the same distinguished node label, we know the root of Q' is the distinguished node of Q'. Therefore $V_i \subseteq V_j$. This proves the following corollary of Theorem 4.

Corollary 1. *Let* $V_1, \ldots, V_n$ *be compatible views. If* $V_1, \ldots, V_k$ *(k < n) are redundant, then for every* $i \in [1, k]$, *there exists* $j \in [k+1, n]$ *such that* $V_i \subseteq V_j$.

Note that in Definition 1, and so in Theorem 4, we haven't taken into account the possibility of rewriting a query using multiple views. We are only considering rewritings using single views. For example, the query $Q = a[b][c]/d$ does not have a CR using either $V_1 = a[b]/d$ or $V_2 = a[c]/d$, but it has a CR using $V_1 \cap V_2$, which is equivalent to Q. We will consider this case in the next section.

Finding Redundant Views. Theorem 4 provides a means to find the redundant views. To see whether V_i is redundant we only need to check whether there exists V_j such that V_i has an equivalent rewriting using V_j. To do so we can use Lemma 4.8 of [13].

5   TP Rewriting Using Intersection of Views

As noted in the previous section, given compatible views $V_1, \ldots, V_n$, it is possible for a query Q to have no CRs using any single view but it has a CR using the intersection of these views. In this section we investigate the form of equivalent rewriting problem and redundant views problem, taking into consideration of this possibility. We need to formally define CR and ER using $V_1 \cap \cdots \cap V_n$ first.

Definition 2. *Let $V_1, \ldots, V_n$ be compatible views in $P^{\{/,//,[]\}}$, and Q be a query in $P^{\{/,//,[]\}}$. Suppose $V_1 \cap \cdots \cap V_n$ is not always empty. A contained rewriting (CR) of Q using $V_1 \cap \cdots \cap V_n$ is a TP $Q' \in P^{\{/,//,[]\}}$, such that for any XML tree t, $Q'(V_1(t) \cap \cdots \cap V_n(t)) \subseteq Q(t)$. Q' is said to be an equivalent rewriting (ER) if $Q'(V_1(t) \cap \cdots \cap V_n(t)) \supseteq Q(t)$ also holds. The maximal contained rewriting (MCR) of Q using the intersection is the union of all CRs of Q using the intersection.*

In the presence of DTD G, the CR, MCR and ER are defined similarly, except we consider only XML trees conforming to G.

The next example shows that it is possible for a query to have different CRs using the intersection, thus the union of these CRs produces strictly more answers than any single CR.

Example 3. Consider the views $V_1 = a[x]/b$ and $V_2 = a[y]/b$, and the query $Q = a[x][y][//b/d]//b[c]$. It can be verified that $Q_1 = b[c][d]$, $Q_2 = b[d]//b[c]$, $Q_3 = b[//b/d][c]$ and $Q_4 = b[//b/d]//b[c]$ are all CRs of Q using $V_1 \cap V_2$, and none of them is contained in another.

However, if the union of all CRs becomes equivalent to Q, then one of the CRs is equivalent to Q. In other words, Theorem 1 can be extended to rewritings using the intersection of compatible views.

Theorem 5. *Let $V_1, \ldots, V_n$ be compatible views in $P^{\{/,//,[]\}}$, and Q be a query in $P^{\{/,//,[]\}}$. If the MCR of Q using $V_1 \cap \cdots \cap V_n$ is equivalent to Q, that is, it produces all answers to Q when evaluated on $V_1(t) \cap \cdots \cap V_n(t)$ for all t, then one of the CRs is an ER of Q using $V_1 \cap \cdots \cap V_n$.*

Proof. The proof uses an important property of intersection of TPs: there are TPs $V_1', \ldots, V_k' \in P^{\{/,//,[]\}}$ such that $V_1 \cap \cdots \cap V_n$ is equivalent to $V_1' \cup \cdots \cup V_k'$ (For the proof of this property see the full version of this paper. We omit it because of page limit). Therefore, Q' is a CR of Q using $V_1 \cap \cdots \cap V_n$ if and only if it is a CR of Q using every V_i' for $i = 1, \ldots, k$. Suppose $Q_1, \ldots, Q_m$ are all of the CRs of Q using $V_1 \cap \cdots \cap V_n$. Then the MCR of Q using the intersection is equivalent to Q implies

$$Q \subseteq \bigcup_{i=1}^{m} \bigcup_{j=1}^{k} (V_j' \oplus Q_i).$$

Since all TPs involved are in $P^{\{/,//,[]\}}$, by Theorem 2, we know there is an i and a j such that $Q \subseteq V_j' \oplus Q_i$. Therefore, the CR Q_i is an equivalent rewriting of Q using $V_1 \cap \cdots \cap V_n$.

Theorem 5 still holds in the presence of non-recursive DTDs. This is because, under a non-recursive DTD G, $V_1 \cap \cdots \cap V_n$ is equivalent to a single TP, say V, in $P^{\{/,//,[]\}}$. As given in [7], the MCR of Q using V under G is contained in a single CR of Q using V under G. Therefore, the MCR of Q using $V_1 \cap \cdots \cap V_n$ under G is contained in a single CR, say Q', of Q using $V_1 \cap \cdots \cap V_n$ under G. If the MCR produces all answers to Q, namely Q is contained in the MCR, then Q' is an ER of Q.

Theorem 5 can also be extended to the case where rewritings using both individual views and using intersections of views are considered together. Let $V = \{V_1, \ldots, V_n\}$ be a set of compatible views. We use MCR(Q, V) to denote the union of all CRs of Q using a single view in V or using the intersection of a subset of views in V. Similar to the proof of Theorem 5, we can prove the theorem below.

Theorem 6. *If* MCR(Q, V) *is equivalent to* Q, *then there exists a single* CR, Q', *of* Q *using either a single view, or using the intersection of some of the views in* V, *such that* Q' *is equivalent to* Q.

Next we re-examine the redundant views problem, taking into consideration of rewritings using intersections of views as well as individual views. The new meaning of redundant views is as follows.

Definition 3. *Let* $V = \{V_1, \ldots, V_n\}$ *be a set of compatible views and* V' *be a proper subset of* V. *We say that* V' *is strongly redundant if for every query* Q, MCR$(Q, V) \subseteq$ MCR$(Q, V - V')$.

The following theorem can be proved in a way similar to the proof of Theorem 4.

Theorem 7. *Let* $V = \{V_1, \ldots, V_n\}$ *be a set of compatible views and* V' *be a proper subset of* V. V' *is strongly redundant iff for every view* $V \in V'$, $V =$ MCR$(V, V - V')$.

6   Other Related Work

Besides the papers on rewriting XPath queries using views [7,13,8] discussed in Section 1, our work is closely related to [9], which studies the the containment problem of different classes of TPs and provides the basis for TP rewriting. For XQuery rewriting using views, [11] deals with equivalent rewriting of XQuery queries using XQuery views, and [4] translates XQuery rewriting to relational rewriting problems. Query rewriting using views has also been extensively studied in the relational database context, see [6] for a survey and [12,3,5,10] for more recent developments.

7 Conclusion

We presented some interesting results on the form of equivalent rewritings, and identified necessary and sufficient conditions for a subset of views to be redundant, in two scenarios. The first is when the rewritings are using a single view, and second is when the rewriting can use the intersection of compatible views in addition to individual views. These results are of theoretical and practical importance to TP query processing.

References

1. Amer-Yahia, S., Cho, S., Lakshmanan, L.V.S., Srivastava, D.: Minimization of tree pattern queries. In: SIGMOD Conference, pp. 497–508 (2001)
2. Balmin, A., Özcan, F., Beyer, K.S., Cochrane, R., Pirahesh, H.: A framework for using materialized XPath views in XML query processing. In: VLDB, pp. 60–71 (2004)
3. Cohen, S., Nutt, W., Sagiv, Y.: Rewriting queries with arbitrary aggregation functions using views. ACM Trans. Database Syst. 31(2) (2006)
4. Deutsch, A., Tannen, V.: Reformulation of XML queries and constraints. In: Calvanese, D., Lenzerini, M., Motwani, R. (eds.) ICDT 2003. LNCS, vol. 2572, pp. 225–241. Springer, Heidelberg (2002)
5. Gou, G., Kormilitsin, M., Chirkova, R.: Query evaluation using overlapping views: completeness and efficiency. In: SIGMOD Conference, pp. 37–48 (2006)
6. Halevy, A.Y.: Answering queries using views: A survey. VLDB J. 10(4), 270–294 (2001)
7. Lakshmanan, L.V.S., Wang, H., Zhao, Z.J.: Answering tree pattern queries using views. In: VLDB, pp. 571–582 (2006)
8. Mandhani, B., Suciu, D.: Query caching and view selection for XML databases. In: VLDB, pp. 469–480 (2005)
9. Miklau, G., Suciu, D.: Containment and equivalence for an XPath fragment. In: PODS, pp. 65–76 (2002)
10. Nash, A., Segoufin, L., Vianu, V.: Determinacy and rewriting of conjunctive queries using views: A progress report. In: Schwentick, T., Suciu, D. (eds.) ICDT 2007. LNCS, vol. 4353, pp. 59–73. Springer, Heidelberg (2006)
11. Onose, N., Deutsch, A., Papakonstantinou, Y., Curtmola, E.: Rewriting nested XML queries using nested views. In: SIGMOD Conference, pp. 443–454 (2006)
12. Wang, J., Topor, R.W., Maher, M.J.: Rewriting union queries using views. Constraints 10(3), 219–251 (2005)
13. Xu, W., Özsoyoglu, Z.M.: Rewriting XPath queries using materialized views. In: VLDB, pp. 121–132 (2005)

Querying Capability Modeling and Construction of Deep Web Sources

Liangcai Shu[1], Weiyi Meng[1], Hai He[1], and Clement Yu[2]

[1] Department of Computer Science, SUNY at Binghamton,
Binghamton, NY 13902, U.S.A
{lshu, meng, haihe}@cs.binghamton.edu
[2] Department of Computer Science, University of Illinois at Chicago,
Chicago, IL 60607, U.S.A
yu@cs.uic.edu

Abstract. Information in a deep Web source can be accessed through queries submitted on its query interface. Many Web applications need to interact with the query interfaces of deep Web sources such as deep Web crawling and comparison-shopping. Analyzing the querying capability of a query interface is critical in supporting such interactions automatically and effectively. In this paper, we propose a querying capability model based on the concept of atomic query which is a valid query with a minimal attribute set. We also provide an approach to construct the querying capability model automatically by identifying atomic queries for any given query interface. Our experimental results show that the accuracy of our algorithm is good.

Keywords: Deep Web, query interface, querying capability modeling.

1 Introduction

It is known that public information in the deep Web is 400 to 550 times larger than the so-called surface Web [1]. A large portion of the deep Web consists of structured data stored in database systems accessible via complex query interfaces [3] (also called search interfaces). Many Web applications such as comparison-shopping and deep Web crawling [2] require interaction with these form-based query interfaces by program. However, automatic interaction with complex query interfaces is challenging because of the diversity and heterogeneity of deep Web sources and their query interfaces. Automatic interaction includes automatically identifying search forms, submitting queries, and receiving and processing result pages. Some related work that has been carried out by different researchers includes: source extraction and description [4, 8] and deep Web crawling [2, 7, 11, 12].

The focus of this paper is on the automatic analysis of the *querying capability* of complex query interfaces, i.e., we want to find out what kinds of queries are *valid* (acceptable) by any given interface. A typical complex query interface consists of a series of attributes, each of which has one or more control elements like textbox, selection list, radio button and checkbox. A query is a combination of pairs, each consisting of an attribute and its assigned value. Those combinations that are accepted by the query interface syntax are *valid queries*; others are *invalid queries*.

B. Benatallah et al. (Eds.): WISE 2007, LNCS 4831, pp. 13–25, 2007.

Determining whether a query is valid is an important aspect of querying capability analysis. Consider the query interface in Fig. 1. Query {<*Author*, "John Smith">} is valid but query {<*Published date*, (2000, 2006)>} is invalid because this interface requires that at least one of the attributes with "*" must have values. The query with both conditions {<*Author*, "John Smith">, <*Published date*, (2000, 2006)>} is valid.

In this paper, we define a *query* as a set of attributes that appear in the query conditions. Traditional query conditions that include both attributes and values are defined as *query instances*. We propose the concept of *atomic query* to help describe querying capability. An atomic query is a set of attributes that represents a minimum valid query, i.e., any proper subset of an atomic query is an invalid query. For the query interface in Fig. 1, query {*Author*} is an atomic query because any query formed by filling out a value in the *Author* textbox is accepted by the search engine (note that a valid query does not guarantee any results will be returned). We can determine if a given query is valid if we identify all atomic queries for a query interface in advance. We propose a method to identify atomic queries automatically.

The work that is most closely related to our work is [8]. In [8], source descriptions are discussed. The focuses are on what kind of data is available from a source and how to map queries from the global schema to each local schema. In contrast, we are interested in describing the full querying capability of a query interface (i.e., what queries are valid) and constructing the querying capability model automatically. Both the issues studied and the solutions provided in these two works are different.

Fig. 1. Query interface of abebooks.com   **Fig. 2.** Query interface of aaronbooks.com

The paper makes the following contributions:
(1). Make a classification of attributes. The attributes are classified into four types: functional attribute, range attribute, categorical attribute and value-infinite attribute.
(2). Propose the concept of atomic query (AQ) and a querying capability model.
(3). Present an algorithm to construct the AQ set that represents querying capability for a given query interface.
(4). Compare different classifiers' performance on result page classification.

The rest of this paper is organized as follows. Section 2 introduces attribute types. Section 3 presents a querying capability model of query interfaces and introduces the concept of atomic query. Section 4 discusses constructing querying capability of a query interface. Section 5 reports experimental results. Section 6 concludes the paper.

2 Attribute Types

We first classify attributes on query interfaces. Different types of attributes may have different functions and impacts on query cost and need to be coped with differently.

On a query interface, an attribute usually consists of one or more control elements (*textbox, selection list, radio button* and *checkbox*) and their labels (in this paper we refer to an attribute by its primary label), and it has a specific semantic meaning. For example, in Fig. 1, attribute *Author* has one textbox element, whereas attribute *Price* has two textbox elements. Many attributes on a query interface are directly associated with the data fields in the underlying database.

There are also *constraints* on a query interface. Constraints are not directly associated with data in the underlying database but they pose restrictions to the attributes involved. For example, *All/Any* in Figure 2 poses restrictions to *Keyword(s)*.

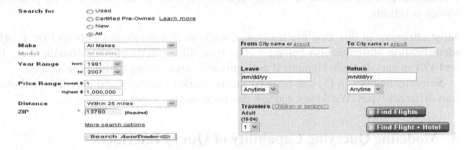

Fig. 3. Query interface of autotrader.com   **Fig. 4.** Query interface of orbitz.com

Definition 2.1. An attribute is a *functional attribute* if its values, when used in a query, affect only how query results are displayed but not what results are retrieved.

For a functional attribute, different values assigned to it yield the same set of search results, assuming the underlying database is unchanged. Functional attribute *Sort results by* (Fig. 1) determines the order of results while *Results per page* (Fig. 1) and *Display* (Fig. 2) decide how many result records are displayed per page.

From our observation, for a given domain, the total number of functional attributes is a small constant if attributes with the same function but different names are considered to be the same (e.g., *Results per page* in Figure 1 and *Display* in Figure 2). Attribute matching techniques (e.g., [6]) can be used to identify such attributes.

Definition 2.2. If any value assignment of an attribute implies a range of values, the attribute is a *range attribute*.

A range attribute could have one or two control elements. For example, *Published date (min, max)* and *Price (min, max)* in Fig. 1 have two elements. Generally, if an element of a range attribute has value *Null* (no value) or *All*, it means no restriction on this attribute, i.e., there is no restriction on the returned results from this attribute.

Definition 2.3. An attribute is a *categorical attribute* if it has a fixed, usually small, set of distinct values that categorize all entities of the site.

For a categorical attribute, we can usually identify a complete set of its distinct values. Typically, this kind of attributes involves selection lists, such as *Make* and *Model* in Fig. 3, and *Binding* in Fig. 1. A group of radio buttons is equivalent to a selection list if only one of the radio buttons can be selected at a time, e.g., *Search for* in Fig. 3. A categorical attribute may be implemented as a textbox like *ZIP* in Fig.3.

A group of checkboxes can be considered as a special categorical attribute with special semantics. Take *Attributes* in Figure 1 as an example. It has three checkboxes - *First edition, Signed,* and *Dust jacket*. If none of them is checked, it means no restriction with the attribute. If one is checked, say, *First edition*, it means that only *first-edition* books are selected. But if all three checkboxes are checked, it means that *first-edition* or *signed* or *dust-jacket* books are selected, i.e., they are *disjunctive* rather than *conjunctive* whereas most attributes in a query interface are conjunctive.

Definition 2.4. An attribute is a *value-infinite attribute* if the number of its possible values is infinite.

Most attributes in this kind are textboxes, such as *Keywords* and *Title* in Fig. 1. All identification attributes also belong to this type, like *ISBN* in books domain (Fig. 1) and *VIN* in cars domain. Numerical and date attributes belong to this type as well.

Some techniques in [5] can be used in our work to identify and classify attributes into different types. We do not address this issue in this paper.

3   Modeling Querying Capability of Query Interfaces

3.1   Query Expression

In this paper, we assume that a *query* is a conjunctive query over a query interface. In other words, the conditions for all attributes in a query are conjunctive. Based on our observation, almost all query interfaces support only conjunctive queries.

Traditionally, a query condition contains values that are assigned to attributes, like {*<Author,* "Meng">, *<Keywords,* "Databases">}. There are three special values - *Null, All,* and *Unchecked. Null* is for textboxes, *All* for selection lists and radio buttons, and *Unchecked* for checkboxes. They have the same function – applying no restriction to the attribute. Thereinafter we use *Null* to represent any of these values.

In this paper, we focus on querying capability and each attribute's function in a query condition. We do not care about what value is assigned to attribute unless the value is *Null*. Thus, we do not include values but attributes in our *query expression*. An attribute is precluded from our query expression if its assigned value is *Null*.

Definition 3.1. A *query expression* (or *query* for short) is a set of attributes that appear in some query condition. A *query instance* is a specific query condition consisting of attribute-value pairs with the value valid for the corresponding attribute.

An attribute is assigned with a valid non-*Null* value if and only if it appears in a query expression. For example, query instance {*<Author,* "Meng">, *<Keywords,* "Databases">} is represented by query expression {*Author, Keywords*} on the query interface in Fig. 1. Therefore, each query instance corresponds to one query expression. But one query expression may represent a group of query instances, with the same attributes but not totally the same valid values.

Definition 3.2. A query is *valid* if at least one of its query instances is accepted by the query interface, regardless of whether or not any records are returned.

That a query instance is accepted means that it is executed by the underlying database system. When a query instance is rejected by the query interface, an error message will be returned usually.

3.2 Conditional Dependency and Attribute Unions

Dependency may exist between a constraint and an attribute, e.g., constraint *All/Any* depends on attribute *Keyword* in Fig. 2. We name this dependency as *conditional dependency*. It is a restriction between a non-*functional* attribute and a constraint.

Definition 3.3. Given an attribute A and an constraint C, if any value of C is insignificant unless a non-*Null* value is assigned to A, we say a *conditional dependency* exists between A and C, denoted as $A{\rightarrow}C$.

Generally the attribute and its dependent constraints appear next to each other. Conditional dependency does not involve functional attributes. We use conditional dependency to define *attribute union* below.

Definition 3.4. Attribute A, its value type T, and constraints $C_1, ..., C_n$ ($n>0$) form an *attribute union*, denoted as $AU(A,T,C_1,...,C_n)$, if and only if $A{\rightarrow}C_i$ ($i=1,...,n$) and no other conditional dependency involves A. In $AU(A,T,C_1,...,C_n)$, T and C_i are optional. And attribute A is named the *core attribute* of the *attribute union*, denoted as CA.

Note that attribute A alone forms an *attribute union* iff there is no conditional dependency involving A. In this case, the *core attribute* is A and the attribute union is denoted by the attribute name A for convenience. In Fig. 2, $AU(Keyword, All/Any)$ is an attribute union. Attribute unions do not involve functional attributes.

3.3 Atomic Query

Our observations show that some attributes are necessary to form valid queries.

Example 1. For the query interface in Fig. 1, any valid query must include at least one of the attributes *Author, Title, ISBN, Keywords* and *Publisher*. Other attributes are not required for a valid query.

Example 2. For the query interface in Fig. 4, all of the attributes *From, To, Leave, Return* and *Travelers* are necessary for any valid query.

To describe the function of these attributes in a query, we propose the concept of *atomic query*, a key concept in our querying capability model. If a query is valid, its superset query is also valid. For example, in Fig. 1, if query {*Author*} is valid, query {*Author, Price*} is also valid.

We have stated that a query is a set of attributes. Actually, it also can be considered as a set of *attribute unions*. An *atomic query* is an irreducible set of *attribute unions* that forms a valid query. All supersets of an atomic query are valid and all valid queries are supersets of some atomic query. Therefore, the identification of atomic queries is the most important part of querying capability analysis.

Definition 3.5. Given a set of attribute unions $S=\{AU_1,...,AU_n\}$ that represents a valid query. S is an *atomic query* iff the query represented by any $T \subset S$ (here $\subset$ denotes proper subset) is not valid. *Atomic query* S is also denoted as $AQ(AU_1,...,AU_n)$.

A query interface may have more than one atomic query. The interface in Fig. 1 has five atomic queries, each consisting of a single attribute: $\{Author\}$, $\{Title\}$, $\{ISBN\}$, $\{Keywords\}$ and $\{Publisher\}$. The interface in Fig. 4 has one atomic query: $\{From,$ $To, Leave, Return, Travelers\}$ with five attribute unions.

3.4 Describing Querying Capability

We describe querying capability of a query interface using a model similar to the Backus-Naur Form (BNF). Different from BNF, the order of attributes in the query expression is not significant. Any attribute in the expression has a non-*Null* value. Any attribute not in the expression has value *Null*.

There are three special functions used in BNFs in this paper: *AQ, AU* and *ADJ*.

- Function $AQ(O_1, ..., O_n)$: if objects $O_1, ..., O_n$ form an *atomic query*, the function returns the atomic query. Otherwise, it returns an invalid symbol (note that a derivative rule in BNF does not apply to any invalid symbol).
- Function $AU(O_1, ..., O_n)$: if objects $O_1, ..., O_n$ form an *attribute union*, the function returns the attribute union; else, it returns an invalid symbol.
- Function $ADJ(A, B)$: if the control elements (or control element groups) A and B are *adjacent* (both left-right and up-down are considered) on the query interface, the function return the combination of A and B. Otherwise, it returns an invalid symbol. Adjacency is determined by some threshold: if the shortest distance between A and B is within the threshold, A and B is considered to be adjacent.

Symbols used and their meanings are as follows:

Q: Query
AQ: Atomic Query, or function *AQ*
AM: Attribute Module, an attribute union or a functional attribute or exclusive attribute (see below)
OQ: Optional Query, a group of attribute unions whose values are optional for a query
AU: Attribute Union, or function *AU*
CoA: Core Attribute, an attribute other than functional attribute
VT: Value Type, the value type of core attribute
C: Constraint
EA: Exclusive Attribute, one of a group of attributes that is exclusive to each other (i.e., only one of them can be used in any query)

ANO: Attribute Name Option, one of the options in a list of attribute names
RA: Range Attribute
CA: Categorical Attribute
VA: Value-infinite Attribute
FA: Functional Attribute
RBG: Radio Button Group, a group of radio buttons
CBG: Checkbox Group, a group of checkboxes
TB: Textbox
SL: Selection List
RB: Radio Button
CB: Checkbox

```
<Query>::= <AtomicQuery> [<OptionalQuery>]
<AtomicQuery>::=AQ(<AttributeModule> [, <AttributeModule>]*)
```

<OptionalQuery>::=<AttributeModule> [<AttributeModule>]*
<AttributeModule>::=<AttributeUnion> | <FunctionalAttribute> | <ExclusiveAttribute>
<AttributeUnion>::=AU(<CoreAttribute>[<ValueType>][<Constraint>]*)
<ExclusiveAttribute>::=<CoreAttribute> <AttributeNameOption>
<AttributeNameOption>::=<SelectionList> | <RadioButtonGroup>
<CoreAttribute>::=<RangeAttribute> | <CategoricalAttribute> | <Value-infiniteAttribute>
<ValueType>::= char | number | Boolean | date | time | datetime | currency
<Constraint>::=<SelectionList> | <RadioButtonGroup> | <CheckboxGroup>
<RangeAttribute>::=<SelectionList> | ADJ(<Textbox>,<Textbox>)
      | ADJ(<SelectionList>, <Textbox>) | ADJ(<SelectionList>, <SelectionList>)
<CategoricalAttribute>::=<SelectionList> | <RadioButtonGroup> | <CheckboxGroup>
      | <Textbox>
<Value-infinite Attribute>::= <Textbox>
<FunctionalAttribute>::=<SelectionList> | <RadioButtonGroup> | <CheckboxGroup>
<RadioButtonGroup>::= <RadioButton> | ADJ(<RadioButton>, <RadioButtonGroup>)
<CheckboxGroup>::=<Checkbox> | ADJ(<Checkbox>, <CheckboxGroup>)

The grammar above can be used to determine if a query is valid.

Example 3. The query interface in Fig. 1 has five atomic queries: $AQ(Author)$, $AQ(Title)$, $AQ(ISBN)$, $AQ(Keywords)$ and $AQ(Publisher)$. They are defined by the logic of the query interface. There are five *attribute unions* that can be part of any *optional query* - Published date, Price, Bookseller country, Binding, and *Attributes*. And there are two functional attributes that can be part of any *functional query* - Sort results by and Results per page. Consider a query Q = {*Author, Title, Binding, Sort results by, Results per page*}. This query contains two atomic queries. Interpretation of the query is not unique. Both *Author* and *Title* are atomic queries. Either of them can be treated as part of an optional query. Fig. 5 is one interpretation tree. A query is valid if there exists at least one way to interpret it.

Several observations:

1. Any attribute module that is part of an atomic query can also be part of an optional query if two atomic queries appear in a valid query.
2. An attribute with a default value *All* is not part of an *atomic query*.
3. For a range attribute, if it reaches its largest range or its value is empty, it has a value *All*, so it is not part of an atomic query. For example, *Year range* in Figure 3, *from 1981 to 2007* means *All*.
4. A non-functional attribute that always has a non-*Null* value automatically becomes part of *atomic query*. An example is *Travelers* in Fig. 4.

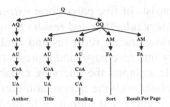

Fig. 5. An interpretation tree

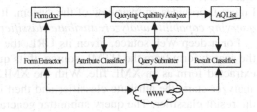

Fig. 6. Overview of the system

4   Querying Capability Construction

4.1   Attributes, Constraints and Control Elements

Constructing the querying capability model of a query interface relies on attribute analysis. Some attributes' functionalities are impacted by constraints.

Constraints and hidden attributes

Constraints generally appear as a radio button group. They are always assigned their default values because different values will not impact querying capability construction, especially for the identification of atomic queries. Hidden attributes are invisible on a query interface but have fixed values assigned to them in each query. The values are included in the form tag on the query interface and can be identified.

Values *Null* and *All*

Values *Null* (for textbox and checkbox) and *All* (for selection list and radio button) mean no constraint on their attributes. An empty textbox means value *Null*.

For attributes with multiple elements, *Null* or *All* values may be assigned in different ways. For example, for attributes with two textboxes to have value *Null*, both textboxes should be assigned *Null* (e.g., *Price* in Fig. 1). To assign an *All* value to an attribute with two selection lists of numerical values (e.g., *Year Range* in Fig. 3), select the smallest value in the first selection list and the largest value in the second one, namely, assign the maximum range.

Type I and type II attributes

From the angle of values *Null* (for a selection list, *Null* implies *All*), any visible non-*functional* attribute can be classified into either of two types as defined below.

Definition 4.1. If a non-functional attribute has no value *Null* in its possible values, it is a *type I attribute* (in most cases, they are selection lists.). If it can be assigned *Null*, it is a *type II attribute* (this type includes textboxes/selection lists with value *Null*.)

Exclusive attributes

Exclusive attributes are generally included in a selection list or a radio button group. Only one of them can be used in any given query. Techniques for identifying exclusive attributes are discussed in [5].

4.2   Constructing Querying Capability Model

4.2.1   Overview

Figure 6 provides an overview of the system. It consists of five parts: *form extractor, querying capability analyzer, attribute classifier, query submitter* and *result classifier.*

For a deep Web source, given its URL, the form extractor retrieves and analyzes the search form from the HTML source of its query interface page, and represents the extracted form as an XML file. With the XML file as input, the querying capability analyzer starts the attribute classifier and then interacts with the query submitter and the result classifier. The query submitter generates a query and submits it to the deep Web source while the result classifier receives result page, analyzes whether the query is accepted, and reports its decision to the querying capability analyzer. Through a series of query submissions and result analyses, the querying capability analyzer determines the list of the atomic queries for the query interface.

4.2.2 An Algorithm for Searching Atomic Queries

A brute force method for finding atomic queries is to submit a query for every combination of all attributes on the query interface. This is obviously not feasible due to its huge query cost. We now describe a more efficient method.

Definition 4.3. For a query interface, a query including exactly n type II attributes is called a *level n query*.

A query interface with m type II attributes has $\binom{m}{k}$ level k queries, including one level 0 query and m level 1 queries. All type I attributes are included in any query with their default values on the query interface. When a type II attribute is included in a query, it is assigned a non-*Null* value. Fig. 7 shows our algorithm.

The algorithm first recognizes each exclusive attribute as an AQ. If there are no exclusive attributes, it searches for AQs by submitting queries. For any level k query q, if a query that is a proper subset of q (we treat a query as an attribute set) is already in the AQ set, then q cannot be an AQ and should be discarded. Otherwise, submit q, download the result page, and analyze the page. If the page represents acceptance, the query is added to the AQ set of the query interface.

```
Algorithm: identifyAQs(I)
Input: I - a query interface,
       A₁ = {a | a is a type I attribute on interface I},
       A₂ = {a | a is a type II attribute on interface I}.
Output: L - a set of AQs for I; L's initial value is ∅.
for each a ∈ A₁ ∪ A₂ do
       if the value set of a represents a set of exclusive attributes E
              A₁ := A₁ − {a};
              for each e ∈ E do
                     s := A₁∪ {e};   // s represents an AQ
                     L := L ∪ {s};
              return L ;
for k = 0 to |A₂| do
       Qₖ := {q | q is a level k query on interface I};
       for each q ∈ Qₖ do
              if k = 0
                     R := ∅;
              else
                     R := {r | r is a level k−1 query, r ⊆ q, r ∈ L};
              if R = ∅
                     submit query q;                    (1)
                     download result page P;
                     if P is an acceptance page           (2)
                            L := L ∪ {q}
Return L
```

Fig. 7. An algorithm for searching atomic queries

In the following, we discuss two important steps in the algorithm: submit a query at line (1) and decide if a result page is an acceptance page at line (2).

4.2.3 Automatic Query Submission

To submit a query is actually to submit a form on a web page. We use the following method to submit a query automatically:

1. Identify attributes to which non-*Null* value will be assigned, and assign values. To assign a non-*Null* value to a selection list is quite straightforward – just select a non-*Null* value from its value list. To do so for checkbox is easy too – just check it. But for textboxes, giving a proper value needs careful consideration. The following two methods can be used to assign non-*Null* value to a textbox.
    *   For every domain, keep a table of attributes and some corresponding values
    *   Keep a table of value types and some corresponding values. For example, value 1998 is for year or date, and value 13902 is for zip code.
2. Assign value *Null* to other attributes. For textboxes and checkboxes, just leave them blank or unchecked. For selection lists, select the *Null (All)* value.
3. Assign values to constraints and hidden attributes. Constraints generally are radio button groups; assign default values to them. Hidden attributes have fixed values.
4. Construct URL string with value assignments. A value assignment is a control-name/current-value pair.

4.2.4 Result Page Classification

To construct the querying capability model, we need to submit queries and decide whether the query is accepted. Therefore we must know whether the result page of the query represents acceptance or rejection. The performance of the querying capability construction depends on the accuracy of result page classification.

A typical acceptance page is like Fig. 8 and a rejection page is like Fig. 9.

Fig. 8. Part of an acceptance result page of amazon.com

Fig. 9. Part of a rejection result page of abebooks.com

Since a result page is a document, the problem can be treated as a document classification problem. Let's first identify the special features that appear on result pages that represent acceptance and rejection below.

Evidences for Query Acceptance

1. Sequences exist in the result page. Two kinds of sequences may exist with order numbers in a result page, one for result records and the other for result pages (see Fig. 8).
2. Large result page size. A large result page size means this query is probably accepted. From observation, a query is probably accepted if its result page is five times larger than the smallest of a group of result page from the same site.

3. Result patterns for query acceptance. Some patterns imply that the query is accepted, for example, *Showing 1 - 24 of 9,549,732 Results* in title (see Fig. 8).

Evidences for query rejection

1. Error messages for query rejection. When a query is rejected, frequently an error message appears on the result page, such as *"please enter at least one search term"* (Fig. 9), and *"no valid search words"*, etc.
2. The result page is similar to its query interface page. The result page sometimes has the same interface as the query interface page to allow users to enter another query.
3. Small result page size. Rejection pages generally have a relative small size.

In Section 5, we discuss how to build a result page classifier based on the observations above.

5   Experiments

The experiment consists of two parts: (1) build a result page classifier; (2) use the classifier to identify atomic queries.

We use two datasets of deep Web sources from three domains: books, music and jobs. Dataset 1 is for training, i.e., build a classifier. Dataset 2 is for identification using the built classifier. These sources are mainly from the TEL-I dataset at [13].

*   Dataset 1: 11 book sources, 10 music sources, and 10 job sources.
*   Dataset 2: 12 book sources, 11 music sources, and 10 job sources.

5.1   Build a Result Page classifier

We have discussed evidences for query acceptance and rejection in Section 4.2.4. Based on the evidences, we select five features: length_of_max_sequence (numeric), pagesize_ratio (numeric), contain_result_pattern (boolean), contain_error_message (boolean), similar_to_query_interface (boolean). The class label has two values: *accept* and *reject*.

From Dataset 1, we generated 237 result pages. For each result page, we compute the values for the five features above, and manually assign the class label. The five values plus the class label for the same result page construct one training record. Based on the 237 training records, we build a result page classifier using Weka [14].

We tried four different classifiers, aiming to select the one with the best performance. The four classifiers are C4.5 algorithm of decision tree (C4.5), *k*-nearest neighbors (KNN), naïve Bayes (NBC), support vector machines (SVM).

For training records, we use one part for training, remaining part for validation. If the training ratio (TR) is 0.6, then 60% are used for training, and 40% for validation.

We experimented with four different TRs from 0.5 to 0.8 and the four classifiers. The results are shown in Fig. 10. Classifier C4.5 has the best overall performance. When TR = 0.7, its accuracy is 98.6%. We notice that for all four classifiers, the accuracy reaches their highest when TR = 0.7. When TR is changed to 0.8, the

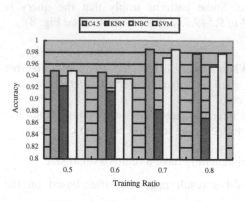

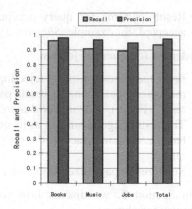

Fig. 10. Comparison for different classifiers

Fig. 11. Average recall and precision

accuracy gets lower due to overfitting. The C4.5 classifier built with TR = 0.7 will be used in the experiment in the later part.

5.2 Metrics for Atomic Query Identification

For a query interface I, we determine its actual atomic queries (AQ) by manually submitting queries. After the system finishes AQ identification, we compare the system identified AQs with actual AQs to determine its performance. In this paper, the performance is measured using average *recall* and *precision* over all query interfaces tested.

5.3 Experimental Results

All the results are based on Dataset 2. The accuracy of result page classification affects the recall and precision of AQ identification. Fig. 11 shows the average recalls and precisions for three domains as well as the overall recall and precision. The recall ranges from 89.5% for jobs domain to 95.7% for the books domain while the precisions are significantly higher. This indicates that our method can identify AQs accurately. For every domain, the recall is lower than the precision. This indicates our method for deciding if queries are accepted is conservative. In other words, if there is no strong evidence that a query is accepted, the system tends to treat it as rejected.

6 Conclusions

This paper proposed a method to model the querying capability of a query interface based on the concept of atomic queries. We also presented an approach to construct querying capability automatically. The experimental results show that the approach is practical. The constructed querying model can be useful for all applications that need to interact with a query interface.

Acknowledgment. This work is supported in part by the following NSF grants: IIS-0414981, IIS-0414939 and CNS-0454298.

References

1. Bergman, M.: The Deep Web: Surfacing Hidden Value (September 2001), http://www.BrightPlanet.com
2. Raghavan, S., Garcia-Molina, H.: Crawling the Hidden Web. In: VLDB, pp. 129–138 (2001)
3. Chang, K.C-C, He, B., Li, C., Patel, M., Zhang, Z.: Structured Databases on the Web. Observations and Implications. SIGMOD Record 33(3), 61–70 (2004)
4. Zhang, Z., He, B., Chang, K.C-C: Understanding Web Query Interfaces: Best-Effort Parsing with Hidden Syntax. In: SIGMOD Conference, pp. 107–118 (2004)
5. He, H., Meng, W., Yu, C., Wu, Z.: Constructing Interface Schemas for Search Interfaces of Web Databases. In: Ngu, A.H.H., Kitsuregawa, M., Neuhold, E.J., Chung, J.-Y., Sheng, Q.Z. (eds.) WISE 2005. LNCS, vol. 3806, pp. 29–42. Springer, Heidelberg (2005)
6. He, H., Meng, W., Yu, C., Wu, Z.: Automatic integration of Web search interfaces with WISE-Integrator. VLDB J. 13(3), 256–273 (2004)
7. Wu, P., Wen, J., Liu, H., Ma, W.: Query Selection Techniques for Efficient Crawling of Structured Web Sources. In: ICDE (2006)
8. Levy, A., Rajaraman, A., Ordille, J.: Querying Heterogeneous Information Sources Using Source Descriptions. In: VLDB, pp. 251–262 (1996)
9. Ipeirotis, P., Agichtein, E., Jain, P., Gravano, L.: To Search or to Crawl? Towards a Query Optimizer for Text-Centric Tasks. In: SIGMOD Conference (2006)
10. BrightPlanet.com, http://www.brightplanet.com
11. Bergholz, A., Chidlovskii, B.: Crawling for Domain-Specific Hidden Web Resources. In: WISE 2003, pp. 125–133 (2003)
12. Arasu, A., Garcia-Molina, H.: Extracting Structured Data from Web Pages. In: SIGMOD Conference, pp. 337–348 (2003)
13. Chang, K.C.-C., He, B., Li, C., Zhang, Z.: The UIUC web integration repository. CS Dept., Uni. of Illinois at Urbana-Champaign (2003), http://metaquerier.cs.uiuc.edu/repository
14. Witten, I., Frank, E.: Data Mining: Practical machine learning tools and techniques, 2nd edn. Morgan Kaufmann, San Francisco (2005)

Optimization of Bounded Continuous Search Queries Based on Ranking Distributions

D. Kukulenz, N. Hoeller, S. Groppe, and V. Linnemann

Luebeck University, Institute of Information Systems
Ratzeburger Allee 160, 23538 Luebeck, Germany
{kukulenz,hoeller,groppe,linnemann}@ifis.uni-luebeck.de

Abstract. A common search problem in the World Wide Web concerns finding information if it is not known when the sources of information appear and how long sources will be available on the Web, as e.g. sales offers for products or news reports. Continuous queries are a means to monitor the Web over a specific period of time. Main problems concerning the optimization of such queries are to provide high quality and up-to-date results and to control the amount of information returned by a continuous query engine. In this paper we present a new method to realize such search queries which is based on the extraction of the distribution of ranking values and a new strategy to select relevant data objects in a stream of documents. The new method provides results of significantly higher quality if ranking distributions may be modeled by Gaussian distributions. This is usually the case if a larger number of information sources on the Web and higher quality candidates are considered.

1 Introduction

An important problem concerning the search for information in the World Wide Web is to find information objects if it is not known when the information appears on the Web and how long the information will be available or is possibly overwritten by new information. If a user wants e.g. to obtain offers for sale for a specific product it is important for him to obtain the respective Web pages containing the offers for sale immediately in order to be able to purchase the products. If a scientist works on a specific research project it is important for him to obtain contributions in his field immediately after publication in order to take advantage of related research as soon as possible and furthermore to avoid the risk of working on scientific problems which have already been solved. In both examples it is usually not known when the information appears on the Web. Also in particular in the case of sales offers information may disappear from the Web after an unknown period of time. Furthermore in the case of the World Wide Web the information usually has to be extracted from a very large number of candidates.

A query language to specify similar search requests is a *bounded continuous search query* [10]. In this query language a user may specify a query profile, e.g. provided by a number of query terms, an information source, as e.g. a fraction of the Web and a bounding condition containing the maximal information load the user is willing to accept. Furthermore a user may specify a trigger condition which determines the points in time when the query is executed and a freshness value which indicates the maximal

B. Benatallah et al. (Eds.): WISE 2007, LNCS 4831, pp. 26–37, 2007.

period of time a user is willing to accept between the detection of a relevant information object on the Web and the time of a respective notification message.

The optimization of such queries is basically twofold. A first problem is to find an adequate relevance estimation method for data objects on the Web, which includes a means to state query profiles and a distance measure between query profiles and data objects on the Web. A second problem concerns the filtering of relevant information according to the relevance estimation. At the end of a query period the items (documents) observed so far may be compared to each other and high quality objects may easily be extracted. In order to provide fresher results a strategy has to be conceived to find or estimate relevant objects at an earlier stage in time where the entire sequence of ranking values is not known yet. A respective selection strategy may only be based on previous observations.

A well-known strategy [7], [13] is simply to divide the query period into smaller time intervals ('evaluation periods'), to store document ratings in each evaluation period and to compare the ratings at the end of each evaluation period, e.g. once a day. Thereby the best documents may be found in each evaluation period. It may easily be seen that in this strategy there is a tradeoff between freshness and quality of selected query results. The fresher query results are requested, the shorter the size of the evaluation periods has to be chosen. The shorter the evaluation periods, the less information the selection process in each evaluation period is based on. This results in a decreasing quality of the selection strategy with a decreasing size of evaluation periods. A strategy to deal with this freshness/quality tradeoff is to apply statistical decision theory, in particular optimal stopping in order to find high-quality information immediately based only on previous observations [4], [16].

Previous methods, in particular the statistical decision method proposed in [10] don't take advantage of the actual distributions of quality values. It is shown in this paper that in many cases the distribution of ranking values may be approximated by a normal distribution. In these cases (denoted as 'informed' selection problem [5]) selection strategies may be conceived and are presented in this paper which provide results of significantly higher quality than realized by uninformed selection.

In summary, our contributions in this paper are:

- Based on a well-known relevance estimation for data objects on the Web, the distributions of relevance values are examined by considering real data sources on the Web.
- Different classes for relevance values of information objects are extracted and it is shown that relevance distributions of a significant part of information on the Web may be modeled by normal distributions.
- Based on this observation the main contribution of this paper is a new selection strategy for data objects on the Web which is based on estimating the distribution of ranking values and selecting high-quality results according to this distribution.
- The proposed selection method is compared to previous methods based on an archive of real data on the Web which was recorded over 303 days.

In the next section we describe the theoretical background. In sections 3 and 4 we present the new algorithm for the processing of bounded continuous search queries and evaluation results. We conclude with an overview of related research and a summary.

2 Theoretical Background

In this section bounded continuous search (BCS) queries are introduced, the optimization problem concerning BCS queries is defined and previous processing methods are described that are used as reference strategies in order to evaluate the new method in section 3.

2.1 Definition of Bounded Continuous Search Queries

Continuous search queries are a means to specify search requests that are not processed immediately but that exist over a period of time and continuously provide a user with query results or respective alerts. In the continuous query language OpenCQ [11] a continuous query is a triple $(Q, T_c, Stop)$ consisting of a normal query Q (e.g. written in SQL), a trigger condition T_c and a termination criterion $Stop$. In this work we consider a similar structure of a continuous query. We consider a query component Q that includes a query profile which consists of a number of query terms. A source specification defines the information sources on the Web which should be considered for the query execution. The trigger condition T_c basically defines the query execution time period for the continuous query and the execution frequency of the query. In addition to this in the considered query language a user may specify a 'bounding condition' that defines the maximal amount of information a user is willing to accept, as e.g. a maximal number of Web pages over the course of the query execution time. An example for a bounded continuous search (BCS) query is the following:

Query: SELECT ESTIMATED BEST 10 FROM PAGE www.ebay.de
WHERE query='dvd player' Trigger=9h and 17h, Start=now, Stop=10days

This query selects the (estimated) 10 best Web pages at the site www.ebay.de with respect to the query terms 'dvd player' that appear during the next 10 days. Further examples are provided in [10].

2.2 Definition of the Optimization Problem

The basic problem concerning the optimization of bounded continuous search queries is to return high quality results to a user that appear on the Web at a specific point in time with only the knowledge about ranking values of documents that appeared prior to a specific evaluation time.

A first question concerning the processing of BCS queries is the estimation of the relevance of information sources given a query profile provided by a user. A well-known distance measure to compute the distance between a document d_j and a query profile (i.e. a set of query terms) Q is the *tf-idf* measure [15]:

$$R_{j,Q} = \sum_{\text{term: } w_k \in Q} tf_{j,k} idf_k \tag{1}$$

In the formula the term $tf_{j,i}$ (*term frequency*) is the frequency of a term w_i in the document d_j. N is the number of documents in the collection, $idf_i := \log(N/df_i)$ is the *inverse document frequency* where df_i is the number of documents in the collection that contain w_i. However, in our context the entire set of documents that appear during the considered query period is not known at a specific point in time. During the query execution time new documents have to be processed by the query engine which may change the idf component of the quality measure. The relative ranking values of documents observed at previous points in time may be modified if documents at subsequent points in time are considered. However in order to compare the presented query processing method to previous methods, in particular the optimal stopping methods [5], [6], [14], the relative ranking values of documents have to be stable over time. For this reason in this work we consider only the tf component in formula 1, i.e. $R'_{j,Q} = \sum_{\text{term: } w_k \in Q} tf_{j,k}$.

A second quality measure is the delay between the time a high quality object is detected on the Web and the time a respective alert message is sent to a user. The larger the delay, the more objects may be compared to each other and high quality objects may be extracted. In this work however, an immediate delivery of alert messages is considered.

The considered optimization problem in this paper is therefore to extract high quality results estimated by $R'_{j,Q}$ and to return these documents to a user immediately after detection on the Web.

2.3 Previous Query Processing Methods

A well-known method to process bounded-continuous search queries, which is similar to methods applied by current continuous query systems [7], [13], is to evaluate information objects after constant periods of time (*evaluation periods*). Data objects are collected in each evaluation period and objects are compared and filtered at the end of each evaluation period. By using this method best objects may be extracted in each evaluation period. It is however shown in [10] that if very fresh results are required and therefore very small evaluation periods have to be considered, the quality of returned results decreases.

In order to obtain higher quality results, statistical estimation methods may be applied which estimate a ranking value of a document in the entire set of documents, based however only on previous observations. This estimation problem is known from the field of optimal stopping as the Secretary-Selection-problem (SSP).

A possible optimal stopping strategy in order to find the best candidate in a sequence of candidates is first to reject a number of k candidates [8]. Then the next candidate is selected which has a higher ranking than the first rejected k candidates. It may be shown that an optimal value for k, the number of rejected candidates, is $\frac{n}{e}$, where n is the number of candidates and e is the Euler-number. More general cases of the Secretary-selection-problem concern e.g. finding the k best candidates by k choices, denoted as k-SSP. Optimization methods were proposed in [5], [6] and [14]. In [10] optimal stopping methods are applied to the optimization of BCS queries and compared to the evaluation period method. In this work we apply the respective method as a reference strategy in order to evaluate the new method presented in the subsequent section.

3   Query Optimization Based on Ranking Distributions

Following the idea in [5] and [17], in this paper the actual ranking distribution of retrieved documents is applied to improve the quality of the document set returned by a

BCS query. The method proposed in this paper consists of two separate tasks. First, the distribution of ranking values of documents in a stream of documents with respect to a given BCS query has to be estimated. Obviously the ranking distribution is dynamic and depends on the BCS query (i.e. in particular the query profile or query terms) and also depends on the considered information source (i.e. fraction of the Web) and may change over time. The second task is to find a strategy to select relevant documents that maximizes the probability to select the k best documents given the estimated ranking distribution.

3.1 Estimation of Ranking Distributions

The exact distribution of ranking values is only known at the end of the query period if all ranking values of documents were recorded during the query execution time. Since results for a BCS query (or respective alerts) should be presented to a user immediately, the distribution of ranking values has to be estimated at a previous point in time.

One basic question is the knowledge (i.e. the set of documents) the distribution estimation should be based on. There are two basic possibilities for the selection of this training set. First, the estimation of the ranking distribution can be based on historic ranking scores with respect to the considered query that were recorded previously (offline). The problem with this approach is that the ranking distribution may change over time. For example the distribution of ranking values of the query 'World Cup 2006 Germany' may change significantly over the course of the year 2006, considering in particular ranking distributions before and after the event.

Therefore, a second possibility is the estimation of the ranking distribution based on (a certain number of) the ranking values that are returned at the (beginning of the) query execution time. This estimation is based on more current training data and therefore supposed to be more appropriate for ranking distributions that change over time. In the estimation method applied in this paper the documents used in the training set are not sent to the user. Therefore, if training data are extracted during the actual query execution time a user may not be informed about high quality documents that appear during the distribution estimation phase. However the more documents are used as training data, the better the distribution estimation is expected to represent the actual distribution of ranking values. An optimal size of the training set with respect to the retrieval quality is found empirically in this paper.

The problem to estimate distributions of ranking values can be simplified by assuming standard distributions. If e.g. a ranking distribution has to be estimated that is similar to the Gaussian distribution, only two parameters have to be estimated, the mean and the standard deviation. The quality of the estimation depends on how the assumed distribution conforms to the real ranking distribution of the query result set.
In section 4 we motivate the use of specific standard distributions by considering distributions of ranking values empirically.

3.2 A Selection Method Based on Ranking Distributions

In the presented method the task is to find a score threshold S that divides the retrieved N documents into two sets, k accepted best documents and N-k rejected, less relevant documents. This threshold can be computed if the distribution of ranking values is known as described below.

In the following we assume that the ranking distribution is available, as e.g. obtained by the procedure in section 3.1. The task is to compute a ranking threshold S with the goal "k score values of the given N are better than S". Thereby the best k candidates may be extracted simply by applying a strategy that selects every document with a ranking value above the threshold S. In order to determine the threshold S in this paper we consider the expectation value of the number of score values larger than S and choose it to be k in the following. If the score values of subsequent documents are as denoted as

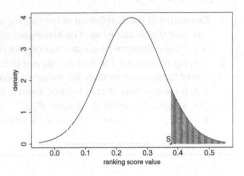

Fig. 1. Estimation of the optimized threshold S for a given gaussian distribution

$x_1, x_2, ..., x_N$, the threshold S should fulfill the equation $E(|\{i \text{ with } x_i > S\}|) = k$. Based on this equation we can reformulate:

$$E(|\{i \text{ with } x_i > S\}|) = k \Leftrightarrow E(\sum_{i=1}^{N} 1_{\{x_i > S\}}) = k \Leftrightarrow \sum_{i=1}^{N} E(1_{\{x_i > S\}}) = k \quad (2)$$

$$\overset{equal\ dist.}{\Longleftrightarrow} N * E(1_{\{x > S\}}) = k \Leftrightarrow N * P(x > S) = k \Leftrightarrow P(x > S) = \frac{k}{N} \quad (3)$$

Here x is an arbitrary ranking score at an arbitrary point of time, where the point of appearance of each ranking score has equal likelihood. The characteristic function $1_C(x)$ based on a condition 'C' is defined to be 1 if the condition is true and 0 otherwise. If S is chosen according to equation 2 the probability of any value being higher than S is k/N. Based on this result and the distribution function of ranking values, the threshold S can directly be determined. Denoting the distribution function of ranking values as $F(x)$ with the density $f(x)$ it follows:

$$P(x > S) = \int_{S}^{x_{max}} f(x)dx = F(x_{max}) - F(S) = \frac{k}{N} \quad (4)$$

This means that for the density function f the area of the graph between the threshold S and the maximum score value has to be equal to $\frac{k}{N}$. In order to find S we have to solve the equation with respect to S, which may be realized numerically in logarithmic complexity.

Figure 1 shows an example where f is a Gaussian density function. In this example the goal is to find the best $k = 10$ out of $N = 100$ documents. Therefore we need to solve the equation $\int_{S}^{x_{max}} f(x)dx = \frac{1}{10} = 0.1$. The derived value for S is approximately $S = 0.3782$. The expectation value for the number of documents with a score value better than S is then given by the requested number of 10 documents. Therefore S may be applied as a selection threshold in order to retrieve the best 10 documents.

Distribution-optimized method for answering BCSQ (Input: candidates N, 'number of best choices' k, Query Q)

1. Estimation of the distribution of the ranking score values
   a) global estimation based on historical ranking distributions
   a') local estimation based on a training set obtained during the query period
2. Solving of equation 4 to find the optimal threshold S
3. Apply the decision strategy for the subsequent documents
   (a) if the user was already notified about k documents stop process
   (b) accept a document d_j, if score $R_{j,Q} > S$
   (c) accept a document d_j, if $k - t$ documents were accepted and only t candidates are left
   (d) reject a document d_j, if $R_{j,Q} \leq S$

Fig. 2. Optimized method for answering BCSQ by using ranking distributions

The presented BCS query processing method is summarized in figure 2. In the method first the distribution of ranking values is estimated in step 1, based on the query profile Q and one of the two choices for the training data as described in section 3.1. In step 2 the threshold value S is estimated according to equation 4 based on the distribution function of ranking values and the query specification by a user, in particular k and N. Third, in step 3 a document in the stream is selected if the ranking value is higher than S until k documents have been selected. The selection procedure in step 3 guarantees that in any case k documents are selected.

4 Experiments

In this section we apply the new selection method to real information on the Web. The respective Web archive used for testing was recorded over a period of 303 days and consists of 26 Web sites which were reloaded twice a day. [1] The new selection method is compared to the optimal stopping method (k-SSP) based on relative ranking values described in section 2.3.

For the new method the first task is to extract classes of distributions of ranking values for different BCS queries and information sources in section 4.1. Second, based on these distribution classes we apply the new selection strategy and compare it to the reference strategy in section 4.2.

4.1 Empirical Estimation of Ranking Distributions

The estimation of ranking distributions may be simplified, if ranking distributions approximately conform to standard distributions with only a few parameters. In this section we extract different classes of ranking distributions by an empirical evaluation of BCS queries with respect to our Web archive. In particular we motivate the use of Gaussian

---

[1] The list of evaluated Web sites is available at http://www.ifis.uni-luebeck.de/projects/smartCQ/WISE07.html

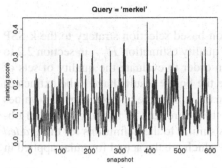

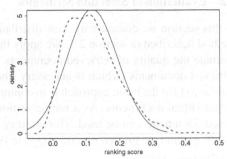

(a) Document score chart (estimated relevance) for the query term 'merkel' over a period of 303 days (2006).

(b) Approximation of the ranking distribution of query 'merkel' by a Gaussian distribution.

Fig. 3. An example for the estimation of the distribution of ranking values

distributions for a large variety of BCS queries. The applied query profiles consist of (single) query terms that are selected randomly from the entire set of terms in the archive. Only query terms with a minimal 'support' (number of appearances) are selected.

An example for the estimated relevance (section 2.2) with respect to the query term 'merkel', the name of the German chancellor Angela Merkel, and with respect to the entire archive over a period of 303 days (year 2006) is shown in figure 3a.

Figure 3b shows the fitted distribution of ranking values (dashed line). The drawn-through line in figure 3b shows a fit assuming a Gaussian distribution. This distribution of relevance values is an example for a significant fraction of BCS queries where the ranking distribution may be modeled by Gaussian distributions. We evaluate the quality of this fit implicitly by applying the selection strategy in section 3.2 based on this distribution fit and by measuring the quality of the retrieved results in the subsequent section.

In the empirical evaluation of ranking distributions we observed that not all distributions appear to conform to Gaussian distributions. A second common class of distributions may be denoted as *bursts*. The respective ranking scores are non-zero at only a few points in time and non-zero scores usually appear at nearby points in time. Examples for such queries are e.g. the query terms 'gun rampage' or 'earthquake'. The media cover the respective (rare) events usually for a few days. Therefore the score distribution is characterized by an extreme maximum at zero and only a few score values at non zero values.

Fortunately, the new selection strategy in section 3.2 may also be applied to burst-like ranking distributions. Since the ranking distribution has a mean and deviation of approximately zero, the presented selection strategy will usually estimate a threshold near zero, $S \approx 0$. Applying this threshold, every zero scored document is rejected and every positively scored document is accepted. Since ranking score bursts are rare and happen infrequently (not many more than the requested number of k documents are non-zero) the new selection method will answer those queries fairly.

4.2   Evaluation of Selection Strategies

In this section we compare the new distribution-based selection strategy to the k-SSP method described in section 2.3. We apply the quality estimation $R'_{j,Q}$ in section 2.2 to evaluate the quality of retrieved documents. In order to evaluate the quality of sets of retrieved documents (which is necessary since the best k results have to be extracted) similar to [10] the basic approach is to compute the sum of the respective quality values of individual documents. As a normalization the entire sum of ranking values of obtained documents can be used. This strategy is similar to the definitions for the *graded recall (gr)* and the *graded precision (gp)* in [9] which are the actual formulas applied in the evaluation process:

$$
\begin{aligned}
gr : &= \sum_{x \in \text{retr}} \text{relevance}(x) / \sum_{x \in D} \text{relevance}(x) \\
gp : &= \sum_{x \in \text{retr}} \text{relevance}(x) / |retr| ,
\end{aligned}
\tag{5}
$$

where D denotes the entire set of documents and $relevance{:=}R'_{j,Q}$ is the function providing relevance values (ranking scores) for documents, $retr$ is the set of retrieved documents. This evaluation has to be computed at the end of the query period, where the entire set of documents is known to the evaluation algorithm.

In the experiments we consider different subsets (baskets) of the entire set of Web sites in the archive. In particular we consider single Web pages, as the pages 'www.welt.de' and 'www.spiegel.de', a basket containing the 8 English pages in the archive, a basket of the 16 German pages and finally the entire archive containing 24 pages.

In each experiment we evaluate and compute the average over 500 BCS queries containing a single query term. The respective query terms are generated by considering the entire set of terms in the archive and by selecting random query terms with a minimal number of occurences in the archive. In the experiments the number of requested 'best' documents k is varied ($k \in \{5, 10, 20, 40\}$). The applied selection strategies are the k-choice SSP (k-SSP) in section 2.3 and the new distribution-based selection strategy (section 3). In the new selection strategy we consider different kinds of training data (section 3.1). First the training data for the distribution estimation is taken from the actual sequence of documents that appear during query execution time. In different experiments the first $10, 20, \ldots 100$ documents are used for the distribution estimation. Further, we consider the offline case in section 3.1 where the training data is taken from a different set of documents. This case is simulated in the experiments by considering 300 BCS queries similar to those considered in the evaluation process, however with distinct query terms.

Results. Figure 4a shows the evaluation results for the page www.welt.de. The x-axis denotes the different applied selection strategies. First, the new selection strategy is considered with a different number of training data of $10, 20, \ldots, 100$. K-SSP denotes the k-Secretary strategy, 'fix-Dis' denotes the new strategy where training data are taken

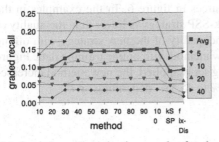

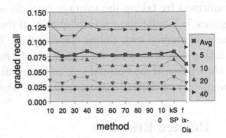

(a) Graded recall evaluation results for the website www.welt.de.

(b) Graded Recall Evaluation results for the website www.spiegel.de.

Fig. 4. Evaluation results for two single Web pages

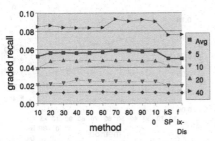

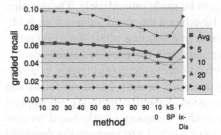

(a) Graded recall evaluation results for the German web archive.

(b) Graded recall evaluation results for the English web archive.

Fig. 5. Evaluation results for two baskets of English and German pages

from separate sequences of documents. The y-axis denotes the obtained retrieval quality (gr in (5)).[2] In figure 4 different evaluation results are connected by lines in order to distinguish different choices for $k \in \{5, 10, 20, 40\}$, the number of best items to be returned.

Figure 4a shows that for most values of k the optimal size of the training set is about 40 candidates. The retrieval quality stays approximately constant for larger training sets. The retrieval quality of the k-SSP method is significantly lower for all values of k and similar to the retrieval quality of the new method ('fix-Dis' in figure 4a) with offline training data. The results for a different data source (www.spiegel.de) are shown in figure 4b. In this example the retrieval quality of the new method is maximal for a training set of 10 documents. The retrieval quality is slightly lower for the k-SSP strategy and is also lower for the offline training set. In the next experiments we computed the average over all 16 German and the 8 English data sources in figures 5a,b. For the German test pages, the optimal size of the training set depends on k and maximizes quality e.g. at a training size of 40 documents if $k = 10$. For the English pages in figure 5b the optimal size of the training set is 10 for all values of k. This result is also

---

[2] In the experiments we do not consider the graded precision (gp) in (5), which is linear dependant of gr if $k = |retr|$ is constant.

confirmed by taking the average over all sources in figure 6. In the examples in the figures 5a,b and 6 the retrieval quality of the k-SSP strategy is lower than the quality of the new method with online training data. The quality of the new strategy with offline training data is between the k-SSP and the new method with online training data. Figure 6 shows that, if the optimal size of the training set is applied, the new method increases the quality of retrieved documents from 44.4% to 45.1%, depending on k.

5 Related Research

The problem to extract relevant documents from a stream of documents is a research topic in the field of information retrieval, in particular information filtering [18], [3], [2], [19], [12] and the field of topic detection and tracking [1], [20]. In the filtering track of the TREC conference the binary decision if a document is relevant or not has to be made immediately. The decision is based on a threshold value which is obtained in a learning phase based on training data and the provided query profile with respect to the maximization of a utility measure, usually based on recall and precision. Adaptive methods are applied to modify the query profile based on user feedback in order to improve the threshold value. In topic tracking usually a significantly smaller number of training data is considered.

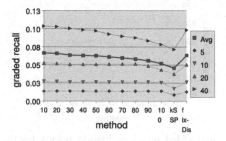

Fig. 6. Graded recall evaluation results for the entire web archive

A disadvantage of these methods is that the information load of a user may not be controlled without further user-interaction (feedback). In many cases too much information is provided by a query engine that may hardly be processed by a user. In contrast to the above strategies in this paper we focus on bounded search queries where the amount of returned information may be controlled by a user. We present a new approach based on learning of ranking distributions, following ideas from the field of mathematical statistics [5], [17] and compare the new approach to previous optimal stopping methods based only on relative ranking values [8], [6], [14], [16].

6 Conclusion

In this paper we present a new method to process bounded continuous search (BCS) queries. In contrast to previous methods in the field of information filtering and topic tracking, BCS queries provide a means to restrict the amount of information returned to a user. Further user interaction in order to adjust threshold values for the selection of relevant documents is not required in contrast to the field of adaptive filtering.

The presented processing method for BCS queries includes an estimation step where the distribution of ranking values is estimated. Based on this distribution we propose a selection strategy in order to provide high quality documents to a user immediately. We evaluate the presented method by comparing it to previous optimal stopping methods that are based on relative ranking values. The new method provides results of significantly higher quality, if information sources of a larger size (i.e. number of Web pages) are considered. In this case usually many high quality documents are available, where the ranking distribution may be modeled by Gaussian distributions.

References

1. Allan, J., Carbonell, J., Doddington, G., Yamron, J., Yang, Y.: Topic detection and tracking pilot study: Final report. In: Proc. of the DARPA Broadcast News Transcription and Understanding Workshop, pp. 194–218 (1998)
2. Arampatzis, A., van Hameran, A.: The score-distributional threshold optimization for adaptive binary classification tasks. In: SIGIR 2001. Proc. of ACM SIGIR conf. on Research and development in IR, pp. 285–293. ACM Press, New York, NY, USA (2001)
3. Collins-Thompson, K., Ogilvie, P., Zhang, Y., Callan, J.: Information filtering, novelty detection, and named-page finding. In: TREC 2002, Gaithersburg (2002)
4. DeGroot, M.H: Optimal Statistical Decisions. Wiley Classics Library (2004)
5. Gilbert, J.P., Mosteller, F.: Recognizing the maximum of a sequence. Journal of the American Statistical Association 61(313), 35–73 (1966)
6. Glasser, K.S., Holzsager, R., Barron, A.: The d choice secretary problem. Comm. Statist. -Sequential Anal. 2(3), 177–199 (1983)
7. Google alert (2006), http://www.googlealert.com
8. Kadison, R.V.: Strategies in the secretary problem. Expo. Math. 12(2), 125–144 (1994)
9. Kekalainen, J., Jarvelin, K.: Using graded relevance assessments in IR evaluation. J. of the American Society for Information Science and Technology 53(13) (2002)
10. Kukulenz, D., Ntoulas, A.: Answering bounded continuous search queries in the world wide web. In: Proc. of WWW-07, World Wide Web Conf., ACM Press, Banff, Canada (2007)
11. Liu, L., Pu, C., Tang, W.: Continual queries for internet scale event-driven information delivery. Knowledge and Data Engineering 11(4), 610–628 (1999)
12. Liu, R.-L., Lin, W.-J.: Adaptive sampling for thresholding in document filtering and classification. Inf. Process. Manage. 41(4), 745–758 (2005)
13. Windows live alerts (2006), http://alerts.live.com/Alerts/Default.aspx
14. Praeter, J.: On multiple choice secretary problems. Mathematics of Operations Research 19(3), 597–602 (1994)
15. Salton, G., Buckle, C.: Term-weighting approaches in automatic text retrieval. Information Processing Management 24(5), 513–523 (1988)
16. Shiryaev, A., Peskir, G.: Optimal Stopping and Free-Boundary Problems (Lectures in Mathematics. ETH Zürich). Birkhauser (2006)
17. Stewart, T.J.: Optimal selection from a random sequence with learning of the underlying distribution. Journal of the American Statistical Association 73(364) (1978)
18. Text retrieval conf. (TREC) (2006), http://trec.nist.gov/
19. Yang, Y.: A study on thresholding strategies for text categorization. In: Proc. of SIGIR-2001, Int. Conf. on Research and Development in IR, New Orleans, US, pp. 137–145. ACM Press, New York (2001)
20. Yang, Y., Pierce, T., Carbonell, J.: A study of retrospective and on-line event detection. In: SIGIR 1998. Proc. of the ACM SIGIR conf. on Research and development in IR, pp. 28–36. ACM Press, New York, NY, USA (1998)

Evaluating Rater Credibility for Reputation Assessment of Web Services

Zaki Malik[1] and Athman Bouguettaya[1,2]

[1] Department of Computer Science, Virginia Tech, VA. 24061 USA
[2] CSIRO-ICT Center, Canberra, Australia
{zaki,athman}@vt.edu
http://www.eceg.cs.vt.edu/

Abstract. This paper investigates the problem of establishing trust in service-oriented environments. We focus on providing an infrastructure for evaluating the credibility of raters in a reputation-based framework that would enable trust-based Web services interactions. The techniques we develop would aid a service consumer in assigning an appropriate weight to the testimonies of different raters regarding a prospective service provider. The experimental analysis show that the proposed techniques successfully dilute the effects of malicious ratings.

Keywords: Reputation, Credibility, Quality.

1 Introduction

With the introduction of *Web services*, the Web has started a steady evolution to become an environment where applications can be automatically invoked by other Web users or applications. A Web service is a set of related functionalities that can be programmatically accessed and manipulated over the Web [5]. More precisely, a Web service is a self-describing software application that can be advertised, located, and used across the Web using a set of standards such as WSDL, UDDI, and SOAP [15] [16]. The service-oriented Web represents an attractive paradigm for tomorrow's interactions spanning a wide range of domains from e-economy to e-science and e-government.

The ultimate goal of Web services is to serve as independent components in an organization of services that is formed as a result of consumer demand and which may dissolve post-demand completion. As services proliferate on the Web, several services may provide similar functionalities. Thus, it will be necessary to select a Web service that provides the "best" service compared to other candidates. However, the selection of the best service is not based only on what (or how) the service offers. Web services may make promises about the provided service and its associated quality but may fail partially or fully to deliver on these promises. Thus, the major challenge lies in providing a *trust* framework for enabling the *selection* of Web services based on trust parameters. The fundamental rationale behind the need for trust is the necessity to interact with unknown entities [2] [10]. The lack of a *global* monitoring system for the service-oriented environment exacerbates the problem of trust.

B. Benatallah et al. (Eds.): WISE 2007, LNCS 4831, pp. 38–49, 2007.

Research results show that reliable *reputation* systems increase users' trust in the Web [8] [18] [8]. Many online businesses have recognized the importance of reputation systems in improving customers' trust and consequently stimulating sales [18]. Examples include *Ebay* and *Amazon* [1]. In fact, managing reputation has itself become a Web business. Examples of online companies whose prime vocation is reputation collection and dissemination include *Bizrate*, *Epinions*, *Citysearch* [8]. Reputation systems are particularly vital in online marketplaces, also known as C2C e-auctions. Several studies have investigated and generally confirmed that in these types of business environments, reputation systems benefit both sellers and buyers [11] [12]. We anticipate that the deployment of reputation systems on the service Web (where services would interact with each other directly much like P2P systems) [16], will have a significant impact on the growth of the different emerging applications.

A number of research works have recognized the importance of reputation management in online systems and a number of solutions have been proposed (as [9] [10] [2] [23] [3] etc). The proposed solutions use different techniques, and study different aspects of online systems to establish trust among participants. For instance, [21] [22] have focused on protecting reputation systems from *unfair* or malicious raters' input (aka. votes or ratings). In this paper, we also focus on the problem of protecting the reputation system from malicious raters. Unlike previous solutions that may require special statistical distributions, may use *screening* to drop some ratings, or involve extensive probabilistic computations [21] [22], we provide easy-to-use techniques for a service-oriented environment that consider all the submitted ratings, can be implemented in a straight forward manner, and which do not involve extensive statistical computations. The proposed techniques are based on the premise that honest and *credible* raters' testimonies should be given more weight as compared to unknown or dishonest raters. The experimental evidence suggests that the proposed techniques are fairly successful in diluting the effects of malicious ratings, when various rating feedbacks are aggregated to derive a service provider's reputation. We provide the details of our proposed techniques in the next section. In Section 3, we present an experimental analysis of the proposed work. This is followed by Section 4, which concludes the paper with directions for future work.

2   Credibility of Raters

We assume a reputation model that is distributed in nature. In contrast to third-party-based traditional approaches for reputation management, no single entity is responsible for collecting, updating, and disseminating the reputation of Web services. Each service consumer records its own perceptions of the reputation of only the services it actually invokes. This perception is called *personal evaluation*. For each service s_j that it has invoked, a service consumer t_i maintains a p-element vector $PerEval^{ij}$ representing t_i's perception of s_j's behavior. Thus, $PerEval^{ij}$ reflects the performance of the provider in consumer's views. However, other service consumers may differ or concur with t_i's observation of s_j. A service consumer that inquires about the reputation of a given service provider from its peers may get various differing personal evaluation "feedbacks." To get a correct assessment of the service provider's behavior, all the personal evaluations for s_j need to be aggregated. Formally, the reputation of s_j, as viewed by a consumer i is defined as:

$$Reputation(s_j, i) = \bigwedge_{x \in L} (PerEval_k^{xj})$$ (1)

where L denotes the set of service consumers which have interacted with s_j in the past and are willing to share their personal evaluations of s_j, and $\bigwedge$ represents the aggregation function. It can be as simple as representing the union of personal evaluations where the output is a real number, or an elaborate process that considers a number of factors to assess a fairly accurate reputation value. Equation 1 provides a first approximation of how the assessed reputation may be calculated. However, the assessed reputation calculation involves various factors that need to be precisely defined and measured. As mentioned earlier, we focus only on one such factor in this paper, i.e., evaluating the credibility of raters, to assign proper weights to their testimonies for reputation aggregation.

Credibility Evaluation Requirement

A service provider that provides satisfactory service (in accordance with its promised quality agreement [10]), may get incorrect or false ratings from different evaluators due to several malicious motives. In order to cater for such "bad-mouthing" or collusion possibilities, a reputation management system should weigh the ratings of highly credible raters more than consumers with low credibilities [9] [6] [17] [23]. Thus, Equation 1 can be re-written as a weighted mean where the rater's credibility defines the weight of its testimony:

$$Reputation(s_j, i) = \frac{\sum_{t_x=1}^{L} (PerEval_k^{xj} * C_r(t_x, i))}{\sum_{t_x=1}^{L} C_r(t_x, i)}$$ (2)

where $Reputation(s_j, i)$ is the assessed reputation of s_j as calculated by the service consumer i and $C_r(t_x, i)$ is the credibility of the service rater t_x as viewed by the service consumer i. The credibility of a service rater lies in the interval [0,1] with 0 identifying a dishonest rater and 1 an honest one.

There are a few existing online systems such as eBay, Amazon, Yahoo! Auctions, Auction Universe, Edeal, etc. that use a centralized reputation system. eBay uses a three point scale with with +1 for a positive rating, 0 for neutral and -1 for a negative rating. A user's reputation is a mere aggregation of all the feedbacks received. Thus, a user with 50 positive feedback ratings will have a reputation value equaling one with 300 positive and 250 negative feedback ratings [14]. This clearly does not provide an ideal solution. Some other online businesses use an average over all ratings to compute the reputation of a user. For instance, Amazon's auction site uses this method. It allows transaction participants to rate on a scale from 1 to 5. Then an average of all feedback ratings to date is calculated to compute an overall reputation score. Thus, a user with ratings of 1, 1, 9, 1, 1, 1, 9, 9, and 1 would have an overall reputation score of 3.7. Clearly, this score is also not in accordance with the ratings received. Thus, designing a ratings system that is robust enough to detect and mitigate the effects of disparate ratings is a fundamental issue [7] [22].

Evaluating Rater Credibility

To overcome the above mentioned problems, several methods have been proposed in literature that screen the ratings based on their deviations from the majority opinion. Examples include the Beta Deviation Feedback [3], Beta Filtering Feedback [22], Likemindedness [20], and Entropy-Based Screening [21]. We adopt a similar notion to dilute the effects of *unfair* or inconsistent ratings. We use a majority rating scheme, in which the "uniformity of ratings" indicates their accuracy. The basic idea of the proposed method is that: if the reported rating agrees with the majority opinion, the rater's credibility is increased and decreased otherwise. Unlike previous models, we do not simply disregard/discard the rating if it disagrees with the majority opinion but consider the fact that the rating's inconsistency may be the result of an actual experience. Hence, only the credibility of the rater is changed, but the rating is still considered.

We use a data clustering technique to define the majority opinion by grouping similar feedback ratings together [7] [19]. We use the k-mean clustering algorithm [13] on all current reported ratings to create the clusters. The most densely populated cluster is then labelled as the "majority cluster" and the centroid of the majority cluster is taken as the *majority rating* (denoted M): $M = centroid(max(\Re_k)) \ \forall k$, where k is the total number of clusters, $max(x)$ gives the cluster $\Re$ with the largest membership and $centroid(x)$ gives the centroid of the cluster x.

The Euclidean distance between the majority rating (M) and the reported rating (V) is computed to adjust the rater credibility. The change in credibility due to majority rating, denoted by Mf is defined as:

$$Mf = \begin{cases} 1 - \frac{\sqrt{\sum_{k=1}^{n}(M_k - V_k)^2}}{\sigma} & \text{if } \sqrt{\sum_{k=1}^{n}(M_k - V_k)^2} < \sigma \\ 1 - \frac{\sigma}{\sqrt{\sum_{k=1}^{n}(M_k - V_k)^2}} & \text{otherwise} \end{cases} \quad (3)$$

where σ is the standard deviation in all the reported ratings. Note that Mf does not denote the rater's credibility (or the weight), but only defines the effect on credibility due to agreement/disagreement with the majority rating. How this effect is applied will be discussed shortly. There may be cases in which the majority of raters collude to provide an incorrect rating for the provider Web service. Moreover, the outlier raters (ones not belonging to the majority cluster) may be the ones who are first to experience the deviant behavior of the providers. Thus, a majority rating scheme 'alone' is not sufficient to accurately measure the reputation of a Web service.

We supplement the majority rating scheme by adjusting the credibility of a service rater based on its past behavior as well. The historical information provides an estimate of the *trustworthiness* of the service raters [22]. The trustworthiness of the service is computed by looking at the 'last assessed reputation value', the present majority rating and that service consumer's provided rating. It is known that precisely defining what constitutes a *credible* rating is an interesting and hard research problem by itself [23]. However, we have attempted to define the credibility of Web services in a practical manner according to the information available to the service consumer. We define a credible rater as one which has performed consistently, accurately, and has proven to be useful (in terms of ratings provided) over a period of time.

Consistency is the defined behavior of a service that exhibits similar results under standard conditions. We believe that under controlled situations (i.e., other variables being the same), a service consumer's perception of a Web service should not deviate much, but stay consistent over time. We assume the interactions take place at time t and the service consumer already has record of the previously assessed reputations (denoted A), which is defined as:

$$A = \bigsqcup_{t-1}^{t-k} Reputation(s_j, i)^t \qquad (4)$$

where $Reputation(s_j, i)$ is as defined in Equation 1 for each time instance t, $\bigsqcup$ is the aggregation operator and k is the time duration defined by each service consumer. It can vary from one time instance to the complete past reputation record of s_j. Note that A is not the "personal evaluation" of either the service rater or the service consumer but is the "assessed reputation" calculated by the service consumer at the previous time instance(s). If the provider behavior does not change much from the previous time instance, then A and the present rating V should be somewhat similar. Thus, the effect on credibility due to agreement/disagreement with the last assessed reputation value (denoted Af) is defined in a similar manner as Equation 3 (by replacing M_k with A_k).

In real-time situations it is difficult to determine the different factors that cause a change in the state of a Web service. A rater may rate the same service differently without any malicious motive, i.e., accurately (but not consistent with the last reporting). Thus, the credibility of a rater may change in a number of ways, depending on the values of V, Mf, and Af. The equivalence of the majority rating Mf, submitted personal evaluation rating V and the assessed reputation at the previous time instance Af is used in adjusting the service rater's credibility C_r. The general formula is:

$$C_r(t_x, i) = C_r(t_x, i) \pm \aleph * (|Mf \pm Af|) \qquad (5)$$

where $\aleph$ is the credibility adjustment normalizing factor, Mf, and Af are as defined above. The signs $\pm$ indicate that either $+$ or $-$ can be used, i.e., the increment or decrement in the credibility depends on the situation. These situations are described in detail in the upcoming discussion. Note that Equation 5 is stated such that a service consumer that wishes to omit the Af effect, and only use the majority rating as basis of credibility calculation, can do so with ease (by setting $Af = 0$). Also, the definition of Af can be extended to interpolated or forecasted reputation values instead of past assessed reputation value.

As mentioned earlier, the equivalence or difference of V with M and A directly affects the credibility of a service rater. Similar to previous works [3] [22] [20] [21], we place more emphasis on the ratings received in the current time instance than the past ones. Thus, equivalence or difference of V with M takes a precedence over that of V with A. This can be seen from Equation 5, where the $+$ sign with ℵ indicates $V \simeq M$ while $-$ sign with ℵ means that $V \neq M$. ℵ is defined as:

$$\aleph = \begin{cases} \frac{C_r(t_x, i)}{\alpha} & \text{if } V \simeq M \\ C_r(t_x, i) \times \beta & \text{otherwise} \end{cases} \tag{6}$$

where α is a natural number with $\alpha \geq 1$, and β is a rational number in the range $[0, 1]$. The exact values of α and β are left at the discretion of the service consumer. However, through experimental evidence we have found that $\alpha \geq max.rep.val.$, and $\beta < 0.5$ (where $max.rep.val$ is the maximum reputation value on the scale: 10 in our case) provide the most accurate results. α and β directly effect the value of ℵ and in turn influence the rater credibility. Note that in Equation 5, a higher value of α results in low increment of the rater credibility (as ℵ becomes low), a higher value of β results in a larger decrement of the rater credibility (as ℵ becomes large) and vice versa. In essence, α and β only influence the value of ℵ, but the similarity or difference of V, M, and A determines whether rater credibility is incremented or decremented.

Adjusting Rater Credibilities

V, M, and A can be related to each other in *one of four* ways. In the following, we provide an explanation of each and show how credibilities are updated in our proposed model.

1. The local reported reputation value is similar to both the majority rating and the previously assessed reputation, i.e., $(V \simeq M \simeq A)$. The equality $M \simeq A$ suggests that the majority of raters believe the quality of s_j has not changed. The service rater's credibility is updated as:

$$C_r(t_k, i) = C_r(t_k, i) + \aleph * (|Mf + Af|) \tag{7}$$

   Equation 7 states that since all factors are equal, the credibility is incremented.

2. The individual reported reputation rating is similar to the majority rating but differs from the previously assessed reputation, i.e., $(V \simeq M)$ and $(V \neq A)$. In this case, the change in the reputation rating could be due to either of the following. First, the rater may be colluding with other service consumers (raters) to increase/decrease the reputation of s_j. Second, the quality of s_j may have actually changed since A was last calculated. The service rater's credibility is updated as:

$$C_r(t_k, i) = C_r(t_k, i) + \aleph * (|Mf - Af|) \tag{8}$$

Equation 8 states that since $V \simeq M$, the credibility is incremented, but the factor $V \neq A$ limits the incremental value to $|Mf - Af|$ (not as big as the previous case).

3. The individual reported reputation value is similar to the previously assessed reputation but differs from the majority rating, i.e., $(V \neq M)$ and $(V \simeq A)$. The individual reported reputation value may differ due to either of the following. First, V may be providing a rating score that is out-dated. In other words, V may not have the latest score. Second, V may be providing a "false" negative/positive rating for s_j. The third possibility is that V has the correct rating, while other consumers contributing to M may be colluding to increase/decrease s_j's reputation. Neither of these three options should be overlooked. Thus, the service rater's credibility is updated as:

$$C_r(t_k, i) = C_r(t_k, i) - \aleph * (|Mf - Af|) \tag{9}$$

Equation 9 states that since $V \neq M$, the credibility is decremented. And to cater for the above mentioned possibilities brought in due to the factor $V \simeq A$, the value that is subtracted from the previous credibility is adjusted to $|Mf - Af|$.

4. The individual reported reputation value is not similar to either the majority rating or the calculated reputation, i.e., $(V \neq M)$ and $(V \neq A)$. V may differ from the majority rating and the past calculated reputation due to either of the following. First, V may be the first one to experience s_j's new behavior. Second, V may not know the actual quality values. Third, V may be lying to increase/decrease s_j's reputation. The service rater's credibility is updated as:

$$C_r(t_k, i) = C_r(t_k, i) - \aleph * (|Mf + Af|) \tag{10}$$

Equation 10 states that the inequality of all factors means that rater's credibility is decremented, where the decremented value is the combination of both the effects (Mf and Af).

Even with the above mentioned techniques in place, every ratings submission that a service consumer receives from service raters may not prove useful. In other words, the consumer's own experience (denoted OE) with the provider may differ from the rater's feedback (denoted RF). We propose that after each interaction, apart from rating the provider s_j, the service consumer also evaluate the usefulness of the raters that provided a rating for s_j. If the Euclidean distance between OE and RF falls below a predefined threshold, RF is deemed useful, otherwise it is not. We assume that raters and consumers have similar attribute preferences, and thus no philosophical differences exist between them while rating the same provider, i.e., under normal (and non-malicious) cases, both should have similar experiences.

The usefulness of a service is required to calculate a service rater's "propensity to default," i.e., the service rater's tendency to provide false/incorrect ratings. There may also be cases where raters alternate between being useful and not useful, over a period of time. Thus, to get a correct estimate of the rater's propensity to default, we compute the *ratio* of the total number of times the ratings submission was useful (k) over the total number of submissions (n). The usefulness factor (u_f) is:

$$u_f = \frac{\sum_{i=1}^{k} U_i}{\sum_{x=1}^{n} V_x} \qquad (11)$$

where U_i is the submission where the rater was termed 'useful' and V_x denotes the total number of ratings submissions by that service. The rater's credibility (calculated using either of Equations 7 through 10) is then adjusted as:

$$C_r(t_k, i) = C_r(t_k, i) * u_f \qquad (12)$$

We need to make a few observations for the above mentioned techniques. First, the consumer can base his decision only on the information he has (from the past), and the information he gathers (in form of feedback ratings). Second, the credibility of the raters is directly influenced by the number of ratings that are similar to each other, and previously assessed provider reputation. The proposed techniques emphasize the "majority rating" where the agreement with the majority results in a credibility (and hence weight) increment. We believe this is a valid assumption as malicious raters are likely to be scattered, and an attempt to gain a majority (through collusion) would prove too costly [19]. Third, even if the large majority in one round wrongfully alters a provider's reputation (and rater credibilities), the consequent rounds will detect malicious rating anomalies. If a large number of raters continue to act maliciously for extended periods of time, then such anomalies are hard to detect as a service consumer cannot decide on the actual honesty of the majority of raters. This is also in accordance with the real life social networks phenomenon [4]. However, the consumer's own personal experience aids in detecting such malicious raters. We believe that the strength of our proposed techniques lies in the ability of a service consumer to identify malicious raters (either in the present round or in consequent ones), and assess the provider reputation accordingly. This is substantiated through a series of experiments that follow.

3   Experimental Evaluations

We have conducted preliminary experiments to assess the robustness of the proposed techniques. Here we report on the service environment created with interactions spanning over 1000 time-instances for a set of hundred (100) Web services. The number of services and time iterations are not fixed, and can both be extended to produce similar results. The service providers are divided into five groups of twenty members each. The first group of providers behave consistently with high quality (different attributes are pre-defined). The next group performs with consistently low quality. The third group performs with high values for the first 500 time instances but then suffer a performance degradation. The fourth group acts in an opposite manner to the third group where providers perform with low values in the beginning. After the tenth time instance, they ameliorate their behavior and start performing with high quality. The final group of providers perform in a random manner, switching between high and low values. We have also divided the group of service raters to simulate actual rating behavior. The hundred service consumers have been divided into three groups. Each group represents members with high credibility in terms of rating providers, members with low rating credibility and members with varied (random) rating credibility respectively.

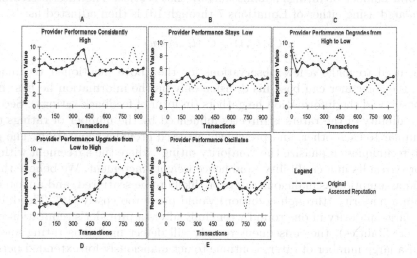

Fig. 1. Reputation Assessment when Low Credibility Raters Out-Number Others

In our experiments, service raters with varying credibilities can exist in the environment in one of four ways: honest raters (high credibility) can out number dishonest raters (low credibility), the number of honest and dishonest raters can center around a "mean credibility value," the number of honest and dishonest raters can be equal, or the dishonest raters can out number honest raters. We acknowledge the fact that rater credibilities can change over a period. Due to space restrictions, we show results from the *worst* case, i.e., where low credibility raters out number their honest peers. Figure 1 shows the reputation assessment procedure in the situation where consumer credibilities are low. Though comparable to the actual provider behavior, these ratings show slight variation. This is due to the *loaded* reporting of non-credible raters. Since consumer reporting is not uniform and is dishonest, the assessed reputations show variance from the actual provider performance values. The majority ratings calculated in any instance are used directly to compute rater credibilities and provider reputations. In the following, we list the results of reputation evaluation using different β (for punishment) values (with only a constant α/reward value due to space restrictions) to compute rater credibility and consequently provider reputations.

Reputation Evaluation with Low β Values: The reputations calculated using low β values in Equation 6 are compared with the original provider performance and the results are shown in Figure 2. Figure 2-A shows the results for

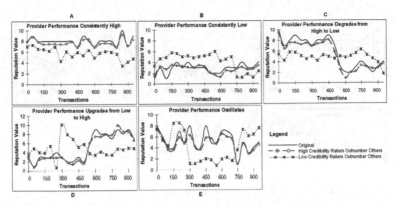

Fig. 2. Original Performance and Assessed Reputation Comparison Using Low β Value

a "high performing" provider, i.e., the provider performs consistently with high quality values. The assessed reputations are shown for two main scenarios. In the first scenario, the majority of raters have high credibility. In the second scenario, malicious raters out-number honest raters. Since low β values are chosen, rater credibility suffers low decrement in case of a dishonest rating report. The first scenario results in the calculated reputations being very close to the original provider performance (shown by a dashed line) since dishonesty is minimal. However, in the second scenario, the large number of malicious raters directly affects the majority rating and hence the final assessed reputation. Therefore, the assessed reputation (shown by a dotted line) is not close to the original performance. Similarly graphs in Figure 2-B through E show the comparison for other service provider groups.

Reputation Evaluation with High β Values: The comparison between original provider performance and assessed reputation using high β values in Equation 6 is shown in Figure 3. Figure 3-C shows the results for a provider that performs with high quality values in the beginning and then its performance drops. Similar to the the previous case, the assessed reputations are shown for two main scenarios. In the first scenario, the majority of raters have high credibility. In the second scenario, malicious raters out-number honest raters. In our proposed model, any deviation from either M or A negatively effects the rater's credibility. The results in the first scenario are not much different from the previous case and assessed reputations are very close to the original provider performance (shown by a dashed line). This is due to the manner in which credibilities are evaluated. However, the results in the second scenario using a high β value differs from the previous case, and the assessed reputations are relatively closer to the original provider performance. This is due to the 'punishing' behavior when rater's evaluation differs from the majority rating and the previous assessed reputation (Equation 7 through Equation 10). Although the assessed reputation (shown by a dotted line) is not so close to the original performance. But the calculated values mostly lie within an acceptable two-threshold. Similarly graphs in Figure 3-A through E show the comparison for other service provider groups.

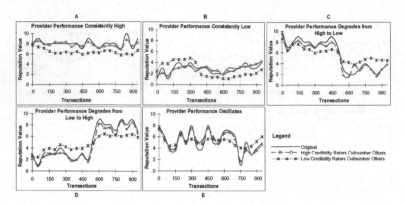

Fig. 3. Original Performance and Assessed Reputation Comparison Using High β Value

4   Conclusion and Future Work

We have presented a framework to evaluate the credibility of service raters for feedback based service-oriented reputation systems. However, the proposed approach could easily be extended to any kind of P2P interactions involving unknown peers. We conducted some experiments to verify the presented framework and some of the results are included in this paper. In the future, we aim to incorporate techniques to address *bootstrapping* issues and provide methods for automatic rating of services. Another important research direction that we intend to undertake is to define a reputation model for *composed* Web services. We anticipate that solving trust related issues this will provide arguments to introduce fundamental changes in today's approaches for Web services selection and composition.

Acknowledgement. This work is supported by the National Science Foundation grant number 0627469.

References

1. Aberer, K., Despotovic, Z.: On Reputation in Game Theory - Application to Online Settings. Working Paper (2004)
2. Bertino, E., Ferrari, E., Squicciarini, A.C.: Trust-X: A Peer-to-Peer Framework for Trust Establishment. IEEE Transactions on Knowledge and Data Engineering 16(7), 827–842 (2004)
3. Buchegger, S., Le Boudec, J.-Y.: A robust reputation system for p2p and mobile ad-hoc networks. In: Proceedings of the Second Workshop on the Economics of Peer-to-Peer Systems (2004)
4. Buskens, V.: Social Networks and the Effect of Reputation on Cooperation. In: Proc. of the 6th Intl. Conf. on Social Dilemmas (1998)
5. Casati, F., Shan, E., Dayal, U., Shan, M-C.: Business-Oriented Management of Web services. ACM Communications (October 2003)

6. Delgado, J., Ishii, N.: Memory-Based Weighted-Majority Prediction for Recommender Systems. In: ACM SIGIR 1999 Workshop on Recommender Systems: Algorithms and Evaluation (1999)
7. Dellarocas, C.: Immunizing online reputation reporting systems against unfair ratings and discriminatory behavior. In: 2nd ACM Conference on Electronic Commerce (2000)
8. Dellarocas, C.: The Digitalization of Word-of-Mouth: Promise and Challeges of Online Feedback Mechanisms. Management Science (October 2003)
9. Huynh, T.D., Jennings, N.R., Shadbolt, N.R.: Certified reputation: how an agent can trust a stranger. In: AAMAS 2006. Proceedings of the fifth international joint conference on Autonomous agents and multiagent systems, pp. 1217–1224. ACM Press, New York, NY, USA (2006)
10. Benatallah, B., Skogsrud, H., Casati, F.: Trust-Serv: A Lightweight Trust Negotiation Service. In: VLDB Demo, Toronto, Canada (August 2004)
11. Kesler, C.: Experimental Games for the Design of Reputation Management Systems. IBM Systems Journal 42(3) (2003)
12. Li, Z., Wang, S.: The foundation of e-commerce: Social reputation system–a comparison between american and china. In: ICEC 2005. Proceedings of the 7th international conference on Electronic commerce, pp. 230–232. ACM Press, New York, NY, USA (2005)
13. MacQueen, J.B.: Some methods for classification and analysis of multivariate observations. In: Proceedings of 5-th Berkeley Symposium on Mathematical Statistics and Probability, pp. 281–297. University of California Press, Berkeley (1967)
14. Malaga, R.: Web-Based Reputation Management Systems: Problems and Suggested Solutions. Electronic Commerce Research 1(1), 403–417 (2001)
15. Papazoglou, M.P., Georgakopoulos, D.: Serive-Oriented Computing. Communcications of the ACM 46(10), 25–65 (2003)
16. Papazoglou, M.P.: Web Services and Business Transactions. World Wide Web: Internet and Web Information Systems 6(1), 49–91 (2003)
17. Park, S., Liu, L., Pu, C., Srivatsa, M., Zhang, J.: Resilient trust management for web service integration. In: ICWS 2005. Proceedings of the IEEE International Conference on Web Services, pp. 499–506. IEEE Computer Society, Washington, DC, USA (2005)
18. Resnick, P., Zeckhauser, R., Friedman, E., Kuwabara, K.: Reputation Systems. Communication of the ACM 43(12) (December 2000)
19. Vu, L.-H., Hauswirth, M., Aberer, K.: QoS-based Service Selection and Ranking with Trust and Reputation Management. In: CoopIS 2005. 13th International Conference on Cooperative Information Systems (October 31 - November 4, 2005)
20. Walsh, K., Sirer, E.G.: Fighting peer-to-peer spam and decoys with object reputation. In: P2PECON 2005. Proceedings of the 2005 ACM SIGCOMM workshop on Economics of peer-to-peer systems, pp. 138–143. ACM Press, New York, NY, USA (2005)
21. Weng, J., Miao, C., Goh, A.: Protecting online rating systems from unfair ratings. In: Katsikas, S.K., Lopez, J., Pernul, G. (eds.) TrustBus 2005. LNCS, vol. 3592, pp. 50–59. Springer, Heidelberg (2005)
22. Whitby, A., Jøsang, A., Indulska, J.: Filtering Out Unfair Ratings in Bayesian Reputation Systems. The Icfain Journal of Management Research 4(2), 48–64 (2005)
23. Xiong, L., Liu, L.: PeerTrust: Supporting Reputation-based Trust for Peer-to-Peer Electronic Communities. IEEE Trans. on Knowledge and Data Engineering (TKDE) 16(7), 843–857 (2004)

An Approach to Trust Based on Social Networks

Vincenza Carchiolo, Alessandro Longheu, Michele Malgeri, Giuseppe Mangioni,
and Vincenzo Nicosia

Dipartimento di Ingegneria Informatica e delle Telecomunicazioni
Facoltà di Ingegneria - Università degli Studi di Catania
{vcarchiolo,alongheu,mmalgeri,gmangioni,vnicosia}@diit.unict.it

Abstract. In the era of Internet-based communities, trusting has become a crucial matter in communities creation and evolution. In this
paper we introduce an approach to the question of trusting inspired
by social, real world networks. We aim at reproducing the behaviour
individuals adopt in their life when they establish and maintain trust
relationships, gathering information and opinions from their friends to
assign trust to a new acquaintance they need to contact. We show that
preliminary results are promising for building an effective and efficient
trusting management in Internet communities.

1 Introduction

The notion of trust has been considered just a matter of sociology up to recent
years. Many persons use the concept of trust as roughly equivalent to solidarity,
meaning and participation: this leads to *unconditional trust*, as generated in
families and small-scale societies and cannot be easily transferred to complex
society; trust instead needs social relationship. Trust is closely related to "truth"
and "faithfulness", indeed in Webster dictionary[8] trust is defined as:

> *"assured resting of the mind on the integrity, veracity, justice, friend-
> ship, or other sound principle, of another person; confidence; reliance;"*

The question of trust is still important in Internet-based communities due
to the spread of networks and sharing of information over last decade; trust
evaluation should be provided both in traditional client-server applications as
well as in distributed environments, e.g. P2P applications.

Reproducing the real world behaviour individuals adopt in order to exploit
the properties of the corresponding network of people relationships seems to be
a promising idea for addressing trust management inside the Internet communities. Thus, here we consider how individual solves the problem in the real life,
emulating the behaviour of a single person that asks his acquaintances for their
judgement about the person to be trusted, in order to get a first estimation
for trust; in this sense, our proposal falls within the context of *reputation-based*
systems, building trust estimation from other people's opinions.

The model we present can be considered as an underlying layer useful for solving several collaborative tasks inside Internet communities, e.g. spam filtering,
P2P message routing and so on.

B. Benatallah et al. (Eds.): WISE 2007, LNCS 4831, pp. 50–61, 2007.

In section 2 we consider related work, section 3 addresses the question of trusting in real life, in section 4 we formalize real world trusting behaviour, also presenting first promising simulation results. Finally, in section 5 we present conclusions and future works.

2   Related Work

In the last decades, the recommender system is gaining an important role in today's networked worlds because they provide tool to support decision helping in selecting reliable from unreliable. Recommender systems help users to identify particular items that best match their tastes or preferences. Reliability is often expressed through a trust value with which each agent labels its neighbors. Thus a model of trust in the collaborative world must be provided and [9,6] explore the use of agent-based system managing trust, but they does not investigate in the topic of formation of trust based on real-world social studies.

Trust plays a central role not only in forming reputation but also supporting decision. Some recent works have suggested to combine distributed recommendation systems with trust and reputation mechanisms [9,7].

Since either building expertise or testing items available on the market are costly activities, individuals in the real world attempt to reduce such costs through the use of their social/professional networks. Social networks have received special attention and it has been shown that their structure plays an important role in decision making processes.

People surfing the Web has already faced the matter to form opinion and rate them against trust, as far as the well-known reviewers' community *Epinions (http://www.epinions.com)* which collects reviews from the community members, any members can also decide to "trust" or "block" (formerly known as "distrust") each other. All the trust and block relationships interact and form a hierarchy known as the Web of Trust (WOT). WOT is combined with rating to determine in what order opinions are shown to readers. However trusting model remains centralized thus not being democratic.

Ziegler and Golbeck [4,12] believe that computational trust model bear several favorable properties from social filtering than they expect notions of trust must reflect user similarity. Therefore a reputation system is an important tool in every network, but assume a central role in emerging P2P networks where many people interacts with many others in a sort of "democratic" fashion. Clearly a centralized trust system which stores the history of each peer and ranks the peer according with their past behaviours if not feasible. Some author discusses decentralized methods that calculates an approximate ranks according to local history and a notion of neighborhood [1] where trust is calculated trying advantage of small-world properties often emerging in networks that mimic real world. In P2P area EigenTrust [5] propose a distributed implementation of PageRank [10] that needs also a distributed structure to store data and imposes to pre-trust all nodes belonging to the net thus reducing the "de-centralisation".

3 Trusting in Real World

To build and analyze the behaviour of a trusting network, we first consider real world interaction using an informal approach, i.e. describing how a person interacts with an existing trust-based (i.e. friendship) community with whom he is initially unacquainted, subsequently formalizing our discussions.

3.1 Definition of Trust

We can say that Alice (A) trusts Bob (B) if A is persuaded that information coming from B are true, for example, if A needs B to help him in finding a job, and B says that he will help A, A is guaranteed that B says the truth, hence he will actually help A. This simple definition expresses the same concepts introduced in[2], where trust is intended as the choice A makes when he perceives an ambiguous path and B's actions determine how the result of following the path will be; a similar definition for trust is given in[11], where "trust is a bet about the future contingent actions of others"; In[3] authors state that "trust in a person is a commitment to an action based on a belief that the future actions of that person will lead to a good outcome".

How much A trusts B is simply quantified using a number in the range [-1,1], where -1 means A completely distrusts B, 1 indicates that A completely trusts B, and values included in this range represent different (and more realistic) values. The presence of both positive and negative values let us to model both individuals A trusts in, i.e. whose associated trust value belongs to $]0,1]$ range, as well as persons A distrusts, i.e. those with a strictly negative value. A value of 0 models the indifference, due to either a lack of information about B or when exactly opposite judgements about B are available, hence A is unable to decide whether trust or not in B.

3.2 Resumes

An additional factor to consider is that trust is strictly related to acquiring information, i.e. a trust relationship in a social network is established between A and B if A collects some information about B, either with a direct interaction with B or even getting such information from other sources, e.g. friends, acquaintances, web sites, e-mails, documents, papers, books, contact board, and so on. Indeed, if A does not know anything about the existence B, A also cannot assign any trust value to B. We name these information A owns about B as B's "resume" denoted as $C_B^A = \{B, \overline{C_B^A}\}$, where B is the identifier for the node B the resume is about, and $\overline{C_B^A}$ represent information about B as study degree, working experiences, personal skills and attitudes, and so on; at this stage, we do not impose any constraint on what information should be present or which structure the resume should have (it can be viewed a flat text file). Note that such information is about B, but different nodes actually can know different things about him, as in real world occur, thus we indicate the owner node A in the notation $\overline{C_B^A}$. Resumes will be also used when a person needs to trust an unknown person, as explained in detail in the following section. Note that in general for a given person more resumes are available, each actually owned by one of his acquaintances, differing because some are incomplete and/or out of date; the only updated and complete resume is the one provided by the person it concerns.

3.3   The Mechanism of Trusting

To model trusting mechanisms inside a community, we use an oriented labeled graph, where each node represent a person, and an oriented labeled arc from A to B denotes that A trusts B, being the arc label the pair (trust value, resume).

To show how links are established, let us consider Alice (A), an unacquainted person who wants to establish some contact with an existing community, usually to satisfy some need in her everyday life. For instance A may be a new student that needs to know where classroom are located, hence she asks for some help to other students she meets in a corridor, or she can search for a computer science professor, finding some name on the University website. In the first case, A directly contacts another individuals, say Bob (B) and Carl (C), thus having links form A to B and from A to C, whereas in the last case A collects information without a direct interaction with individuals, but she is still collecting resumes about some professors, say Danny (D) and Eve (E), so links from A to D and E are present too. Note that arcs are oriented since mutual trust levels are not necessarily the same, or one of them could also be not present. Referring to the example above, A could completely trust in B and C, since she is in a new, unknown context, but B and C might assign a low trust level to A, being an acquaintance they are meeting for the first time; moreover, A will assign a trust level to professor D and E based on their resumes stored in a web page, but D and E actually do not know anything about the existence of A, hence neither arc from D to A nor from E to A will be present. All these initial trust values are refined over time, as soon as other information (resumes) is available or new interactions (personal experiences) occur. The described situation is represented in figure 1, where thinner edges indicates other trust relationships owned by nodes, being A the new node in the network.

When a person X does not belong to the set of A's neighbors (i.e. nodes directly connected to A), A cannot exploits her personal interactions with X; in this case, trusting evaluation should be performed (as in real world occurs) by simply asking to X's neighbors, i.e. to those people who directly know X and can provide a judgement about him. These judgements should be collected and mediated by A, so she can refine her knowledge and assign a proper trust value about X exploiting other people's personal experience with X. Here we focus on this mechanism, since it better represents social network interactions, allowing to assign a more reliable trust value based on several, possibly different opinions rather than just the personal one. Since in the worst case A does not know anyone of X's neighbors, to actually implement this mechanism A asks *her* personal neighbors whether they know X, or whether neighbors of A's neighbors know X and so on, actually exploiting social (trusting) network relationships.

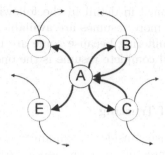

Fig. 1. Node A joins an existing trust network

Referring to the figure 1, A consider all her acquaintances (B,C,D and E) and send them a request (also referred as "query" in the following) for their opinion about X. The query is forwarded through the network until one of X's neighbors is (possibly) reached, then propagating back properly mediated opinions, until A gets answers and build her mediated opinion, as in real world occur. Note that here the terms neighbors and acquaintances here are equivalent. The steps the trusting mechanism consists of are:

1. A initially establish how many opinions (say, n_r) about X she needs to assign X a trust value, the more opinions will ask, the more important the query is, the more accurate the trust value to assign will be, the drawback will be the flooding of queries and related answers traversing the network. n_r is less or equal to the total number of A's acquaintances (but X). The n_r value can be less or equal to the total number of A's acquaintances, but in real world this value is generally significantly less than the total number; indeed we usually know many people, but rarely we will contact *all* of them each time we need an opinion about someone.

2. in real world interactions, A establishes an order according to which acquaintances are contacted. In particular, A first considers acquaintances she trusts more, at the same time having a resume similar to B's resume; to show why we consider resumes similarity in addition to high trust values, suppose for instance that from X's resume, A knows that X studies nuclear physics at the university, and A also suppose that trusts B and C the same, but B also is a nuclear physics student. To assign a trust value to X, in the real world A will first ask B, because A supposes that similarity in resumes increases the possibility that B directly knows X, whereas C will be contacted at a second stage, i.e. if B is not able to give any opinion about X; note that this last situation can still happen since it is possible that no path from B to X actually exists, hence acquaintances with different resumes and/or lower trust values should be exploited when no "qualified" acquaintances are available. To quantify these considerations, we first define the correspondence (i.e. similarity) between resumes associated to nodes X (to be trusted) and I (one of A's acquaintances) using classical information retrieval techniques aiming

at extracting inverse index vector for both resumes, then we define the *relevance* as the product between the similarity and trust value A assigned to I; the relevance measures how much A consider her acquaintance I *qualified* for asking him an opinion about X. To separate *qualified* from *unqualified* acquaintances, a threshold is established for relevance.

3. A sends the query about X to qualified acquaintances first, randomly choosing among them, in order to avoid always contacting the same acquaintances in the same order. If A cannot get n_r answers from them, also unqualified will be contacted in a further step. Note that any query sent to an acquaintance does not guarantee to provide significant results since we do not know whether that acquaintance will lead to X, i.e. whether a path from the acquaintance to X actually exists, as in real life occur. In particular, when each of A's acquaintances receives the query, his task is to answer the query with his opinion about X, but the current acquaintance could still get into the same trouble of A, i.e. he does not know directly X; similarly to A, such node will forward the request to his acquaintances (but A, obviously), waiting for a set of trust values about X to be mediated, hence the resulting mean trust will be assigned by the node to X and it will be also delivered to A (or generally to the requesting neighbor node). The query propagation will occur until one of X's acquaintances neighbors is reached; such a node can give his opinion about X without forwarding the query, since he directly knows X. Similarly to the first node that generated the query (A), an *intermediate* node needs n_r opinions about X, but differently from A, if n_r is not achieved exploiting his qualified acquaintances, no more answers will be asked to acquaintances having a relevance below the threshold, modeling the absence of a personal interest the intermediate node has in determining an opinion about X. Then, in the worst case, the intermediate node will give back a trust value of 0 (i.e. he will answer he cannot give a significant opinion about X). The number n_r of requested opinions and the threshold are the same A established; this models the propagation of the *importance* A assigned to the query. Finally, note that in mediating opinions coming from acquaintances, each intermediate should also add a distance factor that models the fact that in the real world we consider more significant an opinion if it comes from a person that directly knows X, decreasing the value of this opinion as soon as increases the number of individuals to be contacted to get to this opinion.

4. As soon as A gets all n_r answers, she uses them to build a mediated opinion about X. Actually, this is evaluated as a weighted mean value, where opinions are weighted according to trust level A assigned to acquaintances that provided that opinion, reflecting the real world behaviour. A similar formula is adopted by intermediate nodes, where average is multiplied by the distance factor cited above.

3.4   Weak and Strong Links

Each time a node A collects information about an unacquaintance node X and consequently assigns X a (first) trust value without involving X's acquaintances, we name the arc between A and X as a *weak* link, since the value of trust has not been verified neither by A's personal experience, nor by collecting acquaintances opinions; As soon as A has a personal experience with X or he gets other opinions and build a refined trust value using the mechanism discussed previously, the arc from A to X will be named as *strong* link. A weak link is used to model the first contact a person has with one another, a contact that implies the assignment of a first, temporary and possibly wrong trust level, as in real world actually occurs. A strong link can be lowered to a weak one when its associated trust level has been cvaluated a very long time ago, hence that link is "old" and it should be not considered affordable anymore (a new evaluation should be performed). This aging mechanism is described in section 3.5

During the trusting process, each node forwarding the query actually learns about the existence of node X being trusted, hence a weak link is established between such node and X. At the same time, each node learns about the existence of the node that is forwarding him the query, hence a weak link between these node is established. Note that both weak links here described could be already present in the network; in this case, nothing is done.

Note that the mechanism of evaluating trusting by collecting opinions from acquaintances mainly relies on strong links, i.e. qualified acquaintances are just with strong links, whereas others includes both strong and weak links.

3.5 Links Aging

In real social networks, links have a lifecycle: they are established, exploited whenever trusting opinions are required, but they can also become *obsolete*, or finally they can be removed from the network. We distinguish two different situations, weak and strong links, detailed in the following:

- if a link between two nodes A and B is weak, it simply represents the fact that A knows that B exists; if such a link does not become strong (either through personal interaction between A and B or through the mediated opinions mechanism) after a long period it will be removed from the network, modeling the fact that A forgets about B's existence if he never considers B during his life. This is also useful to avoid a network with an excessive (and probably useless) links;
- as soon as a link between A and B becomes strong, it can be considered reliable for trusting evaluation, but this assumption will be probably wrong after a very long period. In real world networks indeed, if A has a good opinion about B, but over time A has no contacts with B and B turns into a "bad guy", our opinion about B is invalid. Similarly, if the trust assigned by A to B was obtained through mediated opinions, it could happen that acquaintances changed their opinion about B, or networks changes might occur (e.g. an acquaintances dies or leave his country), in both cases actually still invalidating our mediated trust level. All such situations can be modeled by assigning an *age* to the strong link. As time goes by without link *updates*, that is neither further personal interactions between A and B occur nor re-evaluation requests for B's trust level by A to his acquaintances are sent,

the link becomes old; after a very long period, this link is degraded to a weak one, meaning that A still knows that B exists, but he should no more trust him neither considering his opinion from trusting someone else until A's personal opinion about B is refreshed, either with a new interaction or by re-evaluating mediated opinion.

4 Formalization of Trusting Mechanism

To formalize the mechanism introduced in the previous section we first model the trusting network as an oriented labeled graph $\mathcal{G}$, defined as a pair $(\mathcal{N}, \mathcal{E})$ where $\mathcal{N}$ is a set of nodes and $\mathcal{E}$ is a set of edges (or links) between vertices, i.e. $\mathcal{E} = \{(u,v)|u,v \in \mathcal{N} \wedge u \neq v\}$; a link from node A to node B is labelled with the pair (t_B^A, C_B^A) where $t_B^A \in [-1;1]$ represents A's trust about B and C_B^A represents the knowledge (resume) A owns about B. In the following we introduce and comment the algorithm used when a node A trusts a node X:

---

trustNode

Require: $\mathcal{G}$
Require: A, X $\in \mathcal{N}$ where A is a node that wishes to trust node X
Require: C_X
Require: τ_{good} that is threshold used to select node for query forwarding
Require: n_r maximum number of request, id_q is a query identifier
 1: **if** X $\notin \mathbf{N}^A$ **then**
 2:     $addToNet$(A,X)
 3: **end if**
 4: $\overline{\mathbf{R}} = \{r_i \in \mathbf{R}_X^A | r_i > \tau_{good}\}$
 5: $\mathbf{T}_X^A = \emptyset$
 6: **for all** $(I \in \overline{\mathbf{R}}) \wedge (| \mathbf{T}_X^A | < n_r)$ **do**
 7:     $t_I^X = forward(I, C_X, id_q)$
 8:     $\mathbf{T}_X^A = \mathbf{T}_X^A \bigcup \{I, t_X^I\}$
 9: **end for**
10: **if** $| \mathbf{T}_X^A | < n_r$ **then**
11:     **for all** $(I \in (\mathbf{N}^A - \overline{\mathbf{R}})) \wedge (| \mathbf{T}_X^A | < n_r)$ **do**
12:         $t_X^I = forward(I, C_X, A, id_q)$
13:         $\mathbf{T}_X^A = \mathbf{T}_X^A \bigcup \{I, t_X^I\}$
14:     **end for**
15: **end if**
16: $t_X^A = trust(\mathbf{T}_X^A, \mathbf{N}^A)$

---

1. the *addToNet* function (line 2) creates the arc from node A to X (randomly assigns an initial trust value) whenever X does not already belong to the set of her acquaintances (denoted as $\mathbf{N}^A$);
2. we then define the correspondence and relevance functions, used the former to evaluate resumes similarity, and the latter to sort acquaintances (based on correspondence and assigned trust level). Specifically, the *correspondence*

is defined as $corr : C \times C \rightarrow [0;1]$, also noted with $corr(C_I, C_X)$; we call $c_{I,X}$ the result of correspondence between nodes I and X. The implementation of this function simply adopt classical information retrieval techniques aiming at extracting inverse index vector for both resumes in order to compare their contents. The *relevance* function is defined as $r_I^{A,X} = c_{X,I} \cdot t_I^A$, where A is the node that ask an opinion about X to one of her neighbors I; we denote the set of all relevance values as $\mathbf{R}_X^A$. The $\overline{\mathbf{R}}$ set in line 4 is the set of *qualified* acquaintances, i.e. those whose relevance is greater than the threshold τ_{good}

3. when the n_r answers are collected, a mediated opinion is evaluated, actually using the following formula:

$$t_X^A = trust(\mathbf{T}_X^A, \overline{\mathbf{N}}_A) = \frac{\sum t_X^i \cdot t_i^A}{\sum t_i^A} \tag{1}$$

In this weighted average formula main terms are t_X^i, i.e. acquaintances opinions about X, weights are t_i^A, that is trust values assigned by A to his acquaintances. To understand why t_i^A is also present at the denominator, we impose that acquaintances A assigned low trust values should have less influence over the resulting mean value, in order to reflect the real world behaviour adopted by a person, i.e. he tends to neglect the opinion of an acquaintances he trusts in with a low degree. To do this, at the denominator we place the sum of *contributions*, being each acquaintance's contribution the trust value A assigned him, so it tends to 0 when A does not trust into him, and tends to 1 when A completely trust in him. At the limit, if A assigned 1 to all acquaintances, the sum of contribution is equal to their number, having in this case the standard weighted average formula, whereas at the opposite if A trusts with a low degree someone (i.e. trust value tends to 0), his contribution as well as also the corresponding term at the numerator will be neglected; if trust value would be exactly 0, contribution at the denominator and weighted opinion about X at the numerator will both removed, reflecting the fact that in real world we do not consider the opinion of an acquaintance we do not trust in. The t_X^A is associated (line 16) to the arc from A to X.

4. the *forward* function invoked in line 7 and 12 allows the query to be forwarded through the network; actually this function is similar to the *trustNode* algorithm, except for two mechanisms, already introduced in the previous section: first, the generic intermediate node that performs the *forward* function will just contact relevant nodes since he is not personally interested in answering the query; second, the formula adopted for evaluating a mediated trust value will also take into account how distance exist from A to X, actually decreasing the average trust using a multiplying distance factor in the range $]0,1]$.

To implement links lifecycle described in section 3.5, we invoke a *linksAging* function (not illustrated here) whenever a given number of queries has been requested, modeling global time expiration by query computation.

Note that a trade-off should be achieved between links creation, actually performed by the trusting algorithm shown before, and links removal executed by linksAging function, in order to both avoid an excessive number of (probably useless) links and also to avoid too few links in the network, which could prevent an effective trusting network exploitation. This trade-off is implemented inside the linksAging function by using an adjustable threshold, so as long as the number of links for a given node is over this threshold, weak links are removed; if the threshold is still exceeded, then also older strong links are degraded to weak; note that these new weak links are not removed in the current call of linksAging function.

4.1 Preliminary Results

A first set of simulation has been performed using the trusting algorithm described previously, in addition with aging mechanism, in order to test the convergence of the network Simulation have been performed considering networks with different number of nodes (i.e. 100, 200 and 400), with initially no links; the trusting algorithm is invoked when queries are generated, and it allows the growing of links number, hence the network evolves through a transient state until a stable set of connection is achieved.

Input parameters we assigned for trusting algorithm are n_r=3 (in order to avoid the flooding of the network), and distance factor=0.9 (to allow long paths for query forwarding to be considered).

Results are quantified by introducing two measures:

1. average trusting requests deepness, denoted as $ATRD$ (where a TR is a query), indicating the deepest path from the requesting node to the node to be trusted, mediated among all TRs; this property is used to express the *efficiency* of the trusting process
2. average number of failed TRs ($AFTR$), where a TR is considered *failed* when no answer are collected to build mediated opinion, for instance when no relevant nodes are found for TR forwarding. This property is used to express the *effectiveness* of the trusting process

In fig. 2 and 3 running $ATRD$ and $AFTR$ are represented, respectively. These express how many TR are needed for the network to converge, i.e. to provide stable values for both properties. This two experiments clearly point out that the proposed trusting algorithm converge and, in particular, TRs deepness reaches stable values when about 1500–3000 (for networks of 100 to 400 nodes, respectively) TRs are performed. Note that TRs deepness initially increases since when the network is created, too few links are present, hence long paths are used to reach a node to be trusted; as soon as links number increases, i.e. best paths are available, the deepness decreases. The number of failed TRs rapidly decreases to zero when the network converges, meaning that when a trusting request is issued, exactly n_r opinions are received from the source node allowing it the evaluation of the mediated opinion using the equation 1.

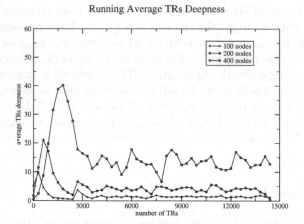

Fig. 2. Average TRs deepness

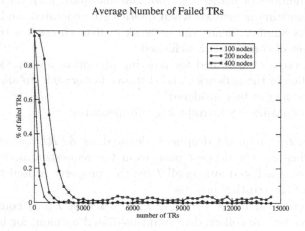

Fig. 3. Average number of failed TRs

5   Conclusions and Future Work

In this paper we introduced the idea of emulating human behaviour in social networks to address the question of trust evaluation; an algorithm has been presented, together with preliminary results, showing that average query deepness tends to decrease, as well as the number of failed queries, thus the network converges. Further works we planned are:

- investigation on the role of resumes, in particular how communities based on nodes resumes are created and evolve over time; resumes should play a central role in the improvement of queries routing, as when the "small world" effect emerges

- analysis on network robustness, for instance considering typical question as empowering and whitewashing
- positive and negative feedback, i.e. in our proposal nodes are someway *passive* since they provide an opinion when contacted, but they do not actively propagate trust values, for instance informing acquaintances whose opinion about a node is very different from a mediated one

References

1. Dell'Amico, M.: Neighbourhood maps: Decentralised ranking in small-world p2p networks. In: 3rd International Workshop on Hot Topics in Peer-to-Peer Systems (Hot-P2P), Rhodes Island, Greece (April 2006)
2. Deutsch, M.: Cooperation and trust, some theoretical notes. In: Nebraska University Press, editor, Nebraska Symposium on MOtivation (1962)
3. Golbeck, J.: Trust and nuanced profile similarity in online social networks
4. Golbeck, J., Hendler, J.: Reputation network analysis for email filtering (2004)
5. Kamvar, S.D., Schlosser, M.T., Garcia-Molina, H.: The eigentrust algorithm for reputation management in p2p networks (2003)
6. Kinateder, M., Pearson, S.: A Privacy-Enhanced Peer-to-Peer Reputation System. In: Bauknecht, K., Tjoa, A.M., Quirchmayr, G. (eds.) E-Commerce and Web Technologies. LNCS, vol. 2738, pp. 206–215. Springer, Heidelberg (2003)
7. Massa, P., Bhattacharjee, B.: Using trust in recommender systems: An experimental analysis. In: Jensen, C., Poslad, S., Dimitrakos, T. (eds.) iTrust 2004. LNCS, vol. 2995, pp. 221–235. Springer, Heidelberg (2004)
8. Merriam-Webster, http://www.m-w.com
9. Montaner, M., López, B., de la Rosa, J.L.: Opinion-based filtering through trust. In: Klusch, M., Ossowski, S., Shehory, O.M. (eds.) CIA 2002. LNCS (LNAI), vol. 2446, pp. 164–178. Springer, Heidelberg (2002)
10. Page, L., Brin, S., Motwani, R., Winograd, T.: The pagerank citation ranking: Bringing order to the web. Technical report, Stanford Digital Library Technologies Project (1998)
11. Sztompka, P.: Trust: A sociological theory (1999)
12. Ziegler, C.-N., Golbeck, J.: Investigating correlations of trust and interest similarity - do birds of a feather really flock together? Decision Support System (2005) (to appear)

A New Reputation-Based Trust Management Mechanism Against False Feedbacks in Peer-to-Peer Systems

Yu Jin[1], Zhimin Gu[1], Jinguang Gu[2], and Hongwu Zhao[2]

[1] School of Computer Science and Technology,
Beijing Institute of Technology, Beijing 100081, China
[2] College of Computer Science and Technology,
Wuhan University of Science and Technology, Wuhan 430081, China
wustjy@bit.edu.cn, zmgu@263.net

Abstract. The main challenge in reputation system is to identify false feedbacks. However current reputation models in peer-to-peer systems can not process such strategic feedbacks as correlative and collusive ratings. Furthermore in them there exists unfairness to blameless peers. We propose a new reputation-based trust management mechanism to process false feedbacks. Our method uses two metrics to evaluate peers: feedback and service trust. Service trust shows the reliability of providing service. Feedback trust can reflect credibility of reporting ratings. Service trust of sever and feedback trust of consumer are separately updated after a transaction, furthermore the former is closely related to the latter. Besides reputation model we also propose a punishment mechanism to prevent malicious servers and liars from iteratively exerting bad behaviors in the system. Simulation shows our approach can effectively process aforesaid strategic feedbacks and mitigate unfairness.

Keywords: trust, reputation, feedback, peer-to-peer. security.

1 Introduction

In P2P systems interactions often occur among previously unknown parties, and no authority dictates the rules for peers' interactions. Peers might maliciously behave and harm others in the system[1][2]. Therefore, one of the fundamental challenges for open and decentralized P2P systems is the ability to manage risks involved in interacting and collaborating with previously unknown and potentially malicious peers. Reputation-based trust management systems can successfully mitigate this risk by computing the trustworthiness of a certain peer from that peer's behavior history.

The efficiency of reputation system depends on the quality of feedback. Therefore the main challenge in reputation system is to identify false feedbacks. Currently most of reputation systems for peer-to-peer applications assume a correlation between service and feedback reputation [3][4][5][6][7], i.e., they assume that peers providing good services also report truthful feedback and peers

B. Benatallah et al. (Eds.): WISE 2007, LNCS 4831, pp. 62–73, 2007.

with poor service performance tend to submit false feedbacks. These reputation systems are vulnerable when facing correlation feedback attack: Strategic peers can heighten their reputation by providing good services, and then destroy the reputation system by reporting false feedbacks. Moreover poor performance and lying are not necessarily related.

Currently a new based on bi-ratings feedback processing mechanism is presented in which after a transaction both service consumer and provider report ratings about the outcome of this transaction [3][8]. If the two ratings are consistent, the reputation value of server will be increased, while if they are inconsistent both the consumer and server will be punished by decreasing their reputation value or trans-acting cycles. However, this consistent mechanism is weak that can not resist collusion. For example, two members of a malicious league can always report consistent ratings about a failing or even an inexistent transaction, therefore the reputation of one member will continuously increase. Furthermore it is unfair for the blameless partner in this mechanism when the ratings disagree with each other.

This paper proposes a new reputation management system in which the feedback trust is separate from service trust. Our method is also based on bi-ratings. However, our mechanism uses both credibility and reliability metrics when evaluating a peer, namely, credibility can reflect the capability of reporting truthful feedbacks while reliability can evaluate the competence of providing services. In our method the update of reputation of server is related to the credibility of the reporter, therefore the consistent mechanism in our approach is powerful to resist collusion when the ratings agree and mitigate the unfairness when they are inconsistent. Furthermore in order to prevent malicious servers and liars from iteratively exerting malicious behaviors in the systems, we introduce a punishment mechanism complementary to the reputation model, namely, the request cycles of feedback reporter and service cycles of server are accordingly decreased when ratings are inconsistent.

2   Related Work

Cornelli et al. propose a reputation-based approach for P2P file sharing systems (called P2PRep) [9]. In P2PRep reputation sharing is based on a distributed polling protocol. Damiani et al. present a similar approach, called XRep, which considers the reputations of both peers and resources [10]. Both of them do not give any metrics to quantify the credibility of voters. More currently Mari et al.[11] discuss the effect of reputation information sharing on the efficiency and load distribution of a peer-to-peer system, in which peers only have limited (peers share their opinions) or no information sharing (peers only use their local ratings). Similarly this approach does not distinguish the ratings for service (reliability) and ratings for voting (credibility). While our mechanism separates feedback trust from service trust.

Yu et al. [4]address credibility and performance in order to prevent strategic behavior. However, this approach has no explicit mechanism for assessing the credibility of the rating; This issue is dealt together with a trust metric regarding behavior. This argument also applies to the approach of Kamvar et al. [6] where, a global reputation metric regarding performance of each peer is calculated. To this end, each peer's local beliefs on the performance of other peers are weighted by the others' beliefs on his own performance. Y. Wang et al. [7] propose a Bayesian network-based trust model in peer-to-peer networks that provides a flexible method to represent differentiated trust and combine different aspects of trust. However when computing integrated trust value in a given peer, other peers' ratings are weighted by their own service performance. These models assume there is a correlation between service performance with credibility. Therefore none of them can process correlation feedback attack.

L.Xiong et al. present a based P-Grid [12] reputation framework that includes an adaptive trust model (called PeerTrust [13]. In PeerTrust in order to separate feedback trust from service trust, peers use a personalized similarity measure to more heavily weigh opinions of peers who have provided similar ratings for a common set of past partners. In a large P2P system, however, finding a statistically significant set of such past partners is likely to be difficult. Aberer et al. [5] present an approach to evaluate trustworthiness (i.e. the combination of credibility and performance) of peers based on the complaints posed for them by other peers following transactions. The main idea is that a peer is considered less trustworthy the more complaints he receives or files. However this approach does not examine the effectiveness in the case of collaborated liars.

R. Gupta et al. [3] have proposed a reputation management framework based trust group for large-scale peer-to-peer networks. In the proposed framework the reputation of nodes is a measure of their wealth–a form of virtual currency. The framework is based on bi-feedbacks, when two ratings are consistent the reputation of server will be increased; while if they are inconsistent the reputation of two transacting peers will be decreased. Clearly this model can not resist collusion when ratings are consistent and has unfairness to blameless partner when ratings are inconsistent. Furthermore this model also assumes there is a correlation between service performance and credibility. Similar to [3], T.G.Papaioannou et al. [8] present a based on bi-feedback mechanism in peer-to-peer reputation systems. It uses an ncr factor to evaluate the credibility of both rating reporters. When two ratings are consistent, in addition to increase the reputation of server, the credibility of both reporters will also increase; while if not, the credibility of them will decrease, and both of them will be restrained from transacting with others, namely, neither responding requests nor launching queries are allowed for them . Therefore this mechanism has also limitation like [3].

3   Reputation Management Mechanism

3.1   Trust Model

Service and feedback trust metrics. Our mechanism uses two metrics to evaluate peers: service and feedback trust. Service trust shows the reliability of providing service. Feedback trust can reflect credibility of reporting ratings. In other words, peers that have low feedback trust can also be service providers as long as their service trust are high , while peers that have high feedback trust but low service trust can not be candidates for providing services. Similarly peers that have a low service trust value can exert more important influence on reputation system if only they have high credibility; contrarily, the ratings from those peers that have very high service trust but low feedback trust are paid less importance by reputation system.

In our mechanism service trust in peer is normalized values ranging from -1.0 to 1.0, the initial service trust is set to 0 that indicates a neutral trust. If the service trust of a server is negative, its service competence will be under suspicion; while feedback trust range from 0 to 1.0, and 0.5 shows a moderate credibility. In our mechanism the initial value is set to 1.0 that signifies a highest credibility. If the feedback trust of a peer is less than a given threshold T_θ, we think it may be incredible.

Updating service and feedback trust. After a transaction, both transacting partners are required to submit rating about this transaction. Besides voting the transaction as successful or not each rating also contains a quantifiable performance metric, e.g. the number of transferred bytes of useful content. We assume that the observed performance is with high probability the same with that actually offered. (The opposite may only occur due to unexpected events during a transaction like network congestion etc.) Thus, if ratings for a transaction disagree (either in their performance metric or in their vote), then, with high probability, at least one of the transacted peers is lying. Suppose R_s is the rating of this transaction observed by service provider, and R_c is the rating of service consumer in this transaction. We have

$$f = C\{R_s, R_c\} \tag{1}$$

Where C is a consistent function of R_s and R_c. Given a rating-inconsistence tolerance threshold t, if $|R_s - R_c| < t$, we think R_s and R_c agree, f equal 1, otherwise f is -1. In our mechanism we take into consideration malicious servers as well as liars, therefore there may be following four circumstances appearing in the system. Accordingly f will be as follows.

a) Virtuous server provides good service, honest consumer report truth rating; $f = 1$

b) Virtuous server provides good service, lying consumer report false rating; $f = -1$

c) Malicious server provides bad service, honest consumer report truth rating; $f = -1$

d) Malicious server provides bad service, lying consumer report false rating. $f = 1$

Clearly only a) and c) are real and useful to the reputation system, while b) has unfairness to good service provider, and collusion attack may appear in d).

When receiving these two ratings, the service trust of server is updated as follows.

$$ST_s^{new} = \begin{cases} \frac{v}{FT_c} \times ST_s^{old} + (1 - \frac{v}{FT_c}) \times f \ if FT_c \geq T_\theta \\ ST_s^{old} \qquad\qquad\qquad\qquad else \end{cases} \tag{2}$$

Where, ST_s^{new} denotes the new service trust of service provider; While ST_s^{old} denotes the accumulative value. FT_c is current feedback trust of consumer. v is learning rate-a real number in the interval $(0, T_\theta)$, and the smaller v is, the more quickly is the ST_s^{old} change (increase or decrease). In order to limit unfairness when two ratings are inconsistent and resist collusion when they agree, we take the feedback trust of reporter into consideration when updating the service trust of server. As shown in equation (2) when $f = 1$, and if the feedback trust of reporter is more than T_θ, we think consumer may report truthful rating and the service trust of good server will increase like a); while if the credibility of the reporter is less than T_θ, the authenticity of its rating will be under suspicion and the service trust of malicious server will not increase. Therefore this can resist collusion that may appear in d); while when $f = -1$, and if reporter has a high feedback trust, we will decrease the service trust of malicious server like c); while if the reporter is thought it may be a liar, the service trust of good server will not decrease. That can mitigate the unfairness for good provide like b).

Subsequently according to above two ratings we have

$$FT_c = \frac{N_c^{con}}{N_c^{total}} \tag{3}$$

where N_c^{con} is the number of consistencies where $f = 1$, namely two ratings agree each other. N_c^{total} is the number of consuming services. Equation (3) shows the larger the number of consistencies with other peers, the more this peer is credible because after all most of peers are virtuous and honest in the system.

3.2   Punishment Mechanism

Although our trust model can resist collusion and mitigate unfairness to a certain extent, iterative malicious behaviors of malicious servers and liars would damage greatly the credibility of honest reporters and reputation of good providers. In order to reduce this risk we combine a punishment mechanism with our trust model. As described in algorithm 1 and algorithm 2 we introduce six variables to implement this approach. That is NCS (Non-Consistent Service), NCR (Non-Consistent Reporting), PCS (Punishment Cycles for Service), PCR (Punishment Cycles for Requesting), SPS (Service Punishment State), and RPS (Requesting Punishment State). NCS and NCR are integers that are no less than zero. They record the number of inconsistencies, namely NCS represents the number of server receiving inconsistent service ratings, while NCR indicates the number of consumer reporting inconsistent service ratings. Clearly the larger NCS of a peer is, the less reliable its service is; similarly the larger NCR is, the less credible its rating is. PCS is the duration of limiting providing service that equals $base^{NCS}$, $base > 1$, and PCR is the duration of restraining from sending requests that equals $base^{NCR}$. When upon disagreement, two transacting partners are punished, but malicious servers (or liars) will be punished to a much more extent than honest reporters (or virtuous servers) because the NCS (or NCR) of the former must be larger than the NCR (NCS) of the latter. Take another way, the duration of punishment of the false partner is much longer

---

Algorithm 1. Feedback Processing Mechanism for Server

---

if *SPS and server receiving a new request* **then**
    // SPS ← 1 when a new peer joining system
    Wait() ;   // server providing a service
    Receiving two ratings;
    if $f == 1$ **then**
        $ST_s^{new} \leftarrow$ equation (2);
    else
        $ST_s^{new} \leftarrow$ equation (2);
        $NCS \leftarrow NCS + 1$;
        $SPS \leftarrow 0$;
        $startTime \leftarrow currentTime()$;
    end
    Discarding this request ;
else
    if $SPS == 0$ *and* $currentTime() - startTime \geq PCS$ **then**
        $SPS \leftarrow 1$;
    else
        Discarding this request ;   // *SPS==0*,not allowed to serve this
        request
    end
end

---

than that of sincere one because PCS (or PCR) is the exponential in NCS (or NCR). SPS and RPS are binary state variables that are used to mark whether a peer is under punishment. If SPS is 0 that indicates this peer is restrained from providing service but can request service within its PCS. Similarly if RPS is 0 that shows this peer can not be allowed to launch a service request but can provide service within its PCR. NCS, PCS, and SPS can be helpful to purify the service trust of good service providers, and NCR, PCR, and RPS can cleanse the feedback trust of honest reporters.

In our reputation management mechanism, malicious servers and liars must suffer from great punishment for their iterative bad behaviors, so our mechanism can encourage peers to provide good services and report truthful ratings in the system.

4 Experiments

In order to evaluate our approach, we develop a series of simulations in a file sharing P2P system. The simulations are developed in Java based on the Peersim 1.0 platform.

4.1 Simulation Setup

Peer behavior categorization. In a P2P system, we consider two kinds of general behaviors of peers: good and malicious. Good peers are those that send

---

Algorithm 2. Feedback Processing Mechanism for Consumer

---

if *RPS and consumer having a new request* **then**
    `// RPS ← 1 when a new peer joining system`
    Wait() ;   `// consumer consuming a service`
    Receiving two ratings;
    if $f == 1$ **then**
        $N_c^{total} \leftarrow N_c^{total} + 1$;
        $N_c^{con} \leftarrow N_c^{con} + 1$;
        $FT_c \leftarrow$ equation (3);
    else
        $N_c^{total} \leftarrow N_c^{total} + 1$;
        $FT_c \leftarrow$ equation (3);
        $NCR \leftarrow NCR + 1$;
        $RPS \leftarrow 0$;
        $startTime \leftarrow currentTime()$;
    end
    Discarding this request ;
else
    if $RPS == 0$ *and* $currentTime() - startTime \geq PCR$ **then**
        $RPS \leftarrow 1$;
    else
        Holding this request ;   `// RPS==0,not allowed to send this`
        `request.`
    end
end

---

authentic files and do not lie in their feedbacks (Type $P1$). Malicious peers can be divided into three categories: 1) peers that send authentic files and lie in their feedbacks (Type $P2$), 2) peers that send inauthentic files and do not lie in their feedbacks (Type $P3$), and 3) peers that send inauthentic files and lie in their feedbacks (Type $P4$). A lying peer is one that gives ratings not according to the real equality of receiving file but to a given lying strategy.

System and parameters configuration. In the experiments T_θ is 0.5 and v is 0.3. In the punishment mechanism base can help to restrain malicious behaviors in the system. We must think over to assign it appropriate value. If it is too small, we will not achieve the goal to punish the malicious peers; while if it is too big this will enhance the unfairness to blames partners. In our simulations it is set to 7, and the number of peers in our system is 1000.

4.2 Simulation Results

In order to evaluate the efficiency of proposed mechanism, we compare our approach with other existing models against different lying strategies mentioned below. For purpose of clear depiction we call our mechanism SCM (Strong Consistence Model), reputation systems like[9][10] as WithoutFM (Without

Feedback Mechanism), reputation systems like [4][6][7] as Correlation, and those reputation system in [3] [8]as WCM (Weak Consistence Model).

Destructive lying. Destructive lying is a complementary strategy. That is liars always give ratings contrary to the reality. The goal of this part is to examine the efficiency of our mechanism when there is only destructive lying attack in the system. Suppose R_{P1}, R_{P2}, R_{P3} and R_{P4} are separately the ratio of peers belonging to $P1$, $P2$, $P3$ and $P4$ to the size of system. In this lying model R_{P3} and R_{P4} constantly are 0.1, while R_{P2} changes.

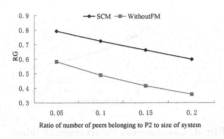

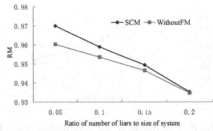

Fig. 1. Comparison for correct identifying good peers in destructive lying

Fig. 2. Comparison for correct identifying malicious peers in destructive lying

The main task of feedback processing mechanism is to filtrate false ratings and correctly identify good or bad providers. Suppose RG represents the ratio of number of good providers correctly identified to total good providers in the system, and RM is the ratio of number of malicious providers correctly identified to total malicious providers in the system. Correct identifying a peer means that the service trust of a good provider is higher than zero, while that of a malicious server is less than 0. As shown in Fig.1 and Fig.2, with the increase of number of liars, RG and RM are descending in SCM and WithoutFM, while both RG and RM in SCM are higher than that in WithoutFM because our reputation system adopts a credibility mechanism that can reduce the damage brought by false feedbacks. In our mechanism change of service trust of providers depends on two ratings as well as credibility of reporters. Incredible peers can exert little bad influence on the reputation of other good providers even if they report negative ratings, while reporters with high credibility can play an important role in decreasing the reputation of malicious providers. Therefore with our credibility mechanism reputation system can effectively and correctly differentiate good peers from bad ones.

Then when R_{P2} is 0.1, we examine the efficiency of our mechanism in restraining iterative malicious behaviors. Suppose SM represents the ratio of number of services provided by malicious servers to total service, and RL is the ratio of number of requests of liars to total requests. As Fig.3 and Fig.4 describe, with the running of simulation malicious behaviors gradually are limited by punishment mechanism. However in our mechanism the extent to which malicious behavior is restrained is larger than that in reputation systems without feedback processing mechanism. For example SM and RL in our approach are smaller than that in WithoutFM. In our punishment mechanism when upon disagree two transacting partners will be punished, but the malicious peers are punished more severely than blameless ones. Therefore false service providers will gradually disappear from service responders. Similarly peers that often lie will be allowed to send less and less service requests.

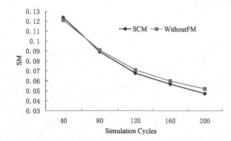 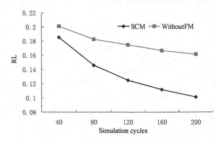

Fig. 3. Comparison for restraining malicious services in destructive lying

Fig. 4. Comparison for restraining lying requests in destructive lying

Above two experiments show the efficiency of credibility mechanism in correct identifying peers and limiting malicious behaviors. However the final goal of feedback processing mechanism is to enhance the average service performance of peers. Suppose R_{succ} is the percentage of successful services. In this experiment R_{P2} also is set to 0.1. As shown in Fig.5 although with the running of simulation R_{succ} in two reputation systems are increasing, R_{succ} in our approach is much more than that in WithoutFM. The essential reason for it is that our reputation system introduces a powerful credibility mechanism.

Subsequently we discuss, when two ratings are inconsistent, the efficiency of our mechanism in reducing unfairness compared with WCM in which both two partners will be punished in reputation or transacting cycles. Therefore successful transaction volume will decline due to unfair punishment for good peers. As described in Fig.6 when number of liars increase numbers of successful transactions in two systems are on the decline, but they in our mechanism are more than that in WCM. Furthermore the R_{succ} in our approach also is bigger than that of WCM. For example when R_{P2} is 0.1, R_{succ} is 82.14% in our method while is 81.69% in WCM; when R_{P2} is 0.2, R_{succ} is 74.20% in our method while is 73.81% in WCM. This shows peers can receive more and better services in our reputation system. The reasons for reducing unfairness for good peers in our mechanism are as follows.

1) Only the service trust of provider may be decreased but its feedback trust keeps constant, and feedback trust of reporter will be decreased while its service trust does not change when two ratings disagree.

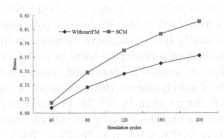
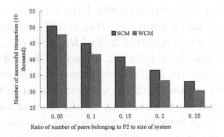

Fig. 5. Average service performance comparison in destructive lying

Fig. 6. Unfairness comparison in destructive lying

2) Service provider will be limited to respond request, but it can continuously send service requests under punishment. Punished consumer can not launch requests, but it can provide services as long as it has high service trust.

Correlative lying. This attack model emerges due to the weakness of the existing reputation systems where the credibility of peers is assumed to have much to do with their service performance. This experiment aims to analyze the efficiency of our mechanism against correlation lying attack compared with those reputation systems (Correlation) in which the credibility of reporters has a correlation with their service performance. Correlation lying peers' target is to damage the reputation of others by their high service trust, accordingly reduce the successful transaction volume in the system. As Fig.7 describes when size of peers who adopt correlation lying strategy when submitting ratings becomes larger and larger, numbers of successful transactions in both approaches decrease. However there are more successful transactions in our mechanism than in Correlation system. Moreover the R_{succ} in our approach is also little higher than that in Correlation. Take an example when the ratio of size of correlation lying peers to the system is 0.3, R_{succ} in our approach is 82.8%, while it is 82.6% in Correlation; when the ratio of size of correlation lying peers to the system is 0.5, R_{succ} in our approach is 66.0%, while it is 65.5% in Correlation. The reason for this is that our mechanism separate feedback trust from service trust, and reputation of service provider is updated according to the credibility of reporter other than service trust. So correlation liars can not succeed in attacking others' reputation by using their high service trust at all.

Collusive lying. In this lying model liars always report positive ratings about their cahoots even if the transactions fail or are nonexistent. However when transacting partners are peers out of their leagues, Collusion liars always give negative ratings no matter what is the real outcome of the transactions. The objective of this experiment is to evaluate the efficiency of our mechanism against collusive attack compared with WCM in which the service trust of provider will raise when two ratings are upon agreement. Our mechanism can resist collusive lying strategy to a certain extent. In this experiment besides collusive lying we also take destructive lying into consideration. In other words liars out of collusive league adopt destructive lying strategy when submitting ratings. R_{P4} keeps constant and is set to 0.1, R_{P2} and R_{P3} change from 0.05 to 0.2. The size of collusion league is separately half of R_{P2} and R_{P3}. Namely half of collusive members help to raise the reputation of another half of malicious service providers by collusive ratings. As Fig.8 shows when size of collusive league is increasing, successful transactions become less and less in both of mechanisms. However number of successful transactions in our mechanism is larger than in WCM. Furthermore R_{succ} is raised by our approach. For instance, when the ratio of size of collusion to the system is 0.1, R_{succ} in our approach is 78.6%, while it is 78.3% in WCM. When the ratio of size of collusion to the system is 0.15, R_{succ} in our approach is 71.0%, while it is 70.7% in WCM. This experimental result shows our mechanism can effectively identify collusive ratings and resist collusion attack. The reason for this is, unlike WCM, our mechanism considers the feedback trust of reporters when updating service trust of providers. If collusive reporter has a low feedback trust he will fail in raising the service trust of his cahoots.

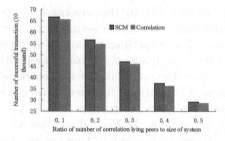
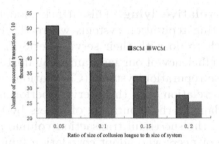

Fig. 7. Correlation attack comparison in correlation lying

Fig. 8. Collusive attack comparison in collusion lying

5   Conclusion

In this paper, we propose a new reputation-based trust management mechanism against strategic ratings like destructive, correlative and collusive lying. Our scheme separates feedback from service trust, and feedback trust of reporter determines how to update the service trust of provider. Furthermore we adopt a new punishment mechanism to prevent bad peers from iteratively exerting malicious behaviors in the system. In this way our mechanism is more robust to peers' maliciousness and less unfair to their goodness.

Acknowledgment

This work was partially supported by China Postdoctoral Science Foundation (20060400275) and NSF of Hubei Educational Agency (Q200711004).

References

1. Chien, E.: Malicious threats of peer-to-peer networking. Technical report, Symantec corporation (2003)
2. Liang, J., Kumar, R., Xi, Y., Ross, K.: Pollution in P2P File Sharing Systems. In: Proceeding of the IEEE Inforcom (Inforcom 2005), pp. 1174–1185. IEEE Computer Society Press, Miami, USA (2005)
3. Gnupta, R., Somani, A.K.: Reputation Management Framework and its Use as Currency in Large-Scale Peer-to-Peer Networks. In: P2P 2004. Proceeding of the 4th International Conference on Peer-to-Peer Computing, pp. 124–132. IEEE Computer Society Press, Zurich, Switzerland (2004)
4. Yu, B., Singh, M.P.: An Evidential Model of Distributed Reputation Management. In: Proceeding of the first of International Joint Conference on Autonomous Agents and Multi-agent Systems, pp. 294–301. IEEE Computer Society Press, Bologna, Italy (2002)
5. Aberer, K., Despotovic, Z.: Managing Trust in a Peer-2-Peer Information System. In: CIKM 2001. Proceeding of the 10th International Conference on Information and Knowledge Management, pp. 310–317. ACM Press, New York, USA (2001)
6. Kamvar, S.D., Schlosser, M.T., Garcia-Molina, H.: EigenRep: Reputation Management in Peer-to-Peer Networks. In: ACM WWW 2003. Proceeding of the Twelfth International World Wide Web Conference, pp. 123–134. ACM Press, Budapest, Hungary (2003)
7. Wang, Y., Vassileva, J.: Bayesian Network Trust Model in Peer-to-Peer Networks. In: Moro, G., Sartori, C., Singh, M.P. (eds.) AP2PC 2003. LNCS (LNAI), vol. 2872, pp. 23–34. Springer, Heidelberg (2004)
8. Papaioannou, T.G., Stamoulis, G.D.: An IncentivesMechanism Promoting Truthful Feedback in Peer-to-Peer Systems. In: CCGrid 2005. Proceeding of the 2005 IEEE International Symposium on Cluster Computing and the Grid, pp. 275–283. IEEE Computer Society Press, Cardiff, UK (2005)
9. Cornelli, F., Damiani, E., Capitani, S.D.: Choosing Reputable Servents in a P2P Network. In: WWW 2002. Proceeding of the 11th International World Wide Web Conference, pp. 441–449. ACM Press, Hawaii, USA (2002)
10. Damiani, E., Vimercati, S., Paraboschi, S.: A Reputation-based Approach for Choosing Reliable Resources in Peer-to-Peer Networks. In: CCS 2002. Proceeding of the 9th ACM conference on Computer and communications security, pp. 207–216. ACM Press, Washington, USA (2002)
11. Marti, S., Garcia-Molina, H.: Limited reputation sharing in P2P systems. In: Proceedings of the 5th ACM Conference on Electronic Commerce, pp. 91–101. ACM Press, New York, USA (2004)
12. Aberer, K.: P-Grid:A self-organizing access structure for P2P information systems. In: Proceeding of the ninth international conference on cooperative information systems, New York, USA, pp. 91–101 (2001)
13. Xiong, L., Liu, L.: Peertrust:a trust mechanism for an open peer-to-peer information system. Technical report, College of Computing, Georgia Institute of Technology (2002)

Freshness-Aware Caching
in a Cluster of J2EE Application Servers

Uwe Röhm[1] and Sebastian Schmidt[2]

[1] The University of Sydney, Sydney NSW 2006, Australia
[2] University of Karlsruhe, Karlsruhe, Germany

Abstract. Application servers rely on caching and clustering to achieve high performance and scalability. While queries benefit from middle-tier caching, updates introduce a distributed cache consistency problem. The standard approaches to this problem, cache invalidation and cache replication, either do not guarantee full cache consistency or impose a performance penalty.

This paper proposes a novel approach: *Freshness-Aware Caching* (FAC). FAC tracks the freshness of cached data and allows clients to explicitly trade freshness-of-data for response times. We have implemented FAC in an open-source application server and compare its performance to cache invalidation and cache replication. The evaluation shows that both cache invalidation and FAC provide better update scalability than cache replication. We also show that FAC can provide a significant better read performance than cache invalidation in the case of frequent updates.

1 Introduction

Large e-business systems are designed as n-tier architectures: Clients access a webserver tier, behind which an application server tier executes the business logic and interacts with a back-end database. Such n-tier architectures scale-out very well, as both the web and the application server tier can be easily clustered by adding more servers into the respective tier. However, there is a certain limitation to the performance and scalability of the entire system due to the single database server in the back-end. In order to alleviate this possible bottleneck, it is essential to keep the number of database calls to a minimum.

For this reason, J2EE application servers[1] have an internal EJB cache (EJB = enterprise java bean). One type of EJBs, entity beans, encapsulate the tuples stored in the back-end database. Their creation is especially costly, because their state has first to be loaded from the database. Hence, the typical caching strategy of an application server is to cache entity beans as long as possible and hence to avoid calling the database again for the same data.

In the presence of clustering, things become more complex. The separate EJB caches of the application servers in the cluster form one distributed EJB cache,

---

[1] This paper concentrates on the J2EE standard and entity beans; the principles of FAC are however applicable to any middle-tier application server technology.

B. Benatallah et al. (Eds.): WISE 2007, LNCS 4831, pp. 74–86, 2007.

where a copy of the same entity bean can be cached by several nodes. Distributed caching involves managing a set of independent caches so that it can be presented to the client as a single, unified cache. While queries clearly benefit from this caching at the middle-tier, updates introduce a distributed cache consistency problem. Two standard approaches to solve this problem are *cache invalidation* and *cache replication*. But both either do not guarantee full cache consistency or impose a performance penalty.

This paper presents a new approach to distributed cache management that guarantees clients the freshness and consistency of cached data: *Freshness-aware Caching (FAC)*. The core idea is that clients can explicitly ask for stale data up to a specific freshness limit. FAC keeps track of the freshness of data in all caches and provides clients with a consistent and staleness-bound view of data. Stale objects can be cached as long as they meet the clients' freshness requirements. This allows clients to trade freshness of data for improved cache utilisation, and hence shorter response times. The main contributions of this paper are as follows:

1. We present two novel freshness-aware caching algorithms, FAC and FAC-δ, that guarantee clients the freshness and consistency of cached data.
2. We give an overview of our implementation of FAC in the JBoss open-source J2EE application server.
3. We present results of a performance evaluation and quantify the impact of the different parameters of FAC on its performance and scalability.

We evaluated our approach versus JBoss' standard cache invalidation algorithm and a cache replication solution. In these experiments with varying cluster sizes up to 7 nodes, both cache invalidation and FAC provided similar update scalability much better than cache replication. It also showed that FAC can provide a significant better read performance than cache invalidation in the case of frequent updates.

The remainder of this paper is organised as follows: In the subsequent section, we give an overview of related work. We formally define freshness-aware caching in Section 3, and briefly discuss its implementation in Section 4. Section 5 presents the results of an experimental evaluation of our approach with other J2EE caching solutions. Section 6 concludes.

2   Distributed Cache Management for J2EE Clusters

State-of-the-art is an asynchronous cache invalidation approach, as used in, e.g., the Bea WebLogic and JBoss application servers [1,2]: When an entity bean is updated in one application server node, that server multicasts a corresponding invalidation message throughout the cluster after the commit of the update transaction. Due to this invalidation message, any old copy of the updated bean is removed from the other caches. JBoss sends its invalidation messages using a group communication protocol which provides guaranteed delivery [2]. Cache invalidation falls short with regard to two important aspects:

1. It cannot give any consistency guarantees for the whole distributed cluster cache, because the invalidation is done asynchronous and decoupled from the commit at the originating cache node.
2. Cache invalidation leads to more cache misses. After an update, all copies of the updated bean get invalidated in the remaining cluster nodes. Hence, the next access to that bean will result in a cache miss.

This has triggered interest on cache replication which gives strong consistency, but for the price of reduced update scalability (e.g. [3]). There are several third party in-process caching solutions for J2EE servers [4,5,6,7]. They typically provide no transparent caching for entity beans, but rather expose special caching data structures which have to be explicitly used in an J2EE application. BEA WebLogic provides in-memory replication for stateful session beans, but not for persistent entity beans which are the scope of this work [1]. Hence we did not compare to those solutions; rather, in our evaluation, we used our own cache replication framework for JBoss that is completely transparent to the application programmer and that supports entity beans [8].

In recent years, middle-tier database and query caching has gained some attention in database research [9,10,11,12]. In a nutshell, a middle-tier database cache is a local proxy DBMS that is deployed on the same machine as the application server and through which all data requests are routed. The whole database is still maintained in a central back-end DBMS server. From the client-view, the access is transparent because the middle-tier database cache has the same schema as the back-end database. But it does cache only a subset of the back-end data. Queries are answered from the local cache as often as possible, otherwise they are routed automatically to the back-end system. Updates are always executed on the back-end database, and later propagated with an asynchronous replication mechanism into the middle-tier database caches. All these approaches such as IBM's DBCache [9] or MTCache from Microsoft [11] are out-of-process caching research prototypes with an relatively heavy-weight SQL interface. Due to the lazy replication mechanisms, the solutions cannot guarantee distributed cache consistency, although some recent work around MTCache started at least specifying explicit currency and consistency constraints [13].

There are also some commercial products available: TimesTen offers a mid-tier caching solution built on their in-memory database manager [14]. Oracle was also offering a database cache as part of its Oracle9i Application Server [15], but has dropped support for this database cache in its latest version Oracle 10g.

3   Freshness-Aware Caching

The design goal of FAC is a fast and consistent caching solution with less cache misses than cache invalidation, and with better update scalability than cache replication. The core idea is that FAC neither immediately evicts nor replicates updated objects, but each cache keeps track of how stale its content is, and only returns data that is fresher than the freshness limits requested by clients.

3.1 Assumptions

Our approach is based on the following assumptions that represent the typical behaviour and setup of today's business applications:

Firstly, we assume a cluster of application servers with a single, shared database server in the back-end that is owned by the application servers. In other words, the database will be only accessed and modified via the application servers — no external transaction can affect the consistency of the database.

Secondly, our approach assumes an object cache inside the application server that caches all data accessed from the back-end database. Because of the exclusive database access, it is safe for application servers to cache persistent objects across several transactions. In the context of the J2EE standard, this is called *Commit Option A* when using container managed persistent (CMP). Our implementation specifically assumes CMP entity beans with Commit Option A.

Further, we assume that one client request spans exactly one transaction and no transaction spans more than one client request. A client request in our case is a method invocation issued by the client that requests or updates one or more entity bean values. We further assume that each such transaction remains local to one cluster node, i.e. we do not allow distributed transactions (transactions that are executed on more than one application server in their lifetime).

We finally assume that no bean will be updated and then read within the same transaction, i.e. we either have update or read-only transactions.

3.2 Freshness Concept

Freshness of data is a measure on how outdated (stale) a cached object is as compared to the up-to-date master copy in the database. There are several approaches to measure this: *Time-based* metrics rely on the time duration since the last update of the master copy, while *value-based* metrics rely on the value differences between cached object and its master copy. A time-based staleness metric has the advantages that it is independent of data types and that it does not need to access the back-end database (a value-based metric needs the up-to-date value to determine value differences). Hence, FAC uses a time-based metric, and the freshness limit is a requirement for staleness to be less than some amount.

Definition 1 (Staleness Metric). *The* staleness *of an object o is the time duration since the object's stale-point, or 0 for freshly cached objects. The stale-point $t_{stale}(o)$ of object o is the point in time when the master copy of o got updated while the cached object itself remained unchanged.*

$$stale(o) := \begin{cases} (t_{now} - t_{stale}(o)) & | \; if \; master(o) \; has \; been \; updated \; at \; t_{stale}(o) \\ 0 & | \; otherwise \end{cases}$$

3.3 Plain FAC Algorithm

We first present a limited algorithm with exact consistency properties, called plain FAC algorithm. This guarantees clients that they always access a consistent (though potentially stale) snapshot of the cache. The algorithm keeps track of the freshness of all cached objects in the middle-tier to be able to always meet the clients' freshness requirements. Clients issue either read-only or update transactions; read-only transactions include an additional parameter, the *freshness limit*, as central QoS-parameter for FAC.

Definition 2 (Freshness-Aware Caching). *Freshness-aware caching (FAC) is an object cache that keeps metadata about the freshness of each cached object. Let $t_{read} = (\{r(o_1), ..., r(o_n)\}, l)$ be a read-only transaction that reads a set of objects o_i with freshness limit l. FAC gives the following guarantees to t_{read}:*

$$\text{(freshness)} \qquad stale(o_i) \leq l \quad | \text{ for } \quad i = 1, ..., n$$

$$\text{(consistency)} \quad stale(o_i) = stale(o_j) \mid \text{if } stale(o_i) \neq 0, \text{ for } i, j = 1, ..., n$$

The core idea of FAC is to allow clients to trade freshness-of-data for caching performance (similar to the rationale of [16]). The staleness of data returned by FAC is bound by the client's freshness limit. The rationale is to achieve higher cache-hit ratios because stale cache entries can be used longer by FAC; if the cached objects are too stale to meet the client's freshness limit, the effect is the same as with a cache miss: the stale object is evicted from the cache and replaced by the latest state from the back-end database.

The plain FAC algorithm guarantees clients to always access a consistent (although potentially stale) snapshot of the cache. This means that for read-only transactions, all accessed objects are either freshly read from the back-end database or they have the same stale point. Update transactions only update fresh objects. If one of these consistency conditions is violated, FAC aborts the corresponding transaction.

FAC Algorithm. Our FAC algorithm extends the object caches of an application server with an additional staleness attribute: a freshness-aware cache is a set of tuples $Cache := \{(o, s)\}$, where o is a cached object and s is the *stale point* of object o, or 0 for freshly cached objects. Furthermore, to provide consistent

---

Algorithm 1. Freshness-aware caching algorithm.

```
 1   function read(t_read, oid, l)
 2   if ( ∃(o_oid, s) ∈ Cache :  (now − s) > l ) then
 3        Cache ← Cache \ (o_oid, s)          // evict too stale o_oid
 4   fi
 5   if ( ∄(o_oid, s) ∈ Cache ) then
 6        o_oid ← master(o_oid)               // load from database
 7        Cache ← (o_oid, 0)
 8   fi
 9   (o, s)  ←  (o_oid, s)  ∈  Cache :  (now − s) ≤ l
10   if ( s = 0 ∨ s = s_{t_read} ) then        // consistency check
11        if ( s ≠ 0 ) then s_{t_read} ← s fi
12        return o
13   else  abort t_read  fi
```

access to several cached objects, FAC keeps track of the first staleness value > 0 seen be a read-only transaction $(s_{t_{read}})$. If FAC cannot guarantee that all accessed objects have the same staleness value, it aborts the request. Algorithm 1 shows the pseudo-code of our algorithm.

3.4 Delta-Consistent FAC

The plain FAC algorithm will give full consistency for cache accesses, but high numbers of updates can lead to high abort rates of readers. Following an idea from [17], we are introducing *delta-consistent FAC* to address this problem.

This approach, called FAC-δ, is to introduce an additional *drift* parameter to FAC that allows clients to read data from different snapshots within the same transaction – as long as those snapshots are within a certain range.

Definition 3 (FAC-δ). *Let* $t = (\{r(o_1), ..., r(o_n)\}, l, \delta)$ *be a read-only transaction that reads a set of objects* o_i *with a freshness-limit* l *and drift percentage* δ *($0 \leq \delta \leq 1$); let* C_{FAC} *be a freshness-aware cache. FAC-δ is a freshness-aware caching algorithm that gives client transactions the following two guarantees:*

$$\text{(freshness)} \qquad stale(o_i) \leq l \qquad |\text{ for } \quad i = 1, ..., n$$
$$\text{(delta} - \text{consistency)} \quad |stale(o_i) - stale(o_j)| \leq \delta \times l \,|\, \text{ for } i, j = 1, ..., n$$

Delta-consistent FAC will abort transactions if they are accessing objects that come from different snapshots which are too far apart as compared to the δ drift value (percentage of the maximum staleness). This requires only two changes to plain FAC: Firstly, our FAC-δ algorithm has to keep track of two stale-points per transaction (the oldest and the newest accessed). Secondly, line 10 of Algorithm 1 changes to check delta-consistency as defined in Definition 3.

4   Implementation

We prototypically implemented freshness-aware caching into the JBoss open source J2EE server [18] for container-managed persistent (CMP) entity beans. Our framework is totally transparent to the CMP entity beans that are deployed into the custom EJB container: there is no special caching API to be called and the whole caching behaviour can be changed just in the deployment descriptor. In the following, we give an overview of our implementation; for space reasons, we restrict the overview to the main concepts of the FAC framework.

FAC maintains *freshness* values for each cache object. Our implementation extends the J2EE container's *entity enterprise context* so that it additionally keeps a timestamp with each entity bean. This timestamp is the bean's stale point, i.e. the point in time when the cached object got updated somewhere in the cluster and hence its cached state became stale. Newly loaded objects have an empty timestamp as by definition fresh objects have no stale point. The timestamp is checked by the *freshness interceptor* to calculate the staleness of a cache object as the time difference between the current time and the stale-timestamp associated with each object.

The *freshness interceptor* is an additional trigger in the interceptor chain of the J2EE container that contains the FAC caching logic. It checks each objects freshness against the current freshness limit and evicts too stale objects before the request is executed; such objects will be automatically freshly loaded by the next interceptor in the chain (the `CachedConnectionInterceptor`). The freshness interceptor also detects state changes of an entity bean and sends those state changes to the *FAC collector*. The *FAC collector* groups all state changes of one or more entity beans in a given transaction.

When a transaction commits, this collection is forwarded to the *freshness cache manager* that informs all caches in the cluster about the stale point of the changed objects. Our implementation uses JBoss' group communication toolkit [2] for all communication between the independent caches. This toolkit guarantees atomicity and message ordering and hence offloads this functionality from our implementation. Each node checks which of the modified beans it also caches and whether they have no stale point associated with them so far, and if both is the case, sets their corresponding stale point.

5   Evaluation

In the following, we present the results of a performance evaluation of FAC versus JBoss' standard cache invalidation and our own cache replication solution [8]. We are in particular interested on the impact of the different configuration parameters on the performance and scalability of FAC.

5.1   Experimental Setup

We used a simple J2EE test application that allows us to concentrate on the application server tier to clearly identify any caching effects. The benchmark application consists of two components: a J2EE application and a test driver in the form of a standalone Java client. The application consists of two entity beans and one stateless session bean. Each entity bean has 10 integer attributes that can be both read and updated; there are corresponding getter and setter methods. The session bean provides two methods to update and to read a several beans. The back-end database consists of the two tables with 2000 tuples each.

All experiments have been conducted on an application server cluster consisting of eight nodes, each with a 3.2GHz Intel Pentium IV CPU and 2 GBytes memory under Redhat Enterprise Linux version 4. One node was used as dedicated database server running Oracle 10.1g. The application server was JBoss 4.0.2 with container managed persistence. The different tests for FAC, invalidation and replication were run using separate JBoss configurations.

The middle-tier database caching solution was TimesTen 5.1 with Oracle as back-end. In TimesTen, we configured the cache schema as one user-defined cache group with incremental auto-refresh every 5 seconds, and automatic propagation of updates from TimesTen to the back-end server at commit time. The distributed cache was managed using the JBoss invalidation method.

5.2   Evaluation of Cache-Miss Costs

First of all, we want to quantify the costs of a cache-miss at the J2EE server, and the improvements possible by using a freshness-aware cache. To do so, we measured inside JBoss the duration of accessing 1000 individual CMP-based entity beans in different system configurations:

1. cache-hit in the EJB cache of JBoss,
2. cache-miss with state loading from a local middle-tier database cache,
3. cache-miss with state loading from the remote back-end database (without any middle-tier database cache).

This gives us the costs at the middle-tier for accessing a cached entity bean or for creating a new CMP-based entity bean instance in case of a cache miss. In the case of a cache hit, the call corresponds to a lookup in the internal EJB cache and returning a pointer to an already existing bean instance. In the case of a cache miss, the requested entity bean must first be created and its state loaded from the underlying relational database.

The resulting access time distributions, as shown in Figure 1, clearly demonstrate the benefits of caching at the middle-tier. The access latency for the internal EJB cache hit is about 0.117 ms on average (median). This is factor 23 times faster than a cache-miss with loading of the state of the CMP entity bean from the remote back-end database (with a median of 2,723 ms and for a warm database cache). Moving the persistent data nearer to the application server on the

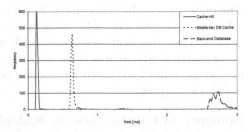

Fig. 1. Histogram of server-internal access times to access one EJB

same middle-tier machine reduces those cache-miss costs drastically. However, having an 'out-of-process cache-hit' in the middle-tier database cache is still about 5 times slower than a cache-hit inside the J2EE process. This is the motivation behind FAC: to improve cache usage by allowing clients to access stale cached data rather than to experience an up-to 23-times slower cache-miss.

5.3   Evaluation of Update Scalability

Next, we are interested in quantifying the scalability of FAC for updates with regard to varying update complexities and varying cluster sizes.

Influence of Update Complexity on Scalability. We start by evaluating the efficiency of FAC with regard to varying update complexity. We do so by

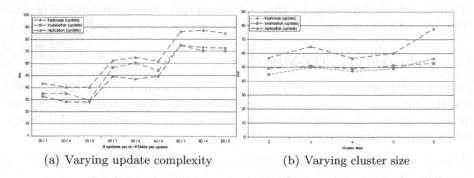

(a) Varying update complexity   (b) Varying cluster size

Fig. 2. Influence of update complexity and cluster size on update response times

varying the number of updates per update transaction and the complexity of updates in terms of how many attributes are modified on a fixed size cluster with 3 nodes. Figure 2(a) shows the results.

We see only a slight variance between the response times for the different number of fields for cache invalidation and FAC, as both send only one message per updated object. On the other side, replication has to send one message per field, which should show up with a higher response time for a growing number of fields. The only reason why this can not be seen in Figure 2(a) is that replication uses a hashmap to save the (field, value) pairs and the minimum size for a hashmap is 12 entries. The figure also shows higher response times for transactions with more updates. All three approaches show a linear update scalability, with cache invalidation and FAC around the same performance, while cache replication is about 15% slower.

Influence of Cluster Size on Scalability. Next, we are interested in the influence of the cluster size on the update performance of the different caching approaches. This will tell us the costs of the overhead of FAC. We ran a number of update transactions of fixed complexity consisting of 40 update operations each using a J2EE cluster of varying size between two and six nodes. The results are shown in Figure 2(b).

All three methods, replication, invalidation, and freshness-aware caching, show a linearly increasing update response time with increasing cluster size. Cache replication shows the slowest update performance and the worst scalability, with a 37% increase of its update time from two to six nodes. In contrast, FAC and cache invalidation show a similar update performance with a very good scalability. This is to be expected as our implementation of freshness-aware caching uses the same update-time propagation mechanism than cache invalidation.

5.4   Influence of Read/Write Ratio

In the following, we concentrate on a comparison between cache invalidation and FAC with regard to multiple concurrent read and update transactions. The next

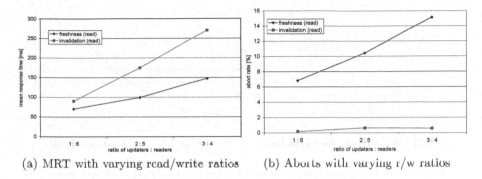

(a) MRT with varying read/write ratios   (b) Aborts with varying r/w ratios

Fig. 3. MRT and abort rates with varying ratios of read- and update-transactions

test is designed to show the influence of higher cache hit rates in freshness-aware caching and in replication with increasing ratio of updaters to readers. This test runs on a cluster of 7 nodes and simulates 7 concurrent clients, some of which issue read-only transactions, some of which issue update transactions (from 6 readers and 1 updater to 4 readers and 3 updates). Each read-only transaction accessed 100 objects, while each update transaction modified 500 objects.

Figure 3(a) shows how the varying read-write ratios affect the cache performance. The more updaters, the slower the mean response times for readers because more cache misses occur. However, FAC is always faster than cache invalidation, in particular with higher number of updaters (up to 80% faster response time with 3 writers and 4 readers). This is due to the fact that with increasing number of updates cache invalidation evicts more and more objects from the cache. In contrast, FAC can still use many stale objects in the cache due to the freshness limits given by the readers (max staleness was 2.5s).

As Figure 3(b) shows, the performance of FAC has also a drawback: higher abort rates. With increasing number of updaters, the cache contains objects with more different staleness values. This increases the chance, that FAC cannot provide a client with a consistent set of stale objects from the cache and hence must abort the reader. Cache invalidation is not affected by this, and consequently shows much lower abort rates due to conflicting reads and writes.

5.5   Influence of Freshness Limit

Next, we want to explore the effect of trading freshness of data for read performance. The experiment was done on a cluster of six nodes and with 60 concurrent clients consisting of 85% readers and 15% updaters. Each read-only transaction was accessing 10 objects, while each update transaction was changing 50 objects. We varied the freshness limit of the readers and measured the effect of the different freshness limits on the mean response times for readers.

The results in Figure 4(a) show how applications which tolerate a higher staleness of the cached data can yield faster response times. There is however a

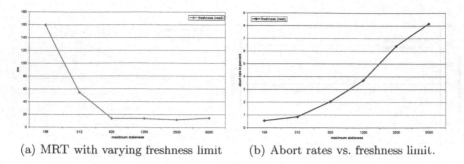

(a) MRT with varying freshness limit   (b) Abort rates vs. freshness limit.

Fig. 4. Mean response times and abort rates with varying freshness limits

limitating effect on the performance by the number of aborts due to inconsistent objects within the same read transaction. This is illustrated in Figure 4(b).

The more readers accept to access outdated data, the higher is the probability that they access objects of different staleness within the same transaction. If those different staleness values exceed the drift value tolerated by the application, the corresponding transaction is aborted. In our case, we configured a maximum drift value of 80%. As shown in Figure 4(b), the abort rate increases from below 1% up to 8% with increasing staleness value. From a maximum staleness of about 625ms onwards, the number of aborted readers exceeds 2% and the throughput in the system actually decreases from now on.

5.6   Influence of Drift Constraint

The previous experiments showed that the costs of lower response times via wider freshness limits are more aborts. In Section 3.4, we had hence introduced FAC-δ that does only provide delta consistency to clients. In the last experiment, we are now investigating the influence of the drift constraint on the read performance with FAC-δ. The following test uses again a cluster of seven nodes and simulates 60 concurrent clients, of which 15% are updaters. Each read-only transaction was accessing 100 objects, while each update transaction was changing 500 objects.

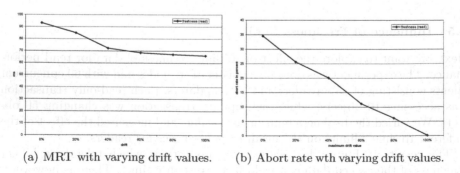

(a) MRT with varying drift values.   (b) Abort rate wth varying drift values.

Fig. 5. Mean response times and abort rates with varying drift values

The results in Figure 5 show that the drift parameter of FAC-δ has the desired performance effect: larger drift values result in lower abort rates and subsequently to faster mean response times. A drift value of 0% represents the basic FAC algorithm, from where the abort rate decreases linearly with increasing drift percentage (cf. Figure 5(b)). With a drift of 100, the abort rate is finally zero, as reads are allowed to read any stale data below the freshness limit.

6   Conclusions

This paper presented a novel approach to caching in a cluster of application servers: *freshness-aware caching* (FAC). FAC tracks the freshness of cached data and allows clients to explicitly trade freshness-of-data for response times. We have implemented FAC in the JBoss open-source application server. In an experimental evaluation, we studied the influence of the different parameters of FAC on its performance and scalability.

It showed that there are significant performance gains possible by allowing access to stale data and hence avoid cache misses. However, because FAC also guarantees clients a consistent cache snapshot, higher freshness limits increase the abort rates. We addressed by introducing *delta-consistent FAC* that allows clients to tolerate stale data from different snapshots within a maximum delta. When comparing our approach with JBoss' standard cache invalidation, it showed that both have a similar scalability and update performance. However, with higher read-write ratios, FAC clearly outperforms cache invalidation.

We are currently working on extending FAC to support consistency constraints to provide scalable full consistency for multi-object requests.

References

1. BEA: BEA WebLogic Server 10.0 Documentation (2007), edocs.bea.com
2. Stark, S.: JBoss Administration and Development. JBoss Group, 3rd edn. (2003)
3. Wu, H., Kemme, B., Maverick, V.: Eager replication for stateful J2EE servers. In: Proceedings of DOA2004, Cyprus, pp. 1376–1394 (October 25-29, 2004)
4. JBoss Cache: A replicated transactional cache, http://labs.jboss.com/jbosscache/
5. SwarmCache: Cluster-aware caching for java, swarmcache.sourceforge.net
6. ehcache: ehcache project (2007), ehcache.sourceforge.net
7. Progress: DataXtend CE (2007), http://www.progress.com/dataxtend/
8. Hsu, C.C.: Distributed cache replication framework for middle-tier data caching. Master's thesis, University of Sydney, School of IT, Australia (2004)
9. Luo, Q., Krishnamurthy, S., Mohan, C., Pirahesh, H., Woo, H., Lindsay, B., Naughton, J.: Middle-tier database caching for e-business. In: SIGMOD (2002)
10. Altinel, M., Bornhövd, C., Krishnamurthy, S., Mohan, C., Pirahesh, H., Reinwald,: Cache tables: Paving the way for an adaptive database cache. In: VLDB (2003)
11. Larson, P.Å., Goldstein, J., Zhou, J.: MTCache: Transparent mid-tier database caching in SQL Server. In: Proceedings of ICDE2004, pp. 177–189. Boston, USA (2004)

12. Amza, C., Soundararajan, G., Cecchet, E.: Transparent caching with strong consistency in dynamic content web sites. In: Proceedings of ICS 2005, pp. 264–273 (2005)
13. Guo, H., Larson, P.Å., Ramakrishnan, R., Goldstein, J.: Relaxed currency and consistency: How to say 'good enough' in SQL. In: SIGMOD 2004, pp. 815–826 (2004)
14. TimesTen Team: Mid-tier caching: The TimesTen approach. In: Proceedings of ACM SIGMOD 2002, pp. 588–593 (June 3-6, 2002)
15. Oracle: Oracle 9i application server: Database cache. White paper (2001)
16. Röhm, U., Böhm, K., Schek, H.J., Schuldt, H.: FAS – a freshness-sensitive coordination middleware for a cluster of OLAP components. In: VLDB, pp. 754–765 (2002)
17. Bernstein, P., Fekete, A., Guo, H., Ramakrishnan, R., Tamma, P.: Relaxed-currency serializability for middle-tier caching & replication. In: SIGMOD (2006)
18. JBoss Group: JBoss (2007), http://www.jboss.org

Collaborative Cache Based on Path Scores*

Bernd Amann and Camelia Constantin

LIP6, Université Paris 6, France
{bernd.amann,camelia.constantin}@lip6.fr

Abstract. Large-scale distributed data integration systems have to deal with important query processing costs which are essentially due to the high communication overload between data peers. Caching techniques can drastically reduce processing and communication cost. We propose a new distributed caching strategy that reduces redundant caching decisions of individual peers. We estimate cache redundancy by a distributed algorithm without additional messages. Our simulation experiments show that considering redundancy scores can drastically reduce distributed query execution costs.

Keywords: Distributed query evaluation, caching policy.

1 Introduction

Many large-scale distributed data integration and data sharing systems are built on top of a federation of autonomous distributed data sources defining distributed queries on local and remote data. The evaluation of such queries recursively triggers the evaluation of sub-queries resulting in the generation of a large number of messages and a high communication cost. This overhead could often drastically be reduced by the temporary caching of frequently used data.

Caching is a widely adopted optimization technique for reducing computation and communication cost in client-server web applications. On the server's side, caching avoids the recomputation of dynamically generated web pages (query result), whereas on the client's side it reduces communication cost and increases data availability. In a more distributed setting where the query workload is distributed on a network of collaborating peers, efficient caching policies are attractive for several reasons. First, peer-to-peer query workloads seem to follow a similar Zipf-like distribution as standard web server workloads [16] (the relative probability of a request for the ith most popular page is proportional to $1/i$) where caching has been shown to be useful. Secondly, and more important, caching a query result avoids the evaluation of all its direct and indirect sub-queries, which multiplies the achieved cost benefit. Finally, collaborative caching strategies can federate storage and computation capacities of different peers.

Finding an optimal cache replacement strategy in a distributed setting is difficult for several reasons. First, as for more traditional client-server settings, the workload and the ordering of the incoming queries at each peer is unknown in

---

* This work was funded by the French national research grant ACI MD SemWeb.

B. Benatallah et al. (Eds.): WISE 2007, LNCS 4831, pp. 87–98, 2007.

advance and all peers exploit certain statistical workload properties (temporal locality, access frequency) for estimating the *usage* of materialized data. Second, even in the case of a known/estimated workload with well defined data access frequencies, finding an optimal data-placement in a distributed setting has been shown NP-complete [8]. Moreover a straightforward caching approach consisting in deploying standard client-server caching strategies on each peer provides limited performance since it does not explicitly take into account possibly redundant caching behaviors. Knowledge about the caching decisions of other peers could help peers to take better cache decisions and avoid redundant caching.

In this article, we propose a new cache replacement strategy that caches query results depending on their usage and on their *cache redundancy scores* reflecting caching decisions made by other peers *using* these results. The contributions of this paper are the following: (*i*) a cache-based query processing and cost model for statically distributed defined queries, (*ii*) a collaborative cache replacement policy based on data usage and cache redundancy estimations, (*iii*) an efficient distributed on-line algorithm exploiting existing query execution messages for estimating global cache redundancy based on results presented in [7], and (*iv*) simulation experiments evaluating the performance of our approach.

The article is organized as follows. Section 2 describes related work. Section 3 presents an abstract cache-based distributed query processing model. Section 4 formally defines our cache replacement strategy. Section 5 presents experiments and Section 6 concludes.

2   Related Work

Web cache replacement policies [15] can be distinguished by the heuristics for estimating the cache benefit of web pages. A successful family of replacement heuristic are based on the Greedy-Dual (*GD*) algorithm [19] using a cost function for estimating the cache benefit of web pages. Greedy-Dual-Size (*GDS*) [5] materializes pages based on their size, the acquisition cost and the cache age. Extensions of *GDS* use recent access-frequency (GDSP, [3]) and document popularity (*GreedyDual∗*, [10]) to estimate the utility of caching some given web page. Our method deploys *GDSP* in a distributed context and compares its performances to the ones of our algorithm considering cache redundancy scores.

Collaborative web cache proxies improve cache effectiveness by sharing distributed cache space and bandwidth. [12] introduces a hierarchical cooperative cache placement algorithm minimizing access cost by caching web pages close to clients. Other distributed web proxies like *Squirrel* [9] and [18] pool the cache space and bandwidth of local web browsers without needing additional dedicated hardware. Our solution can be seen as a form of collaborative caching strategy, since the cache made by a peer considers the decisions made by other peers.

More and more web contents is generated dynamically by structured queries on relational or XML databases. This allows to implement more powerful caching strategies by extending existing database technologies. For example, *Ace-XQ* [6] is a client-server XML query engine which caches dynamically generated results and integrates cache decisions into the query evaluation process. Middle-tier database caching systems like *DBCache* [4] and *DBProxy* [2] improve access latency and scalability by advanced view materialization and query matching techniques for frequently generated query results. *PeerOLAP* [11] caches OLAP query results obtained from a centralized warehouse on a large number of clients connected through an arbitrary P2P network.

Caching also has been applied for P2P query optimization. In *PeerDB* [13] data and queries are distributed and each peer can locally cache the answers returned from remote nodes. Cache replacement is decided locally by applying a least-recently used (LRU) benefit metrics. Cache redundancy is minimized by caching on each peer a copy of maximally one answer set for each source node. [17] proposes a replication method for an unstructured P2P network, that creates or deletes on demand replicas on peers in a decentralized manner. Similarly to [11], peers might cache query results in behalf of other peers by generating additional cache request messages. In our cache model each peer only caches results it needs for evaluating local queries. This has the advantage that it is possible to obtain a certain kind of collaboration without generating additional messages on the network.

3   Cache-Based Query Processing

We consider a distributed system of peers evaluating queries on local and remote data. Each peer is identified by a unique global identifier p (e.g. a URL) and publishes a set of queries $\mathcal{Q}(p)$ on locally stored data and on remote query results. In the following, $q@p \equiv q \in \mathcal{Q}(p)$ denotes that query q is published by peer p and $\mathcal{S}ub(q@p)$ is the set of queries on neighbor peers of p whose results are necessary for evaluating q. Each peer p allocates a fixed space of memory for storing remote query results and $\mathcal{C}ache(p, t)$ denotes the set of all remote queries materialized by peer p at time t. $\mathcal{C}ache(p, t)$ is called the *local cache configuration* of p at time t. By definition, $\mathcal{M}at(q@p, t) = \mathcal{S}ub(q@p) \cap \mathcal{C}ache(p, t)$ is the set of materialized query results necessary for evaluating q at time t and $\mathcal{V}(q@p, t) = \mathcal{S}ub(q@p) \setminus \mathcal{M}at(q@p, t)$ is the set of queries which must be called by peer p at time t for answering q. Virtual and materialized query results are identified by the corresponding query identifier q.

The evaluation of a query q_0 at some instant t_0 recursively triggers the execution of other queries q_i whose results have to be materialized for answering q_0, representing the finite set of *evaluation paths* $\mathcal{P}(q_0, t_0)$. Each path $\pi \in \mathcal{P}(q_0, t_0)$ is a sequence of queries $\pi = q_0(t_0) \rightarrow q_1(t_1) \rightarrow \ldots q_n(t_n)$ where $q_{i+1}(t_{i+1})$ is triggered by query $q_i(t_i)$ $(q_{i+1} \in \mathcal{V}(q_i, t_i))$ and $\mathcal{V}(q_n, t_n)$ is empty. The set $\mathcal{P}(q_0, t_0)$ can be defined recursively: $\mathcal{P}(q_i, t_i) = \{q_i(t_i) \rightarrow \pi | \pi \in \cup_{q_{i+1} \in \mathcal{V}(q_i, t_i)} \mathcal{P}(q_{i+1}, t_{i+1})\}$

Figure 1-a) shows five peers publishing queries on local (not shown in the figure) and remote data and each arc from q_i to q_j signifies that q_i triggers q_j (timestamps were removed for clarity). q_1 generates the paths $q_1 \rightarrow q_4 \rightarrow q_6$ and $q_1 \rightarrow q_3$. Caching obviously can reduce the number of queries used in distributed evaluation. Figure 1-b) shows a cache configuration where p_a cached the result

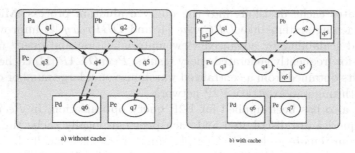

a) without cache   b) with cache

Fig. 1. Distributed query processing with caching

of q_3, p_b cached the result of q_5 and p_c materialized the result of q_6. q_1 and q_2 then only generate the paths $q_1 \rightarrow q_4$ and $q_2 \rightarrow q_4$.

The *evaluation cost* of some query $q@p$ at some instant t is denoted $cost(q, t)$ and obviously depends on the cache configurations of the peers participating to the evaluation of q:

$$cost(q, t) = \alpha \times comp(q, t) + \sum_{q' \in \mathcal{V}(q,t)} (comm(p, q', t') + cost(q', t')) \qquad (1)$$

where $comp(q, t)$ denotes the local computation cost of q on peer p in CPU cycles and $comm(p, q', t')$ is the communication cost of sending query q' and receiving its results at p. Weight α expresses the relative importance of the computation cost of p compared to the cost to obtain the result of q' [1].

It is easy to see that the cache decisions of each peer strongly influences the cost of a given query workload W. The evolution of the local cache configurations of each peer p depends on the queries executed by p: each time p receives the result of some query q', it applies a *cache replacement policy* $\mathcal{R}$ which decides about the materialization of the result of q'. If this is the case, $\mathcal{R}$ identifies a subset $Evicted \in \mathcal{C}ache(p, t)$ of already cached query results that are replaced by the result of q'. In this article we consider cost-aware cache replacement policies [5] which minimize the total execution cost of a *query workload* W. The major issue of cost-based policies is to estimate so-called *cache benefit scores* which reflect the future cost savings obtained by replacing already materialized results by a new result. [3] proposes the following definition for estimating the benefit of caching the result of some query q at time t:

Cache benefit and utility score: The *cache benefit score* $\beta(q', t')$ for the result of some query $q' \in \mathcal{V}(q, t)$ received at some instant t' by p is defined as:

$$\beta(q', t') = v(q', t') * \frac{cost(q', t') + comm(p, q', t')}{size(q')} \qquad (2)$$

This definition is based on the heuristics that the benefit of caching some result is proportional to its computation and acquisition cost and caching a result of large size is less beneficial than caching many results of smaller size. These heuristics are completed by a *utility score* $v(q', t')$ taking into account more statistical information about the query workload for improving the cost of a given query workload W. As shown in the following section, this score might for example reflect the recent usage of a query result [3] (results which have been accessed recently and frequently have higher benefit).

4   Cache Utility Scores

In this section we present a new distributed cache replacement policy for minimizing the cost of distributed query workloads. We extend an existing usage-based utility score [3] by so-called *cache redundancy scores* estimating the "redundancy" of caching some query result received by some peer p with respect to the cache decisions of the other peers in the system.

Local usage-based utility scores: We first describe the usage-based utility score defined in [3] consisting in materializing data that were used frequently and recently. Peer p estimates the access-frequency of the virtual or materialized result of some remote query q by a counter which is decreased by an *aging* function to de-emphasize the significance of past accesses. Let $v_l(q, t^k)$ denote the local utility score computed at the kth access to the result of q at instant t^k. Then, the *local utility score* $v_l(q, t^{k+1})$ of the $(k+1)^{th}$ access to the result of q is computed from previous utility score $v_l(q, t^k)$ as follows:

$$v_l(q, t^{k+1}) = v_l(q, t^k) \times 2^{-(t^{k+1} - t^k)/T} + 1 \qquad (3)$$

where $v_l(q, t^0)$ is initialized to 0 for all q, and T is the period after which the utility is divided by 2 if the results of q are not used.

Path cache redundancy scores: As shown in section 3, if some peer p_{i-1} decides to cache the result of a query $q_i@p_i$, all future evaluation paths $\pi = q_0 \rightarrow \ldots q_{i-1}@p_{i-1} \rightarrow q_i@p_i \rightarrow q_{i+1}@p_{i+1} \rightarrow \ldots q_n$ will "stop" at query $q_{i-1}@p_{i-1}$ and caching the result of $q_{k+1}@p_{k+1}$ where $k \geq i$ on peers p_k will have less or no benefit in future executions of q_0 on path π. This is illustrated in figure 2 showing the evaluation of queries $q_1@p_a$ and $q_2@p_b$ of figure 1 with a cache configuration obtained when all peers locally applied a cache replacement strategy $\mathcal{R}$. Cache of p_c is redundant with respect to the cache of p_a and p_b, since it optimizes the cost of q_4 (by caching q_6) which will not be called anymore during the evaluation of q_1 and q_2. After a certain time period the benefit of caching q_6 will decrease compared to the benefit of q_7 which continues to be used and p_c will finally cache the result of q_7 instead of q_6 (figure 2-b). The following strategy reduces redundant caches by exploiting explicit knowledge on the caching decisions of other peers in the system.

  We estimate for each remote query q_i called by peer p_{i-1} for evaluating query $q_{i-1}@p_{i-1}$ on some evaluation path π, a *path cache redundancy score* $\rho(\pi, q_i) \in [0, 1]$ reflecting the probability that p_{i-1} or some peer p_k before p_{i-1} on π will

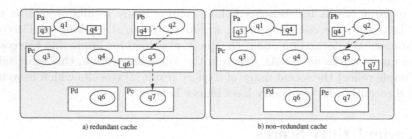

a) redundant cache   b) non–redundant cache

Fig. 2. Redundant cache configurations

cache the result of q_{k+1} on π. More formally, each peer p_k on path π locally estimates the probability that it will cache the result of q_{k+1} received from p_{k+1}, it combines this *local materialization score* with the path redundancy score $\rho(\pi, q_k)$ received from p_{k-1} and propagates it to $q_{k+1}@p_{k+1}$. Then, if p_i receives through $\pi = q_0 \rightarrow q_1 \rightarrow \ldots q_{i-1}@p_{i-1} \rightarrow q_i@p_i$ a path redundancy score $\rho(\pi, q_i)$ that is close to 1 (or 0), it estimates as probable that no (at least one) peer p_k "before" it on π caches the intermediary result of q_{k+1} and that the future evaluation of q_0 will (not) reach p_i on this path. Therefore p_i estimates as (not) "useful" to cache q_{i+1} on path π.

Global redundancy score : Let $\pi = q_0@p_0(t_0) \rightarrow \ldots \rightarrow q_n@p_n(t_n)$ be an evaluation path in $\mathcal{P}(q_0, t_0)$ for some workload query q_0 and $\mathcal{M}(q_i, t_i)$ be the estimated probability that p_{i-1} caches the result of q_i executed at t_i (we show below how to estimate this probability). The *global redundancy score* $\rho(\pi, q_i)$ of q_i on π is sent by p_{i-1} to $q_i@p_i$ at the same time as the call parameters and is defined by the product of the local materialization scores of all queries q_k, $k \leq i$:

$$\rho(\pi, q_i) = (1 - \mathcal{M}(q_0, t_0)) \times \ldots \times (1 - \mathcal{M}(q_i, t_i)) \tag{4}$$

Global utility score : Let $v_g(q_{i+1}, t^k)$ denote the global utility score of caching the results of q_{i+1} on peer p_i computed by p_i at the kth access to the result of q_{i+1} at instant t^k. Let $\rho(\pi, q_i) \in [0, 1]$ be the global redundancy score of query $q_i@p_i$ on path π received by p_i at the $k + 1^{th}$ access to the results of q_{i+1}. The *global utility score* of caching the result of q_{i+1} on p_i at time t^{k+1} is defined as :

$$v_g(q_{i+1}, t^{k+1}) = v_g(q_{i+1}, t^k) \times 2^{-(t^{k+1} - t^k)/T} + \rho(\pi, q_i) \tag{5}$$

Note that the global utility score $v_g(q_{i+1}, t^{k+1})$ of caching the result of some query q_{i+1} at peer p_i depends on its usage (access-frequency and recency) *and on the cache redundancy scores received in the past*. It represents the total estimated number of query evaluation paths which use the result of q_{i+1} until t^{k+1} and did not cache some query result before q_{i+1}. Local utility scores are equivalent to global utility scores with a cache redundancy score equal to 1 on all paths.

Estimating local materialization scores : The *materialization score* $\mathcal{M}(q_i, t_i) \in [0, 1]$ estimates the probability that peer p_{i-1} *will* materialize the result of $q_i(t_i)$

during a certain time period. A precise estimation should take into account different properties (size, access frequency) of q_i and of other queries at p_{i-1}, but also the current and future cache configurations on p_{i-1}. However, the estimation function should also be computed efficiently without additional query overload.

In our experiments we used a simple size-based heuristics for estimating the materialization score of some query $q_i(t_i)$ at some peer p_{i-1}. First, peer p_{i-1} obviously can only cache query results whose size is smaller than the maximal cache size and $\mathcal{M}(q_i, t_i) = 0$ for all queries q_i whose results are larger than the total cache size, denoted by $Total(p_{i-1})$. Second, $q_i(t_i) \in \mathcal{V}(q_{i-1}, t_{i-1})$ is a sub-query of some query $q_{i-1}@p_{i-1}(t_{i-1})$ and *all* the sub-queries in $\mathcal{V}(q_{i-1}, t_{i-1})$ are candidates for being cached by p_{i-1}. Thus the materialization score of $q_i(t_i)$ should take into account the total size of all virtual results $\mathcal{V}(q_{i-1}, t_{i-1})$ for reflecting this conflict.

Materialization score: Let $Occup(p_{i-1}, t_{i-1})$ be the total cache space occupied at p_{i-1} at time $t-1$, $Vsize(q_{i-1}, t_{i-1}) = \sum_{q_k \in \mathcal{V}(q_{i-1}, t_{i-1}) \wedge size(q_k) \leq Total(p_{i-1})} size(q_k)$ be the size of all virtual query results whose estimated size is smaller than the total cache size $Total(p_{i-1})$. The *materialization score* of a query $q_i \in \mathcal{V}(q_{i-1}, t_{i-1})$ called at instant t_i is defined as:

$$\mathcal{M}(q_i, t_i) = \begin{cases} 0 & if\ size(q_i) > Total(p_{i-1}) \\ min(1, \frac{Total(p_{i-1})}{Occup(p_{i-1}, t_{i-1}) + Vsize(q_{i-1}, t_{i-1})}) & otherwise \end{cases}$$

(6)

Note that the above computation supposes that $\mathcal{V}(q_{i-1}, t_{i-1})$ is known before evaluating q_i. If this information is not available, the estimation might replace $Vsize(q_{i-1}, t_{i-1})$ by $size(q_i)$ and consider each sub-query q_i independently. We also suppose that $size(q_i)$ of the result of q_i is known by the peer p_{i-1} when it calls q_i. When q_i is called by p_{i-1} for the first time, $size(q_i)$ is estimated by the average size of the cached data.

5 Experimental Evaluation

This sections presents several experiments conducted on top of a simulation environment for large-scale distributed query processing networks. We consider a network of $P = 1000$ peers and a set of workload queries W uniformly distributed on all peers. Each workload query generates an acyclic graph of sub-queries. The maximal width and height of a workload query are distributed uniformly in $[1, M]$ and $[1, H]$, respectively (triggers on average $(M/2)^{H/2}$ sub-queries). All sub-queries of some query $q@p$ are remote queries on a different peer p' which is chosen randomly among all the peers (clique network). The probability of choosing an existing query $q'@p'$ or of creating a new query $q_{new}@p'$ is 0.5. Unless specified otherwise, all experiments use a workload of $N = 10,000$ queries with maximal width $M = 3$ and maximal height $H = 10$. Each query $w \in W$ has a frequency $f(w) \in [1, F]$ measured in hours (it is executed each hour with probability $\frac{1}{f(w)}$). The minimal query frequency F is set to 8 hours. Frequency values are uniformly distributed on all queries in W, the total number of workload queries executed each hour is on average $(N/F) \times \sum_{k=1}^{F}(1/k) \sim (N/F) \times ln(F)$. For facilitating the interpretation of results, we assume that all query results have the same size. The default cache size of a peer p is 25% of the total number of sub-queries called by p.

We measured the performance of each cost-based cache replacement at t by the sum of execution costs of all queries in the workload each hour until t. We simulated the execution of the query workload during $t = 48$ hours. Transfer rate is chosen randomly for each couple of peers (p, p') in the interval $[0.01, 1.00]$. The computation cost is the same for all queries on all peers and is set to 0.001 (query evaluation cost is mainly influenced by communication cost of its sub-queries). We implemented and compared three utility score measures for estimating the cache benefit of the result of some query $q(t)$:

- *local* [11] : constant utility score $v(q, t) = 1$; the cost benefit only depends on the local query cost and size.
- *local+* [3] : local utility score $v_l(q, t)$ which also takes into account usage estimations (equation 3).
- *global* (our contribution): global utility $v_g(q, t)$ which takes into account usage and *cache redundancy* estimations (equation 5).

Figures 3 and 4 compare the cost of the workload depending on the maximal query width M (Figure 3) and maximal number of evaluation hops per workload query H (Figure 4). Increasing the average height and/or width of all workload queries increases the total number of sub-queries triggered by each workload query and, without caching, the evaluation cost would increase. However all caching policies *decrease* the query execution cost. This is because the average query frequency at each peer increases with the average query height and/or width (the number of peers and workload queries is fixed) and it is more probable that the materialized data is useful for its neighbor peers. The peers take advantage of the cache of their neighbors, and materialize data whose cost was not improved by their neighbors. This "collaborative behavior" is equivalent to a growth in the cache capacity of each peer, thus reducing the *global* cost of the workload. This is also confirmed by Figure 5 showing that the workload cost decreases when the average number of neighbor peers increases.

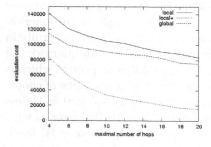

Fig. 3. Variable number of hops H

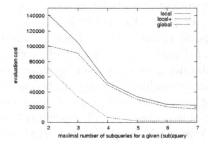

Fig. 4. Variable number of neighbors M

The influence of a caching decision of some peer p on some other peer p' decreases with the distance between p and p' : caching some data at some peer has greater probability of being useful for "near" peers than for distant ones (if the peer is farther it is possible that other peer between them has made a cache for it). This is confirmed experimentally in Figures 3 and 4 where the average query width (Figures 3) has a stronger impact on the workload cost saving than the average query height (Figure 4). For example, the number of sub-queries generated with configuration $c_1 : (M = 3, H = 20)$ is the same than the number of sub-queries of configuration $c_2 : (M = 4.5, H = 10)$. Nevertheless, the cost achieved by the *local+* strategy is twice bigger with c_1 than with c_2 (the same is true for *local*). *Global* improves the cost in configuration c_2 by 75% compared to configuration c_1.

Figure 5 shows the cost depending on the maximal number of neighbors per peer. The total cost decreases with the increasing number of neighbors and stabilizes at 100 neighbors/peer (*i.e* when each peer knows 10% of all peers). When the number of neighbors is limited to 1, each peer receives, on average, calls from a single other peer and all evaluation paths on which a peer contributes to the workload mainly visit the same peers. In this case the improvement achieved by the *global* compared to *local+* is only about 20%. By increasing the average number of neighbor peers, the relative improvement between the *global* and the *local+* policy reaches 65% at 10 neighbors/peer.

Figure 6 shows the cost for different number of workload queries. For each strategy the cost increases linearly with the number of workload queries, but this increase is much slower for the *global* strategy. The *global* caching thus

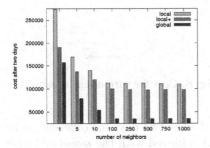

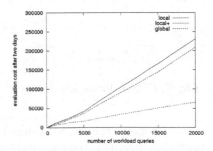

Fig. 5. Allowed neighbors per peer   **Fig. 6.** Variable workload queries (N)

performs well for a large number of workload queries. The *local+* strategy is better than *local* since it takes into consideration the usage of the cached data. But it also creates more redundant cache than the *global* strategy since the scores of redundant data increase. Therefore *local+* evicts them from the cache after a longer period.

Figure 7 shows the *Detailed Cost Saving Ratio(DCSR)* [11] at the end of each day during 7 days for two query tree configurations. The *DCSR* corresponds to the relative gain of a given strategy (*local+* or *global*) compare to the *local* strategy. The first configuration, c_1, is the randomly-generated one used above. In configuration c_2, a subset of queries (the popular ones) on each peer has many more clients than the other queries of the peer. Each peer has a subset of popular results (produced by popular neighbor queries) which are more used than the others by the queries of the peer. Since many queries of each peer share the same results, the peer makes few different calls. The difference between the costs with strategies *local+* and *global* in configuration c_1 decreases at the end of the seventh day, but remains significant (22% vs. 60%). This is because the *local+* and *local* strategies also evict redundantly cached data after some time period. In configuration c_2, the cost improvements of *global* and *local+* do not change in time. Both strategies cache from the beginning the data for the most popular services and do not change their decisions in the following hours when other queries of the workload are also executed. In this configuration the cost improvement of *global* is less significant compared to *local+* (45% vs. 60%). The *global* strategy does not cache some (unpopular) data which might improve the workload cost since their estimated utility is too low compared to the utility of the popular ones. The *local+* caches more unpopular data than *global* (each call has an estimated utility of 1) and finally performs better since it only considers their usage.

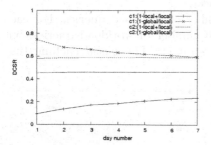

Fig. 7. $DCSR$ during seven days   **Fig. 8.** Cost for various cache capacities

Figure 8 shows the $DCSR$ values computed after two days for different cache capacities. Obviously, the cost of the workload achieved with the *global* and with the *local+* strategies is equal to the cost with the local strategy when the size of the cache is 0% or 100%. When all peers have a small or high cache capacity, the absolute cost improvement is small compared to the total workload cost for both strategies. *Global* has a maximal relative cost improvement for a cache size of 40%, whereas *local+* follows a more flat curve (maximal relative gain is 20%).

Figure 9 shows the effect of additional queries generated by an expiration based cache refresh policy. We assume that all cached data expires after some fixed time period and is removed from the cache. In general, this increases the workload cost since even popular query results have to be rematerialized periodically which generates additional queries. The cost of evaluating these queries is higher with *global* than with *local+*, since it cannot benefit from redundantly

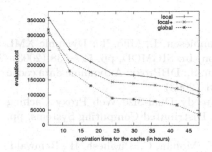

Fig. 9. Expiration of the cached data

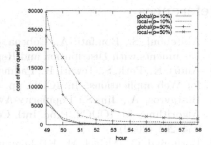

Fig. 10. Cost of 10%/50% new queries

cached query results. In consequence, for small expiration periods, the cost improvement achieved by minimizing redundancy is erased by the additional cost of recomputing expired query results.

Figure 10 shows the behavior of the system for a sudden surge in the workload (unexpected number of queries for which no data was cached). We executed $(100-p)\%$ of the workload during two days, with frequencies as described above. Then we sent the additional $p\%$ queries each hour, during 10 hours. We measured the sum of the costs of the *new* queries each hour during 10 hours. During the first hour of the third day the total cost of the new queries is higher with *global*, since *local+* makes redundant caches on the peers on which the new queries are executed, and a part of the new queries takes advantage of those caches. Nevertheless during the following hours the *global* strategy outperforms *local+*.

6   Conclusion and Future Work

We present a distributed cache replacement policy for optimizing distributed query evaluation. We extend the query evaluation process by an efficient distributed cache replacement policy deciding which peer should materialize which query result by using local access frequency information and global caching estimations. We have shown how our path materialization scores can be used for implementing a collaborative cache replacement strategy without generating any additional messages. We evaluated the qualitative performance of our strategy by experimentation on simulated query workloads.

We intend to extend an existing P2P XML query engine in the context of a distributed data warehouse application (WebContent) in order to verify if our dynamic caching strategy is still useful in a more complex environment where it interacts with other sophisticated query optimization techniques based on query rewriting and data indexing. We concentrated on the problem of cache replacement and did not consider data updates and cache synchronization. Our experiments confirmed that all caching strategies are sensitive to cache refresh strategies. We believe that our framework can efficiently be combined with advanced cache synchronization protocols like proposed in [14].

References

1. Abiteboul, S., Bonifati, A., Cobena, G., Manolescu, I., Milo, T.: Dynamic XML Documents with Distribution and Replication. In: SIGMOD, pp. 527–538 (2003)
2. Amiri, K., Park, S., Tewari, R., Padmanabhan, S.: DBProxy: A dynamic data cache for Web applications.. In: ICDE, pp. 821–831 (2003)
3. Bestavros, A., Jin, S.: Popularity-Aware Greedy Dual-Size Web Proxy Caching Algorithms. In: ICDCS. Proc. Intl. Conf. on Distributed Computing Systems, pp. 254–261 (2000)
4. Bornhövd, C., Altinel, M., Krishnamurthy, S., Mohan, C., Pirahesh, H., Reinwald, B.: DBCache: Middle-tier Database Caching for Highly Scalable e-Business Architectures. In: SIGMOD, p. 662 (2003)
5. Cao, P., Irani, S.: Cost-Aware WWW Proxy Caching Algorithms. In: USENIX Symposium on Internet Technologies and Systems (1997)
6. Chen, L., Rundensteiner, E.A.: ACE-XQ: A CachE-aware XQuery Answering System. In: WebDB. Proc. Intl. Workshop on the Web and Databases, pp. 31–36 (June 2002)
7. Constantin, C., Amann, B., Gross-Amblard, D.: A Link-Based Ranking Model for Services. In: CoopIS. Proc. Intl. Conf. on Cooperative Information Systems (2006)
8. Gribble, S.D., Halevy, A.Y., Ives, Z.G., Rodrig, M., Suciu, D.: What Can Database Do for Peer-to-Peer?. In: WebDB. Proc. Intl. Workshop on the Web and Databases, pp. 31–36 (2001)
9. Iyer, S., Rowstron, A.I.T., Druschel, P.: Squirrel: a Decentralized Peer-to-Peer Web Cache. In: ACM Intl. Symp. on Principles of Distributed Computing (PODC), pp. 213–222 (2002)
10. Jin, S., Bestavros, A.: GreedyDual* Web Caching Algorithm: Exploiting the Two Sources of Temporal Locality in Web Request Streams. Computer Communications 24(2), 174–183 (2001)
11. Kalnis, P., Ng, W.S., Ooi, B.C., Papadias, D., Tan, K.-L.: An Adaptive Peer-to-Peer Network for Distributed Caching of OLAP Results. In: SIGMOD, pp. 480–497 (2002)
12. Korupolu, M.R., Dahlin, M.: Coordinated Placement and Replacement for Large-Scale Distributed Caches. TKDE 14(6), 1317–1329 (2002)
13. Ng, W.S., Ooi, B.C., Tan, K.-L., Zhou, A.: PeerDB: A P2P-based System for Distributed Data Sharing. In: ICDE, pp. 633–644 (2003)
14. Olston, C., Widom, J.: Best-Effort Cache Synchronization with Source Cooperation. In: SIGMOD, pp. 73–84 (2002)
15. Podlipnig, S., Böszörményi, L.: A survey of web cache replacement strategies. ACM Comput. Surv. 35(4), 374–398 (2003)
16. Sripanidkulchai, K.: The popularity of gnutella queries and its implications on scalability (February 2001), http://www.cs.cmu.edu/ kunwadee/research/p2p/ gnutella.html
17. Tsoumakos, D., Roussopoulos, N.: An Adaptive Probabilistic Replication Method for Unstructured P2P Networks. In: CoopIS. Proc. Intl. Conf. on Cooperative Information Systems, pp. 480–497 (2006)
18. Xiao, L., Zhang, X., Andrzejak, A., Chen, S.: Building a Large and Efficient Hybrid Peer-to-Peer Internet Caching System. TKDE 16(6), 754–769 (2004)
19. Young, N.E.: On-line caching as cache size varies. In: SODA. Proc. of the Second Annual ACM-SIAM Symposium on Discrete Algorithms, pp. 241–250 (1991)

Similarity-Based Document Distribution for Efficient Distributed Information Retrieval

Sven Herschel

Humboldt-Universitt zu Berlin
Unter den Linden 6, 10099 Berlin, Germany
herschel@informatik.hu-berlin.de

Abstract. Performing information retrieval (IR) efficiently in a distributed environment is currently one of the main challenges in IR. Document representations are distributed among nodes in a manner that allows a query processing algorithm to efficiently direct queries to those nodes that contribute to the result. Existing term-based document distribution algorithms do not scale with large collection sizes or many-term queries because they incur heavy network traffic during the distribution and query phases.

We propose a novel algorithm for document distribution, namely *distance-based document distribution*. The distribution obtained by our algorithm allows answering any IR query effectively by contacting only a few nodes, independent of both document collection size and network size, thereby improving efficiency. We accomplish this by linearizing the information retrieval search space such that it reflects the ranking formula which will be used for later retrieval.

Our experimental evaluation indicates that effective information retrieval can be efficiently accomplished in distributed networks.

1 Introduction

Distributed information retrieval systems have attracted much research interest lately. Applications for reliable, content-based access to a potentially very large, distributed collection of documents can be found in ad-hoc sharing of personal libraries as well as in larger collaboration scenarios of publicly available document collections. [1]

These applications not only require effective IR, meaning a satisfactory result quality to user queries. Indeed, they also require distributed IR to be efficient so that satisfactory results are retrieved in a reasonable amount of time. This is especially challenging in distributed scenarios without a centralized index: document representations must be distributed among the nodes in such a manner that any node on the network may issue a query and identify nodes that potentially contribute to its answer[1]. Of special interest in this context are distributed IR systems built on structured overlay networks, since they provide key-based

---

[1] For obvious reasons of scalability, we omit the theoretical alternative to query all nodes within the network.

B. Benatallah et al. (Eds.): WISE 2007, LNCS 4831, pp. 99–110, 2007.

routing and a scalable, self-maintaining, and reliable network infrastructure for distributed applications.

Key-based routing (KBR)[2] provides the means to reach a node by an application-specific identifier, which potentially carries semantic meaning. This identifier is translated by the network overlay into a node address (nodeId), allowing us to reliably address nodes by arbitrary keys. Current distributed IR systems built on top of structured overlays naively distribute the inverted index—they use the index entries (or: *terms*) as the application-specific identifier. This approach, however, fails to scale with increasing document collection size, because (1) each document in the entire collection must be indexed under all its terms—a potentially very expensive process—and (2) a query consisting of more than one term must contact more than one node—with potentially large intermediate results.

We take an entirely new perspective at distributing documents within a structured overlay: instead of naively distributing the inverted index, we already include the ranking calculations into our distribution strategy. This gives us the opportunity to cluster documents onto nodes based on document similarity, so that the documents that constitute the answer to a query will be located on the same node.

Contribution. While we build our research on the routing infrastructure of a structured overlay, we still face the challenge of distributing an inherently centralized application like an information retrieval system among the nodes within this overlay.

We propose a novel approach of mapping documents to nodes in a structured overlay. Instead of a term-based distribution strategy, we use an *similarity-based* distribution strategy that builds linear identifiers [3] for documents based on the ranking function of the IR model. This results in:

– A similarity-based clustering of documents onto nodes, allowing direct routing of a query to relevant nodes according to the ranking function. This strategy is obviously very efficient in identifying nodes within the network that answer a given query.
– A retrieval process that only transfers document representations over the network that contribute to the query result—our precision is always 100%. As opposed that, term-based approaches also return results that only match part of the query and will not be part of the final result.
– Incremental result improvement: it is not necessary to retrieve the entire result set at once. Instead—comparable to the "next"-button of search engines—results can be requested incrementally.

Our preliminary experimental evaluation shows promising results which indicate that effective information retrieval can be efficiently accomplished in structured overlay networks.

Structure of this paper. Sec. 2 introduces structured overlays and our terminology. Sec. 3 introduces our algorithm for distributing document representations and querying a collection and Sec. 4 evaluates our approach regarding precision, recall and efficiency. Sec. 5 covers related work and Sec. 6 concludes the paper and gives an outlook on future research on this subject.

2 Overlay Networks in a Nutshell

In this section, we give a brief introduction to structured overlay networks and introduce the terminology used in this paper. In order to provide a consistent path through our paper, we will work with the following example.

Example 1. Consider a document collection D of three documents d_1, d_2, d_3. We have two terms $T = t_1, t_2$ that exhaustively represent the contents of our documents as follows: $d_1 = t_1, d_2 = t_1, t2, d_3 = t_1, t_2, t_2$. We now face the challenge of distributing these three documents among our network of nodes $N = (n_1, n_2, n_3)$. We target much larger collections, but we choose this small example for better comprehension. □

2.1 Overlay Networks

When we talk about networks of nodes without central administration that aim to achieve a common goal, we talk about so-called *overlay networks*. In general, an overlay network is built on top of an existing network, in our case the internet. This abstraction allows us to route messages without being concerned with establishing or maintaining the logical connections of the distributed network or ensuring message routing between nodes of the overlay. Overlay networks have received much research attention [2] due to some compelling properties:

- They distribute processing tasks among all nodes of the network and are therefore able to grow organically as supply (i.e., available storage, network bandwidth, etc.) increases together with demand (queries to be processed).
- They don't require a systems administrator and can therefore decentrally be deployed and maintain themselves.
- They are robust in the sense that there is no single point of failure which takes the entire network down. Usually, processing results degrade gracefully with failing nodes.

2.2 Structured Overlay Networks

Structured overlay networks[4,5,6,7] give us the additional benefits of a well-defined network structure in which the nodes are arranged, reliable routing and application-specific keys. To construct a structured overlay network, let us assume that each node of the network has a unique *nodeId* from a very large id space, i.e. $0 \ldots 2^{64} - 1$, nodes are randomly distributed among this id space and the ends of the id space are connected, so nodes are arranged in the structure of a circle.

Let us further assume that each node n_x is *responsible* for all ids between its own id and the id of the following node: $nodeId(n_x) < id <= nodeId(n_{x+1})$.

Existing routing algorithms for structured overlays are capable of routing any message that has been adressed to an *id* to the responsible node, even under adverse conditions like failing nodes and changing network setups. Applications based on these networks provide a hash function $hash : object \rightarrow id$ that converts an application-specific key (*object*) into a valid ID for the networks (*id*).

The challenge for researchers has now been to model complex applications such as a distributed IR system on top of these structured overlays.

Consider an application-specific key filename and a hash function $hash : filename \rightarrow id$ that converts any filename into an *id* as required by our structured overlay. We then partition our collection as follows: each document d is stored at the node responsible for the id $hash(filename(d))$. This is called *document-based* partitioning and has been successfully implemented in distributed file systems, for example.

The drawback of using the filename as the key is that it is necessary to know a document's filename in order to find it. Information retrieval requires *content-based search*, i.e., search within the content of documents. In order to support content-based search, we need to distribute the inverted index of the collection among the nodes of our network. This approach is called *term-based* partitioning and distributed information retrieval systems (e.g., [8]) have been proposed using this approach. The following example illustrates term-based partitioning for our running example.

Example 2. To support content-based search, we tokenize each document into terms and distribute the document's representation to each node responsible for one of the contained terms. Let us assume that—by an appropriate *hash* function—n_1 is responsible for t_1, and n_2 is responsible for t_2. Since $d_1 = t_1, d_2 = t_1, t_2, d_3 = t_1, t_2, t_2$, we distribute d_1, d_2, and d_3 to node n_1, and d_2 and d_3 to node n_2. See also Fig. 1 (a).   □

Term-based distribution has the big advantage that it is very easy to route a query: the query engine splits the query into its individual terms $T(q)$ and contacts the nodes responsible for these terms. Systems based on term-based partitioning have proven to work quite effectively, but their major drawback is efficiency. Each document is indexed under *all* of its terms, resulting in a lot of indexing network traffic. Similarly, once we consider queries with more than one term, we need to contact more than one node and each node sends back all results that contain the single term it has received. This results in lots of unneccesary network traffic because a node can neither decide whether the query consisted of additional terms, nor whether these are part of the indexed document, nor whether the final ranking will include the document in the result list.

As we can see, both the filename-based partitioning as well as the term-based partitioning have substantial drawbacks. In this paper, we introduce a new partitioning approach: *similarity-based* partitioning. We split the document collection along document similarity, so similar documents will be clustered to the same node. The interesting question is "similar compared to what"? We will address this question in Sec. 3.

2.3   Terminology

We consider a document collection D of size $p = ||D||$, consisting of individual documents $d_1 \ldots d_p$. Documents are distributed among nodes $N = (n_1 \ldots n_q)$ with $q = ||N||$ being the number of nodes in the network. Each node has a unique id $nodeId(n)$. Similarly, each document stored at a node has a unique id $docId(n,d)$, incorporating the nodeId and a (node-unique) document id. We refer to the set of all documents stored at a node as $D(n)$ and we assume that $\bigcup D(n) = D$. For the purpose of our paper it does not make a difference whether a document is actually stored at a node or whether it is just a document representation. For clarity of presentation we have decided to speak of documents being stored at nodes.

Documents consist of terms t from a term universe T [2]. Therefore, each document can be represented as a vector $\boldsymbol{d}$ with dimensionality $||T||$. The individual vector values represent for each term the weight each term has within the document. A term that never occurs within the document has a weight of 0.

A user queries the system by means of a query q, consisting of terms from T.

Finally, we have a function $hash$ that maps a set of terms to a set of $nodeIds$: $h : \mathcal{P}(T) \rightarrow \mathcal{P}(nodeId(n))$. We use this function to map documents as well as queries (both being a set of terms) onto the nodes that are responsible for storing the documents and processing the queries, respectively.

3   Similarity-Based Document Partitioning

As we have seen in Sec. 2, one challenge of information retrieval in structured overlays is how document representations are partitioned among the nodes. In this section, we describe similarity-based document partitioning, a novel partitioning strategy that allows efficient and effective IR. We first provide an overview of our approach before demonstrating its application for indexing documents and processing queries.

3.1   Quick Overview

Our algorithm is based on *document similarity*: instead of storing a document representation at all nodes responsible for one of the terms in the document, as term-based approaches do, we store the document at the *same* location as similar documents. This allows both for very efficient indexing of the collection as well as for very efficient retrieval as a query can be routed to the node storing documents that are most similar to the query.

Example 3. Fig. 1 shows the differences between term-based partitioning and similarity-based partioning. In Fig. 1(a) node n_1 is responsible for term t_1. It therefore indexes all documents containing t_1, i.e., d_1, d_2 and d_3. Similarly, n_2 indexes d_2 and d_3. Obviously, d_2 and d_3 are indexed at multiple nodes. In Fig. 1(b)

---

[2] We assume that common information retrieval optimizations always take place so, for example, terms are stemmed and stopwords are not part of the term universe.

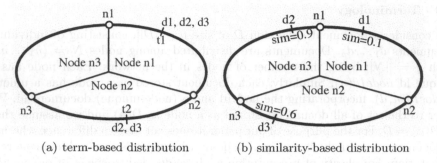

(a) term-based distribution   (b) similarity-based distribution

Fig. 1. Comparison of distribution strategies

documents are indexed according to our approach (see Sec.3.2). This results in each document being indexed at exactly one node.   □

Essentially, for each document we calculate its distance from a virtual landmark document LD in the center of the vector space. This distance value is our application-specific key for structured overlays: it is mapped into the network's ID space using *hash* and therefore tells us at which node to store the document representation. Queries then have the same distance from the landmark document as the expected result documents, allowing us to route queries directly to the responsible nodes.

Another advantage of similarity-based document distribution is that the user may decide herself how many documents she cares to retrieve. For a quick result, only the responsible node is contacted. For more exhaustive results she directs the query to the neighbors of the responsible node. As soon as her information need is fulfilled, she stops query processing. Compared to term-based partitioning approaches, which provide an all-or-nothing alternative (since all responsible nodes must be contacted before a result can be computed), our approch provides gradually improving query results. Speed of query processing can easily be traded against result quality. This means that precision of our results is always 100% while recall can be increased incrementally. Thinking of user experience on for instance search engines such as Google or Yahoo, we believe that this incremental solution is acceptable and avoids network traffic during query execution. Indeed, users of search engines do not need all results returned by a search engine, they are usually satisfied with the first few results.

3.2   Indexing Documents

Having discussed the main idea of similarity-based distribution of documents, we now provide details on both indexing and querying. To index documents, similarity-based distribution uses a fixed location within the vector space which we name the *landmark document LD*.

The landmark document. As stated before, the vector space in our system has $||T||$ dimensions, one dimension for each term of the term universe. Each

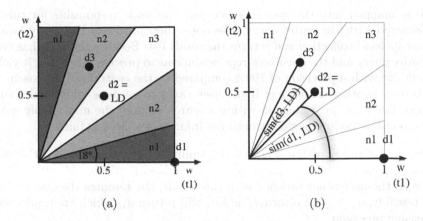

Fig. 2. (a) shows a three documents in a two-dimensional term vector space, (b) shows how their angle distance from the landmark document determines their node location

document is represented by a vector d, the vector values represent the importance of the specific term within the document. Importance values come from the interval $[0 \ldots 1]$.

We define the *landmark document* to be the center of this vector space. Therefore, LD has a value of 0.5 in all its vector dimensions: $LD = (0.5, 0.5, \ldots)^T$. Any other fixed point within the vector space can be chosen as a landmark document, as long as all nodes in the system use the same LD for their calculations.

Distributing a document. To find the node responsible for a specific document, we calculate the document's similarity to the landmark document $sim(d, LD)$ (Equ. 1) and map the score into the id space using Equ. 2.

$$sim(d, LD) = 1 - \left| \cos^{-1} \frac{d \cdot LD}{||d|| \cdot ||LD||} \right| \tag{1}$$

$$docId = (2^{64} - 1) \cdot sim(d, LD) \tag{2}$$

Example 4. Fig. 2 illustrates document indexing for our example with two terms and three documents. We calculate the vector weights as the normalized term frequencies: $w_{t,d} = \frac{TF(t,d)}{|d|}$. Since d_1 only consists of one term t_1, it has a document vector of $(d_1) = (1, 0)^T$. Correspondingly, $(d_2) = (0.5, 0.5)^T$ and $(d_3) = (0.333, 0.667)^T$.

The landmark document vector in this two-dimensional scenario is $LD = (0.5, 0.5)^T$. Therefore, $sim(d_2, LD) = 1$ and $docId(d_2) = 2^{64} - 1$. Since n_3 is responsible for this id space, d_2 is distributed to n_3. Similarly, d_1 and d_3 are distributed to n_1 and n_2, respectively. □

3.3 Processing Queries

Querying follows the same principles as distribution: given a query q, we calculate the angular distance from the landmark document LD using Formula 1. This value is mapped into the *nodeId* space and the node responsible for this *id* is contacted with the entire query. The node then processes the query locally against its local collection and returns the result list. Because the node has both the entire query and the document representations to process this query, it yields a result list with a precision of 100% compared to the centralized approach.

There is another advantage: If the user cares to retrieve additional results (comparable to a "next" button within a search engine) she may simply query the responsible node's neighbors until her information need is fulfilled.

Example 5. A query with $q = t_1, t_2$ has a similarity of 1 to LD and is therefore routed to node n_3, which is the node where the best result for this query, d_2, is stored. If the user is not satisfied with this result, she reroutes the query to n_2, and, possibly, n_1. n_2 will return d_3 and n_1 will return d_1, which are results with decreasing precision.   □

4 Evaluation

In this Section, we present initial results we have obtained using similarity document distribution. Our evaluation is split in two: (1) We evaluate the effectiveness of our algorithm against a centralized approach using the same retrieval model. (2) We evaluate the reduction in network traffic we achieve compared to the standard approach for distributed IR in structured networks.

4.1 Experimental Setup

Inspired by the research of [9] for a standardized, open data set for distributed information retrieval systems, we built our document collection from the Wikipedia[3] as follows: We split the wikipedia dump from January 25th, 2006, into individual files, named according to their page title. We then removed wikipedia special pages, i.e., pages starting with either "Image:", "Wikipedia:", "Template:", or "Category:", because these documents do not contribute to our document retrieval. Similarly, all "redirect" documents were removed. While being named like normal wikipedia documents, these documents only contain a "redirection" instruction, pointing to a different document within the same collection, i.e., for resolving synonym conflicts. Due to the large number of individual experiments, we opted for a document corpus of moderate size and therefore considered each 1,200th document from the remaining file list, resulting in a document corpus D consisting of 2,000 documents.

For our network, we experimented with two network setups, a relatively small one, arbitrarily chosen to consist of 100 nodes, and a relatively large one, chosen to consist of 30,000 nodes. In both setups, we evenly distributed nodes across the *nodeId* space.

Our approach works best with longer queries, i.e., queries consisting of multiple terms. We therefore posed queries of 15 to 20 terms each to the system. The

---

[3] http://en.wikipedia.org

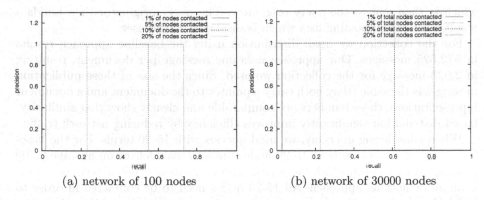

(a) network of 100 nodes   (b) network of 30000 nodes

Fig. 3. Precision and recall for query processing using the similarity-based algorithm

queries were generated by randomly choosing terms from the term universe T. We ran 1,000 of these queries against the centralized index and both the distributed indexes. The result from the centralized index served as our gold standard for evaluating the results from distributed indexes. The resulting recall/precision values were averaged among all queries.

4.2 Relative Recall

We evaluate our approach against a centralized search engine using the same retrieval model and measure its *relative recall*. That means, that if our approach yielded the same documents, the relative recall would constantly stay at "1".

Fig. 3 shows the results of our experiments. We would like to emphasize on the following points: (1) the precision is always 100%, i.e., only document representations that are part of the result set, are actually transferred. (2) Recall improves when additional nodes are contacted: While it is sufficient to only contact a few nodes, it is possible to further expand the result set by contacting additional nodes. We believe, however, that there's a maximum number of documents a user would like to see—common search engines limit the pagesize of result sets to 10.

4.3 Efficiency Estimation

Our efficiency estimation considers both distributing the documents and querying the network. We assume a naive term-based distribution algorithm as our baseline and use the same workload as before. The naive term-based approach works as follows: During distribution, each document is distributed to each node that is responsible for one of its term. During querying, for each term in the query, the responsible node is contacted for documents matching this term. The node collects all matching document representations in a so-called posting list which is then sent to the query originator. The query originator then builds a result list from all posting lists which is displayed to the user.

For the collection we used, distribution using the baseline approach results in 572,525 messages. Our approach uses one message per document, resulting in 2,023 message for the collection we used. Since the size of these publication messages is the same (they both carry a pointer to the document and a document representation), these numbers are comparable and clearly show that similarity-based distribution significantly improves efficiency by reducing network traffic.

When considering querying, we used queries with 15-20 terms. For the baseline approach, this results in 15-20 node contacts that deliver on average 2,200 posting list items from which the matching documents are then determined. This is an all-or-nothing approach: All 15-20 nodes need to be contacted in order to build the result set. Our approach only retrieves document postings that contribute to the result set. The user therefore decides herself how many documents she cares to retrieve. This approach follows the user interface pattern of search engines which only present few results and deliver additional results on demand.

5   Related Work

Our work is based on three main areas of research.

Linearization of vector spaces. Our work is about similarity-preserving linearization of high-dimensional vector spaces for efficiency reasons. While vector linearization has been explored for quite some time[3], most fail to be similarity-preserving because of the "curse of dimensionality". A recent work exploring vector linearization for nearest neighbor search in vector spaces is [10]. To our knowledge, no work exists taking advantage of similarity-preserving vector linearization for distributed information retrieval.

Information retrieval in distributed networks. The foundations for information retrieval using the vector space model were laid by Salton et. al. in [11]. Since then centralized information retrieval has advanced and inspired research in distributed information retrieval [1]. A general architecture for distributed information retrieval comes from [12]. Our work in inspired by theirs.

Keyword search as a subset to the distributed information retrieval problem has been addressed in [13]. [14] introduces a straightforward approach for implementing standard information retrieval algorithms within a structured overlay network, but does not address efficiency issues. [15] et. al. focus on information retrieval in unstructured overlay networks, combining collection structure with information retrieval. Because of reliability concerns[4] we focussed our research on structured overlays. Interestingly, IR researchers within unstructured networks start incorporating structure into their networks in order to be more efficient [16].

---

[4] Unstructured P2P networks are unable to efficiently answer the "non-existence" question, i.e., provide the assurance that an object does *not* exist in the network.

Structured overlay networks. The algorithms that served as the starting points for the structured overlay revolution were Pastry [4], CAN [5], Chord [6] and Tapestry [7]. The group around Karl Aberer proposed a general reference architecture for overlay networks [2]. Our work relies on structured overlays as the routing substrate for our similarity-based IR system.

6   Conclusion

We introduced *similarity-based* distribution of documents among nodes within a distributed information retrieval system. Similarity is based on a *landmark document* in the center of the vector space. This approach allows us to forestall the ranking function and distribute documents according to their similarity to the landmark document. Since queries use the same ranking function, it is sufficient for query processing to calculate the query's distance to the landmark document and start retrieving documents at the responsible node.

We experimented with our approach within different network setups and the results of our experiments support our approach:

Only meaningful documents. Our approach ensures that only documents that contribute to the result set are transferred. Precision remains at 100%. This is contrary to term-based approaches where intermediate posting lists are transferred and only integrated and pruned at the querying node.

Incremental result improvements. Our approach allows to incrementally contact additional nodes if the result set does not satisfy the user's information need. A new node contact only yields and transfers results contributing to the result set.

Outlook and future research. The work presented in this paper provides some promising results showing that similarity-based document placement can support efficient and effective distributed IR. However, there ist still room for improvement, especially on the following two issues:

Improving document placement. While promising, we are not satisfied with the recall after contacting only a few nodes. We will therefore further improve the forestalled ranking function to even better place documents on responsible nodes. An especially interesing question is the fact that documents' distances from the landmark document are not evenly distributed. We will therefore experiment with irregularly distributed node ranges.

Dealing with dynamics. Both dynamics in the document collection as well as the node configuration have not been adressed in this paper and will be the subject of further research. This includes documents added to and removed from the document collection as well as nodes added to and removed from the network.

References

1. Callan, J.: Distributed Information Retrieval. In: Bruce Croft, W. (ed.) Advances Information Retrieval: Recent Research from the CIIR, Ch. 5, pp. 127–150. Kluwer Academic Publishers, Dordrecht (2000)
2. Aberer, K., Alima, L., Ghodsi, A., Girdzijauskas, S., Haridi, S., Hauswirth, M.: The essence of p2p: a reference architecture for overlay networks. In: P2P 2005. Fifth IEEE International Conference on Peer-to-Peer Computing, pp. 11–20 (August 31-September 2, 2005)
3. Samet, H.: The Design and Analysis of Spatial Data Structures. Addison-Wesley, Reading (1989)
4. Rowstron, A.I.T., Druschel, P.: Pastry: Scalable, decentralized object location, and routing for large-scale peer-to-peer systems. In: Guerraoui, R. (ed.) Middleware 2001. LNCS, vol. 2218, pp. 329–350. Springer, Heidelberg (2001)
5. Ratnasamy, S., Francis, P., Handley, M., Karp, R., Schenker, S.: A scalable content-addressable network, pp. 161–172 (2001)
6. Stoica, I., Morris, R., Karger, D., Kaashoek, M.F., Balakrishnan, H.: Chord: A scalable peer-to-peer lookup service for internet applications, pp. 149–160 (2001)
7. Zhao, B.Y., Kubiatowicz, J.D., Joseph, A.D.: Tapestry: An Infrastructure for Fault-tolerant Wide-area Location and Routing, University of California at Berkeley (2001)
8. Tang, C., Xu, Z., Dwarkadas, S.: Peer-to-peer information retrieval using self-organizing semantic overlay networks. In: SIGCOMM 2003. Proceedings of the 2003 conference on Applications, technologies, architectures, and protocols for computer communications, pp. 175–186. ACM Press, New York, NY, USA (2003)
9. Neumann, T., Bender, M., Michel, S., Weikum, G.: A reproducible benchmark for p2p retrieval. In: Bonnet, P., Manolescu, I. (eds.) ExpDB, pp. 1–8. ACM, New York (2006)
10. Aghbari, Z.A., Makinouchi, A.: Linearization approach for efficient KNN search of high-dimensional data. In: Li, Q., Wang, G., Feng, L. (eds.) WAIM 2004. LNCS, vol. 3129, pp. 229–238. Springer, Heidelberg (2004)
11. Salton, G., Wong, A., Yang, C.S.: A vector space model for automatic indexing. Commun. ACM 18(11), 613–620 (1975)
12. Aberer, K., Klemm, F., Rajman, M., Wu, J.: An architecture for peer-to-peer information retrieval [17]
13. Reynolds, P., Vahdat, A.: Efficient peer-to-peer keyword searching. In: Endler, M., Schmidt, D.C. (eds.) Middleware 2003. LNCS, vol. 2672, pp. 21–40. Springer, Heidelberg (2003)
14. Tang, C., Xu, Z., Mahalingam, M.: psearch: information retrieval in structured overlays. SIGCOMM Comput. Commun. Rev. 33(1), 89–94 (2003)
15. Nottelmann, H., Fischer, G., Titarenko, A., Nurzenski, A.: An integrated approach for searching and browsing in heterogeneous peer-to-peer networks. In: Heterogeneous and Distributed Information Retrieval (2005)
16. Bender, M., Michel, S., Weikum, G., Zimmer, C.: Bookmark-driven query routing in peer-to-peer web search [17]
17. Callan, J., Fuhr, N., Nejdl, W. (eds.): Proceedings of the SIGIR Workshop on Peer-to-Peer Information Retrieval, 27th Annual International ACM SIGIR Conference, Sheffield, UK (July 29, 2004). In: Callan, J., Fuhr, N., Nejdl, W. (eds.): Peer-to-Peer Information Retrieval (2004)

BEIRA: A Geo-semantic Clustering Method for Area Summary

Osamu Masutani and Hirotoshi Iwasaki

Research and Development Group, Denso IT Laboratory, Inc.
3-12-22 Shibuya Shibuya-ku, Tokyo, 150-0002, Japan
{omasutani,hiwasaki}@d-itlab.co.jp

Abstract. This paper introduces a new map browser of location based contents (LBC) that summarizes area characteristics. Recently various web map services have been widely used to search web contents. As LBC increase, browsing a number of LBC which are viewed as POI (point of interest) on a geographical map becomes inefficient. We tackle this issue by using AOI (area of interest) instead of POI. With the AOI a user can instantly find area characteristics without viewing each content of POI. We assume that semantically homogeneous and geographically distinguishable areas are suitable for the AOI. The AOI is formed by geo-semantic clustering which is a co-clustering that takes into account both geographical and semantic aspects of POI information. By the experiment using real LBC on the web, we confirmed our method has potential to extract good AOI.

Keywords: Map interface, web mining, co-clustering.

1 Introduction

Various web map services (WMSs) come into wide use recently. WMS is more suitable than an ordinary search engine to find location based contents (LBCs) which are contents associated with some location. For example, to find a favorite restaurant in specific area by WMS is much easier than by search engines. The WMS has function to display the distribution of search results on a geographic map and specify geographic area to exclude results on irrelevant areas.

WMSs have started to be used not only with on desktop PCs, but also on mobile devices. On a driving situation, which is our main target, some considerations are needed to simplify the user interface of map viewing. For example, the result of restaurant search in Tokyo includes thousands of candidates to choose. Only within 500m from Shibuya station, there exist about 1000 restaurants (Fig. 1). It may be feasible to choose among the candidates on a PC, however it should be unfeasible on a mobile device because of its limited user interface. Moreover, on a mobile device, users often have less time to find destination than on a PC.

Of course, filtering by a genre of restaurants can reduce the number of candidates, but on a driving situation filtering cause some problems. For example, a user chooses Italian restaurant which is favorite for him, but there might be no Italian restaurant

B. Benatallah et al. (Eds.): WISE 2007, LNCS 4831, pp. 111–122, 2007.

near current location. Even if there are some Italian restaurants, the user might miss to choose other genre of restaurant which is popular and only find in the area. Such situations often occur in an unfamiliar area.

These issues about limited user interface and improper information retrieval are caused by lack of the overlooking view of entire contents. The clustering and summarization techniques are often used to overlook data. Then we also use clustering and summarization of POI (point of interest) data.

Some studies propose methods to browse and retrieve POIs, by linking neighbor contents with geographic similarity [1], by showing textual area summary for each predefined area (ex. city) [2]. Using predefined area is not best solution for area summary because users would like to know which area exactly corresponds to the summary. In [3], the area is automatically extracted by geographic association of POIs. However, they don't take semantic (textual) aspect into account. We assume that partitioning area should be refined by taking semantic aspect into account.

We introduce AOI (area of interest) instead of POI to represents such area summary (Fig. 1). An AOI consists of an area surrounded by irregular shaped boundary and some summarization labels. We use clustering of POI contents to define the area boundary. The clustering is performed by "geo-semantic" co-clustering which take both geographic and semantic aspects into account. To summarize an AOI, we propose "location aware" summarizing that emphasizes local terms. We think the AOI view efficiently reduces search processes, as long as users prefer representative and unique POI in a certain area. We developed a map based contents browser BEIRA that extracts and displays AOI. We evaluate our geo-semantic clustering and location aware summarization by using real restaurant contents on web.

POI   AOI

Fig. 1. The map user interfaces with POI (left) and with AOI (right)

2   Related Works

Major web search engines recently begin to introduce geographical or local search service, for example Google Maps[1]. Only the POI address in the contents is used to bridge contents with a map. They use geo-coding that covert address to geographical coordinates (latitude and longitude). However other features of contents are used as only accompanying information for POI.

---

[1] http://maps.google.com/

Dealing more contents information associated with geographic features is also studied in web mining or geographic information system/science (GIS) field. At first, there have been a number of geographical web search studies. Some of them are based on geo-coding of a general location name with information extraction technique [4][5]. Studies such as [2] propose a comprehensive map-based digital library for location based contents using general name geo-coding. Some researches address the geographical query categorization or result ranking [6,7,8,9] that is geographically aware search engine. Their purpose is eliminating unrelated query results according to geographical scope of a user request. In another type of studies, they partition web crawling targets according to geographical location of web pages or web sites [10]. Their focus is to enhance the web search performance.

There also exist studies using GIS to deal location based contents [1,3,11]. They use various geographic manipulations over POI datasets which is extracted from a web. In [1], GIS is used to introduce "generic link" which links geographically close contents. Using the generic link users can find geographically similar contents without actual web link. In [11] construct versatile contents management system GeoWorlds on GIS. It handles not only ordinary geographic data (such as remote sensing data) but also location based contents. The user can analyze textual data in a certain area with some NLP techniques. However textual data is dealt almost separately with geographic data. Geographic features of contents are used only to filter texts within specified area. In [3] they address extracting spatial knowledge from location based contents. The system finds optimal region by spatial distributions of POI and then extract textual summary within the region. In [12] they use regional co-occurrence summation to rank place names to highlight place name on the map. They employ a major POI for area summary in each area. These studies still use geographic data and semantic data in separate processes. Our main contribution is to extract area summary by combining them seamlessly.

Combining semantic data and other type of data is also studied. One of major field of such study is multimedia information retrieval such as image retrieval. The image retrieval is to search images from a textual query from database. Most of the commercial services such as Google Image[2] use only textual information surrounding the image on the homepage. Some advanced systems use both textual and image features to characterize each image on web. Vector space model of text and image features are combined by linear combination [13] or joint distribution model [14].

Furthermore there exist some web mining approaches which use semantic features with other type features such as user characteristics or query log. Various types of co-clustering which are able to handle heterogeneous data are summarized in [15]. One of approach is tensor-based representation of heterogeneous data [16]. Each feature of data composes a product space of some heterogeneous dimensions. The other approach is based on unified relationship matrix (URM) [17,18]. The approach integrates pair wise relationship matrices into a unified matrix. We use both of these approaches to combine geographic and semantic features.

---

[2] http://images.google.co.jp/

3 BEIRA – Bird's Eye Information Retrieval Application

Our application BEIRA (Bird's Eye Information Retrieval Application) is a prototype of a WMS. BEIRA has two pane of POI for information browsing (Fig.2). The left pane is a geographic map. A user can use same functions of typical WMSs, such as searching POI by name or keyword, filtering candidates by some attributes such as genre, or focus candidates according to their locations. Furthermore a user can see AOIs on the pane. An AOI is drawn as a colored contour with some label. By using AOI a user can understand area characteristics at a glance, and AOI helps to narrow candidates of POIs. In driving situation, AOI reduce a load factor of operation of car navigation when the user must specify a destination from a number of candidates (AOI also can be used as a destination).

The right pane is a semantic map which shows a distribution of POIs on semantic space. The semantic space is a multidimensional space of aspects described in latter part of this paper. They help comparing candidate POIs on (selected) 2 dimensional aspect space. Because the location on a semantic map indicates respective characteristics with other POIs, a user is able to specify a favorite POI without reading POI contents. The semantic map also has (purely semantic) AOIs.

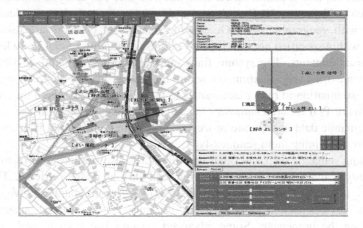

Fig. 2. BEIRA user interface

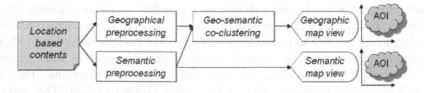

Fig. 3. BEIRA data process flow

The location based contents of POIs are extracted from some web sites. The contents are processed through geographic and semantic preprocessing (Fig.3). Both geographic and semantic representations of POI are used to make clusters by geo-semantic co-clustering. Then the clusters or AOI on geographic and semantic map are drawn by contour graphics. The label of AOI is summarization of each cluster.

3.1 Preprocessing

We use vector space model (VSM) to represent semantic space. The terms used for our VSM are adjective, noun and verb in location based text. We employ tf-idf weighting. Then we perform dimension reduction by latent semantic indexing (LSI). We call the LSI space as the semantic (aspect) space. The semantic map browser on right side of BEIRA simply represents a selected two dimensional view of the space. The AOI on semantic map view is provided by k-means clustering on the two dimensional aspect space. The concept and methodology of semantic map is almost same with the ones in [19]. A geographic space is simply a space of geographic coordinates. The addresses in contents are geo-coded into the geographic coordinates.

3.2 Labeling and Drawing of AOI

The label that summarizes AOI is top ranked list of terms in each POI text in the AOI cluster. The term ranking is according to tf-idf weighting or its refined version as described in following part of this paper. We employed convex hull to draw AOI. Each POI member of AOI is surrounded by convex hull. Additionally the result of convex hull polygon is smoothed by some smoothing technique of GIS.

3.3 Geo-semantic Co-clustering

In this section we explain necessity of co-clustering on the geo-semantic space and also explain some co-clustering methods applied in BEIRA.

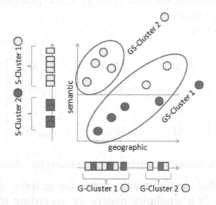

Fig. 4. Explanation of geo-semantic clustering

Fig.4 shows how the geo-semantic blending works to extract better cluster than either geographic clustering or semantic clustering. The geo-semantic space is represented as 2 dimensional space here. Each POI is mapped onto the space and 2 semantic characteristics are represented as black and white for example. Both two mapping to semantic and geographic dimension are shown as 1 dimensional distributions of POI. Three types of clustering are demonstrated on the POI distribution. Each cluster is labeled according to the density of type of POIs.

By semantic clustering (left side), 2 clusters are purely homogeneous in semantic aspect, but the distribution on geographic dimension of two clusters is heavily overlapped. The purpose of clustering is extracting visually distinguishable area, therefore this result isn't appropriate. By geographic clustering (down side), 2 clusters make very distinguishable areas. However both 2 clusters isn't homogeneous. Furthermore the black is hidden behind white characteristics. By geo-semantic clustering (circles), moderately homogeneous and geographically distinguishable cluster can be extracted. The result seems to be compromised in both aspect, however it meets user requirement in an aspect in finding "rough" summary of each area.

Some variations of co-clustering of two heterogeneous data are proposed [15]. Simple one is tensor based co-clustering which combines two heterogeneous data spaces into one product space of each dimension. The concept of the tensor based geo-semantic co-clustering is almost same with Fig. 4. The combined data is shown as a matrix L_{tensor} :

$$L_{tensor} = [\lambda_G D_G \quad \lambda_S D_S] \tag{1}$$

where D_G is a geographic data matrix and D_S is a semantic data matrix. We use simple k-means clustering on the geo-semantic tensor space L_{tensor}.

Another type of co-clustering is URM (Unified Relationship Matrix) based co-clustering which combines multiple relational data into one matrix. For URM based clustering we employ M-LSA (Multi-type Latent Semantic Analysis) which constructs multiple inter-relational data [18]. M-LSA uses eigenvalue decomposition on URM to cluster each objects.

URM L_{URM} is defined as :

$$L_{URM} = \begin{bmatrix} \lambda_{11} M_1 & \lambda_{12} M_{12} & \cdots & \lambda_{1N} M_{1N} \\ \lambda_{21} M_{21} & \lambda_{22} M_2 & \cdots & \lambda_{2N} M_{2N} \\ \vdots & \vdots & \ddots & \vdots \\ \lambda_{N1} M_{N1} & \lambda_{N2} M_{N2} & \cdots & \lambda_{NN} M_N \end{bmatrix} \tag{2}$$

where M_i is intra-type adjacency matrix, M_{ij} is intra-type adjacency matrix. We use 2 types of data geographic and semantic data. Geographic data defines intra-type relationship of POI by POI similarity matrix M_G according to geographic closeness between two POIs. The other relationship matrix M_{GS} is semantic relationship between POI and semantic concepts which is the result of LSI. The geo-semantic URM L_{GS} is shown as follows:

$$L_{GS} = \begin{bmatrix} \lambda_{GG}M_G & \lambda_{GS}M_{GS} \\ \lambda_{SG}M_{SG} & 0 \end{bmatrix} \qquad (3)$$

A bias parameter $R = \lambda_{GG} / \lambda_{GS}$ is left to choose. This is a geo-semantic ratio to balance geographic and semantic aspects of clustering. Tensor based clustering has also similar definition of geo-semantic ratio $R = \lambda_G / \lambda_S$.

3.4 Location Aware Weighting of Terms

We use term weighting to construct VSM and labeling of AOI. Normally TF/IDF is used to represent term significance. However we would like to define term significance as not only term rarity but also locality of term. The geographical distribution of term is defined by term occurrence in POI documents. Fig. 5 shows distributions of some terms. A general term "onion" distributes widely and uniformly. On the other hand, the location name "Dogenzaka" has heavily clustered distribution. The other term having similar clustered distribution is "wedding". Wedding party tends to be held in a calm sophisticated area which is often separated with congested area of a city. Therefore the term "wedding" is more clustered than general terms. The term that has high locality (the value of K here) is regarded as suitable label for AOI than low locality term, because users would like to know local and unique characteristics.

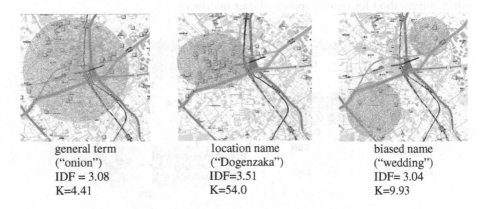

general term	location name	biased name
("onion")	("Dogenzaka")	("wedding")
IDF = 3.08	IDF=3.51	IDF= 3.04
K=4.41	K=54.0	K=9.93

Fig. 5. Some type of noun distributions

We employ location aware TF/IDF (L-TF/IDF) to calculate term importance. L-TF/IDF of term t on document d is d_t defined as follows:

$$d_t = L(t)TF(d,t)IDF(t) \qquad (4)$$

where $L(t)$ is a value of Repley's K-function which is one of point distribution analysis index. K-function means expectation of the density of data points in the circle around the randomly selected points. Its estimate is as follows:

$$L(t) = K(t, dist) = \frac{A}{N^2} \sum_{i \neq j} I(\|x_i - x_j\| < dist)$$ (5)

where dist is distance parameter, A is a area of target region, N is the number of points in the area and I() is a delta function.

In Fig. 5, we also described the value IDF and K(1km) value of each term. We chose three terms having similar IDF value. K value of "wedding" is higher than "onion", therefore "wedding" is regarded as more important than "onion". Such group of terms that has highly condensed distribution are, for example, "brand" (K=26.1), "department store" (K=26.1), "foreigner"(K=30.0), "underground"(K=28.6).

3.5 System Architecture

BEIRA is a .NET application with GIS component SIS[3](See Fig.6). Drawing AOI is performed by SIS's convex hull drawing function with smoothing. Most of backend processes are written in Java with some commercial or open source libraries.

We use NQL[4] to extract web pages on POI information sites. We use GATE[5] and Weka[6], Sen[7] as basic Japanese language analysis. Note that most of our technique are language-independent though we evaluated only by data in Japanese. We employed TCT[8] as a high performance Java based co-clustering tool with enhancement to enable higher-order co-clustering. TCT use ARPACK for sparse matrix computation which is suitable for natural language processing. We also use CUDA[9] to accelarate matrix computation for rapid response of user interface.

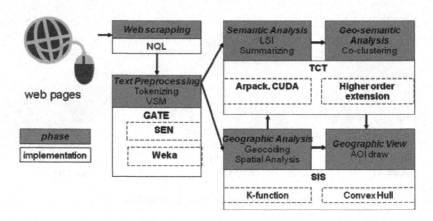

Fig. 6. System architecture of BEIRA

[3] SIS / Informatix Inc. / Cadcorp Ltd. (http://www.cadcorp.com/)
[4] NQL / NQL tech (http://www.nqltech.com/)
[5] GATE(A General Architecture for Text Engineering) / Sheffild University (http://gate.ac.uk/)
[6] Weka / The University of Waikato (http://www.cs.waikato.ac.nz/ml/weka/)
[7] Sen (Japanese morphological analyzer) (http://ultimania.org/sen/)
[8] TCT (Text Clustering Toolkit) / University College Dublin (http://mlg.ucd.ie/)
[9] CUDA (Compute Unified Device Architectur) / NVidia (http://developer.nvidia.com/)

4 Evaluation

To evaluate our application we use real world data. One of most popular POI is restaurant. There are massive numbers of restaurants in Tokyo and finding the best favorite restaurant is very time consuming. We use reputation text on the restaurant reputation sites asku[10] which mainly focus on reputation. Each article in the web site has address, so geo-coding is not a problem here. Over 30,000 restaurants in Tokyo area are registered and commented on the site. We choose 289 cafes in Shibuya-ward, because café seems to reflect area characteristics rather than all types of restaurant, and Shibuya-ward has various characteristics of sub-areas.

We prepare correct data in the dataset. We carefully choose cluster member by reading each reputation of cafés. Only the semantically and geographically condensed cluster is chosen. The dimension of result semantic space of LSI is 200. LSI is performed on a VSM of about 30,000 restaurant texts. Geographical coordinate is converted into rectangular coordinates system and its unit is kilo meter. K-function for L-TF/IDF is calculated at distance 1km.

4.1 Geo-semantic Ratio

We confirmed clustering result using a ratio R between 1.0E-04 (semantic) and 1.0E-06. Fig.7 shows how AOI is drawn on each R. By semantic clustering (R=1.0E-04) the drawing contour is failed because members of a cluster is spread over whole area. By geographic clustering (R-1.0E-06) the AOI becomes circular area around POI cluster. By geo-semantic clustering extracts some non-circular areas whose members are semantically homogeneous (ex. the contour in the center area).

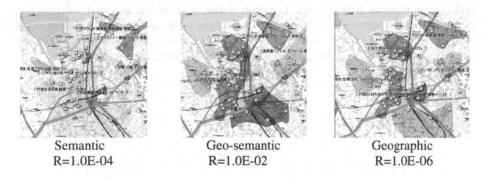

| Semantic | Geo-semantic | Geographic |
| R=1.0E-04 | R=1.0E-02 | R=1.0E-06 |

Fig. 7. AOI by each geo-semantic ratio R

We also performed sensitivity analysis over R (Fig.8). The evaluation index is F-measure if precision and recall on the correct data. We confirmed there was an optimal R between geographic and semantic extremes. The optimal R was around 0.01 for both clustering. It tells us geo-semantic clustering is better than purely geographic or semantic clustering though the optimal ratio here might not be general

---

[10] http://www.asku.com

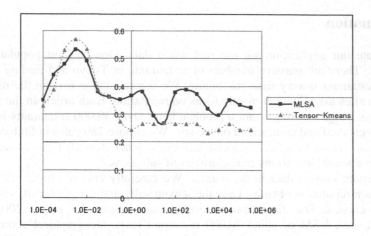

Fig. 8. F-value for each R for M-LSA and tensor-based co-clustering

value. However no significant differences are found between MLSA and Tensor-based clustering in this examination.

Furthermore the result confirms geo-semantic ratio R has potential to be calibrated by some method. This will be our next target to reveal.

4.2 Location-Aware Summarizing

Next we evaluated our location-aware summarizing which is calculated by location-aware weighting of terms. We evaluate term weighting method by performance to rank location name higher rank. Location name is regarded as good estimator of locally important term. Of course part of speech or term category information is not used to extract them. We evaluate performance of term weighting methods by density of location names in top 1,000 ranked term list by each weighting method.

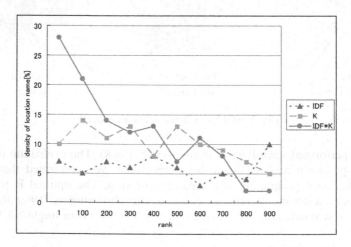

Fig. 9. Density of location name in top 1,000 rank of each term weighting method

This result on Fig.9 shows L-IDF (K-function value by IDF) is best method to weight term importance. High occurrence terms tend to have high K-function value so only by K-function location names aren't extracted efficiently. The list of the "local terms" which are extracted by this method, might also be a good estimator of dialect.

5 Conclusion

In this paper we proposed a new map based browser of web contents. The core concept is the AOI which has semantically homogeneous and geographically distinguishable cluster of POIs. A geo-semantic clustering is assumed to be able to extract AOI by POI data. The geo-semantic ratio R is only free parameter and we evaluate it on real restaurant reputation web pages. The result tells geo-semantic clustering extracts more suitable border of AOI than geographic clustering or semantic clustering. Additionally we proposed a location aware summarizing method which is take term locality into account. We confirmed our location aware term weighting is good index for location aware summarizing.

For future work, we will evaluate our method with other types of location based contents such as restaurants, shops, and sightseeing spots. And we will also attempt usability evaluation to confirm what the best AOI for users is. An optimal ratio R for AOI on every type of region or genre should be estimated.

References

1. Hiramatsu, K., Ishida, T.: An Augmented Web Space for Digital Cities. In: SAINT 2001. Proceedings of the International Symposium on Applications and the Internet, p. 105 (2001)
2. Lim, E., Goh, D., Wee-Keong, N., Khoo, C., Higgins, S.: G-Portal: a map-based digital library for distributed geospatial and georeferenced resources. In: Proceedings of the 2nd ACM/IEEE-CS joint conference on Digital libraries, pp. 351–358 (2002)
3. Morimoto, Y., Aono, M., Houle, M., McCurley, K.: Extracting Spatial Knowledge from the Web. In: SAINT 2003. Proceedings of the International Symposium on Applications and the Internet, pp. 326–333 (2003)
4. Kanada, Y.: A Method of Geographical Name Extraction from Japanese Text for Thematic Geographical Search. In: CIKM 1999. 18th International Conference on Information and Knowledge Management, pp. 46–54 (1999)
5. Amitay, E., Har'El, N., Sivan, R., Soffer, A.: Web-a-where: geotagging web content. In: Proceedings of the 27th annual international ACM SIGIR conference on Research and development in information retrieval, pp. 273–280 (2004)
6. Buyukokkten, O., Cho, J., Garcia-Molina, H., Gravano, L., Shivakumar, N.: Exploiting Geographical Location Information of Web Pages. In: Proceedings of Workshop on Web Databases WebDB 1999 (1999)
7. Ding, J., Gravano, L., Shivakumar, N.: Computing Geographical Scopes of Web Resources. In: 26th International Conference on Very Large Databases (2000)
8. Silva, M.J., Martins, B., Chaves, M., Cardoso, N., Afonso, A.: Adding geographic scopes to web resources. In: ACM SIGIR 2004 Workshop on Geographic Information Retrieval (2004)

9. Asadi, S., Xu, J., Shi, Y., Diederich, J., Zhou, X.: Calculation of Target Locations for Web Resources. In: Aberer, K., Peng, Z., Rundensteiner, E.A., Zhang, Y., Li, X. (eds.) WISE 2006. LNCS, vol. 4255, pp. 277–288. Springer, Heidelberg (2006)

10. Exposto, J., Macedo, J., Pina, A., Alves, A., Rufino, J.: Geographical partition for distributed web crawling. In: Proceedings of the 2005 workshop on Geographic information retrieval, pp. 55–60 (2005)

11. Neches, R., Yao, K., Ko, I., Bugacov, A., Kumar, V., Eleish, R.: GeoWorlds: Integrating GIS and Digital Libraries for Situation Understanding and Management. The New Review of Hypermedia and Multimedia NRHM 7, 127–152 (2001)

12. Tezuka, T., Kurashima, T., Tanaka, K.: Toward tighter integration of web search with a geographic information system. In: WWW 2006. Proceedings of the 15th International Conference on World Wide Web, pp. 277–286. ACM Press, New York (2006)

13. Sclaroff, S., Cascia, M., Sethi, S.: Unifying textual and visual cues for content-based image retrieval on the World Wide Web. Computer Vision and Image Understanding 75(1-2), 86–98 (1999)

14. Barnard, K., Forsyth, D.: Learning the Semantics of Words and Pictures. In: International Conference on Computer Vision, vol. 2, pp. 408–415 (2001)

15. Liu, T.Y.: High-order Heterogeneous Object Co-clustering. In: The 4th Chinese workshop on Machine Learning and Application (2006)

16. Sun, J.-T., Zeng, H.-J., Liu, H., Lu, Y., Chen, Z.: CubeSVD: A Novel Approach to Personalized Web Search. In: WWW 2005. Proceedings of the 15th International Conference on World Wide Web, pp. 382–390 (2005)

17. Xi, W., Zhang, B., Chen, Z., Lu, Y., Yan, S., Ma, W., Fox, E.: Link fusion: a unified link analysis framework for multi-type interrelated data objects. In: WWW 2004. Proceedings of the 13th international conference on World Wide Web, pp. 319–327. ACM Press, New York (2004)

18. Wang, X., Sun, J., Chen, Z., Zhai, C.: Latent Semantic Analysis for Multiple-Type Interrelated Data Objects. In: SIGIR 2006, pp. 236–243 (2006)

19. Yoshida, N., Zushi, T., Kiyoki, Y., Kitagawa, T.: Context Dependent Dynamic Clustering and Semantic Data Mining Method for Document Data. In: IPSJ, vol. 41(SIG1), pp. 127–139 (2000)

Building the Presentation-Tier of Rich Web Applications with Hierarchical Components

Reda Kadri[1,2], Chouki Tibermacine[3], and Vincent Le Gloahec[1]

[1] Alkante SAS, Cesson-Sévigné, France
[2] VALORIA, University of South Brittany, Vannes, France
[3] LIRMM, University of Montpellier II, Montpellier, France
r.kadri@alkante.com, Chouki.Tibermacine@lirmm.fr,
v.legloahec@alkante.com

Abstract. Nowadays information systems are increasingly distributed and deployed within the Internet platform. Without any doubt, the World Wide Web represents the *de facto* standard platform for hosting such distributed systems. The use of a multi-tiered architecture to develop such systems is often the best design decision to reach scalability, maintainability and reliability quality goals. Software in the presentation-tier of this architecture needs in practice to be designed with structured and reusable library modules. In this paper, we present a hierarchical component model which allows developers to build (model, generate code and then reuse) this software level of rich Web applications. In this model, components can be connected via their interfaces to build more complex components. These architecture design models can be reused together with their corresponding code using an association mechanism. As shown in this paper this is a valuable feature in assisting developers to position their developed documents within the overall software design and thus enable maintaining the consistency between artifacts of these two stages of the development process.

1 Introduction and Background

There are already a few years that the debate exists about rich Web applications. However it is the article published by Jesse James Garrett[1], co-founder of *Active Path*, on his blog on February 2005, which seems to have started the awakening of developers. From Google (Gmail and Google Sugest) to Yahoo (Flickr), a dozen of general public Web sites already adopted *Ajax* [1] which becomes a frightening competitor for existing rich client technologies like Flash or those emerging such as XUL and Eclipse RCP. Indeed Ajax provides the same advantages as its competitors (advanced ergonomy, etc.) but it does not impose the installation of a plugin in the Web browser.

In our research, we are interested in the development of such rich Web applications in a context of multi-tiered architectures of Web information systems. Our work aims at introducing new high-level languages, methods and tools in this

---

[1] http://www.adaptivepath.com/publications/essays/archives/000385.php

B. Benatallah et al. (Eds.): WISE 2007, LNCS 4831, pp. 123–134, 2007.

field. In that context we propose in this paper a hierarchical reusable component model to design the architecture of these applications. The proposed model is hierarchical because rich Web applications are by definition pieces of software based on elements that are organized hierarchically. A simple HTML page can include a form, which can be composed of many `input`, `select` or `text area` components. All of these components are reusable assets that can be used in order to compose other client applications. For instance, an existing component representing an e-mail user interface can be reused, customized and composed with other components to build a more complex client application, such as a user agenda. In the following section, we present the proposed component model. An example defined using this model is then illustrated in section 3. The components that we deal with vary from simple HTML elements (forms, frames, hypertext links, etc) to more complex components such as authentication, e-mail, editorial management components. We make use of reusable software modules because we argue that functional requirements evolution of these components is rare, as discussed in [9]. Starting from applications designed with this component model, we can then generate PHP and Ajax code.

Existing composition techniques of Web-based components address only the static and structural aspect to generate interfaces of applications, and much work remains to do for developers in order to implement collaborations between components. The challenge that we raise in our work is to allow developers to define compositions of interactive components in order to build straightforwardly and hierarchically rich Web applications. In section 4, we present the tools developed for implementing our proposals. Before concluding this paper and presenting the future work, we provide a comparison between the proposed work and the related one.

2   Component-Based Web Application Architecture Model

Instead of proposing a new component model and introducing another encumbering architecture modeling language, we chose to reuse an existing well-known standard which is the UML notation. Indeed, UML is a modeling language which has been adopted by many software development teams in industry and academia. Proposing UML extensions, even standard ones (like UML profiles), did not appear as a good solution, because we found in the version 2.0 of the UML specification all abstractions needed in modeling rich Web applications with hierarchical entities.

2.1   Architectural Elements

The component model proposed here (referred to as AlCoWeb) introduces a set of architectural abstractions, such as components, interfaces, connectors and ports.

– **Components.** Components represent Web elements at different levels of abstraction. They can be either atomic or hierarchical. Atomic components are black-boxes (which do not have an explicit internal structure) or basic components which do not have an internal structure at all. Hierarchical components, however have an explicit internal structure. They are described using component assemblies. Components or component assemblies are modeled using UML 2 components. Examples of components include HTML text fields, authentication forms, auto-complete text boxes and check box lists.
– **Interfaces.** Interfaces represent public services defined by components. They are modeled using UML interfaces and can be of different kinds. **Synchronous Interfaces** are interfaces that contain traditional object-oriented operations. They are decomposed into two kinds: **Provided Interfaces** define provided services to other components. For instance, a component HTMLTextField can define an interface which provides services like getFormattedValue() which returns the formatted value of the text field, or getPage() which returns the page that contains the text field. **Required Interfaces** declare the dependencies of the component in terms of required services, which should be provided by other components. A component CheckBox can for example define a required interface that declares services like setExternalValue(). This service allows a given component to set the value of another component attribute (value of a text field component, for example). **Event-based Interfaces** represent asynchronous operations which are based on events. Each service is executed only if a particular event occurs. The implementation of such operations is mainly defined using a client-side scripting language such as JavaScript. Examples of these services include HTML button onClick, text onSelect, HTML form object onFocus, or mouseOver operations. **Checking Interfaces** group some operations which are invoked to validate contents of components. As for the previous interfaces, the implementation of these operations is frequently defined using client-side scripting languages. For instance, in an HTML form component, we could perform some checking to see whether all mandatory items in the form are completed. We could also add a checking interface to a TextField component so that we can make some format checking (well-formed dates, valid URLs by querying a DNS server component, etc). In AlCoWeb, all kinds of interfaces are modeled using UML interfaces. No extensions are needed for this purpose. We argue this is sufficient to build components without ambiguity and reuse them afterwards.
– **Ports.** Ports represent a set of interfaces of the same kind and related to the same functionality. They can represent provided, required, checking or event interfaces. They are modeled with the traditional UML ports.
– **Connectors.** Connectors are interaction-oriented architectural elements. They link interfaces of different components and encapsulate interaction protocols. Examples are provided in Section 3. These connectors are modeled using UML connector abstractions and can be of different kinds. **Hierarchical Connectors** are bindings between a hierarchical component and its

sub-components. **Assembly Connectors** are bindings between components
at the same level of hierarchy.

2.2 Assembling Architectural Elements

Configurations of the elements introduced above can be described using component assemblies. These assemblies allow developers to build applications starting from components by linking them through connectors. Additional architecture constraints can be described to formalize design decisions.

Component Assemblies. Assemblies represent configurations of whole applications. Components can be modeled from scratch or reused and customized after checking them out from repositories. These components can then be bound together, using newly modeled connectors. An illustrative example is presented in Section 3.

Architecture Constraints. In order to describe architecture constraints, we introduced a constraint language which accompanies this component model. This language is an ACL profile for Web development. As introduced in [13], ACL is a multi-level language. It separates predicate-level from architecture-level expression. Predicate-level concepts, like quantifiers and set operations, are described using an OCL (Object Constraint Language) [12] dialect, and the architecture-level expression concepts are encapsulated in MOF [11] metamodels. The metamodel defined for this ACL profile summarizes the concepts introduced in the previous sections.

For instance, we may need to define a constraint which states that the component of name `TextField` should not be connected to more than two different components. This constraint could be defined using the ACL profile as follows:

```
context TextField:Component inv:
TextField.interface.connectorEnd.connector.connectorEnd
.interface.component->asSet()->size() <= 3
```

This constraint navigates to all connectors to which are attached the interfaces of TextField. It then gets all components whose interfaces are linked to the connector ends of all obtained connectors. The resulting collection (Bag) is then transformed into a set to remove duplicates. The obtained set contains even the component TextField, this is the reason why its size should be less than or equals 3 (instead of 2, as stated in the constraint of the previous paragraph).

2.3 Component Deployment

Once the development of an application is finished, we can proceed to its deployment. An application is characterized by the description file of the component assembly. There are three possible kinds of deployment. The first is called *Evaluation Deployment*, whose purpose is the internal deployment of the application within the development team. It serves for component testing. The second kind of deployment is called *Remote Qualification Deployment*, which aims at deploying the application in the customer environment. This is performed in order to test the application by the customers before validation. The last kind of deployment is said *Production Deployment*, which corresponds to the final product delivery.

2.4 Component Evolution and Reuse

The development process which is used with AlCoWeb introduces two professions. On the one hand, *component developers* model and code components, and put them into the repository. On the other hand, *component assemblers* checkout existing components from the repository in order to build larger applications. The two professions work on separate environments and the repository constitutes the bridge between the two professions. When component assemblers need new components in order to satisfy a particular requirement, they ask component developers to develop them and put them in the repository, or enhance existing ones and add them as new component versions.

2.5 Association of Design Artifacts to Code

Every entity in the Web application model is associated to some implementation elements in the code. When we navigate hierarchically in the model, we go through the implementation code in the same manner. We thus introduce an association link between design models and the code. These associations link a given entity in the model to the code, which is marked by the entity identifier as a comment. The associations can also reference files or directories. We argue that this mechanism is a good practice in making relationships between views [2] of the Web application architecture (relationships between structural and physical views).

3   Illustrative Example

We developed using AlCoWeb a large set of components that implement a plethora of technologies. Examples of these components include directory access systems (Active Directory, OpenLdap, etc.), password-based cryptographic system (MD5, DES, etc.), database access systems (MySQL, PostgreSQL/PostGIS, etc.), geographical web service access (WFS, WMS, etc.), AJAX widgets based on Dojo, Scriptaculous, Rico, Google and Yahoo Ajax APIs. For reasons of brevity and space limitation, we prefer not to detail some of these components and present below a simple example.

The left side of Figure 1 depicts an example of a `TextField` component. This component provides and requires a number of interfaces and defines some event interfaces. Provided and required interfaces defined for this component are separated into PHP-specific (`PhpTxtfPrdInterface` and `PhpTxtfReqInterface`) and JSP-specific (`JspTxtfPrdInterface` and `JspTxtfReqInterface`). Event interfaces are decomposed into two kinds: events whose source is an application user (`ClientTxtfSideEvents`) and events whose source is the server where the application is hosted (`ServerTxtfSideEvents`).

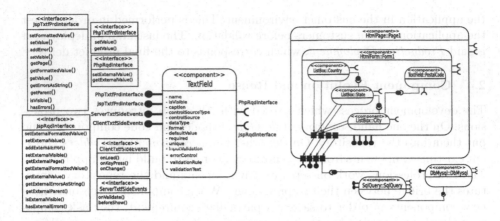

Fig. 1. A simple example of a component assembly

The right side of the figure illustrates an HTML page of a Web application. This page is represented by a component (HTMLPage) which defines multiple required, provided and event-based interfaces. This component contains another component HTMLForm. Hierarchical connectors bind interfaces of the super-component to the sub-component. The latter component contains four sub-components. The first sub-component (at the right) represents architecturally the TextField component introduced previously, and functionally its value represent the postal code of a city. The other sub-components are ListBox components, and represent lists of countries, states and cities. Suppose that all information about these geographic places is stored in a database, represented by the two components SqlQuery and DbMysql. Components are bound together through connectors defined by the developer by gluing interfaces (sockets and lollipops). The component ListBox representing countries connects to the database in order to get the list of available countries. As soon as this component receives a change event from the user, it executes the operation onChange() which updates the second ListBox component. The latter connects to the database using XMLHttpRequest in order to get the states of the chosen country. The same events occur for the last ListBox component. This Ajax-based functioning of the application is illustrated here using hierarchical components.

4   AlCoWeb-Builder: A Tool for Component-Based Web Development

The environment we developed rely on several frameworks offered by the Eclipse platform. AlCoWeb-Builder has been designed as a set of plug-ins that allows to separate the underlying component model from the graphical editor itself. The following section presents which frameworks have been used to develop this tool, and how they communicate to process from model design to code generation.

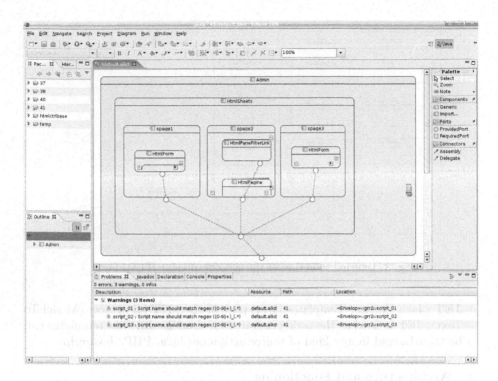

Fig. 2. Screen-shot of AlCoWeb-Builder

4.1   Underlying Technologies

Eclipse is now a mature and productive platform. It is composed of various projects that provide frameworks for software development; such projects are for example Eclipse RCP (Rich Client Platform), BIRT (Business Intelligence & Reporting) or WST (Web Standard Tools), which provide support for EJB and Ajax development. Three main frameworks have been used to implement AlCoWeb-Builder:

- **GMF** (Graphical Modeling Framework [4]) offers a generative component and runtime infrastructure to produce a full-feature graphical editor based on EMF (Eclipse Modeling Framework) and GEF (Graphical Editing Framework). On the one hand, the component model is designed using EMF functionalities and provides a set of Java classes which represents that model. On the other hand, GMF enriches GEF with purely graphical functionalities.
- **MDT** project (Model Development Tools [4]) provides two frameworks: UML2 and OCL. The UML2 framework is used as an implementation of the UML^{TM} 2.0 specification. In the same way, the OCL framework is the implementation of the OMG's OCL standard. It defines an API for parsing and evaluating OCL constraints on EMF models.

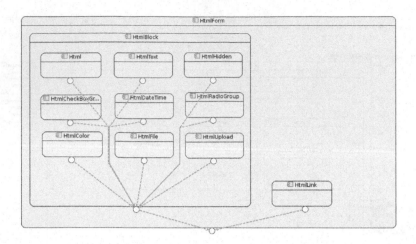

Fig. 3. Internal structure of the sub-component HTMLForm

- **JET** (Java Emitter Templates [4]) is part of the M2T [4] project (Model To Text). JET is used as the code-generator for models. JSP-like templates can be transformed in any kind of source artifacts (Java, PHP, Javascript ...).

4.2   Architecture and Functioning

The first prototype of AlCoWeb-Builder has been split into four main plug-ins : modeling, editing, constraints and transforming. The modeling and editing plug-ins represent the graphical editor itself, with all basic UML modeling features to design, edit and save modeling artifacts (as shown in Figure 2). It represents the generic part of the architecture which is mapped to the component model described before. Constraints plug-in allow the validation of modeling diagrams. For instance by ensuring that an `HtmlTextField` component cannot be added into an `HtmlForm` component. Code generation is represented also as a separate plug-in in order to allow multiple target languages (JSP, ASP, ...).

In a first time, our component model has been designed using EMF facilities. This model is an instance of the EMF metamodel (Ecore) which is a Java implementation of a core subset of the OMG MOF (Meta Object Facility). In further releases, we plan to use directly the Java implementation of the UML2 meta-model, provided by the Eclipse MDT UML2 framework, as our component model.

From this model definition described in XMI (XML Meta-data interchange), EMF produces a set of Java classes. Thoses classes serve as a domain model in the GMF architecture, mainly build upon the MVC (Model View Controller) pattern. The GMF runtime offers a set of pre-integrated interesting features, such as diagram persistence, validation and OCL. In AlCoWeb-Builder, the persistence feature allows to save the diagram into two separated XMI resources : a

Fig. 4. Overview of the example application interface

domain file – the component model instance – and a diagram file – the notation model for graphical elements –.

EMF also adds support for constraint languages. In our tool, OCL constraints are used to validate the diagram and ensure model integrity. An OCL editor has been developed into a larger plug-in which implements evolution contracts and a quality oriented evolution assistance of the designed component diagrams [14]. This OCL editor is based on the OCL framework proposed by the Eclipse platform, but adds some auto-completion capabilities to make easier the edition of constraints for developers.

Finally, the JET2 framework provides the code generation facility. It consists of two sets of files : an input model and templates files. The input file, given in XML, is in our case a component model instance previously modelled with the GMF editor. Templates use the XPath language to refer to nodes and attributes of the input model, and generate text of any kind. The first prototype of AlCoWeb-Builder uses those templates to generate PHP code.

Figure 3 shows the detailed internal structure of the component HTMLForm shown in the previous figure. Components are designed hierarchically and incrementally. A double click on a given component allows either to access to the architecture of this component if this one exists, or to define a new internal structure for this component. The interpretation of the generated PHP, HTML, JavaScript (Ajax) code for the whole application is depicted in Figure 4.

5   Related Work

Modeling software architectures using the UML language has already been discussed in multiple works, such as [10,8]. In these works different types of extensions has been proposed to deal with the lack of expressiveness of UML in describing some aspects of software architectures. As in these approaches, in this paper we showed how we can use UML to model component-based architectural abstractions, but within a particular application domain which is Web software engineering. In addition, we make use of UML as it is in version 2.0 and not 1.5.

In the literature, many works contributed in the modeling of Web applications with UML. The most significant work in this area is Jim Conallen's one in [3]. The author presents an approach which makes a particular use of UML in modeling Web applications. Web Pages are represented by stereotyped classes, hyperlinks by stereotyped associations, page scripts by operations, etc. Additional semantics are defined for the UML modeling elements to distinguish between server-side and client-side aspects (page scripts, for instance). While Conallen's approach resembles to the approach presented here, it deals with traditional Web applications and not with rich Web applications which enable in some situations direct communication between presentation-tier components and data-tier components (as illustrated in section 3). The author proposes an extension to the UML language through stereotypes, tagged values and constraints, while what we propose in this work is simply the use of UML as it is. Indeed, the distinction between server-side and client-side elements is not of great interest in our case. Moreover, hierarchical representation of these elements is not considered in his work, while the presentation-tier of Web applications is by nature hierarchical.

In [6], Hennicker and Koch present a UML profile for Web Application Development. The work presented by these authors is more process-oriented than product-centric. Indeed, they propose a development method which starts by defining use cases as requirement specifications and then, following many steps, deduce presentation models. These models are represented by stereotyped UML classes and composite objects, and contain among others fine-grained HTML elements. As stated above, the work of Hennicker and Koch is process-oriented. It shows how we can obtain a Web application starting from requirement specifications. In our work we focus on the modeling of presentation models they discuss. We do not deal with the method used to obtain these models. In addition, we consider hierarchical elements in Web applications like their hierarchical presentation elements represented by stereotyped composite objects. However, presentation elements that we deal with in our work are interactive and collaborative (like forms and form objects); their presentation elements are navigation-specific (HTML pages that contain images and texts).

In [5], the authors present an approach for end users to develop Web applications using components. The proposed approach focuses on the building of high level generic components by developers, which can be reused by end users to assemble, deploy and run their applications. Components in this approach vary from form generators to database query engines. The concern is thus with whole Web applications, all the tiers are targeted here and their running environment.

In the work we presented in this paper, components are presentation-tier specific and focuses on the modeling of architectures of applications based on collaborating library components. As stated previously, our approach is more product-centric than organizational-oriented like in [5].

6   Conclusion and Future Work

Alkante develops cartography-oriented rich Web applications for regional and local communities in Brittany (France)[2]. The work presented in this paper is the continuation of a former work [7] realized in this company which targeted the business-tier of these applications, and provided a new development process based on building and reusing software components. In Alkante's development team, reusing hierarchical modules is a recurrent aspect when producing geographical web information systems. In order to make easy this task we developed a component model which enables architecture modeling of rich Web applications and reusing at the same time these design models and their corresponding code. In addition, we defined a version management mechanism for AlCoWeb components. This is based on a repository which uses, as in traditional version systems, like CVS, branches and tags for their organization.

In the near future, we plan to make available AlCoWeb-Builder and our component repository in order to be enriched by the community (in an open-source development perspective) and to serve as a testbed database for the component-based software engineering community. AlCoWeb offers to the Web information systems community a lightweight language and an environment which is convivial (allowing hierarchical, incremental and reuse-based development of Web information systems) and transparent (with simple navigation between models and code).

References

1. Alliance, O.A.: Open ajax alliance web site (Last access: February 2007), http://www.openajax.org/
2. Clements, P., Bachmann, F., Bass, L., Garlan, D., Ivers, J., Little, R., Nord, R., Stafford, J.: Documenting Software Architectures, Views and Beyond. Addison-Wesley, Reading (2003)
3. Conallen, J.: Modeling Web Applications with UML, 2nd edn. Addison-Wesley Professional, Reading (2002)
4. Eclipse. Eclipse web site (Last access: June 2007), http://www.eclipse.org/
5. Ginige, J.A., De Silva, B., Ginige, A.: Towards end user development of web applications for smes: A component based approach. In: Lowe, D.G., Gaedke, M. (eds.) ICWE 2005. LNCS, vol. 3579, pp. 489–499. Springer, Heidelberg (2005)
6. Hennicker, R., Koch, N.: Systematic design of web applications with uml. In: Unified Modeling Language: Systems Analysis, Design and Development Issues, pp. 1–20. Idea Group Publishing, Hershey, PA, USA (2001)

---

[2] Alkante Website: www.alkante.com

7. Kadri, R., Merciol, F., Sadou, S.: Cbse in small and medium-sized enterprise: Experience report. In: Gorton, I., Heineman, G.T., Crnkovic, I., Schmidt, H.W., Stafford, J.A., Szyperski, C.A., Wallnau, K. (eds.) CBSE 2006. LNCS, vol. 4063, Springer, Heidelberg (2006)
8. Kandé, M.M., Strohmeier, A.: Towards a uml profile for software architecture descriptions. In: Evans, A., Kent, S., Selic, B. (eds.) UML 2000. LNCS, vol. 1939, Springer, Heidelberg (2000)
9. Larsson, M.: Predicting Quality Attributes in Component-based Software Systems. PhD thesis, Mälardalen University, Sweden (2004)
10. Medvidovic, N., Rosenblum, D.S., Redmiles, D.F., Robbins, J.E.: Modeling software architectures in the unified modeling language. ACM Transactions On Software Engineering and Methodology 11(1), 2–57 (2002)
11. OMG. Meta object facility (mof) 2.0 core specification, document ptc/04-10-15 (2004), Object Management Group Web Site: http://www.omg.org/cgi-bin/apps/doc?ptc/04-10-15.pdf
12. OMG. Object constraint language specification, version 2.0, document formal/2006-05-01(2006), Object Management Group Web Site: http://www.omg.org/cgi-bin/apps/doc?formal/06-05-01.pdf
13. Tibermacine, C., Fleurquin, R., Sadou, S.: Preserving architectural choices throughout the component-based software development process. In: WICSA 2005. Proceedings of the 5th IEEE/IFIP Working Conference on Software Architecture, November 2005, pp. 121–130. IEEE Computer Society Press, Pittsburgh, Pennsylvania, USA (2005)
14. Tibermacine, C., Fleurquin, R., Sadou, S.: On-demand quality-oriented assistance in component-based software evolution. In: Gorton, I., Heineman, G.T., Crnkovic, I., Schmidt, H.W., Stafford, J.A., Szyperski, C.A., Wallnau, K. (eds.) CBSE 2006. LNCS, vol. 4063, Springer, Heidelberg (2006)

WeBrowSearch: Toward Web Browser with Autonomous Search

Taiga Yoshida, Satoshi Nakamura, and Katsumi Tanaka

Department of Social Informatics, Graduate School of Informatics, Kyoto University
Yoshida-Honmachi, Sakyo, Kyoto 606-8501 Japan
{yoshida,nakamura,tanaka}@dl.kuis.kyoto-u.ac.jp

Abstract. In this paper, we propose a Web browser that has an autonomous search capability for complementary information related to a currently browsed page. The system automatically searches for pages having the complementary information, and shows a keyword map, in which each keyword is a type of hyperlink anchor. When a user moves or double clicks a keyword in the keyword map, the system enables users to navigate from the browsed page to the complementary page just as if navigating by ordinary hyperlinks. The proposed Web browser is particularly useful for navigating Web pages that are not connected by ordinary hyperlinks, to compare them.

Keywords: information retrieval, visualization, query-free search, interactive operation.

1 Introduction

The usage of search engines and Web browsers to obtain necessary information from the WWW has become popular. Web browsers basically provide users with a facility to navigate among Web pages based on predefined hyperlinks in Web pages. On the other hand, search engines enable users to search for Web pages independently from their predefined hyperlinks. In this sense, browsing by hyperlinks and searching by search engines have been achieved independently from each other.

The problems arising in browsing and searching are as follows:

(1) Navigation from page to page is restricted to navigation by the hyperlinks predefined by the authors of the Web pages. For example, when a user is reading a certain company's page and he/she wishes to know its rival company's information, hyperlinks to those rival companies are seldom predefined.
(2) When search engines are invoked, they do not know what pages have been visited by Web browsers.

The above problems become more crucial when a user feels the need to obtain complementary information when he/she is browsed a certain Web page.

In this paper, we propose a new Web browser to cope with the above problem. The major features and contribution of our Web browser are summarized as follows:

B. Benatallah et al. (Eds.): WISE 2007, LNCS 4831, pp. 135–146, 2007.

(1) Browsing with autonomous search
(2) Searching for complementary information
(3) Visualization of complementary information by using the keyword map
(4) Dynamic generation of hyperlinks to complementary information

The system searches for related pages automatically in conjunction with a page browsed by a user. When a user browses a certain page with a browser, the system constructs a query according to terms that appear in the browsed page and executes the query. The system visualizes terms in the browsed page on the keyword map and searches related pages according to the keyword map. Each term plotted on the keyword map can be operated by a user with a mouse. The system presents pages containing complementary information. The user can browse other related pages with a simple mouse operation. We call our system "WeBrowSearch". WeBrowSearch is a coined word from a combination of the words "Web," "browse," and "search."

In this paper, first, we describe the motivation and concept of our system. Then, we discuss related works. After that, we explain the design and implementation of our system and describe the evaluation of our system. Finally, we conclude this paper and describe future work.

2   Motivation

Our research group surveyed Web search activity [1]. In this survey, we created an online questionnaire consisting of 26 questions that were answered by 1,000 Internet users. Users were divided into four categories depending on their age: 20-29, 30-39, 40-49, and 50-59 years old. Each group consisted of 250 respondents; half were males and half were females. In this survey, there are two questions that are related to our research. One question is about the situation of using Web search engines. The other question is about the reason for searching the Web.

The results to the question about situations of using Web search engines are shown in Fig. 1. We found that users decide to use search engines when they want to research particular information or browse the Web. Searching the Web without any particular reason is also common. Two other common situations in which searches were performed are when watching TV and reading e-mail.

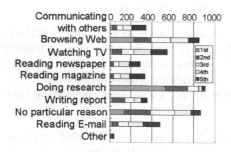

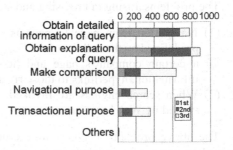

Fig. 1. Situations in which users search Web   **Fig. 2.** Reasons for searching Web

The results of answers to the question about reasons for searching the Web are shown in Fig. 2. Users search the Web mostly because they require basic (46.0% of respondents selected it as a first reason) or detailed (36.8%) information about particular things. Another motivation for searching the Web is to do some comparison in shopping for example (7.4%). Few users chose other reasons for searching the Web.

These results mentioned that the Web search while browsing is very important and required by many people. However, searching related pages while browsing the Web is not easy for people. Users have to open the page of a Web search service and input a proper query to find related Web pages.

3 Related Work

A query-free search is the concept that a system automatically searches for related information without any user operation. Hart et al.[2] made a system that enables a user to search a manual of a printer without inputting a query. The system that Henzinger et al.[3] made automatically presents Web pages related to a program that a user is watching according to captions of the watched program. Our system is also based on a query-free search, but we adopted the keyword map and accepted some user interactions. By some interactions, the system can understand a user's intention further.

There are some research studies that enable a user to search related pages while doing something else.

WebTelop[4] is a system that searches pages related to a TV program that a user has browsed. The system automatically presents a page that supplements the TV program that is being watched. In this system, a user cannot specify what aspect of the topic to look at. Our system presents related pages while browsing and enables a user to change an aspect of related pages by operating the keyword map.

Nadamoto et al.[5] made a system that enables a user to compare a web page with similar web pages. In using the system named *CWB*, when a user browses a certain news page in a news site, the system enables users to search similar news pages in another news site. A user can obtain related news using the system, but he/she had to know the URL of the target news site.

The system made by Yumoto et al. [6, 7], and the system named *CWS*[8] enable users to compare two objects by analyzing pages obtained using names of these objects. Initially, these systems require names of objects being compared.

Our system constructs a query and searches related pages without requiring a user to input the name of the comparison target.

There are many systems that can visualize web pages or information.

KeyGraph[9] is a system that can visualize important words in text data using a network graph in which important words as nodes are connected to each other. However, *KeyGraph* is intended for one text file and not for all Web search results. This system does not consider interactive network operation.

Yahoo! Mindset[10] is a system that reflects user's intentions. This system determines scores of Web pages according to whether they are more commercial or more informational. If a user wants to search for commercial pages, the system returns commercial pages by dragging a slider to the "shopping" side. If a user wants to search for informational pages, then, only dragging a slider to the "researching" side is necessary.

This system is the same as our system with respect to re-ranking search result pages according to user operation. However, this system does not generate new axes dynamically. In our system, important terms are extracted from result pages, and they are used for generating new axes for re-ranking on the keyword map.

WebGlimpse[11] is a system that combines the two paradigms of searching and browsing. A user can find an ideal page reachable by browsing hyperlinks of the current document. Searched pages are confined to pages that can be browsed from a browsed page with this system. However, a user can find any page with our system because the system searches related pages by constructing many flexible queries.

4   WeBrowSearch

4.1   Concept

The concept of WeBrowSearch is to support users who browse the Web using an autonomous search for complementary information. Mainly, our system is used as a Web browser. In a background process, the system automatically searches the related pages, which include complementary information according to users' Web browsing. Then, the system shows topic terms about complementary information on the keyword map. In this work, we classify topic terms as follows:

- Internal topic terms: Topic terms are extracted from a browsed page
- External topic terms: Topic terms are extracted from surrounding pages of a browsed page except for terms in a browsed page

A detailed image of the keyword map is shown in Fig. 3. Terms extracted from a web page and related pages are plotted on the keyword map. Positions of term nodes are determined as described below:

- Internal topic terms are plotted in a circle
- External topic terms are plotted around a circle

Users can understand the topic in a browsed page by viewing terms on the keyword map and find an ideal related page by clicking and dragging and dropping on the keyword map.

An image of the system is shown in Fig. 4. When a user visits a new page on the browser, which is shown on the left side of the window, the system extracts internal topic terms from a browsed page. The system composes a search query that consists of internal topic terms to obtain related pages and external topic terms. These terms are mapped on the keyword map, which is shown in the upper right of the window. Concurrently, the system presents a list of related pages in the lower right of the window. On the keyword map, a user can click the term node and move the position of the term node by using a drag-and-drop operation. If there is a term node that matches the user's intention, and he/she wants to know more about the term, a user can browse a related page about a topic of the term by moving the term node toward the center of the keyword map. In the case where the user does not want to browse pages that contain certain terms, he/she can indicate his/her intention by moving the nodes toward the outer side of the keyword map.

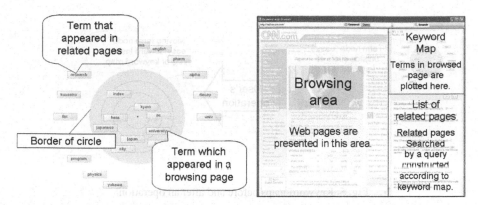

Fig. 3. Keyword map   **Fig. 4.** Image of system

When a user drags or double clicks a term node, the system chooses and presents a new related page in a browsed area according to a layout of the keyword map. If the user double clicks a term node, the system composes a new query and searches for related pages again, and the system presents a new related page found by the query in the browsed area. A user can browse another related page by operating on terms on the keyword map or by clicking the "next related page" button.

4.2 Usage Scenario

We explain the usage of the system by giving an example of searching for a recipe on the Internet. We assume that a user has a lot of beef, and wants to cook with beef. However, his/her cooking skill is not high.

1. First, a user connects to a recipe site and searches with the query "beef".
2. He/she finds some recipe pages about a beef stew in a recipe site.
3. However, he/she felt the recipe is difficult to cook.
4. Then, he/she moves the term node "easy" toward the inside of a circle on the keyword map.
5. The system automatically presents a related page about a beef stew recipe that is easy to cook.
6. He/she decided to make a beef stew following a recipe on the related page.

The keyword map of a recipe page about a beef stew is shown in Fig. 5. The left side of the figure is an initial keyword map, and the right side of the figure is the keyword map after a user's operation.

In Fig. 6, a page that a user initially browsed is shown on the left side, and a related page searched by a user's operation is shown on the right side.

5 Design and Implementation

To implement our system, some functions are required. The required functions are as follows:

- Extracting topic terms from a browsed Web page
- Mapping extracted terms on the keyword map

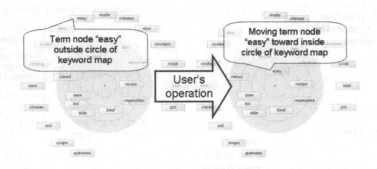

Fig. 5. Keyword maps before and after an operation

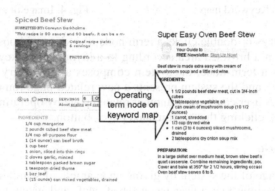

Fig. 6. Initial page and relative page

- Constructing a query according to the keyword map
- Searching related pages, extracting topic terms, and plotting them on the keyword map
- Selecting presented related pages according to the keyword map

We explain these functions in this section.

5.1   Design Based on a Requirement

When using this system, first, a user inputs a URL of a page or a query. If he/she wants to browse a page using a URL, the page is loaded by clicking the "Connect" button. He/she can execute the query by clicking on the "Search" button. When a user clicks the "Search" button, a search results page of Google is displayed. The user can browse the page in the browsing area.

When the page is loaded, the system extracts some terms from the text of the browsed page. These terms are extracted by choosing terms that occur frequently in the browsed page. Some stop words like "about" or "here" are excluded from the extracted terms.

The ten most frequently emerging terms are plotted as term nodes on the keyword map. Each term node is plotted based on the frequency of the term. If a term occurred frequently in a browsed page, the term node is plotted close to the center of the keyword map. The less frequently a term occurs in a browsed page, the farther the term node is plotted from the center of the keyword map.

Then, related pages are searched automatically by the system. Initially, the system constructs a query according to positions of term nodes on the keyword map. The query is composed of terms that are near the center of the keyword map. The system executes the query and obtains 100 result pages.

On the keyword map, term nodes are plotted reflecting the frequency of terms in a browsed page. Therefore, we can regard a page that contains many terms plotted in the vicinity of the center of the keyword map as strongly related to a browsed page. The system calculates scores of every searched related page based on the frequency of each term and the distance from the position of each term to the center of the keyword map. Searched related pages are re-ranked in descending order of the calculated score. In this way, searched related pages are re-ranked in order from a page strongly related to a browsed page to a page that is not related to a browsed page. Some of the highly ranked related pages are presented as a list of related pages in the window.

When pages are searched with a search engine, those pages are expressed as a set of components. Title, URL, and Snippet are contained in the set of components. Snippet is a summary of the pages made by extracting sentences surrounding the query term. The system counts the number of times every term occurs in 100 pages, and finds the 20 most frequently occurring terms. These terms are plotted around the keyword map.

When terms are plotted on the keyword map initially, terms extracted from a browsed page are plotted inside a border of a circle, and terms extracted from related pages are plotted outside a border of a circle. The circle can be thought of as a user's intention about a page which he/she wants to browse. The system cannot know a user's intention adequately until a user performs an operation on the keyword map, so the system attaches a high value to a page that is similar to a browsed page.

A user can move term nodes on the keyword map with a mouse operation. By moving a term node, the score of the term is recalculated. If a user moves the term node toward the center of the keyword map, the score of the term is increased, and if he/she moves the term toward the outside of the keyword map, the score of the term is decreased. When a user moves a term node, the system automatically calculates scores of related pages and re-ranks these pages.

Terms that occur frequently in a browsed page are plotted inside a border of a circle, and terms extracted from related pages are plotted outside a border of a circle. Terms that have already been extracted from a browsed page are not extracted from related pages. Therefore, we can consider that terms extracted from searched related pages are not contained in a browsed page or that there are few occurrences if the terms are contained. If a user drags a certain term node from outside a border of the circle to inside a border of the circle, he/she probably thinks a browsed page does not have sufficient information about the term and wants to browse a page that contains the term. The system supposes that the term is an important term and assigns a high score to the term. On the contrary, if a user drags a certain term node from inside a border of the circle to outside a border of the circle, he/she probably does not want to browse a page that contains the term even though that browsed page contains the term. The system assigns a minus score to the term.

A list of related pages is created when a browsed page is loaded. That means a user's intention about a target page is not regarded apart from information that he/she decided to see the page. Therefore, the query that was composed according to the initial layout of the keyword map may not indicate the user's intention and searched related pages may not contain any page he/she wants to browse. If a user feels that searched related pages do not contain interesting pages, he/she can transform a query and search again. By double clicking a term node, the system constructs a query using the term and executes the query, and a list of related pages is obtained again.

A related page that is chosen based on a layout of the keyword map is navigated in a browsing area synchronized with a user's operation performed on the keyword map. In addition, a list of searched related pages is shown in the lower right of the page. There are a few pages that are displayed in the list of related pages. However, a user can browse various pages by performing operations on terms on the keyword map or clicking the "next related page" button.

5.2 Implementation

5.2.1 Extracting Topic Terms from Browsed Web Page

First, html code of a browsed page is extracted from the browser and html tags are removed from the html code. Then, the system tokenizes the text and obtains terms from the browsed page. Terms that are not considered part of topic are removed from the list of terms as stop words. Several hundred words including prepositions and pronouns are removed from the extracted terms. The frequencies of all remained terms are counted and the terms are sorted by their frequencies. The top ten terms are chosen as topic terms.

5.2.2 Mapping Extracted Terms on Keyword Map

The system plots extracted terms on the keyword map according to frequencies of these terms. A term that has a high frequency is plotted near the center of the keyword map. The system decides the distance from the center of the keyword map to a term node according to the formula shown below.

$$d(t) = \frac{r}{\sqrt[4]{freq(t)+1}} \tag{1}$$

In formula (1), t means each term, d(t) means the distance between the center of the keyword map and the term node, r means the radius of the circle in the keyword map, and freq(t) means the frequency of a term in a browsed page. This distance varies from 0 to 0.841 times of the radius of the circle. The system plots terms according to the calculated distance from the center of the keyword map.

5.2.3 Constructing Query According to Keyword Map

The system chooses three terms for constructing a query by selecting the nearest three terms from the center of the keyword map. Calling these terms T_1, T_2, and T_3, the system constructs a query Q like below.

$$Q = (T_1 \text{ OR } T_2) \text{ AND } (T_2 \text{ OR } T_3) \text{ AND } (T_3 \text{ OR } T_1)$$

Pages that contain two of these three terms can be found by this query. The reason why we do not construct a query $Q = T_1$ AND T_2 AND T_3 is that there is a risk the number of searched result will be too few if there is a term which have no relation with the topic of a browsed page among these term.

5.2.4 Search Related Pages, Extract Topic Terms, and Plot Them on the Keyword Map

A query constructed on the basis of frequencies of terms in browsed pages is executed with a search engine. We used the Yahoo! Web Search API to search for pages.

The system obtains 100 result pages using a search engine and extracts 20 terms that emerge frequently in titles and snippets of searched result pages. The method of extraction is the same as that in 5.2.1. Extracted terms are plotted around the circle.

5.2.5 Selecting Presenting Relative Pages According to the Keyword Map

Each term on the keyword map has its own score and is calculated according to the distance between the term node and the center of the keyword map. The score of a page $S(p)$ is calculated as the total score of terms contained in the relative page. The formula is shown below.

$$S(p) = \sum_{t=1}^{N} x_{pt} (r - d(t))$$ (2)

In formula (2), p means each searched result page, N means the number of the topic terms on the keyword map, and t, r, and d have the same meaning as those in 5.2.2. If a page p contains term t, $x_{pt} = 1$; otherwise $x_{pt} = 0$.

If a term node is placed outside a border of a circle, the score of a term will be calculated as a minus score. However, if a term outside a border of a circle is extracted from a related page, the system gives the term a score of 0.

Searched related pages are sorted in descending order by $S(p)$., and some pages that have a high value of $S(p)$ are presented in the window as a list of related pages.

When a user moves a term on the keyword map, the page that has the highest $S(p)$ value is navigated in the browsing area

6 Evaluation

To validate a feasibility of the system, we browse some test pages using the system. Browsed pages are shown in Table 1.

We examined what query were constructed by the system (A_1), how many terms were extracted from relative pages (A_2), how many different pages were searched by moving each term from outside a circle to inside (A_3), examples of a moved term and a searched related page (A_4).

In this experimentation, the system initially loaded a page. The system constructed a query (A_1) and searched relative pages. Next, we counted terms extracted from relative pages and validated that whether the terms are related with the topic of the browsed

Table 1. Titles and URLs of pages for experimentation

Page1	Title	Strong evidence of wet past on Mars
	URL	http://edition.cnn.com/2007/TECH/space/05/22/ mars.rovers.ap/index.html
Page2	Title	Apple - iPod nano
	URL	http://www.apple.com/ipodnano/
Page3	Title	Kyoto travel guide – Wikitravel
	URL	http://wikitravel.org/en/Kyoto

page or not (A_2). Then, we moved each term from outside a circle of the keyword map to inside. And we counted how many pages we were able to find using the system and also counted how many pages which corresponded to the topic of the browsed page were among those pages (A_3).

Table 2 shows the result of the experiment.

We could find relative pages for all these three pages using our system. However, in the case of page 1, we could get pages about "silica" but we could not get pages about "mars". This is because the query which the system constructed did not contain "mars". So, we double clicked the node "mars" and the system searched related pages again with a new query. Then, we could get pages about "mars" for example a page with the title "Broken Wheel Reveals Water On Mars."

Table 2. Attributes of pages for experimentation

Page1	A_1	(silica I water) (water I deposit) (deposit I silica)
	A_2	17 (7 terms were related with the topic)
	A_3	12 (6 pages were related with the browsed page)
	A_4	surface : What Utah gemstone is rarer than diamond? - Utah Geological Survey kitchen : a kitchen witches work is never done cleaning : Yahoo! Answers - Household cleaning tips?
Page2	A_1	(ipod I nano) (nano I apple) (apple I ipod)
	A_2	13 (11 terms were related with the topic)
	A_3	6 (6 pages were related with the browsed page)
	A_4	buy : Apple Ipod :: Apple Ipod Online generation : Apple iPod nano 4GB (2nd Generation) - Articles / Reviews store : Apple ipod nano video: The apple store (u.s.).
Page3	A_1	(kyoto I station) (station I bus) (bus I kyoto)
	A_2	15 (15 terms were related with the topic)
	A_3	10 (9 pages were related with the browsed page)
	A_4	hotel : Kyoto Brighton Hotel - kyoto hotels - kyoto accommodations japan : japan-guide.com forum - Trains + kyoto 10 days plan map : Direction to the hotel and Kyoto University Clock Tower

From this experimentation, it can be seen that constructing a proper query is a very important task for searching relative pages. To improve the accuracy of the system, we plan to develop another algorithm for constructing a query which is not limited to extracting terms that emerge frequently in a browsed page.

This experimentation shows that the system can provide users with related pages of a browsed page by simple operations.

7   Conclusion

In this paper, we proposed a new browser. The system searches related pages automatically while a user browses web pages. We conclude this paper and discuss future work in this section.

When we browse web pages, finding a target page that has ideal information is difficult.

First, judging whether a browsed page matches a user's intention is difficult. In this point, our system presents topics in a browsed page by plotting terms that frequently emerge in a browsed page.

Second, if a browsed page does not have a topic that a user wants to browse, he/she has to think about a proper query and execute that. However, thinking about a proper query is difficult work. This system supports a user to construct a query by presenting related pages and terms extracted from those pages. A user only has to choose which term matches his/her intention, without thinking about what is an appropriate term that expresses the intention.

When a user searches pages by a query, many pages are presented as result pages. And it is hard to find a page which contains topics he/she wants. This system supports a user to find a page that contains these topics.

However, our system has some problems. One problem is the time for searching related pages automatically. When a user browses a web page, he/she probably pays attention to the browsed page itself. Therefore, a user would not want to wait for searching after a browsed page is loaded. Maintaining an original browser in operation so that a user does not have to wait until a search for related pages is completed would be necessary.

Using this system, a user can understand topics of a browsed page and related pages. The system presents which topics are in a browsed page and which topics are not. However, a user cannot know whether each topic is a major or minor topic. We will improve the system to be able to inform a user of the generality of a topic by changing the size of a term node for example.

In addition, there is a situation in which a term that a user wants to emphasize or disregard is not plotted on the keyword map. Therefore, the system must have a method of adding more term nodes to the keyword map by means of a user operation. We will also implement a system that finds terms related to a term that a user has added and plots such terms.

When a user browses a long page, there is a situation when he/she wants to customize the layout of that page. We will extend the system to enable users to remake a layout of a page like resizing the font size of a paragraph or highlighting some terms by performing operations on a keyword map.

Acknowledgments

This work was supported in part by "Informatics Education and Research Center for Knowledge-Circulating Society" (Project Leader: Katsumi Tanaka, MEXT Global COE Program, Kyoto University), and by #18049041, #18049073 and #18700129.

References

1. Nakamura, S., Konishi, S., Jatowt, A., Ohshima, H., Kondo, H., Tezuka, T., Oyama, S., Tanaka, K.: Trustworthiness Analysis of Web Search Results. In: ECDL. Proceedings of the 11th European Conference on Research and Advanced Technology for Digital Libraries (2007)
2. Hart, P., Graham, J.: Query-free information retrieval. IEEE Expert 12(5), 32–37 (1997)
3. Henzinger, M., Chang, B.-W., Milch, B., Brin, S.: Query-Free News Search. In: WWW. Proceedings of theTwelfth International World Wide Web Conference (2003)
4. Ma, Q., Tanaka, K.: WebTelop: Dynamic tv-content augmentation by using web pages. In: Proceedings of IEEE International Conference on Multimedia and Expo (ICME) (II), pp.173–176 (2003)
5. Nadamoto, A., Tanaka, K.: A Comparable Web Browser (CWB) for Browsing and Comparing Web Pages. In: WWW. Proceedings of the 12th International World Wide Web Conference, pp. 727–735 (2003)
6. Yumoto, T., Tanaka, K.: Finding Pertinent Page-Pairs from Web Search Results. In: Fox, E.A., Neuhold, E.J., Premsmit, P., Wuwongse, V. (eds.) ICADL 2005. LNCS, vol. 3815, Springer, Heidelberg (2005)
7. Yumoto, T., Tanaka, K.: Page Sets as Web Search Answers. In: Sugimoto, S., Hunter, J., Rauber, A., Morishima, A. (eds.) ICADL 2006. LNCS, vol. 4312, pp. 244–253. Springer, Heidelberg (2006)
8. Sun, J.-T., Wang, X., Shen, D., Zeng, H.-J., Chen, Z.: CWS: a comparative web search system. In: WWW. Proceedings of the 15th international conference on World Wide Web (2006)
9. Ohsawa, Y., Benson, N.E., Yachida, M.: KeyGraph: automatic indexing by co-occurrence graph based onbuilding construction metaphor, Research and Technology Advances in Digital Libraries. In: ADL 1998. Proceedings. IEEE International Forum on Volume, pp. 12–18 (1998)
10. Yahoo! Mindset, http://mindset.research.yahoo.com/
11. Manber, U., Smith, M., Gopal, B.: WebGlimpse - Combining Browsing and Searching. In: Proceedings of the Usenix Technical Conference (1997)

A Domain-Driven Approach for Detecting Event Patterns in E-Markets: A Case Study in Financial Market Surveillance

Piyanath Mangkorntong[1] and Fethi A. Rabhi[2]

[1] School of Computer Science and Engineering
The University of New South Wales, Sydney, NSW 2052, Australia
pman@cse.unsw.edu.au
[2] School of Information Systems, Technology and Management
The University of New South Wales, Sydney, NSW 2052, Australia
f.rabhi@unsw.edu.au

Abstract. An e-market can be thought of as a distributed event system where an event is generated every time the market's state changes in response to a number of human or computing agents. The paper describes a practical application of event processing in an e-market context through conventional Event Processing Systems (EPSs). A new EPS architecture that allows an integration of several existing EPSs under a unified domain-specific user interface and execution environment is proposed. We assess the performance of the system for a case study in financial market surveillance and its ability to provide a common interface for two existing EPSs – SMARTS and Coral8. A discussion on the experimental results and the issues arising from the proposed EPS architecture are also provided.

Keywords: e-markets, event-driven architecture, event pattern model.

1 Introduction

E-markets[1] can be thought of as trading environments over computer networks. Since millions of transactions typically occur on e-markets everyday, the motivation behind our research arises from the increasing need to analyse this transactional data for use in many business processes such as market strategy evaluation and illegal market activity detection.

Building analysis systems for e-markets is a challenging task. One of the biggest challenges is to deal with a large amount of data produced in real-time. This is because e-markets involve a large number of users (possibly from all around the world) and have been growing in size rapidly over the last decade. Other challenges include developing software components that interface effectively with the market feeds and handling the different types of data formats used to encode e-market transactions.

B. Benatallah et al. (Eds.): WISE 2007, LNCS 4831, pp. 147–158, 2007.
© Springer-Verlag Berlin Heidelberg 2007

Our approach considers an e-market as a distributed event-driven system where an event is generated when the market state changes. This allows for the analysis of e-market data to be performed efficiently using event processing concepts[7]. The paper presents an architecture that reuses existing Event Processing Systems (EPSs) and is structured as follows. The next section discusses the role and limitations of existing EPSs in analysing events arising from e-markets. Section 3 proposes an EPS framework dedicated for analysing e-markets. Section 4 describes a case study involving financial market surveillance and Section 5 concludes this paper.

2   Background

An Event Processing System (EPS) is a system that can be used to search for a set of events that match a given event pattern in real-time. The four main reasons for introducing an EPS into organisations are surveillance, fault diagnostics, security protection and performance monitoring[7, 11]. At present, there are several EPSs available commercially. For example, Ipswitch WhatsUp Professional[6] is an EPS that has been designed for a network management domain. Coral8[2], Apama[12] and StreamBase[14] are examples of EPSs that can be customised to fit any application domain.

For the purpose of this paper, we carried out a comparison between a general-purpose (Coral8) and a dedicated EPS (SMARTS):

- **Coral8[2]:** Coral8 has been developed based around a Service Oriented Architecture (SOA). It has been designed to support many different types of input/output formats. The Event Pattern Language (EPL) used is called Continuous Computation Language (CCL). The majority of the features in CCL have been derived from the Standard Query Language (SQL). However, the workflow of CCL and the environments that it runs in are significantly different from traditional relational databases.

- **SMARTS[13]:** The Securities Markets Automated Research Trading and Surveillance (SMARTS) is known as one of the world-leading commercial financial market surveillance software systems. SMARTS provides a software component to interface with the transactions coming from the financial market feed. The transactions collected are stored in the SMARTS database which can only be accessible by SMARTS software modules. Event patterns used in SMARTS are described in ALICE which is the SMARTS proprietary EPL.

Using selected comparison criteria from [3, 5] which are adaptability, time usage, memory usage, scalability and flexibility, both SMARTS and Coral8 offer some different system characteristics in processing and expressing event patterns. For example, SMARTS has been specifically designed to fit the requirements of the financial market domain and therefore is much more scalable than Coral8 in terms of handling large amounts of transactions from different financial market feeds. However, in terms of adaptability and flexibility, Coral8 provides some features including publish-subscribe and customised data format support which makes it more adaptable and flexible than SMARTS. The idea behind our work is to propose an approach that makes it possible to use several existing EPSs (called slave EPSs) to support an efficient event detection process for complex patterns. Our approach is described in the next section.

3 Proposed Approach

The main ingredients of our solution consist of a unified event pattern representation which can be used to represent event patterns for different slave EPSs (Sections 3.1 and 3.2) and a domain-driven architecture which relies on the proposed event pattern representation (Sections 3.3, 3.4 and 3.5).

3.1 Representing Event Patterns

We use events to represent any changes to the state of an e-market. In previous work[10], we have defined a number of data types which are used to describe the entities that comprise the market. This will be referred to as e-market data model in the rest of this paper. Figure 1 shows our e-market data model that fits over our financial market case study.

We have classified event patterns into primitive and complex event patterns. Complex event patterns can be described as combinations of primitive event patterns. Our primitive event patterns comprises of the Coral8 event design patterns[4] which include most common event patterns such as Filtering, Aggregation over Windows and Event Pattern Matching.

The diagram in Figure 2 illustrates the different ways event patterns are represented in this paper. Each event pattern has a unique model and different code representations associated with each slave EPS. The event pattern model, which is generic and independent from any slave EPS, is represented by parameters and constraints. They are stored in the Event Pattern Definition Database. For example, the 'Model' section of Figure 2 shows the model of Event Pattern-1 consisting of its parameters and constraints.

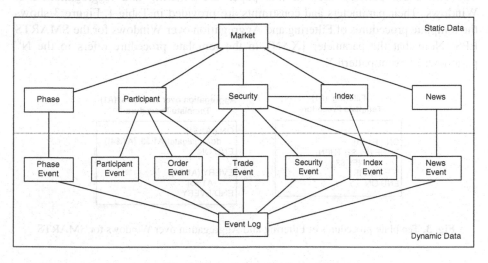

Fig. 1. Financial market data model

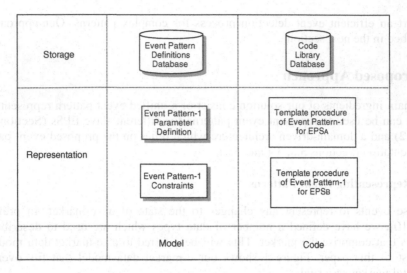

Storage	Event Pattern Definitions Database	Code Library Database
Representation	Event Pattern-1 Parameter Definition Event Pattern-1 Constraints	Template procedure of Event Pattern-1 for EPSA Template procedure of Event Pattern-1 for EPSB
	Model	Code

Fig. 2. Event pattern representation

The event pattern code is EPS-dependent and represents the event pattern implementation which we refer to as a "template procedure". The event pattern template procedures are stored in the Code Library Database. For example, the 'Code' section of Figure 2 shows two template procedures of Event Pattern-1 for EPS$_A$ and EPS$_B$. Examples of event patterns are provided in the next section. More details on the event pattern representation used in this paper can be found in [9].

3.2 Event Pattern Examples

The two event patterns used in this paper are Filtering and Aggregation over Windows. Their parameters and constraints are provided in Table 1. Figure 3 shows the template procedures of Filtering and Aggregation over Windows for the SMARTS EPS. Note that the parameter {X.$N} in the template procedure refers to the Nth parameter of event pattern X.

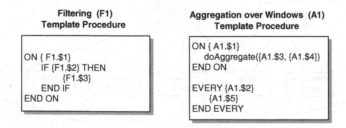

Filtering (F1) Template Procedure

```
ON { F1.$1}
    IF {F1.$2} THEN
        {F1.$3}
    END IF
END ON
```

Aggregation over Windows (A1) Template Procedure

```
ON { A1.$1}
    doAggregate({A1.$3}, {A1.$4})
END ON

EVERY {A1.$2}
    {A1.$5}
END EVERY
```

Fig. 3. Template procedures of Filtering and Aggregation over Windows for SMARTS

Table 1. Event pattern examples

Event Pattern Name	Parameters	Description	Constraints
F1 (Filtering)	1. Input Event Stream 2. Filtering Condition 3. Output Event Stream	Returns a stream of input events whose property matches the 'Filtering Condition' specified.	1. The type of F1.$1 can only be one of the event types defined in our e-market data model.
A1 (Aggregation over Windows)	1. Input Event Stream 2. Window Size 3. Aggregation Operator 4. Aggregated By 5. Output Event Stream	Subdivides a stream of input events into groups of objects according to the 'Aggregated By' parameter and then combines them into a single object using the 'Aggregated Operator'.	1. Same as previous with A1.$1 2. The type of aggregated objects is constrained by the type of A1.$1.

3.3 Proposed EPS Architecture

Our EPS architecture is an improvement over the architecture described in [10]. It is a 3-tier architecture with a user interface, business logic and data layers (see Figure 4). The Graphical User Interface (GUI) supports the two main tasks involved in an event pattern detection process which are event pattern design and event pattern execution. The Technical GUI component provides graphical interfaces for constructing new event patterns. The Domain GUI component provides graphical interfaces for event pattern execution. A user can select a previously created event pattern, customise its execution by supplying some parameters and then start the event pattern execution process.

The business logic layer has two main software components – Design Manager and Execution Manager. The Design Manager is responsible for the event pattern design process. This includes retrieving, storing and manipulating event pattern definitions, construction algorithms and template procedures. The Execution Manager is responsible for the event pattern compilation and execution processes which involve accessing event pattern definitions and template procedures as well as instantiating template procedures. It also facilitates access to financial transactions for the relevant slave EPSs, monitors execution and reports back the complex events detected to the user through the Domain GUI component in real-time.

The data layer comprises of the Event Pattern Definitions Database, the Code Library Database and the Event Database. The Event Pattern Definitions Database and the Code Library Database store representations associated to event patterns

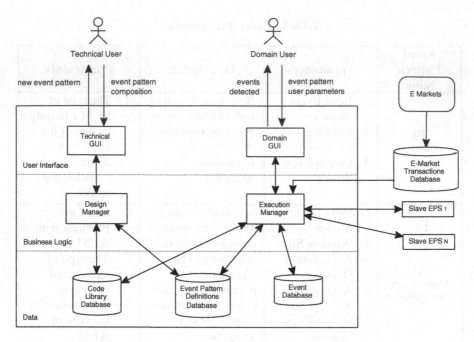

Fig. 4. Proposed EPS architecture

(see Section 3.1). The Event Database is used to store instances of event patterns detected during the execution process.

Currently, a system prototype based on the proposed EPS architecture is being developed using C# .NET and SQL Server 2005. Two slave EPSs used in the prototype are Coral8 and SMARTS.

3.4 Event Pattern Construction Process

The event pattern construction process is handled by the Design Manager. We demonstrate this process through the use of an example (see [9] for more details).

The Pair Trading event pattern is used in financial market surveillance to count the number of trades arising from particular subset of market participants (called traders) across a period of time. This event pattern is a simplified form of the event pattern used for detecting cyclic trading. This type of behaviour, where a group of traders manipulate the market by trading amongst themselves, is considered illegal in some financial markets. The Pair Trading event pattern can be expressed as a combination of two primitive event patterns – Filtering and Aggregation over Windows (called children event patterns).

There are a number of pre-defined ways for composing children event patterns, each of which has a set of rules associated with it. The rules specify how to establish the new event pattern's parameter definitions and construct its template procedure from its children event patterns. For example, to construct the Pair Trading event pattern, Filtering and Aggregation over Windows are connected in a pipeline fashion as illustrated in Figure 5. Table 2 shows the resulting parameters and constraints of the Pair Trading event pattern which are derived from its children event patterns.

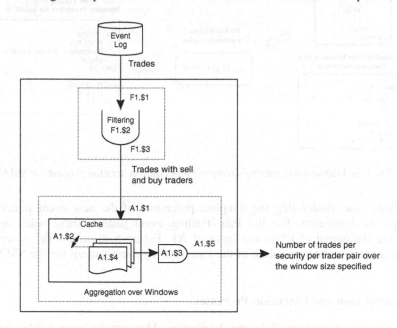

Fig. 5. Representation of the Pair Trading event pattern

Table 2. Parameters of the Pair Trading event pattern

Event Pattern Name	Parameters	Description	Constraints
P1 (Pair Trading)	1. Input Event Stream 2. Filtering Condition 3. Window Size 4. Aggregation Operator 5. Aggregated By 6. Output Stream Event	It derives parameters 1-2 from F1 (Filtering) and parameters 3-6 from A1 (Aggregation over Windows). It uses F1 to extract the trades that satisfy the Filtering Condition specified and then A1 groups the trades and counts them by the specified trader pair.	It derives the three constraints from F1 and A1. The additional constraint is that the value of F1.$3 must be the same as the value of A1.$1.

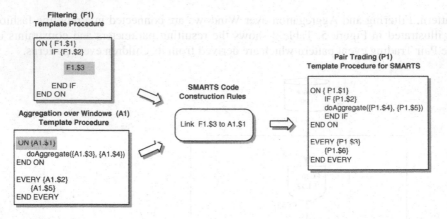

Fig. 6. The Pair Trading event pattern's template procedure generation process for SMARTS

The rules for constructing the template procedure of the new event pattern are based on its constraints. For the Pair Trading event pattern, these rules involve redirecting the output of F1 to the input of A1. Figure 6 illustrates the process of generating the template procedure of the Pair Trading event pattern for the SMARTS EPS.

3.5   Compilation and Execution Processes

As mentioned in Section 3.3, the Execution Manager is responsible for the compilation and execution processes of the proposed EPS. These processes are now described in more detail.

The compilation process is triggered by the user after selecting the event pattern and a suitable slave EPS to process the event pattern. It involves retrieving the template procedures that correspond to the selected event pattern and slave EPS from the Code Library Database. Some event patterns are not supported by some of the EPSs so the compilation option for this event pattern/EPS pair will not be available. The retrieved template procedures are instantiated with user-supplied parameter values. This is followed by invoking the compiler of the selected EPS on the instantiated procedures to build the final executable files.

When a user issues a command to execute the complex event pattern, the Execution Manager notifies the selected EPS with the location of the executable, the dataset selected and the destination of the output file. In this research, we assume that the Financial Market Transactions database consists of pre-processed datasets[1] for use with different EPSs. Our system also integrates different ways it can monitor the event pattern execution process performed by the selected slave EPS (e.g. standard output redirection and publish-subscribe method) as well as allow real-time notifications when an event pattern instance is detected. When the execution finishes, the event pattern detection results are written to the output file which is a part of the Event Database.

---

[1] The data pre-processing step is not required when selecting SMARTS since it has the ability to process real-time market data as well.

4 Evaluation

In this research, the two criteria used to select the most suitable EPS to process the event pattern are:

- Expressive Power: The ability of an EPS to express event patterns of varying complexity and can be assessed through the EPL associated with an EPS.
- System Performance: The performance of the proposed EPS when selecting a particular slave EPS to process a complex event pattern. This can be assessed by measuring the execution time required in processing a complex event pattern.

We have performed an experiment which involved selecting the most suitable EPS based on the expressive power of slave EPSs (Coral8 and SMARTS). The assessment results have been published in [10]. In this paper, we will be focusing on selecting the slave EPS based on the performance of the proposed EPS through the system prototype developed. Two slave EPSs used in the experiments are Coral8 and SMARTS. The experiment and results are provided in Section 4.1 followed by a discussion of the issues arising from the implementation in Section 4.2.

4.1 Experiments

The purpose of the experiments is to demonstrate the ability to switch between different EPSs through the same user interface and study the performance implications. In this case study, the time usage required in to execute the Pair Trading event pattern (described earlier in Section 3.4) across different time window sizes has been measured. The graphs in Figure 7 show a comparison of the execution time when using different window sizes for both SMARTS and Coral8. The vertical axis represents the amount of time required in processing the event pattern by an EPS and the horizontal axis represents the window size (in minutes). The experiments were carried out by using the real historical data from the Stock Exchange of Thailand (SET)[2]. For each window size, we have taken the average execution time of ten experiments where each experiment is run against a dataset collected from different normal trading days (i.e. no suspension). The experiments have been conducted using an Intel Centrino Duo Core Processor 1.83 GHz with the 1 GB of memory running on the Windows XP platform.

From the graphs, we can see clearly that the performance of the proposed EPS when selecting SMARTS as the EPS to process the Pair Trading event pattern requires less time than when selecting Coral8 for the same pattern.

4.2 Discussion

The experimental results show that by far, the Pair Trading event pattern can be processed faster using SMARTS than Coral8. This suggests that if an event pattern can be executed by many EPSs then the most suitable EPS should be the one with the best performance. However, this conclusion can only be true if there are no other factors that can affect the system performance.

---

[2] Data supplied by Securities Industry Research Centre of Asia-Pacific (SIRCA) on behalf of Reuters.

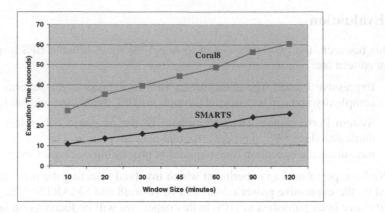

Fig. 7. Comparing SMARTS and Coral8 performance for the Pair Trading pattern using Stock Exchange of Thailand (SET) data

In this case study, the performance when selecting Coral8 to process the Pair Trading event pattern is affected by three other known factors which include the data pre-processing step, delays in starting up the event pattern execution and publishing input data. The data pre-processing step is required in order to extract the raw e-market data into the format supported in Coral8. The delays in starting the event pattern execution process are due to the way Coral8 operates. The Coral8 executable must be registered with the server before starting the event pattern execution process and unregistered once the process is finished. The proposed system also has to publish the input data to the Coral8 server, therefore adding to the overheads of the execution process. This is unlike the SMARTS server which, given the dataset location, has the ability to access and process the dataset automatically.

In order to clearly demonstrate this argument, we have measured the time required in the data pre-processing step and the time required for starting up the event pattern execution process in Coral8. The graphs in Figure 8 compare between Coral8 performance (shown as several solid lines) and SMARTS performance (shown as a dotted line) depending on whether several additional overheads associated with Coral8 are considered or not. From the graphs, we can see that after removing the three factors affecting the proposed EPS performance, Coral8 execution time is very close to when SMARTS is selected. In fact, if the time used to publish input data took 20% of the execution time, using Coral8 to process the Pair Trading pattern is actually slightly faster than using SMARTS.

Apart from the issues arising from the experiments discussed above, another limitation on the system performance are the sets of Application Program Interface (API) provided by the slave EPSs for facilitating the integration with external systems. SMARTS does not provide any API to connect to other systems and therefore redirecting the standard output into the system is the only option. This makes the performance of the proposed EPS when selecting SMARTS less efficient than expected. Coral8 provides a set of libraries for integration with other systems in many languages. However, the C# libraries are still not mature because some of APIs are only partially completed.

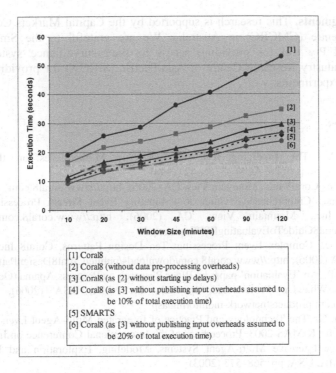

Fig. 8. Decomposing Coral8 performance according to different overheads for the Pair Trading pattern using Stock Exchange of Thailand (SET) data

The issues and limitations discussed here show that selecting an EPS based on its performance is not trivial. More experiments are required in order to investigate other factors affecting the system's performance. This includes examining the system with other event patterns of varying complexities as well as integrating other existing EPSs into the system.

5 Conclusions and Future Work

In this paper, we have applied the concept of distributed event systems to e-markets. A case study in financial market surveillance has been used to demonstrate our approach for representing event patterns independently of the underlying EPS. Our contribution is a proposed event pattern model and an architecture that can support an integration of different EPSs. We concentrated on evaluating the system performance of the proposed EPS through the system prototype developed.

Future work of this research include testing the developed prototype against a comprehensive set of event patterns used in financial market surveillance, collecting feedback from industry partners on the system usability and performance and investigating issues arising due to the integration of existing EPSs. We are also planning to investigate the automatic selection of slave EPSs according to the complexity of an event pattern as well as the characteristics of its constituent EPSs.

Acknowledgments. This research is supported by the Capital Markets Collaborative Research Centre (CMCRC) in Australia. We are grateful to the Smarts Group International Pty Ltd for providing access to their surveillance system and to Securities Industry Research Centre of Asia-Pacific (SIRCA) for providing the data used in the experiments.

References

1. Bakos, Y.: The Emerging Role of Electronic Marketplaces on the Internet. Communications of the ACM 41, 35–42 (1998)
2. Coral8 Inc., Coral8 Inc., Mountain View CA (2006), http://www.coral8.com/
3. Coral8 Inc., Comprehensive Guide to Evaluating Event Stream Processing Engines, Coral8 Inc., Mountain View, CA (2006), http://www.coral8.com/downloads/public/Coral8GuideToEvaluatingESPEngines.pdf
4. Coral8 Inc., Complex Event Processing: Ten Design Patterns, Coral8 Inc., Mountain View, CA (2006), http://www.coral8.com/downloads/public/Coral8DesignPatterns.pdf
5. Howard, P.: An 'Evaluation' paper by Bloor Research on Progress Apama (October 2006)
6. Ipswitch WhatsUp Professional, Ipswitch, Inc., MA, USA (2006), http://www.ipswitch.com/products/network-monitoring.asp
7. Jakobson, G.: The Technology and Practice of Integrated Multi-Agent Event Correlation Systems. In: KIMAS 2003. Proceedings of the International Conference on Integration of Knowledge Intensive Multi-Agent Systems, Modeling, Exploration and Engineering, Boston, MA, USA, pp. 568–573 (2003)
8. Luckham, D.: The Power of Events: An Introduction to Complex Event Processing in Distributed Enterprise Systems. Addison Wesley Professional, MA, USA (2002)
9. Mangkorntong, P., Rabhi, F.A.: A High-Level Approach for Defining & Composing Event Patterns and Its Application to E-Markets. In: Proceedings of The Second International Workshop on Event-driven Architecture, Processing and Systems (EDA-PS 2007) at the 33rd International Conference on Very Large Data Bases (VLDB 2007), Vienna, Austria (2007)
10. Mangkorntong, P., Rabhi, F.A.: Detecting Event Patterns in E-Markets: A Case Study in Financial Market Surveillance. In: Proceedings of the IADIS International Conference on E-Commerce, Barcelona, Spain, pp. 112–119 (2006)
11. Palmer, M.: Event Stream Processing - A New Physics of Software. DM Direct Newsletter, Issue. DM Review and SourceMedia, Inc. (July 29, 2005), http://www.dmreview.com/article_sub.cfm?articleID=1033537
12. Progress Apama Algorithmic Trading Platform (2006), http://www.apama.com
13. SMARTS, Smarts Group International Pty Ltd. (2006), http://www.smartsgroup.com
14. StreamBase, StreamBase Systems, Inc. (2006), http://www.streambase.com

Adaptive Email Spam Filtering Based on Information Theory

Xin Zhang, Wenyuan Dai, Gui-Rong Xue, and Yong Yu

Department of Computer Science and Engineering
Shanghai Jiao Tong University, Shanghai 200240, China
{zhangxin, dwyak, grxue, yyu}@apex.sjtu.edu.cn

Abstract. Most previous email spam filtering techniques rely on traditional classification learning which assumes the data from training and test sets are drawn from the same underlying distribution. However, in practice, this *identical-distribution* assumption often violates. In general, email service providers collect training data from various public available resources, while the tasks focus on users' individual inboxes. Topics in the mail-boxes vary among different users, and distributions shift as a result. In this paper, we propose an *adaptive* email spam filtering algorithm based on information theory which relaxes the identical-distribution assumption and adapts the knowledge learned from one distribution to another. Our work focuses on the content analysis which minimizes the *loss in mutual information* between email instances and word features, before and after classification. We present theoretical and empirical analyses to show that our algorithm is able to solve the adaptive email spam filtering problem well. The experimental results show that our algorithm greatly improves the accuracy of email filtering, against the traditional classification algorithms, while scaling very well.

Keywords: adaptive, email spam, information theory.

1 Introduction

Email becomes increasingly popular in people's daily life while spam is more and more severe. As a result, people spend increasing amount of time for reading emails and deciding whether they are spam or non-spam. In this situation, a robust junk mail filter is highly needed. Usually, email service provides some kinds of spam filters to help users detect spam. In general, it is not able to train such filters based on labeled messages from individual users, but the available labeled data are often public, e.g. newsgroups messages or emails received through "spam traps"[1]. Although the spam emails do not much differ among various users' inboxes since spammers often send spam emails non-targetedly, the non-spam emails in the training set and users' inboxes could be rather diversity due to topics drifting among different users. In the case, a basic assumption of most machine learning techniques violates, which is the training examples should be drawn from the same underlying distribution as the test ones. If

---

[1] Spam traps are email addresses published visually invisible for humans but get collected by the web crawlers of spammers.

B. Benatallah et al. (Eds.): WISE 2007, LNCS 4831, pp. 159–170, 2007.

we use the public data to train a traditional classifier for predicting the class labels of the emails from individual users, the performance could be rather limited. Therefore, there is a critical email spam filtering problem how to classify emails under different distributions, which is known as *transfer* or *adaptive learning*.

In this paper, we focus on the problem of identifying spam emails. Different from traditional machine learning tasks, in this work, the training and test data come from different distributions. An adaptive email spam filtering algorithm based on information theory is proposed in this paper. We define an objective function (or evaluation criterion) for the categorization focusing on both training and test data. The objective function is formulated by the *loss in mutual information* between email instances and word features, before and after prediction. By optimizing the objective function, test emails should be well self-organized, while the labels in the training set could be considered as a kind of constraints to help the predictions. This classification method does not make the *identical-distribution* assumption, and hence is capable for the adaptive email spam filtering problem. The theoretical analysis shows that our algorithm is able to monotonically optimize the value of objective function. The experimental results support our theory and demonstrate that our algorithm is effective in adaptively filtering spam emails with a high rate of convergence.

The rest of paper is organized as follows. In Section 2, we introduce some preliminary concepts from information theory. The problem is presented in Section 3. Our AdaFilter algorithm based on information theory is proposed in Section 4. Section 5 presents the experiments and empirical analysis. In Section 6, we review some related works. We conclude the whole paper and give the future work in Section 7.

2   Preliminaries

We briefly introduce some preliminary concepts in information theory which will be used frequently in this paper. For more details, please refer to [7]. Let $\mathcal{X}$ and $\mathcal{Y}$ be two random variable sets with a joint distribution $p(x, y)$ and marginal distributions $p(\mathcal{X})$ and $p(\mathcal{Y})$. The *mutual information* $I(\mathcal{X}; \mathcal{Y})$ is defined as

$$I(\mathcal{X}; \mathcal{Y}) = \sum_{x \in \mathcal{X}} \sum_{y \in \mathcal{Y}} p(x, y) \log \frac{p(x, y)}{p(x)p(y)}. \tag{1}$$

Mutual information indicates a kind of dependency between two variables. Larger mutual information value means more certainty that random variables depend on each other. *Kullback-Leibler* (KL) *divergence* [15] or *relative entropy* measures the distance between the two probability distributions. Let $p(x)$ and $q(x)$ be two probability mass functions where $x \in \mathcal{X}$. The KL-divergence is defined as

$$D(p\|q) = \sum_{x \in \mathcal{X}} p(x) \log \frac{p(x)}{q(x)}. \tag{2}$$

KL-divergence is a kind of distance between two different distributions, although it is not a real distance measure because it is not symmetric that $D(p\|q)$ is usually unequal to $D(q\|p)$. Besides, KL-divergence is always non-negative.

3   Problem Formulation

We formally define the problem as follows. Let $\mathcal{E}_{tr}$ be the (training) set of *labeled emails* using for training, $\mathcal{E}_{ad}$ be the (adaptive or test) set of *unlabeled emails* to be predicted. As we have discussed in Section 1, $\mathcal{E}_{tr}$ and $\mathcal{E}_{ad}$ are about the different topics under different distributions. We denote $\mathcal{E}$ as the set of all the labeled and unlabeled emails, so that $\mathcal{E} = \mathcal{E}_{tr} \cup \mathcal{E}_{ad}$. The word feature set of $\mathcal{E}$ is denoted by $\mathcal{W}$. For each email $e \in \mathcal{E}$, there is a class-label $c \in C = \{spam, ham\}$ associated with it. Our objective is to estimate a hypothesis $h: \mathcal{E} \to C$ which predicts the labels of spam/ham emails in $\mathcal{E}$ as accurately as possible.

4   The Adaptive Email Spam Filtering Algorithm

4.1   Objective Function

Let $\hat{\mathcal{E}} = \{\hat{e} | \hat{e} = spam \vee \hat{e} = nonspam\}$ be a prediction to $\mathcal{E}$ given by some hypothesis h. We consider $\mathcal{E}$, $\mathcal{W}$ and $\hat{\mathcal{E}}$ as the random variable sets which take emails, words, and predictions (*spam* or *nonspam*) as random variables. The *mutual information* $I(\mathcal{E}; \mathcal{W})$ measures the amount of information email instances $\mathcal{E}$ contain about their word features $\mathcal{W}$ [7]. Likewise, $I(\hat{\mathcal{E}}; \mathcal{W})$ measures the amount of information predictions $\hat{\mathcal{E}}$ contain about the word features $\mathcal{W}$. Since the emails are presented by word features, the mutual information $I(\mathcal{E}; \mathcal{W})$ could be considered as an information theoretic property about the data set $\mathcal{E}$. A good prediction $\hat{\mathcal{E}}$ should retrain this property. That is, $I(\hat{\mathcal{E}}; \mathcal{W})$ should close to $I(\mathcal{E}; \mathcal{W})$. In this paper, we use the *loss in mutual information* before and after prediction as the objective function which is formulated as

$$I(\mathcal{E}; \mathcal{W}) - I(\hat{\mathcal{E}}; \mathcal{W}). \tag{3}$$

The objective function is to be minimized. As we will show later in Lemma 1, the objective function is always non-negative.

In order to calculate Equation (3), let us first define two probability distributions, and then the approach for calculation will be presented in Lemma 1. Let $p(\mathcal{E}, \mathcal{W})$ denote the joint probability distribution of the emails $\mathcal{E}$ and the word features $\mathcal{W}$. The joint distribution under a prediction $\hat{\mathcal{E}}$ is denoted by $\hat{p}(\mathcal{E}, \mathcal{W})$ which is defined as

$$\hat{p}(e, w) = p(\hat{e}, w)p(e|\hat{e}) = p(\hat{e}, w)\frac{p(e)}{p(\hat{e})}, \tag{4}$$

where $e \in \hat{e}$. Note that $p(e|\hat{e}) = \frac{p(e)}{p(\hat{e})}$ since e totally depends on $\hat{e}$. Lemma 1 gives an alternative calculation approach for Equation (3). Furthermore, it also builds a connection between *loss in mutual information* and *KL-divergence*.

Lemma 1. *For a fixed prediction* $\hat{\mathcal{E}}$, *we can express the objective function in Equation (3) as*

$$I(\mathcal{E}; \mathcal{W}) - I(\hat{\mathcal{E}}; \mathcal{W}) = D(p(\mathcal{E}, \mathcal{W}) \| \hat{p}(\mathcal{E}, \mathcal{W})), \tag{5}$$

where $D(\cdot \| \cdot)$ *is the KL-divergence defined in Equation (2).*

The proof of Lemma 1 is omitted due to the limitation of space. It can be easily derived by the definitions of the mutual information and KL-divergence. From Equation (5), we can find that the loss in mutual information before and after prediction equals in value to the KL-divergence between $p(\mathcal{E}, \mathcal{W})$ and $\hat{p}(\mathcal{E}, \mathcal{W})$. As a consequence, we can minimize $D(p(\mathcal{E}, \mathcal{W}) \| \hat{p}(\mathcal{E}, \mathcal{W}))$ instead of minimizing the objective function in Equation (3).

4.2 Optimization

Equation (5) is a function based on joint probability, which is difficult to optimize. In Lemma 2, we will convert the calculation of the objective function into the form of conditional probability. Then, the objective function can be optimized by minimizing the KL-divergence between two conditional probability distributions.

Lemma 2. *The objective function in Equation (3) can be rewritten into a form of conditional probability*

$$D(p(\mathcal{E}, \mathcal{W}) \| \hat{p}(\mathcal{E}, \mathcal{W})) = \sum_{\hat{e} \in \hat{\mathcal{E}}} \sum_{e \in \hat{e}} p(e) D(p(\mathcal{W}|e) \| \hat{p}(\mathcal{W}|\hat{e})). \tag{6}$$

Proof

$$D(p(\mathcal{E}, \mathcal{W}) \| \hat{p}(\mathcal{E}, \mathcal{W})) = \sum_{\hat{e} \in \hat{\mathcal{E}}} \sum_{w \in \mathcal{W}} \sum_{e \in \hat{e}} p(e, w) \log \frac{p(e, w)}{\hat{p}(e, w)}.$$

Since

$$\hat{p}(e, w) = p(\hat{e}, w) \frac{p(e)}{p(\hat{e})} = p(e) \frac{p(\hat{e}, w)}{p(\hat{e})} = p(e) p(w|\hat{e}),$$

and

$$\hat{p}(w|e) = \frac{\hat{p}(\hat{e}, w)}{\hat{p}(\hat{e})} = \frac{\sum_{e \in \hat{e}} p(\hat{e}, w) \frac{p(e)}{p(\hat{e})}}{\sum_{e \in \hat{e}} p(\hat{e}) \frac{p(e)}{p(\hat{e})}} = \frac{p(\hat{e}, w) \sum_{e \in \hat{e}} \frac{p(e)}{p(\hat{e})}}{p(\hat{e}) \sum_{e \in \hat{e}} \frac{p(e)}{p(\hat{e})}} = \frac{p(\hat{e}, w)}{p(\hat{e})} = p(w|\hat{e}),$$

$D(p(\mathcal{E}, \mathcal{W}) \| \hat{p}(\mathcal{E}, \mathcal{W}))$ can be expressed by

$$D(p(\mathcal{E}, \mathcal{W}) \| \hat{p}(\mathcal{E}, \mathcal{W})) = \sum_{\hat{e} \in \hat{\mathcal{E}}} \sum_{w \in \mathcal{W}} \sum_{e \in \hat{e}} p(e) p(w|e) \log \frac{p(e) p(w|e)}{p(e) \hat{p}(w|\hat{e})}$$

$$= \sum_{\hat{e} \in \hat{\mathcal{E}}} \sum_{e \in \hat{e}} p(e) \sum_{w \in \mathcal{W}} p(w|e) \log \frac{p(w|e)}{\hat{p}(w|\hat{e})}$$

$$= \sum_{\hat{e} \in \hat{\mathcal{E}}} \sum_{e \in \hat{e}} p(e) D(p(\mathcal{W}|e) \| \hat{p}(\mathcal{W}|\hat{e})).$$

□

From Lemma 2, we can find that optimizing $D(p(\mathcal{E}, \mathcal{W}) \| \hat{p}(\mathcal{E}, \mathcal{W}))$ is equivalent to optimizing $\sum_{\hat{e} \in \hat{\mathcal{E}}} \sum_{e \in \hat{e}} p(e) D(p(\mathcal{W}|e) \| \hat{p}(\mathcal{W}|\hat{e}))$. Thus, for a single email instance e, if we want to decrease $D(p(\mathcal{W}|e) \| \hat{p}(\mathcal{W}|\hat{e}))$, an alternative approach is to assign each e to a better $\hat{e}$ which is able to reduce the value of $D(p(\mathcal{W}|e) \| \hat{p}(\mathcal{W}|\hat{e}))$. As a result, the objective function can be reduced through assigning better $\hat{e}$ to e. Then, the algorithm is derived as follows. For each email instance e, the algorithm chooses the best $\hat{e}$ to minimize the value of $D(p(\mathcal{W}|e) \| \hat{p}(\mathcal{W}|\hat{e}))$, and then assigns e to $\hat{e}$. Based on Lemma 2, we know that this process can reduce the value of the objective function. The detailed description of the algorithm is presented in Fig. 1.

The Adaptive Email Spam Filtering Algorithm

Input: a labeled training set $\mathcal{E}_{tr}$; an unlabeled test (or adaptive) set $\mathcal{E}_{ad}$; an initial prediction $h^{(0)}$; the number of iterations T.

Output: the final prediction $h_f : \mathcal{E} \to \mathcal{C}$

1. Estimate the probability distribution p based on $\mathcal{E} = \mathcal{E}_{tr} \cup \mathcal{E}_{ad}$.

2. Initialize the probability distribution $p^{(0)}$ based on $h^{(0)}$ and Equation (4).

3. For $t = 1, \dots, T$

4.   Update the adaptive emails $\mathcal{E}_{ad}$ based on
$$h^{(t)} = \mathrm{argmin}_{c \in \mathcal{C}} D(p(\mathcal{W}|e) \| \hat{p}^{(t-1)}(\mathcal{W}|c))$$
where $\hat{p}^{(t-1)}(\mathcal{W}|c) = \hat{p}^{(t-1)}(\mathcal{W}|\hat{e})$, and $\forall e' \in \hat{e}, h^{(t-1)}(e) = c$

5.   For $e \in \mathcal{E}_{tr}$, $h^{(t)}(e) = h^{(t-1)}(e)$.

6.   Update $\hat{p}^{(t)}$ based on $h^{(t)}$ and Equation (4).

7. End For

8. Return $h^{(T)}$ as the final prediction h_f.

Fig. 1. The description of the Adaptive Email Spam Filtering (AdaFilter) algorithm

In Fig. 1, in each iteration, the algorithm AdaFilter chooses the best label c for each email instance e in the adaptive set $\mathcal{E}_{ad}$ to minimize the function $D(p(\mathcal{W}|e) \| \hat{p}^{(t-1)}(\mathcal{W}|\hat{e}))$. It has been proven that this process is able to reduce the value of the objective function. Note that, the predictions for all the $e \in \mathcal{E}_{tr}$ stay unchanged throughout the whole procedure, since these emails are already labeled.

4.3   Convergence

We have already presented the algorithm details. Since our algorithm AdaFilter is iterative, an important issue is to prove its convergent property. In the following theorem, we will show that AdaFilter could monotonically decrease the objective function, and then prove the termination property of AdaFilter.

Theorem 1. *The algorithm* AdaFilter *monotonically decreases the objective function in Equation (6). That is,*

$$D\left(p(\mathcal{E}, \mathcal{W}) \| \hat{p}^{(t)}(\mathcal{E}, \mathcal{W})\right) \geq D\left(p(\mathcal{E}, \mathcal{W}) \| \hat{p}^{(t+1)}(\mathcal{E}, \mathcal{W})\right) \tag{7}$$

Proof

Following Lemma 2, we have

$$D\left(p(\mathcal{E}, \mathcal{W}) \| \hat{p}^{(t)}(\mathcal{E}, \mathcal{W})\right) = \sum_{\hat{e}:h^{(t)}} \sum_{e \in \hat{e}} p(e) \sum_{w \in \mathcal{W}} p(w|e) \log \frac{p(w|e)}{\hat{p}^{(t)}(w|\hat{e})}.$$

Based on the Steps 4 and 5 in Fig. 1,

$$\sum_{\hat{e}:h^{(t)}} \sum_{e \in \hat{e}} p(e) \sum_{w \in \mathcal{W}} p(w|e) \log \frac{p(w|e)}{\hat{p}^{(t)}(w|\hat{e})}$$

$$\geq \sum_{\hat{e}:h^{(t)}} \sum_{e \in \hat{e}} p(e) \sum_{w \in \mathcal{W}} p(w|e) \log \frac{p(w|e)}{\hat{p}^{(t)}(w|h^{(t+1)}(e))}$$

$$= \sum_{\hat{e}:h^{(t+1)}} \sum_{e \in \hat{e}} p(e) \sum_{w \in \mathcal{W}} p(w|e) \log \frac{p(w|e)}{\hat{p}^{(t)}(w|\hat{e})}$$

$$\geq \sum_{\hat{e}:h^{(t+1)}} \sum_{e \in \hat{e}} p(e) \sum_{w \in \mathcal{W}} p(w|e) \log \frac{p(w|e)}{\hat{p}^{(t+1)}(w|\hat{e})}$$

$$= D\left(p(\mathcal{E}, \mathcal{W}) \| \hat{p}^{(t+1)}(\mathcal{E}, \mathcal{W})\right).$$

Note that, the third inequality follows by

$$\sum_{\hat{e}:h^{(t+1)}} \sum_{e \in \hat{e}} p(e) \sum_{w \in \mathcal{W}} p(w|e) \log \frac{p(w|e)}{\hat{p}^{(t)}(w|\hat{e})}$$

$$= \sum_{\hat{e}:h^{(t+1)}} \sum_{e \in \hat{e}} p(e) \sum_{w \in \mathcal{W}} p(w|e) \log p(w|e)$$

$$+ \sum_{\hat{e}:h^{(t+1)}} \sum_{e \in \hat{e}} p(e) \sum_{w \in \mathcal{W}} p(w|e) \log \frac{1}{\hat{p}^{(t)}(w|\hat{e})},$$

and

$$\sum_{\hat{e}:h^{(t+1)}} \sum_{e \in \hat{e}} p(e) \sum_{w \in \mathcal{W}} p(w|e) \log \frac{1}{\hat{p}^{(t)}(w|\hat{e})}$$

$$= \sum_{\hat{e}:h^{(t+1)}} \left(\sum_{e \in \hat{e}} p(e) \sum_{w \in \mathcal{W}} p(w|e) \right) \log \frac{1}{\hat{p}^{(t)}(w|\hat{e})}$$

$$= \sum_{\hat{e}:h^{(t+1)}} \hat{p}^{(t+1)}(\hat{e}) \sum_{w \in \mathcal{W}} \hat{p}^{(t+1)}(w|\hat{e}) \log \frac{1}{\hat{p}^{(t)}(w|\hat{e})}$$

$$\geq \sum_{\hat{e}:h^{(t+1)}} \hat{p}^{(t+1)}(\hat{e}) \sum_{w \in \mathcal{W}} \hat{p}^{(t+1)}(w|\hat{e}) \log \frac{1}{\hat{p}^{(t+1)}(w|\hat{e})}$$

Note that, the last inequality follows by the non-negativity of the Kullback-Leibler divergence.   □

Theorem 1 proves that AdaFilter decreases the objective function monotonically. From Theorem 1, we know that AdaFilter is guaranteed to converge in a finite number of iterations. Note that, AdaFilter can only find a locally optimal solution, and finding the global optimal solution is NP-hard.

Regarding the computational complexity of AdaFilter, suppose the non-zeros in $p(\mathcal{E}, \mathcal{W})$ is N. In each iteration, AdaFilter calculates $h^{(t)}$ in $O(N)$ and update $\hat{p}^{(t)}(\mathcal{W}|\hat{e})$ in $O(|\mathcal{W}|)$. In general, $|\mathcal{W}|$ is not larger than N. Thus, the time complexity of AdaFilter is $O(N)$, which can be considered as good scalability.

5   Experiments

In this section, we empirically evaluate our algorithm AdaFilter. Our focus is the superior of the algorithm, so we did not pay much attention on optimization by feature selection or considering some structure features.

5.1   Data Set

In order to evaluate the properties of our algorithm, we conducted our experiments on the three data sets from ECML/PKDD discovery challenge 2006[2].

In this corpus, for each training data set, 50% of the labeled training data are assigned to spam which were sent by blacklisted servers of the Spamhaus project[3]. 40% of the labeled data are non-spam from the SpamAssassin corpus and the other 10% are non-spam which were sent by about 100 different subscribed English and German newsletters. The composition of the labeled training is summarized in Table 1.

Table 1. Composition of labeled train data

	#documents
emails sent from blacklisted severs	2000
SpamAssassin emails	1600
newsletters	400
total	4000

Three users' inboxes with the size of 2500 are tested in the experiments. Table 2 summarizes the composition of the evaluation inboxes consisting of 50% spam and 50% non-spam emails. For more details about the data sets, please refer to [3].

The test (or adaptive) data were collected from different users' inboxes using real but open messages. The non-spam part of the inboxes consists of ham-messages received by distinct Enron employees from the Enron corpus [14] cleaned from spam. The spam part of the inboxes consists of spam-messages from distinct spam sources. Since users' topic interests differ from each other, the distributions of emails are different among all the inboxes.

---

[2] http://www.ecmlpkdd2006.org/challenge.html
[3] http://www.spamhaus.org

Table 2. Composition of adaptive evaluation inboxes for data sets

Inbox ID	Non-spam	Spam source
0	Beck	Spam trap of Bruce Gunter (www.em.ca/~bruceg/spam)
1	Kaminski	Spam collection of SpamArchive.org
2	Kitchen	Personal spam of Tobias Scheffer (www.em.ca/~bruceg/spam)

Fig. 2 shows the instance-term co-occurrence distribution on the first data set. In this figure, instances 1 to 4000 are from the training data $\mathcal{E}_{tr}$ while instances 4001 to 6500 are from the user's inbox $\mathcal{E}_{ad}$. The instances are ordered first by their source ($\mathcal{E}_{tr}$ or $\mathcal{E}_{ad}$, and second by their categories (*spam* or *nonspam*). The terms (or words) are sorted by $n_+(t) / n_-(t)$, where $n_+(t)$ and $n_-(t)$ represent the number of occurrence the feature term t appears in spam and non-spam emails, respectively. From Fig. 2, it can be found that the distributions of training and test data are somewhat different, whereas the figure also shows amount of commonness exists between the two data. The common part ensures that feasibility for training data $\mathcal{E}_{tr}$ from public available source to help categorization on test data $\mathcal{E}_{ad}$ from users' inboxes.

5.2   Evaluation Metrics

The performance of the proposed methods was evaluated using two metrics. The first metric is accuracy, which expresses the proportion of email instances that are predicted correctly. The second metric is AUC value [16]. The AUC value is the area under the ROC curve (Receive Operating Characteristic curve). The AUC value specifies the confidence that the decision function assigns higher values to positive instances than to negative ones.

5.3   Comparison Methods

In order to show the superiority of our algorithm AdaFilter, we compare it with several traditional classification algorithms. We take Naïve Bayes classifier (NBC) [17] and Support Vector Machines (SVM) [5] as the baseline methods. Transductive Support Vector Machines (TSVM) is also introduced as a comparison semi-supervised

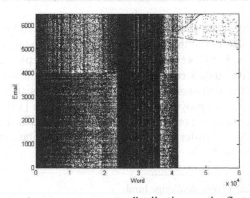

Fig. 2. Instance-term co-occurrence distribution on the first data set

learning methods. SVM and TSVM were implemented by SVM^{light} [12] with default parameters (linear kernel). The initial hypothesis $h^{(0)}$ of AdaFilter is given by NBC. All the comparison methods are implemented in the way that the classifier is trained using $\mathcal{E}_{tr}$ as labeled data and $\mathcal{E}_{ad}$ as unlabeled data, and then predict the labels of the emails in $\mathcal{E}_{ad}$.

5.4   Performance

Table 3 presents the performance in accuracy on each data set given by NBC, SVM, TSVM and our algorithm AdaFilter. We can see from the table that AdaFilter always provides the best results, compared with traditional classification methods. Besides, NBC gives better performance than SVM. In our opinion, although SVM is known as a much stronger classifier than NBC, NBC is more general for adaptive learning. Moreover, TSVM outperforms SVM, because it utilizes the information given by unlabeled data from the adaptive data set.

Table 4 presents the AUC values by NBC, SVM, TSVM and AdaFilter for each data set (each user's inbox).   The AUC value specifies the confidence that the decision function assigns higher values to positive instances than negative ones. Thus, the decisions made by AdaFilter are more confident than NBC, SVM and TSVM.

Fig. 3 presents the accuracy on different sizes of labeled training data. We randomly choose the labeled data from $\mathcal{E}_{tr}$ of the first data set by different proportion. It can be seen that our AdaFilter algorithm always gives the best performance, while the proportion of the training data set changes from 1% to 64%.   Furthermore, when the size of the labeled data decreases, the performance of AdaFilter drops much slower than SVM and NBC. We believe that the performance of AdaFilter is not much sensitive to the size of the training data set. NBC and SVM quickly get worse when the proportion of training data set is less than 16%. Meanwhile, our AdaFilter algorithm stays relatively more stably.

Table 3. Accuracy for each classifier on each data set

Data Set	NBC	SVM	TSVM	AdaFilter
0	0.764	0.713	0.752	**0.830**
1	0.774	0.726	0.770	**0.835**
2	0.861	0.856	0.929	**0.955**

Table 4. AUC for each classifier on each data set

Data Set	NBC	SVM	TSVM	AdaFilter
0	0.765	0.738	0.799	**0.875**
1	0.769	0.779	0.851	**0.854**
2	0.926	0.925	0.979	**0.992**

5.5   Convergence

Since our algorithm AdaFilter is iterative, the convergence property becomes an important issue. Theorem 3 has already proven the convergence of AdaFilter theoretically. Now, we empirically show the convergence property of AdaFilter. Fig. 4 gives the accuracy curves as functions measure the performances after each iteration on all the

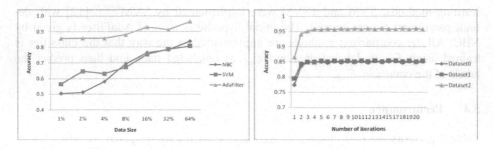

Fig. 3. Accuracy curves on different size of training data on the first data set

Fig. 4. Accuracy curves after each iteration on all the data sets

data sets. It can be seen in the figure clearly that AdaFilter always reaches almost convergent points within 5 iterations. As a result, we believe that AdaFilter converges very fast, and 5 iterations are enough in practice.

6   Related Works

In this section, we review some prior works mostly related to our work. Addressing on the email spam filtering problem, [22] proposed a Bayesian method to classify spam/non-spam emails. Three feature regimes were considered in their works, i.e. words only, words + phrases, words + phrases + domain specific features. [1] presented a throughout evaluation of the Bayesian methods for email spam filtering on an email spam corpus *Ling-Spam*[4]. [11] evaluated several machine learning techniques for detection spam emails, including C4.5 [20], Naïve Bayes Classifier [17], PART [10], Support Vector Machines [5], Rocchio [21]etc. All the above works assumes the training examples are under the same distribution from which the test data are drawn. However, as we have discussed in the previous sections, this assumption does not hold in general. Several other researches addressing on the same or similar work are [2, 6, 25, 19, 18] etc., to be mentioned.

Recently, email spam filtering has been recognized as an adaptive learning problem that the underlying distributions of the training and test data are different. Several heuristic approaches have been proposed during the last year's ECML/PKDD Discovery Challenge[5]. For example, the first place algorithm [13] proposed to use a simple statistical classifier which detects spam emails based on *strong* spam/non-spam words. Self-training using the statistical classifier as the basic learner is applied to improving the performance further. [4] addressed the same problem using a nonparametric hierarchical Bayesian model to learn a common prior and impose on a new email account. In contrast, we proposed a new method under the information theoretic framework. This framework, with well theoretic supporting, does not make any assumptions on the specific underlying distributions, while Bayesian model usually requires the distribution be Gaussian.

---

[4] http://www.aueb.gr/users/ion/data/lingspam_public.tar.gz
[5] http://www.ecmlpkdd2006.org/challenge.html

Another related research is information theory based learning. Information theory is widely used in machine learning, e.g. Decision Tree [20], feature selection [24], etc. Recently, mutual information has been applied to improving clustering [23]. [8] proposed a word clustering method which minimizes the loss in mutual information between words and class-labels before and after clustering. Using the similar strategy, mutual information based co-clustering was proposed [9]. In contrast to these works, we try to design an information theoretic approach to solve the adaptive email spam filtering problem.

7   Conclusions and Future Work

Detecting email spam is challenging in both research and practice. In this paper, we focus on the email spam filtering problem that labeled and unlabeled data are under different distributions. Usually, this situation comes true, because email service providers gather training data from public available sources, but the test data are from users' individual inboxes. Due to the topic drift among different users, the underlying distributions from which the training and test data are drawn should be different as a consequence. In our work, an adaptive classification algorithm, called AdaFilter, is proposed based on information theory. The algorithm is motivated by minimizing the *loss in mutual information* between email instances and word features, before and after prediction. An iterative approach was designed to achieve the goal. Our theoretical analysis demonstrates that AdaFilter monotonically optimizes the value of the objective function. The experimental results support our theory and present the superior performance and scalability of AdaFilter, comparing with several traditional classification algorithms.

In our work, we focus on filtering email spam in one individual user's inbox by using global training data. In the future, we want to modify our algorithm in order to deal with multiple users' inboxes simultaneously. We believe the relations between different users' inboxes are able to help the prediction. Moreover, we also want to deal with the case that there are some but not sufficient labeled data in each user's inbox. Collaboratively using these labeled data could be a challenging and exciting task.

References

1. Androutsopoulos, I., Koutsias, J., Chandrinos, K.V., Paliouras, G., Spyropoulos, C.D.: An Evaluation of Naive Bayesian Anti-Spam Filtering. In: Proceedings of the Workshop on Machine Learning in the New Information Age, 11th European Conference on Machine Learning (2000)
2. Androutsopoulos, I., Koutsias, J., Chandrinos, K.V., Spyropoulos, C.D.: An Experimental Comparison of Naive Bayesian and Keyword-Based Anti-Spam Filtering with Encrypted Person-al E-mail Messages. In: Proceedings of the 23rd Annual International ACM SIGIR Conference on Research and Development in Information Retrieval (2000)
3. Bickel, S.: ECML-PKDD Discovery Challenge 2006 Overview. In: Proceedings of the ECML/PKDD Discovery Challenge Workshop (2006)

4. Bickel, S., Scheffer, T.: Dirichlet-Enhanced Spam Filtering based on Biased Samples. Advances in Neural Information Processing Systems (2006)
5. Boser, B.E., Guyon, I., Vapnik, V.: A Training Algorithm for Optimal Margin Classifiers. In: Proceedings of the Fifth Annual Workshop on Computational Learning Theory (1992)
6. Carreras, X., Mrquez, L.: Boosting Trees for Anti-spam Email Filtering. In: Proceedings of the 2001 International Conference on Recent Advances in Natural Language Processing (2001)
7. Cover, T.M., Thomas, J.A.: Elements of information theory. Wiley-Interscience, New York, NY, USA (1991)
8. Dhillon, I.S., Mallela, S., Kumar, R.: Enhanced Word Clustering for Hierarchical Text Classifi-cation. In: Proceedings of the Eighth ACM SIGKDD International Conference on Knowledge Dis-covery and Data Mining (2002)
9. Dhillon, I.S., Mallela, S., Modha, D.S.: Information-Theoretic Co-clustering. In: Proceedings of the Ninth ACM SIGKDD International Conference on Knowledge Discovery and Data Mining (2003)
10. Frank, E., Witten, I.H.: Generating accurate rule sets without global optimization. In: Proceedings of the Fifteenth International Conference on Machine Learning (1998)
11. Hidalgo, J.G.: Evaluating cost-sensitive unsolicited bulk email categorization. In: Proceedings of 17th ACM Symposium on Applied Computing (2002)
12. Joachims, T.: Learning to classify text using support vector machines. Dissertation, Kluwer (2002)
13. Junejo, K., Yousaf, M., Karim, A.: A Two-Pass Statistical Approach for Automatic Persona-lized Spam Filtering. In: Proceedings of the ECML/PKDD Discovery Challenge Workshop (2006)
14. Klimt, F., Yang, Y.: The Enron corpus: A new dataset for email classification research. In: Proceedings of the European Conference on Machine Learning (2004)
15. Kullback, S., Leibler, R.A.: On information and sufficiency. Annals of Mathematical Statis-tics 22(1), 79–86 (1951)
16. Hanley, J., McNeil, B.: A Method of Comparing the Areas under Receiver Operating Characteristic Curves Derived from the Same Cases. Radiology 148, 839–843 (1983)
17. Lewis, D.D.: Representation and Learning in Information Retrieval. Doctoral dissertation, Amherst, MA, USA (1992)
18. Metsis, V., Androutsopoulos, I., Paliouras, G.: Spam Filtering with Naive Bayes? Which Naive Bayes? In: Proceedings of the 3rd Conference on Email and Anti-Spam (2006)
19. Michelakis, E., Androutsopoulos, I., Paliouras, G., Sakkis, G., Stamatopoulos, P.: Filtron: A Learning-Based Anti-Spam Filter. In: Proceedings of the 1st Conference on Email and Anti-Spam (2004)
20. Quinlan, J.R.: C4.5: Programs for Machine Learning. Morgan Kaufmann, San Mateo, CA (1993)
21. Rocchio, J.J.: Relevance Feedback in Information Retrieval. In: The SMART Retrieval System: Experiments in Automatics Document Processing (1971)
22. Sahami, M., Dumais, S., Heckerman, D., Horvitz, E.: A Bayesian Approach to Filtering Junk E-mail. In: AAAI 1998 Workshop on Learning for Text Categorization (1998)
23. Slonim, N., Tishby, N.: Document Clustering using Word Clusters via the Information Bottle-neck Method. In: Proceedings of the Twenty-Third Annual International ACM SIGIR Conference on Research and Development in Information Retrieval (2000)
24. Yang, Y., Pedersen, J.O.: A Comparative Study on Feature Selection in Text Categorization. In: Proceedings of Fourteenth International Conference on Machine Learning (1997)
25. Zhang, L., Yao, T.: Filtering Junk Mail with a Maximum Entropy Model. In: Proceedings of the 20th International Conference on Computer Processing of Oriental Languages (2003)

Time Filtering for Better Recommendations with Small and Sparse Rating Matrices*

Sergiu Gordea and Markus Zanker

University Klagenfurt
9020 Klagenfurt, Austria
{sergiu,markus}@ifit.uni-klu.ac.at

Abstract. The recommendation technologies are used as viable solutions for advertising the best products and for helping users to orientate themselves in large e-commerce platforms offering various product assortments. Despite their popularity they still suffer of cold start and sparse data matrices limitations, which affect seriously the effectiveness of recommenders employed in applications with less user-system interaction. Having the aim to improve the quality of recommendation lists in such systems we introduce time heuristics into the recommendation process and propose two new variants of collaborative filtering algorithms for solving these problems. A time aware method is proposed for making more correct evaluations of recommenders used in domains with strong time dependencies.

1 Introduction

Recommendation Technologies became very popular since very large and successful e-commerce platforms like amazon.com "conquered" the electronic market. They support online customers when navigating through rich product assortments and facilitate the access to products which fulfill user requirements.

In general, collaborative filtering (CF) algorithms provide better recommendations than the alternative content based (CB) solutions and have lower implementation and maintenance costs than knowledge based deployments, therefore this is the most used technology in commercial applications [1]. Despite their successful employment in large online-shops, CF recommenders still suffer from a set of limitations:

– **Cold start**, also known as the rump-up problem, refers to the fact that it is impossible to learn preferences of a new user, or to recommend a new product, until a certain amount of ratings is available in the system (for the new user, new item respectively) [2], [3].

---

* This work is carried out with financial support from the EU, the Austrian Federal Government and the State of Carinthia in the Interreg IIIA project Software Cluster South Tyrol - Carinthia.

B. Benatallah et al. (Eds.): WISE 2007, LNCS 4831, pp. 171–183, 2007.

- The **Sparsity** problem typically occurs in systems with large product offers, in which there are plenty of items rated only by few users, and many users which rated few items. In this situations the members of user neighborhood (e.g. users with similar tastes, which provided similar rates for items rated in common) provided rates for many different products, that will come into recommendation list, but with low recommendation scores. These are low trust recommendations [4], [5].
- Having a **Small number of users** in combination with the sparsity problem will affect the performance of CF recommenders very much. There are many web applications which are selling products to a very restricted section of online customers (e.g. fine cigars shop, financial services, guided city tours, etc.) [6]. These online applications have, typically, less than thousand customers per year, being hard to identify very similar neighbors that make good recommendations.

Few user-system interaction is found usually in small applications, that have small number of users. This situation can be met in larger applications, too, if they sell products that are less frequently purchased (e.g. in average, regular customers change their digital camera once in 3 years). Especially in this kind of context, the upper enumerated problems lead to the situation when recommender systems are not able to identify highly correlated user neighborhoods. Therefore the recommendation lists are populated with items having relative low scores, which means not very trusted recommendations. In contrast to other approaches, which try to eliminate the sparsity problem by populating the rating matrix with artificial ratings, we aim at improving the recommendation lists by constructing additional filters which have the job of cleaning them from bad or unexpected items. The contribution of this paper lays in the introduction of time aware recommendation solutions, with Time Decay and Time Window Filtering algorithms which improve the prediction accuracy of recommender systems by incorporating usability heuristics into the classification model. Another new concept presented in this paper is the time aware evaluation methodology which measures the effectiveness of recommenders by reconstructing the application context in experimental simulations. We consider this method to be better suited for off-line experimentations than the classic cross validation method, especially in domains with time dependencies (e.g. tourism domain, where accommodation bookings are dependent on the time when the customers have holiday and the opening period of hotels).

2   Related Work

According to Burke's classification, there are three main categories of recommendation technologies: content based filtering, collaborative filtering and hybrid recommenders [1]. There are basically two paradigms that inspired the construction of recommenders, and there is also the possibility to combine them in hybrid algorithms. The first paradigm *'get more of the same'* is used in Content Based (CB) and Knowledge Based (KB) algorithms. It makes recommendations by searching commonalities between product descriptions (e.g. item content) and user tastes [7], [8], [9], [10]. The second one *'get most popular items'* exploits the fact that users having similar tastes will buy the same products. This idea is implemented in Collaborative Filtering (CF) applications [11], [12].

Many of recommendation algorithms are evaluated in laboratory conditions using large and cleaned up data-sets like the movielens one [1]. Zanker et al. show the differences between laboratory and commercial data-sets, pointing out that recommenders have more problems and lower prediction accuracy when employed in middle and small sized commercial applications [6]. The authors also present a comparative evaluation of CB, CF, KB and hybrid algorithms in a commercial context.

The sparsity problem was identified since the beginning of this research field as being one of the recommenders' greatest limitations [11], [5]. There are several approaches that try to reduce the sparsity of the rating matrix by filling the missing values with artificial rates. The content boosted collaborative algorithm proposed by Melville et al. uses content based item similarities as artificial rates [13]. This method increases the number of users that get recommendations (better user coverage) but it leads to a small degradation of the overall prediction accuracy [6]. Other solutions use usability heuristics, user clustering or implicit rates generated by filterbots in order to overcome the inconvenient of sparse rating matrices [5], [14].

The effectiveness of recommenders can be improved by integrating more knowledge into the classification models. Adamovicius et al. proposed an approach based on a multidimensional input matrix. One dimension in the matrix is filled with the regular ratings and the other ones with user context information (e.g. place, time, company, etc.) [4]. The authors present experimental results that indicate an improvement of recommendation lists when different projections of user context are used in the recommendation process.

The time is for sure an important component of the user and application context, which was used in some approaches to infer artificial ratings, to improve the quality of user profiles, or to weight the existing ratings when computing recommendations [14], [15]. In our work we use time heuristics for filtering recommendations, and for generating deterministic simulations when evaluating recommendation algorithms.

3   Time Aware Recommendations

The explicit rating matrix is replaced or derived from transaction data in many online applications. This is the case of the amazon.com website, for example, on which the recommendations are accessible under the section "Customers who bought this item also bought". This replacement is made because buying a product is the most powerful indicator that the user likes the given item. A second reason for using transaction information for making recommendation is the lack

---

[1] http://www.movielens.umn.edu

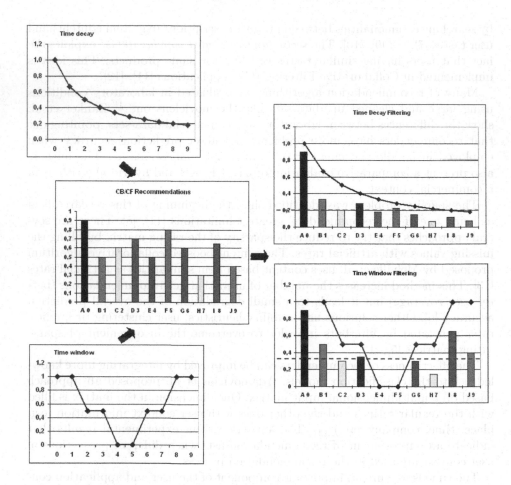

Fig. 1. Time Filtering

of explicit ratings. Many users buy products without rating any of them, or provide their opinion only after testing the products. In the great majority of cases the distribution in time of product purchases draw some particular patterns. For example, the users are interested to buy the newest models of electronic products (e.g. digital cameras, mp3 players), or they buy certain products only in particular seasons of the year (e.g. skiing articles are both in late autumn and winter, T-shirts and sport shoes are both in spring and summer). We integrate this information into recommendation algorithms through the usage of time filters as shown in Figure 1.

The time filters are used to attenuate the recommendation scores for items that have a small chance to be bought at the moment in time when the customers are surfing on the online shop's website. The two variants, time decay filtering and time window filtering are built by combining the appropriate time filter with an classic CB or CF recommendation algorithm. The recommendation scores are multiplied with the time function representing the filter, and the items that fall under the recommendation threshold are eliminated from recommendation list (e.g. the score lay under the predefined threshold, or under the nth position in the recommendation list). We propose two types of time filters which model the following situations:

- *Time decay filter*, modeling the high interests for items that are recently launched on the market.
- *Time window filter*, modeling the periodical increase of interest for certain categories of products.

3.1   Time Filter Selection

Time decay filtering. The identification of the appropriate filtering technique is done buy analyzing the distribution in time of the ratings for each item (see Figure 2). The time decay filtering is efficient when the ratings distribution looks like in Figure 2 a). It can be observed that the greatest amount of ratings is concentrated in a relative short time interval after the item's creation date. The decay function which models this behavior has the following expression:

$$h_u(t) = \frac{1}{1 + DIN_u * Age(t)} \tag{1}$$

The slope of the function is controlled by two parameters which are learned from historical data, the size of the aging interval and user's degree of interest for novel information (DIN_u). The size of the aging interval is learned from rating distribution as being the time interval which concentrates 90% of the rates (see Figure 2 a)). The configuration of this filter parameter is crucial for the performance of the recommendation algorithm. Therefore, it is advised to perform a tuning of this parameter using experimental simulations (see Section 4). The age of each item is computed in aging intervals. For example, if the size of the aging interval was identified to be one month, all items introduced in the system in the last month have the age of 0, other items made available two months ago have the age of 1 and so on.

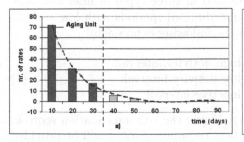 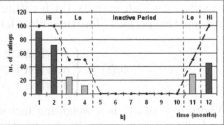

Fig. 2. Filter selection and configuration

The time decay function can be also used to push the newest items into the recommendation list. There is also known that online shops have the interest to advertise and promote their newest products, that represent new market trends. In this case the aging interval can be set by a domain expert (e.g. the shop manager) who will set for how long the items should be favored in the recommendation list.

The DIN_u is a component of the user profile, being the personalization element of the decay function. There can be noticed differences between user behaviors when analyzing the logs of the online systems. There are users that are interested to buy the newest products all the time, and there are users that want to buy good products but not very expensive ones. The second category of users will carefully evaluate the characteristics of older and newer products, and will pursue the evolution of prices. Therefore the item access logs are a good indicator to which category a user belongs. The DIN_u parameter is computed as follows:

$DIN_u = \frac{\#u_{dacc}}{\#u_{tacc}}$,

where $u_{\#dacc}$ is the number of distinct items accessed by user u, and $u_{\#tacc}$ is the total number of access logs of user u (including duplicates). For example, a user that accessed the description of only 3 distinct items 6 times, has the $DIN = 0.5$. The value of the DIN parameter characterizes the customer's behaviour, and it is automatically updated after each user-system interaction session.

Time window filtering. The time window filtering solution will be chosen in the case when the purchases are not uniformly distributed over the whole period of the year (see Figure 2). Such a time dependency can be met in the tourism domain, when the hotel resorts at the sea side are booked in summer and the ones in skying areas are popular in winter. There are also differences between the high season periods of the items belonging to the same category. For example the opening season is longer for the sea side hotels in the south of Spain or Italy, than the ones situated at Baltic Sea. Also, the skying season lasts longer on the skying resorts in high mountains (e.g over 3000 meter altitude) than for the ones situated in lower mountains (e.g. between 1000 and 2000 meters). Therefore, the time window used for recommendations filtering is item specific (e.g. can be seen as a product property). When learning window filters from the rating distribution, the time ax is divided in three types of intervals: time periods with high activity (high rating distribution), periods with low activity and inactive periods. The time filter has the following expression:

$$w_i(t) = \begin{cases} 1, & \text{when } t \in \text{high activity interval;} \\ \alpha, & \text{when } t \in \text{low activity interval;} \\ 0, & \text{when } t \in \text{inactive interval.} \end{cases} \tag{2}$$

Where $\alpha \in (0...1)$ is a weighting factor reducing the recommendation score in the low season. In most of the cases, the best recommendations will be provided using a value of α between 0.5 and 0.8. The fine tuning of this parameter is made with the help of experimental simulations (see Section 4).

In many application domains the filter configuration can be directly derived from product description. For example, the hotel descriptions indicate different prices for high season and extra season, and they are closed in the rest of the year.

If the information regarding the duration of the *high activity*, *low activity* and *inactive intervals* is not available for the given application domain, they are learned from rating distribution as shown in Figure 2 b). Using historical data, the *inactive interval* is identified as the period of time in which no rates are provided for a given product (e.g. the months in which no customer bought this product). The *low activity interval* is defined as the time periods when only a few rates are provided (lower than a given threshold), and the rest of the time is the *high activity interval*. In our evaluation we analyzed the monthly rating distribution in a time frame of one year. Depending on the application domain, different time frames have to be used for deriving the filter configuration. For example, a time frame of a week has to be used when deriving the filter configuration for TV programs recommendation. In this case, the precision of the active/inactive time intervals must be defined in terms of day plus hour.

In rich product offers, where many items get very similar recommendation scores, the attenuation produced by the α parameter may completely eliminate related items from the recommendation list. In these cases, the filter configuration can be reduced to a step function, which allows items to be recommended only in high season.

Collaborative Filtering. The time filtering technique can be used to improve the recommendations of each type of recommender, being it CB, CF or KB. For our experimental evaluation we chose to use the Resnick's CF algorithm, because it has better prediction accuracy than the CB solutions, is independent from the availability of domain knowledge and necessitate lower implementation costs maintenance than the KB Recommenders. The algorithm uses Pearson's correlation to compute the similarity between users and to identify the user neighborhood:

$$sim_{uv} = \frac{\sum_i r_{ui} * r_{vi}}{\sqrt{\sum_i r_{ui}^2 * \sum_i r_{vi}^2}} \tag{3}$$

Afterwards, the recommendation scores are computed using the following weighed sum:

$$score_{ui} = \frac{1}{N} * \sum_{v=1}^{N} sim_{uv} * r_{vi}, \tag{4}$$

Where:

- sim_{uv} is the similarity between user u and user v
- r_{ui}, r_{vi} are the ratings of user u, respectively v for item i
- $score_{ui}$ is the computed utility of item i for user u
- N is the number of members of user neighborhood

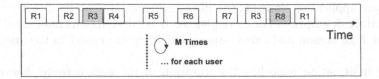

Fig. 3. M fold Cross validation. *All but 2* simulation.

4   Evaluation

4.1   Evaluation Methods

M-Fold Cross-Validation. The effectiveness of recommendation algorithms is usually measured through the Recall metric, which measures the ratio of items correctly suggested by recommenders from the total number of items rated by users as good items. The evaluation is typically performed off-line by separating the data set in two parts, one set containing rates used for model building (training or learning set) and a second set of rates used for model validation (test or validation set). The size and the content of these two sets is given by the type of the simulation and evaluation strategy. The *All but N* simulation type (also known as *leave N out*) uses a validation set of N ratings, the rest of the ratings being used in the training set.

In the case of M Fold Cross Validation method, the N items of the learning set are randomly selected. Therefore, the simulation results are indeterministic and the recommendation performance is measured over M simulation repetitions. This process is sketched in Fig 3, where R_i represents the rate of the current user for the item i, and the marked items (i.e. R_3, R_8) compose the randomly selected validation set. The overall prediction accuracy is computed as the average recall over all simulations of each user.

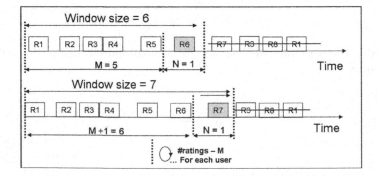

Fig. 4. Varying window experimentation. *All but 2* simulation.

Varying Window Experimentation. The main inconvenience of cross validation method is the fact that it completely ignores time dependencies when selecting the training set and computing the recommendation lists. The list of recommendations is built by using all information found in the system at experimentation time (i.e. this is different from the rating time). Therefore, in the recommendation list may come items that had no rates, or even items that were not available in the system at the moment in time when the user evaluated a given product.

In order to overcome these problems and to make a more correct evaluation of the recommender systems we introduce the varying window experimentation method, which is time aware. The ratings (R_i) are ordered by their timestamp, and the training set is selected by reconstructing the application context at rating time (see Figure 4). If we consider the case of an initial window size of 6 in an *All but 1* simulation, the first 5 rates (in the chronological order) will compose the learning set and the next one the test set. The rest of the ratings are ignored in the first iteration. In the next run the window size will be incremented by including the next rate into the training set, and replacing it with a new one in the validation set. This step is repeated until the window size will reach the number of user's rates. The same simulation process is executed for each system user. In this kind of experimentation, the recommendation lists are built as the system would have computed them at the rate time (i.e. the moment in time when the user has rated the each individual object).

4.2   Evaluation Context

For the experimental evaluation we used two commercial data-sets from domains with strong time dependencies. KMPortal[2] is a knowledge management tool used as knowledge exchange platform by different partners involved in an European research project in public administration domain. The interaction logs of this system show a high interest of users for the newest information available in the system, therefore we use this data-set for evaluating the Time Decay Filtering algorithm. The system contains 424 information objects introduced into the system during a period of time of two years, starting with the end of 2004. There are 3524 document access logs registered into the system for 89 users. Many users prove to be inactive users, only 30 of them have accessed more than 5 different information objects. The 5 users that are members of the project management group have a very different behavior than the regular users and they were not taken in account in the experimental evaluation.

The second data-set contains binary ratings representing transaction data from consumer services domain. This second data-set is used for evaluating the Time Window Algorithm. It contains the user-system interaction logs registered over a period of 3 years (starting with January 2004). 3853 users bought 18825 items from a list of 1387 product alternatives, and more than 2000 users have bought at least 5 distinct products in this time interval.

---

[2] see www.kmportal.net

4.3   Experimental Results

Using the data-sets described in the previous subsection we implemented two experiments in which we are looking forward for answering the following questions:

- How good is the recommendation accuracy of CF algorithms used in commercial applications with less user system interaction?
- Do the time filtering solutions improve effectiveness of recommenders employed in those systems?
- Is there a significant difference between the experimental results measured with cross validation method and the ones computed with time aware evaluation method?

Table 1. Time decay filtering recommendation performance (Recall)

	M	CF_{cv}	CF_{vw}	$CFTD_{vw}$		
Aging Unit	-	-	-	20	30	60
ABN 1	3	0.20	0.64	0.47	0.52	0.47
ABN 1	4	0.23	0.35	0.50	0.47	0.50
ABN 1	5	0.33	0.66	0.35	0.67	0.42
ABN 1	6	0.27	0.27	0.30	0.45	0.30
AVG		0.258	0.480	0.405	0.527	0.422

We used *All but 1* simulations for measuring the prediction accuracy of the three variants of collaborative filtering recommenders: Resnicks's algorithm (CF), Time Decay Filtering (CFTD) and Time Window Filtering (CFTW) algorithms. The results are presented in Table 1 and Table 2, where cv and vw indices denote the evaluation method: cross-validation or varying window. The **M** parameter sets the minimum number of ratings used for model learning (i.e. the size of the training set). For the varying window evaluation method, **M**+1 represents the initial size of the ratings selection window (see Figure 4). The CF_{vw} experiment reported an average recall of about 45-48%, which was improved by the time filtering techniques to ∼53%, while cross validations simulations indicated poor prediction accuracy of ∼ 25%, and ∼ 17% respectively, for the second data set.

Table 2. Time window filtering recommendation performance (Recall)

	M	CF_{cv}	CF_{vw}	$CFTW_{vw}$			
α	-	-	-	0	0.5	0.8	1
ABN 1	4	0.17	0.556	0.564	0.597	0.599	0.610
ABN 1	6	0.18	0.496	0.510	0.535	0.583	0.524
ABN 1	7	0.15	0.431	0.478	0.511	0.528	0.498
ABN 1	9	0.19	0.419	0.481	0.497	0.509	0.491
AVG		0.172	0.473	0.508	0.535	0.554	0.530

In order to find the best filter configuration we performed a variation of the filter parameters. For the Time Decay Filter, we found that the ideal value of the *Aging Unit* lays between 30 and 45 days. Using a value outside this interval affects seriously the performance of the filter. For this data set, it reduces the overall performance of the recommender below the one of the classic CF Algorithm. For the Time Window Filter, the best results were obtained with a value of the α parameter set to 0.8. The special situations, $\alpha = 1$ and $\alpha = 0$ transform the window filter into a step function. This doesn't change the recommendation score provided by CB/CF algorithms, but stops items to be recommended in the inactive interval, and inactive + low activity interval, respectively. Even these simple filters were able to improve the recommendation score with 5 to 10 %.

Discussion. Using the cross validation method to evaluate the recommenders in domains with strong time dependencies can lead to false conclusions as shown in the experimental results. The cross validation results lay far away from the values computed with time aware evaluation methods on both data sets. Given the relative small number of users available in the KMPortal dataset, the experiment results have a higher variance than in the case of Consumer Services data-set. The time decay and time window filtering algorithms proved to bring an improvement of 10 and 17% respectively, over the classic CF algorithm. Anyway, if the filters are bad configured, they can negatively affect the quality of the recommendations (see Table 1).

In the results of the Time Window filtering experiment (Table 2), it is interesting to notice a slightly negative correlation between the size of the training set (M) and the recall metric. This may suggest a degradation of the classification model that can be explained through changes of user preferences, given the distance in time between the first and the last ratings. In other words, customers that buy more services are interested to get a higher product diversification than the regular users (e.g. heavy customers buy products from 2-3 or more different categories).

The Time Filtering is used in collaboration with content based and collaborative filtering techniques, which are the core of the recommendation algorithms. The time filters do not suggest new items, they mainly improve the accuracy of the recommendations (i.e. Precision metric) by changing the positions of the items in the recommendation list. Anyway, given the fact that the recommendation lists are truncated in real applications (i.e. only the top 5 or top 10 recommendations are presented to the user), the prediction precision (i.e. Recall metric) of the algorithms is also improved. In our experimentation we used the Recall metric, since this is the standard measure for the quality of the recommenders.

The experimental evaluation of the time filtering approaches was made individually using two different data-sets. At this time we are not in the possession of a data-set which presents both types of preferences, decayed and periodical. Therefore, we were not able to evaluate how effective the combination of the two recommendation approaches can be.

5   Conclusions

In this paper we presented the time filtering technology which was proved to improve the quality of recommendation lists in domains with time dependant user preferences and time dependant item availability, such as knowledge management domain and customer services domain, respectively. Both time decay and time window filters have proved to be effective when used in the right context with the correct configuration. The time aware evaluation method using varying window simulations reconstructs the application context for computing the recommendation list. This method is more appropriate to be used for evaluating the effectiveness of recommenders in time dependant domains, than the classic cross validation method. As future work we plan to build more personalized time filters by incorporating more information from user profile into filter configuration. For example, the age of the client is a relevant factor for selecting the time period and the destination of their journey. Also we plan to evaluate the performance of the proposed algorithms with public data-sets (e.g. movielens).

References

1. Burke, R.: Hybrid recommender systems: Survey and experiments. User Modeling and User-Adapted Interaction 12(4), 331–370 (2002)
2. Adamavicius, G., Tuzhilin, A.: Towards the next generation of recommender systems: A survey of the state-of-the-art and possible extensions. IEEE Transactions on Knowledge and Data Engineering 17(6) (2005)
3. Herlocker, J.L., Konstan, J.A., Terveen, L.G., Riedl, J.T.: Evaluating collaborative filtering recommender systems. ACM Trans. Inf. Syst. 22(1), 5–53 (2004)
4. Adomavicius, G., Sankaranarayanan, R., Sen, S., Tuzhilin, A.: Incorporating contextual information in recommender systems using a multidimensional approach. ACM Transactions on Information Systems 23(1), 103–145 (2005)
5. Sarwar, B., Karypis, G., Konstan, J., Riedl, J.: Recommender systems for large-scale e-commerce: Scalable neighborhood formation using clustering. In: International Conference on Computer and Information Technology (2002)
6. Zanker, M., Jessenitschnig, M., Jannach, D., Gordea, S.: Comparing recommendation strategies in a commercial context. IEEE Intelligent Systems 22 (2007)
7. Balabanovic, M., Shoham, Y.: Fab: Content-based, collaborative recommendation. Communications of the ACM 40(3), 66–72 (1997)
8. Mooney, R.J., Roy, L.: Content-based book recommending using learning for text categorization. In: Proceedings of DL-00, 5th ACM Conference on Digital Libraries, San Antonio, US, pp. 195–204. ACM Press, New York (2000)
9. Burke, R.: Knowledge-based recommender systems. Encyclopedia of Library and Information Systems 69(2) (2000)
10. Felfernig, A., Gordea, S.: AI Technologies Supporting Effective Development Processes for Knowledge Based Recommender Applications. In: SEKE 2005. 17th International conference on Software Engineering and Knowledge Engineering, ACM, New York (2005)
11. Konstan, J., Miller, B., Maltz, D., Herlocker, J., Gordon, L., Riedl, J.: Grouplens: Applying collaborative filtering to usenet news. Communications of the ACM 3(40), 77–87 (1997)

12. Karypis, G.: Evaluation of item-based top-n recommendation algorithms. In: CIKM 2001: Proceedings of the tenth international conference on Information and knowledge management, pp. 247–254. ACM Press, New York (2001)
13. Melville, P., Mooney, R., Nagarajan, R.: Content-boosted collaborative filtering. In: Eighteenth national conference on Artificial intelligence, American Association for Artificial Intelligence, Menlo Park, CA, USA, pp. 187–192 (2001)
14. Middleton, S.E., Shadbolt, N.R., de Roure, D.C.: Ontological user profiling in recommender systems. ACM Transactions on Information Systems 22(1), 54–88 (2004)
15. Ding, Y., Li, X.: Time weight collaborative filtering. In: 14th ACM international conference on Information and knowledge management, pp. 485–492. ACM Press, New York (2005)

A Survey of UML Models to XML Schemas Transformations*

Eladio Domínguez[1], Jorge Lloret[1], Beatriz Pérez[1],
Áurea Rodríguez[1], Ángel L. Rubio[2], and María A. Zapata[1]

[1] Dpto. de Informática e Ingeniería de Sistemas,
Facultad de Ciencias, Edificio de Matemáticas,
Universidad de Zaragoza, 50009 Zaragoza, Spain
noesis,jlloret,beaperez,arv852,mazapata@unizar.es
[2] Dpto. de Matemáticas y Computación, Edificio Vives,
Universidad de La Rioja, 26004 Logroño, Spain
arubio@unirioja.es

Abstract. UML is being increasing used for the analysis and design of Web Information Systems. At the same time, many XML–based languages are cornerstones in the development of this kind of system. As a consequence of the predominance of these languages, there are many works in the literature devoted to exploring the relationships between UML and XML. In this paper we present a survey of current approaches to the transformation of UML models into XML schemas. The study is focused on the case of transformation of UML class diagrams to XML schemas, since we have not found any proposal regarding other kinds of UML diagrams.

Keywords: UML model, XML Schema, Model Transformation.

1 Introduction

Several recent papers (see [5, 24, 25, 35]) address the increasing interest in using UML in order to analyze and design Web Information Systems. The use of UML within this context has some limitations [24] that have been partially resolved by means of the definition of some UML extensions for web modeling [10]. Nevertheless, the advantages of using a well–established, tool–supported and standard language are beyond doubt. At the same time, more and more XML–based technologies are being used in the implementation and deployment of Web Applications. Therefore, it is not surprising that there are many works in the literature devoted to studying the different relationships between UML and XML. This number will probably grow in the foreseeable future due to the proliferation of Model–Driven Development approaches, which will be applied to the development of Web Information Systems in a natural way.

* This work has been partially supported by DGI (project TIN2005-05534 and FPU grant AP2003-2713), by the Government of La Rioja (project ANGI 2005/19), by the Government of Aragon and by the European Social Fund.

B. Benatallah et al. (Eds.): WISE 2007, LNCS 4831, pp. 184–195, 2007.
© Springer-Verlag Berlin Heidelberg 2007

In the present paper we investigate numerous approaches in the literature for the transformation of UML models to XML schemas. One result of our research is that, to our knowledge, there are no approaches devoted to the transformation of UML models to XML schemas other than UML class diagrams. On the other hand, there are many papers that deal with the transformation of UML class diagrams to XML schemas. Because of this, the main contribution of the paper is to provide an exhaustive revision and comparison of such transformations.

The comparison of the different approaches has been realized in a systematic way. On the one hand, and taking as a base the framework proposed in [11], we have compared the different features of transformations proposed by each approach. One conclusion regarding this comparison is that only one approach [15] considers traceability properties, which are considered as desirable features in transformation frameworks such as MDA [1]. On the other hand, we have analyzed which UML class diagram elements are considered by each proposal. A conclusion of this analysis is that the proposal of [26, 27] is the most complete but it does not provide general transformation rules for UML models including the application of some profiles. However, we would stress the necessity for such rules, since the use of profiles is the most common method for tailoring UML to the requirements of specific domains, as for instance Web Information Systems Development. For this reason, we have made a proposal including profiles, with a practical application in a medical context [13].

The paper is structured as follows: in the following section we present an overview of the different UML to XML Schema transformation proposals. A statement about the method of comparison that has been used is presented in Section 3. Sections 4 and 5 are devoted, respectively, to the presentation of the feature–based comparison and the element–based comparison. Finally, Section 6 includes some conclusions and future work.

2   UML Models to XML Schemas Proposals

Due to the widespread use of UML and XML Schema languages, it is no wonder that there are many works in the literature whose goal is to propose a transformation of UML diagrams into XML schemas.

These transformations are far from being trivial, and one reason for this complexity is that UML and XML Schema have several important differences [42]. For instance, UML is primarily used in the early stages (analysis, design) of the development process. However, XML Schema is more implementation oriented and includes programmatic and syntactic details (such as ordering or scoping) that are not generally captured in UML models. Another (almost obvious) difference is that UML is mainly a graphical, visual language whereas XML Schema has essentially a textual representation. Moreover, the most important difference is that while XML Schema is concerned with the *structure* of XML data representing entities and their components and relations, UML allows us to represent not only the *structure* but also the *behavior* of a system, by means of different kinds of diagrams. Because of this, it is not surprising that all the papers we have found regarding the transformation of UML to XML Schema only deal with UML class diagrams (the most important part of UML devoted to the representation of structural aspects). In the rest of the paper we focus our attention on this kind of transformation, carrying out an exhaustive revision and comparison. Let us make now a brief comment about the transformations of UML diagrams different from class diagrams to XML.

The main result in this respect is that all the proposals translate the other kinds of UML diagrams to *XML documents* not to XML schemas. On the one hand, the OMG's XML Metadata Interchange (XMI) [38] defines rules to map MOF [37] metamodels and metadata to XML documents. In particular, it can be used to transform any type of UML diagram into an XMI document. Although XMI is widely used as an interchange format, some works claimed its verbosity, poor readability and compactness [11]. On the other hand, several works are concerned with the transformation of particular kinds of diagrams to XML documents. With respect to *UML sequence* and *collaboration diagrams*, the paper [43] defines a previous XML schema for UML sequence and collaboration diagrams in order to be used as a base for the definition of XML documents. Regarding *UML activity diagrams*, several papers [18, 34] use the XML Process Definition Language (XPDL) to transform activity diagrams to XML documents. As for *UML state machines*, StateChart XML (SCXML) [9] has been proposed by the W3C in order to provide a generic state–machine based execution environment, and XTND (XML Transition Network Definition) [33] is a notation for simple finite state machines.

3 Comparison Method Statement

As we have mentioned previously, henceforth we only deal with the transformation of UML class diagrams to XML schemas. In particular, in this section we explain the method followed to select the papers to be compared, as well as the decisions we have taken to determine the type of comparison to be performed.

With regard to the selected papers, firstly we made a bibliographic search of papers proposing UML to XML Schema transformations, without restricting the search to the area of web information systems. Sometimes the proposals of different authors were very similar, in which case we have selected only one proposal representing all of them. In these cases, we have considered that the papers not included in the comparison do not provide any additional information. Finally, we have analyzed 22 papers in all, grouped into 13 different groups, which provide a general vision of the different published proposals.

It should be noted that the goal of some papers we have analyzed [2, 3, 21, 26, 27, 28, 32, 41, 44] is the definition of a UML to XML Schema transformation, whereas other papers [4, 6, 7, 8, 12, 13, 14, 15, 16, 17, 19, 20, 30, 36, 39, 40] form part of more general projects that do not have this transformation as the only goal. Also, it is worth mentioning that, although several of the analyzed papers explicitly indicate the application domain (database design [27], construction industry [44], e-learning [4, 30], e-business [6, 7, 8, 12, 19, 36], evolution architecture [14, 15, 16, 17], life sciences [13, 20]), most of the works present their transformation approach in a general way without specifying a specific domain where it could be applied. We have not included a summary of individual works for space reasons.

As for the type of comparison, we have decided to compare the selected papers according to two different dimensions. On the one hand, we have analyzed the *features* of the approach followed to define the UML to XML Schema transformation. In order to do this, we have considered several proposals of model transformation classifications [11, 22, 31]. Of these, we have decided that the taxonomy proposed in [11] is the most rigorous and complete approach. For this reason, we have chosen it to carry out our feature–based comparison.

On the other hand, we have considered it appropriate to compare the papers in terms of the specific UML elements that are taken into account by each proposal. Besides, the strategy followed in order to translate each UML element into XML Schema elements is also analyzed.

We also want to note that, beyond the specific papers we compare in this work, the comparison method we have followed can be used as a general framework to compare other UML to XML Schema transformation proposals.

4   Feature–Based Comparison

Due to the wide range of varied model transformations proposed in the literature, several recent papers [11, 22, 31] have tried to determine taxonomies or frameworks that help to analyze and compare different approaches. We have considered it appropriate to use these papers in order to perform our comparison. In particular, we have performed the comparison mainly on the basis of the feature model proposed in [11] which makes explicit several possible design choices for a model transformation approach.

Since the type of transformation proposals we want to analyze is not general, some of the features proposed in [11] are not really possible choices. To be precise, in all the analyzed papers:

(1) the proposed transformation is *exogenous*, since the source and target models are not the same,

(2) the proposed transformation involves two *different technological spaces* [31],

(3) regarding the resulting XML schema, although it has a tree–based structure, it is a textual representation of the UML class model, so that we have considered that the type of transformation to be defined is a *model-to-text* transformation.

Aside from these features shared by all the analyzed works, other features that we have used to compare the papers are proposed in [11]. Results of the comparison are summarized in Table 1 which shows the characteristics of each work according to nine different features. In the remaining part of this section, we explain each feature and we comment on the conclusions that can be deduced from each of them.

Table 1. Feature–based comparison

	[2,3]	[4,30]	[6-8]	[12,19]	[13,15,16]	[20]	[21]	[26,27]	[32]	[36]	[39,40]	[41]	[44]
Directionality	Bid.	Unid.	Bid.	Unid.	Unid.	Unid.	Unid.	Unid.	Unid.	Unid.	Unid.	Unid.	Unid.
Traceability					Yes								
Incrementality					Target								
Metamodel Approach	UML + Profile	UML + Profile	UML + Profile		UML		UML	UML	UML	UML + Profile	UML + Profile	UML	UML
Category	M2T	M2T2T	M2T2T	M2T	M2M2T	M2T	M2T2T	M2T2T	M2T2T	M2T	M2T2T	M2M2T	M2T
Number of Metamodels	1	1	1	1	3	1	1	1	1	1	1	2	1
Kind of Transformation	Transf. pattern	XSLT	XSLT	—	XSLT	—	XSLT	XSLT	Rules	Production Rules	—	—	Transf. Rule
Paradigm	—	Decl.	Decl.	—	Decl.	—	Decl.	Decl.	Hybrid	Decl.	—	—	Decl.
Tool		U2XAptly	Hypermodel						UML-XML Converter	—			

Directionality. The majority of papers only propose transformations from UML class diagrams to XML schemas (unidirectional), only two approaches, [2,3] and [6,7,8], consider the possibility of executing the transformations in both directions (bidirectional), so that the XML to UML case is also considered.

Traceability and Incrementality. Traceability is concerned with maintaining an explicit link between the source and the target models of a model transformation [11,31]. On the other hand, incrementality refers to the ability to update the target model based on changes in the source model [11]. To our knowledge, the proposal developed within our research group [15] is the only work that deals with traceability, maintaining explicit links between the UML and XML models. The aim of this proposal is to use the automatic generated trace for keeping the consistency between the models when evolution tasks are carried out. In this way, target incrementality is achieved since the target model is updated according to the evolution changes performed in the source model.

Metamodel Approach. A metamodel approach is followed in most cases. In these cases, all the papers define UML metamodels and in five cases the metamodel is defined together with a UML profile.

Category. Among the analyzed papers a distinction can be defined on the basis of the fact whether they perform directly a *Model-to-Text* (M2T) transformation from UML Class Diagram to XML Schema or they accomplish some intermediate transformation. In this second case, the analysis of the works has led us to consider two different categories. *Model-to-Text-to-Text* (M2T2T): the UML class diagram is translated first into another representation using some intermediate textual language (XMI in the majority of cases) and then another transformation is performed to obtain the XML schema. *Model-to-Model-to-Text* (M2M2T): an intermediate metamodel is determined so that the UML class diagram is first translated into another model which is an instance of the intermediate metamodel. This second model is then transformed into an XML schema. Two works ([15] and [41]) propose this type of transformation. In both cases, the intermediate metamodel represents a metamodel of the XML Schema language. Let us note that [29] also proposes to use an M2M2T approach but we have not included it in the table because the paper is very short and we do not have enough information to describe it according to the different criteria.

Number of Metamodels. This feature depends on the three previous ones. The M2T and M2T2T papers with a metamodel approach use only one metamodel for the UML class models. The paper [41] with an M2M2T approach defines two metamodels, which are used, respectively, at the conceptual and logical levels. Finally, the M2M2T proposal with traceability, developed within our research group [15], defines three metamodels: two of them for the conceptual and logical levels and a third metamodel representing modeling knowledge with regard to the way in which UML elements are translated into XML Schema elements.

Kind of Transformation and Paradigm. These features are difficult to determine because some authors explain their proposals by means of examples but they do not explain the way in which the transformation is implemented. In these cases, this information cannot be extracted from the papers (the symbol '–' is used to represent this fact). When this feature is mentioned, the great majority of authors implement the transformation by means of XSLT. We have considered, as it is stated in [23], that XSLT is a declarative language, in the sense that the transformation you require is described, rather than providing a sequence of procedural instructions to achieve it. The rest of the papers use different approaches: (1) transformation patterns [2], (2) declarative approach by means of mapping rules [36, 44] and (3) a hybrid approach (declarative and imperative) by means of rules [32].

Tool. According to the information we have gathered from the papers, few works develop a tool based on their proposals. Only three of them indicate the name and the characteristics of the tool, and one work explains that they have a tool but they do not give its name (which is shown by means of the symbol '–').

5   Element–Based Comparison

The goal of this comparison is to analyze which UML Class Diagram elements are considered by each proposal and the strategy followed in order to translate them into XML Schema elements.

The results are summarized in Table 2 indicating in each cell the proposal of each author. Let us note that (1) an empty cell represents that the authors do not comment anything about the element in question, (2) the symbol '–' indicates that the authors propose to translate the corresponding element but they do not explain their strategies and, (3) the symbol 'X' means that the authors explicitly state that they do not translate that element.

The first conclusion that can be extracted from Table 2 is that the proposal of [26, 27] is the most complete approach. Furthermore, in order to compare the way in which each element is considered by each author, we have gathered together the rows in five groups (represented by alternating grey/white colors

Table 2. Element–based comparison

	[2,3]	[4,30]	[6-8]	[12,19]	[13,15,16]	[20]	[21]	[26,27]	[32]	[36]	[39,40]	[41]	[44]
Class	complex Type	element	complex Type + elem.	element	complex Type + elem.	complex Type + elem.	elem.	complex Type + elem.	complex Type + elem.	elem.	complex Type	complex Type	element
Attribute	elem. or attrib.	element	elem. or attrib.	element	element	elem. or attrib.	elem.	element	element	attrib.	elem. or attrib.	elem. or attrib.	elem. or attrib.
Associat.	—	referenc. via associat. element	referenc. via associat. element	referenc. via associat. element	key / keyref referenc.	referenc. with XLink	referenc. via associat. element	key/keyref referenc.	referenc. via associat. element	referenc. via associat. element	key / keyref referenc.	nested element	
Generalizat.	G2	G2, G4	G2, G4	G2	G2	G2	G1	G2	G3	G1	G4		
Built-in data type	simple-Type	simple-Type	simple-Type			simple-Type	simple-Type	simple-Type	simple-Type	simpleT. or string	simpleT. or string	simple-Type	
Abstract class	abstract elem.		abstract elem.	abstract elem.		abstract elem.		abstract elem.	abstract elem.	x			
Multiplicity of attribute		Min/Max Occurs attrib.	Min/Max Occurs attrib.	Min/Max Occurs attrib.		Min/Max Occurs attrib.		Min/Max Occurs attrib.				Min/Max Occurs attrib.	
Multiplicity of associat.		Min/Max Occurs attrib.	Min/Max Occurs attrib.		Min/Max Occurs attrib.	Min/Max Occurs attrib.	Min/Max Occurs attrib.	Min/Max Occurs attrib.	Min/Max Occurs attrib.		Min/Max Occurs attrib.	Min/Max Occurs attrib.	
Model	root elem.	root elem.	root elem.		root elem.			root elem.			root elem.	root elem.	
Note	—	documentation element	documentation element	documentation element				documentation element	documentation element	documentation element			
User defined data type		complex-Type elem.	derived through generaliz. from simpleT.			complex-Type elem.	simple-Type elem.	complex-Type elem.	derived through generaliz. from simpleT.		derived through generaliz. from simpleT.	derived through generaliz. from simpleT.	
Composit.	child elem.		child element	child element				nested elem. + referenc. elem.	nested elem. + referenc. attrib.	child element	child element		
Restriction in the value of attribute	RVA1, RVA2, RVA3	RVA1, RVA2, RVA3	RVA2, RVA3			RVA1, RVA3		RVA3		RVA3	RVA1, RVA2, RVA3	RVA1, RVA2	
Aggregat.		Ag2	Ag1, Ag2	Ag1				Ag1	Ag3	Ag1			Ag4
Complex attribute		element	complex-Type + elem.			complex-Type + elem.		element					
Package	—		XML namespace					elem. whithout attribute or XML namespace			XML namespace		
Associat. class						complexT. + elem. + references		complexT. + elem. + references			complexT. + elem. + references	Different options nesting based on navigation	
Dependency		Type's attrib. content						x	nested elem. + referenc. attrib.				attrib. relation ="deleg ation"
Derived attribute								attrib.+ elem. with name constraint +condition as fixed.	elem. with name, type attrib. for data type				
Stereotype					S1			attrib					
Method								x	x				

in Table 2) using a mixture of two criteria. The first and the last group are determined according to the number of works that deal with each element. In particular, the first group includes the elements for which the majority of works propose a translation. In contrast, the elements of the last group are analyzed by either few works or none. The rest of the groups have been determined according to the similarities or differences among the translation strategies followed by the authors. Next, we explain in detail each group.

As we have said, the *first group* includes the elements that are considered by almost every proposal. It is no wonder that the elements of this group are the most widely used UML Class Diagram elements: class, attribute, association, generalization and built–in data type. With regard to the strategy followed to translate classes, attributes and built–in data types, the authors more or less agree. This is not the case of the associations and the generalizations.

As for associations, a study of the possible translation strategies is presented in [26,27] proposing four different options: (1) nested element, (2) key/key references of elements, (3) references via association element and (4) references with XLink and XPointer. These authors, like our proposal [13,15], decide to use the second option, but, as can be seen in Table 2, the other three options are more frequently used by the other authors.

Something similar occurs with regard to generalizations. A study of possible translation strategies is presented in [28] proposing three different options (which we have called G1, G2 and G3) and, besides, a fourth option (G4) is proposed in [6, 7, 8] and [39, 40]. Of these, the option G2 is the most frequently used by the authors. The meaning of each option is the following:

- G1: All classes are transformed into `complex types` and `elements` of these types. The generalization relation between a class and its subclasses is mapped to containment between the corresponding elements.
- G2: The root class is mapped to an `element` and a `complex type`. The subclasses are mapped to `complex types` derived by extension from the `complex type` of the superclass.
- G3: `Elements` are declared for every class and they are organized in a substitution group.
- G4: The root class is mapped to an `element` and a `complex type`. Specialization classes are mapped to types derived by restriction from the type of the superclass.

The *second group* includes the UML Class Diagram elements for which all the proposals that deal with them propose the same transformation. This is the case of: abstract class, multiplicity of attribute, multiplicity of association, model and note. The elements of the *third group*, user defined data type and composition, are not translated in the same way by the authors but there are very few differences among the proposals.

Different approaches are proposed with regard to the two elements of the *fourth group*. As for the restriction in the value of attribute, in general, the authors decide to propose the use of several options based on the kind of restriction. In the case of aggregation the option we have called Ag1 is the most frequently used. The meanings of the abbreviations used to express the proposed translation strategies of these two elements are the following.

Restriction in the value of attribute

- RVA1: Range restrictions are declared using `min/max Exclusive` and `min/max Inclusive` properties.
- RVA2: Pattern restrictions are mapped to templates for the instances of an attribute.
- RVA3: Enumeration restrictions are translated as enumeration of possible values of attributes.

Aggregation is rendered by

- Ag1: a nested child `element` linked to its remote `element` definition.
- Ag2: a proxy `element` linked to its remote `element` definition.
- Ag3: an XML `element` with `ref attribute` pointing to the referencing class.
- Ag4: an `attribute` named 'relation' with value 'aggregation'.

As we have said before, the elements of the *last group* are considered by none or almost none of the authors. Specially remarkable is the fact that methods are not translated by any author, and some authors even state that methods of a class do not have an equivalent representation in the XML Schema [32]. We also want to note that only two proposals ([26, 27] and [13, 15]) take into account UML stereotypes. The proposal of [26, 27] translates stereotypes without considering their properties, whereas the proposal developed within our research group [13, 15] is the only work that specifies transformation rules for the general application of UML profiles. According to this proposal (denoted as S1 in Table 2), firstly, an XML schema is created for each UML profile and, later, the stereotyped elements are translated into XML elements of a specific complex type (the definition of the complex types is based on the XML schemas previously created).

Finally, we want to note that, due to space reasons, we have not included in Table 2 elements that appear only in one paper. They are: qualified association [27], XOR constraint [27], n-ary association [27], primary key of a class [41], identity constraint (such as unique, pk, exists or isa constraints) [16] and null, non-nillable or not-absence constraints [16].

6   Conclusions and Future Work

In this paper we have presented a survey of current approaches to the transformation of UML models to XML schemas. A conclusion is that there are numerous papers that propose transformations of UML class diagrams into XML schemas. The rest of the UML diagrams are translated in the papers into XML documents but not into XML schemas. We have compared the UML Class Diagram to XML Schema transformations from two different viewpoints: first, considering the features of each one of the transformations and, second, checking what UML Class Diagram elements each proposal takes into account.

With regard to the feature–based comparison, the majority of authors propose a unidirectional transformation (from UML to XML Schema), implemented by means of XSLT, making use of an intermediate structure (instead of performing a direct translation). Generally a UML metamodel for the representation of the

UML class diagrams is defined, and some authors propose to use other meta-models for achieving a transformation with better features. In this respect, we consider that the proposal developed within our research group [13, 15, 16] is a noteworthy work, with some remarkable features (traceability, target incrementality and M2M2T approach).

As for the element–based comparison, the majority of works propose a translation for the most widely used UML Class Diagram elements (class, attribute, association, generalization and built–in data type). Nevertheless, there is no agreement about the way of translating associations and generalizations. As another conclusion, the proposal of [26, 27] is the most complete approach, but a weakness of this proposal is that it does not specify complete transformation rules for the general application of UML profiles. However, based on the widespread use of UML profiles as a way to add domain specific semantics to UML class models, we consider it necessary to have at our disposal a general way for translating stereotyped class models. For this reason, in [13] we have made a proposal in this sense.

There are several lines of future work. One important issue is the transformation of other kinds of UML diagrams to XML schemas. For instance, the representation of a UML state machine by an XML schema would allow us to store the execution traces of the state machine in an XML document conforming with that XML schema. We will incorporate the results of this research to our general framework of *evolution* of UML/XML models [13, 15, 16]. In particular, the development of an evolution tool is an ongoing project.

References

1. Bast, W., Kleppe, A., Warmer, J.: MDA explained. The Model Driven Architecture: Practice and Promise. Addison–Wesley, London (2003)
2. Bernauer, M., Kappel, G., Kramler, G.: Representing XML Schema in UML -An UML Profile for XML Schema. Technical report, Business Informatics Group, Ins. of Soft. Tech. and Inter. Sys., Vienna University of Technology (November 2003) (Last visited: June 2007), Available at http://www.big.tuwien.ac.at/research/publications/papers03.html
3. Bernauer, M., Kappel, G., Kramler, G.: Representing XML Schema in UML - A Comparison of Approaches. In: Koch, N., Fraternali, P., Wirsing, M. (eds.) ICWE 2004. LNCS, vol. 3140, pp. 440–444. Springer, Heidelberg (2004)
4. Bertolino, A.: Initial Recommendations on Advantage Testing Technologies. Technical report, D09 (November 2004) (Last visited: June 2007), Available at http://www.imsglobal.org/telcert/D09_Testing_Research_v1.0.pdf
5. Caceres, P., Marcos, E., Vela, B.: A MDA–Based Approach for Web Information System Development. In: Stevens, P., Whittle, J., Booch, G. (eds.) UML 2003 - The Unified Modeling Language. Modeling Languages and Applications. LNCS, vol. 2863, Springer, Heidelberg (2003)
6. Carlson, D.: Modeling XML Vocabularies with UML: Part II (Last visited June 2007), Available at http://www.xml.com/pub/a/2001/09/19/uml.html
7. Carlson, D.: Modeling XML Applications with UML: practical e-business applications. Addison-Wesley, Reading (2001)

8. Carlson, D.: Modeling XML Vocabularies with UML: Part III (2001) (Last visited June 2007), Available at http://www.xml.com/pub/a/2001/10/10/uml.html
9. Carter, J., Barnett, J., Bodell, M., Hosn, R., Burnett, D.: State chart XML (SCXML): State Machine Notation for Control Abstraction. W3C working draf, (February 2007) (Last visited June 2007), Available at http://www.w3.org/TR/2007/WD-scxml-20070221/
10. Conallen, J.: Building Web Applications with UML. Addison–Wesley, London (2000)
11. Czarnecki, K., Helsen, S.: Feature-based Survey of Model Transformation Approaches. IBM Systems Journal 45(3), 621–646 (2006)
12. Damodaran, S.: RosettaNet: Adoption Brings New Problems, New Solutions. In: Proceedings of the XML 2005 Conference and Exhibition, Atlanta (November 2005)
13. Domínguez, E., Lloret, J., Pérez, B., Rodríguez, A., Rubio, A.L., Zapata, M.A.: MDD-based Transformation of Stereotyped Class Diagrams to XML Schemas in a Healthcare Context. 2007. Accepted for Publication in CMLSA (2007)
14. Domínguez, E., Lloret, J., Rubio, A.L., Zapata, M.A.: An MDA-Based Approach to Managing Database Evolution. In: Rensink, A. (ed.) Proceedings of the Workshop Model Driven Architecture: Foundations and Applications, CTIT Technical Report, vol. TR-CTIT-03-27, pp. 97–102 (2003)
15. Domínguez, E., Lloret, J., Rubio, A.L., Zapata, M.A.: Evolving XML Schemas and Documents Using UML Class Diagrams. In: Andersen, K.V., Debenham, J., Wagner, R. (eds.) DEXA 2005. LNCS, vol. 3588, pp. 343–352. Springer, Heidelberg (2005)
16. Domínguez, E., Lloret, J., Rubio, A.L., Zapata, M.A.: Validation of XML Documents: From UML Models to XML Schemas and XSLT Stylesheets. In: Yakhno, T., Neuhold, E.J. (eds.) ADVIS 2006. LNCS, vol. 4243, pp. 48–59. Springer, Heidelberg (2006)
17. Domínguez, E., Lloret, J., Zapata, M.A.: An Architecture for Managing Database Evolution. In: Olivé, À., Yoshikawa, M., Yu, E.S.K. (eds.) ER 2002 Workshops. LNCS, vol. 2784, pp. 63–74. Springer, Heidelberg (2003)
18. Guelfi, N., Mammar, A.: A Formal Framework to Generate XPDL Specifications from UML Activity Diagrams. In: Proceedings of the 2006 ACM symposium on Applied computing, pp. 1224–1231 (2006)
19. Heikkinen, B.: Component-based Modelling with UML and XML-Schemas in RosettaNet (2002) (Last visited June 2007), Available at http://smealsearch2.psu.edu/95558.html
20. Hucka, M.: SCHUCS: An UML-Based Approach for Describing Data Representations Intended for XML Encoding. Sys. Biol. Workbench Develop. Group (2000)
21. Jeckle, M.: Practical Usage of W3C's XML-Schema and a Process for Generating Schema Structures from UML Models. In: Proceedings of the 2nd International Conference of Advances in Infrastructure for E-Business, Science and Education on the Internet, Rome, Italy (August 2001)
22. Jouault, F., Kurtev, I.: Transforming Models with ATL. In: Bruel, J.-M. (ed.) MoDELS 2005. LNCS, vol. 3844, pp. 128–138. Springer, Heidelberg (2006)
23. Kay, M.: XSLT Programmer's Reference, 2nd edn. Wrox Press Ltd., Birmingham (2003)
24. Koch, N., Kraus, A.: The Expressive Power of UML-based Web Engineering. In: IWWOST 2002. Second International Workshop on Web-oriented Software Technology, Malaga, Spain, pp. 105–119 (2002)
25. Kraus, A., Koch, N.: Generation of Web Applications from UML Models Using an XML Publishing Framework. In: Proc. of IDPT 2002, Pasadena, USA (2002)
26. Krumbein, T.: Logical Design of XML Databases by Transformation of a Conceptual Schema. Master's Thesis (in German), HTWK Leipzig (2003)

27. Krumbein, T., Kudrass, T.: Rule-Based Generation of XML Schemas from UML Class Diagrams. In: WebDB 2003. Proceedings of the XML Days at Berlin, Workshop on Web Databases, pp. 213–227 (2003)
28. Kurtev, I., Berg, K.V., Aksit, M.: UML to XML-Schema Transformation: a Case Study in Managing Alternative Model Transformations in MDA. In: FDL 2003. Proceedings of the Forum on specification and Design Languages, European Electronic Chips & Systems design Initiative, Frankfurt, Germany, (September 2003)
29. Liu, H., Lu, Y., Yang, Q.: XML Conceptual Modeling with XUML. In: Osterweil, L.J., Rombach, H.D., Soffa, M.L. (eds.) ICSE 2006. International Conference on Software Engineering, pp. 973–976 (2006)
30. Marchetti, E.: Automatic XML Schema Generation from UML Application Profile. Elektrotechnik und Informationstechnik (e&i) Journal of Springer Verlag 122(12), 485–487 (2005)
31. Mens, T., Van Gorp, P., Varró, D., Karsai, G.: Applying a Model Transformation Taxonomy to Graph Transformation Technology. Electronic Notes in Theoretical Computer Science 152, 143–159 (2006)
32. Narayanan, K., Ramaswamy, S.: Specifications for Mapping UML Models to XML Schemas. In: WiSME 2005. Proceedings of the 4th Workshop in Software Model Engineering, Montego Bay, Jamaica (2005)
33. Nicol, G.T.: XTND - XML Transition Network Definition (November 2000) (Last visited June 2007), Available at http://www.w3.org/TR/2000/NOTE-xtnd-20001121/
34. Noh, H.M., Wang, B., Yoo, C.J., Chang, O.B.: An Extension of UML Activity Diagram for Generation of XPDL Document. In: Zhang, Y., Tanaka, K., Yu, J.X., Wang, S., Li, M. (eds.) APWeb 2005. LNCS, vol. 3399, pp. 164–169. Springer, Heidelberg (2005)
35. Novikov, B., Gorshkova, E.: Exploiting UML Extensibility in the Design Phase of Web Information Systems. In: BalticDB&IS 2002. Proceedings of the Baltic Conference, Tallinn, Estonia, pp. 49–64 (2002)
36. OASIS. ebXML Business Process Specification Schema v1.01 (May 2001) (Last visited June 2007), Available at http://www.ebxml.org/specs/ebBPSS.pdf
37. OMG. MOF 2.0 Core Final Adopted Specification Document, ptc/03-10-04 (2004), Available at http://www.omg.org/
38. OMG. MOF 2.0 XMI Mapping Specification, v2.1, Document formal/05-09-01 (2005), Available at http://www.omg.org/
39. Provost, W.: Enforcing Association Cardinality (2002) (Last visited June 2007), Available at http://www.xml.com/lpt/a/2002/06/26/schema_clinic.html
40. Provost, W.: UML for W3C XML Schema Design (2002) (Last visited June 2007), Available at http://www.xml.com/lpt/a/2002/08/07/wxs_uml.html
41. Routledge, N., Bird, L., Goodchild, A.: UML and XML Schema. In: Zhou, X. (ed.) ADC 2002. Thirteenth Australasian Database Conference, ACS, Melbourne, Australia (2002)
42. Salim, F.D., Price, R., Krishnaswamy, S., Indrawan, M.: UML Documentation Support for XML Schema. In: Australian Software Engineering Conference, pp. 211–220 (2004)
43. Singh, J.: Mapping UML Diagrams to XML. Master's Thesis, Jawaharlal Nehru University, New Delhi (2003)
44. Wu, I.C., Hsieh, S.H.: An UML-XML-RDB Model Mapping Solution for Facilitating Information Standardization and Sharing in Construction Industry. In: Proceedings of the 19th International Symposium on Automation and Robotics in Construction, Gaithersburg, Maryland, pp. 317–321 (September 2002)

Structural Similarity Evaluation Between XML Documents and DTDs

Joe Tekli, Richard Chbeir, and Kokou Yetongnon

LE2I Laboratory UMR-CNRS, University of Bourgogne
21078 Dijon Cedex France
{joe.tekli, richard.chbeir, kokou.yetongnon}@u-bourgogne.fr

Abstract. The automatic processing and management of XML-based data are ever more popular research issues due to the increasing abundant use of XML, especially on the Web. Nonetheless, several operations based on the structure of XML data have not yet received strong attention. Among these is the process of matching XML documents with XML grammars, useful in various applications such as documents classification, retrieval and selective dissemination of information. In this paper, we propose an algorithm for measuring the structural similarity between an XML document and a Document Type Definition (DTD) considered as the simplest way for specifying structural constraints on XML documents. We consider the various DTD operators that designate constraints on the existence, repeatability and alternativeness of XML elements/attributes. Our approach is based on the concept of tree edit distance, as an effective and efficient means for comparing tree structures, XML documents and DTDs being modeled as ordered labeled trees. It is of polynomial complexity, in comparison with existing exponential algorithms. Classification experiments, conducted on large sets of real and synthetic XML documents, underline our approach effectiveness, as well as its applicability to large XML repositories and databases.

Keywords: Semi-structured XML-based data, XML grammar, DTD, structural similarity, tree edit distance.

1 Introduction

Computing similarity between the ever-increasing number of XML documents is becoming essential in several scenarios such as in change management and data warehousing [7], [8], [9], XML query systems (finding and ranking results according to their similarity) [28], [36] as well as the clustering of similar XML documents gathered from the web [10], [24]. Similarly, estimating similarity between XML grammars is useful for data integration purposes, in particular the integration of Document Type Definitions/schemas that contain nearly or exactly the same information but are constructed using different structures [12], [13], [21], [22].

On the other hand, evaluating similarity between documents and grammars can also be beneficial in various applications, e.g. for classifying XML documents against

B. Benatallah et al. (Eds.): WISE 2007, LNCS 4831, pp. 196–211, 2007.

a set of DTDs/schemas declared in an XML database (just as DB schemas are necessary in traditional DBMS for the provision of efficient storage, retrieval, protection and indexing facilities, the same is true for DTDs and XML repositories) [3], [4], [24], XML document retrieval via structural queries [4], [15] (a structural query being represented as a DTD in which some additional constraints on data content could be posed), as well as in the selective dissemination of XML documents (user profiles being expressed as DTDs against which the incoming XML data stream is matched. Hence, an XML document would be distributed to the users whose profiles – DTDs – are similar enough to the document) [3], [4].

In this study, we focus on the *document/grammar* comparison issue, introducing a novel method for evaluating the structural similarity between an XML document and a DTD. Different from previous heuristic approaches, e.g. [3], [4], our method makes use of the well know techniques for finding the edit distance between tree structures, XML documents and DTDs being modeled as ordered labeled trees (OLTs). The contributions of the paper can be summarized as follows: i) introducing a tree representation model, that copes with the expressive power of XML DTD grammars, ii) introducing a polynomial tree edit distance algorithm for evaluating the structural similarity between an XML document and a DTD, iii) developing a prototype to evaluate and validate our approach.

The rest of the paper is organized as follows. Section 2 reviews background in XML/grammar similarity approaches and related problems, such as approximate pattern matching and tree edit distance. Section 3 presents preliminary definitions and provides the tree representations of XML documents and DTDs adopted in this study. In Section 4, we develop our XML/DTD structural similarity approach. Section 5 presents our experimental tests. Section 6 concludes the paper and outlines some future research directions.

2 Related Work

In the following, we briefly review the problem of approximate pattern matching with *variable length don't cares (VLDC)*, which is seemingly comparable to the issue of XML/DTD matching. Then, we present a glimpse on the state of the art concerning tree edit distance algorithms, widely employed to compare semi-structured data, XML documents in particular. Subsequently, we go over existing XML/Grammar similarity computing algorithms.

2.1 Approximate Pattern Matching with Variable Length Don't Cares

An intuitive XML/DTD comparison approach could be that of approximate matching with the presence of *Variable Length Don't Cares (VLDC)*. In strings, a *VLDC* symbol (e.g. $\wedge$) in a given pattern may substitute for zero or more symbols in the data string [2] [17]. Approximate *VLDC* string matching means that after the best possible substitutions, the pattern still does not match the data string and thus a matching distance is computed. For example, "*comp $\wedge$ ng*" matches "*commuting*" with distance 1 (representing the cost of removing the "*p*" form "*comp $\wedge$ ng*" and having the "$\wedge$" substitute for "*mmuti*"). The string *VLDC* problem has been generalized for trees [35], introducing *VLDC* substitutions for whole paths or sub-trees.

Nonetheless, despite being comparable, it is clear that the repeatability and alternativeness operators in DTDs are fairly different from the *VLDC* symbols. *VLDC* symbols can replace any string (w.r.t. string matching) or sub-tree (w.r.t. tree matching) whereas the DTD operators specify constraints on the occurrence of a particular and well known node (and consequently the sub-tree rooted at that node). For example, in DTD of Figure 3.a, the operator '?' associated with element *length* (*length?*) designates that the specific node entitled *length* (and not any other node) can appear *0* or *1* time. Likewise for all DTD constraint operators.

2.2 Tree Edit Distance

In the past few years, tree edit distance algorithms, derived from edit distance between strings [20] [31] [32], have been extensively studied and exploited to compare semi-structured data, e.g. XML documents. Various methods for determining the structural similarities between Ordered Labeled Trees (OLTs) have been proposed. In essence, all these approaches aim at finding the cheapest sequence of edit operations that can transform one tree into another. Early approaches [29] [34] allow insertion, deletion and relabeling of nodes anywhere in the tree. However, they are relatively complex. For instance, the approach in [29] has a time complexity $O(|A||B| depth(A) depth(B))$ where $|A|$ and $|B|$ denote tree cardinalities while $depth(A)$ and $depth(B)$ are the depths of the trees. In [7] [9], the authors restrict insertion and deletion operations to leaf nodes and add a move operator that can relocate a sub-tree, as a single edit operation, from one parent to another. However, corresponding algorithms do not guaranty optimal results. Recent work by Chawathe [8] restricts insertion and deletion operations to leaf nodes, and allows the relabeling of nodes anywhere in the tree, while disregarding the move operation. The overall complexity of [8]'s algorithm is of $O(N^2)$. Nierman and Jagadish [24] extend the approach in [8] by adding two new operations: insert tree and delete tree to allow insertion and deletion of whole sub-trees within in an OLT. Experimental results provided in [24] show that their algorithm outperforms that of Chawathe [8], as well as [29]'s algorithm. The approach in [24] is of $O(N^2)$ complexity. A specialized version of [24]'s is provided in [10], where tree insertion/deletion costs are computed as the sum of the costs of inserting/deleting all individual nodes in the considered sub-trees (in comparison with [24]'s, where certain sub-tree similarities are considered while assigning tree operations costs).

2.3 Similarity Between XML Documents and XML Grammars

Very few approaches have been developed to measure the structural similarity between XML documents and grammars. To our knowledge, the only methods are provided by Grahne and Thomo [15] and Bertino *et al.* [3] [4]. In [15], the authors address the problem of determining whether semi-structured data conform to a given data-guide, in the context of approximate querying. Note that this approach is not developed for XML documents and DTDs/schemas but for generic semi-structured data and data-guides (i.e. there are no constraints on the repeatability and alternativeness of elements). In addition, no experimental results assessing the performance of the approach were reported. Another approach provided by Bertino *et al.* [4] was dedicated to measuring the structural similarity between an XML document and a DTD. The proposed algorithm considers the level (i.e. depth) in which the elements occur in the hierarchical structure of the XML and DTD tree representations (in [3] [4], DTDs are modeled as labeled trees, similarly to XML documents) as well as element complexity (i.e. the cardinality of the sub-tree rooted at the element) when computing similarity values. The authors state that their approach is of exponential complexity but show that it becomes polynomial when the following assumption holds: *in the declaration of an element, two sub-elements with the same tag are forbidden.* In short, although the approach in [4] is proven effective in various situations [3], it is heuristic and time consuming.

3 Preliminaries

In this study, we focus on the XML/DTD comparison problem while aiming to achieve two specific targets: i) First of all, we intend to develop a fine-grained and precise method to compare XML documents and DTDs, w.r.t. the generic approach in [15] (disregarding DTD constraints) and heuristic method in [3], [4]; ii) Second, we aim to obtain an algorithm able to compare XML documents and DTDs in polynomial time, in comparison with the exponential method in [3], [4].

Making use of tree edit distance would help provide a structural comparison method that is more precise, more natural in the XML context, and that could be executed in polynomial time when comparing XML documents and DTDs, edit distance based methods being generally of quadratic $(O(N^2))$ or quadric $(O(N^4))$ complexity. In addition, a prominent advantage of using the edit distance as a similarity measure is that along with the distance value, a mapping between the nodes in the compared trees is provided in terms of the edit script (cf. Definition 4). The mapping can be visualized and can serve as an explanation of the similarity measure.

Several algorithms [5] have been developed to compute an edit distance, as the sum of a sequence of elemental edit operations (mainly node insertion, deletion and update) that can transform one tree structure into another (cf. Section 2). However, with DTD tree structures, constraints on the existence, repeatability and alternativeness of nodes come to play (mainly via the ?, *, +, *Or* operators). As a result, the problem of comparing an XML document to a DTD comes down to computing the edit distance between two trees, taking into account the various DTD constraint operators. In the following, we present some definitions and basic notions prior to developing our XML/DTD structural similarity approach.

3.1 Basic Definitions

Definition 1 - Ordered Labeled Tree: It is a rooted tree in which the nodes are ordered and labeled. We denote by $R(T)$ the root node of tree T.

Definition 2 - Constraint operators: These are operators utilized in DTDs to specify constraints on the existence, repeatability and alternativeness of elements/attributes. With constraint operators, it is possible to specify whether an element is optional ('?'), an element may occur several times ('*' for 0 or more times and '+' for 1 or more times), or some sub-elements are alternative w.r.t. each other ('|' representing the *Or* operator) or are grouped in a sequence (',' representing the *And* operator). It is also possible to specify whether an attribute is optional (*"Implied"*) or mandatory (*"Required"*). An element/attribute with no constraints (*null*) is mandatory and should appear exactly once.

Note that in this study, similarly to that of Bertino *et al.* in [3] [4], we do not associate constraint operators to expressions (e.g., *(a, b)**, *(a | c)*+, ...) and focus on the case where operators are applied to elements only. This issue will be considered in upcoming studies.

Definition 3 - Lc-pair representation of a node: We define the lc-pair representation of a node as the pair *(l, c)* where: *l* is the node's label and *c* the associated constraint operator. We use *p.l* and *p.c* to refer to the label and the constraint operator of an *lc-pair* node *p* respectively.

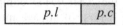

Fig. 1. Lc-pair graphical representation of an XML node *p*

On the other hand, our edit distance XML structural similarity approach utilizes five edit operations, adopted from [8], [24]: *leaf node insertion, leaf node deletion* and *node update*, as well as *tree insertion* and *tree deletion*. Formal definitions are here omitted due to lack of space.

Definition 4 - Edit script: It is a sequence of edit operations $op_1, op_2, ..., op_k$. When applied to a tree *T*, the resulting tree *T'* is obtained by applying edit operations of the *Edit Script (ES)* to *T*, following their order of appearance in the script. By associating costs with each edit operation, $Cost_{Op}$, the cost of an *ES* is defined as the sum of the costs of its component operations: $Cost_{ES} = \sum_{i=1}^{|ES|} Cost_{Op_i}$.

Definition 5 - Edit distance: It is defined as the minimum cost edit script that transforms the tree *A* to the tree *B*: $Dist(A, B) = Min\{Cost_{ES}\}$.

Consequently, the problem of comparing two trees *A* and *B*, i.e. evaluating the structural similarity between *A* and *B*, is defined as computing the corresponding tree edit distance [34].

3.2 Tree Structures

3.2.1 Tree Representation of XML Documents

An XML document is represented as a rooted ordered labeled tree (OLT) [33], where nodes stand for XML elements and are labeled with corresponding element tag names. Attributes appear as children of their encompassing elements, sorted by attribute name and appearing before all sub-element siblings. Similarly to [14], [24], element/attribute values are ignored since we are only interested in the structure of the document (cf. Figure 2).

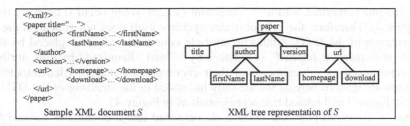

Fig. 2. Tree representation of an XML document

3.2.2 Tree Representation of DTDs

As for XML documents, DTD can also be represented as a single tree or as a series of trees, depending on element constraints. In our approach, we represent a DTD lacking the '|' constraint operator (i.e. *Or*) as a single rooted ordered labeled tree (OLT), so as to attain a structure similar to that of XML document trees. The ',' operator (i.e. *And*) is mapped to its parent node in the document tree. Nodes grouped via the ',' operator are represented as the roots of the paths that are conjunctively connected. Constraint operators '?' and '*' are denoted via the *Lc-pair* representation of corresponding nodes (cf. Definition 3).

```<!DOCTYPE paper[1```
```<!ELEMENT paper ((author+ | publisher), version, length?, url*)>```
```<!ATTLIST paper title CDATA #REQUIRED```
```   category CDATA #IMPLIED]>```
```<!ELEMENT author (firstName, middleName?, lastName)   5```
``<!ELEMENT publisher (#PCDATA
	(firstName,```
	``` middleName?, lastName)>```
	```<!ELEMENT length (#PCDATA)>```
	```<!ELEMENT version (#PCDATA)>```
	```<!ELEMENT url (homepage, download+)> 10```
	```<!ELEMENT homepage (#PCDATA)>```
	```<!ELEMENT download (#PCDATA)>```
	```<!ELEMENT firstName (#PCDATA)>```
	```<!ELEMENT middleName (#PCDATA)>```
	```<!ELEMENT lastName (#PCDATA)>]> 15```
		```<!DOCTYPE paper[1```
		```<!ELEMENT paper (author+, version, length?, url*)>```
		```<!ATTLIST paper title CDATA #REQUIRED```
		``` category CDATA #IMPLIED >```
		```<!ELEMENT author (firstName, middleName?, lastName) 5```
		```<!ELEMENT version (#PCDATA)>```
		```<!ELEMENT length (#PCDATA)>```
		```<!ELEMENT url (download+, homepage)>```
		```<!ELEMENT download (#PCDATA)>```
		```<!ELEMENT homepage (#PCDATA)> 10```
		```<!ELEMENT firstName (#PCDATA)>```
		```<!ELEMENT middleName (#PCDATA)>```
		```<!ELEMENT lastName (#PCDATA)]>```

a. Sample DTD *D* with two *Or* ('|') operators	**b.** First conjunctive DTD corresponding to *D*, *D₁*

```<!DOCTYPE paper[1```
```<!ELEMENT paper (publisher, version, length?, url*)>```
```<!ATTLIST paper title CDATA #REQUIRED```
```   category CDATA #IMPLIED >```
```<!ELEMENT publisher (#PCDATA)>   5```
```<!ELEMENT version (#PCDATA)>```
```<!ELEMENT length (#PCDATA)>```
```<!ELEMENT url (homepage, download+)>```
```<!ELEMENT homepage (#PCDATA)>```
```<!ELEMENT download (#PCDATA)>   ]>   10```
	```<!DOCTYPE paper[1```
	```<!ELEMENT paper (publisher, version, length?, url*)>```
	```<!ATTLIST paper title CDATA #REQUIRED```
	``` category CDATA #IMPLIED >```
	```<!ELEMENT publisher (firstName, middleName?, lastName)> 5```
	```<!ELEMENT version (#PCDATA)>```
	```<!ELEMENT length (#PCDATA)>```
	```<!ELEMENT url (download+, homepage)>```
	```<!ELEMENT homepage (#PCDATA)>```
	```<!ELEMENT download (#PCDATA)> 10```
	```<!ELEMENT firstName (#PCDATA)>```
	```<!ELEMENT middleName (#PCDATA)>```
	```<!ELEMENT lastName (#PCDATA)>]>```

c. Second conjunctive DTD *D₂* corresponding to *D*	**d.** Third conjunctive DTD *D₃* corresponding to *D*

Fig. 3. A sample DTD including *Or* operators and its *original* and *disjunctive normal* forms

An element assigned a '+' constraint operator (i.e. which would occur 1 or more times in the XML documents conforming to the DTD at hand) is represented as two consecutive nodes: one is mandatory (constraint = *null*, which would occur exactly once) and one is marked with a '*' constraint (which would occur 0 or many times, cf. Figure 4). Therefore, the '+' constraint operator itself does not appear in the tree representation of the DTD. Similarly, attribute constraints can be replaced by those applied on elements (*"Implied"* is replaced by '?' and *"Required"* by *null*, attributes being represented as sub-elements of their encompassing nodes). Element/attribute data types are ignored here as we are only interested in the structure of the DTD (cf. DTDs in Figure 3 and related tree representations in Figure 4).

When '|' comes to play, the DTD is decomposed into a *disjunctive normal form* [28]. The disjunctive normal form of a DTD is a set of conjunctive DTDs. Consider for example the DTD in Figure 3.a; it encompasses two *Or* operators. Thus, it can be represented as three separate conjunctive DTDs (cf. Figures 3.b, 3.c and 3.d) underlining the three possible structural configurations that could be attained via the different combinations of the *Or* operators. Consequently, the resulting conjunctive DTDs, that do not encompass *Or* ('|') operators, would be represented as OLTs, as discussed in the previous paragraph (cf. Figure 3).

Now that XML documents and DTDs are represented as OLTs, the problem of comparing an XML document and a DTD comes down to comparing the corresponding trees.

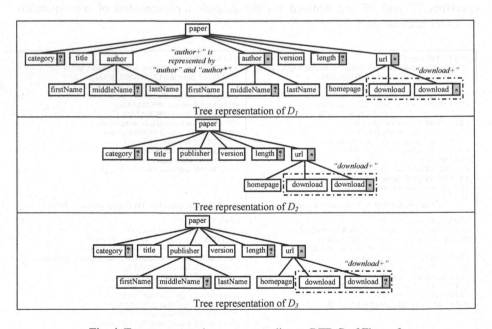

Fig. 4. Tree representations corresponding to DTD *D* of Figure 3

4 Proposal

Our XML/DTD structural similarity approach consists of two phases: i) a pre-processing phase for transforming documents and DTDs into OLTs, ii) and an tree edit distance phase for computing the structural similarity between an XML document and a DTD. The overall algorithm, entitled *XD_Comparison*, is presented in Figure 5.

```
Algorithm XD_Comparison()
Input: DTD D and XML document S
Output: Similarity value between D and S ∈ [0,1]
Begin
    Begin Pre-processing phase                                                      1
        DTreeSet = DTD_to_Tree(D)      // Generating the set of trees representing D
        STree = XML_to_Tree(S)         // Generating the tree representation of S
    End Pre-processing phase                                                        5
    Begin Edit distance phase                                                       7
        Dist[] = new [1... |DTreeSet|]    // |DTreeSet| designates the number of trees
                                          // representing D (cf. Section 3.2)       10
        For (i=1 ; i ≤ |DTreeSet| ; i++)
        {
            Dist[i] = XD_EditDistance(DTreeSet[i] , STree)    // Edit distance algorithm
        }
        Return _____1_____            // Structural similarity value between D and S
               1 + min(Dist[i])
    End Edit distance phase                                                        15
End
```

Fig. 5. The overall algorithm of our approach

After transforming the DTD and the XML document into their tree representations (Figure 5, lines 1-5), the edit distance between each DTD tree (recall that multiple trees are obtained when the *Or* operator is utilized, otherwise, the DTD is represented by one single tree, cf. Section 3.2) and the XML document tree is computed (Figure 5, lines 7-13). Subsequently, the minimum distance attainable is taken into account when computing the overall similarity value (Figure 5, line 14). Note that similarity measures based on edit distance are generally computed as *1/(1+Dist)* and vary in the *[0, 1]* interval. As a result, the maximum similarity *(=1)* in our case indicates that the concerned XML document is valid for the DTD to which it is compared.

While DTD/XML tree representations are central in our approach, the corresponding construction process is undertaken in average linear time. It is achieved by performing a pre-order traversal of the XML element/attributes in the file at hand, adding the corresponding tree nodes respectively. As for DTDs in particular, an additional tree instance is created every time an '|' operator (*Or*) is encountered, the ',' operators (*And*) being mapped to their parent nodes. In addition, DTD tree nodes would encompass '?' and '*' constraint operators, the '+' operator being treated as a special case of '*' (detailed algorithms are disregarded here).

4.1 The Algorithm

Our edit distance algorithm for comparing a DTD tree and an XML document tree is given in Figure 6. It is based on a dynamic programming formulation, which is frequently used to solve minimum edit distance problems, and employs the edit operations mentioned in Section 3.1 (i.e. node insertion/deletion/update, as well as tree insertion/deletion). When the DTD tree is free of constraint operators (i.e. when all elements are mandatory), the algorithm comes down to a classical approximate tree matching process [24].

Note that when employing dynamic programming to compute the distance between a source tree (in our case the DTD tree) and a destination tree (in our case the XML document tree), a key issue is to first determine the cost of inserting every sub-tree of the destination tree and that of deleting every sub-tree of the source tree. An intuitive way of determining the cost of inserting (deleting) a sub-tree usually consists of computing the sum of the costs of inserting (deleting) all individual nodes in the considered sub-tree [10]. As for single node operations, unit costs are naturally utilized ($Cost_{Ins} = Cost_{Del} = 1$. $Cost_{Upd}(a, b) = 1$ when $a.l \neq b.l$, otherwise, $Cost_{Upd} = 0$, underlining that no changes are to be made to the label of node a). In this study, we restrict our presentation to the basic cost schemes above. Note that the investigation of alternative operations cost models, and how they affect our tree edit distance measure, is out of the scope of this paper. Such issues will be addressed in future studies.

Comparing an XML document and a DTD, in our approach, amounts to computing the edit distance between the corresponding trees, taking into account the '?', '*' and '+' DTD constraint operators. Recall that the 'l' and ',' operators are handled via our DTD tree model (cf. Section 3.2).

In order to consider, in our edit distance computations, that a DTD element with constraint operator '?' can only appear 0 or 1 time in the XML trees conforming to the DTD, a null cost is induced when the corresponding element node (as well as the sub-tree rooted at that node) is to be deleted (cf. Figure 6, lines 9-11 and 22-29). In other words, when an element assigned the '?' operator is to be deleted, the algorithm identifies the fact that its presence is optional and does not consider the corresponding deletion operation in the edit distance process.

In order to consider that a DTD element with the '*' constraint can appear 0 or many times in the XML trees conforming to the DTD, we proceed as follows:

- A null cost is induced when the corresponding element node (as well as the sub-tree rooted at that node) is to be deleted (cf. Figure 6, lines 9-11, 30-37), considering the fact that its presence is optional.
- When combining *insert tree*, *delete tree* and the recursive edit distance computations costs in order to determine the minimum cost script to be executed (cf. Figure 6, lines 30-37), the algorithm propagates the minimum distance value along with the recursive process (line 33). This allows to take into account the repeatability of the element that is assigned the '*' constraint without increasing the overall distance value.

As for the '+' operator, it is treated via the same processes dedicated to the '*' constraint operator. Since the '+' operator designates that the associated element can occur 1 or multiple times in the XML document, the designated element could be represented as two consecutive nodes: one that is mandatory (which will certainly appear in the document) and one marked with a '*' constraint (which would appear 0 or multiple times). Thus, as stated previously, the '+' constraint itself is disregarded in the tree representation of the DTD.

```
Algorithm XD_EditDistance()
Input: DTD tree D and XML tree S
Output: Edit distance between D and S
Begin                                                                              1
    M = Degree(D)                        // The number of first level sub-trees in A.
    N = Degree(S)                        // The number of first level sub-trees in B.
    Dist [][] = new [0...M][0...N]
    Dist[0][0] = Cost_Upd(R(D).l, R(S).l)    // R(D).l and R(S).l are the labels of the roots   5
                                             // of trees D and S

    For (i = 1 ; i ≤ M ; i++)
    {
        If (R(D_i).c = "?" or R(D_i).c = "*")    // R(D_i) is the root of 1^st level sub-tree D_i in D
        {                                                                          10
            Dist[i][0] = Dist[i-1][0]     // no cost is added when ? and * are encountered
        Else
            Dist[i][0] = Dist[i-1][0] + Cost_DelTree(D_i)    // Traditional tree edit distance.
        }
    }                                                                              15
    For (j = 1 ; j ≤ N ; j++) { Dist[0][j] = Dist[0][j-1] + Cost_InsTree(S_j) }
    For (i = 1 ; i ≤ M ; i++)
    {
        For (j = 1 ; j ≤ N ; j++)                                                  20
        {
            If (R(D_i).c = "?")                  // The ? operator is encountered
            {
                Dist[i][j] = min {
                    Dist[i-1][j-1] + XD_EditDistance(D_i, S_j),    // Dynamic programming.  25
                    Dist[i][j-1] + Cost_InsTree(S_j),              // Simplified tree
                    Dist[i-1][j]                                   //operation syntax.
                }
            }
            Else If (R(D_i).c = "*")             // The * operator is encountered           30
            {
                Dist[i][j] = min{
                    min{ Dist[i-1][j-1], Dist[i][j-1] } + XD_EditDistance(D_i, S_j),
                    Dist[i][j-1] + Cost_InsTree(S_j),
                    Dist[i-1,j]                                                     35
                }
            }
            Else                    // no constraint operators (mandatory element/sub-tree)
            {
                Dist[i][j] = min{                        // Traditional tree edit distance.  40
                    Dist[i-1][j-1] + XD_EditDistance(D_i, S_j),    // Dynamic programming.
                    Dist[i][j-1] + Cost_InsTree(S_j),             // Simplified tree
                    Dist[i-1][j] + Cost_DelTree(D_i),             // operations syntaxes.
                }
            }                                                                      45
        }
    }
    Return Dist[M][N]
End
```

Fig. 6. Edit distance algorithm for comparing a DTD tree and an XML document tree

4.2 Computation Example

We present the result of comparing sample XML document S in Figure 2 to the complex DTD D in Figure 3 in which the 'l' operator is utilized. Following our approach, comparing XML document S to DTD D comes down to comparing S to trees D_1, D_2 and D_3 which represent the *disjunctive normal form* of DTD D

(D encompassing two *Or* operators). Distance values are reported below (detailed computations are omitted due to space limitations):

- Dist(D_1, S) = 0. *S* is valid for the DTD tree D_1
- Dist(D_2, S) = 3. Node *publisher* in D_2 is to be replaced by *author* in *S*. In addition, nodes *firstName* and *lastName* in *S* are not defined in D_2
- Dist(D_3, S) = 1. Node *publisher* in D_3 is to be replaced by *author* in *S*.

As stated previously, when a DTD corresponds to multiple conjunctive DTD trees, the minimum edit distance between those trees and the XML tree is retained. Consequently, the structural similarity between the DTD *D* and XML document *S* is computed as follows:

$$Sim(D, S) = \frac{1}{1 + min(Dist(D, S))} = \frac{1}{1 + min(Dist(D_1, S))} = 1 \quad \text{(maximum similarity value)}$$

which indicates that *S* conforms to *D*.

4.3 Overall Complexity

The computational complexity of our XML/DTD structural similarity approach simplifies to a polynomial formulation: $O(N^3)$. Let $|D|$ be the number of nodes (elements/attributes) in the DTD tree *D* considered (i.e. cardinality/size of *D*), $|D'|$ the cardinality of the largest conjunctive DTD corresponding to *D*, N_D the maximum number of conjunctive DTDs corresponding to *D*, and $|S|$ the cardinality of the XML tree *S* considered. The algorithm *XD_Comparison* (cf. Figure 5) is of complexity $O(|D|+|S| + (N_D|D'||S|)$:

- The pre-processing phase of *XD_Comparison* (cf. Figure 5, lines 2-5) is of average linear complexity and simplifies to $O(|D|+|S|)$.
- *XD_EditDistance*, which is of $O(|D'||S|)$ complexity, is executed for each conjunctive form of the DTD *D* (cf. Figure 5, lines 10-13). Thus, the complexity of the edit distance phase comes down to $O(N_D|D'||S|)$.

Note that the maximum number of *conjunc*tive DTDs, making up the *disjunctive normal form* of a generic DTD *D*, is $|D|-1$. This is obtained when the DTD has $(|D|-2)\times Or$ operators which is the maximum number of operators that can occur in an XML definition. Therefore, in the worst case scenario, $O(|D|+|S|+N_D|D'||S|) = O(|D|+|S|+(|D|-1)|D'||S|)$ which simplifies to $O(N^3)$, where *N* is the maximum number of nodes in the DTD/XML trees being compared.

5 Experimental Evaluation

5.1 Evaluation Metrics

In order to validate our XML/DTD structural similarity approach, we make use of structural classification. In addition, we adapt to XML classification the *precision (PR)*, *recall (R)*, *noise (N)* and *silence (S)* measures, widely utilized in information retrieval evaluations, and propose a novel method for their usage in order to attain consistent experimental results (in the same spirit as Dalamagas *et al.*'s approach [10] w.r.t. XML document clustering). For an extracted class C_i that corresponds to a given DTD_i:

- a_i is the number of XML documents in C_i that correspond to DTD_i (correctly classified documents, i.e. documents that conform to DTD_i).
- b_i is the number of documents in C_i that do not correspond to DTD_i (mis-classified).
- c_i is the number of XML documents not in C_i, although they correspond to DTD_i (documents that conform to DTD_i and that should have been classified in C_i).

Therefore, $PR = \dfrac{\sum_{i=1}^{n} a_i}{\sum_{i=1}^{n} a_i + \sum_{i=1}^{n} b_i}$, $R = \dfrac{\sum_{i=1}^{n} a_i}{\sum_{i=1}^{n} a_i + \sum_{i=1}^{n} c_i}$, $N = \dfrac{\sum_{i=1}^{n} b_i}{\sum_{i=1}^{n} a_i + \sum_{i=1}^{n} b_i}$ and $S = \dfrac{\sum_{i=1}^{n} c_i}{\sum_{i=1}^{n} a_i + \sum_{i=1}^{n} c_i}$

where n is the total number of classes, which corresponds to the total number of DTDs considered for the classification task ($PR, R, N, S \in [0, 1]$ interval).

Thus, as with traditional information retrieval evaluation, high *precision/recall* and low *noise/silence* (indicating in our case high classification quality) characterize a good similarity method (structure-based XML/DTD similarity).

5.2 Classifying XML Documents

In our experiments, we undertook a series of multilevel classification tasks, varying the classification threshold in the *[0, 1]* interval:

- For the starting classification threshold ($s_1=0$), all XML documents appear in all classes, the starting classes.
- For the final classification threshold ($s_n=1$), with n the total number of classification levels (i.e. number of classification sets identified in the process), each class will only contain XML documents which actually conform to the DTD identifying the class.
- Intermediate classification sets will be identified for levels s_i / $s_1 < s_i < s_n$.

Then, we compute *precision, recall, noise* and *silence* values for each of the classification sets identified in the multilevel classification phases, thus constructing *PR, R, N* and *S* graphs that describe the system's evolution throughout the classification process.

5.3 Results

We conducted experiments on real and synthetic XML documents. Two sets of 1000 documents were generated from 20 real-case[1] and synthetic DTDs, using an adaptation of the IBM XML documents generator[2]. We varied the *MaxRepeats* parameter to determine the number of times a node will appear as a child of its parent node. For a real dataset, we considered the online version of the ACM SIGMOD Record[3]. *Precision, recall, noise and silence* graphs are depicted in Figure 7.

---

[1] *http://www.xmlfiles.com* and *http://www.w3schools.com*
[2] *http://www.alphaworks.ibm.com*
[3] *http://www.acm.org/sigmod/xml*

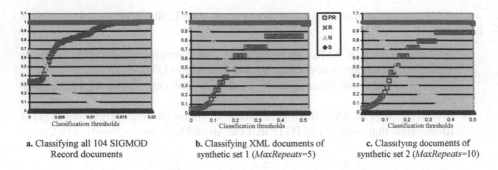

a. Classifying all 104 SIGMOD Record documents

b. Classifying XML documents of synthetic set 1 (*MaxRepeats*=5)

c. Classifying documents of synthetic set 2 (*MaxRepeats*=10)

Fig. 7. PR, R, N and S graphs for classifying real and synthetic XML documents

One can clearly realize that *recall* (*R*) and *silence* (*S*) are always equal to *1* and *0* respectively. These values reflect the fact that our XML/DTD comparison algorithm constantly identifies in the DTD classes, regardless of the classification threshold, the XML documents that actually conform to the DTDs considered (i.e. documents and DTDs having *Sim(DTD, XML) =1*). On the other hand, *precision* (*PR*) gradually increases towards *1*, whereas *noise* (*N*) decreases to *0*, while varying the classification threshold from *0* to *1*:

- When the classification threshold is equal to 0, all documents in the XML repository are considered in each and every one of the classes corresponding to the DTDs at hand. That is underlined by minimum *precision* and maximum *noise* values.
- Then, as the classification threshold increases, inconsistent documents are gradually filtered from the DTD classes, ultimately yielding classes that only encompass documents conforming to the considered DTDs.

PR and *N* results in Figure 7 show that our algorithm yields optimal classification quality, attaining optimal classes at a very early stage of the multilevel classification process (with classification thresholds < *0.5*). Note that we do not experimentally compare our method to that of Bertino *et al* [4] since the authors do not provide the detailed algorithms of their approach.

5.4 Timing Analysis

We have shown, in Section 4.3, that the complexity of our XML/DTD structural comparison algorithm is polynomial following the size of the XML/DTD trees being compared, $O(|D|+|S|+|+(|D|-1)|D||S|)$ (which simplifies to $O(N^3)$, *N* being the maximum number of nodes in the XML/DTD trees *S* and *D*). This polynomial dependency on the size of each tree is experimentally verified, timing results being presented in Figure 8. The timing experiments were carried out on a PC with a Xeon 2.7 GHz processor (1GB RAM). Figure 8.a shows that the time to identify the structural similarity between XML trees and conjunctive DTD trees of various sizes grows in a linear fashion with tree size (underlining the complexity of *XD_EditDistance*, $O(|D'||S|)$). On the other hand, Figure 8.b shows that the time to

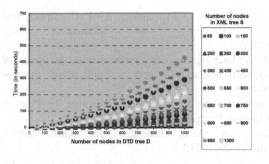
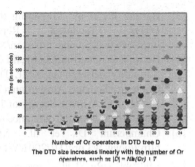

a. Timing results to compute pair-wise distances for XML/DTD trees (conjunctive DTDs) of various sizes.

b. Varying the number of *Or* operators (i.e. the number of conjunctive DTDs) in the DTD.

Fig. 8. Timing Results

execute our *XD_Comparison* algorithm grows in a polynomial fashion with DTD tree size when varying the number of *Or* operators (i.e. the number of conjunctive DTDs) in the DTD.

6 Conclusion

In this paper, we proposed a structure based similarity approach for comparing XML documents and DTDs. Based on the tree edit distance concept, our approach takes into account the various DTD constraints on the existence, repeatability and alternativeness of XML elements/attributes. Our theoretical study as well as our experimental evaluation showed that our approach is of polynomial complexity, in comparison with existing exponential algorithms, and that it yields optimal comparison results.

As continuing work, we are exploring the use of our approach in order to address, not only the structural aspect of XML documents (element/attribute labels), but also their information content (element/attribute values). By adding additional constraints on the data content of elements/attributes, DTDs could be exploited as *content-and-structure* queries, taking into account the structure of XML data in the search process, and returning ranked answers in the spirit of information retrieval (IR). Note that combining database (DB) structural "binary answer" XML search (e.g. XML-QL [11], XQuery [6]) and information retrieval query result ranking, is one of the most recent and promising trends in both DB and IR research. On the other hand, we plan to study XML/DTD comparison with DTDs including repeatable expressions (expressions assigned '?', '*' and '+' constraint operators) as well as recursive definitions. In this context, it would be interesting to extend the tree model to a more general graph model for DTDs, and adapt (if possible) our tree edit distance algorithm accordingly.

References

[1] Aho, A., Hirschberg, D., Ullman, J.: Bounds on the Complexity of the Longest Common Subsequence Problem. ACM J. 23(1), 1–12 (1976)
[2] Akatsu, T.: Approximate String Matching with Don't Care Characters. Information Processing Letters 55, 235–239 (1995)

[3]   Bertino E., et al.: Measuring the Structural Similarity among XML Documents and DTDs, Technical Report, University of Genova (2002), http://www.disi.unige.it/person/MesitiM

[4]   Bertino, E., Guerrini, G., Mesiti, M.: A Matching Algorithm for Measuring the Structural Similarity between an XML Documents and a DTD and its Applications. Elsevier Computer Science, Amsterdam (2004)

[5]   Buttler, D.: A Short Survey of Document Structure Similarity Algorithms. In: Proc. of the 5th International Conference on Internet Computing, pp. 3–9 (2004)

[6]   Chamberlin, D., et al.: XQuery: A Query Language for XML (2001), http://www.w3.org/TR/2001/WD-xquery-20010215

[7]   Chawathe, S., et al.: Change Detection in Hierarchically Structured Information. In: SIGMOD, pp. 493–504 (1996)

[8]   Chawathe, S.: Comparing Hierarchical Data in External Memory. In: Proceedings of the 25th Int. Conf. on Very Large Data Bases, pp. 90–101 (1999)

[9]   Cobéna, G., et al.: Detecting Changes in XML Documents. In: Proc. of the 18th ICDE Conference, pp. 41–52 (2002)

[10] Dalamagas, T., Cheng, T., Winkel, K., Sellis, T.: A methodology for clustering XML documents by structure. Inf. Syst. 31(3), 187–228 (2006)

[11] Deutsch, A., et al.: XML-QL: A Query Language for XML. Computer Networks 31(11-16), 1155–1169 (1999)

[12] Do, H.H., Rahm, E.: COMA: A System for Flexible Combination of Schema Matching Approaches. In: The 28th VLDB Conf., Honk Kong, pp. 610–621 (2002)

[13] Doan, A., Domingos, P., Halevy, A.Y.: Reconciling Schemas of Disparate Data Sources: A Machine Learning Approach. In: ACM SIGMOD, pp. 509–520 (2001)

[14] Flesca, S., et al.: Detecting Structural Similarities between XML Documents. In: WebDB, pp. 55–60 (2002)

[15] Goldman, R., Windom, J.: Dataguides: Enabling Query Formulation and Optimization in Semi-structured Databases. In: Proc. of the 23rd VLDB Conference, pp. 436–445 (1997)

[16] Grahne, G., Thomo, A.: Approximate Reasoning in Semi-structured Databases. In: Proc. of the 8th International Workshop on Knowledge Representation meets Databases, vol. 45 (2001), http://CEUR-WS.org/Vol-45/03-thomo.ps

[17] Landau, G.M., Vishkin, U.: Fast Parallel and Serial Approximate String Matching. Journal of Algorithms 10, 157–169 (1989)

[18] Lee, J.H., Kim, M.H., Lee, Y.J.: Information Retrieval Based on Conceptual Distance in IS-A Hierarchies. Journal of Documentation 49(2), 188–207 (1993)

[19] Lee, M., et al.: XClust: Clustering XML Schemas for Effective Integration. In: CIKM, pp. 292–299 (2002)

[20] Levenshtein, V.: Binary Codes Capable of Correcting Deletions, Insertions and Reversals. Sov. Phys. Dokl. 6, 707–710 (1966)

[21] Madhavan, J., et al.: Generic Schema Matching With Cupid. In: 27th VLDB Conference, pp. 49–58 (2001)

[22] Melnik, S., Garcia-Molina, H., Rahm, E.: Similarity Flooding: A Versatile Graph Matching Algorithm and its Application to Schema Matching. In: Proc. of the 18th ICDE Conference, pp. 117–128 (2002)

[23] Myers, E.: An O(N D) Difference Algorithm and Its Variations. Algorithmica 1(2), 251–266 (1986)

[24] Nierman, A., Jagadish, H.V.: Evaluating structural similarity in XML documents. In: Proc. of the 5th Int. Workshop on the Web and Databases, pp. 61–66 (2002)

[25]   Pereira, F.: Technologies for Digital Multimedia Communications: An Evolution Analysis of MPEG Standards. China Communications Journal, 8–19 (2006)
[26]   van Rijsbergen, C.J.: Information Retrieval. Butterworths, London (1979)
[27]   Sanz, I., Mesiti, M., Guerrini, G., Berlanga Lavori, R.: Approximate Subtree Identification in Heterogeneous XML Documents Collections. In: Bressan, S., Ceri, S., Hunt, E., Ives, Z.G., Bellahsène, Z., Rys, M., Unland, R. (eds.) XSym 2005. LNCS, vol. 3671, pp. 192–206. Springer, Heidelberg (2005)
[28]   Schlieder, T.: Similarity Search in XML Data Using Cost-based Query Transformations. In: WebDB, pp. 19–24 (2001)
[29]   Shasha, D., Zhang, K.: Approximate Tree Pattern Matching. In: Pattern Matching in Strings, Trees and Arrays, Oxford University Press, Oxford (1995)
[30]   Tai, K.C.: The Tree-to-Tree correction problem. ACM J. 26, 422–433 (1979)
[31]   Wagner, J., Fisher, M.: The String-to-String correction problem. J. of the ACM (21), 168–173 (1974)
[32]   Wong, C., Chandra, A.: Bounds for the String Editing Problem. Journal of the Association for Computing Machinery 23(1), 13–16 (1976)
[33]   WWW Consortium, The Document Object Model (August 10, 2006), http://www.w3.org/DOM
[34]   Zhang, K., Shasha, D.: Simple Fast Algorithms for the Editing Distance between Trees and Related Problems. SIAM J. 18(6), 1245–1262 (1989)
[35]   Zhang, K., Shasha, D., Wang, J.: Approximate Tree Matching in the Presence of Variable Length Don't Cares. J. of Algorithms 16(1), 33–66 (1994)
[36]   Zhang, Z., Li, R., Cao, S., Zhu, Y.: Similarity Metric for XML Documents. In: Knowledge Management and Experience Management Workshop, Karlsruhe, Germany, pp. 255–261 (2003)

Using Clustering and Edit Distance Techniques for Automatic Web Data Extraction*

Manuel Álvarez, Alberto Pan**, Juan Raposo, Fernando Bellas, and Fidel Cacheda

Department of Information and Communications Technologies
University of A Coruña, Campus de Elviña s/n. 15071. A Coruña, Spain
{mad,apan,jrs,fbellas,fidel}@udc.es

Abstract. Many web sources provide access to an underlying database containing structured data. These data can be usually accessed in HTML form only, which makes it difficult for software programs to obtain them in structured form. Nevertheless, web sources usually encode data records using a consistent template or layout, and the implicit regularities in the template can be used to automatically infer the structure and extract the data. In this paper, we propose a set of novel techniques to address this problem. While several previous works have addressed the same problem, most of them require multiple input pages while our method requires only one. In addition, previous methods make some assumptions about how data records are encoded into web pages, which do not always hold in real websites. Finally, we have tested our techniques with a high number of real web sources and we have found them to be very effective.

1 Introduction

In today's Web, there are many sites providing access to structured data contained in an underlying database. Typically, these sources, known as "semi-structured" web sources, provide some kind of HTML form that allows issuing queries against the database, and they return the query results embedded in HTML pages conforming to a certain fixed template. For instance, Fig. 1 shows a page containing a list of data records, representing the information about books in an Internet shop.

Allowing software programs to access these structured data is useful for a variety of purposes. For instance, it allows data integration applications to access web information in a manner similar to a database. It also allows information gathering applications to store the retrieved information maintaining its structure and, therefore, allowing more sophisticated processing.

Several approaches have been reported in the literature for building and maintaining "wrappers" for semi-structured web sources ([2][9][11][12][13]; [7] provides a brief survey). Although wrappers have been successfully used for many

---

* This research was partially supported by the Spanish Ministry of Education and Science under project TSI2005-07730.
** Alberto Pan's work was partially supported by the "Ramón y Cajal" programme of the Spanish Ministry of Education and Science.

B. Benatallah et al. (Eds.): WISE 2007, LNCS 4831, pp. 212–224, 2007.

web data extraction and automation tasks, this approach has the inherent limitation that the target data sources must be known in advance. This is not possible in all cases. Consider, for instance, the case of "focused crawling" applications [3], which automatically crawl the web looking for topic-specific information.

Several automatic methods for web data extraction have been also proposed in the literature [1][4][5][14], but they present several limitations. First, [1][5] require multiple pages generated using the same template as input. This can be inconvenient because a sufficient number of pages need to be collected. Second, the proposed methods make some assumptions about the pages containing structured data which do not always hold. For instance, [14] assumes the visual space between two data records in a page is always greater than any gap inside a data record (we will provide more detail about these issues in the related work section).

In this paper, we present a new method to automatically detecting a list of structured records in a web page and extract the data values that constitute them. Our method requires only one page containing a list of data records as input. In addition, it can deal with pages that do not verify the assumptions required by other previous approaches. We have also validated our method in a high number of real websites, obtaining very good effectiveness.

The rest of the paper is organized as follows. Section 2 describes some basic observations and properties our approach relies on. Sections 3-5 describe the proposed techniques and constitute the core of the paper. Section 3 describes the method to detect the data region in the page containing the target list of records. Section 4 explains how we segment the data region into data records. Section 5 describes how we extract the values of each individual attribute from the data records. Section 6 describes our experiments with real web pages. Section 7 discusses related work.

Fig. 1. Example HTML page containing a list of data records

2 Basic Observations and Properties

We are interested in detecting and extracting lists of structured data records embedded in HTML pages. We assume the pages containing such lists are generated according to the page creation model described in [1]. This model formally captures the basic observations that the data records in a list are shown contiguously in the page and are formatted in a consistent manner: that is, the occurrences of each attribute in several records are formatted in the same way and they always occur in the same relative position with respect to the remaining attributes. For instance, Fig. 2 shows an excerpt of the HTML code of the page in Fig. 1. As it can be seen, it verifies the aforementioned observations.

HTML pages can also be represented as DOM trees as shown in Fig. 3. The representation as a DOM tree of the pages verifying the above observations has the following properties:

- **Property 1**: Each record in the DOM tree is disposed in a set of consecutive sibling subtrees. Additionally, although it cannot be derived strictly from the above observations, it is heuristically found that a data record comprises a certain number of *complete* subtrees. For instance, in Fig. 3 the first two subtrees form the first record, and the following three subtrees form the second record.
- **Property 2:** The occurrences of each attribute in several records have the same path from the root in the DOM tree. For instance, in Fig. 3 it can be seen how all the instances of the attribute *title* have the same path in the DOM tree, and the same applies to the remaining attributes.

3 Finding the Dominant List of Records in a Page

In this section, we describe how we locate the data region of the page containing the main list of records in the page.

From the property 1 of the previous section, we know finding the data region is equivalent to finding the common parent node of the sibling subtrees forming the data records. The subtree having as root that node will be the target data region. For instance, in our example of Fig. 3 the parent node we should discover is n_1.

```
<html><body>
  <div> ... </div>
  <div> ... </div>
  <div>
    <table> ... </table>
    <table>
      <tr><td><table>
        <tr><td>
          <span><a>Head First Java. 2nd Edition</a></span>
          <br>by <span>Kathy Sierra and Bert Bates</span>
          <br><span>Paperback</span> - Feb 9. 2005</td>
        <td><img></td></tr></table></td></tr>
      <tr><td>Buy new: <span>29.67€</span>
        Price used: <span>20.00€</span></td></tr>
```

```
  <tr><td><table>
    <tr><td>
      <span><a>Java Persistence with Hibernate</a></span>
      <br>by <span>Christian Bauer and Gavin King</span>
      <br><span>Paperback</span> - Nov 24. 2006</td>
      <td><img></td></tr></table></td></tr>
    <tr><td>Buy new: <span>37.79€</span></td></tr>
    <tr><td>Other editions: <span>e-Books & Docs</span>
      </td></tr>
    ...
  </table>
  ...
  </div>
</body></html>
```

Fig. 2. HTML source code for page in Fig. 1

Our method for finding the region containing the dominant list of records in a page p consists of the following steps:

1. Let us consider N, the set composed by all the nodes in the DOM tree of p. To each node $n_i \in N$, we will assign a score called s_i. Initially, $\forall_{i=1..|N|} s_i = 0$.
2. Compute T, the set of all the text nodes in N.
3. Divide T into subsets $p_1, p_2,...,p_m$, in a way such that all the text nodes with the same path from the root in the DOM tree are contained in the same p_i. To compute the paths from the root, we ignore tag attributes.
4. For each pair of text nodes belonging to the same group, compute n_j as their deepest common ancestor in the DOM tree, and add 1 to s_j (the score of n_j).
5. Let n_{max} be the node having a higher score. Choose the DOM subtree having n_{max} as root of the desired data region.

Now, we provide the justification for this algorithm. First, by definition, the target data region contains a list of records and each data record is composed of a series of attributes. By property 2, we know all the occurrences of the same attribute have the same path from the root. Therefore, the subtree containing the dominant list in the page will typically contain more texts with the same path from the root than other regions. In addition, given two text nodes with the same path in the DOM tree, the following situations may occur:

1. By property 1, if the text nodes are occurrences of texts in different records (e.g. two values of the same attribute in different records), then their deepest common ancestor in the DOM tree will be the root node of the data region containing all the records. Therefore, when considering that pair in step 4, the score of the correct node is increased. For instance, in Fig. 3 the deepest common ancestor of d_1 and d_3 is n_1, the root of the subtree containing the whole data region.
2. If the text nodes are occurrences from different attributes in the same record, then in some cases, their deepest common ancestor could be a deeper node than the one we are searching for and the score of an incorrect node would be increased. For instance, in the Fig. 3 the deepest common ancestor of d_1 and d_2 is n_2.

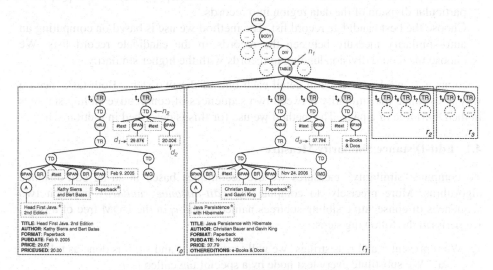

Fig. 3. DOM tree for HTML page in Fig. 1

By property 2, we can infer that there will usually be more occurrences of the case 1 and, therefore, the algorithm will output the right node. Now, we explain the reason for this. Let us consider the pair of text nodes (t_{11}, t_{12}) corresponding with the occurrences of two attributes in a record. (t_{11}, t_{12}) is a pair in the case 2. But, by property 2, for each record r_i in which both attributes appear, we will have pairs (t_{11}, t_{i1}), (t_{11}, t_{i2}), (t_{12}, t_{i1}), (t_{12}, t_{i2}), which are in case 1. Therefore, in the absence of optional fields, it can be easily proved that there will be more pairs in the case 1. When optional fields exist, it is still very probable.

This method tends to find the list in the page with the largest number of records and the largest number of attributes in each record. When the pages we want to extract data from have been obtained by executing a query on a web form, we are typically interested in extracting the data records that constitute the answer to the query, even if it is not the larger list (this may happen if the query has few results). If the executed query is known, this information can be used to refine the above method. The idea is very simple: in the step 2 of the algorithm, instead of using all the text nodes in the DOM tree, we will use only those text nodes containing text values used in the query with operators whose semantic be *equals* or *contains*. For instance, let us assume the page in Fig. 3 was obtained by issuing a query we could write as *(title contains 'java') AND (format equals 'paperback')*. Then, the only text nodes considered in step 2 would be the ones marked with an '*' in Fig. 3.

4   Dividing the List into Records

Now we proceed to describe our techniques for segmenting the data region in fragments, each one containing at most one data record.

Our method can be divided into the following steps:

- Generate a set of candidate record lists. Each candidate record list will propose a particular division of the data region into records.
- Choose the best candidate record list. The method we use is based on computing an auto-similarity measure between the records in the candidate record lists. We choose the record division lending to records with the higher similarity.

Sections 4.2 and 4.3 describe in detail each one of the two steps. Both tasks need a way to estimate the similarity between two sequences of consecutive sibling subtrees in the DOM tree of a page. The method we use for this is described in section 4.1.

4.1   Edit-Distance Similarity Measure

To compute "similarity" measures we use techniques based in string edit-distance algorithms. More precisely, to compute the *edit-distance similarity* between two sequences of consecutive sibling subtrees named r_i and r_j in the DOM tree of a page, we perform the following steps:

1.   We represent r_i and r_j as strings (we will term them s_i and s_j). It is done as follows:
     a.   We substitute every text node by a special tag called *text*.

    b. We traverse each subtree in depth first order and, for each node, we generate a character in the string. A different character will be assigned to each tag having a different path from the root in the DOM tree. Fig. 4 shows the strings s_0 and s_1 obtained for the records r_0 and r_1 in Fig. 3.

2. We compute the *edit-distance similarity* between r_i and r_j, denoted as *es* (r_i, r_j), as the string edit distance between s_i and s_j $(ed(r_i, r_j))$ calculated using a variant of the Levenshtein algorithm [8], which does not allow substitution operations (only insertions and deletions are permitted). To obtain a similarity score between 0 and 1, we normalize the result using the equation (1). In our example from Fig. 4, the similarity between r_0 and r_1 is $1 - (2 / (26 + 28)) = 0.96$.

$$es(r_i, r_j) = 1 - ed(s_i, s_j) / len(s_i) + len(s_j) \qquad (1)$$

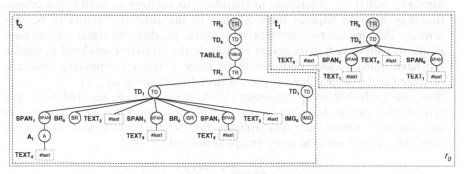

$s_0 \rightarrow t_0$: TR$_0$ TD$_0$ TABLE$_0$ TR$_1$ TD$_1$ SPAN$_1$ A$_1$ TEXT$_4$ BR$_0$ TEXT$_2$ SPAN$_1$ TEXT$_3$ BR$_0$ SPAN$_1$ TEXT$_3$ TEXT$_2$ TD$_1$ IMG$_0$
    t_1: TR$_0$ TD$_0$ TEXT$_0$ SPAN$_0$ TEXT$_1$ TEXT$_0$ SPAN$_0$ TEXT$_1$

$s_1 \rightarrow t_2$: TR$_0$ TD$_0$ TABLE$_0$ TR$_1$ TD$_1$ SPAN$_1$ A$_1$ TEXT$_4$ BR$_0$ TEXT$_2$ SPAN$_1$ TEXT$_3$ BR$_0$ SPAN$_1$ TEXT$_3$ TEXT$_2$ TD$_1$ IMG$_0$
    t_3: TR$_0$ TD$_0$ TEXT$_0$ SPAN$_0$ TEXT$_1$
    t_4: TR$_0$ TD$_0$ TEXT$_0$ SPAN$_0$ TEXT$_1$ $len(s_0) = 26$ $len(s_1) = 28$ $\boxed{ed(r_0, r_1) = ed(s_0, s_1) = 2}$

Fig. 4. Strings obtained for the records r_0 and r_1 in Fig. 3

4.2 Generating the Candidate Record Lists

In this section, we describe how we generate a set of candidate record lists inside the data region previously chosen. Each candidate record list will propose a particular division of the data region into records.

By property 1, every record is composed of one or several consecutive sibling sub-trees, which are direct descendants of the root node of the data region. We could leverage on this property to generate a candidate record list for each possible division of the subtrees verifying it. Nevertheless, the number of possible combinations would be too high: if the number of subtrees is n, the possible number of divisions verifying property 1 is 2^{n-1} (notice that different records in the same list may be composed of a different number of subtrees, as for instance r_0 and r_1 in Fig. 3). In some sources, n can be low, but in others it may reach values in the hundreds (e.g. a source showing 25 data records, with an average of 4 subtrees for each record). Therefore, this exhaustive approach is not feasible. The remaining of this section explains how we overcome these difficulties.

Our method has two stages: clustering the subtrees according to their similarity and using the groups to generate the candidate record divisions.

Grouping the subtrees. For grouping the subtrees according to their similarity, we use a clustering-based process we describe in the following lines:

1. Let us consider the set $\{t_1,...,t_n\}$ of all the subtrees which are direct children of the node chosen as root of the data region. Each t_i can be represented as a string using the method described in section 4.1. We will term these strings as $s_1,...,s_n$.
2. Compute the *similarity matrix*. This is a *nxn* matrix where the *(i,j)* position (denoted m_{ij}) is obtained as $es(t_i, t_j)$, the *edit-distance similarity* between t_i and t_j.
3. We define the *column similarity* between t_i and t_j, denoted $cs(t_i,t_j)$, as the inverse of the average absolute error between the columns corresponding to t_i and t_j in the similarity matrix (2). Therefore, to consider two subtrees as similar, the column similarity measure requires their columns in the similarity matrix to be very similar. This means two subtrees must have roughly the same *edit-distance similarity* with respect to the rest of subtrees in the set to be considered as similar. We have found *column similarity* to be more robust for estimating similarity between t_i and t_j in the clustering process than directly using $es(t_i, t_j)$.
4. Now, we apply bottom-up clustering [3] to group the subtrees. The basic idea behind this kind of clustering is to start with one cluster for each element and successively combine them into groups within which inter-element similarity is high, collapsing down to as many groups as desired.

$$cs(t_i,t_j)=1-\sum_{k=1..n}|m_{ik}-m_{jk}|/n \qquad (2)$$

$$s(\Phi)=2/|\Phi|\,(|\Phi|-1)\sum_{t_i,t_j\in\Phi}cs(t_i,t_j) \qquad (3)$$

Fig. 5 shows the pseudo-code for the bottom-up clustering algorithm. Inter-element similarity of a set Φ is estimated using the *auto-similarity* measure ($s(\Phi)$), and it is computed as specified in (3).

```
1. Let each subtree t be in a singleton group {t}
2. Let G be the set of groups
3. Let Ωg be the group-similarity threshold and
       Ωe be the element-similarity threshold
4. While |G| > 1 do
      4.1 choose Γ,Δ ∈ G, a pair of groups which maximize the auto-similarity measure s(Γ∪Δ) (see equation 3).
          The set Γ∪Δ must verify:
            a)  s(Γ∪Δ)>Ωg
            b)  ∀i∈Γ∪Δ, j∈Γ∪Δ, cs(i,j)>Ωe
      4.2 if no pair verifies the above conditions, then stop
      4.3 remove Γ and Δ from G
      4.4 let Φ=Γ∪Δ
      4.5 insert Φ into G
5. End while
```

Fig. 5. Pseudo-code for bottom-up clustering

We use *column similarity* as the similarity measure between t_i and t_j. To allow a new group to be formed, it must verify two thresholds:

- The global auto-similarity of the group must reach the *auto-similarity threshold* Ω_g. In our current implementation, we set this threshold to 0.9.
- The column similarity between every pair of elements from the group must reach the *pairwise-similarity threshold* Ω_e. This threshold is used to avoid creating groups that, although showing high overall auto-similarity, contain some dissimilar elements. In our current implementation, we set this threshold to 0.9.

Generating the candidate record divisions. For generating the candidate record divisions, we assign an identifier to each of the generated clusters. Then, we build a sequence by listing in order the subtrees in the data region, representing each subtree with the identifier of the cluster it belongs to. For instance, in our example of Fig. 3, the algorithm generates three clusters, leading to the string $c_0c_1c_0c_2c_2c_0c_1c_2c_0c_1$.

The data region may contain, either at the beginning or at the end, some subtrees that are not part of the data. For instance, these subtrees may contain information about the number of results or web forms to navigate to other result intervals. These subtrees will typically be alone in a cluster, since there are not other similar subtrees in the data region. Therefore, we pre-process the string from the beginning and from the end, removing tokens until we find the first cluster identifier that appears more than once in the sequence. In some cases, this pre-processing is not enough and some additional subtrees will still be included in the sequence. Nevertheless, they will typically be removed from the output as a side-effect of the final stage (see section 5).

Once the pre-processing step is finished, we proceed to generate the candidate record divisions. By property 1, we know each record is formed by a list of consecutive subtrees (i.e. characters in the string). From our page model, we know records are encoded consistently. Therefore, the string will tend to be formed by a repetitive sequence of cluster identifiers, each sequence corresponding to a data record. The sequences for two records may be slightly different. Nevertheless, we will assume they always either start or end with a subtree belonging to the same cluster (i.e. all the data records always either start or end in the same way). This is based on the following heuristic observations:

- In many sources, records are visually delimited in an unambiguous manner to improve clarity. This delimiter is present before or after every record.
- When there is not an explicit delimiter between data records, the first data fields appearing in a record are usually *key* fields appearing in every record.

Based on the former observations, we will generate the following candidate lists:

- For each cluster c_i, $i=1..k$, we will generate two candidate divisions: one assuming every record starts with c_i and another assuming every record ends with c_i. For instance, Fig. 6 shows the candidate divisions obtained for the example of Fig. 3.
- In addition, we will add a candidate record division considering each record is formed by exactly one subtree.

This reduces the number of candidate divisions from 2^{n-1}, where n is the number of subtrees, to $1+2k$, where k is the number of generated clusters, turning feasible to evaluate each candidate list to choose the best one.

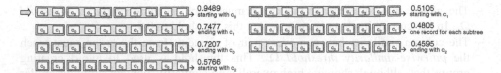

Fig. 6. Candidate record divisions obtained for example page from Fig. 3

4.3 Choosing the Best Candidate Record List

To choose the correct candidate record list, we rely on the observation that the records in a list tend to be similar to each other. Therefore, we will choose the candidate list showing the highest auto-similarity.

Given a candidate list composed of the records $<r_1, ..., r_n>$, we compute its auto-similarity as the weighted average of the *edit-distance similarities* between each pair of records of the list. The contribution of each pair to the average is weighted by the length of the compared registers. See equation 4.

$$\sum_{i=1..n,\, j=1..n, i\neq j} es(r_i, r_j) \left(len(r_i) + len(r_j)\right) \; / \; \sum_{i=1..n,\, j=1..n, i\neq j} len(r_i) + len(r_j) \qquad (4)$$

For instance, in Fig. 6, the first candidate record division is chosen.

5 Extracting the Attributes of the Data Records

In this section, we describe our techniques for extracting the values of the attributes of the data records identified in the previous section.

The basic idea consists in transforming each record from the list into a string using the method described in section 4.1, and then using *string alignment* techniques to identify the attributes in each record. An alignment between two strings matches the characters in one string with the characters in the other one, in such a way that the edit-distance between the two strings is minimized. There may be more than one optimal alignment between two strings. In that case, we choose any of them.

For instance, Fig. 7a shows an excerpt of the alignment between the strings representing the records in our example. Each aligned text token roughly corresponds with an attribute of the record. Notice that to obtain the actual value for an attribute we may need to remove common prefixes/suffixes found in every occurrence of an attribute. For instance, in our example, to obtain the value of the *price* attribute we would detect and remove the common suffix "€". In addition, those aligned text nodes having the same value in all the records (e.g. "*Buy new:*", "*Price used:*") will be considered "labels" instead of attribute values and will not appear in the output.

To achieve our goals, it is not enough to align two records: we need to align all of them. Nevertheless, optimal multiple string alignment algorithms have a complexity of $O(n^k)$. Therefore, we need to use an approximation algorithm. Several methods have been proposed for this task [10][6]. We use a variation of the *center star* approximation algorithm [6], which is also similar to a variation used in [14] (although they use tree alignment). The algorithm works as follows:

1. The longest string is chosen as the "master string", m.
2. Initially, S, the set of "still not aligned strings" contains all the strings but m.
3. For every $s \in S$, align s with m. If there is only one optimal alignment between s and m and the alignment matches any null position in m with a character from s, then the character is added to m replacing the null position (an example is shown in Fig. 7b). Then, s is removed from S.
4. Repeat step 3 until S is empty or the master string m does not change.

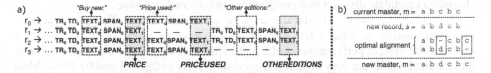

Fig. 7. (a) Alignment between records of Fig. 1 (b) example of alignment with the master

6 Experience

This section describes the empirical evaluation of our techniques. During the development phase, we used a set of 20 pages from 20 different web sources. The tests performed with these pages were used to adjust the algorithm and to choose suitable values for the used thresholds.

For the experimental tests, we chose 200 new websites in different application domains (book and music shops, patent information, publicly financed R&D projects, movies information, etc). We performed one query in each website and collected the first page containing the list of results. Some queries returned only 2-3 results while others returned hundreds of results. The collection of pages is available online[1].

While collecting the pages for our experiments, we found three data sources where our page creation model is not correct. Our model assumes that all the attributes of a data record are shown contiguously in the page. In those sources, the assumption does not hold and, therefore, our system would fail. We did not consider those sources in our experiments. In the related work section, we will further discuss this issue.

We measured the results at three stages of the process: after choosing the data region containing the dominant list of data records, after choosing the best candidate record division and after extracting the structured data contained in the page. Table 1 shows the results obtained in the empirical evaluation.

In the first stage, we use the information about the executed query, as explained at the end of section 3. As it can be seen, the data region is correctly detected in all pages but two. In those cases, the answer to the query returned few results and there was a larger list on a sidebar of the page containing items related to the query.

In the second stage, we classify the results in two categories: *correct* when the chosen record division is correct, and *incorrect* when the chosen record division contains some incorrect records (not necessarily all). For instance, two different records may be concatenated as one or one record may appear segmented into two.

---

[1] http://www.tic.udc.es/~mad/resources/projects/dataextraction/testcollection_0507.htm

As it can be seen, the chosen record division is correct in the 93.50% of the cases. It is important to notice that, even in incorrect divisions, there will usually be many correct records. Therefore, stage 3 may still work fine for them. The main reason for the failures at this stage is that, in a few sources, the auto-similarity measure described in section 4.3 fails to detect the correct record division, although it is between the candidates. This happens because, in these sources, some data records are quite dissimilar to each other. For instance, in one case where we have two consecutive data records that are much shorter than the remaining, and the system chooses a candidate division that groups these two records into one.

In stage 3, we use the standard metrics *recall* and *precision*. These are the most important metrics in what refers to web data extraction applications because they measure the system performance at the end of the whole process. As it can be seen, the obtained results are very high, reaching respectively to 98.55% and 97.39%. Most of the failures come from the errors propagated from the previous stage.

7 Related Work

Wrapper generation techniques for web data extraction have been an active research field for years. Many approaches have been proposed [2][9][11][12][13]. [7] provides a brief survey.

All the wrapper generation approaches require some kind of human intervention to create and configure the wrapper. When the sources are not known in advance, such as in focused crawling applications, this approach is not feasible.

Several works have addressed the problem of performing web data extraction tasks without requiring human input. IEPAD [4] uses the Patricia tree [6] and string alignment techniques to search for repetitive patterns in the HTML tag string of a page. The method used by IEPAD is very probable to generate incorrect patterns along with the correct ones, so human post-processing of the output is required.

RoadRunner [5] receives as input multiple pages conforming to the same template and uses them to induce a union-free regular expression (UFRE) which can be used to extract the data from the pages conforming to the template. The basic idea consists in performing an iterative process where the system takes the first page as initial UFRE and then, for each subsequent page, tests if it can be generated using the current template. If not, the template is modified to represent also the new page. The proposed method cannot deal with disjunctions in the input schema and it requires receiving as input multiple pages conforming to the same template.

As well as RoadRunner, ExAlg receives as input multiple pages conforming to the same template and uses them to induce the template and derive a set of data extraction rules. ExAlg makes some assumptions about web pages which, according to the own experiments of the authors, do not hold in a significant number of cases: for instance, it is assumed that the template assign a relatively large number of tokens to each type constructor. It is also assumed that a substantial subset of the data fields to be extracted have a unique path from the root in the DOM tree of the pages. It also requires receiving as input multiple pages.

Table 1. Results obtained in the empirical evaluation

Stage 1	# Correct	# Incorrect	% Correct
	198	2	99.00
Stage 2	# Correct	# Incorrect	% Correct
	187	13	93.50
			Precision
Stage 3	# Records to Extract	3557	98.55
	# Extracted Records	3515	Recall
	# Correct Extracted Records	3464	97.39

[14] presents DEPTA, a method that uses the visual layout of information in the page and tree edit-distance techniques to detect lists of records in a page and to extract the structured data records that form it. As well as in our method, DEPTA requires as input one single page containing a list of structured data records. They also use the observation that, in the DOM tree of a page, each record in a list is composed of a set of consecutive sibling subtrees. Nevertheless, they make two additional assumptions: 1) that exactly the same number of sub-trees must form all records, and 2) that the visual gap between two data records in a list is bigger than the gap between any two data values from the same record. Those assumptions do not hold in all web sources. For instance, neither of the two assumptions holds in our example page of Fig. 3. In addition, the method used by DEPTA to detect data regions is considerably more expensive than ours.

A limitation of our approach arises in the pages where the attributes constituting a data record are not contiguous in the page. Those cases do not conform to our page creation model and, therefore, our current method is unable to deal with them. Although DEPTA assumes a page creation model similar to the one we use, after detecting a list of records, they try to identify these cases and transform them in "conventional" ones before continuing the process. These heuristics could be adapted to work with our approach.

References

1. Arasu, A., Garcia-Molina, H.: Extracting Structured Data from Web Pages. In: Proc. of the ACM SIGMOD Int. Conf. on Management of Data (2003)
2. Baumgartner, R., Flesca, S., Gottlob, G.: Visual Web Information Extraction with Lixto. In: VLDB. Proc. of Very Large DataBases (2001)
3. Chakrabarti, S.: Mining the Web: Discovering Knowledge from Hypertext Data. Morgan Kaufmann Publishers, San Francisco (2003)
4. Chang, C., Lui, S.: IEPAD: Information extraction based on pattern discovery. In: Proc. of 2001 Int. World Wide Web Conf., pp. 681–688 (2001)
5. Crescenzi, V., Mecca, G., Merialdo, P.: ROADRUNNER: Towards automatic data extraction from large web sites. In: Proc. of the 2001 Int. VLDB Conf., pp. 109–118 (2001)
6. Gonnet, G.H., Baeza-Yates, R.A., Snider, T.: New Indices for Text: Pat trees and Pat Arrays. In: Information Retrieval: Data Structures and Algorithms, Prentice Hall, Englewood Cliffs (1992)
7. Laender, A.H.F., Ribeiro-Neto, B.A., da Silva, A.S., Teixeira, J.S.: A Brief Survey of Web Data Extraction Tools. ACM SIGMOD Record 31(2), 84–93 (2002)

8. Levenshtein, V.I.: Binary codes capable of correcting deletions, insertions, and reversals. Soviet Physics Doklady 10, 707–710 (1966)

9. Muslea, I., Minton, S., Knoblock, C.: Hierarchical Wrapper Induction for Semistructured Information Sources. In: Autonomous Agents and Multi-Agent Systems, pp. 93–114 (2001)

10. Notredame, C.: Recent Progresses in Multiple Sequence Alignment: A Survey. Technical report, Information Genetique et. (2002)

11. Pan, A., et al.: Semi-Automatic Wrapper Generation for Commercial Web Sources. In: EISIC. Proc. of IFIP WG8.1 Conf. on Engineering Inf. Systems in the Internet Context (2002)

12. Raposo, J., Pan, A., Álvarez, M., Hidalgo, J.: Automatically Maintaining Wrappers for Web Sources. Data & Knowledge Engineering 61(2), 331–358 (2007)

13. Zhai, Y., Liu, B.: Extracting Web Data Using Instance-Based Learning. In: Ngu, A.H.H., Kitsuregawa, M., Neuhold, E.J., Chung, J.-Y., Sheng, Q.Z. (eds.) WISE 2005. LNCS, vol. 3806, pp. 318–331. Springer, Heidelberg (2005)

14. Zhai, Y., Liu, B.: Structured Data Extraction from the Web Based on Partial Tree Alignment. IEEE Trans. Knowl. Data Eng. 18(12), 1614–1628 (2006)

A Semantic Approach and a Web Tool for Contextual Annotation of Photos Using Camera Phones

Windson Viana*, José Bringel Filho**, Jérôme Gensel, Marlène Villanova-Oliver, and Hervé Martin

Laboratoire d'Informatique de Grenoble, équipe STEAMER
681, rue de la Passerelle, 38402 Saint Martin d'Hères, France
{carvalho, bringel, gensel, villanov, martin}@imag.fr

Abstract. The increasing number of personal digital photos on the Web makes their management, retrieval and visualization a difficult task. To annotate these images using Semantic Web technologies is the emerging solution to decrease the lack of the description in photos files. However, existing tools performing manual annotation are time consuming for the users. In this context, this paper proposes a semi-automatic approach for annotating photos and photo collections combining OWL-DL ontologies and contextual metadata acquired by mobile devices. We also describe a mobile and Web-based system, called PhotoMap, that provides automatic annotation about the spatial, temporal and social contexts of a photo (i.e., where, when, and who). PhotoMap uses the Web Services and Semantic Web reasoning methods in order to infer information about the taken photos and to improve both browsing and retrieval.

Keywords: Semantic Web, ontologies, context sensing, mobile devices, photo annotation, and spatial Web 2.0.

1 Introduction

One of the main motivations of people for taking personal photos is to keep a trace of a situation that they can share afterwards, with their families and friends by using Web photo gallery systems (e.g., Picasa). Shooting new pictures gets easier with the modern digital cameras and the storage costs in these devices has decreased enormously. Hence, the amount of photos in personal digital collections has grown rapidly and looking for a specific photo to be shared has become a frustrating activity [5][7][10]. A similar difficulty is found in off-the-shelf Web image search engines. Generally, a keyword search results in providing the user with images that she/he did not expected [7]. Both problems are related to the lack of description in images files. One solution to facilitate organization and retrieval processes is to have annotations about the photos in a machine-readable form. This image metadata describes both the

---

* Supported by CAPES – Brasil.
** Supported by the Programme Alban, the European Union Programme of High Level Scholarships for Latin America, scholarship no. E06D104158BR.

B. Benatallah et al. (Eds.): WISE 2007, LNCS 4831, pp. 225–236, 2007.
© Springer-Verlag Berlin Heidelberg 2007

photo context (e.g., where and when the picture has been taken) and information about the photo content (e.g., who are the people in the photo, what are they doing) [3][4][6]. The annotation can be exploited by Web image search engines in order to allow users to retrieve their photos by searching on content and context information, instead of only querying filenames. In addition, context annotation of a photo provides a support to conceive better browsing and management image systems [6]. Several Web applications and research works provide tools to help the creation of manual annotation. For instance, Flickr allows the association of spatial information (location tags) and free-text keywords with photos. Other applications, such as [2] and [3], offer a more powerful metadata description by using conceptual graphs and spatial ontologies in RDF. Despite the efforts of innovative annotation tools, such as EspGames[1], manual annotation is still a time consuming and a boring activity. The multimedia research community proposes systems that automatically suggest image annotations. These systems employ the low-level visual features of the image (e.g., colors, textures, shapes) for indexing the photos and extracting annotations by using algorithms to identify similarity correspondences between a new image and pre-annotated pictures. However, there is a semantic gap between the recommended annotations of these systems and the user desired annotations [6][10].

In the near future, the use of new mobile devices can change the photo annotation radically. These devices will have high-resolution built-in cameras and the majority of the users will progressively substitute their traditional digital cameras by this new generation of mobile devices. Besides that, the progressive addition of sensors to mobile devices (e.g., GPS, RFID readers, and temperature sensors) allows the acquisition of a vast number of contextual information about the user's situation when she uses her cameraphone for taking a photo [9]. The interpretation and inference about this context data generate new information that can be employed for annotating photos automatically [5]. In addition, using mobile applications the users can assign metadata at the capture point of their photos, avoiding the so-called "time lag problem" of desktop postponed annotations [5]. In this context, we propose a semi-automatic approach for annotating photos by using OWL-DL ontologies and mobile devices. In this article, we present an ontology called ContextPhoto that allows manual and automatic context annotations of photos and photo collections. In order to validate the context annotation process we propose, we have also designed and developed a mobile and Web Information System. This novel system, called PhotoMap, is an evolution of the related mobile annotation systems since it provides automatic annotation about the spatial, temporal and social contexts of a photo (i.e., where, when, and who was nearby). PhotoMap also offers a Web interface for spatial and temporal navigation in photo collections. This interface exploits spatial Web 2.0 services for showing where the user takes her photos and the followed itinerary for taking them. In addition, users can look into the inferred and manual annotations of their photos and exploit them for retrieval purposes.

This article is organized as follows: section 2 presents our ontology for photo annotation; section 3 gives an overview of the proposed Web and mobile system; section 4 discusses the related works in image annotation and photos visualisation; and, finally, we conclude in section 5, and outline potential future works.

---

[1] http://www.espgame.org/

2 ContextPhoto Ontology

Annotation is the main tool to associate semantics with an image. Annotation highlights the significant role of photos in restoring forgotten memories of visited places, party events, and people. In addition, the use of annotation allows the development of better organization, retrieval and visualization processes for personal image management. The existing approaches for multimedia annotation employ several representation structures to associate metadata with images. These structures may be attribute-value pairs inserted in the header of image files (e.g., EXIF, and IPTC formats), or more expressive representations such as the MPEG-7 standard and the RDF/OWL ontologies. In the context of the Semantic Web, the use of ontologies for annotation representation is more suitable for making the content machine-understandable. Ontologies can reduce the semantic gap between what image search engines find and what the people expect to be retrieved when they make a search. Moreover, with a description in a formal and explicit way, reasoning methods can be employed in order to infer about the content of a photo and its related context. In the field of multimedia annotation, different ontologies for image metadata representation have been proposed [11][12][2][3]. These ontologies are well suited for extensive content-oriented annotation. In our approach, however, we are more interested in representing what was the context of the user (social, spatial, temporal) when she took her photos. We claim, as the authors in [6] and [7], that contextual metadata are particularly useful to photos organization and retrieval, since, for instance, knowing location and time of a photo often describes the photo itself a lot even before a single pixel is shown. A graphical representation of our ontology, called ContextPhoto, is presented in Fig. 1. ContextPhoto is an OWL-DL ontology to annotate photos and also photo collections with contextual metadata and content textual descriptions.

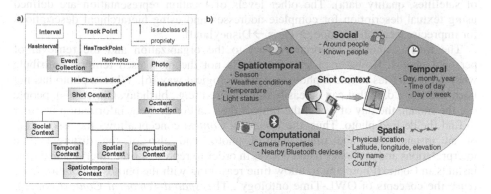

Fig. 1. ContextPhoto Ontology

The main idea of ContextPhoto is to associate a spatial itinerary and a temporal interval with a photo collection. The authors [6] and [4] have pointed out that grouping photos in events (e.g., a vacation, a tourist visit, and a party) is one of the most common way people use to recall and to organize their photos. The concept EventCollection of ContextPhoto represents the idea of a photo collection associated with an event. A time interval and an ordered list of track points (timestamp and geographic coordinates) are linked to EventCollection (see Fig. 1a). The EventCollection property hasPhotos describes the photos (Photo) related to the collection. The Photo concept contains the basic image properties (e.g., name, size, width and height). Besides these properties, each photo has two types of annotation: the content annotation (Content Annotation) and the contextual annotations (Shot Context). An annotation system can exploit our ontology for manual content annotation and they can suggest keyword tags derived from the shot context annotation. The integration of ContextPhoto with other image annotation ontologies such as the Visual Descriptor Ontology (VDO) [12] can also be envisioned to offer wider description possibilities. ContextPhoto supports five types of contextual metadata: spatial (Spatial Context), temporal (Temporal Context), social (Social Context), computational (Computational Context), and spatiotemporal (SpatioTemporal Context). These concepts correspond to the major elements for describing a photo (i.e., where, when, who, with what) [6][4][7].

Location is the most useful information to recall personal photos. A person can remember a picture using an address ("Champs Élysées avenue") or a more imprecise description of a place ("Disneyland"). However, systems to acquire location information (GPS, A-GPS) describe location in terms of georeferenced coordinates (latitude, longitude, elevation, coordination system). Hence, the Spatial Context concept incorporates different semantic representation levels of location description. For instance, in order to describe location places in a geometric way (polygons, lines, points), ContextPhoto imports the NeoGeo2 ontology. NeoGeo is an OWL-DL representation of the core concepts of GML (Geographic Markup Language), an open interchange format defined by the Open Geospatial Consortium (OGC) to express geographical features. Using the gml:Point concept, ContextPhoto define a subclass of Spatial Context to represent GPS coordinates. We have also added properties to store the elevation data and the imprecision information of the GPS receiver (number of satellites, quality data). The other levels of location representation are defined using textual description for complete addresses and using hierarchical descriptions for imprecise location (Europe → Paris →Disneyland).

The time is another important aspect to the organization and the retrieval of personal photos. However, a specific date is not the most used temporal attribute when a person tries to find a photo in a chronological way [6]. When a photo has not been taken at a date that can be easily remembered (e.g., birthday, Christmas), people use the month, the day of week, the time of day, and/or the year information in order to find the desired photo. Thus, the *Temporal Context* concept allows the association of an instant (date and time) with a photo, and also of the different time interpretations and attributes listed above. In order to represent time intervals and time instants in ContextPhoto and to allow time reasoning with the photos annotations, we reuse the concepts of OWL-Time ontology[3]. The *Spatial-Temporal Context* concept contains attributes of the shot context of a photo that depends of a time and a location data to be calculated. In this first version of ContextPhoto, we have defined only one spatiotemporal class: the physical environment. This concept has properties that describe how were the season, the temperature, the weather conditions, and the day light status (e.g., day, night, after sunset) when the user took her photo. These

---

[2] http://mapbureau.com/neogeo/neogeo.owl
[3] http://www.w3.org/TR/owl-time/

spatiotemporal attributes are both employed to search purposes and to increase the described information of a photo. The *Computational Context* concept describes all the digital devices present at the time of the photo shot (i.e., the camera, and the surrounding Bluetooth devices). This concept groups two other classes: *Camera*, and *BluetoothDevice*. The *Camera* class describes the characteristics of the digital camera and the camera settings used to take the photograph (e.g., aperture, shutter speed, and focal length). This concept integrates the core attributes of the EXIF format. The *Bluetooth Device* class contains the Bluetooth addresses of the nearby devices. This concept plays a central role in the inference process of the photo social context.

One of the innovating characteristics of ContextPhoto ontology is the ability to describe the social context of a photo. The description of social context is based on the proposal of [5] and FoafMobile[4] to use the Bluetooth address of personal devices in order to detect the presence of a person. The main idea is to associate a hash value of Bluetooth addresses with a unique Friend-Of-A-Friend[5] (FOAF) profile. FOAF is a RDF ontology that allows the description of a person (name, email, personal image) and of her social networks (the person's acquaintances). Thus, we can use the nearby Bluetooth addresses stored in the *Computation Context* element in order to calculate if these addresses identify a user's friend, and afterward, to annotate the photo with the names of the nearby acquaintances. In order to represent acquaintances and the user, we define in ContextPhoto two classes that can be elements of the *Social Context* concept of a photo: the *Person*, and the *Owner* classes. The *Person* is defined as a subclass of *foaf:Person* that contains a Bluetooth device (see Formula 1). The *Owner* describes the owner of the photo (the photographer and user of our system) (see Formula 2). The inference process to calculate the nearby people is presented in details in the section 3.2.

$$Person \subseteq foaf:Person \qquad (1)$$
$$\equiv \exists hasBTDevice \; ctxt:BTDevice$$

$$Owner \subseteq Person \qquad (2)$$
$$\equiv \exists hasEventCollection \; ctxt:EventCollection$$

3 PhotoMap

PhotoMap is a mobile and Web Information System for contextual annotation of photos taken by mobile users. The three main goals of PhotoMap are: (1) to offer a mobile application enabling users to take pictures and to group theirs photos in event collections; (2) to propose a Web system that organizes the user photos using the acquired spatial and temporal data; and (3) to improve the users recall of their photos showing the inferred spatial, temporal, and social information.

3.1 General Overview

The PhotoMap system is structured according to a client-server model. The PhotoMap client is a mobile application running on J2ME[6]-enabled devices. Using this mobile

---

[4] http://jibbering.com/discussion/Bluetooth-presence.1
[5] http:// www.foaf-project.org/
[6] http://java.sun.com/javame/

application, users can create collections of photos representing events (parties, concerts, tourist visits). The user can give a name and a textual description when she starts a collection. **Fig. 2a** illustrates a scenario of use of the PhotoMap mobile application. The PhotoMap client runs a background process that monitors the current physical location of the mobile device. The mobile application accesses the device sensors (e.g., built-in GPS, Bluetooth-enabled GPS receiver) via the Location API[8] (i.e., JSR 179) or via the Bluetooth API[8] (i.e., JSR 82). The gathered coordinates (latitude, longitude, and elevation) are stored in order to build a list of track points. This list represents the itinerary followed by the user to take the pictures. The acquired track list is associated with the metadata of the current photo collection.

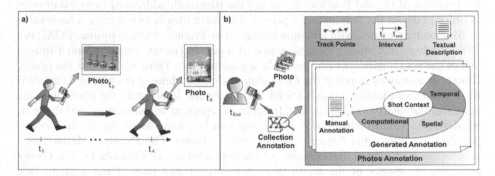

Fig. 2. Photomap use and the generated annotation

In addition, the PhotoMap client captures the photo context when a user takes a picture with her camera phone. The mobile client gets the device geographic position, the date and time information, the bluetooth addresses of the nearby devices, and the configuration properties of the digital camera. All this metadata is stored for each photo of the collection. After she has taken a picture, the user can add manual annotations for each photo. This feature reduces the time lag problem that occurs in desktop annotations tools [1] since the user is still involved in the shot situation when she writes the photo textual metadata. The user indicates the end of the current photo collection with two clicks in the interface of the PhotoMap client. Thus, the mobile application generates the collection annotation by using the ContextPhoto ontology. The textual description of the collection, its start and end instants, and the track points list are added to the annotation. For each photo, the gathered context (i.e.; computational, spatial, and temporal contexts) and the possible manual annotation are stored in the ontology instantiation. The taken photos and the generated annotation are stocked in the mobile device file system (see Fig. 2b). Afterward, the user uploads her photos and the annotation metadata of the collection to the PhotoMap Web server. The user can execute this upload process from her mobile phone directly (e.g., via a HTTP connection). PhotoMap offers an alternative in order to avoid payment of data service charges. The user can transfer to a desktop computer (e.g., via a USB connection) her photo collections and their annotations. Thus, the user executes the Web interface of PhotoMap to finish the upload process. The PhotoMap server is a

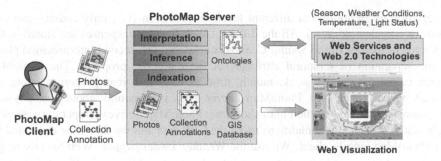

Fig. 3. Overview of the PhotoMap architecture

J2EE Web-based application. Besides acting as an upload gateway, the PhotoMap server is in charge of the photos indexing and of the inference process of associated contextual annotations. **Fig. 3** shows an overview of the PhotoMap architecture.

After the transmission of a collection and its semantic metadata, the PhotoMap server reads the annotations associated with each photo. The spatial and temporal information are translated to a more useful representation. For example, geographical coordinates are transformed in a textual representation like city and country names. In addition, PhotoMap accesses to off-the-shelf Web Services in order to get information about the physical environment where each photo was taken (e.g., temperature, weather conditions). Furthermore, PhotoMap uses Semantic Web technologies in order to infer information about the social context of each photo. Automatically, PhotoMap identifies the user's friends that were nearby during the shot time of each photo. The inference information about each photo is added to the ContextPhoto instantiation. The generated annotations are exploited by the visualization tool of PhotoMap. After an indexing process, the PhotoMap Web site allows the user to view her photo collections and the itinerary followed to take them. The user can see on a map, using spatial Web 2.0 services, the places where she took her photos. She also has access to the rich generated annotation.

3.2 Interpretation, Inference and Indexing Processes

Fig. 4 shows the sequence of three fundamental processes performed by the PhotoMap server in order to increase knowledge about the shot context of a photo and to optimize the mechanisms of spatial and temporal queries. When a user sends a photo collection to PhotoMap, the annotation contains only information about the computational, spatial and temporal contexts of a photo. Thus, PhotoMap accesses off-the-shelf Web Services in order to augment the description of these context annotations. This approach, proposed by [6], reduces the development cost and profits from the advantages of the Web Services technology (i.e., reutilization, standardization). First, PhotoMap executes an **interpretation** of the gathered spatial metadata. The PhotoMap server uses a Web Service to transform GPS data of each photo into physical addresses. The *AddressFinder* Web Service[7] offers a hierarchic

---

[7] http://ashburnarcweb.esri.com/

description of an address at different levels of precision (i.e., only country and city name, a complete address). All the different Web Service responses are stored in the Datatype properties of the *Spatial Context* subclasses. The second interpretation phase is the separation of temporal attributes of the date/time property. The PhotoMap server calculates day of week, month, time of day, and year properties using the instant value. Later, the PhotoMap server gets information about the physical environment as it was at the photo shot time. PhotoMap derives temperature, season, light status, and weather conditions using the GPS data and the date/time annotated by the PhotoMap mobile client. We use the *Weather Underground*[8] Web Service to get weather conditions and the temperature information. In addition, the PhotoMap server uses the *Sunrise and Sunset Times*[9] Web Service to get the light status. The season property is easily calculated using the date and GPS data.

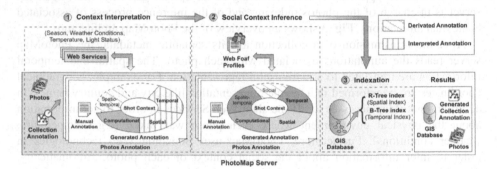

Fig. 4. Interpretation, inference and indexing process

After the interpretation process, the PhotoMap server execute the **inference** process in order to derivate the social context of a photo. This process gets the FOAF profiles of the user friends and tries to identify who was present at the moment of a photo shot. In our approach, we use the FOAF profile of the photo's owner is a start point of a search. We navigate through the properties "*foaf:Knows*" and "*rdf:seeAlso*" to get the FOAF profiles of people that she knows. After that, we repeat the same process with the friend profiles in order to find the friends of the owner's friends. All the found profiles are used to instantiate individuals of the *Person* class. After the instantiation of the *Person* and *Owner* individuals, we use a rule-based engine in order to infer which acquaintances were present at the moment of the photo shot. Formula 3 shows the SWRL rule used to infer the presence of an owner's friend. After the end of the inference process, individuals representing the nearby acquaintances are associated with the *Social Context* of the current photo. Hence, the final OWL annotation of a photo contains spatial, temporal, spatial-temporal, computational and social information. Then, the PhotoMap server executes an **indexing** process in order to optimize browsing and interaction methods. The number

---

[8] www.weatherunderground.com/
[9] http://www.earthtools.org/

of photo collections annotations should increase in our system quickly. To avoid problems of sequential searches in these OWL annotations, spatial and temporal indexes are generated for each collection using the PostgreSQL database extended with the PostGIS module.

$$
\begin{aligned}
&Owner(?owner) \wedge Person(?person) \wedge SocialContext(?scctxt) \\
&ComputationalContext(?compctxt) \wedge BTDevice(?btDv) \\
&foaf:knows(?person, ?owner) \wedge hasBTDevice(?person, ?btDv) \\
&hasContextElement(?scctxt, ?owner) \wedge \\
&hasContextElement(?compctxt, ?btDv) \\
&\rightarrow hasContextElement(?scctxt, ?person)
\end{aligned}
\qquad(3)
$$

3.2 Browsing and Querying Photos

The PhotoMap Web site offers graphical interfaces for navigation and query over the users captured collections of photos. We have decided to design PhotoMap Web site using map-based interfaces for browsing photos purposes. Usability studies [8] show that map interfaces present more interactivity advantages than browsing photos only with location hierarchical links (e.g.; Europe -> France -> Grenoble). Furthermore, with a map-based interface, we can easily represent the itineraries followed by the users for taking their photos. **Fig. 5** shows a screen shot of the PhotoMap Web site.

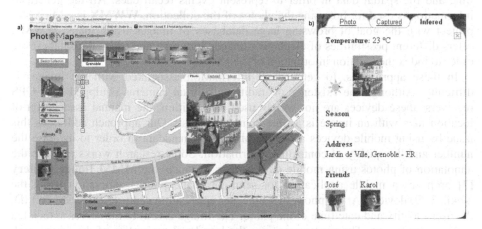

Fig. 5. Screen shot of the PhotoMap Web Site

The rectangular region positioned at the left side of the Web Site is called the *menu-search view*. This window shows the PhotoMap main functionalities, the social network of the user, and a keyword search engine for her photo collections. On the right side, from top to bottom, are presented the *event-collection view*, the *spatial view*, and the *temporal query window*. When a user enters in the PhotoMap Web site, the PhotoMap uses the temporal index to determinate the ten latest collections. PhotoMap then shows, in the *event-collection view,* thumbnails of the first photos of each collection and the collection names annotated by the user. The latest collection is selected and the itinerary followed by the user is displayed in the *spatial view*. Placemarks are also inserted in the map for each photo, and the user can click to view the photo and the generated annotation. **Fig. 5** b shows the contextual information of a selected photo.

The *spatial view* displays maps using the Google Maps API. The navigation on the map, in a spatial query mode, changes the displayed map and also the visible collections in the *event-collection view*. In order to perform this operation, PhotoMap queries the spatial database that is constrained to return only the collections intersecting a defined view box. PhotoMap uses the bounding box coordinates of the *spatial view* and the current zoom value in order to calculate the view box coordinates. The temporal query window can be used to restraint the displayed collections (e.g., to show only the collections intersecting the view box and taken in January). A mouse click in a placemark shows the photo and the context annotation. Photomap reads the collection and photos annotation using the Jena API.

4 Related Works

The possibility to have access to both spatial and temporal data of an image is exploited by the tools **PhotoCompas** [6] and **WWMX** [8]. **PhotoCompas** use Web Services to derivate information about the location and physical environment (e.g., weather conditions, temperature). PhotoCompas calculates image clusters using the time and the spatial data in order to represent events recall cues. All the generated information is used for browsing the photos of the collection. **WWMX** is a Microsoft project with the goal of browsing large databases of images using maps. WWMX offers different possibilities of displaying images in a map and uses GIS databases in order to index the location information of the photos.

In these approaches, to get context information associated with the images is difficulty. Although the existence of traditional digital cameras equipped with GPS receivers, these devices are not widely used. In addition, the manual association of location tags with an image is a time consuming task. Our approach addresses this issue by using mobile devices as source of context metadata in order to increase the number and the quality of contextual information. Some research works propose the annotation of photos using mobile devices. For instance, the **MMM Image Gallery** [7] proposes a mobile application for semi-automatic photos annotation using the Nokia 3650 devices. At the photo shot time, the application captures the GSM cell ID, the user identity and date/time of the photo. The image is sent to the server associated with its annotation. The server combines the low level properties of the image with location information in order to derivate information about its content (e.g., is the Eiffel Tower the photo subject?). After that, the user uses a mobile XHTML browser in order to read and validate the generated annotation. The imprecision of the GSM cell ID and the slow response time of the server during the image upload and validation processes were identified as the major usability problems of MMM [7].

Zonetag [1] is a Yahoo! research prototype allowing user to upload their photos to Flickr from their mobile phones directly. ZoneTag leverages location information from camera phones (GSM cell ID and GPS coordinates) and suggests tags to be added to the photo. ZoneTag derives its suggestions from the tags used by the community for the same location, and from the tags assigned to the user photos.

In [5], the design of a mobile and Web system combines Bluetooth addresses acquisition, FOAF profiling and face detection tools in order to identify people in a photo. The authors suggest that the generated annotation is represented in a RDF file and embedded in the header of the image. They propose to use the Bluetooth addresses of the nearby devices as key inputs to a FOAF Query Service that is able to deliver the requested FOAF profiles. This approach seems to us not easily feasible. First, this proposition does not take into account the distributed characteristics of FOAF profiles. Moreover, to conceive a service that indexes all the FOAF profiles of the Web is not realistic. In our approach, we use the FOAF profile of the photo owner as the starting point of a search from which we access to profiles that are useful to annotate the photo (i.e., the profiles of friends, of friends of friends).

The major difference between the other works and our approach is that PhotoMap is both a mobile application for semi-automatic image annotation and a Web system for organization and retrieval of personal image collections. Besides the ability of acquire context automatically, the PhotoMap mobile client allows user to create their event collections and to insert textual annotation to them. The PhotoMap Web site allows the user to view her photos collections and also the itinerary she has followed to take them.

5 Conclusion and Future Works

Context information can be useful for organization and retrieval of images on the Web. We have described, in this article, a novel mobile and Web System that captures and infers context metadata of photos. We have proposed an OWL-DL ontology to annotate event collections of photos and to be employed for time, spatial, and social reasoning. PhotoMap Web site offers interfaces for spatial and temporal queries over the user photo collections. In future works, we will define other query modes for the PhotoMap Web site. Instead of only query over their collections, users will be able to search public photos using all the contextual attributes (e.g., show photos near to the Eiffel Tower in winter before the sunset). Usability tests will be performed in order to evaluate the PhotoMap mobile application and the visualization Web site. Finally, PhotoMap will be extended towards a Web recommendation system of tourist itineraries using a community collaboration model.

References

1. Ames, M., Naaman, M.: Why We Tag: Motivations for Annotation in Mobile and Online Media. In: CHI 2007. Proc. of Conference on Human Factors in computing systems (2007)
2. Hollink, L., Nguyen, G., Schreiber, G., Wielemaker, J., Wielinga, B., Worring, M.: Adding Spatial Semantics to Image Annotations. In: Proc. of 4th International Workshop on Knowledge Markup and Semantic Annotation (2004)
3. Lux, M., Klieber, W., Granitzer, M.: Caliph & Emir: Semantics in Multimedia Retrieval and Annotation. In: Proc. of 19th International CODATA Conference 2004: The Information Society: New Horizons for Science, Berlin, Allemane (2004)

4. Matellanes, A., Evans, A., Erdal, B.: Creating an application for automatic annotations of images and video. In: SWAMM. Proc. of 1st International Workshop on Semantic Web Annotations for Multimedia, Edinburgh, Scotland (2006)
5. Monaghan, F., O'Sullivan, D.: Automating Photo Annotation using Services and Ontologies. In: Mdm 2006. Proc. of 7th International Conference on Mobile Data Management, pp. 79–82. IEEE Computer Society, Washington, DC, USA (2006)
6. Naaman, M., Harada, S., Wang, Q., Garcia-Molina, H., Paepcke, A.: Context data in geo-referenced digital photo collections. In: MULTIMEDIA 2004. Proc. of 12th ACM international Conference on Multimedia, pp. 196–203. ACM, New York, NY, USA (2004)
7. Sarvas, R., Herrarte, E., Wilhelm, A., Davis, M.: Metadata creation system for mobile images. In: MobiSys 2004. Proc. of 2th International Conference on Mobile Systems, Applications, and Services, pp. 36–48. ACM, Boston, MA, USA (2004)
8. Toyama, K., Logan, R., Roseway, A.: Geographic location tags on digital images. In: MULTIMEDIA 2003. Proc. of 11th ACM international Conference on Multimedia, pp. 156–166. ACM Press, Berkeley, CA, USA (2003)
9. Yamaba, T., Takagi, A., Nakajima, T.: Citron: A context information acquisition framework for personal devices. In: Proc. of 11th International Conference on Embedded and real-Time Computing Systems and Applications (2005)
10. Wang, L., Khan, L.: Automatic image annotation and retrieval using weighted feature selection. Journal of Multimedia Tools and Applications, 55–71 (2006)
11. Schreiber, A.,Th., D.B., Wielemaker, J., Wielinga, B.J.: Ontology-based photo annotation. IEEE Intelligent Systems (2001)
12. Athanasiadis, Th., Tzouvaras, V., Petridis, K., Precioso, F., Avrithis, Y., Kompatsiaris, Y.: Using a Multimedia Ontology Infrastructure for Semantic Annotation of Multimedia Content. In: SemAnnot 2005. Proc. of 5th International Workshop on Knowledge Markup and Semantic Annotation, Galway, Ireland (November 2005)

Formal Specification of OWL-S with Object-Z: The Dynamic Aspect

Hai H. Wang, Terry Payne, Nick Gibbins, and Ahmed Saleh

University of Southampton, Southampton SO17 1BJ, UK
{hw, trp, nmg, amms}@ecs.soton.ac.uk

Abstract. OWL-S, one of the most significant Semantic Web Service ontologies proposed to date, provides Web Service providers with a core ontological framework and guidelines for describing the properties and capabilities of their Web Services in unambiguous, computer-interpretable form. To support standardization and tool support of OWL-S, a formal semantics of the language is highly desirable. In this paper, we present a formal Object-Z semantics of OWL-S. Different aspects of the language have been precisely defined within one unified framework. This model not only provides a formal unambiguous model which can be used to develop tools and facilitate future development, but as demonstrated in the paper, can be used to identify and eliminate errors in the current documentation.

1 Introduction

The Semantic Web and Semantic Web Services have been recognized as promising technologies that exhibit huge commercial potential. Current Semantic Web Service research focuses on defining models and languages for the semantic markup of all relevant aspects of services, which are accessible through a Web service interface [1, 9]. OWL-S is one of the most significant Semantic Web Service frameworks proposed to date [1]. It provides Web Service developers with a core ontological model which can be used for describing the domain-independent properties and capabilities of their Web services in an unambiguous and computer-interpretable form.

In the linguistics of computer languages, the terms syntax, static semantics and dynamic semantics are used to categorize descriptions of language characteristics. To achieve a consistent usage of a language, all these three aspects must be precisely defined. The syntax and semantics of OWL-S are defined in terms of its metamodel – as a set of OWL ontologies complemented with supporting documentation in English[1]. Although OWL has a well-defined formal meaning (as a small fragment of FOL), OWL lacks the expressivity to define all the desired properties of OWL-S. Despite its aim of providing an unambiguous model for defining services, OWL-S users and tool developers have had to rely heavily on English-language documentation and comments to understand the language and interpret its models. However, this use of natural language is ambiguous and can be interpreted in different ways. This lack of precision in defining

---

[1] There has been some published work on the definition of the dynamic semantics of OWL-S [8], as well as its precursor [2].

B. Benatallah et al. (Eds.): WISE 2007, LNCS 4831, pp. 237–248, 2007.

the semantics of OWL-S can result in different usage, as Web service providers and tool developers may have a different interpretation and understanding of the same OWL-S model. Another major problem with the current OWL-S definition is that the three components of OWL-S (i.e. the *syntax*, *static semantics* and *dynamic semantics*) have been separately described in various formats (such as natural language, Ontologies and Petri Nets etc.). These different descriptions contain redundancy and sometimes contradiction in the information provided. Furthermore, with the continuous evolution of OWL-S, it has been very difficult to consistently extend and revise these separate descriptions. To support a common understanding, and facilitate standardization for OWL-S, a formal semantics of its language is highly desirable. Also, if we are to reason about OWL-S and ultimately formally verify, or even adequately test the correctness of a Semantic Web Service system, we need to have a precise specification of the OWL-S model. In addition, being a relatively young field, research into Semantic Web services and OWL-S is still ongoing, and therefore a semantic representation of OWL-S needs to be reusable and extendable in a way that can accommodate this evolutionary process.

In this paper we present a formal denotational model of OWL-S using the Object-Z (OZ) specification language [6]. Object-Z [6] is an extension of the Z formal specification language to accommodate object orientation. The main reason for this extension is to improve the clarity of large specifications through enhanced structuring. A denotational approach has been proved to be one of the most effective ways of defining the semantics of a language, and has been used to define the formal semantics for many programming and modeling languages [7]. OZ has been used to provide one single formal model for the syntax, the static semantics and the dynamic semantics of OWL-S. Because these different aspects are described within a single framework, the consistency between these aspects can be easily maintained. In our previous work, we have presented the formal model for the syntax and static semantics of OWL-S [12]. This paper focuses on the dynamic semantics of OWL-S.

We chose Object-Z over other formalisms to specify OWL-S because:

- The object-oriented modelling style adopted by Object-Z has good support for modularity and reusability.
- The semantics of Object-Z itself is well studied. The denotational semantics and axiomatic semantics of Object-Z are closely related to Z standard work [13]. Object-Z also has a fully abstract semantics.
- Object-Z provides some handy constructs, such as *Class-union* [3] etc., to define the polymorphic and recursive nature of language constructs effectively. Z has previously been used to specify the Web Service Definition Language (WSDL)[2]; however, as Z lacks the object-oriented constructs found in OZ, a significant portion of the resulting model focused on solving several low level modeling issues, such as the usage of free types, rather than the WSDL language itself. Thus, using OZ can greatly simplify the model, and hence avoid users from being distracted by the formalisms rather than focusing on the resulting model.
- In our previous work [11], OZ has been used to specify the Semantic Web Service Ontology (WSMO) language, which is another significant Semantic Web Service alternative. Modeling both OWL-S and WSMO in the same language provides an

---

[2] http://www.w3.org/TR/wsdl20/wsdl20-z.html

opportunity to formally compare the two approaches and identify possible integration and translation between the two languages.

The paper is organized as follows. Section 2 briefly introduces the notion of OWL-S and Object-Z. Section 3 is devoted to a formal Object-Z model of OWL-S dynamic semantics. Section 4 discusses some of the benefits of this formal model. Section 5 concludes the paper and discusses possible future work.

2   Background

OWL-S. OWL S [1], formally known as DAML-S, originated from a need to define Web Services or Agent capabilities in such a way that was semantically meaningful (within an open environment), and also to facilitate meaningful message exchange between peers. Essentially, OWL-S provides a service model based on which an abstract description of a service can be provided. It is an upper ontology whose root class is the Service class, that directly corresponds to the actual service that is described semantically (every service that is described maps onto an instance of this concept). The upper level Service class is associated with three other classes: *ServiceProfile*, *ServiceModel* and *ServiceGrounding*. In detail, the OWL-S *ServiceProfile* describes *what the service does*. Thus, the class SERVICE *presents* a *ServiceProfile*. The service *profile* is the primary construct by which a service is advertised, discovered and selected. The OWL-S *ServiceModel* tells *how the service works*. Thus, the class SERVICE is *describedBy* a *ServiceModel*. It includes information about the service inputs, outputs, preconditions and effects (IOPE). It also shows the component processes for a complex process and how the control flows between the components. The OWL-S *grounding* tells *how the service is used*. It specifies how an agent can pragmatically access a service.

Object-Z (OZ). The essential extension to Z in Object-Z is the *class* construct, which groups the definition of a state schema with the definitions of its associated operations. A class is a template for *objects* of that class: the states of each object are instances of the state schema of the class, and its individual state transitions conform to individual operations of the class. An object is said to be an instance of a class and to evolve according to the definitions of its class. Operation schemas have a Δ-list of those attributes whose values may change. By convention, no Δ-list means that no attribute changes value. OZ also allows composite operations to be defined using different operation operators, such as conjunction operator '$\wedge$', parallel operator '$\|$', sequential operator '$\S$', choice operator '$[]$' and etc. The standard behavioral interpretation of Object-Z objects is as a transition system [10]. A behavior of a transition system consists of a series of state transitions each affected by one of the class operations.

3   Formal Object Model of OWL-S

The existing specification of OWL-S informally or semi-formally describes the language from three different aspects: *syntax* (an OWL-S model is well-formed), *static semantics* (an OWL-S model is meaningful) and *dynamic semantics* (how is an OWL-S model interpreted and executed). We propose the use of OZ to provide a formal specification of all aspects of OWL-S in one single unified framework, so that the semantics of the language can be more consistently defined and revised as the language evolves. The formal models of syntax and static semantics have been addressed in a separate paper [12], whereas this paper extends the static model and focuses the dynamic semantics of OWL-S [3]. Our model is based on the latest version of OWL-S (1.2 Beta)[4].

As the model of the dynamic behaviours of OWL-S requires the understanding of some basic OWL-S entities, such as values, variables and expressions etc, we will start from these basic constructs. The general approach of the framework is that the OWL-S elements are modeled as different OZ classes. The syntax of the language is captured by the attributes of an OZ class. The predicates are defined as *class invariant* used to capture the static semantics of the language. The class operations are used to define OWL-S's dynamic semantics, which describe how the state of a Web service changes.

3.1 Types, Values and Variables

Types. OWL-S includes two kinds of value types – simple types from XML Datatype Schema and RDFS/OWL classes. The XML datatype types include *boolean*, *integer*, *string*, *datetime* etc. Users can also define their new datatypes. Each of these datatypes are modeled as OZ classes, such as:

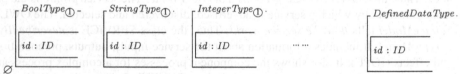

Note that the subscript $\textcircled{1}$ in OZ indicates that a class can only have one instance, which means that there is only one *BoolType*. It is a syntactic sugar for a system constraint $\#BoolType = 1$. *ID* is an OZ class which has been defined in the static model.

Everything defined in the Semantic Web is a *Resource* (*Resource* is an OZ class which will be defined in the next subsection). This includes OWL classes, properties, instances, ontologies itself and etc. Each of them is modeled as a subclass of *Resource*. Among them, OWL class can also be used as a value type. *DataType* and *Type*, defined using OZ class union, denote all kinds of datatype and OWL-S value types.

$Class, Instance, Property, Ontology :$
$\qquad \mathbb{P}\ Resource$

$\varnothing \quad \text{....}$

$DataType \cong BoolType \bigcup ... \bigcup StringType$

$Type \cong DataType \bigcup Class$

Values. OWL-S values include datatype values and web resources. OWL-S variables can be bound to a value of their types, or can be *nil*, which means *unavailable* or *non_value*. We define two special value types *nil* and *void* value, which are modeled as classes *Nil* and *VoidVal*. They are used to define how the variables are bound. A variable equating to *nil* means that the variable has no value (not being bound yet), while a variable that equates to a *void* value means that it has a value but we do not care

---

[3] Because of the limited space, we only present a partial model here and a more complete model can be found at http://www.ecs.soton.ac.uk/ hw/OWLS.pdf

[4] http://www.ai.sri.com/daml/services/owl-s/1.2/

what it is. The meaning of a void value is simply just the value itself. When an object of classes *Nil* or *VoidVal* is instantiated, the identifier *self* will denote the object self.

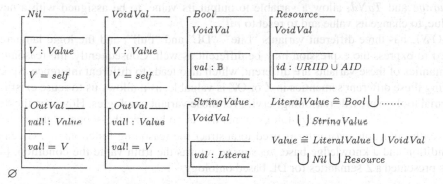

The class *Bool* and *StringValue* denote the boolean and XML string values. The definitions for the other types of XML datatypes are omitted here. *LiteralValue* denotes all datatype values. As we mentioned before, everything defined in the Web is a *Resource*. *Resource* is one kind of *values* as well. The OWL-S *value* in general is modeled as a class union.

$$type\_of : Value \rightarrow Type; \ compatible : Type \times Type \rightarrow \mathbb{B}$$
$$reasoning : (Condition \times Ontology) \rightarrow Bool \cup Nil$$

___ProcessVar_____

$$id : VID; \ parameterType : URIID; \ V : Value$$
$$\Delta type : Type$$

$$\exists_1 t : Type \bullet t.id = parameterType \wedge type = t$$
$$V \notin Nil \Rightarrow compatible(type\_of(V), type)$$

Change	_OutVal_	_ToNil_	_Assign_
$\Delta(V)$	$val! : Value$	$\Delta(V)$	$\Delta(V)$
	$val! = V$	$V' \in Nil$	$val? : Value$
			$V' = val?$

∅

Variables. *ProcessVar* denotes the variables used in an OWL-S service process. It has the attribute *id* denoting its name and *parameterType* denoting the type of the variable which is specified using a URI. *VID* and *URIID* are classes denoting variable id and URI id (details are omitted here). We also introduce an attribute *V* which denotes the value a variable bounded to and a secondary attribute [5] (denoted by Δ) *type* to denote the value type represented by the URI address – *parameterType*. The first predicate defined in the class invariant indicates that there exists only one 'type' for a given URI and the second predicate shows that a variable can only be bound to a value which type is compatible with (i.e, same or sub-typed) the declared variable type. The function *type_of* denotes the relation between values and its type. *Compatible* shows if two types are same or sub-typed. We only provide an abstract view of these two functions and details of them go beyond the scope of this paper. [2] gives some formal type rules for OWL-S. Furthermore, four operations defined in *ProcessVar* – *OutVal*, *Assign*, *Change* and *ToNil*, allow a variable to output its value, to be assigned with a new value, to change its value and to reset to *nil*.

OWL has three different variants "Lite", "DL" and "Full", and the logic language used to express the expression may be different as well. Consequently, the reasoning semantics of these variants are different, which also leads to different inference. Specifying these different semantics in Z or OZ is valuable, as it allows us to reuse existing formal tools to provide reasoning service for those variant ontologies. However, the detailed specification of these ontology semantics is go beyond the scope of this paper. Here, the function *reasoning* is used to abstract the reasoning functionality. Given a condition and a knowledge base, *reasoning* returns the truth value the condition. [4] has presented a Z semantics for DL based ontology.

3.2 Process Behavior Model

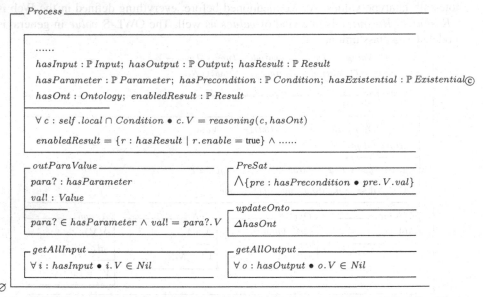

In OWL-S, *process model* is the key element to represent the functional (and behavioural) aspects of a service. It tells *how the service works*, and includes information about the service inputs, outputs, preconditions and effects (*IOPE*). It also shows the component processes for a complex process and how the control flows between the components. In our static model of OWL, *Process* has been defined to capture the common characters of an OWL-S process and we will extend this definition here. We define an attribute *hasOnt* in a *Process* to denote the knowledge base (KB) a service agent has. We assume that each service agent maintains its own KB. Some OWL-S implementation assumes that there exists one unified global KB sheared by all agents. To model that we simply assert the axiom $\forall p_1, p_2 : Process \bullet p_1.hasOnt = p_2.hasOnt$. We also define an attribute *enabledResult* to denote those results which can be performed. The first predicate shown in the class invariant shows that for all the *conditions* contained in the service (such as preconditions and result conditions), their truth values will respect to the local KB at all the time. Note that *self.local* is a predefined OZ term denoting the directly or indirectly local objects of *self*. Operation *outParaValue* is used to output the service's parameters' value. *PreSat* denotes that all the preconditions of a *service* are satisfied. *GetAllInput* and *getAllOutput* are used to check if all inputs of a service are initialized before the execution and all output variables are assigned after the execution.

Result

$inCondition : \mathbb{P}\ Condition;\ withOutput : \mathbb{P}\ OutputBinding$
$hasEffect : \mathbb{P}\ Expression,\ hasResultVar : \mathbb{P}\ ResultVar;\ \ldots\ldots$
$\Delta enable : \mathbb{B}$

$enable \Leftrightarrow (\exists\ rvb : hasResultVar \rightarrow Value \bullet$
$\qquad \forall\ rv : hasResultVar \bullet rv.V = rvb(rv) \Rightarrow (\forall\ ic : inCondition \bullet ic.value.val \Leftrightarrow \text{true}))$

$E \cong \left[\ enable \Leftrightarrow \text{true}\ \right] \bullet$
$\qquad (\bigwedge rv : hasResultVar \bullet rv.update \ {}^{\circ}_{9} \left[\ \forall\ ic : inCondition \bullet ic.value.val \Leftrightarrow \text{true} \right] {}^{\circ}_{9}$
$\qquad inProcess.updateOnt \ {}^{\circ}_{9} \left[\ \forall\ ef : hasEffect \bullet ef.value.val \Leftrightarrow \text{true} \right] {}^{\circ}_{9}(\bigwedge ob : withOutput \bullet ob.E))$
$\qquad [] \left[\ \neg\ (enable \Leftrightarrow \text{true}) \right]$

$\varnothing$

Result in OWL-S model specifies under what conditions the outputs are generated as well as what domain changes are produced during the execution of the service. In *Result*, *InCondition* denotes the conditions under which the result occurs. *WithOutput* denotes the output bindings of output parameter of the process to a value form. *HasEffect* is a set of expressions that captures possible effects to the context. *HasResultVar* declares variables that bound in the *inCondition*. Details about the static definition of *Result* are omitted. To define the execution of *Result*, we introduce a secondary attribute *enable* to denote if a *Result* is ready to be performed. A *Result* is ready to be performed if there exist bindings for its *ResultVars* so that all its *inConditions* are satisfied (.*value.val* attribute defined on *Condition* class is true). The execution of a *Result* is defined as follows. If *enable* is true, the result variables (*hasResultVar*) are first bound to concrete values which make all the conditions in *inConditions* true. After that, the service's (denoted by the attribute *inProcess*) KB is updated based on *hasEffects*. Finally the outputs of the *Result* are processed. On the other hand, if *enable* is not true, the behavior of *result* is undefined. The tool developers has the freedom to implement their own scenarios to handle this situation, such as terminate the execution and report an error message or simply ignore it.

AtomicProcess

$Process;\ \ldots\ldots$

$updateExi \cong \bigwedge ev : hasExistential$
$\qquad \bullet ev.Update$
$E \cong getAllInput \ {}^{\circ}_{9} updateExi \ {}^{\circ}_{9} PreSat$
$\qquad {}^{\circ}_{9}(\bigwedge er : enabledResult \bullet er.E)$

$\varnothing$

CompositeProcess

$Process$

$composeOf : ControlConstruct;\ \ldots\ldots$

$E \cong composeOf.E$

AtomicProcess is a kind of *Process* denoting the actions a service can perform by engaging in a single interaction. The execution of an *AtomicProcess* is defined as follows. Firstly, all the inputs of the *AtomicProcess* must be initialized (*getAllInput*).

Then the existential variables are bounded so that all the pre-condidtions of the process are true (*updateExi* and *PreSat*). Finally, the enabled results are executed.

Composite processes are processes which are decomposable into other processes. A *CompositeProcess* must have a *composedOf* attribute which indicates the control structure of the composite (*ControlConstruct*). The execution of a composite process is determined by the different kinds of the control constructs.

```
┌─ Perform ──────────────────────────────
│ ControlConstruct
│  ┌─────────────────────────────────────
│  │ performProcess : Process
│  │ hasDataFrom : ℙ InputBinding; ........
│  │
│  │ E ≘ (⋀ df : hasDataFrom • df.E)⁰₉
│  │     performProcess.getAllInput
│  │     ⁰₉performProcess.E⁰₉
│  │     performProcess.getAllOutput
│  └─────────────────────────────────────
│ │ ThisPerform : Variable
∅
```

```
┌─ Sequence ─────────────────────────────
│ ControlConstruct
│  ┌─────────────────────────────────────
│  │ components : SeqControlConstruct; .....
│  │
│  │ E ≘ ⁰₉ i : 1...#components • components(i).E
```

```
┌─ Split ────────────────────────────────
│ ControlConstruct
│  ┌─────────────────────────────────────
│  │ components : ℙ ControlConstruct; .....
│  │
│  │ E ≘ ⋀ c : components • c.E
```

Perform is a special kind of *ControlConstruct* used to refer a process to a composite process. It has two attributes – *performProcess* which denotes the process to be invoked and *hasDataFrom* which denotes the necessary input bindings for the invoked process, where *InputBinding* is a subset of *Binding* with *toVar* to *Input* (*Binding* will be defined in the next subsection). OWL-S also introduces a standard variable, *ThisPerform*, used to refer, at runtime, to the execution instance of the enclosing process definition. The static model of *Perform* is omitted here. The execution of a *Perform* is defined as follows. Firstly, all the necessary input bindings for the invoked process are executed. If all the inputs of the invoked process have been initialized, the process is executed. We also need to insure that all the output values are produced.

```
┌─ Choice ───────────────────────────────
│ ControlConstruct
│  ┌─────────────────────────────────────
│  │ components : ℙ ControlConstruct; .....
│  │ ┌─ selectComp ──────────────────────
│  │ │ comp! : components
│  │
│  │ E ≘ selectComp • comp!.E
∅
```

```
┌─ IfThenElse ───────────────────────────
│ ControlConstruct
│  ┌─────────────────────────────────────
│  │ ifCondition : Condition
│  │ then, else : ControlConstruct; .....
│  │
│  │ E ≘ [ condition.value.val ⇔ true ] • then.E []
│  │     [ condition.value.val ⇔ false ] • else.E []
│  │     [ ¬ (condition.value.val ⇔ true ∨ condition.value.va
```

Sequence, as a subclass of *ControlConstruct*, denotes a list of control constructs to be done in order, while the *ControlConstruct Split* has a set of process components and they are executed together. *Choice* calls for the execution of a single control construct from a given set of control constructs. Any of the given control constructs may be chosen for execution. *If-Then-Else* class is a control construct that has properties *ifCondition*, *then* and *else* holding different aspects of the If-Then-Else. The execution for other kinds of control constructs can be defined as well.

3.3 Data Flow and Variable Bindings Execution

In OWL-S, there are two complementary ways to specify data flow between steps: *consumer-pull* and *producer-push*. The *consumer-pull* approach specifies the source of a datum at the point where it is used. OWL-S implements this convention by providing a notation for arbitrary terms as the values of input or output parameters of a process step, plus a notation for subterms denoting the output or input parameters of prior process steps. *Bindings* are used to specify the *consumer-pull* data flow. The attribute *to Var* of *Binding* denotes the variable to be bound. *ValueType*, *valueSource* and etc. are different kinds of possible value sources it receives from. They are also called *valueSpecifier*. The execution of a *Binding* is to assign to *to Var* with the value of *valueSpecifier*.

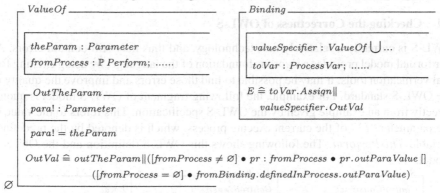

For example, *Binding*'s attribute *ValueSource* and its type *ValueOf* are one sort of data source. It is used to describe the source of a value that is a parameter (*theParam*) of another process within a composite process (*fromProcess*). We model that type of *fromProcess* as a set of composite processes with a predicate of cardinality less then one to show that there is at most one *fromProcess* is defined. If the *fromProcess* corresponds to the empty set, then *theParam* must be a different parameter from the same process. The static model of *ValueOf* is omitted here.

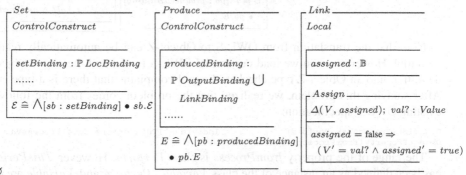

Producer-Push is another ways in OWL-S to specify how data flows. It is used when the output of a process depends on the branch of conditions the agent takes. So at the point in a branch where the data required to compute the output are known, we insert a *Produce* pseudo-step to say what the output will be. *Producer*, modeled as a subclass of *ControlConstruct*, 'pushes' a value available at run time to an Output. The attribute *producedBinding* specify the target of data flow, which specifies either a *LinkBinding* or an *OutputBinding*. *LinkBinding* is one kind of *Binding* such that the variable to be bound must be a *Link* variable.

Link is a special kind of local variable which is used as a communication channel that may be written exactly once. We define an attribute *assigned* to denote if a link variable has been written before. Some initial and reset operations have been omitted.

4 Discussions

The formal specification of OWL-S can be beneficial to the Semantic Web service communities in many different ways, as discussed in the following Subsections.

4.1 Checking the Correctness of OWL-S

OWL-S is currently a relatively new technology, and thus may still contain errors. As our formal model provides a rigorous foundation of the language, by using existing formal verification tools, it may be possible to find those errors and improve the quality of the OWL-S standard. For example, the following fragment of OWL-S model is quoted directly from an example given by the OWL-S specification. This refers to the value of the parameter "*I*1" of the current execute process, which is denoted by the predefined variable *ThisPerform*. The following shows this OWL-S definition and the Object-Z model.

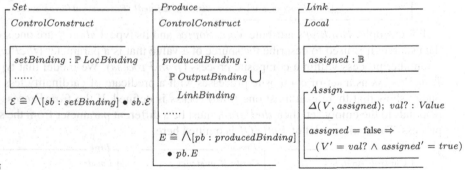

Note that the translation from OWL-S to Object-Z can be automatically realized by a tool. However, when we load our formal OWL-S model and the above Object-Z definition into an Object-Z type checker, the tool complains that there is a type error. After studying this problem, we realized that the problem comes from the following existing OWL-S specification:

```
Disjoint(ProcessVar,Perform)              Individual(ThisPerform type(ProcessVar))
ObjectProperty(fromProcess} range(Perform))
```

The range of the property *fromProcess* is class *Perform*. However *ThisPerform* has been defined as an instance of the class *Variable*. *Perform* and *Variable* are considered disjointed in many Semantic Web languages. Briefly speaking, the value of property *fromProcess* is expected to be an instance of *Perform*, while it has been assigned to an instance of variable (even its value denotes an instance of *Perform*). Thus, the OWL-S standard should be revised as follows:

```
ObjectProperty(fromProcess   range(unionOf(Perform ProcessVar)))
```

4.2   Providing a Unified and Precise Description of OWL-S

The syntax, static semantics and dynamic semantics of OWL-S are described separately using different formats, such as English, the Ontology itself, and some simple axioms. It has been very difficult to consistently extend and revise these descriptions with the continuous evolution of OWL-S. For example the text definition of *AtomicProcess* is described as "*.... for an atomic process, there are always **only** two participants, TheClient and TheServer.*" On the other hand, the following shows the corresponded ontological definition of *AtomicProcess*.

```
AtomicProcess
    (restriction(hasClient hasValue TheClient)  restriction(performedBy hasValue TheSever))
```

According the OWL semantics, the ontological definition requires that for an atomic process, there are **always** two participants, TheClient and TheServer. However, there could exist some other participants. Those inconsistent definitions can lead to many difficulties when OWL-S is used in practice. Furthermore, large sections of the OWL-S document are in normative text, which could result in several divergent interpretations of the language by different users and tool developers. Furthermore, the documentation makes many assumptions and implications, which are implicitly defined. This could lead to inconsistent conclusions being drawn. Our formal OWL-S model can be used to improve the quality of the normative text that defines the OWL-S language, and to help ensure that: the users understand and use the language correctly; the test suite covers all important rules implied by the language; and the tools developed work correctly and consistently.

4.3   Reasoning the OWL-S by Using Exiting Formal Tools Directly

Since research into Semantic Web Services in general, and OWL-S in particular is still evolving, current verification and reasoning tools (though rudimentary) for validating OWL-S models are also evolving. In contrast, there have been decades of development into mature formal reasoning tools that are used to verify the validity of software and systems. By presenting a formal semantic model of OWL-S, many Object-Z and Z tools can be used for checking, validating and verifying the OWL-S model. For example, in our previous work, we have applied Z/EVESto reason over Web ontologies. We also applied an Object-Z type checker to validate an OWL-S model. Instead of developing new techniques and tools, reusing existing tools provides a cheap, but efficient way to provide support and validation for standards driven languages, such as OWL-S.

4.4   The Ease of Extendibility

As OWL-S is still evolving, an advantage of using an object-oriented approach in the language model is to achieve the extendibility of the language model. Suppose that later a new kind of OWL-S *ControlConstruct* was defined. Then in our model it is necessary to add a subclass of *ControlConstruct* with proper defined '*enable*' condition and execution operation '*E*' to include this aspect: The introduction of the extension do not involve any changes to the classes defined in the previous section. Validation tools can then be used to confirm the validity of the extended model.

5   Conclusion

This paper has presented an Object-Z formal operational semantics of OWL-S. Together with our previous work [12], we have presented a complete formal model of OWL-S. The advantage of this approach is that the abstract syntax, static and dynamic semantics of each OWL-S construct are grouped together and captured in one single Object-Z class; hence the language model is more structural, concise and easily extendible. We believe this OZ specification can provide a useful document for developing support tools for OWL-S.

References

1. Ankolekar, A., Burstein, M., Hobbs, J., Lassila, O., McDermott, D., Martin, D., McIlraith, S., Narayanan, S., Paolucci, M., Payne, T., Sycara, K.: DAML-S: Web Service Description for the Semantic Web. In: First International Semantic Web Conference (ISWC) Proceedings, pp. 348–363 (2002)
2. Ankolekar, A., Huch, F., Sycara, K.P.: Concurrent execution semantics of daml-s with sub-types. In: Horrocks, I., Hendler, J. (eds.) ISWC 2002. LNCS, vol. 2342, pp. 318–332. Springer, Heidelberg (2002)
3. Dong, J.S., Duke, R.: Class Union and Polymorphism. In: Mingins, C., Haebich, W., Potter, J., Meyer, B. (eds.) Proc. 12th International Conference on Technology of Object-Oriented Languages and Systems. TOOLS 12, pp. 181–190. Prentice-Hall, Englewood Cliffs (1993)
4. Dong, J.S., Lee, C.H., Li, Y.F., Wang, H.: Verifying DAML+OIL and Beyond in Z/EVES. In: ICSE 2004. Proc. The 26th International Conference on Software Engineering, Edinburgh, Scotland, pp. 201–210 (May 2004)
5. Dong, J.S., Rose, G., Duke, R.: The Role of Secondary Attributes in Formal Object Modelling. Technical Report 95-20, Software Verification Research Centre, Dept. of Computer Science, Univ. of Queensland, Australia (1995)
6. Duke, R., Rose, G.: Formal Object Oriented Specification Using Object-Z. Cornerstones of Computing. Macmillan (March 2000)
7. Kim, S.K., Carrington, D.: Formalizing UML Class Diagram Using Object-Z. In: France, R.B., Rumpe, B. (eds.) UML 1999. LNCS, vol. 1723, Springer, Heidelberg (1999)
8. Narayanan, S., McIlraith, S.A.: Simulation, verification and automated composition of web services. In: WWW 2002. Proceedings of the 11th international conference on World Wide Web, pp. 77–88. ACM Press, New York, NY, USA (2002)
9. Roman, D., Keller, U., Lausen, H., de Bruijn, J., Lara, R., Stollberg, M., Polleres, A., Feier, C., Bussler, C., Fensel, D.: Web services modeling ontology. Journal of Applied Ontology 39(1), 77–106 (2005)
10. Smith, G.: A fully abstract semantics of classes for Object-Z. Formal Aspects of Computing 7(3), 289–313 (1995)
11. Wang, H.H., Gibbins, N., Payne, T., Saleh, A., Sun, J.: A Formal Semantic Model of the Semantic Web Service Ontology (WSMO). In: The 12th IEEE International Conference on Engineering Complex Computer Systems, Auckland (July 2007)
12. Wang, H.H., Saleh, A., Payne, T., Gibbins, N.: Formal specification of owl-s with object-z. In: OWL-S: Experiences and Directions, Innsbruck, Austria (June 2007)
13. Woodcock, J.C.P., Brien, S.M.: W: A logic for Z. In: Proceedings of Sixth Annual Z-User Meeting, University of York (December 1991)

An Approach for Combining Ontology Learning and Semantic Tagging in the Ontology Development Process: eGovernment Use Case

Ljiljana Stojanovic, Nenad Stojanovic, and Jun Ma

FZI at the University of Karlsruhe, Haid-und-Neu Strasse 10-14,
76131 Karlsruhe, Germany
{Ljiljana.Stojanovic, Nenad.Stojanovic, Jun.Ma}@fzi.de

Abstract. In this paper we present a novel method for ontology development that combines ontology learning and social-tagging process. The approach is based on the idea of using tagging process as a method for refinement (pruning) of the ontology that has been learned automatically from available knowledge sources. In the nutshell of the approach is a model for the conceptual tag refinement, which basically searches for terms that are conceptually related to the tags that are assigned to an information source. In that way the meaning of the tags can be disambiguated, which support better usage of the tagging process for the ontology pruning. We have developed a software tool, an annotation framework, which realizes this idea. We present results from the first evaluation studies regarding the application of this approach in the eGovernment domain.

Keywords: social tagging, tagging modeling, ontology development.

1 Introduction

One of the main hindrances for an efficient ontology development process is the common agreed nature of an ontology. Indeed, an ontology provides a view on a domain that is common for a community of interest, what requires an active involvement of many individuals in the ontology development process. However, people are usually reluctant to such an involvement, what makes ontology development a difficult task. From this reason lots of methods have been developed for the automatic generation of an ontology from a set of relevant knowledge sources, known as ontology learning methods. However, although they can be used as a useful starting point for structuring a domain conceptually, these automatically learned ontologies required substantial effort for pruning, in order to be conceptually sound (e.g. lots of noisy concepts are usually produced). Moreover, due to its "automatic" nature, the produced ontology structures lack the real community flavour.

On the other side, recent popularity of collaborative community portals (e.g. flickr or del.icio.us) in which web users provide simple metadata about the published content, shows that users are willing to participate in a light-weight knowledge acquisition process (so called social tagging). Moreover, it seems that this process can be easily used for ontology development, since it simulates exactly the creation of a

B. Benatallah et al. (Eds.): WISE 2007, LNCS 4831, pp. 249–260, 2007.

common-shared vocabulary for a domain of interest. However, there are many challenges in translating "tags clouds", which are results of social tagging process, into a well-defined structure, which is in the nutshell of a formal ontology.

In this paper we present a novel method that supports an efficient usage of the tagging process in the ontology development task. The approach support users in the tagging process by providing them with the context of the tags they have assigned to an information resource. By considering that context, the user can easily extend/refine defined tags, or determine how/where to add these tags in the ontology. The approach is based on our previous work in the query refinement [1]. Indeed, we treat this challenge for tagging using the experience from the information retrieval (search in particular): very short queries are ambiguous and an additional refinement step is required in order to clarify what a user meant with a query, i.e. what is his information need. In an analogous way we are talking about the annotation (tagging) need, i.e. what did the user consider as relevant in a document he has tagged: if the tag consists of only 1-2 keywords, then it is very difficult to conclude properly what the user meant with that tagging. Consequently, the usage value of tags decreases. For example, due to this ambiguity in the meaning, they are not very suitable for the development of the conceptual model of the target domain (but they could be used). Therefore, our approach introduces tag refinement as a method for the disambiguation of the meaning of a tagging in order to define its conceptualisation.

We have developed an annotation framework that realizes this idea and we have used it for creating ontologies in three real use-case studies from the eGovernment domain. The evaluation results are very promising and we present some results related to the quality of the developed ontologies.

The paper is organised as follows: In the second section we analyse the problems in the ontology learning process. In third section we present our approach in details. Section four contains evaluation details. In section five we give some concluding remarks.

2   Ontology Learning: Problems in Current Approaches

Ontology learning is defined as methods and techniques used for constructing an ontology from scratch, enriching, or adapting an existing ontology in a semi-automatic fashion using several sources. The terms generation, ontology mining, ontology extraction are also used to refer the concept of semi-automatic ontology construction.

Many existing ontology learning environments focus either on pure machine learning techniques [2] or rely on linguistic analysis ([3], [4]) in order to extract ontologies from natural language text. Shallow NLP (natural language processing) is very frequently used as the backbone of the ontology learning process. Hearst patterns (e.g., NP_0 such as NP_1, NP_2, and | or NP_n) are frequently used in ontology learning approaches. For example, the statement "the diseases such as x, y, z" can be used for concluding that x, y, z are types (subclasses) of the concept diseases.

Since building an ontology for a huge amount of data is a difficult and time consuming task a number of tools such as TextToOnto [5], Text2Onto [6] , the Mo'k Workbench [2], OntoLearn [4] or OntoLT [3], have been developed in order to support the user in constructing ontologies from a given set of (textual) data.

In this work we have used Text2Onto tool, although, in principle, any other tool could be applied. Text2Onto (http://sourceforge.net/projects/texttoonto/) is a tool suite for ontology learning from textual data. It represents the learned ontological structures at meta-level. It can be translated to different kind of knowledge representation languages which are commonly used such as OWL, and RDFS. The data-driven discovery capability in Text2Onto keeps the robustness although changes are made in the corpus. Text2Onto based on linguistic analysis tool called GATE framework, combined with machine learning techniques. GATE allows basic linguistic processing such as tokenization, lemmatization, and shallow parsing which is flexible in algorithm. Text2Onto currently only supports ontology learning from English, German and Spanish texts. Text2Onto generates following modelling primitives: concepts, subclass-of relations, mereological relations (part-of) and general relations.

However, it is already shown that the precision of the ontology learning process is rather low, so that usually a very huge amount of ontology entities is discovered. Therefore, although ontology learning promises to learn ontologies automatically, the role of the human in this process is inevitable. For example, after ontology learning phase, Text2Onto approach requires the involvement of a domain expert in the (manual) selection of entities that are suitable for the inclusion in the ontology (it can be treated as an ontology pruning process). Besides requesting an additional (significant) effort from the domain experts, such an approach has the general drawback that the community-flavour of the learned ontology is lost. Indeed, as mentioned above, ontology learning algorithms have statistical nature and do not reflect necessary the community aspects. There is no guarantee that the terms that are relevant for the domain from the ontology point of view will appear in a text corpus very frequently and vice versa. Finally, by reviewing the list of ontology candidates in the ontology pruning process only by one domain expert, the community-driven aspects of an ontology will be discarded.

In the next section we show how the process of social tagging can be used for the efficient involvement of community members in the ontology pruning process.

3   Social Tagging in the Ontology Pruning Process

3.1   The Approach

Tagging seems to be the natural way for people to classify objects as well as an attractive way to discover new material. Tagging services provide users with a repository of tagged resources (a.k.a tagspace) that can be searched and explored in different ways. Indeed, tagging is simple, it does not require a lot of effort and it is very useful to find the tagged objects later. People tag pictures, videos, and other resources with a couple of keywords to easily retrieve them in a later stage. Therefore, social tagging provides a "brilliantly lazy" way of constructing personal (private) context and as such it seems to be a suitable method for expressing community aspects of an ontology development (pruning) process.

However, there are three main challenges for using social tagging process directly for the ontology development (pruning) task:

- First, a user in the tagging process should have a more structured view on the existing tag structure, instead of being provided with the so called "tag cloud" (e.g. del.ico.us)
- Next, instead of just providing several keywords as tags for a document, more structured tagging information is needed to define the suitability of a tag for the inclusion in the ontology.
- Finally, if a term has been selected for the inclusion into the ontology, then the next task is to determine the right position of that term in the ontology.

Figure 1 illustrates how the approach we developed deals with these challenges. Note that one of the crucial requests for the approach was that the main benefits of a tagging process (and Web2.0 technologies in general) – simplicity in usage and low-entry barrier – should be retained.

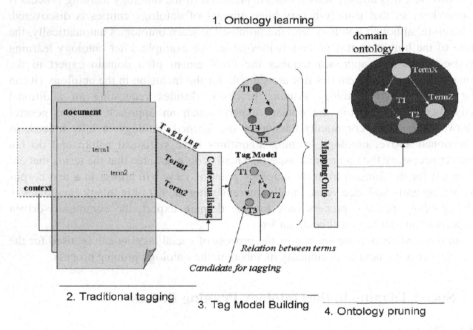

Fig. 1. Main steps in the proposed approach for using social tagging in pruning an ontology

The first task in the approach is the ontology learning process, which an ontology is developed in (c.f. Figure 1).

In the second step the traditional tagging process is performed, i.e. a user should define several keywords that express his/her view on a document (or a part of the document).

In the third step, the contextualisation of the tags is performed: the tags defined by users are put in the context of 1) the text, they are summarizing and 2) already existing domain. The result is the so called Tag Model that formalizes the meaning of the user's tags, by defining the relations between tags and their context. In that way the disambiguation of the meaning of tags is performed.

In the last step the previously derived Tag Model is mapped onto existing ontology in order to define the most appropriate position (if any) for including/placing the user's tag in the ontology.

In the rest of this section we describe the last two steps in more details.

3.2 Contextualisation: Tag Model Building

3.2.1 The Problem

The goal of this step is to transform a set of tags (several keywords), that a user has defined in the tagging process for a document (cf. Figure 1, step 2) into a more structured form, that will enable better understanding of the user's intention in the tagging process. We consider here the analogy to the "problem" of short Boolean queries in the information retrieval [7]: due to the ambiguity in the interpretation of the meaning of a short query, lots of irrelevant document are retrieved. Indeed, keyword-based tags do not define in a disambiguous way the intension of the user in the tagging process. For example, by tagging a page with terms "process" and "knowledge", a user has open a large space of possible interpretations what he meant: 1) "process knowledge" as a type of knowledge, 2) "knowledge process" as a type of process, or 3) "process that is somehow, but not directly related to knowledge". Therefore, the main problem for the effective usage of tags is to determine the context in which the user has used them.

3.2.2 The Conceptual Tagging Model

In order to enable better interpretation of a tagging (i.e. tags, that a user has assigned to a document) we have defined the Conceptual Tagging Model, illustrated in Figure 2. This work leverages our work in the Conceptual query refinement [1].

The interpretation of a tagging depends on the interpretations of the individual tag terms (cf. Figure 2, concept Term). The interpretation of a tag term can be defined through its relations with other terms (concept Relation, cf. Figure 2), whereas a relationship can be established between concepts Relations as well. Therefore, a meaning of a tagging is represented as a context that encompasses nested relations between tag terms (i.e. relations between relations and terms are allowed too). A meaning of a tagging represents so called tagging (annotation) need. For example, if a user has defined the tagging = "knowledge, management, quality" for a document, then the meaning (i.e. tagging need) "quality regarding knowledge management" can be in the given model represented as a relation between the term "quality" and a relation between terms "knowledge" and "management", like rel2("quality", rel1("knowledge", "management")). This is called a Tag model. Note that a tagging might have several meanings, i.e. it can be mapped into several Tag models.

We introduce several types of relation, which a term belongs to. Indeed, from the conceptual point of view, two terms are either in a specialization relation (cf. Figure 2, Specialize) (i.e. a term specializes the meaning of another term, like "process + workflow = process workflow"), in a modification relation (cf. Figure 2, Modify) (i.e. a term modifies the meaning of another term, like e.g. "process executed by a workflow") or in a co-occurrence relation (cf. Figure 2, Co-occurrence) (i.e. two terms just appear together in a context, without a direct influence on each other, like e.g. "… Process is described using a model, which can be found in the literature about

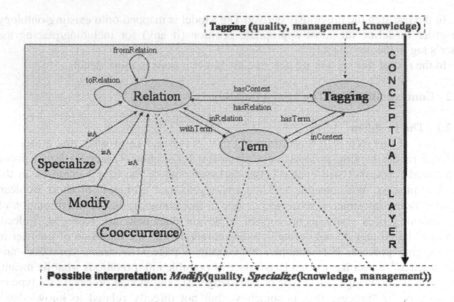

Fig. 2. Conceptual model for interpreting the meaning of a tagging is represented inside the central grey box. An example of interpreting the tagging "quality, knowledge, management" is illustrated as well (dotted box at the bottom).

workflows"). For the completeness of these relation set we rely on our previous work in the query conceptualisation [1].

3.2.3 Building Tag Models

As we already mentioned, a meaning of a tagging (i.e. an interpretation) is defined through the relations that can be established between tag terms, i.e. for a tagging (t_1, t_2, ... t_n) a meaning is defined as $rel_{x1}(t_1, ... (rel_{xz}(t_{n-1}, t_n))...)$, whereas rel_{xi} $\in$ {Specialize, Modify, Cooccurrence}.

Since a user defines a tagging regarding an information resource, an interpretation of the tagging should emerge from that resource. On the other side, the main information that can be derived from an information resource is the linguistic one (shallow NLP): there is a verb, there is a noun, there is a noun phrase, etc. Therefore, in order to build the meaning of a tagging, one has to process this information and derive conceptual relations between tag terms. Figure 3 sketches the process of building tag models.

In the nutshell of the approach is the consideration that the relations between the noun phrases determine the relations between tag terms contained in them. As we already mentioned, structural information are derived from the linguistic processing based on the shallow NPL performed in step 2 (cf. Figure 3). Complete description of linguistic patterns used in the steps 4 and 5 can be found in [1].

Therefore, by mapping the conceptual space to the linguistic space we can build tag models.

buildQueryModels
input:
    Tagging: $t_1, t_2, \ldots t_n$
    Document D
output: ordered list of Tag models
    $rel_1(t_1, rel_2(t_2, \ldots)), rel_i \in \{Specialize, Modify, Cooccurrence\}$.
1. Find the set of sentences from the document D that contain one of query terms (*SetSen*)
2. Perform shallow NLP on the set *SetSen*
3. Find the set of noun phrases from *SetSen* that contain one of tagging terms (*SetNP*)
4. Determine the structure (relation) in which each tagging term appear in the context of another query words, regarding the set *SetNP*
5. Create Tag models

Fig. 3. Procedural description of building query model. Some more explanations are given below in text.

3.2.4 Tag Refinement

By introducing the presented conceptual model of tagging, the tag refinement process can be seen as the refinement of a Tag model. Indeed, for a tag $(q_1, q_2, \ldots q_n)$, where q_i i=1, n are tag terms, several Tag models can be built and for each of them several refinements can be generated. For example: $tag(t_1, t_2, \ldots t_n)$ has several meanings in the form $rel_{x1}(t_1, \ldots (rel_{xz}(t_{n-1}, t_n))\ldots)$, that can be refined in several ways, e.g. $rel_{x1}(t_1, \ldots, rel_{xz-1}(t_{n-2}, rel_{xz}(t_{n-1}, t_n)))\ldots)$
whereas $rel_{xi} \in \{Specialize, Modify, Cooccurrence\}$.

Therefore, by using a conceptual representation of a tag, a refinement is not represented just as a bug of words, but rather as a structure with an implicitly represented meaning. It will enable a user to better understand the refinements and to focus on semantic relevant ones.

3.2.5 Suggesting New Terms for Tagging Using the Tag Model Refinement

Another experience we can reuse form the IR community is that users tend to build short queries and than to expand them incrementally. The same can be valid for the tagging process: user can start with one-two tags (that are easy to define) and then add other ones incrementally.

We assume that, by choosing a term, in the isolation as a tag (annotation) for an information source, a user cannot manage to express his "annotation need" completely. For example, by selecting just the term "expert" as a tag, the user fails to describe clearly what the page is about. What is missing is the context in which this term appears. The context can usually be described as the relation to some other terms, like "domain expert" or "instructions for expert". Obviously, this is much more descriptive than having the terms "expert", "domain" and "instructions" in isolation.

Therefore, in order to support the process of learning an ontology, the tagging process should provide the tags in the context of other terms from that document. In that way, the "material" that can be used for learning is much more useful. Indeed, form the quality of an ontology point of view, it is much better to learn the concept "domain expert" instead of learning concepts "domain" and "expert" in the isolation.

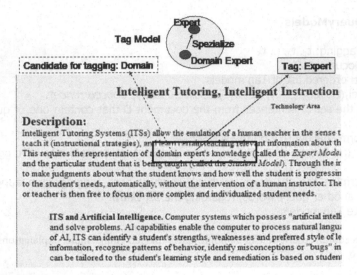

Fig. 4. An example of incremental tagging: User started with the tag "expert" and the system has suggested the refinement "domain expert", since the term "expert" appears in that context in the target document

Another, more challenging, problem is the mechanism for generating these more complex annotations. Indeed, we cannot assume that a user will provide compound annotations on his own (i.e. from scratch), but rather that the system can provide to him such a list of more complex annotations proactively. Obviously, this list should be based on the initial tag given by the user. For example, if the user annotates a page with "expert", the system should provide automatically suggestions like "domain expert".

The presented approach alleviates this process by discovering potential candidates for tagging. Indeed, if we consider the relations of a tag term to all other terms (not only included in tagging) from related noun phrases, then we can discover some potential candidates for tagging. Figure 4 illustrates such a situation: the user annotates the document with the term "expert", but the system discovers the "Specialize" relation to the term "domain", what leads to the tagging "domain expert".

3.3 Ontology Pruning

This phase is related to the extension of the ontology with terms generated in the tagging process. There are two main issues here:

1) how to determine which term should be added to the ontology;

2) if a term has been selected for adding into the ontology, then how to determine the right position of that term in the ontology.

Regarding first issue, we see an ontology as a common shared vocabulary, so that the relevance of a term for the community as the whole plays an important role. It doesn't mean that the terms that have been proposed only by a single person will be not treated as candidates for the inclusion in an ontology, but rather that the confidence that they are important for the whole communities decreases. The heuristics for updating an ontology is statistic-based: Lots of annotations with the same term means that corresponding term is a candidate for adding into the ontology. As the frequency of the same tags increases, in the same way increases the probability that the term will be relevant for the given domain.

In the next subsection we elaborate more on the second issue.

3.3.1 Adding New Ontology Term in the Right Place in the Ontology
Finding the right position of a term (included in taggings) for the placement in the ontology is a difficult task since not the term itself, but moreover the context in which it appears in the tagging(s) determine its role in the ontology.

However, as already elaborated, in the traditional tagging process the tagging terms are added in the isolation. Regarding the example from the Figure 6, without having a Tag model, the concept "domain" and "expert" will be created as sub-concepts of the top concept in the ontology, without having any conceptual relation between them. By using the Tag model the concept "domain expert" can be added as one entity in the ontology. However, the problem of finding the right position remains – without any additional information it will be added on the top of an ontology.

It is clear that from the point of view of the quality of an ontology, the approach to place new concepts very high in the hierarchy has lots of drawbacks: the ontology will be very flat and probably contain redundant parts. A solution could be that a human expert (i.e. an ontology engineer) adds these relations manually. However, for a human expert it is quite difficult to puzzle these isolated terms in a meaningful structure, since she/he don't know the context in which all these terms have been created. In other words, the specific context is which a term appears in a document is missing in order to determine the context (or to find the similar one) in the ontology. Moreover, such an approach of relying on one expert will again discard the community aspects of an ontology.

Therefore, any semi-automatic support would be very useful, not only for the efficiency of the approach, but also for the quality of the learned ontology. As already mentioned, especially important is to know the context in which a candidate for the adding into ontology appears. This problem is related to the section 3.2.5 in which a term appears in an information source.

However, in this case, the complementary information to the information provided in the tag model refinement process is needed. Indeed, instead of searching, like in the model refinement, for the possible refinements of the model, here is the search oriented towards finding model for a given refinement. This process we call inverse-refinement of a tag model.

We explain this process on the example illustrated in Figure 4. Let's assume that there is the concept "knowledge" in the ontology (related to some other concept) and that the term "domain expert" (as explained in Section 3.2.5) has been selected for the addition in the ontology. Instead of adding yet another term, the system will try to find a conceptual relation between that term and concepts from the ontology, but according to the context of the current page. From the current context selected in Figure 4, it is clear that this term appears in the context of the term "knowledge" that represents a concept from the ontology. In that case, if we perform inverse model refinement for the term "domain expert", then we will find the term "knowledge" as a possible model for the "domain expert". Finally, the user will be provided with the suggestion to define "domain expert knowledge" as a specialisation (isA relation) of the concept "knowledge". Therefore, although the user selected only the term "domain expert", the system concludes that the context in which it appears is a concept that already exists in the ontology ("knowledge") and suggests such a subconcept (subsumption) relation. In the case that inverse model refinement found that "domain expert" appears as a modification of the "knowledge", a nontaxonomic-relation between "domain expert" and "knowledge" will be suggested to the user.

In that way the "optimal" position of a new concept will be found in the ontology.

4 Evaluation

The approach presented in the previous section has been implemented in an annotation tool, that has been used for the ontology development in three eGovernment organisations. This work has been performed in the scope of the EU IST project FIT, http://fit-project.org.

The main prerequisite for the presented approach is the existence of digital documents (e.g. a web portal) that describe a domain that will be used for learning the domain ontology and which can be afterwards annotated by the members of the community. In the scope of the project we have applied our approach for the development of domain ontologies in three eGovernment organisations: a city in Austria, a ministry from Greece and a ministry from Serbia. The main sources for ontology learning were the web portals of the corresponding organisation. Text2Onto was used for the ontology learning task. Six experts from the domain (public administrators from the corresponding organisations) were used for the annotation task. Note that we learned one domain ontology pro organisation, since the problem domain is different for each of them (e.g. ministry for internal affairs in Greece vs. ministry for science in Serbia). The developed ontologies have been used as a means for the personalisation of the portal structure, i.e. the user behaviour in the portal is interpreted using the domain ontology in order to discover some potential problems, by analysing the problems users experienced in the past.

In this paper we present the evaluation of the quality of the developed ontologies

Since there is no gold standard ontology for the selected eGovernment domains, we could not evaluate the quality of the resulted ontology per se. Instead, we evaluated the effects of the pruning process, i.e. the added quality of the ontology, introduced by our approach. We compared the ontology produced by our approach with the ontology produced by an expert from the domain (who was not involved in the tagging process). In other words, that expert was provided with the results of the ontology learning process directly, like in the traditional ontology learning process. We are interested in what is the difference between the ontology pruning process performed by one domain expert (single person) and performed by a group of the community members. This could be measured by comparing the newly added concepts and relations introduced by expert and our approach. Results are summarized in Table 1 and Table 2. We present only the average values for all three use case studies. The variations between studies were present but not significant. For this paper we focus only in the validity of our ontology pruning process. It is the task

Table 1. Results from the evaluation study (part 1)

Approach	Different concepts (%)	Different relations (%)
Our	20%	40%
Traditional	5%	3%

Table 2. Results from the evaluation study (part 2). This data is inferred from Table 1 in order to make more clear the effects of our approach.

Approach	Same concepts (%)	Same relations (%)
Our / Traditional	80%	60%

for the future work to go into the quality of our approach regarding a particular situation and to conclude under which circumstances our approach best performs.

Discussion: This evaluation has shown that our approach could replace the human in the ontology learning process. Table 2 and Table 1 show, respectively, that our approach behaves similarly to an expert and adds new quality in the ontology. Indeed, Table 2 shows that 80% (60%) of selected concepts (relations) are the same as selected by an human expert and Table 1 shows that only 5% (3%) of concepts (relations) selected by the human expert are missing. On the other side, the quality added by our approach to the ontology development process is the significant additional number of entities that are selected for the inclusion in the ontology and which are a result of the common agreement between several domain experts (20% additional concepts and 40% additional relations). What we didn't evaluate formally is the quality of common-agreement achieved by our approach. This would be a part of the future work.

5 Conclusion

In this paper we presented a novel approach for combining ontology learning and semantic tagging in the ontology development process. The approach is based on the contextualisation of the tags provided by the user in order to define their ontological meaning (and enable the pruning of an ontology). The contextualisation is treated as a model refinement problem.

We have developed a tool for supporting the whole approach. The tool is realised as an annotation editor, which allows domain experts to tag interesting Web documents with their own personal concepts (folksonomies). By designing this tool we did not try only to exploit the advantages of these processes (easy to use vs. formal semantics). More important we tried to mix them in order to resolve their drawbacks. As already known, tagging systems suffer from having too little formal structure and easily result in "metadata soup". On the other side, annotation user interfaces cannot be based upon a closed, hierarchical vocabulary (i.e. ontology), since they are awkward and inflexible. A strict tree of ontology concepts does not reflect usage and intent.

The tool has been applied in three uses cases in the eGovernment domain. The first evaluation results are very promising. Our approach could support an ontology pruning process very efficiently. However, it is still to be done in evaluating the whole approach, especially the direct impact of the community in the pruning task.

Acknowledgement

The research presented in this paper was partially funded by the EC in the project "IST PROJECT 27090 - FIT".

References

1. Stojanovic, N.: Method and Tools for Ontology-based Cooperative Query Answering, PhD thesis, University of Karlsruhe, Germany (2005)
2. Bisson, G., Nedellec, C., Canamero, L.: Designing clustering methods for ontology building, The Mo'K workbench. In: Proceedings of the ECAI Ontology Learning Workshop, pp. 13–19 (2000)
3. Buitelaar, P., Olejnik, D., Sintek, M.: OntoLT: A protege plug-in for ontology extraction from text. In: ISWC. Proceedings of the International Semantic Web Conference (2003)
4. Velardi, P., Navigli, R., Cuchiarelli, A., Neri, F.: Evaluation of ontolearn a methodology for automatic population of domain ontologies. In: Buitelaar, P., Cimiano, P., Magnini, B. (eds.) Ontology Learning from Text: Methods, Applications and Evaluation, IOS Press, Amsterdam (2005)
5. Maedche, A., Staab, S.: Ontology learning. In: Staab, S., Studer, R. (eds.) Handbook on Ontologies, pp. 173–189. Springer, Heidelberg (2004)
6. Cimiano, P., Völker, J.: Text2Onto -A Framework for Ontology Learning and Data-driven Change Discovery. In: Montoyo, A., Muñoz, R., Métais, E. (eds.) NLDB 2005. LNCS, vol. 3513, pp. 227–238. Springer, Heidelberg (2005)
7. Silverstein, C., Henzinger, M., Marais, H., Moricz, M.: Analysis of a Very Large Alta Vista Query Log., SRC Technical Note, 1998-14 (1998)

Term Rewriting for Web Information Systems – Termination and Church-Rosser Property

Klaus-Dieter Schewe[1] and Bernhard Thalheim[2]

[1] Massey University, Information Science Research Centre
Private Bag 11222, Palmerston North, New Zealand
k.d.schewe@massey.ac.nz

[2] Christian Albrechts University Kiel, Department of Computer Science
Olshausenstr. 40, D-24098 Kiel, Germany
thalheim@is.informatik.uni-kiel.de

Abstract. The use of conditional term rewriting on the basis of Kleene algebras with tests is investigated as an approach to high-level personalisation of Web Information Systems. The focus is on the possible action sequences that can be represented by an algebraic expression called plot. By exploiting the idea of weakest preconditions such expressions can be represented by formal power series with coefficients in a Boolean algebra. This gives rise to a sufficient condition for termination based on well-founded orders on such power series. As confluence cannot be guaranteed, the approach further proposes critical pair completion to be used in order to enforce the desirable Church-Rosser property.

1 Introduction

On a high-level of abstraction a Web Information System (WIS) can be represented by a storyboard, which in an abstract way captures who will be using the system, in which way and for which goals. An important component of a storyboard is the *plot* that is specified by an assignment-free process, thus captures the possible action sequences, each of which is called a *story*. The formalisation of plots has been addressed in [1] using ASMs and in [13,15] using Kleene Algebras with Tests (KATs) [10]. In both cases the motivation for the formalisation was the ability to formally reason about a WIS specification already at a high-level of abstraction in order to capture the anticipated usage of the system.

In particular, the formalisation of a plot by an algebraic expression can be used to apply term rewriting on the basis of equations that may be used to express user preferences. In this way the research contributes to the important personalisation problem for WISs using an inferential approach rather than duplicating its specification. This idea was further investigated in [14] and [16]. In this paper we continue this line of research trying to clarify the relationship between the two approaches, and establish conditions, under which term rewriting can be effectively applied.

A system of equations on a KAT defines a quotient in a natural way. What we are interested in are "simple" representatives of equivalence classes in the case

B. Benatallah et al. (Eds.): WISE 2007, LNCS 4831, pp. 261–272, 2007.

of KATs that are generated out of some finite set of atoms. We derive a canonical representation through formal power series with coefficients in a Boolean algebra, which arises by adopting weakest preconditions. This is the basis for the definition of power series orders on KATs, which will allow us to characterize effective systems of equations by sufficient conditions for terminating term rewriting. In order to circumvent the confluence problems we adopt Critical Pair Completion (CPC) as the key element of the Knuth-Bendix algorithm [9]. We conclude by discussing challenges arising from the research done so far.

2   Personalisation of WISs as Term Rewriting

The more a WIS respects the personal preferences of its users on how to use the system, the higher are its chances of being accepted. Such preferences can refer to the content offered, the available functionality, and the presentation. Quite often the major emphasis of WIS personalisation is the presentation aspect. Our concern here is the functionality in the sense that we want to discard actions from a WIS specification, if the user preferences imply that they will not be executed. The high-level specification of a WIS by means of a storyboard is an open invitation for personalisation with respect to functionality, whereas personalisation with respect to content or presentation requires more details. Therefore, we concentrate on the *plot* that describes the possible sequence of actions [15].

2.1   Plots and Kleene Algebras with Tests

In order to formalise plots we refer to Kleene algebras with tests (KATs) [15] that are known to provide a formalism that exceeds the expressiveness of propositional Hoare logic [11]. While Kleene algebras are known since the cradle days of automata theory – they were introduced by Kleene in [8] – KATs are a more recent extension. Formally, we obtain the following definition.

Definition 1. A *Kleene algebra* (KA) $\mathcal{K} = (K, +, \cdot, ^*, 0, 1)$ consists of

- a carrier-set K containing at least two different elements 0 and 1, and
- a unary operation * and two binary operations $+$ and $\cdot$ on K

such that the following axioms are satisfied (adopting the convention to write pq for $p \cdot q$, and to assume that $\cdot$ binds stronger than $+$):

- $+$ and $\cdot$ are associative, i.e. for all $p, q, r \in K$ we must have $p + (q + r) = (p + q) + r$ and $p(qr) = (pq)r$;
- $+$ is commutative and idempotent with 0 as neutral element, i.e. for all $p, q \in K$ we must have $p + q = q + p$, $p + p = p$ and $p + 0 = p$;
- 1 is a neutral element for $\cdot$, i.e. for all $p \in K$ we must have $p1 = 1p = p$;
- for all $p \in K$ we have $p0 = 0p = 0$;
- $\cdot$ is distributive over $+$, i.e. for all $p, q, r \in K$ we must have $p(q+r) = pq + pr$ and $(p + q)r = pr + qr$;

- p^*q is the least solution x of $q + px \leq x$ and qp^* is the least solution of $q + xp \leq x$, using the partial order $x \leq y \equiv x + y = y$.

Definition 2. A *Kleene algebra with tests* (KAT) $\mathcal{K} = (K, B, +, \cdot, *, \bar{\ }, 0, 1)$ consists of

- a Kleene algebra $(K, +, \cdot, *, 0, 1)$;
- a subset $B \subseteq K$ containing 0 and 1 and closed under $+$ and $\cdot$;
- and a unary operation $\bar{\ }$ on B, such that $(B, +, \cdot, \bar{\ }, 0, 1)$ forms a Boolean algebra.

Example 1. Take a simple example, where the WIS is used for ordering products, which we adopt from [16]. In this case we may define four scenes:

- The scene $s_0 =$ product would contain product descriptions and allow the user to select products.
- The scene $s_1 =$ payment will be used to inform the user about payment method options and allow the user to select the appropriate payment method.
- The scene $s_2 =$ address will be used to require information about the shipping address from the user.
- Finally, scene $s_3 =$ confirmation will be used to get the user to confirm the order and the payment and shipping details.

There are six actions (their names are sufficient to indicate what they are supposed to do)

- $\alpha_1 =$ select_product is defined on s_0 and leads to a transition to scene s_1.
- $\alpha_2 =$ payment_by_card is defined on s_1 and leads to a transition to scene s_2.
- $\alpha_3 =$ payment_by_bank_transfer is defined on s_1 and leads to a transition to scene s_2.
- $\alpha_4 =$ payment_by_cheque is defined on s_1 and leads to a transition to scene s_2.
- $\alpha_5 =$ enter_address is defined on s_2 and leads to a transition to scene s_3.
- $\alpha_6 =$ confirm_order is defined on s_3 and leads to a transition to scene s_0.

and five Boolean conditions:

- The condition $\varphi_1 =$ price_in_range expresses that the price of the selected product(s) lies within the range of acceptance of credit card payment. It is a precondition for action α_2.
- The condition $\varphi_2 =$ payment_by_credit_card expresses that the user has selected the option to pay by credit card.
- The condition $\varphi_3 =$ payment_by_bank_transfer expresses that the user has selected the option to pay by bank transfer.
- The condition $\varphi_4 =$ payment_by_cheque expresses that the user has selected the option to pay by cheque.
- The condition $\varphi_5 =$ order_confirmed expresses that the user has confirmed the order.

Using these actions and conditions, we can formalise the story space by the algebraic expression $(\alpha_1(\varphi_1\alpha_2\varphi_2 + \alpha_3\varphi_3 + \alpha_4\varphi_4)\alpha_5(\alpha_6\varphi_5 + 1) + 1)^*$.

2.2   Quotient Kleene Algebras with Tests

As already stated in the introduction, we are interested in equations on KATs. Taking an interpretation by assignment-free process involving propositional conditions as in [13], such equations can be interpreted as user preferences and general knowledge about the application area. In particular, equations of the following form are of interest:

- If a process p has a precondition φ, then we obtain the equation $\bar{\varphi}p = 0$.
- If a process p has a postcondition ψ, we obtain the equation $p = p\psi$ or equivalently $p\bar{\psi} = 0$.
- An equation $\varphi(p_1 + p_2) = \varphi p_1$ expresses a conditional preference of p_1 over p_2 in case the condition φ is satisfied.
- An equation $p(p_1 + p_2) = pp_1$ expresses another conditional preference of p_1 over p_2 after the process p.
- An equation $p_1 p_2 + p_2 p_1 = p_1 p_2$ expresses a preference of order.

Thus, let Σ denote a set of equations on a KAT $\mathcal{K} = (K, B, +, \cdot, *, \bar{}, 0, 1)$. Then Σ defines an equivalence relation $\sim$ on K in the usual way, i.e. $p \sim q$ holds iff $p = q$ is implied by Σ and the axioms for KATs in Definitions 1 and 2.

Let K/Σ denote the set of equivalence classes $[p]$ under $\sim$ with $p \in K$, and B/Σ the subset $\{[b] \mid b \in B\}$. Define the KAT operations on K/Σ in the natural way, i.e. $[p] + [q] = [p + q]$, $[p][q] = [pq]$, $[p]^* = [p^*]$, $0 = [0]$, and $1 = [1]$, and the complementation on B/Σ by $\overline{[b]} = [\bar{b}]$. Then the axioms of Kleene algebras in Definitions 1 and 2 are easily verified leading to the following proposition.

Proposition 1. *If* $\mathcal{K} = (K, B, +, \cdot, *, \bar{}, 0, 1)$ *is a KAT and* Σ *a set of equations on* K, *then the quotient* $\mathcal{K}/\Sigma = (K/\Sigma, B/\Sigma, +, \cdot, *, \bar{}, 0, 1)$ *is a KAT.*   □

2.3   Term Rewriting with KAT Expressions

At the level of plots we can address personalisation with respect to functionality, i.e. we restrict the plot according to user preferences and goals. As this leads to equations as outlined above, we are actually looking for a simplified KAT expression p' for a plot p such that p and p' represent the same element in $\mathcal{K}/\Sigma$. In [16] we suggested to use term rewriting for this. As standard in order-sorted algebraic specifications we take two *sorts* B and K ordered by $B \leq K$, and the following (nullary, unary, binary) *operators*:

$$0, 1 :\to B \quad +, \cdot : K\,K \to K \quad * : K \to K \quad \bar{} : B \to B$$

Using these sorts and operators we can define *terms* in the usual way.

A *rewrite rule* is an expression of the form $\lambda \rightsquigarrow \varrho$ with terms λ and ϱ of the same sort, such that the variables on the right hand side ϱ are a subset of the variables on the left hand side λ. A *conditional rewrite rule* is an expression of the form $t_1 = t_2 \to \lambda \rightsquigarrow \varrho$, in which in addition the terms t_1 and t_2 contain the same variables and these form a superset of the set of variables in the left hand side term λ.

The application of a rewrite rule $\lambda \rightsquigarrow \varrho$ to a term t is standard: if t contains a subterm t' that can be matched with λ, i.e. there is a substitution θ such that the application of θ to λ results in t' (denoted $\theta.\lambda = t'$), then replace t' in t by $\theta.\varrho$.

The application of a conditional rewrite rule $t_1 = t_2 \rightarrow \lambda \rightsquigarrow \varrho$ to a term t is defined analogously. Precisely, if t contains a subterm t' that can be matched with λ, i.e. there is a substitution θ such that the application of θ to λ results in t' (denoted $\theta.\lambda = t'$), then replace t' in t by $\theta.\varrho$. However, in this case we have to show that $\theta.t_1 = \theta.t_2$ holds for the substitution θ. For this we start a separate term-rewriting process that aims at showing $\theta.t_1 \rightsquigarrow \cdots \rightsquigarrow \theta.t_2$. We call this separate rewriting process a *co-routine*, because we can run it in parallel to the main rewriting process. The risk is of course that if we fail to verify $\theta.t_1 = \theta.t_2$, then we have to backtrack to t for the main rewriting process.

In order to exploit term-rewriting for the personalisation problem we formulate the axioms of KATs and the personalisation equations as (conditional) rewrite rules, then start with $p\psi$ and apply the rules until we finally obtain a term of the form $p'\psi$ to which no more rule can be applied. Note that $p\psi$ is closed, i.e. it does not contain variables, so during the whole rewriting process we will only have to deal with closed terms.

We use $p, q, r, \ldots$ (if needed with additional indices) as *variables* of sort K, and $a, b, c, \ldots$ (also with indices) as *variables* of sort B. Then we use the following general (conditional) rewrite rules:

$$p + (q + r) \rightsquigarrow (p + q) + r \qquad\qquad p(qr) \rightsquigarrow (pq)r$$
$$p + p \rightsquigarrow p \qquad\qquad p + 0 \rightsquigarrow p$$
$$p1 \rightsquigarrow p \qquad\qquad 1p \rightsquigarrow p$$
$$p0 \rightsquigarrow 0 \qquad\qquad 0p \rightsquigarrow 0$$
$$p(q + r) \rightsquigarrow pq + pr \qquad\qquad (p + q)r \rightsquigarrow pr + qr$$
$$1 + pp^* \rightsquigarrow p^* \qquad\qquad 1 + p^*p \rightsquigarrow p^*$$
$$pq + q = q \rightarrow p^*q + q \rightsquigarrow q \qquad\qquad qp + q = q \rightarrow qp^* + q \rightsquigarrow q$$
$$p + q \rightsquigarrow q + p \qquad\qquad ab \rightsquigarrow ba$$
$$a\bar{a} \rightsquigarrow 0 \qquad\qquad \bar{a}a \rightsquigarrow 0$$
$$a + \bar{a} \rightsquigarrow 1 \qquad\qquad \bar{a} + a \rightsquigarrow 1$$

As an extension we apply these rewriting rules $\lambda \rightsquigarrow \varrho$ not only in one direction, but we also use the reverse rules $\varrho \rightsquigarrow \lambda$ (similar for conditional rewrite rules). In addition, the personalisation equations from above give rise to further rewrite rules:

- A conditional preference equation gives rise to a rule of the form $a(p + q) \rightsquigarrow ap$.
- A precondition gives rise to a rule of the form $\bar{a}p \rightsquigarrow 0$.
- A postcondition gives rise to a rule of the form $p\bar{a} \rightsquigarrow 0$.
- An invariance condition gives rise to a rule of the form $ap\bar{a} + \bar{a}pa \rightsquigarrow 0$.
- An exclusion condition gives rise to a rule of the form $ab \rightsquigarrow 0$.

Example 2. Let us continue Example 1 and localise the plot, i.e. personalise with respect to the location of its users. Consider a European user, for which the option to pay by cheque should be invalidated. So we obtain the equation $\alpha_4 = 0$. Furthermore, assume that this user prefers to pay by credit card, if this turns out to be a valid option. For this we use the equation $\varphi_1(\alpha_3 + \alpha_4) = 0$, which also implies $\varphi_1\alpha_3 = 0$. Let again $\psi = 1$. Now we compute

$$p\psi = (\alpha_1(\varphi_1\alpha_2\varphi_2 + \alpha_3\varphi_3 + \alpha_4\varphi_4)\alpha_5(\alpha_6\varphi_5 + 1) + 1)^*$$
$$\rightsquigarrow (\alpha_1(\varphi_1\alpha_2\varphi_2 + \alpha_3\varphi_3)\alpha_5(\alpha_6\varphi_5 + 1) + 1)^*$$
$$\rightsquigarrow (\alpha_1(\varphi_1\alpha_2\varphi_2 + (\varphi_1 + \bar{\varphi}_1)\alpha_3\varphi_3)\alpha_5(\alpha_6\varphi_5 + 1) + 1)^*$$
$$\rightsquigarrow (\alpha_1(\varphi_1\alpha_2\varphi_2 + \varphi_1\alpha_3\varphi_3 + \bar{\varphi}_1\alpha_3\varphi_3)\alpha_5(\alpha_6\varphi_5 + 1) + 1)^*$$
$$\rightsquigarrow (\alpha_1(\varphi_1\alpha_2\varphi_2 + 0 + \bar{\varphi}_1\alpha_3\varphi_3)\alpha_5(\alpha_6\varphi_5 + 1) + 1)^*$$
$$\rightsquigarrow (\alpha_1(\varphi_1\alpha_2\varphi_2 + \bar{\varphi}_1\alpha_3\varphi_3)\alpha_5(\alpha_6\varphi_5 + 1) + 1)^*$$
$$= p'\psi,$$

i.e. the option to pay by cheque has been removed from the resulting plot p', and the option to pay by bank transfer has become subject to the condition $\bar{\varphi}_1$, i.e. the price must be out of range for credit card payment.

2.4 Formalisation Using Abstract State Machines

In [1] an alternative formalisation of plots using Abstract State Machines was used. Abstract State Machines (ASMs) were originally introduced to capture the notion of sequential and parallel algorithms [2,6]. This leads among others to the definition of a language for the definition of ASM update rules [3]. In doing so, ASMs provide general formal means for developing software systems of all kinds in a systematic and sound way, which has been demonstrated by applications in various areas, e.g. [1,4,12,7].

Let us briefly relate the ASM-based approach to plot formalisation with our approach here, which is based on KATs. As a plot can be described by some assignment-free process, it should correspond to an assignment-free ASM. We will formally establish this relationship.

In a nutshell, we can define the *signature* Σ of an ASM by a finite set of function symbols. Each $f \in \Sigma$ is equipped with an *arity*, so that it can be interpreted by a partial function $A^n \rightarrow A$ for a given super-universe A.

Basic update *rules* are **skip** or assignments $f(t_1, \ldots, t_n) := t$ with terms $t_1, \ldots, t_n, t$ and a (dynamic) function symbol f of arity n. Furthermore, if r_1, r_2, r are rules, φ is a logical expression, and $t, t_1, \ldots, t_n$ are terms, then the following are (complex) rules:

- $r_1; r_2$ (sequence rule)
- $r_1 \| r_2$ (parallel rule)
- **if** φ **then** r_1 **else** r_2 **endif** (if rule)
- **let** $x = t$ **in** r **end** (let rule)
- **forall** x **with** φ **do** r **enddo** (forall rule)

– **choose** x **with** φ **do** r **enddo** (choice rule)
– $r(t_1, \ldots, t_n)$ (call rule)

The semantics of these constructs has been defined formally in [3] via consistent update sets. It is rather easy to see that KATs, if interpreted by processes can be expressed by ASMs. Obviously, 1 can be represented by **skip**, and 0 by the rule **choose** x **with** false **do skip enddo**.

Propositions $b \in B$ are used as a preconditions, such that bp can be represented by the rule **if** $b = 1$ **then** p **else** 0 **endif**. The operator · is the sequence operator, while choice $r_1 + r_2$ can be represented as

> **choose** x **with** $x = 0 \vee x = 1$
> **do if** $x = 0$ **then** r_1 **else** r_2 **endif enddo**

Finally, the Kleene star operator r^* can be represented by the rule **choose** n **with** $n \geq 0$ **do** $\hat{r}(n)$ **enddo** using the rule $\hat{r}(n)$, which is defined as **if** $n = 0$ **then skip else** $r; \hat{r}(n-1)$ **endif**

For an *assignment-free ASM* we start from the request that assignments are absent. Consequently, the forall rule and the let rule become obsolete, and it does no harm to introduce a star rule and a modified choice rule as shortcuts as defined above. Hence the only difference between KATs and assignment-free ASMs would be given by the parallel rule. However, if r_1 and r_2 are assignment-free, $r_1 \| r_2$ could be replaced by $r_1 r_2 + r_2 r_1$, which results in the following proposition.

Proposition 2. *KATs and assignment-free ASMs are equivalent.* □

Note that by using KATs or assignment-free ASMs we make heavily use of the call rule without parameters. Actually, as emphasised in [16] we are primarily interested in KATs $\mathcal{K} = (K, B, +, \cdot, *, \bar{\ }, 0, 1)$, in which the Boolean algebra B is finitely generated out of $\{b_1, \ldots, b_k\}$ using negation, conjunction and disjunction, while K is generated out of B and a set $\{p_1, \ldots, p_\ell\}$ of "elementary" processes, i.e. rules in ASM terminology.

In doing so, we actually hide the effects of the called rules in the sense that we do not know how they affect the values of the propositional variables $b_1, \ldots, b_k$. We may, however, introduce explicit pre- and postconditions for this purpose.

3 Termination and Church-Rosser Property

The outlined term rewriting process is indeed highly non-deterministic, but we wish to obtain a unique final result. For this we first have to ensure that the process terminates. If this can be done, then the second problem is to ensure confluence, i.e. guaranteeing that the result is independent from the choice of rewrite rules and their order. We will investigate these two problems in this section.

3.1   Formal Power Series

In [13] the idea was launched to approach personalisation by term rewriting on KATs. Indeed, the plot of a storyboard can be easily written as a KAT expression, and preference rules give rise to equations as exemplified in the previous section. Then the problem is to find a "simplest" representative $p' \in K$ of an equivalence class $[p] \in K/\Sigma$.

Equations like the ones shown in the previous section show indeed a preferred direction. We would rather see the left hand side replaced by the right hand side than the other way round. That is, we could use equations as rewrite rules $\ell \leadsto r$. This was exemplified in [13,15] and further elaborated in [16]. It may still be necessary to use rules such as $1 \leadsto b + \bar{b}$ for some $b \in B$, but for the moment we are primarily interested in termination.

For this we take up the old idea from [5] to define a well-founded order on terms in such a way that the left hand side of the rewrite rules will always be larger than the right hand side. In particular, we will look at an order defined by formal power series. Thus, our first step will be to look for ways to associate a formal power series with the expressions in a KAT $\mathcal{K} = (K, B, +, \cdot, *, \bar{\ }, 0, 1)$.

In order to so, we start with the assumption that we are given a set $A = \{p_1, \ldots, p_k\}$ of atomic processes and a set $P = \{b_1, \ldots, b_\ell\}$ of propositional atoms with $A \cap P = \emptyset$. Let B be generated out of $P \cup \{0, 1\}$ using conjunction $\cdot$, disjunction $+$, and negation $\bar{\ }$. Then let K be generated out of $A \cup B \cup \{0, 1\}$ using choice $+$, sequence $\cdot$, and iteration $*$. More precisely, we have to consider K/Σ with Σ being the KAT axioms from Definitions 1 and 2, but we ignore this little subtlety.

We even extend the Boolean algebra B by enlarging its base P. The idea is that for $p \in A$ and $b \in P \cup \{0\}$ we add the weakest preconditions $[p]b$ to P. The intended meaning is if p starts in a state satisfying $[p]b$, for which it is defined and terminating, then the resulting state will satisfy b. Formally, $[p]b$ is the largest solution x of $xp\bar{b} = 0$. For simplicity let us stay with the notation P, B and K.

First we have to ensure that these basic preconditions are sufficient. Looking at composed propositions we have

$$[p](b_1 b_2) = ([p]b_1)([p]b_2) \qquad \text{and} \qquad [p]\bar{b} = \overline{[p]0[p]b} \,,$$

which together with $b_1 + b_2 = \overline{\bar{b_1}\bar{b_2}}$ allows us to replace composed propositions by simply ones in P.

For complex processes in K we obtain $[1]b = b$, $[0]b = 0$, $[pq]b = [p][q]b$, and $[p + q]b = ([p]b)([q]b)$. In addition, $[p^*]b$ is the smallest solution x of $x = b[p]x$, but this may not be needed, as we can use the formal power series $p^* = 1 + p + p^2 + \cdots = \sum_{i=0}^{\infty} p^i$ for our purposes, which will result in

$$[p^*]b = \sum_{i=0}^{\infty} [p^i]b = \sum_{i=0}^{\infty} \underbrace{[p] \ldots [p]}_{i \text{ times}} b \,.$$

Definition 3. A *monom* over A of degree m is an expression of the form $q_1 \cdots q_m$ with all $q_i \in A$. A *formal power series* over A and P is an expression of the form

$$\sum_{i=0}^{\infty} \sum_{\sigma:\{1,...,i\}\to\{1,...,k\}} b_{\sigma(1)...\sigma(i)}p_{\sigma(1)} \cdots p_{\sigma(m)}$$

with coefficients $b_{\sigma(1)...\sigma(i)} \in B$. Let $\mathcal{FPS}(K)$ denote the set of formal power series over K.

Note that the Boolean algebra B in this definition contains all the preconditions, and the $p_{\sigma(1)} \cdots p_{\sigma(i)}$ are indeed monoms of degree i. Furthermore, assuming an order $\leq$ on A, this can be canonically extended to a total order on monoms, in which monoms of lower degree precede all those of higher degree.

Now, taking all these equations together with $[p]bp = pb$, which results from the definition of weakest preconditions, we obtain that all propositional conditions can be shifted to the front, which is summarised by the following proposition.

Proposition 3. *With the definitions above each KAT expression $p \in K$ can be represented by a formal power series $[p] \in \mathcal{FPS}(K)$.* □

3.2 Termination

The second step towards our goal of terminating term rewriting consists of defining a well-founded partial order on K such that for rewrite rules $\ell \rightsquigarrow r$ we are interested in we obtain $\ell \geq r$. This suffices to guarantee termination [5]. We will concentrate on fps orders.

Definition 4. An *fps order* $\leq$ on K is defined by a valuation $\nu : B \to \mathbb{N}$.
For KAT expressions $p_1, p_2 \in K$ with

$$[p_j] = \sum_{i=0}^{\infty} \sum_{\sigma:\{1,...,i\}\to\{1,...,k\}} b_{\sigma(1)...\sigma(i)}^{(j)} p_{\sigma(1)} \cdots p_{\sigma(i)}$$

we define $p_1 \leq p_2$ iff $p_1 = p_2$ or there exist some sequence α such that $\nu(b_\alpha^{(1)}) < \nu(b_\alpha^{(2)})$ and $\nu(b_\beta^{(1)}) = \nu(b_\beta^{(2)})$ hold for all $\beta < \alpha$.

For instance, we could take $b \in B$ in disjunctive normal form, and use this to define $\nu(b)$ using the equations

$$\nu(0) = 0 \qquad\qquad \nu(b) = \frac{1}{2} = \nu(\bar{b}) \text{ for } b \in P$$

and

$$\nu(b_1 b_2) = \nu(b_1) \cdot \nu(b_2) \qquad\qquad \nu(b_1 + b_2) = \nu(b_1) + \nu(b_2)$$

The fps order defined by this valuation guarantees that all rewrite rules obtained from the equations in Section 2 will have the desired property. More generally, we can now characterise some classes of rewrite rules that are of particular interest by the property that we obtain a descending sequence of KAT expressions with respect to some fps order.

Definition 5. A set $\mathcal{R}$ of rewrite rules $\ell \leadsto r$ derived from a set Σ of equations $\ell = r$ on K is *safe* iff there exists an fps order $\leq$ on K such that $r \leq \ell$ holds for all rewrite rules in $\mathcal{R}$.

Note that we do not require $r < \ell$, because we can only have finitely many equivalent terms. The following proposition states that indeed a safe set of rewrite rules will guarantee termination.

Proposition 4. *If $\mathcal{R}$ is a safe set of rewrite rules on K, then term rewriting with $\mathcal{R}$ will always terminate.*   □

Note that our definition of fps-order does not require the full expansion of the power series representation, which would be infeasible.

3.3 Confluence and Critical Pair Completion

Besides termination we are interested in a unique result of the rewriting process, for which we have to investigate confluence. Formally, (conditional) term rewriting defines a binary relation $\leadsto$ on terms. Let $\leadsto^*$ denote its reflexive, transitive closure.

Definition 6. A (conditional) term rewriting system is *confluent* iff for all terms t, t_1, t_2 whenever $t \leadsto^* t_1$ and $t \leadsto^* t_2$ hold, there exists a term t' with $t_1 \leadsto^* t'$ and $t_2 \leadsto^* t'$.

A (conditional) term rewriting system is *locally confluent* iff for all terms t, t_1, t_2 whenever $t \leadsto t_1$ and $t \leadsto t_2$ hold, there exists a term t' with $t_1 \leadsto^* t'$ and $t_2 \leadsto^* t'$.

A terminating and confluent (conditional) term rewriting system satisfies the *Church-Rosser property*, i.e. for each term t there exists a unique term t^* with $t \leadsto^* t^*$ such that no rewrite rule is applicable to t^*. The term t^* is usually called the *normal form* of t. In order context of WIS personalisation, in which t and t^* would represent plots, we call t^* the *personalisation* of t with respect to the preference rules used for term rewriting.

For terminating term rewriting systems the notions of confluence and local confluence coincide. Therefore, assuming that we use an fps-order we may restrict our attention to local confluence. Nevertheless, it is easy to see that the application of the rewrite rules as defined in the previous section will not be locally confluent. In order to circumvent this problem we suggest to adopt critical pair completion from the approach of Knuth and Bendix [9].

Definition 7. In case $t \leadsto^* t_1$ and $t \leadsto^* t_2$ hold and neither t_1 nor t_2 permit a further application of rewrite rules, the pair (t_1, t_2) is called *critical*.

The idea of critical pair completion is to add a new rewrite rule $t_1 \rightsquigarrow t_2$ (and its reverse) and continue the rewriting process. In the presence of an fps-order, however, we would check first if $t_1 \leq t_2$ holds or $t_2 \leq t_1$. In the former case we would add $t_2 \rightsquigarrow t_1$, in the latter one $t_1 \rightsquigarrow t_2$ to the set of rewrite rules. Obviously, in this way we enforce confluence and thus the Church-Rosser property.

Proposition 5. *If a safe set $\mathcal{R}$ of rewrite rules on K is combined with critical pair completion, term rewriting with $\mathcal{R}$ will satisfy the Church-Rosser property.*
□

Example 3. In Example 2 we actually applied the rewrite step $\alpha_3 \rightsquigarrow (\varphi_1 + \bar{\varphi}_1)\alpha_3$. This can be rewritten to $\varphi_1\alpha_3 + \varphi_1\alpha_3$ and further to $\varphi_1\alpha_3$ using the preference rule $\varphi_1\alpha_3 = 0$. On the other hand we could undo the introduction of the disjunction and rewrite the term to α_3. Critical pair completion would thus add the rewrite rule $\alpha_3 \rightsquigarrow \bar{\varphi}_1\alpha_3$ and thus enforce the Church-Rosser property.

4   Conclusion

In this paper we continued our work on Kleene Algebras with Tests (KATs) as an algebraic approach to deal with WIS personalisation. Under the assumption of weakest preconditions we developed a representation of certain KAT expressions by formal power series with coefficients in a Boolean algebra, which permits an effective, terminating approach to term rewriting for certain classes of equations on KATs. When combined with critical pair completion we can actually obtain a rewrite system satisfying the Church-Rosser property. We expect that the idea can be fruitfully exploited in practice, maybe even in areas beyond web information systems.

We further demonstrated that KATs are equivalent to assignment-free ASMs. There is of course one fundamental difference between ASMs in general and KATs, as besides assignments the former ones permit parallelism. As long as we do not take assignments into account, the absence of parallelism in personalisation is no loss, as in a KAT $p\|q$ can be represented as $pq + qp$, i.e. parallelism is treated merely as a choice of order. This is obviously false, if assignments are considered. As the move to lower levels of abstraction in WISs leads to replacement of assignment-free processes by database update operations and of propositions by (at least) first-order formulae, which could easily be formalised by ASM refinements, the challenge is to combine our results on rewriting on a high level of abstraction with ASM refinements.

A natural idea is to add a parallelism operator to KATs disregarding the fact that this could be represented by a choice of order. In order to generalise our approach to term rewriting we would need an extension to the idea of precondition. If such an extension is possible, we could still obtain a formal power series representation and hence guarantee that term rewriting will be terminating. Alternatively, we could try to shift the rewriting idea to the level of ASMs. We intend to explore both ideas in our future work.

References

1. Binemann-Zdanowicz, A., Thalheim, B.: Modeling information services on the basis of ASM semantics. In: Börger, E., Gargantini, A., Riccobene, E. (eds.) ASM 2003. LNCS, vol. 2589, pp. 408–410. Springer, Heidelberg (2003)
2. Blass, A., Gurevich, J.: Abstract state machines capture parallel algorithms. ACM Transactions on Computational Logic 4(4), 578–651 (2003)
3. Börger, E., Stärk, R.: Abstract State Machines. Springer, Heidelberg (2003)
4. Börger, E., Stärk, R., Schmid, J.: Java and the Java Virtual Machine: Definition, Verification and Validation. Springer, Heidelberg (2001)
5. Dershowitz, N.: Termination of rewriting. Journal of Symbolic Computation 3(1/2), 69–116 (1987)
6. Gurevich, J.: Sequential abstract state machines capture sequential algorithms. ACM Transactions on Computational Logic 1(1), 77–111 (2000)
7. Gurevich, J., Sopokar, N., Wallace, C.: Formalizing database recovery. Journal of Universal Computer Science 3(4), 320–340 (1997)
8. Kleene, S.C.: Representation of events in nerve sets and finite automata. In: Shannon, McCarthy (eds.) Automata Studies, pp. 3–41. Princeton University Press (1956)
9. Knuth, D.E., Bendix, P.B.: Simple word problems in universal algebras. In: Computational Problems in Abstract Algebra, pp. 263–297. Pergamon Press, Oxford, UK (1970)
10. Kozen, D.: Kleene algebra with tests. ACM Transactions on Programming Languages and Systems 19(3), 427–443 (1997)
11. Kozen, D.: On Hoare logic and Kleene algebra with tests. In: Logic in Computer Science, pp. 167–172 (1999)
12. Prinz, A., Thalheim, B.: Operational semantics of transactions. In: Schewe, K.-D., Zhou, X. (eds.) Database Technologies 2003: Fourteenth Australasian Database Conference. Conferences in Research and Practice of Information Technology, vol. 17, pp. 169–179 (2003)
13. Schewe, K.-D., Thalheim, B.: Reasoning about web information systems using story algebras. In: Benczúr, A.A., Demetrovics, J., Gottlob, G. (eds.) ADBIS 2004. LNCS, vol. 3255, Springer, Heidelberg (2004)
14. Schewe, K.-D., Thalheim, B.: An algorithmic approach to high-level personalisation of web information systems. In: Fan, W., Wu, Z., Yang, J. (eds.) WAIM 2005. LNCS, vol. 3739, pp. 737–742. Springer, Heidelberg (2005)
15. Schewe, K.-D., Thalheim, B.: Conceptual modelling of web information systems. Data and Knowledge Engineering 54(2), 147–188 (2005)
16. Schewe, K.-D., Thalheim, B.: Personalisation of web information systems - a term rewriting approach. Data and Knowledge Engineering 62(1), 101–117 (2007)

Development of a Collaborative and Constraint-Based Web Configuration System for Personalized Bundling of Products and Services

Markus Zanker[1,2], Markus Aschinger[1], and Markus Jessenitschnig[3]

[1] University Klagenfurt, Austria
markus.zanker@uni-klu.ac.at
[2] ConfigWorks Informationssysteme & Consulting GmbH, Austria
masching@edu.uni-klu.ac.at
[3] eTourism Competence Center Austria (ECCA)
markus.jessenitschnig@etourism-austria.at

Abstract. The composition of product bundles like tourism packages, financial services portfolios or compatible sets of for instance skin care products is a synthesis task that requires knowledgeable information system support. We present a constraint-based Web configurator capable of solving such tasks in e-commerce environments. Our contribution lies in hybridizing a knowledge-based configuration approach with collaborative methods from the domain of recommender systems in order to guide the solving process by user preferences through large product spaces. The system is implemented on the basis of a service-oriented architecture and supports a model-driven approach for knowledge acquisition and maintenance. An evaluation gives acceptable computation times for realistic problem instances.

1 Introduction

In many e-commerce situations consumers are not looking for a single product item but they require a set of several different products. For instance on e-tourism platforms online users may either selectively combine different items like accommodation, travel services, events or sights - or they might choose from an array of pre-configured packages. When bundling different items on their own users are performing a synthesis task comparable to configuration.

Mittal and Frayman [1] defined configuration as a special type of design activity, with the key feature that the artifact being designed is assembled from a set of pre-defined components. When bundling different products with each other, product categories represent these pre-defined components. Additional knowledge is required that states which combinations of components are allowed and which restrictions need to be observed. For instance, proposed leisure activities should be within reach from the guest's accommodation or recommended sights need to be appropriate for kids if the user represents a family. Nevertheless, the problem of computing product bundles that are compatible with a set of domain constraints differs from traditional configuration domains in the sense that

B. Benatallah et al. (Eds.): WISE 2007, LNCS 4831, pp. 273–284, 2007.

fewer restrictions apply. For instance a car configurator computes a valid vehicle variant satisfying the user's requirements and all applicable commercial and technical restrictions that derive from the manufacturer's marketing and engineering experts. Contrastingly, when configuring a product bundle relatively fewer strict limitations will apply, because not a new materialized artifact is created but an intangible composition of products or services is defined. As a consequence an order of magnitude more combinations of components are possible and the question of finding an optional configuration becomes crucial.

Optimality can be either interpreted from the provider perspective, e.g. the configuration solution with the highest profit margin, or from the customer perspective. In the latter case the product bundle that best fits the customer's requirements and preferences is proposed. Recommender systems are intended to derive a ranked list of product instances following an abstract goal such as maximizing user's utility or online conversion rates [2]. For instance collaborative filtering is the most common recommendation technique. It exploits clusters of users that showed similar interest in the past and proposes products that their statistically nearest neighbors also liked [3]. We exploit this characteristic of recommender systems for deriving a personalized preference ordering for each type of product in our configuration problem. Our contribution thus lies in integrating a constraint-based configuration approach with soft preference information that derive from recommender systems based on for instance the collaborative filtering paradigm. This way we can guide the solving process towards finding optimal product bundles according to the users assumed preference situation.

Contrasting the work of Ardissono et al. [4] and of Pu and Faltings [5] we do not require explicit preference elicitation by questioning or example-critiquing, but depending on the underlying recommendation paradigm community knowledge and past transaction data are utilized.

The paper is structured as follows: First we give an extensive elaboration on related work and then present a motivating example for illustration purposes in Section 3. Furthermore, we give details on the system development in Section 4 and finalize with practical experiences and conclusions.

2   Related Work

Configuration systems are one of the most successful applications of AI-techniques. In industrial environments they support the configuration of complex products and services and help to reduce error rates and throughput time significantly compared to manual processes. Depending on the underlying knowledge-representation mechanism rule-based, model-based and case-based frameworks for product configuration exist [6]. Configurators that build on the constraint satisfaction problem (CSP) paradigm are within the family of model-based approaches [7,8]. They clearly separate between an explicitly represented knowledge base and a problem solving strategy. In technical domains such as telephone switching systems large problem instances with tens of thousands of components exist. Efficient solving strategies exploit the functional decomposition of the product structure to determine valid interconnections of the different components with each other [7]. Pure sales configuration systems such as online car or pc configuration systems[1] are much simpler from the computational point of view. They allow their users to explore the variant space of different options and add-ons and ensure that users place orders that are technically feasible and have a correct price. However, these systems are typically not personalized, i.e. they do not adapt their behavior according to their current user.

The CAWICOMS project was among the first to address the issue of personalization in configuration systems [4]. They developed a framework for a personalized, distributed Web-based configurator. They build on dynamic user interfaces that adapt their interaction style according to abstract user properties such as experience level or needs situation. The system decides on the questioning style (e.g. asking for abstract product properties or detailed technical parameters) and computes personalized default values if the user's assumed expertise is not sufficient for answering. Pu and Faltings [5] present a decision framework based on constraint programming. They show how soft constraints are well-suited for supporting preference models. Their work concentrates on explicitly stated user preferences especially via an example critiquing interaction model. This way tradeoff decisions are elicited from users. Given a specific product instance users may critique on one product property and specify on which other properties they would be willing to compromise. Soft constraints with priority values are revised in such an interaction scenario and guide the solution search.

In contrast to the work in [4,5], we do not solely rely on explicitly stated user feedback, but integrate configuration with recommender systems to include assumed user preferences.

Recommender systems constitute a base technology for personalized interaction and individualized product propositions in electronic commerce [2]. However, they do not support synthesis tasks like configuration. Given sets of items and users, recommender systems compute for each single user an individualized list of ranked items according to an abstract goal such as degree of interest or buying probability [2]. Burke [9] differentiates between five different recommendation paradigms: *collaborative*, *demographic* and *content-based* filtering as well as *knowledge* and *utility-based* recommendation. Collaborative filtering is the most well known technique that utilizes clusters of users that showed similar preferences in the past and recommends those items to a user her cluster neighbors have liked [3,10]. Content-based filtering records those items the user has highly rated in the past and proposes similar ones. Successful application domains are for instance news or Web documents in general, where the system learns user preferences in the form of vectors of term categories [11]. Demographic filtering builds on the assumption that users with similar social, religious or cultural background share similar views and tastes. Knowledge- and utility-based methods rely on a domain model of assumed user preferences that is developed by a human expert. Jannach [12] developed a sales advisory system that maps explicitly stated abstract user requirements onto product characteristics and computes

---

[1] For instance, see http://www.bmw.com or http://store.apple.com

a set of best matching items. Case-based recommender systems exploit former successful user interactions denominated as cases. When interacting with a new user, the system retrieves and revises stored cases in order to make a proposition. Thus, a human expert is required to define efficient similarity measures for case retrieval [13]. Burke [14] and Ricci [15] have researched and developed several systems within this field.

We integrated our configurator with a polymorphic recommendation service that can be instantiated with an arbitrary recommendation paradigm [16]. Details will be given in Section 4.

Preference-based search takes a more interactive and dynamic approach towards personalized retrieval of items. They build and revise the preference model of the user during interaction, instead of having it beforehand. One of the first applications of interactive assessment of user preferences was the Automated Travel Assistant [17]. Further extensive works on interactive preference elicitation has been conducted recently [18,19,20] and [21] includes an extensive overview.

The work on preference-based search is orthogonal to our contribution. Our implementation supports interactivity between the system and the user during exploration of the search space. Therefore, additional preference constraints can be added and revised at each round of interaction (see Subsection 4.4).

3   Motivating Example

We start the description of our approach by giving a motivating example. Figure 1 depicts a service configuration scenario from the e-tourism domain. The user model states a set of specific requirements for *John* like a travel package to the city of Innsbruck or that the solution should be appropriate for a family with children. In addition, contextual parameters like the weather situation or the current season are represented in the system context. Concrete product instances and their evaluations are part of the Product model, e.g. sights or restaurants. The dotted arrows exemplify some constraints of the configuration knowledge base like *The location of the Sight/Restaurant/Event and the location of the Accommodation should be the same* or *If the weather outlook is rainy propose an indoor event*. Additional preference information is included into the configuration knowledge base by integrating recommendation services for each class of products: for instance, when collaborative filtering recommends items from the product class *Event*, transaction records of other users and the interaction history of *John* will be exploited. Consequently, the higher ranked items in the recommendation result are more probable to be included in the configuration solution.

Informally, given a User Model, a System context and a set of constraining dependencies, the task of the system is to find a set of products from different categories that is optimal with respect to the preference information derived from external recommender systems. In the following we will detail the development of the system.

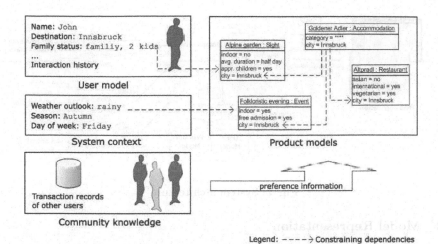

Legend: - - - - → Constraining dependencies

Fig. 1. Example scenario

4 Development

When designing the system we decided on the constraint programming paradigm for knowledge representation and problem solving - comparable to most of the configuration systems referenced in Section 2. The *Choco* Open Source constraint library [22] in Java is the basis for our implementation. In the next subsections we will describe the constraint-representation of the domain model, aspects of knowledge acquisition as well as our system architecture.

4.1 Architecture

In Figure 2 we sketch the system architecture. It consists of a configuration service component and several recommender services delivering personalized rankings of instances from a given class of products. We realized a service-oriented architecture that supports communication via Web services, a php-API and a Java-API. This enables flexible integration with different Web applications, distributed deployment scenarios and ensures the extensibility towards additional recommendation services. The latter requires sharing the identities of users and product instances as well as the semantics of user and product characteristics among all components. This is realized by a central user and product model repository offering service APIs, too.

The user interacts with a Web application, that itself requests personalized product bundles from the configurator via the service-API. The evaluation of contextual parameters can be requested by the same communication means. We have implemented variants of collaborative and content-based filtering recommenders as well as utility and knowledge-based ones as sketched in [16]. A more detailed evaluation of different recommendation strategies on commercial data was done in [23]. Next, we will detail on the domain model for the configuration service itself.

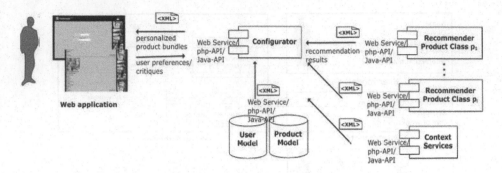

Fig. 2. System architecture

4.2 Model Representation

The constraint satisfaction paradigm is employed for knowledge representation. A Constraint Satisfaction Problem (CSP) is defined as follows [24]:

A CSP is defined by a tuple $\langle X, D, C \rangle$, where $X = \{x_1, \ldots, x_n\}$ is a set of variables, $D = \{d_1, \ldots, d_n\}$ a set of corresponding variable domains and $C = \{c_1, \ldots, c_m\}$ a set of constraints.

Each variable x_i may only be assigned a value $v \in d_i$ from its domain. A constraint c_j further restricts the allowed assignments on a set of variables. On each partial value assignment to variables it can be determined if a constraint is violated or not. In addition, all constraints $c_j \in C$ are defined to be either *hard* ($c_j \in C_{hard}$) or *soft* ($c_j \in C_{soft}$), where $C = C_{hard} \cup C_{soft}$ and $C_{hard} \cap C_{soft} = \emptyset$. Soft constraints may be violated. Each of them is associated with a penalty value and the sum of penalty values of violated constraints has to be minimized when looking for an optimal solution. For further details and definitions of CSPs we refer to [24].

Next, based on this formalization we present our domain model. It consists of a tuple $\langle U, P, X_{UM}, X_{Cx}, X_{PM}, D_{PM}, C_{hard}, C_{soft} \rangle$, where:

- $U = \{u_1, \ldots, u_n\}$ is a set of users,
- $P = \{p_1, \ldots, p_i\}$ a set of product classes - each is associated with a recommendation service that delivers a personalized item ranking upon request,
- $weight(p_j)$ the relative weight of product class p_j used in the overall optimization function,
- $X_{UM} = \{x_1, \ldots, x_j\}$ a set of variables from the User model,
- $X_{Cx} = \{x_1, \ldots, x_k\}$ a set of variables modeling the system context,
- $X_{PM} = \{p_1.x_1, \ldots, p_1.x_m, \ldots, p_i.x_1, \ldots, p_i.x_o\}$ a set of variables modeling product properties, where $p.x$ denotes the product property x of product class p and $p[j].x$ the concrete evaluation of x for product instance j,
- $D_{PM} = \{p_1.d_1, \ldots, p_1.d_m, \ldots, p_i.d_1, \ldots, p_i.d_o\}$ a set of corresponding domains for product properties,
- $C_{hard} = \{c_1, \ldots, c_p\}$ a set of hard constraints on variables in $X = X_{UM} \cup X_{Cx} \cup X_{PM}$,
- $C_{soft} = \{c_1, \ldots, c_q\}$ a set of soft constraints on variables in X and finally
- $pen(c_j)$ the penalty value for relaxing soft constraint c_j.

This domain model is defined and maintained by domain experts themselves in order to reduce the traditional knowledge acquisition bottleneck as outlined in the next subsection.

4.3 Model Definition and CSP Generation

Model definition and maintenance is supported by a modular editor environment based on the Eclipse RCP (Rich-Client Platform) technology[2]. Figure 3 gives an impression of the interaction with the knowledge acquisition workbench. First, those user characteristics that are relevant for solving are retrieved from the User model repository. Second, additional external services providing contextual data such as the current season of the year or weather information are selected. In a third step, the set of product classes P and their associated recommender services are integrated. For each class of products relevant properties are selected from the underlying repository. Finally, hard and soft constraints are defined using a context-sensitive editor.

In Figure 4 we depict the whole process. It is separated into a design phase, where the model is defined and maintained, and an execution phase. During the latter a specific user u is given, when the configurator is invoked. First, product rankings are retrieved from recommendation services and then corresponding product characteristics are requested from the product model repository. In the

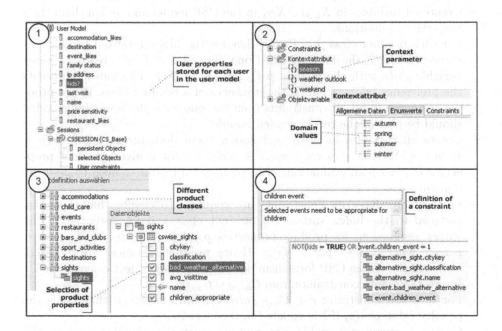

Fig. 3. Knowledge acquisition workbench

---

[2] See http://www.eclipse.org/rcp for reference

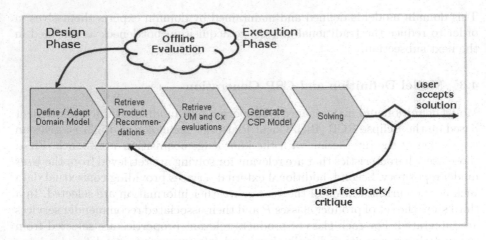

Fig. 4. Design and Execution phase

next step, all variables in $X_{UM} \cup X_{Cx}$ are assigned values by requesting the User model repository and the context services.

Then a CSP model is generated on the fly and the following transformation steps are done:

- Create all variables in $X_{UM} \cup X_{Cx}$ in the CSP model and assign them their respective evaluations.
- For each product class $p \in P$, $p[1]$ denotes the highest ranked product instance and $p[p_n]$ the lowest ranked one. For all $p \in P$ we create an index variable $p.idx$ with the domain $p.d_{idx} = \{1, \ldots, p_n\}$. The index represents the preference information on the instances of a product class. If two product instances fulfill all constraints then the one with the lower index value should be part of the recommended bundle.
- Create all variables in X_{PM} and assign them domains as follows: $\forall p \in P \forall p.x \in X_{PM} \forall i \in p.d_{idx} \quad p[i].x \in p.d_x$, i.e. for a given product property x of class p all evaluations of recommended instances need to be in its domain $p.d_x$.
- Furthermore, integrity constraints are required to ensure that the value assigned to the index variable of product class p is consistent with the values assigned to its product properties $p.x$: $\forall p \in P \forall p.x \in X_{PM} \forall i \in p.d_{idx} \quad p.idx = i \rightarrow p.x = p[i].x$. Hence product instances are modeled, although the *Choco* CSP formalism does not support object orientation.
- Insert all domain constraints from $C_{hard} \cup C_{soft}$.
- For each soft constraint $c \in C_{soft}$, create a variable $c.pen$ that holds the penalty value $pen(c)$ if c is violated or 0 otherwise.
- Create a resource variable res and a constraint defining res as the weighted sum of all variables holding penalty values for soft constraints and all weighted index variables. The tradeoff factor between soft constraints and

product class indexes as well as all weights can be adapted via the knowledge acquisition workbench.

4.4 CSP Solving

Once the CSP model is generated, the *Choco* solver is invoked utilizing its optimization functionality. The goal is to find an assignment to all variables in the CSP model that does not violate any hard constraint and minimizes the resource variable *res*. We extended the branch and bound optimization algorithm [22] to compute the top n solution tuples instead of solely a single product bundle. The solver is capable of following two different strategies for computing n product bundles, namely *1-different* and *all-different*. *1-different* signifies that each tuple of product instances in the set of n solutions contains of at least one product instance that is different in all other solution bundles. In contrast *all-different* means that the intersection between two product bundles from the set of solutions is empty, i.e. a product instance may only be part of at most one solution tuple. In the next section we report on computation times for CSP generation and solving.

5 Evaluation

Based on our e-tourism application domain we developed an example scenario consisting of 5 product classes with a total of 30 different product properties. Their domains are string, bounded integer and boolean. We defined a total of 23 representative domain constraints (13 hard and 10 soft constraints) for configuration knowledge.

We varied the number of the requested product instances from recommender services in 5 different steps between 5 and 100 and denoted the resulting CSP models M1 to M5. Details on problem sizes and times for 'on-the-fly' generation of CSP models are given in Table 1. As can be seen, the number of variables does not depend on the number of recommendations and stays the same for all models. However, the average size of variable domains increases from M1 to M5 due to the higher amount of product instances per product class. The number of constraints depends mainly on the aforementioned integrity constraints to model product instances. Therefore, M5 contains ten times more constraints than M1.

Table 1. Model sizes used in evaluation

Model	Number of Recommendations	Number of Vars	Average Domain Size	Number of Constraints	Generation time in ms
M1	5	58	7,45	206	10
M2	10	58	8,73	374	20
M3	30	58	13,55	1010	60
M4	50	58	16,5	1355	95
M5	100	58	23,23	2093	135

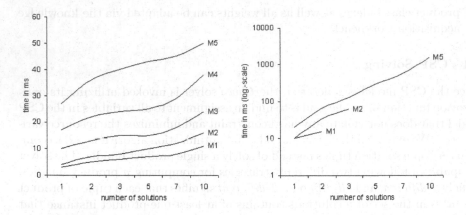

Fig. 5. Results for solution strategies *1-different* (left) and *all-different* (right)

As can be seen in Table 1, the times for generating even the large M5 model are acceptable for interactive applications.

In order to evaluate the performance of our system we ran the experiments on a standard Pentium 4 3GHz processor. The time measurements for the solving step are depicted in Figure 5. For each model we ran the experiment for 100 users and averaged times. We evaluated both strategies to compute a set of top n solution tuples as well as varied the number of n between 1 and 10. The solving times (y-axis) for the *all-different* strategy are on a logarithmic scale. Moreover, we skipped graphs for M3 and M4 for better readability. The *1-different* strategy is less complex and therefore solving time does not exceed 50 ms. To the contrary, *all-different* requires significantly longer computation times. Nevertheless, the performance is still satisfactory. The highest value for model M5 is about 1.5 seconds with 10 solutions computed for a reasonable problem in our domain. Although in typical e-commerce situations at most 10 different product bundles will be required, requests for 100 solution tuples can still be handled within a satisfying time range. For instance, when using the *1-different* solution strategy the solver required around 220 ms to calculate the top 100 solutions for model M5. In case of *all-different* 15 solutions are the maximum amount of solutions we could find within the same model. It took us 6 seconds which indicates the bottleneck of an interactive system.

Nevertheless, we have evidence to believe that our integration of a configurator with different recommender systems is efficient enough to solve standard product bundling tasks in online sales situations. Further experiments in different example domains will be conducted.

6   Conclusion

We presented the development of a generic Web configurator that is realized by integrating recommendation functionality with a constraint solver. Thus, our contribution lies in a novel strategy to personalize configuration results on product bundles. The system observes on the one hand explicit domain restrictions and on the other hand user preferences deriving from recommender systems. The application is developed within the scope of an industrial research project in the e-tourism domain. Our evaluation on realistic problem sizes showed acceptable computation times and system deployment is planned for the next months.

References

1. Mittal, S., Frayman, F.: Toward a generic model of configuration tasks. In: 11th International Joint Conferences on Artificial Intelligence, Menlo Park, California, pp. 1395–1401 (1989)
2. Adomavicius, G., Tuzhilin, A.: Towards the next generation of recommender systems: A survey of the state-of-the-art and possible extensions. IEEE Transactions on Knowledge and Data Engineering 17(6) (2005)
3. Resnick, P., Iacovou, N., Suchak, N., Bergstrom, P., Riedl, J.: Grouplens: An open architecture for collaborative filtering of netnews. In: Computer Supported Collaborative Work (CSCW), Chapel Hill, NC (1994)
4. Ardissono, L., Felfernig, A., Friedrich, G., Goy, A., Jannach, D., Petrone, G., Schäfer, R., Zanker, M.: A framework for the development of personalized, distributed web-based configuration systems. AI Magazine 24(3), 93–108 (2003)
5. Pu, P., Faltings, B.: Decision tradeoff using example-critiquing and constraint programming. Constraints 9, 289–310 (2004)
6. Sabin, D., Weigel, R.: Product configuration frameworks - a survey. IEEE Intelligent Systems 17, 42–49 (1998)
7. Fleischanderl, G., Friedrich, G., Haselböck, A., Schreiner, H., Stumptner, M.: Configuring large systems using generative constraint satisfaction. IEEE Intelligent Systems 17, 59–68 (1998)
8. Mailharro, D.: A classification and constraint-based framework for configuration. Artificial Intelligence for Engineering Design, Analysis and Manufacturing 12, 383–397 (1998)
9. Burke, R.: Hybrid recommender systems: Survey and experiments. User Modeling and User-Adapted Interaction 12(4), 331–370 (2002)
10. Pazzani, M.: A framework for collaborative, content-based and demographic filtering. Artificial Intelligence Review 13(5/6), 393–408 (1999)
11. Balabanovic, M., Shoham, Y.: Fab: Content-based, collaborative recommendation. Communications of the ACM 40(3), 66–72 (1997)
12. Jannach, D.: Advisor suite - a knowledge-based sales advisory system. In: de Mantaras, L.S.L. (ed.) PAIS. 16th European Conference on Artificial Intelligence - Prestigious Applications of AI, pp. 720–724. IOS Press, Amsterdam (2004)
13. O'Sullivan, D., Smyth, B., Wilson, D.: Understanding case-based recommendation: A similarity knowledge perspective. International Journal of Artificial Intelligence Tools (2005)
14. Burke, R.: The Wasabi Personal Shopper: A Case-Based Recommender System. In: IAAI. 11th Conference on Innovative Applications of Artificial Intelligence, Trento, IT, pp. 844–849. AAAI, USA (2000)
15. Ricci, F., Werthner, H.: Case base querying for travel planning recommendation. Information Technology and Tourism 3, 215–266 (2002)

16. Zanker, M., Jessenitschnig, M.: Iseller - a generic and hybrid recommendation system for interactive selling scenarios. In: ECIS. 15th European Conference on Information Systems, St. Gallen, Switzerland (2007)
17. Linden, G., Hanks, S., Lesh, N.: Interactive assessment of user preference models: The automated travel assistant. In: UM. 5th International Conference on User Modeling, Lyon, France (1997)
18. Shimazu, H.: Expert clerk: Navigating shoppers' buying process with the combination of asking and proposing. In: IJCAI. 17th International Joint Conference on Artificial Intelligence, pp. 1443–1448 (2001)
19. Smyth, B., McGinty, L., Reilly, J., McCarthy, K.: Compound critiques for conversational recommender systems. In: IEEE/WIC/ACM International Conference on Web Intelligence (WI), pp. 145–151. IEEE Computer Society, Washington, DC, USA (2004)
20. Viappiani, P., Faltings, B., Pu, P.: Evaluating preference-based search tools: a tale of two approaches. In: AAAI. 22th National Conference on Artificial Intelligence (2006)
21. Viappiani, P., Faltings, B., Pu, P.: Preference-based search using example-critiquing with suggestions. Artificial Intelligence Research 27, 465–503 (2006)
22. Laburthe, F., Jussien, N., Guillaume, R., Hadrien, C.: Choco Tutorial, Sourceforge Open Source, http://choco.sourceforge.net/tut_base.html
23. Zanker, M., Jessenitschnig, M., Jannach, D., Gordea, S.: Comparing recommendation strategies in a commercial context. IEEE Intelligent Systems 22, 69–73 (2007)
24. Tsang, E.: Foundations of Constraint Satisfaction. Academic Press, London, UK (1993)

SRI@work: Efficient and Effective Routing Strategies in a PDMS [*]

Federica Mandreoli[1], Riccardo Martoglia[1], Wilma Penzo[2],
Simona Sassatelli[1], and Giorgio Villani[1]

[1] DII - University of Modena e Reggio Emilia, Italy
{federica.mandreoli, riccardo.martoglia, simona.sassatelli,
giorgio.villani}@unimo.it
[2] DEIS - University of Bologna, Italy
wpenzo@deis.unibo.it

Abstract. In recent years, information sharing has gained much benefit by the large diffusion of distributed computing, namely through P2P systems and, in line with the Semantic Web vision, through Peer Data Management Systems (PDMSs). In a PDMS scenario one of the most difficult challenges is query routing, i.e. the capability of selecting small subsets of semantically relevant peers to forward a query to.

In this paper, we put the *Semantic Routing Index (SRI)* distributed mechanism we proposed in [6] at work. In particular, we present general SRI-based query execution models, designed around different performance priorities and minimizing the information spanning over the network. Starting from these models, we devise several SRI-enabled routing policies, characterized by different effectiveness and efficiency targets, and we deeply test them in ad-hoc PDMS simulation environments.

1 Introduction

In recent years, information sharing has gained much benefit by the large diffusion of distributed computing, namely through the flourishing of Peer-to-Peer systems [1]. On the other hand, in line with the Semantic Web vision, the stronger and stronger need of adding semantic value to the data has emerged. In this view, Peer Data Management Systems (PDMSs) have been introduced as a solution to the problem of large-scale sharing of semantically rich data [5]. In a PDMS, each peer is enriched with a schema (involving a set of concepts and relations among them) representing the peer's domain of interests, and semantic mappings are locally established between peer's schemas. In order to query a peer, its own schema is used for query formulation, and query answers can come from any peer in the network that is connected through a semantic path of mappings [10].

In such a distributed scenario a key challenge is query routing, i.e. the capability of selecting a small subset of relevant peers to forward a query to. Let us consider, for instance, the sample scenario of a portion of network represented

---

[*] This work is partially supported by the FIRB NeP4B Italian project.

B. Benatallah et al. (Eds.): WISE 2007, LNCS 4831, pp. 285–297, 2007.

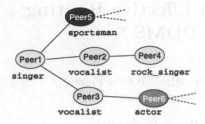

SRI$_{Peer1}$	singer	cd	title	...
Peer1	1.0	1.0	1.0	...
Peer2	0.8	0.6	0.7	...
Peer3	0.6	0.4	0.5	...
Peer5	0.1	0.2	0.3	

(a) The network and concept mappings

(b) A portion of the Peer1's SRI

Fig. 1. The reference example

in Figure1(a). It is a PDMS organized in Semantic Overlay Networks (SONs) [4] where peers of the same color belong to the same SON. In the SON paradigm, nodes with semantically similar contents are clustered together in order to improve data retrieval performances. Let us suppose that a user asks Peer1 for the concept "singer". Figure 1(a) shows an example of a plausible set of pairwise and one-to-one mappings which can be defined from the queried concept (e.g. "singer" is mapped to the Peer2's concept "vocalist" which, in turn, is mapped to the Peer4's concept "rock_singer"). In such a scenario, devising routing policies which are able to exploit the semantic closeness among concepts in order to find the best paths for query forwarding is a fundamental issue. Indeed, notice that a random mechanism could propagate the query to Peer5 and thus ask for "sportsman" which is semantically dissimilar from "singer". On the other hand, a routing strategy limiting itself only to the neighboring peer mappings would prove inadequate in many cases. For instance, though "vocalist" is the corresponding concept for both Peer2 and Peer3 which thus appear equivalent for query forwarding, the path originating at Peer2 is indeed much better, since Peer3's subnetwork includes the Peer6's concept "actor". Therefore, some kind of information about the relevance of the whole semantic paths should be available in the network, maintained up-to-date, and easily accessible for query routing purposes. To this end, in [6], [7] we proposed a PDMS architecture where each peer maintains a *Semantic Routing Index (SRI)* which summarizes, for each concept of its schema, the semantic approximation "skills" of the subnetworks reachable from its neighbors, and thus gives a hint of the relevance of the data which can be reached in each path. A portion of the Peer1's SRI is shown in Figure 1(b).

With SRI at the PDMS's disposal, different routing strategies may be considered, each having effectiveness, efficiency or a trade-off between the two as its priority. For instance, coming back to our example, once the query has reached Peer2, two choices are possible: (a) Going on towards Peer4 which is the nearest unvisited peer (*efficient* strategy, the main objective being to minimize the query path); (b) navigating the way to Peer3, as it provides information more related to the queried concept (*effective* strategy, maximizing the relevance of the retrieved results). In this paper, we present two general SRI-based query execution models, designed around the two different performance priorities, efficiency and effectiveness. Both models are devised in a distributed manner through a protocol of message exchange, thus trying to minimize the information spanning over the network. Starting from these models, we devise several SRI-enabled routing policies (Section 3). Then, we deeply test the proposed approaches in complex PDMS simulation environments (Section 4). Finally, Section 5 relates our work to other approaches in the literature and concludes the paper.

2 Semantic Routing Indices

In our PDMS reference scenario each peer p stores local data, modelled upon a local schema S that describes its semantic contents, and it is connected through semantic mappings to some other neighboring peers p_j. Each of these semantic mappings M_{p_j} defines how to represent p's schema (S) in terms of the p_j' schema (S_j) vocabulary. In particular, it associates each concept in S to a corresponding concept in S_j according to a *score*, denoting the *degree of semantic similarity* between the two concepts. Such similarity is a number between 0 (no similarity) and 1 (identity).

Further, each peer maintains cumulative information summarizing the semantic approximation capabilities w.r.t. its schema of the whole subnetworks rooted at each of its neighbors. Such information is kept in a local data structure which we call *Semantic Routing Index (SRI)*. In particular, a peer p having n neighbors and m concepts in its schema stores an SRI structured as a matrix with m columns and $n + 1$ rows, where the first row refers to the knowledge on the local schema of peer p. Each entry $SRI[i][j]$ of this matrix contains a score expressing how the j-th concept is semantically approximated by the subnetwork rooted at the i-th neighbor, i.e. by each semantic path of mappings originated at p_j. For instance, going back to Figure 1, the score 0.8 in the Peer2 row and the "singer" column is the outcome of the aggregation of the scores associated to the paths Peer2 and Peer2-Peer4.

A formal definition of semantic mappings and SRIs can be found in [6], where the fuzzy theoretical framework of our work is presented.

3 SRI-Based Query Processing and Routing Strategies

Starting from the queried peers, the objective of any query processing mechanism is to answer requests by navigating the network until a stopping condition is reached (see Section 4 for a list of stopping conditions). A query is posed on the schema of the queried peer and is represented as a tuple $q = (id, f, sim, t)$ where: id is a unique identifier for the query; f is a formula of predicates specifying the query conditions and combined through logical connectives; sim is the semantic approximation associated with the semantic path the query is following, i.e. the accumulated approximation given by the traversal of semantic mappings between peers' schemas; t is an optional relevance threshold. When t is set, those path whose relevance sim is smaller than t are not considered for query forwarding.

3.1   Query Execution Models

Query execution can be performed following different strategies, according to two main families of navigation policies: The *Depth First* (*DF*) query execution model, which pursues efficiency as its objective, and the *Global* (*G*), or *Goal-based* model, which is designed for effectiveness. Both approaches are devised in a distributed manner through a protocol of message exchange, thus trying to minimize the information spanning over the network. For both models we experienced various routing strategies, which will be the subject of the next sections. In general, the models work in the following way: Starting from the queried node, a peer p, univocally identified as p.ID, receives query q and a list L of already visited nodes from the calling peer C through an "execute" message. Then, it performs the following steps: 1. Accesses its local repository for query results; 2. decides a neighbor $Next$ among the unqueried ones which forward the query to; 3. reformulates the query q into q_{Next} for the chosen peer, using the semantic mappings M_{Next} towards it; 4. forwards the query q_{Next} to the neighbor $Next$ and waits for a finite response from it. In order to accomplish the forwarding step, the semantic information stored at the peer's SRI is used to rank the neighbors, thus also taking into account the semantic relevance of results which are likely to be retrieved from their own subnetworks. In particular, in order to limit the query path, the DF model only performs a local choice among the p's neighbors, whereas the G model adopts a global ranking policy among the already known and unvisited peers. Once the most promising neighbor has been chosen, the query q is reformulated (through unfolding [10]) into q_{Next}. Then, q_{Next} is assigned the semantic approximation value obtained by composing the semantic approximation obtained so far, i.e. $q.sim$, and the approximation for $q.f$ given by the semantic mapping M_{Next} between the current peer and the neighbor $Next$, thus instantiating $q_{Next}.sim$. Finally, the query execution process can start backtracking, returning the control to the calling peer through a "restore" message. For this reason, NS, an array of navigation states, one for each query processed by p, is maintained at each node for query restoring. As we will see in detail, while the DF model performs backtracking only when a "blind alley" is reached, the G model constantly exploits it in order to reach back potentially distant goals.

3.2   The DF Query Execution Model

The DF query execution model provides an SRI-oriented depth first visiting criteria. The left portion of Figure 2 shows the algorithm in detail. In particular, once a peer receives an `executeDF(`q,C,L`)` message, step 1 is performed through `localExecution(`q,L`)`), step 2 through lines 01-07, step 3 through `prepareNextQuery(`q,q_{Next}`)` and step 4 through lines 11-15. Notice that only L "surfs" the network along with the query which, at each step, is locally instantiated on the peer's dictionary.

As to peer selection (lines 02-05), the evaluation of the semantic relevance of each neighbor N is performed by combining, according to a fuzzy logic approach,

```
executeDF(q,C,L):                        executeG(q,C,L,GL):

00   localExecution(q,L);                00   localExecution(q,L);
01   NeighborList = ∅;                    01   for each neighbor N of p except C
02   for each neighbor N of p except C    02     relevance = fScore(compose(q.sim,
03     relevance = combine(q.f,SRI[N]);                      combine(q.f,SRI[N])));
04     add (N,relevance) to NeighborList; 03     add (p.ID,N.ID,relevance,true) to GL;
05   PQ = order NeighborList in desc      04   order GL in desc order of relevance;
       order of relevance;
                                          restoreG(id,L,GL):
restoreDF(id,L):
                                          05   do
06   do                                   06     NextG = top(GL);
07     (Next,relevance) = pop(PQ);        07   while (NextG not NULL and NextG.active
08     prepareNextQuery(q,q_Next);                 = true and NextG.to ∈ L);
09   while (Next not NULL and             08   if (NextG is NULL)
             (Next ∈ L OR q_Next.sim <= q.t)); 09    stop execution
10   if (Next is NULL)                    10   else if (NextG.active = false
11     delete NS[q.id];                            OR NextG.relevance <= q.t)
12     send message restoreDF(q.id,L) to C; 11    clean(p.ID,GL,NS);
13   else                                 12     send message restoreG(q.id,L,GL) to C;
14     NS[q.id] = (q,PQ,C);               13   else if (NextG.from <> p.ID)
15     send message executeDF(q_Next,p.ID,L) 14   clean(p.ID,GL,NS);
             to Next;                     15     NH = selectNextHop(GL);
                                          16     send message restoreG(q.id,L,GL) to NH;
localExecution(q,L):                      17   else
                                          18     prepareNextQuery(q,q_NextG.to);
00   add p.ID to L;                       19     NS[q.id] = (q,C);
01   execute q on local repository;       20     NextG.active = false;
02   if (stopping condition is reached)   21     send message executeG(q_Next,p.ID,L,GL)
03     stop execution                              to Next;

prepareNextQuery(q,q_Next):               clean(id,GL,NS):

00   q_Next.id = q.id;                    00   for each goal G in GL
01   q_Next.f = reformulate(q.f,M_Next);  01     if (G.from = id and G.active = true)
02   q_Next.sim = compose(q.sim,combine   02       return;
       (q.f, M_Next));                    03     else if (G.to = id)
03   q_Next.t = q.t;                      04       target = G;
                                          05   delete target from GL;
                                          06   delete NS[id];
```

Fig. 2. Query execution algorithms for "DF" (left) and "G" (right) routing models

the similarity values in the SRI's row associated to N, i.e. $SRI[N]$, and corresponding to each concept c in the formula $q.f$: $\text{combine}(q.f, SRI[N])$. For instance, following our reference example in Figure 1 and dealing conjunction with the minimum, the combined score for a query on Peer1 towards Peer2 involving concepts "cd" and "title" connected through AND would be $min(0.8, 0.6) = 0.6$. The peer produces a list NeighborList and such list is then ordered on the basis of the peers' relevance (producing a priority queue PQ, line 05) such that the top neighbor roots the subnetwork with the best approximation skills. L is used for cycle detection, i.e. in order to avoid querying the same peer more times (line 09). The forwarding process starts backtracking when the list of unvisited neighbors for the current peer is empty, returning the control to the calling peer (lines 10-12). When a peer receives the message $\text{restoreDF}(id, L)$, it reactivates the navigation state $NS = (q, PQ, C)$ for the query identifier $q.id$ and then follows lines 06-15 either to select the next peer which to forward the query to, or to continue backtracking. Thus, DF progresses by going deeper and deeper until the stopping condition is verified.

Based on the DF model, we devised the two following routing policies, that will be tested in depth in Section 4:

- **DF policy**: the "standard" depth-first policy, straightly implementing the DF model;
- **DFF (Depth-First Fan) policy**: a variation of DF, performing depth-first visit with an added twist. Specifically, at each node, DFF performs a "fan" by exploring all the neighbors having similarity above threshold t, then it proceeds in depth to the best subnetwork, as DF does. DFF is an attempt to enhance DF, as it tries to capture in less hops more answers coming from short semantic paths and, thus, being potentially more relevant than those retrieved by DF.

Example 1. Let us consider our reference network and see how a query posed on Peer1 and involving the concept "singer" would be routed. We use the following notation: We present the routing sequence of hops as an ordered list, where each entry N means PeerN is accessed and queried, i.e. 3 stands for Peer3, while (N) denotes a backtracking hop through PeerN. We consider the navigation until peers 1 through 4 (the most relevant ones) have been queried. Threshold t is set to 0. For the DF policy this would be the behavior: 1, 2 (most promising subnetwork rooted at Peer1), 4, (2), (1), 3. For DFF: 1, 2, (1), 3, (a fan is performed before exploring the best subnetwork), (1), 5, (1), (2), 4.

3.3   The G Query Execution Model

Along with the DF model, we devised an additional one sharing some common principles but having effectiveness as its primary goal: The right part of Figure 2 shows the G model in detail. Differently from the DF one, in the G model each peer chooses the best peer to forward the query to in a "global" way: It does not limit its choice among the neighbors but it considers all the peers already "discovered" (i.e. for which a navigation path leading to them has been found) during network exploration and that have still not been visited. This is mainly achieved by managing and passing along the network an additional structure, called *Goal List* (*GL*), which is the evolution of DF's NeighborList (and *PQ*) and is a global ordered list of *goals*. Each goal G contains information useful for next peer selection. In particular, it represents an arc in the network topology, starting from an already queried peer and going to a destination (and still unvisited) one, and is structured as a quadruple: (G.from, G.to, G.relevance, G.active), where G.from is the goal's starting peer, G.to is the goal's destination (a neighbor of G.from), G.relevance is the relevance used for ranking and G.active is a boolean value which is used for G's logical deletion. As we will see, inactive goals are used to reconstruct the path to an already queried peer. *GL* is always kept ordered on the basis of the goals' relevances, which are calculated by means of the fScore() function (line 02). Notice that active goals are always kept ahead inactive ones in *GL*'s ordering. Similarly to DF, the fScore() argument represents the semantic relevance of the goal. However, in this case the computation is compose($q.sim$, combine($q.f$, SRI[N]))[1], since, differently

---

[1] compose is a fuzzy composition function. Possible choices are the minimum or the algebraic product; for more details see [6].

from DF, goals must be globally comparable and thus the whole path originating from the querying peer must be taken into account. Further, as we will see later in detail, different `fScore()` functions can be adopted in order to favor different priorities.

Following algorithm `executeG(q,C,L,GL)`: Each Peer, after local execution (line 00), updates GL with the current neighbors' goals (lines 01-03), orders GL (line 04) and simply selects the next goal by accessing the top goal (line 06). As in DF, L is used to avoid cycles (line 07). Then, if NextG starts from the current peer (this means that the destination peer is a neighbor) query forwarding is executed similarly to DF (lines 18-21). Instead, if this is not the case (lines 13-16), the query is sent back one hop at a time to the peer that inserted NextG in GL. Notice that goals are not directly removed from GL after being selected at line 06. Instead, they are set as inactive (line 20) (i.e. already visited) before forwarding the query to their destination peer (line 21). The path to reach a previously queried peer is obtained through the function `selectNextHop(GL)` (line 15), whose code will not be shown in detail for simplicity of presentation; such function exploits GL inactive goals' information in order to identify, hop by hop, the route backtracking to the selected goal. The traversed peers are sent message `restoreG(id,L,GL)`. Obviously going back to potentially distant goals (peers) has a cost in terms of efficiency but always ensures the highest possible effectiveness, since the most relevant discovered peers are always selected. Notice that, in this case, backtracking is not limited to the chain of calling peers, since the selected goal does not necessarily start from one of them.

When, during backtracking, the current peer identifies that all goals starting from itself have been navigated, it can remove unnecessary information from NS and GL. These steps are performed through function `clean(id,GL,NS)`, invoked by peer with identifier p.ID=id at lines 11 and 14. In particular, the function removes the navigation state related to q from NS and the goal G having G.to=p.ID, since the query will not be forwarded again to p. Finally, going back to main execution, if no more goals are available in GL the execution is stopped (line 08-09), while, if only inactive goals are left, final backtracking is executed (lines 10-12).

Based on the G model, we devised two routing policies, which differ on the basis of the `fScore()` function and which will be tested and compared to the DF ones in Section 4:

- **G policy**: The `fScore()` function is the identity and, thus, the goals are selected solely on the basis of their semantic relevance;
- **GH (Global Hybrid) policy**: This "hybrid" policy chooses goals following a trade-off between effectiveness and efficiency. This is achieved by introducing an ad-hoc parameterizable `fScore()` function, which does not only consider a goal G's semantic relevance $semRel$ but also its distance $hops$ (expressed in number of hops) from the current peer: $fScore(semRel) = semRel/(hops)^k$, $k = 0\ldots\infty$. By simply adjusting the value of k, the GH policy can be easily tuned more on efficiency ($k \to \infty$) or on effectiveness ($k \to 0$).

Example 2. Going back to our reference example, this would be the routing sequence for the G policy: 1, 2, (1), 3 (since the relevance of the goal to Peer3 is expected to be higher than the one to Peer4), (1), (2), 4.

4   Experiments

We implemented and put into practice the ideas behind our routing mechanisms in a complete infrastructure, called SUNRISE[2] [7], which completely supports the construction of a PDMS semantic network and its testing through a dedicated *simulation environment*. By means of the SUNRISE simulation module, we were able to reproduce the main conditions characterizing such an environment without using a real P2P system. Further, we could easily abstract from network and communication issues, while maintaining full control on the internal dynamics of the network management. The simulation module is based on SimJava 2.0, a discrete, event-based, general purpose simulator. Through this framework we modelled scenarios corresponding to networks of semantic peers, each with its own schema describing a particular reality. We chose peers belonging to different semantic categories, where the schemas of the peers in the same category describe the same topic from different points of view. As in [10], the schemas are derived from real world-data sets, collected from many different available web sites, such as the DBLP Computer Society Bibliography and the ACM SIGMOD Record and enlarged with new schemas created by introducing structural and terminological variations on the original ones. Then, we distributed these schemas in the network in a clustered way, i.e. the schemas belonging to the same semantic category are located close to each other. This reflects realistic scenarios where nodes with semantically similar content are often clustered together. As to the topology of the semantic network of mappings connecting the peers, we tested our techniques on different alternatives generated with the BRITE topology generator tool. The mean size of our networks was in the order of some hundreds of nodes.

In order to evaluate the effectiveness and efficiency of our routing mechanisms, we simulated the querying process by instantiating different queries on randomly selected peers and propagating them until a stopping condition is reached. We considered two alternatives: stopping the querying process when a given number of hops (*hops*) has been performed and measuring the quality of the results (*satisfaction*) or, in a dual way, stopping when a given satisfaction is obtained and measuring the required number of hops. Satisfaction is a specifically introduced quantity that grows proportionally to the goodness of the results returned by each queried peer. In particular, it is computed as the sum of the $q.sim$ values of the traversed peers. In this section, we compare our routing strategies presented in Section 3 considering two further policies: The Random (R) one, which moves randomly in the network and corresponds to a baseline, and the Global IP-based (GIP) one. The GIP strategy is a variation of the Global (G) mechanism: A direct connection is established between the current peer and the

---

[2] System for Unified Network Routing, Indexing and Semantic Exploration.

peer chosen for the following step, avoiding the hops needed to reach it in the original network topology. This policy involves the modification of the PDMS network (the IP addresses of the visited peers are stored and new connections are created when necessary) and it can not be considered a real P2P strategy, but it is an interesting upper-bound to be shown. Notice that all the results we present are computed as a mean on several hundreds of query executions on several networks.

4.1 Effectiveness and Efficiency Evaluation

We start by studying the effectiveness of the routing strategies with two types of experiments. For the first one, Figure 3 shows the trend of the obtained satisfaction when we gradually vary the stopping condition on hops. As we expected, the Random strategy is outdistanced by the others and the GIP one represents the upper-bound, since all the available hops are exploited for querying peers (there are no backtracking hops). Among the others, the G policy shows the best behavior as it selects for each step the available peer with the higher sim value. Further, the DFF mechanism is initially less effective than the DF one, since it uses a large number of hops for performing its "fan" exploration. Nevertheless, for higher stopping conditions, it becomes increasingly more effective until it outperforms the DF policy and almost reaches the G one: This is due to the fact that it visits nearer peers, which have a higher probability to provide better results. The results for the GH policy are shown in Figure 4, where the trends for different values of the k parameter are represented. Notice that all the curves for the GH strategy are included in the area delimited between the G and the DF ones: The G and DF policies are indeed two particular cases of the GH one, where only the relevance (for G) or hops (for DF) component is considered, i.e. k in the formula is set to 0 (for G) or ∞ (for DF). Further, increasing the k value, the obtained behavior gets closer to the G one, as more importance is given to the relevance component. The results for the second type of experiments are represented in Figure 5, where the percentage of satisfied queries (i.e the queries for

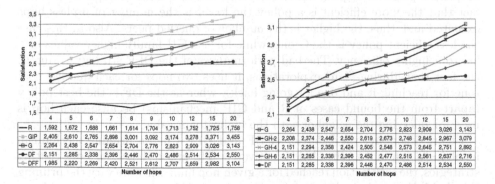

	4	5	6	7	8	9	10	12	15	20
—R	1,592	1,672	1,688	1,661	1,614	1,704	1,713	1,752	1,725	1,758
GIP	2,405	2,610	2,765	2,898	3,001	3,092	3,174	3,278	3,371	3,455
G	2,264	2,438	2,547	2,654	2,704	2,776	2,823	2,909	3,026	3,143
DF	2,151	2,285	2,338	2,396	2,446	2,470	2,486	2,514	2,534	2,550
DFF	1,985	2,220	2,269	2,420	2,521	2,612	2,707	2,859	2,982	3,104

Number of hops

	4	5	6	7	8	9	10	12	15	20
G	2,264	2,438	2,547	2,654	2,704	2,776	2,823	2,909	3,026	3,143
GH-2	2,208	2,374	2,446	2,550	2,619	2,673	2,748	2,845	2,967	3,079
GH-4	2,151	2,294	2,358	2,424	2,505	2,548	2,573	2,645	2,751	2,892
GH-6	2,151	2,285	2,338	2,396	2,452	2,477	2,515	2,561	2,637	2,716
DF	2,151	2,285	2,338	2,396	2,446	2,470	2,486	2,514	2,534	2,550

Number of hops

Fig. 3. Satisfaction reached by routing policies given a maximum number of hops

Fig. 4. Comparison of GH policy varying k w.r.t. G and DF policies

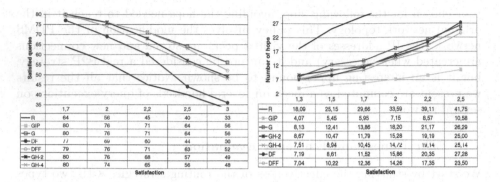

Fig. 5. Percentage of queries that reach a given satisfaction

Fig. 6. Mean number of hops needed to reach a given satisfaction

which the level of satisfaction given as stopping condition is reached) is shown for different levels of satisfaction. The G policy confirms to be the most effective one, while the performances of the GH ones decrease with the value of k. As for the family of the DF strategies, the earlier explained difference between DF and DFF is even further evidenced.

The results of the experiments aiming at verifying the efficiency of the routing strategies are shown in Figures 6 and 7. Figure 6 is the dual perspective of the situation in Figure 3, since it represents the trend of the number of required hops for a given satisfaction goal. As in the previous graph, the R and GIP strategies act as the lower and upper bounds respectively. The G policy is instead the less efficient one, since at each step it chooses the next peer to visit without considering the number of hops needed to reach it. As for the GH strategies, obviously we have that the higher is k the more efficient is the routing policy. It is interesting to note that also from this efficiency perspective, after the initial phase, the DFF policy outperforms the DF one.

Figure 7 shows another measure of the efficiency corresponding to the number of queried peers for a given number of hops: Given a limit of hops, we have in fact that the more efficient is a policy the lower is the number of useless hops executed and, thus, the higher is the number of queried peers. Also in this case the less efficient policy is the G one, while for the GH family higher values of k led to a higher efficiency. The bad performances of the DFF strategy is due to the fact that, during "fan" explorations, it uses two hops for each queried peer (one for reaching it and one for going back), whereas the DF is the best policy (apart from the bound cases of R and GIP mechanisms) because each hop is used for querying the neighbor with the highest sim value.

Figure 8 explores the behavior of DF and DFF when a pruning mechanism involving a non-null $q.t$ threshold is activated. The graph shows, for both routing strategies, the satisfaction reached for a given hop limit when we vary the pruning threshold. Notice that when we use a zero threshold, and consequently apply no pruning mechanism, we obtain the same results presented earlier in the section. As can be seen, increasing the threshold leads initially to an improvement of

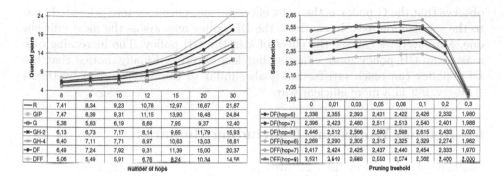

	8	9	10	12	15	20	30
—R	7,41	8,34	9,23	10,78	12,97	16,67	21,87
GIP	7,47	8,39	9,31	11,15	13,90	18,48	24,84
G	5,38	5,83	6,19	6,69	7,95	9,37	12,40
GH-2	6,13	6,73	7,17	8,14	9,65	11,79	15,93
GH-4	6,40	7,11	7,71	8,97	10,63	13,03	16,81
DF	6,49	7,24	7,92	9,31	11,39	15,00	20,37
DFF	5,06	5,49	5,91	6,76	8,24	10,34	14,56

Number of hops

	0	0,01	0,03	0,05	0,08	0,1	0,2	0,3
DF(hop=6)	2,338	2,355	2,393	2,431	2,422	2,426	2,332	1,980
DF(hop=7)	2,396	2,423	2,480	2,511	2,513	2,540	2,401	1,988
DF(hop=8)	2,446	2,512	2,566	2,590	2,598	2,615	2,433	2,020
DFF(hop=6)	2,269	2,290	2,305	2,315	2,325	2,329	2,274	1,962
DFF(hop=7)	2,417	2,424	2,425	2,437	2,440	2,454	2,333	1,970
DFF(hop=8)	2,521	2,540	2,560	2,560	2,574	2,502	2,400	2,000

Pruning treshold

Fig. 7. Mean number of queried peers given a maximum number of hops

Fig. 8. Pruning threshold effect on query satisfaction w.r.t. DF and DFF policies

satisfaction for both strategies. However, further increases of the threshold lead to lower values of satisfaction, signifying that useful subnetworks are pruned. The value of the optimal pruning threshold obviously depends on many factors, however we found out that setting it to 0.1 was a good choice in most tested settings, as Figure 8 shows.

The graph represented in Figure 9 deepens the study of the behavior of the GH strategies, showing the trend of satisfaction (continuous lines) and queried peers (dotted lines) for a given number of hops when we vary the value of k. As we expected, the higher the value of k, the lower is the obtained satisfaction and the higher is the number of queried peers. Increasing k we in fact give more importance to the hops component in the formula and consequently obtain more efficient but less effective strategies.

Finally, Figure 10 shows the relation between the number of queried peers and the satisfaction that every policy reaches given a maximum number of hops. In this way, we are able to visualize how the results obtained by the different policies position themselves in a combined effectiveness/efficiency plane. As expected, we

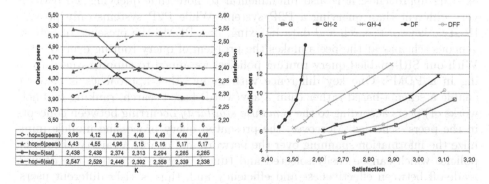

	0	1	2	3	4	5	6
hop=5(peers)	3,96	4,12	4,38	4,48	4,49	4,49	4,49
hop=6(peers)	4,43	4,55	4,96	5,15	5,16	5,17	5,17
hop=5(sat)	2,438	2,438	2,374	2,313	2,294	2,285	2,285
hop=6(sat)	2,547	2,528	2,446	2,392	2,358	2,339	2,338

K

Fig. 9. k influence on efficiency and effectiveness in GH policy

Fig. 10. Effectiveness vs. efficiency of routing policies

observe that the G policy is the most effective one, since its curve is located near to the satisfaction axis. In contrast, the DF policy appears as the most efficient one. Moreover, we can see the effect of k in the GH policy: The increasing of k makes the GH policy more efficient, but less effective. Finally, notice that the DFF policy can reach satisfaction goals similar to the ones reached by the G strategy, but in a more efficient way.

5   Related Works and Concluding Remarks

Much research work in the P2P area has studied query routing issues [9], [3], [2], [8], [11]. Most of these proposals are based on IR-style and machine-learning techniques [3], [2], [8], [11]. However, all of them provide routing techniques either assuming distributed indices which are indeed conceptually global [9], [8], or supporting completely decentralized search algorithms which, nevertheless, exploit information about neighboring peers only. An exception is [3] in which, however, the authors assume a common vocabulary in the network. Also in a P2P scenario, the work [12] is particularly interesting since, differently from many other works, it is completely focused on query routing policies' design and evaluation. Many of the proposed strategies are mainly designed for efficiency purposes, such as "iterative deepening" or "directed BFS", which are completely unaware of data location among peers. Instead, the "local indices" strategy makes use of distributed structures which can be configured to index, at each peer, the location of data on all surrounding peers. In this case, the indexed peers are not only limited to neighbors (an indexing radius r can be specified), however the index size is exponential on r and can not be efficiently maintained for large radii.

As we showed in [6], [7] and through the detailed experiments presented in Section 4, in order to obtain good routing effectiveness without losing network dynamism in difficultly maintainable indexing structures, some kind of summarized information about the data available in the neighbors' rooted subnetworks should be available: Our SRI framework serves this purpose and the SRI size is, for each peer, linear on the number of neighbors. Further, w.r.t. the cited existing approaches, it is also fundamental to note that querying a PDMS is very different than querying a P2P system: While P2P systems only provide IR-style keyword search and data indexing, in a PDMS the presence of heterogeneous schemas at the peers makes the problem of query routing even harder. With our SRI-enabled query routing policies, we aimed to support query routing in a PDMS: One key difference with many other approaches is that we enable a schema-based rather than a key(word)-based search, ranking (subnetworks of) peers according to the *semantic similarity* occurring between concepts in the peers' schemas. Moreover, the presented query processing models minimize the information spanning over the network and enable a variety of routing policies which can be easily selected and tuned in order to obtain the desired trade-off between effectiveness and efficiency and, thus, satisfy different users' needs.

References

1. Gnutella, http://www.gnutella.com/
2. Cooper, B.F.: Using Information Retrieval Techniques to Route Queries in an InfoBeacons Network. In: Ng, W.S., Ooi, B.-C., Ouksel, A.M., Sartori, C. (eds.) DBISP2P 2004. LNCS, vol. 3367, Springer, Heidelberg (2005)
3. Crespo, A., Garcia-Molina, H.: Routing Indices for Peer-to-Peer Systems. In: Proc. of ICDCS (2002)
4. Crespo, A., Garcia-Molina, H.: Semantic Overlay Networks for P2P Systems. In: Moro, G., Bergamaschi, S., Aberer, K. (eds.) AP2PC 2004. LNCS (LNAI), vol. 3601, Springer, Heidelberg (2005)
5. Halevy, A., Ives, Z., Madhavan, J., Mork, P., Suciu, D., Tatarinov, I.: The Piazza Peer Data Management System. IEEE TKDE 16(7), 787–798 (2004)
6. Mandreoli, F., Martoglia, R., Penzo, W., Sassatelli, S.: SRI: Exploiting Semantic Information for Effective Query Routing in a PDMS. In: Proc. of WIDM (2006)
7. Mandreoli, F., Martoglia, R., Penzo, W., Sassatelli, S., Villani, G.: SUNRISE: Exploring PDMS Networks with Semantic Routing Indexes. In: Proc. of ESWC (2007)
8. Michel, S., Bender, M., Triantafillou, P., Weikum, G.: IQN Routing: Integrating Quality and Novelty in P2P Querying and Ranking. In: Ioannidis, Y., Scholl, M.H., Schmidt, J.W., Matthes, F., Hatzopoulos, M., Boehm, K., Kemper, A., Grust, T., Boehm, C. (eds.) EDBT 2006. LNCS, vol. 3896, Springer, Heidelberg (2006)
9. Stoica, I., Morris, R., Karger, D., Kaashoek, F., Balakrishnan, H.: Chord: A Scalable Peer-To-Peer Lookup Service for Internet Applications. In: Proc. of ACM SIGCOMM (2001)
10. Tatarinov, I., Halevy, A.: Efficient Query Reformulation in Peer Data Management Systems. In: Proc. of ACM SIGMOD (2004)
11. Tempich, C., Staab, S., Wranik, A.: REMINDIN': Semantic Query Routing in Peer-to-Peer Networks Based on Social Metaphors. In: Proc. of WWW (2004)
12. Yang, B., Garcia-Molina, H.: Improving Search in Peer-to-Peer Networks. In: Proc. of ICDCS (2002)

Learning Management System Based on SCORM, Agents and Mining

Carlos Cobos[1], Miguel Niño[1], Martha Mendoza[1], Ramon Fabregat[2], and Luis Gomez[3]

[1] Systems Department, University of Cauca, Colombia
{ccobos, manzamb, mmendoza}@unicauca.edu.co
[2] Institute of Informatics and Applications (IIiA), University of Girona, Spain
ramon.fabregat@udg.es
[3] Systems and Computing Engineering School, Industrial University of Santander, Colombia
lcgomezf@uis.edu.co

Abstract. Based on SCORM sequencing and navigation specifications, a learning management system has been developed. The system has intelligent tutoring system capabilities that allow contents, presentation and navigation to be adapted according to the learner's requirements. In order to achieve that development, two concepts were put together: multi-agent systems and data mining techniques (especially the ID3 algorithm). All the implementation code was developed in VS.Net, which implied building a supporting framework for agents. The results of a pilot test were favorable.

Keywords: Learning Management System, Intelligent Tutoring System, Adaptive Hypermedia Systems, SCORM, Agents, Data mining, ID3.

1 Introduction

"Learning Management System (LMS) is a broad term that is used for a wide range of systems that organize and provide access to online learning services for students, teachers, and administrators. These services usually include access control, provision of learning content, communication tools, and administration of user groups. Another term that often is used as a synonym for LMS is learning platform. Two examples of well-known, commercial LMS systems are WebCT and Blackboard." [1]

The current state of development of Learning Management Systems (LMS) is mainly based on Sharable Content Object Reference Model (SCORM) specifications proposed by Advanced Distributed Learning (ADL) [2]. These specifications define a metadata schema for learning objects, a content package structure for courses and some sequencing and navigation rules (SN) [3]. SCORM SN defines both the learning activity concept and the sequencing rules for them. It also defines several smaller cluster units (the basic building blocks of learning activity [4]) and the content organization based on a tree structure called "Activity Tree" (AT) which can be useful as a Learning Instructional Template (LIT). Unfortunately, this AT is used by all learners regardless of particular skills or experience. Each student has his/her unique way of learning. A learning style is defined as characteristic strengths and preferences in the ways people take in and process information [5]. Many different learning style

B. Benatallah et al. (Eds.): WISE 2007, LNCS 4831, pp. 298–309, 2007.

assessment models and instruments are available. In order to build an AT adapted to each learner, a support model for several learning models is essential.

It is extremely ambitious for a teacher to build an AT by hand for each learner. Integrating Adaptive Hypermedia System (AHS) capabilities into an LMS facilitates the generation of personalized AT. The concept of adaptation has been widely investigated in the field of hypermedia systems [6]. "Adaptability" is the system's capacity to adapt itself automatically to the user from its suppositions about the user's necessities [7]. The developed LMS can adapt contents, presentations and navigation [6]. Data mining techniques are useful to analyze the learner profile (LP/user model) and to apply specific SN rules for each learner.

In this paper, agent architecture and a data mining approach are proposed to construct an adapted AT with an associated SN for each learner. Agents work cooperatively to monitor and adapt every activity and content used by the learner based on his/her learning style, while a decision engine based on an ID3 implementation [8] creates a decision tree (DT) and an algorithm called "Delivering" generates an AT for each learner.

In Section 2, related work is described and the added value of this work is explained. In Section 3, the main concepts and the relationships among them are presented in a conceptual model. In Section 4, the LMS architecture is presented to demonstrate the interaction of all agents and SCORM metadata with the decision engine. In Section 5, details of the agent's implementation are shown. In Section 6, the data mining approach with ID3 and Delivering are proposed. In Section 7, the results of a pilot test are presented. Finally, Section 8 presents conclusions and future work.

2 Related Works

Shang et al. [9] proposed an intelligent environment for active learning to support a student-centered, self-paced, and highly interactive learning approach. The learning environment can use the related learning profile (learning style and background knowledge) of students to select, organize, and present customized learning objects.

Triantafillou et al. [10] proposed an adaptive learning system, called AES-CS (Adaptive Educational System based on Cognitive Styles), in which learners are divided into two groups, Field Independent and Field Dependent, according to their cognitive styles. The AES-CS system provides an appropriate strategy and learning objects for each group.

Gilbert et al. [11] applied Case Based Reasoning (CBR) techniques to assign a new learner to the one group out of four with the most similar learning style. The system can offer adaptive learning material based upon the learning experience in the group selected.

In all systems mentioned above, the approaches rely only on the adaptation of learning objects for each group.

Weber and Specht [12] proposed an adaptive learning system called Episodic Learner Adaptive Remote Tutor (ELM-ART). It chooses the next best step according to the curricula presented to the learner. Links to pages are signaled following a "traffic light" metaphor that uses different colors to indicate recommended study sections, learned sections or study sections not recommended because the learner is not ready to learn them yet. However, they do not monitor learner's activities.

Chen et al. [13] apply DT and data cube techniques to analyze the learning behaviors of students and discover the pedagogical rules of students' learning performance from web logs including the amount of time spent reading article, posting article, asking question, login, etc. However, they do not apply educational theory to model the learning characteristics of learners, nor do they apply results to a new learner automatically.

Peña et al. [14] propose an AHS called MAS-PLANG based on learning styles. Here, the adaptation techniques focus on customized learning object selection and navigation tools. MAS-PLANG models a learner with its own monitor agents. However, they use only one model (Felder and Silverman Learning Style Model [5]).

Wang and Shao [15] propose a model called the Hierarchical Bisecting Medoids (HBM) Algorithm which integrates learner clustering, association mining techniques and historical navigation sessions by time. In the same group, the association mining technique was used to analyze navigation sessions and establish a recommendation model. In this approach, however, learning characteristics and student learning sequences were not considered, so personalized recommendation may not be appropriate.

Finally, Su et al. [16] propose an adaptive learning system based on SCORM SN and a four phase Learning Portfolio Mining (LPM) approach, which uses sequential pattern mining, a clustering approach, and DT creation sequentially to extract learning features from learning portfolios and to create a DT to predict which group a new learner belongs to. However, they do not use a multi-agent system for system implementation or adapt learning objects.

In this work, the automatic adaptation of learning objects, sequences and navigation rules classifies students according to their own learning styles while also

Table 1. Differences between this work and related work

Project vs. Characteristic	[9] [10] [11]	[12]	[13]	[14]	[15]	[16]	This work
Adaptation of learning objects	for each group	"traffic light" metaphor	No	object selection and navigation	historical navigation	No	for each learning style
Monitor learner's activities	MAS	No	No	MAS	No	No	MAS
SCORM supported	No	No	No	No	No	SN	metadata, SN
Based on education theory	active learning/ cognitive styles	No	No	learning models	No	No	learning models
Apply results to a new learner automatically	Yes	Yes	No	Yes	Yes	No	Yes
Multiple learning models	No	No	No	No	No	No	Yes
Data mining techniques	No	No	DT, data cube	No	clustering, association	DT, clustering	DT

DT: Decision Tree   SN: Sequencing and Navigation Rules   MAS: Multi Agent System

considering that styles can change over time. A collection of agents was implemented to monitor student activities. Moreover, it allows the institution to base its work on SCORM and teachers to use different learning models according to the experience they have with each one of them. Differences between this work and related works are shown in Table 1.

3 Conceptual Model

To understand all concepts involved in our proposal, we have divided the learning process into four stages. Each stage can be described in terms of roles, users, concepts and their relationships (see Fig.1):

1) A definition of models and learning styles, with questionnaires regarding each model that include questions and answers used by the LMS. Considering that teachers or institutions must have the freedom to choose and use the learning model they consider most appropriate, our work allows administrators and teachers to define the specific model and the questionnaires at the level of course templates or of specific courses.

2) A homogenous SCORM-based course template is defined. This template includes goals, resources and activities, AT, metadata, sequencing and navigation for all learners. Reutilization is an essential attribute promoted by this "homogenization" stage.

3) Using the course template, a teacher can customize a specific course promoting the concept of "free cathedra" that include, among others, activities/resources registry, learning model used by the course, a strategy (support/reject/leveling) depending on the student learning style.

4) Learners develop the course, beginning with a survey that reveals their learning styles according to the selected model, and then a customized AT or LIT is generated by the LMS to monitor and evaluate learning activities (learning based on learning styles).

The Felder-Silverman and the CHAEA (Honey-Alonso Questionnaire about Learning Styles) models are considered in this paper. Felder and Silverman [5] propose a bipolar dimensions model depending on the kind of information perceived by the learner (Active/Reflective, Sensing/Intuitive, Visual/Verbal and Sequential/Global). CHAEA defines four learning styles (Activists, Reflectors, Theorists and Pragmatists - similar to Kolb [17]) according to how information is perceived and processed. Concepts related to learning models are shown at the top of Fig.1.

To achieve content adaptation it was essential to extend the metadata of learning objects to include the relationship of each learning model with styles. For this reason, the concept of SCORM Metadata (plus) is shown in Fig. 1. Although this additional information is stored in a database, it is not present at the XML File of the learning object in order to maintain the original SCORM specification structure.

Therefore, the AT is customized to be used for relevant learning objects inside a specific activity. Even though this is an improvement, it creates a work overload for teachers because more than one learning object is required for each activity.

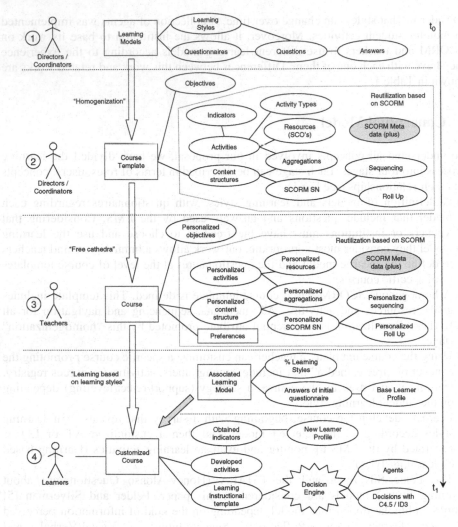

Fig. 1. AHS Conceptual Model. From top to bottom every concept involved in the customization stage for a learner's AT is shown.

Of course, the system can operate with only one learning object per activity, but in that case there is no real content adaptation.

4 LMS Architecture

The developed LMS (plus some standard LMS features such as registry, AT and learning object visualization, chat, forums, etc.) provides the work of a collection of agents cooperatively to support tasks like AT adaptation and learner activity monitoring.

The LMS has three different nodes: the content server and external repositories, the application server (Internet information server) and the Windows client application. The content server stores relational data, learning objects, etc., in a secure network server not accessible from the Internet. The application server stores web services in three layers (facade, business and services logic). Finally, the Windows client application establishes a communication with Web services through a user session with SOAP/XML. This application is based on smart-client architecture (see Fig. 2). Learners interact with the AT viewer, the learning objects viewer, the evaluating services and the anthropomorphic agents, among others.

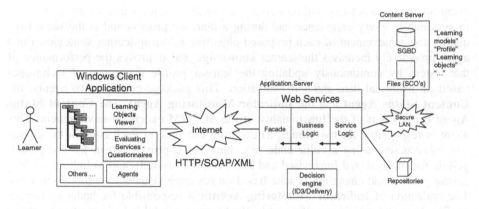

Fig. 2. LMS main components: Content server and repositories, application server and Windows client application

Agents are created in the Windows client application, but they have their logic in the application server (web services) and store all data in the content server. They are grouped according to behavior inside software packages. These software packages interact with the AHS using a communication bridge called **"Agents Coordination"**. Each package is described below.

Interface Agents. This package contains two agents, the **Interface Agent** and the **Anthropomorphic Agent**. The main purpose of the **Interface Agent** is to perform the teacher's role. It is responsible for communication tasks with learners. An **Anthropomorphic Agent** presents an animated anthropomorphic design inside a graphical user interface (GUI) to show that all messages are generated from other agents.

Data Access Agents. This package contains two agents, the **Profile Agent** and the **Learning Styles Agent**. The **Profile Agent** is responsible for managing and storing the user's identity in the content server according to content preferences, consumer settings and the user environment. Moreover, it recovers, modifies, and updates all learner data associated with a content. The **Learning Styles Agent** is responsible for storing the learner's information and associated learning styles in the content server (e.g. in a neuro-linguistic programming model, a learner could be 50% visual, 30% auditory and 20% kinesthetic).

Adapter Agents. These agents are responsible for the creation of the AT and its presentation to the user through the **Interface Agent** inside the **Content Structure Viewer** (a content explorer) according to the learner's specific, predominant learning style. These agents are capable of automatically recommending resources, sequences or navigation strategies according to a specific learning style. This package has two agents: the **Adapter Agent That Supports Style** and is responsible for finding resources related to a predominant learning style, and the **Adapter Agent That Reject Style** and is responsible for finding resources not related with a learner's style.

Monitoring Agents. These agents are responsible for continuously updating learner profiles by monitoring learning activities planned for a course; every learner event related to an activity will be stored in a profile. The main goal of this package is to systematize every experience had during a learning process and at the same time qualify the achievement of each proposed objective to set up learner strategies. Each agent proactively increases the learner knowledge and improves the performance of the system by continuously updating the learner profile using several techniques based on historical data and self evaluation. This package contains two agents, the **Content Status Agent** and the **Indicator Monitoring Agent.** The **Content Status Agent** creates part of the functionality of the SCORM execution environment. This agent is responsible for two things: a) updating the state of each node in the course content structure using a color code (e.g. blue for approved, green for completed, yellow for finished, red for revised and black for not revised), and b) linking itself to course content and changing its state based on resource visualization time indicators. The collection of **Indicator Monitoring Agents** is responsible for updating learner profiles based on activities performed by the learners. Each kind of activity has one or more indicators (comprehension level, number of times that a resource is being used, rest time during the activity and resource use time).

5 Agent's Implementation

The MAS-CommonKADS methodology [18] was used to analyze and design the system, since this methodology is based on knowledge engineering and aspects related to multi-agent systems are covered by it. These aspects are agent knowledge modeling, agent interaction modeling, and the integration of object oriented techniques. MAS-CommonKADS allows the inclusion of UML notation or AUML (Agent Unified Modeling Language) that is focused on the agent's field [19].

On the other hand, it was essential to have a platform or API to create and control the agent's life cycle because the implementation of the system was in VS.NET. A technological exploration was made of the tools used to develop multi-agent systems and only those tools which complied with the FIPA standards were considered. The tools found used Java as a platform making their interoperability with VS.NET technology difficult. Two alternatives were possible. The first was to achieve interoperability between Java and .NET platforms. (Bayo explains [20] how to build and instantiate agents in .NET through JADE.) The second was to develop an API built in .NET but according to FIPA standards.

Initially, the interoperability between the JADE and .NET platforms was tried. However, some JADE services could not be used due to communication problems between those technologies and this option was rejected. Finally, considering some ideas from JADE, a new .NET-based platform to control the agent's life cycle was designed and implemented using C# as the programming language. The implementation of the anthropomorphic agent was developed with Microsoft Agent, a Visual Studio.NET library COM Object.

6 Data Mining Approach with ID3 and Delivering

The agent architecture presented previously and a data mining approach are proposed to construct an adapted AT with an associated SN for each learner. A decision engine based on an ID3 implementation [8] uses a learner's profile (which stores all user experience in the system) to create a unique decision tree (DT). This DT groups all the information about relevant experiences for further use, evaluating new experiences or making a decision (based on an algorithm called "Delivering").

Any kind of information can be stored in a learner profile: for example, gender, marital status, learning styles, hobbies, etc. Developing the decision engine required the use of a knowledge representation mechanism based on a relational model and an inference mechanism that uses DTs. Our learner profile includes the following **learner attributes**: **learning model** (LM), **learning style** (LS) and **style level** (SL, high or low depending on whether the style is predominant or not).

The learner profile also includes **System Attributes** which control how contents are shown to the learner to promote or reject, for example, the learning style of students. These system attributes are the **content structure tree** (CST), the **activity type** (ATy) and an attribute called **strategy,** which has been created from the available learning styles to support or reject the style or leveling between one learning style and its opposite (see Table 2).

Table 2. Learner and System Attributes

From	Learner Attributes	Values
Learner attributes	Learning Model	Felder, CHAEA
	Learning Style	Active, Reflective, Sensing, Intuitive, Visual, Verbal, Sequential, Global
	Style Level	Low, Medium, High
System attributes	Content Structure Tree	Choice, ChoiceExit, Flow, ForwardOnly
	Activity Type	Analysis and Reflection, Demo, Explanation, Enforce, Motivation, Evaluation, Review, Lecture, Simulation, Game
	Strategy	Support, Reject, Leveling

The set of user experiences is registered by monitoring agents. There are three important events to monitor. The first is when the learner does all activities of an aggregation and presents the evaluation. In this case the system stores the experience and its attribute values for the learner and his/her evaluation results. (A Boolean value indicates if the learner approved or not.) The second event to monitor occurs when the learner decides to log off from the system without visualizing the entire number of activities of an aggregation or presenting the evaluation. In this case the system will store the learner's experience, its attribute values and a negative evaluation (indicating that the activities presented were not pleasant for the learner). The third event is when the learner asks the system to change the activities presented in an aggregation (asking the decision engine to generate a new strategy) and the experience report is negative. Those experiences will be presented as a tuple (data row) of the database as follows:

$$S = \text{(Learning-Model, Learning-Style, Style-Level, Content-Structure-Tree, Activity Type, Strategic, Result)}$$
$$S_1 = \text{(Felder, Global, Medium, Choice, Explanation, Support, Positive)}$$
(1)

Result {Positive, Negative}: if experience was appropriate or not.

Next, the ID3 algorithm uses attributes and the set of experiences for its training and generates the DT. ID3 was created by Quinlan in 1986 to make decisions using information gain criteria. Finally, "Delivering" generates a personalized AT for each learner. **Delivering** determines the application order of each strategy (from the most to the least recommended) based on the DT and strategies available in the course (See Fig. 3).

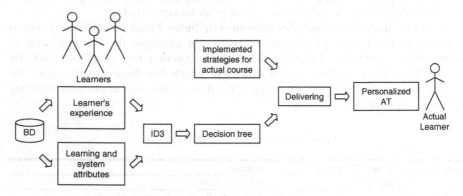

Fig. 3. Components and inference mechanisms. The learners register their experiences. ID3 creates the DT. Finally the system generates the most appropriate AT for a specific learner.

The **Delivering** algorithm finds every route with a positive experience inside the DT and then takes the collection of values that produced a result to separate them into learner and system attributes. The learner attributes are used to compare the values of these same attributes inside the learner profile the system is processing at the time. The goal is to find the same combination and retrieve the collection values of the system that belong to a specific experience in the knowledge base. If there is not an exact combination of values, the case in which there is the greatest number of matched learner attributes is retrieved. If a specific experience for a learner is satisfactory with these system attribute values, there is a high probability that it will be repeated. Fig. 4 provides an example of a DT and shows a set of experiences.

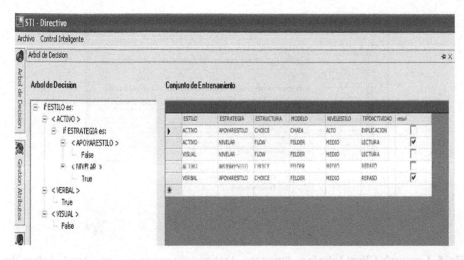

Fig. 4. DT and Set of Experiences. A collection of positive and negative experiences registered by the system (right side) and the DT generated from this data collection (left side).

7 Test Results

To evaluate the proposed model, a test with sixteen (16) learners (with an extensive set of experiences) was performed. The learners were students from the class "Introduction to Computer Science" in the first semester of the System Engineering degree course (computer science) at the University of Cauca. Two topics, "Algorithms" and "Meta-Algorithms", were used. Each topic provided resources for all the learning styles supported. Two main aspects were evaluated on the test. First, the interaction with the LMS (see Fig. 5) and, second, a survey performed for each learner to evaluate that interaction. During the interaction, each learner completed Felder's questionnaire (approximately 15 minutes) and then studied both topics mentioned above (30 minutes) using the adaptation support described previously.

The general results of this test were the following: 1) the predominant style was visual; 2) 87% of the learners agreed with the content presentation; 3) 94% of learners considered that the AHS helps the learning process; and 4) 81% of learners considered that the content presentation related to "Algorithms" helped them pass the exam, while 80% also considered that content presentation related to "Meta-Algorithms" helped them pass the exam.

The functionality test was performed to verify the adaptability process on content presentation, because every content structure presented to learners through the tool of the AHS module was generated/customized according the most predominant learning style of each learner. Of course, this preliminary test highlights the need to do a more formal experiment because the results are closer to a functionality test than a formal experiment.

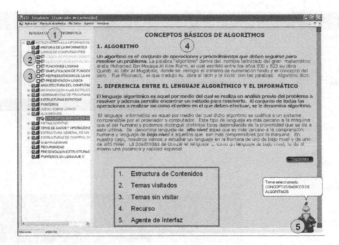

Fig. 5. Learner's Client Interface during test. 1) shows the course studied. 2) and 3) show the aggregations and activities with their different states according to colors previously explained in Section 4. 4) shows the learning object studied. 5) shows the Anthropomorphic Agent.

8 Conclusions and Future Work

In order to model and develop our LMS, an agent-based approach was essential (robustness, scalability and simpler programming). The final product supports several learning models, provides a decision making module that uses the ID3 algorithm and is based on SCORM SN (SCORM Sequencing and Navigation Specification).

The results of the tests are appropriate because the system really adapts contents, presentation and navigation to the user needs. In a near future, we plan to include more attributes in the learner's profile, e.g. gender, learning motivation, media preference, hobbies and social status. In addition, a more controlled and formalized experiment will be designed to determine the effects derived from the use of the LMS in the learning process.

Finally, the implementation of the decision engine using an EC4.5 [21], C4.5 or C5.0 [22] algorithm will manipulate discrete attribute values and in this way eliminate the conversion between a percentage value to a high-medium-low value.

References

1. Paulsen, M.F.: Online Education and Learning Management Systems. Global E-learning in a Scandinavian Perspective. Bekkestua: NKI Forlaget, p. 337 (2003), ISBN: 82-562-5894-2 (printed version), http://www.studymentor.com/studymentor
2. SCORM. Sharable Content Object Reference Model (SCORM) (2004) (retrieved September 22, 2005), http://www.adlnet.org
3. Sharable Content Object Reference Model (SCORM) Sequencing and Navigation (SN) (retrieved September 22, 2005), http://www.adlnet.org/scorm/history/2004/documents.cfm

4. Chang, H.-P., Wang, C.-C., Jan, K.H., Shih, T.K.: SCORM sequencing testing for sequencing control mode. Advanced Information Networking and Applications. In: 20th International Conference on AINA 2006, vol. 2 (April 18-20, 2006), DOI 10.1109/AINA. 2006.295

5. Felder, M.R., Silverman, L.: Learning and Teaching Styles in Engineering Education. Engineering Education 78(7), 674–681 (1988)

6. Brusilovsky, P., Maybury, M.: From Adaptive Hypermedia to the adaptive web. Communications of the ACM 45(5) (2002)

7. Oppermann, R., Rashev, R., Kinshuk: Adaptability and Adaptivity in Learning Systems. In: Behrooz, A. (ed.) Knowledge Transfer, pAce, London, vol. II, pp. 173–179 (1997) ISBN 1-900427-015-X

8. Quinlan, J.R.: C4.5: Programs for Machine Learning. Morgan Kaufmann, San Francisco (1993)

9. Shang, Y., Shi, H.C., Chen, S.S.: An intelligent distributed environment for active learning. ACM Journal of Educational Resources in Computing 1(2), 1–17 (2001)

10. Trantafillou, E., Poportsis, A., Demetriadis, S.: The design and the formative evaluation of an adaptive educational system based on cognitive styles. Computers & Education 41, 87–103 (2003)

11. Gilbert, J.E., Han, C.Y.: Adapting instruction in search of a significant difference. Journal of Network and Computer Application 22, 149–160 (1999)

12. Weber, G., Specht, M.: User modeling and adaptive navigation support in www based tutoring systems. In: Proceedings of User Modeling 1997, pp. 289–300 (1997)

13. Chen, G.D., Liu, C.C., Ou, K.L., Liu, B.J.: Discovering decision knowledge from web log portfolio for managing classroom processes by applying decision tree and data cube technology. Journal of Educational Computing Research 23(3), 305–332 (2000)

14. Peña, C.I., Marzo, J.L., De La Rosa, J.L.L.: Intelligent Agents in a Teaching and Learning Environment on the Web. In: ICALT2002. 2nd IEEE International Conference on Advanced Learning Technologies, Kazan (Russia), pp. 21–27 (September 9-12, 2002)

15. Wang, F.H., Shao, H.M.: Effective personalized recommendation based on time-framed navigation clustering and association mining. Expert Systems with Applications 27, 365–377 (2004)

16. Su, J.-M., Tseng, S.-S., Wang, W., Weng, J.-F., Yang, J.T.D., Tsai, W.-N.: Learning Portfolio Analysis and Mining for SCORM Compliant Environment. Educational Technology & Society 9(1), 262–275 (2006)

17. Kolb, D.: Experiential learning: experience as the source of learning and development. Prentice Hall, Englewood Cliffs, NJ (1984)

18. Iglesias, C.A., Gariji, M., Centeno-González, J., Velasco, J.R.: Analysis and Design of Multiagent Systems Using MAS-Common KADS. In: Rao, A., Singh, M.P., Wooldridge, M.J. (eds.) ATAL 1997. LNCS, vol. 1365, pp. 313–327. Springer, Heidelberg (1998)

19. Odell, J., Muller, P.J., Bauer, B.: Agent UML: A Formalism for Specifying MultiAgent Interaction. In: Ciancarini, P., Wooldridge, M.J. (eds.) AOSE 2000. LNCS, vol. 1957, pp. 91–103. Springer, Heidelberg (2001)

20. Bayo, A.M.: Running Jade over .NET (2003)

21. Ruggieri, S.: Efficient C4.5. IEEE Transactions on Knowledge and Data Engineering 14(2), 438–444 (2002)

22. Larose, D.T.: Discovering Knowledge in Data: An Introduction to Data Mining. John Wiley & Sons, Incorporated, Chichester (2005)

A Web-Based Learning Resource Service System Based on Mobile Agent

Wu Di, Cheng Wenqing, and Yan He

Department of Electronics and Information Engineering, Huazhong University of Science and Technology, PR China
wudi@hust.edu.cn, chengwq@hust.edu.cn, isaacyh@sina.com

Abstract. The quality of learning resource service is a vital point in an E-Learning application system. In this paper, a novel learning resource service model is proposed based on the application of mobile agent and Web Services technology. Unlike the traditional approaches, our system takes into account the digital rights management of learning resource and implements the digital rights validation function with the help of agent communication language and digital rights expression language. In addition, the learning resource transmission strategy based on Web Services and resource search strategy based on mobile agent are also introduced in our model. In this paper, we first introduce the layered-stucture of our learning resource service model. Then we analyze the resource digital rights control process and the resource agent search flow which are all the key business logics in the system. Finally, according to the forenamed technology and model, we implement an applied learning resource service system for a national higher-education E-Learning portal.

Keywords: Learning Resource Service, Mobile Agent, Web Services, Digital Rights Management

1 Introduction

One of the most important materials of E-Learning is the digital learning resource [1]. The quality of learning resource service is a vital problem of E-Learning. However, learning resource is difficult to manage and search, because its format might be quite abundant, such as image, text, audio, video, etc. A learning resource service system should provide integrated service about resource publication, searching, distribution and digital rights protection. To provide high-quality learning resource service for users, the resource service system should make users get the learning resource they need as soon as possible and protect the digital rights of learning resource effectively. In other word, the search effect and the digital rights security of learning resource should be taken into account in a learning resource service system.

Mobile Agent (MA) is an emerging technology. It has been used to design applications ranging from distributed information retrieval to network management [2]. In a learning resource service system, unlike traditional search method, mobile agents can support a flexible autonomous search method. With the help of mobile agents, we can search learning resource more neatly and intelligently.

B. Benatallah et al. (Eds.): WISE 2007, LNCS 4831, pp. 310–321, 2007.

Web services are software components that can be accessed through uniform protocols in a distributed network. A Web service can be defined as a software component, which provides a specific service, and which can be described and advertised, discovered and invoked over the Web, by using standard technologies and protocols [3]. Based on the standard protocols of Web Services, it is more convenient to support learning resource publication, distribution and transmission in a distributed network environment.

In this paper we propose a web-based Learning Resource Service Model (LRSM) based on mobile agent. It makes use of the advantages of mobile agents and Web Services, and can provide an effective way for users to get the learning resource they need. In addition, LRSM uses digital rights expression language (REL) to describe the rights of learning resource and control the rights execution process of learning resource.

2 Related Works

2.1 Learning Resource Standards

Learning resource standards is a precondition for learning resource management and interchange. Many learning resource standards have been proposed by several standardization committees. The most popular standards of learning resource are the learning resource metadata and content packaging. Some main standardization committees, such as IEEE LTSC [4] and IMS [5,6], have proposed specifications about learning resource metadata and content packaging.

Learning object metadata (LOM) proposed by IEEE LTSC is generally accepted as the standard for providing metadata to multimedia learning resources. The aim of using metadata for describing learning objects is to promote the sharing of learning material [7]. LOM can help users know the key information about the learning resource, such as name, authors and version, etc. On the other hand, LOM can be bind with Extensible Markup Language (XML) format which is convenient for content packaging and SOAP binding.

Content packaging (CP) proposed by IMS describes data structures that are used to provide interoperability of web-based content with content creation tools, learning management systems, and run time environments [8]. With the help of CP, users can know the organization structure of a learning resource content package and reorganize the internal structure of a course. On the other hand, LOM can be bind with XML format which is convenient for SOAP binding.

2.2 Simple Object Transmission Protocol

Among distributed technologies, Web Services are beginning to make a significant impact on the design and delivery of e-learning courseware [9]. Web Services are self-contained, self-describing, modular applications that can be published, located, and invoked across the Web [10]. Web Services provides a new framework for distributed application system on Internet. XML is used as the basic bind format in Web Services.

As one of the key standard protocols of Web Services, the XML-based Simple Object Access Protocol (SOAP) has been a vital component that makes Web Services be applied successfully. It provides a standardized mechanism for heterogeneous information systems and applications to communicate with each other regardless of their implementation language and operating platform [11]. In learning resource transmission process, SOAP can be used to transport and interchange operation statement. In addition, based on the SOAP with Attachment (SwA) technology, we can bind learning resource as the SOAP attachment and transmit them together with the operation statement to the target location.

2.3 Agent Communication Language

A mobile agent is a computer program that acts autonomously on behalf of a user and moves through a network of heterogeneous machines [12]. To accomplish a specific task, we usually need more than one agent. Agents in a multi-agent system (MAS) must be able to interact and communicate with each other. This usually needs a common language, an Agent Communication Language (ACL) [13]. One of the most widely used ACLs is Knowledge Query and Manipulation Language (KQML).

KQML is an external high-level language for agents, oriented toward the exchange of messages, independent of the syntax and the ontology of the message content [14]. It is both a message format and a message-handling protocol that supports run-time knowledge sharing among agents. KQML offers a variety of message types (performatives) that express an attitude regarding the content of the exchange. Performatives can also assist agents in finding other agents that can process their requests [15]. KQML can be bind with SOAP messages which also have an XML-based document format.

2.4 Digital Rights Management

Digital Rights Management (DRM) is the common term associated with technologies that are able to protect digital content after the distribution, providing ways to exercise usage control on the content [16]. DRM provides an integrated solution based on information security technology. With the common reusing and sharing of learning resource, a problem emerges that the learning resource distribution lacks sufficient rights management [17]. Therefore, the DRM of learning resource is a key problem that should be solved in the process of learning resource management and distribution.

Right expression language (REL) is a language devised to express conditions of use of digital content. They have been proposed to describe licenses governing the terms and conditions of content access [18]. They can also describe the rights of learning resource effectively. In addition, REL can be integrated into KQML performatives and bind with SOAP messages in an XML format.

3   Resource Service Model

We proposed a distributed learning resource service management model (LRSM) which composes the advantages of learning resource standards, digital rights expression and mobile agent communication. The structure of LRSM includes seven layers (see Figure 1):

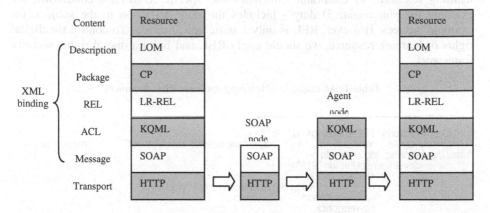

Fig. 1. LRSM structure

1) Transport layer – transport resource packages based on Internet Transmission Control Protocol (TCP), such as Hyper Text Transfer Protocol (HTTP);
2) Message layer – transport and interchange SOAP message, build a communication channel between resource nodes based on Web Services;
3) ACL layer – communicate between mobile agents and interchange control information between search agents and resource agents;
4) REL layer – describe the digital rights of learning resource;
5) Package layer – package learning resource based on learning resource standards, such as CP;
6) Description layer – describe learning resource based on learning resource standards, such as LOM;
7) Content layer – includes the real materials of learning resource, such as physical files.

In LRSM, different layers use different communication languages, a unit in a specific layer only communicate with other units in the same layer. The description layer, package layer, REL layer, ACL layer and message layer all use XML as data format, they bind their data with XML and nest XML nodes layer by layer. So we can get a big XML tree in the description layer and interchange it with other nodes. Layer's same format makes it easy to transport resource information based on mobile agent.

4   Resource Digital Rights Control

REL is a static specification which describes the digital rights of learning resource. Several REL specifications can be used to describe learning resource digital rights. We modify MPEG-21 REL and extend some elements according to the learning resource features. An example of learning resource REL document is shown in Table 1. A learning resource REL document includes five key elements: 1) subject - includes the subject of the digital rights, for example a user; 2) right - includes the basic rights types and definitions, for example the pint right; 3) resource - includes the key metadata info of the target learning resource, for example the id and name of a learning resource; 4) constraint - includes the specific constraint conditions, for example the pint count; 5) duty - includes the duty definition to the subject, for example the fees. However, REL is only a static specification. To control the digital rights of learning resource, we should apply REL and build a digital rights security framework.

Table 1. An example of learning resource REL document

```
<?xml version="1.0" encoding="UTF-8"?>
    <lr-rel:license    xmlns:lr-rel    =    "http://www.celtsc.edu.cn/lr-rel"    xmlns:lr-rdd    =
"http://www.celtsc.edu.cn/lr-rdd">
        <lr-rel:licenseUnit id="012">
            <lr-rel:subject>
                    <lr-rdd:resourceReceiver id="Wu Di"/>
            </lr-rel:subject>
            <lr-rel:right>
                    <lr-rdd:print/>
            </lr-rel:right>
            <lr-rel:resource>
                    <lr-rdd:courseware id="Java Programming Language"/>
            </lr-rel:resource>
            <lr-rel:constraint>
                    <lr-rdd:count>2</lr-rdd:count>
            </lr-rel:constraint>
            <lr-rel:duty>
                    <lr-rdd:payment>
                        <lr-rdd:amount>20.00</lr-rdd:amount>
                    </lr-rdd:payment>
            </lr-rel:duty>
        </lr-rel:licenseUnit>
    </lr-rel:license>
```

A third-party security service framework is a representative structure in DRM system. To control the digital rights of learning resource more effectively, we use REL to uniformly describe the rights of learning resource and use an execution control flow to implement the REL license validation. To control the rights consumption and execution process, we use a third-party object - license manager to manage the digital rights licenses. The license manager is an independent server stores all valid licenses and license consumption log in a license database. The sequence diagram of the whole digital rights control flow includes nine steps (see Figure 2):

1) The user sends a learning resource request message to the border interface. The request message might be heterogeneous message, for example browsing resource or printing resource, etc.

2) Based on the rights description in the request, a license segment is created, which is an XML-based REL document. The document will be sent to the learning resource server.

3) The resource server creates a search request based on the license received from above and sends it to license manager. The search result is used to confirm if the user has the rights or not.

4) The license manager searches from the license database in the license server and gets the corresponding license, then analyze the license and get the digital rights control information.

5) The rights controller matches the license segment send from the user and the segment gotten from the license server, and validate if the user has the rights he wants to get.

6) The validate result is created and sent to the license manager. Based on the result, the digital rights license segment should be modified, so that the consumption info can be included in the license.

7) The license manager tracks the status of license segment sent from the user, modifies and updates the license according to the digital rights consumption info.

8) If the user is permitted to get the rights he wants to get, the encrypted learning resource is returned to the border interface and prepared for the user.

9) The border interface decrypts the learning resource and returns it to the user so that the user can consume the resource based on the digital rights he is permitted.

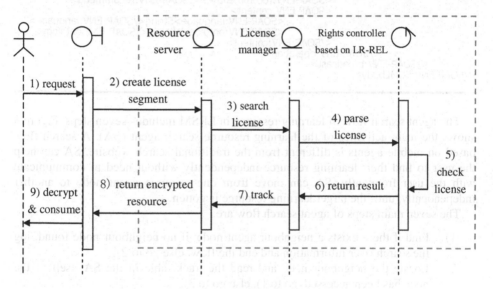

Fig. 2. Learning resource digital rights control flow

5 Resource Agent Search Flow

KQML defines nine key performatives, includes *:sender, :receiver, :from, :to, :in-reply-to, :reply-with, :language, :ontology, :content*, etc. A SOAP message includes three main components: SOAP header, SOAP body and SOAP attachment. When we use a SOAP message to transmit a KQML statement, the KQML statement is bind in the SOAP body component (see Table 2). When the middle node receives the SOAP message, it only parses the SOAP header component and transmits it to the next SOAP node. However, when the final receiver receives the message, it parses the

SOAP body component and get out the KQML statement from the SOAP body. The license manager will check the digital rights according to the REL included in the KQML statement. If the digital rights validate successfully, the target learning resource metadata info will be attached into the SOAP attachment component and the whole SOAP message will be sent back.

Table 2. An examples of KQML SOAP binding segment

```
<SOAP-ENV:KQMLbody>
        <SOAP-ENV:broadcast SOAP-ENV:value="ask-all">
                <SOAP-ENV:operation>
                        <SOAP-ENV:sender>jinjing</SOAP-ENV:sender>
                        <SOAP-ENV:receiver>wangyuan</SOAP-ENV:receiver>
                        <SOAP-ENV:from>wudi</SOAP-ENV:from>
                        <SOAP-ENV:to>yanhe</SOAP-ENV:to>
                        <SOAP-ENV:reply-with>...</SOAP-ENV:reply-with>
                        <SOAP-ENV:in-reply-to>...</SOAP-ENV:in-reply-to>
                        <SOAP-ENV:register>...</SOAP-ENV:register>
                        <SOAP-ENV:recommend>...</SOAP-ENV:recommend>
                        <SOAP-ENV:content>
                                <SOAP-ENV:language>String</SOAP-ENV:language>
                                <SOAP-ENV:ontology>String</SOAP-ENV:ontology>
                        </SOAP-ENV:content>
                </SOAP-ENV:operation>
        </SOAP-ENV:broadcast>
</SOAP-ENV:KQMLbody>
```

The agent search flow of learning resource in LRSM includes seven steps. Figure 3 shows the main activities of the learning resource search agent (SA). A search flow based on mobile agents is different from the traditional search website. SA can help the user to find their learning resource independently without need to communicate with the user frequently. SA can move from one resource agent node to another independently, until the target learning resource is gotten.

The seven main steps of agent search flow are:

1) Find if there exists a neighbour agent node, if no neighbour node found, log the search over information and end the flow, else go to 2);

2) Locate the neighbour node and read the track table in the SA itself, if the node has been accessed, go to 1), else go to 3);

3) Move to the neighbour node and check user account through the REL segment embedded in KQML, if the SA is a valid user of the node, go to 4), else log the lost info and track info, then go to 1);

4) Search the resource database of the node, if the target resource is found, go to 5), else log the lost info and track info, then go to 1);

5) Search the license database of the node, if the relevant license is found, go to 6), else log the lost info and track info, then go to 1);

6) Match the license found in the node with the license segment embedded in REL, if match successfully, go to 7), else log the lost info and track info, then go to 1);

7) SA attaches the learning resource metadata info and return to the user.

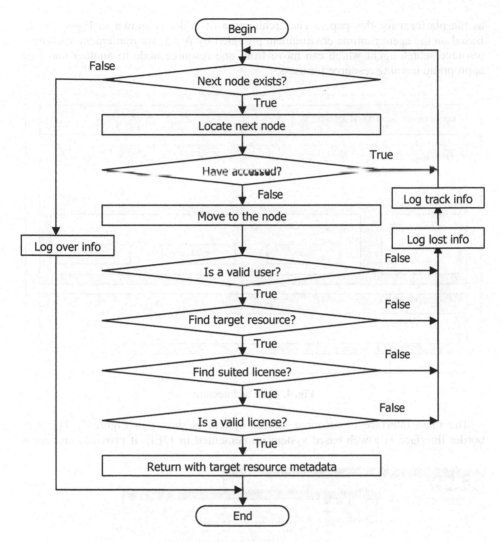

Fig. 3. Learning resource search flow

6   Implementation of LRSM

Based on LRSM, we implement a learning resource service system which is a mobile agent system and manages learning resource in a distributed network environment. Java 2 Enterprise Edition (J2EE) technology provided by SUN corp. is used to implement the system, because it can be deployed in a wide range of configurations with varying performance and scalability needs [19]. In addition, J2EE provides a novel support of Web Services, which is very fitful for learning resource distribution and publication [20].

The software language to construct mobile agent system should be object oriented, platform independent, with communication capability, and implemented with code security [21]. Aglet provided by IBM corp. with Java is a popular tool and is selected as the platform for this paper. The architecture of Aglet is shown in Figure 4 [2]. Based on the agent runtime environment provided by Aglet, we implement a learning resource search agent which can move from one resource node to another and find appropriate learning resource for users.

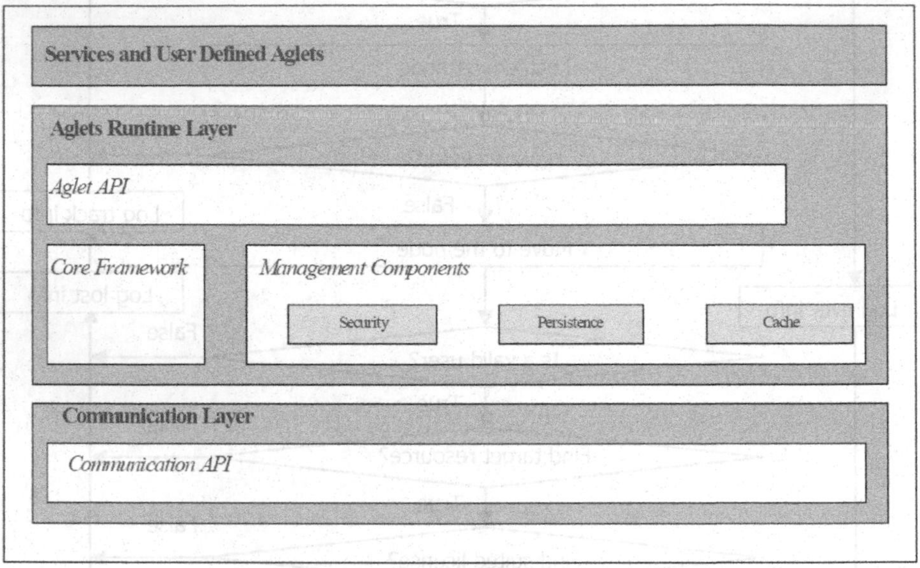

Fig. 4. Aglet architecture

The main interface of resource service system is shown in Figure 5. The user border interface is a web-based system implemented in J2EE, it provides an access

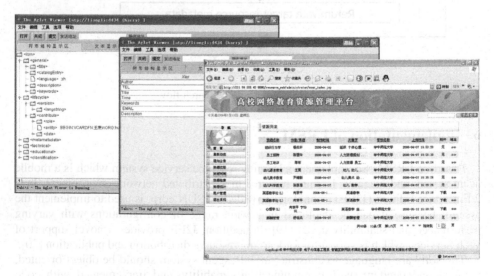

Fig. 5. Learning resource search agent and service system based on LRSM

portal for users, so that users can submit resource search requests and browse learning resource. On the other side, the learning resource search agent is a local application system which is implemented based on Aglet, it creates a resource search agent and communicate with each resource agent node through KQML. The Aglet objects in the system communicate with the resource portal through Java servlet technology. When a user request is submit to the servlet, an Aglet-based mobile agent object is created and move to the distant learning resource servers to find the target resource. When the search agent finds out the target resource, the resource will be return to the servlet and shown in the web for users. In the whole process, users can not see the Aglet interface, the learning resource service process is transparent to users. Therefore, the system is easy for users and novel for resource managers.

7 Conclusion and Future Works

In this paper, we put forward a web-based learning resource service system based on mobile agent, which provides an effective support for learning resource digital rights control. Furthermore, with the help of Aglet and J2EE technology, we implement a resource service system according to the model.

The harvest of this paper has been partially used in China Distance Education and Continuous Education website (CDCE) which is a national higher education E-learning service portal. LRSM is used to support the learning resource service in CDCE.

In the future works, we will try to build a large learning resource service system in a multi-agent environment (MAE), and extend KQML to support more abundant semantic level communication activity about digital rights control and learning resource search.

Acknowledgement

The work described in this paper is fully supported by a grant from the Science and Technology Plan Project, Ministry of Education, China [Project No. 705038-03]. We thank Mr. Liang Li, Mr. Yang Junkai and Ms. Liu Yang for their provocative discussions about the application of the model and their perfect development work in the system implementation.

References

[1] Wu, D., Cheng, W., Yang, Z.: A Mobile Agent Assisted Learning Resource Service Framework Based on SOAP. Web-based Learning: Technology and Pedagogy, 125–134 (July 31- August 8, 2005)

[2] Quah, J.T.S., Leow, W.C.H., Chen, Y.M.: Mobile Agent Assisted E-Learning. In: ICITA 2002. Proceedings of the 1st International Conference on Information Technology and Applications, pp. 595–600 (March 18, 2002)

320   D. Wu, W. Cheng, and Y. He

[3] Anane, R., Bordbar, B., Deng, F., Hendley, R.J.: A Web Services Approach to Learning Path Composition. In: ICALT 2005. Proceedings of 5th IEEE International Conference on Advanced Learning Technologies, pp. 98–102 (July 5-8, 2005)

[4] IEEE Learning Technology Standards Committee, IEEE P1484.12.2/D1, Draft Standard for Learning Technology – Learning Object Metadata – ISO/IEC 11404 Binding (September 13, 2002)

[5] IMS Global Learning Consortium, Inc., IMS Learning Resource Meta-Data Information Model, Version 1.2.1 Final Specification (September 28, 2001)

[6] IMS Global Learning Consortium, Inc., IMS Content Packaging Information Model, Version 1.1.4 Final Specification (October 04, 2004)

[7] Motelet, O., Baloian, N.: Hybrid System for Generating Learning Object Metadata. In: ICALT 2006. Proceedings of the 6th International Conference on Advanced Learning Technologies, pp. 563–567 (July 05-07, 2006)

[8] Wu, D., Yang, Z., Cheng, W.: A Standardized Visual Web-based Courseware Authoring System. In: ICALT 2005. Proceedings of the 5th IEEE International Conference on Advanced Learning Technologies, pp. 78–79 (July 5-8, 2005)

[9] Vossen, G., Westerkemp, P.: E-learning as a Web service. In: IDEAS 2003. Proceedings of the 7th IEEE International Database Engineering and Applications Symposium, pp. 242–249 (July 16-18, 2003)

[10] Chen, W.: Web Services – What Do They Mean to Web-based Education. In: ICCE 2002. Proceedings of the International Conference on Computers in Education, pp. 707–708 (December 3-6, 2002)

[11] Lai, K.Y., Phan, T.K.A., Tari, Z.: Efficient SOAP Binding for Mobile Web Services. In: LCN 2005. Proceedings of the IEEE Conference on Local Computer Networks 30th Anniversary, pp. 218–225 (November 15-17, 2005)

[12] Mohammadi, K., Hamidi, H.: A New Approach for Mobile Agent Fault-Tolerance and Reliability Using Distributed Systems. In: Proceedings of the 1st IEEE and IFIP International Conference in Central Asia on Internet, p. 5 (September 26-29, 2005)

[13] Berna-Koes, M., Nourbakhsh, I., Sycara, K.: Communication Efficiency in Multi-Agent Systems. In: ICRA 2004. Proceedings of the 2004 IEEE International Conference on Robostics & Automation, pp. 2129–2134 (April 26 - May 1, 2004)

[14] Brahimi, M., Boufaida, M., Seinturier, L.: A Federated Agent-based Solution for Developing Cooperative E-business Applications. In: DEXA 2004. Proceedings of the 15th International Workshop on Database and Expert Systems Applications, pp. 289–293 (August 30- September 3, 2004)

[15] Labrou, Y., Finin, T.: A Proposal for a new KQML Specication (TR CS-97-03), p. 2 (February 3, 1997)

[16] Rosset, V., Filippin, C.V., Westphall, C.M.: A DRM Architecture to Distribute and Protect Digital Contents Using Digital Licenses. In: Proceedings of the Advanced Industrial Conference on Telecommunications/Service Assurance with Partial and Intermittent Resources Conference/E-Learning on Telecommunications Workshop, pp. 422–427 (July 17-20, 2005)

[17] Cheng, W., Deng, W., Wu, D.: A Rights Managing Model in Learning Resource Distribution. In: ITHET 2006. Proceedings of the 6th International Conference on Information Technology Based Higher Education and Training (July 7, 2006)

[18] Delgado, J., Prados, J., Rodríguez, E.: Profiles for interoperability between MPEG-21 REL and OMA DRM. In: CEC 2005. Proceedings of the 7th IEEE International Conference on E-Commerce Technology, pp. 518–821 (July 19–22, 2005)

[19] Guo, J., Liao, Y., Parviz, B.: A Performance Validation Tool for J2EE Applications. In: ECBS 2006. Proceedings of the 13th Annual IEEE International Symposium and Workshop on Engineering of Computer Based Systems, p. 10 (March 27-30, 2006)

[20] Wu, D., Yang, Z., Zhao, G.: Learning Resource Registry and Discovery System. In: AINA 2004. Proceedings of the 18th International Conference on Advanced Information Networking and Application, pp. 433–438 (March 29-31, 2004)

[21] James, E.: White Mobile Agents (May 1996), http://www.webtechniques.com/archives/1996/05/ white

Wikipedia Mining for an Association Web Thesaurus Construction

Kotaro Nakayama, Takahiro Hara, and Shojiro Nishio

Dept. of Multimedia Eng., Graduate School of Information Science and Technology
Osaka University, 1-5 Yamadaoka, Suita, Osaka 565-0871, Japan
Tel.: +81-6-6879 4513; Fax: +81-6-6879-4514
{nakayama.kotaro, hara, nishio}@ist.osaka-u.ac.jp

Abstract. Wikipedia has become a huge phenomenon on the WWW. As a corpus for knowledge extraction, it has various impressive characteristics such as a huge amount of articles, live updates, a dense link structure, brief link texts and URL identification for concepts. In this paper, we propose an efficient link mining method pfibf (Path Frequency - Inversed Backward link Frequency) and the extension method "forward / backward link weighting (FB weighting)" in order to construct a huge scale association thesaurus. We proved the effectiveness of our proposed methods compared with other conventional methods such as cooccurrence analysis and TF-IDF.

1 Introduction

A thesaurus is a kind of dictionary that defines semantic relatedness among words. Although the effectiveness is widely proved by various research areas such as natural language processing (NLP) and information retrieval (IR), automated thesaurus dictionary construction (esp. machine-understandable) is one of the most difficult issues. Of course, the simplest way to construct a thesaurus is human-effort. Thousands of contributors have spend much time to construct high quality thesaurus dictionaries in the past. However, since it is difficult to maintain such huge scale thesauri, they do not support new concepts in most cases. Therefore, A large number of studies have been made on automated thesaurus construction based on NLP. However, issues due to complexity of natural language, for instance the ambiguous/synonym term problems still remain on NLP. We still need an effective method to construct a high-quality thesaurus automatically avoiding these problems.

We noticed that Wikipedia, a collaborative wiki-based encyclopedia, is a promising corpus for thesaurus construction. According to statistics of Nature, Wikipedia is about as accurate in covering scientific topics as the Encyclopedia Britannica [1]. It covers concepts of various fields such as Arts, Geography, History, Science, Sports or Games. It contains more than 1.3 million articles (Sept. 2006) and it is becoming larger day by day. Because of the huge scale concept network with a wide-range topic coverage, it is natural to think that Wikipedia can be used as a knowledge extraction corpus. In fact, we already proved that it

B. Benatallah et al. (Eds.): WISE 2007, LNCS 4831, pp. 322–334, 2007.

can be used for accurate association thesaurus construction[2]. Further, several researches have already proved the importance and effectiveness of Wikipedia Mining[3,4,5,6].

However, what seems lacking in these methods is the deep consideration for improving accuracy and scalability. After a number of continuous experiments, we realized that there are possibilities to improve the accuracy because the accuracy changes depending on particular situations. Further, none of previous researches has focused on scalability. WikiRelate [4], for instance, measures the relatedness between two given terms by analyzing (searching) the categories which they belong to. This means that we have to search all combinations of categories of all combinations of terms thus the number of steps for the calculation becomes impossibly huge. To conclude this, we still have to consider the characteristics of Wikipedia and optimize the algorithm in order to extract a huge scale accurate thesaurus.

In this paper, we propose an efficient link structure mining method to construct an association thesaurus from Wikipedia. While almost all researches in this research area analyze the structure of categories in Wikipedia, our proposed method analyzes the link structure among pages because links are explicit relations defined by users.

The rest of this paper is organized as follows. First of all, we introduce a number of researches on automated thesaurus construction in order to make our stance clear. Second, we unveil the characteristics and statistics of Wikipedia in detail. After that, we describe some conventional methods which can be used for Wikipedia mining, and we propose a link mining method "pfibf" and an extension "FB weighting." Then, we describe the results of our experiments. Finally, we draw a conclusion.

2   Wikipedia as a Web Corpus

As we mentioned before, Wikipedia is an attractive Web corpus for knowledge extraction because of the characteristics. In this section, we describe two important characteristics of Wikipedia; a dense link structure and concept identification by URL.

"Dense" means that it has a lot of "inner links," links from pages in Wikipedia to other pages in Wikipedia. Let us show some results of the link structure analysis for Wikipedia (Sept. 2006). Figure 1 shows the distribution of forward and backward links. The statistics unveiled that Wikipedia (Esp. the distribution of backward links) has a typical "power-law" distribution, containing a few nodes with a very high degree and many with a low degree of links. This characteristic, the rich semantic links among concepts, shows us the potential of Wikipedia mining.

Concept identification based on URL is also a key feature on Wikipedia. Ordinary (electric) dictionaries have indexes to find the concepts the user wants to know. However, several concepts are put into one index in most cases. This means that ambiguous terms are listed in one article. This is no problem for humans because it is human readable, but it is not machine understandable. For

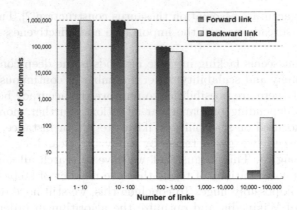

Fig. 1. Distribution of link number

example, if a sentence "Golden delicious is a kind of apple" exists in an article in a dictionary, humans can immediately understand that "apple" means a fruit. However, it is difficult to analyze for a machine because "apple" is an ambiguous term and there is no identification information whether it is a fruit or a computer company. On Wikipedia, almost every concept (article/page) has an own URL as an identifier. This means that it is possible to analyze term relations avoiding ambiguous term problems or context problems.

3   Related Works

In this section, we introduce a number of studies for thesaurus construction which relate to our research. After that, we explain how can we apply/extend conventional methods, cooccurrence analysis and TF-IDF (Term Frequency - Inverse Document Frequency) weighting[7] for thesaurus construction.

3.1   Web Structure Mining

One of the most notable differences between an ordinary corpus and a Web corpus is apparently the existence of hyperlinks. Hyperlinks do not just provide a jump function between pages, but have more valuable information. There are two type of links; "forward links" and "backward links." A "forward link" is an outgoing hyperlink from a Web page, an incoming link to a Web page is called "backward link". Recent researches on Web structure mining, such as Google's PageRank[8] and Kleinberg's HITS[9], emphasize the importance of backward links in order to extract objective and trustful data.

By analyzing this information on hyperlinks, we can extract various information such as topic locality[10], site topology, and summary information. Topic locality is the law that web pages which are sharing the same links have more topically similar contents than pages which are not sharing links.

3.2   Wikipedia Mining

"Wikipedia mining" is a new research area which is recently addressed. As we mentioned before, Wikipedia is an invaluable Web corpus for knowledge extraction. WikiRelate[4] proved that the inversed path length between concepts can be used as a relatedness for two given concepts. However, there are two issues on WikiRelate; the scalability and the accuracy.

The algorithm finds the shortest path between categories which the concepts belong to in a category graph. As a measurement method for two given concepts, it works well. However, it is impossible to extract all related terms for all concepts because we have to search all combinations of category pairs of all concept pairs (1.3 million × 1.3 million). Furthermore, using the inversed path length as the semantic relatedness is a rough method because categories do not represent the semantic relations in many cases. For instance, the concept "Rook (chess)" is placed in the category "Persian loanwords" with "Pagoda," but the relation is not semantical, it is just a navigational relation.

The accuracy problem of WikiRelate is also mentioned in a Gabrilovich's paper[6]. Gabrilovich proposed a TF-IDF based similarity measurement method for Wikipedia and proved that the accuracy is much better than that of WikiRelate, a category based approach.

3.3   Cooccurrence Analysis

Since the effectiveness of the cooccurrence analysis has been widely proved in the thesaurus construction research area[11], it is possible to apply it to relatedness analysis in Wikipedia. A cooccurrence-based thesaurus represents the similarity between two words as the cosine of the corresponding vectors. Term cooccurrence tc between two terms (t_1 and t_2) can roughly be defined by the following formula:

$$tc(t_1, t_2) = \frac{|D_{t_1} \cap D_{t_2}|}{|D_{t_1} \cup D_{t_2}|}. \tag{1}$$

D_{t_1} is a set of documents which contain term t_1. To measure the similarity of two terms, the number of documents which contain the terms is used. Although the effectiveness has been proved, natural language processing has various accuracy problems due to the difficulty of semantics analysis.

We propose a link cooccurrence analysis for thesaurus construction which uses only the link structure of Web dictionaries in order to avoid the accuracy problems of NLP. Since a Web dictionary is a set of articles (concepts) and links among them, it makes sense to use link cooccurrence as a thesaurus construction method. The formula is basically the same as for the term cooccurrences. The difference is that it uses links in documents instead of terms.

3.4   TF-IDF

TF-IDF[7] is a weighting method that is often used to extract important keywords from a document. TF-IDF uses two measurements; tf (Term Frequency) and idf (Inverse Document Frequency). tf is simply the number of appearances of a term in the document. df describes the number of documents containing the term. In short, the importance of the term basically becomes higher according to the term frequency of the term in the document, and it becomes lower according to the inversed document frequency of the term in the whole collection of documents because idf works as a common terms filter.

Since TF-IDF statistically evaluates the importance of a term to a document in a collection of documents, it also can be used for thesaurus construction because a page corresponds to a concept and the links are semantic associations for other concepts in Web dictionaries. The importance of links to a document can be defined as follows:

$$tfidf(l, d) = tf(l, d) \cdot idf(l), \tag{2}$$

$$idf(l) = \log \frac{N}{df(l)}. \tag{3}$$

$tf()$ denotes the number of appearances of the link l in document d. N is the total number of documents and $df(l)$ returns the number of documents containing the link l. In summary, the importance basically increases according to the link frequency of l in the document d and the inversed document frequency of the link l in the whole collection of documents, thus common links will be filtered by idf. Since a page in Wikipedia corresponds to a concept, by calculating TF-IDF for every link in a page, we can extract a vector of weighted links for the concept. After extracting the vectors for each concept, relatedness between two concepts can be calculated comparing their vectors by using correlation metrics such as cosine metrics.

4   pfibf

$pfibf$ (Path Frequency - Backward link Frequency), the method that we are proposing, is a link structure mining method which is optimized for Wikipedia. While TF-IDF analyzes relationships to neighbor articles (1 hop), $pfibf$ analyzes the relations among nodes in n-hop range. $pfibf$ consists of two factors; pf (Path Frequency) and ibf (Inversed Backward link Frequency). The point is that this is a very balanced method in both scalability and accuracy. In this section, we describe $pfibf$ in detail.

4.1   Basic Strategy

Web-based dictionaries such as Wikipedia consist of a set of articles (concepts) and hyperlinks among them, thus they can be expressed by a graph $G = \{V, E\}$ (V: set of articles, E: set of links). Let us consider how we can measure the relativity between any pair of articles (v_i, v_j). The relativity is assumed to be strongly affected by the following two factors:

- the number of paths from article v_i to v_j,

– the length of each path from article v_i to v_j.

The relativity is strong if there are many paths (sharing of many intermediate articles) between two articles. In addition, the relativity is affected by the path length. In other words, if the articles are placed closely together in the graph G and share hyperlinks to same articles, the relativity is estimated to be higher than farther ones.

In addition, the number of backward links on articles is also estimated as a factor of relativity. For instance, assume that there is an article which is referred to from many other articles. This article would have a lot of short paths to many articles. This means that it has a strong relativity to many articles if we used only pf. However, this kind of articles must be considered as a general concept, and the importance of general concepts is not high in most cases. Therefore, we must consider the inversed backward link frequency ibf in addition to the two factors above.

4.2 Dual Binary Tree

The counting of all paths between all pairs of articles in a huge graph is a computational resource consuming work. Thus, making it efficient is a serious issue on Wikipedia mining. Using adjacency matrices and multiplication is not a clever idea because of the low scalability. Wikipedia has more than 1.3 million articles, thus we needs several terabytes just for storing data. Further, we need unimaginably much time to calculate the multiplication because the order is $O(N^3)$. However, a large number of elements in the adjacency matrix of a Web site are zero, thus effective compression data structures and analysis methods are the key to achieve high scalability on Wikipedia mining. Therefore, we propose an efficient data structure named "Dual binary tree" (DBT) and a multiplication algorithm for the DBT.

Since the adjacency matrix of a Web site link structure is a sparse matrix (almost all elements are zero), the DBT stores only the non-zero elements for data compression. Figure 2 shows the image of a DBT. The DBT consists of two types of binary trees; i-tree and j-tree. Each element in the i-tree corresponds to a row in the adjacency matrix and each i-tree element stores a pointer to the root of a j-tree. This means that the DBT consists of totally $N+1$ (1 i-tree and N j-trees) binary trees. The point is that operations for both getting and storing data are very fast because the number of steps is in both cases $O(logN)$.

Next, we define the multiplication algorithm for the DBT as follows:

Algorithm $MultiplyDBT(A)$
1   **for** $i \in$ $i\text{-}Tree$
2   **for** $j \in$ $j\text{-}Tree(i)$
3   **for** $k \in$ $j\text{-}Tree(j)$
4         $R_{i,k} := R_{i,k} + a_{j,k} \cdot a_{i,j};$

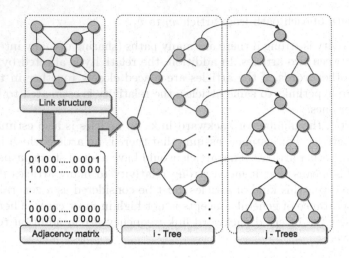

Fig. 2. Dual binary tree for adjacency matrix

The function $j\text{-}Tree(i)$ extracts all elements in the ith row of the adjacency matrix A. $a_{j,k}$ denotes the element in the jth row and kth column of the matrix. The first loop will be executed N times, but the numbers of cycles of the second and third loop depend on the average link number M. Thus the total number of steps is $O(M^2 N log N)$. Further, M is constantly 20 to 40 in Wikipedia in spite of the evolvement of the matrix size N. Finally, the result is stored in another DBT R.

We conducted a benchmark test for DBT and the multiplication algorithm compared with GNU Octave (with ATLAS library), one of the most effective numerical algebra implementations and the result shows the effectiveness of DBT for huge scale sparse matrix multiplication.

pfibf with DBT. In this section, we describe the concrete flow of $pfibf$ calculation by using a DBT. Since $pfibf$ analyzes both forward and backward links of the articles, first we calculate A' by adding A and the transpose matrix A^T as follows:

$$A' = A + A^T. \tag{4}$$

By calculating the power of A', we can extract the number of paths for any pair of articles in n-hop range. An element $a_{i,j}^n$ in matrix A'^n denotes the number of paths from article v_i to article v_j whose length is n. However, before calculating A'^n, each element in A should be replaced by the following formula to approximate ibf:

$$a'_{i,j} = a'_{i,j} \cdot log \frac{N}{|B_{v_j}|}. \tag{5}$$

$|B_{v_j}|$ denotes the number of backward links of article v_j. Finally, we can extract the $pfibf$ for any pair by adding the matrices A'^1, A'^2, ... , A'^n as follows:

$$pfibf(i,j) = \sum_{l=1}^{n} \frac{1}{d(n)} \cdot a_{i,j}^{\prime l}. \tag{6}$$

$d()$ denotes a monotonically increasing function such as a logarithm function which increases the value according to the length of path n.

4.3 FB Weighting

After a number of experiments to evaluate the accuracy of $pfibf$, we realized that the accuracy decreased in particular situations. Then, after having further experiments in order to detect the cause, we finally realized that the accuracy of general term analysis is worse than the accuracy of domain specific terms. General terms have the following characteristics:

- They have a lot of backward links,
- They are referred to from various topic-ranges,
- The content is trustful because it is usually edited by many authorities.

General terms, such as "United states," "Marriage" and "Horse," are referred to from various articles in various topic ranges. This means that the backward link analysis cannot be converged because the topic locality is weaker than in domain-specific terms such as "Microsoft" and "iPod." Although the backward link analysis is not convergent, the forward link analysis is effective because the contents are trustful and usually edited by many authorities.

In contrast to this, domain-specific terms have a much stronger topic locality. Although they have less links from other pages and the contents are sometimes not trustful, each link from other pages is topically related to the content. Therefore, we developed the "FB weighting" method which flexibly changes the weight of the forward link analysis and backward link analysis as follows:

$$W_b(|B_d|) = 0.5/(|B_d|^{\alpha}), \tag{7}$$

$$W_f(|B_d|) = 1 - W_b(|B_d|). \tag{8}$$

$|B_d|$ is the backward link number of article d. The constant α must be optimized according to the environment. After a number of experiments, an α value of about 0.05 was recognized to be suitable for the link structure of Wikipedia. The weight W_b is multiplied for each element on A and W_f for A^T as well. Thus formula (4) must be modified into the following formula (9):

$$A' = W_f \cdot A + W_b \cdot A^T. \tag{9}$$

5 Experiments

To evaluate the advantages of our approach, we conducted several experiments. In this section, we describe these experiments and discuss the results.

5.1 Overview

First of all, we constructed four thesauri from Wikipedia by using the four methods mentioned in this paper; TF-IDF, link cooccurrence analysis, $pfibf$ (2-hop) and $pfibf$ with FB (2-hop) weighting, in order to to evaluate the performance.

After that, we conducted two experiments to evaluate the accuracy of the constructed thesauri. In the first experiment, the accuracy of an association thesaurus extracted by each of the four methods was evaluated by the "WordSimilarity-353 test collection"[12] which has often been used in previous Wikipedia researches[4,6]. The test collection contains 353 word pairs and these pairs have been judged by 13-16 testers to produce a single relatedness score. For each method, we calculated the relatedness for each pair in this test collection. Then, we compared the extracted relatedness with the human judgements by using Spearman's rank correlation coefficient.

In the second experiment, the accuracy for each method was evaluated by CP (Concept Precision)[13]. We developed an evaluation interface which shows the top 30 associated terms for a query for each constructed thesaurus individually. Users evaluated whether the associated terms presented by the system are relevant or not by ranking them into 3 levels (1: irrelevant, 2: Moderate, 3: Relevant). CP is defined by the following formula:

$$CP = \frac{Number\ of\ retrieved\ relevant\ concepts}{Number\ of\ total\ retrieved\ concepts}. \tag{10}$$

"Number of retrieved relevant concepts" means the number of concepts that were scored 3 by users. We randomly gave 10 different queries and 120 associated terms (30 terms / method) from the queries, thus totally 300 term pairs were evaluated. 12 people participated in the experiment and on average 6 testers evaluated one set of pairs.

5.2 Result

Table 1 shows the experimental results.

First, the analysis time for TF-IDF and Cooccurrence was much shorter than that of pfibf because these are sequential approach while pfibf needs a huge scale matrix multiplication. pfibf (FB weighting as well) took totally 62 hours on a single workstation (Pentium 4 2.4 GHz) to extract 300 associated concepts for every concept (1.3 million concepts) in Wikipedia. Thus, by using a single

Table 1. Performance and accuracy for thesaurus construction

Methods	Time / Page	Experiment 1 (Spearman)	Experiment 2 (CP) Top 10	Top 20	Top 30
TF-IDF	0.001 sec.	0.574	69.3%	66.9%	66.3%
Cooccurrence	0.001 sec.	0.538	65.7%	59.8%	55.8%
pfibf	0.30 sec.	0.677	76.3%	71.2%	68.3%
pfibf with FB	0.34 sec.	0.680	81.8%	75.3%	73.2%

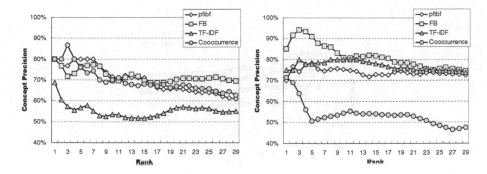

Fig. 3. CP (Concept Precision) of domain specific terms and general terms

workstation, we can extract the thesaurus once per several days. We believe that it is enough scalable for many applications such as IR systems with high-coverage for latest concepts.

However, the analysis time could be reduced by using several workstations because the DBT multiplication is suitable for parallel computing. We used 3 workstations to reconstruct the thesaurus and the result was exactly same but the analysis time become one third of that of a single workstation.

Regarding the two experiments, the results show that both our proposed methods achieved higher accuracy than the other two methods in both CP and Spearman's rank correlation coefficient. This means that the two factors of pfibf (number of paths and length of paths) are helpful in order to construct an accurate thesaurus. According to the CP, the accuracy of pfibf with FB weighting method is better than that of plain pfibf. However, by comparing Spearman's rank correlation coefficient, the accuracy of FB weighting is not much different from that of plain pfibf.

In order to make the effectiveness of FB weighting clear, we compared these four methods in detail. We separated the queries into 2 categories; domain specific terms and general terms. After that, we evaluated the methods in three cases; thesaurus construction for domain specific terms only, general terms only and all terms mixed.

Figure 3 (left) shows a comparison of CP for domain specific terms such as "Microsoft" and "PlayStation." It shows that the link cooccurrence analysis achieves a high precision for the top ranked terms. FB weighting proved to be less effective for the domain specific term analysis than the normal $pfibf$. As we mentioned before, the contents of domain specific terms (articles) are not refined enough compared with general terms, thus irrelevant links occur relatively often. We think that this is the reason why the CP of TF-IDF decreases so drastically for lower ranked pages.

Figure 3 (right) shows a comparison of the CP for general terms such as "Sport" and "Book." It shows that the CP is quite high for the top 10 terms extracted by FB weighting, and the CP decreases softly, but keeping a relatively

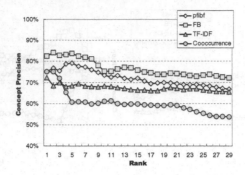

Fig. 4. Concept precision of all terms

Table 2. Sample of queries and the extracted terms by pfibf with FB weighting

Query	Extracted association terms		
Sports	Basketball	Baseball	Volleyball
Microsoft	MS Windows	OS	MS Office
Apple Comp.	Macintosh	Mac OS X	iPod
iPod	Apple Comp.	iPod mini	iTunes
Book	Library	Diamond Sutra	Printing
Google	Search engine	PageRank	Google search
Horse	Rodeo	Cowboy	Horse-racing
Film	Actor	Television	United States
DNA	RNA	Protein	Genetics
Canada	Ontario	Quebec	Toronto

high value even for pages with a very low rank. In contrast to this, the cooccurrence analysis was not effective for general terms except for the top terms. General terms cooccur with various terms in various topics because the topic locality of general terms is not strong. We think that this low topic locality is the cause of the accuracy problem.

Table 2 shows an example of an association thesaurus constructed by *pfibf* with FB weighting.

6   Conclusion

Wikipedia, a very large scale Web-based encyclopedia, is an invaluable Web corpus for knowledge extraction. In this paper, we first unveiled the link structure of Wikipedia in detail to prove that it has a high potential for knowledge extraction. In the next step, we listed up the possible conventional methods that can be used for Wikipedia mining and proposed a link structure mining method *pfibf*, a scalable and high accuracy method for association thesaurus construction. After that, we applied FB weighting as an extension to avoid the accuracy problem on general terms. Finally we confirmed the notable advantage of our proposed methods in a number of experiments.

The constructed thesaurus (pfibf with FB weighting) is accessible under the following URL and it allows users to extract associated terms from any concept in Wikipedia.

http://wikipedia-lab.org:8080/WikipediaThesaurusV2

An association thesaurus is just a first step in our whole project; "Wikipedia mining." Our next step is another project called "Wikipedia Ontology;" a huge scale Web ontology which is automatically extracted by Wikipedia mining. The purpose of this project is to extract not only term associations but also term relations such as "is-a" or "part-of."

We believe that Wiki-based knowledge management in enterprise environments will be popular in near future. This means that the application of our proposed methods, $pfibf$ and FB weighting, are not limited to Wikipedia. These methods also can be applied for extracting organization specific concepts.

Acknowledgment

This research was supported in part by Grant-in-Aid on Priority Areas (18049050), and by the Microsoft Research IJARC Core Project.

References

1. Giles, J.: Internet encyclopaedias go head to head. Nature 438, 900–901 (2005)
2. Nakayama, K., Hara, T., Nishio, S.: A thesaurus construction method from large scale web dictionaries. In: AINA 2007. Proc. of IEEE International Conference on Advanced Information Networking and Applications, pp. 932–939 (2007)
3. Ruiz-Casado, M., Alfonseca, E., Castells, P.: Automatic assignment of wikipedia encyclopedic entries to wordnet synsets. In: Szczepaniak, P.S., Kacprzyk, J., Niewiadomski, A. (eds.) AWIC 2005. LNCS (LNAI), vol. 3528, pp. 380–386. Springer, Heidelberg (2005)
4. Strube, M., Ponzetto, S.: WikiRelate! Computing semantic relatedness using Wikipedia. In: AAAI 2006. Proc. of National Conference on Artificial Intelligence, pp. 1419–1424. Boston, Mass (2006)
5. Milne, D., Medelyan, O., Witten, I.H.: Mining domain-specific thesauri from wikipedia: A case study. In: WI 2006. Proc. of ACM International Conference on Web Intelligence, pp. 442–448 (2006)
6. Gabrilovich, E., Markovitch, S.: Computing semantic relatedness using wikipedia-based explicit semantic analysis. In: IJCAI 2007. Proc. of International Joint Conference on Artificial Intelligence, pp. 1606–1611 (2007)
7. Salton, G., McGill, M.: Introduction to Modern Information Retrieval. McGraw-Hill Book Company, New York (1984)
8. Lawrence, P., Sergey, B., Rajeev, M., Terry, W.: The pagerank citation ranking: Bringing order to the web. Technical Report, Stanford Digital Library Technologies Project (1999)

9. Kleinberg, J.M.: Authoritative sources in a hyperlinked environment. Journal of the ACM (5), 604–632 (1999)
10. Davison, B.D.: Topical locality in the web. In: Proc. of the ACM SIGIR, pp. 272–279 (2000)
11. Schutze, H., Pedersen, J.O.: A cooccurrence-based thesaurus and two applications to information retrieval. International Journal of Information Processing and Management 33(3), 307–318 (1997)
12. Finkelstein, L., Gabrilovich, E., Matias, Y., Rivlin, E., Solan, Z., Wolfman, G., Ruppin, E.: Placing search in context: the concept revisited. ACM Trans. Inf. Syst. 20(1), 116–131 (2002)
13. Chen, H., Yim, T., Fye, D.: Automatic thesaurus generation for an electronic community system. Journal of the American Society for Information Science 46(3), 175–193 (1995)

Economically Enhanced Resource Management for Internet Service Utilities

Tim Püschel[1,2], Nikolay Borissov[2], Mario Macías[1], Dirk Neumann[2],
Jordi Guitart[1], and Jordi Torres[1]

[1] Barcelona Supercomputing Center - Technical University of Catalonia(UPC),
c/ Jordi Girona 29, 08034 Barcelona, Spain
{tim.pueschel,mario.macias,jordi.guitart,jordi.torres}@bsc.es
[2] Institute of Information Systems and Management (IISM)
Universität Karlsruhe (TH), Englerstr. 14, 76131 Karlsruhe, Germany
{pueschel,borissov,neumann}@iism.uni-karlsruhe.de

Abstract. As competition on global markets increases the vision of utility computing gains more and more interest. To attract more providers it is crucial to improve the performance in commercialization of resources. This makes it necessary to not only base components on technical aspects, but also to include economical aspects in their design. This work presents an framework for an Economically Enhanced Resource Manager (EERM) which features enhancements to technical resource management like dynamic pricing and client classification. The introduced approach is evaluated considering various economic design criteria and example scenarios. Our preliminary results, e.g. an increase in achieved revenue from 77% to 92% of the theoretic maximum in our first scenario, show that our approach is very promising.

1 Introduction

Many web applications have strongly varying demand for computing resources. To fulfil these requirements sufficient resources have to be made available. In some cases it is possible to run certain tasks at night to achieve a more even usage, however in many cases it is not feasible to let users wait a long time for results. This leads to a situation in which utilization is very high during certain peak times while many resources lay idle during other times. Due to high competition on global markets many enterprises face the challenge to make use of new applications and reduce process times on one side and cut the costs of their IT-infrastructures on the other side [5].

In light of this challenge the idea of utility computing gained interest. Utility computing describes a scenario where computer resources can be accessed dynamically in analogy to electricity and water [19]. The more resource providers offer their resources or services, the more likely it is they can be accessed at competitive prices. Therefore it is important to attract more providers.

However providers will only offer their services if they can realize sufficient benefit. With state-of-the-art technology, this assimilation is hampered, as the

B. Benatallah et al. (Eds.): WISE 2007, LNCS 4831, pp. 335–348, 2007.

local resource managers facilitating the deployment of the resources are not designed to incorporate economic issues (e.g. price).

In recent times, several research projects have started to develop price-based resource management components supporting the idea of utility computing. Those approaches are devoted to scheduling by utilizing the price mechanism. Clearly, this means that technical issues such as resource utilization are ignored for scheduling. In addition, resource management is much more comprehensive than just scheduling. For example Service-Level-Agreement (SLA) management is also part of resource management that is often omitted in economic approaches. This plays a role when deciding which already ongoing jobs to cancel in overload situations to maintain system stability.

To improve performance in the commercialization of distributed computational resources decisions about the supplied resources and their management should be based on both technical and economic aspects [12]. Hence, this paper is an interdisciplinary work taking into account aspects from computer science and economics. We will explore the use of economic enhancements such as client classification and dynamic pricing to resource management.

Technical resource management systems typically offer the possibility to include priorities for user groups. In purely price-based schedulers it is not possible to distinguish important from unimportant partners, as only current price matters for the allocation. We will motivate that client classification should be integrated into economically enhanced resource management systems. Essentially, there are two main reasons to do so: First, client classification allows the inclusion of long-term oriented relationships with strategically important customers so-called credential components. Second, client classification can be used as an instrument of revenue management, which allows skimming off consumer surplus. The main contribution of this paper is to show how technical parameters can be combined into an economically enhanced resources management that increases revenue for the local resource sites.

2 Motivational Scenarios

2.1 Organisation Selling Spare Resource Capacity

An organisation offers different web services for internal use. However the utilization of the resources is uneven. During certain times there is only a low load, the resources are almost idle, and at other times users have to wait for their results very long. Therefore the organisation decides to buy more capacity on the market when needed and sell its spare capacity when the load is low. When accepting jobs users from within the organisation should always be preferred. To provide and communicate a good dynamic evaluation of resources, market mechanisms are also used for internal users. This also gives internal users an incentive to run their jobs during times of low utilization. Internal users also receive a significant discount compared to external users.

2.2 Resource/Service Provider with Preferred Customers

A big service provider maintains a data center whose resources are sold. The service provider already has a number of clients but still has a high spare capacity. Therefore he joins a marketplace to find new clients and optimize capacity utilization. To maintain the good relations with the current clients and encourage regular use of its services the provider offers a preferred client contract. Preferred clients receive a discount on the reservation price and soft preference when accepting jobs. Soft preference means that their bids are increased by a certain amount for the winner determination. When offering the same price the preferred client is chosen over the external client, but the standard client has the chance to outbid the preferred client. Prices should be calculated dynamically based on utilization, client classification, estimated demand and further pricing policies of the provider.

3 Economically Enhanced Resource Management

To improve performance in the commercialization of distributed computational resources and increase the benefit of service providers we propose the introduction of various economic enhancements into resource management. This chapter first describes the objectives and requirements for the enhancements, then follows a description of the key mechanisms that are to be integrated in the EERM. The chapter ends with a description of the architecture and the components of the EERM.

3.1 Objectives and Requirements

The main goals of these enhancements are to link technical and economical aspects of resource management and strengthen the economic feasibility. This can be achieved by establishing more precise price calculations for resources, taking usage of the resources, performance estimations and business policies into account. The introduced mechanisms should be be able to deal efficiently with the motivational scenarios given earlier. This means they have to feature *client classification*, different types of *priorities* for jobs from certain clients, *reservation* of a certain amount *of resources* for important clients, and *dynamic calculation of prices* based on various factors. In addition to these requirements the system should also offer quality of service (QoS) and be able to deal with situations in which parts of the resources fail. To adapt to different scenarios and business policies of different situations it should be highly flexible and configurable via policies.

When designing mechanisms various economic design criteria [4], [20] should be considered. These following criteria apply to the respective features as well as the overall system and the market mechanisms it is embedded in.

Individual Rationality. An important requirement for a system is that it is individual rational on both sides, i.e. both providers and clients have to have a benefit from using the system.

Simplicity and Computational Costs. While the enhancements introduce some additional factors and they should not introduce any unnecessary complexity. Similarly client classification, quality of service and dynamic pricing add some additional computational complexity, however they should not add any intractable problems and its benefits should outweigh its costs.

Revenue Maximization. A key characteristic for resource providers is revenue maximization or more general utility maximization.

Incentive Compatibility. Strategic behaviour of clients and providers can be prevented if a mechanism is incentive compatible. Incentive compatibility means that no other strategy results in a higher utility than reporting the true valuation.

Efficiency. There are different types of efficiency. The first one considered here is pareto optimimality. An allocation is considered pareto optimal if no participant can improve its utility without reducing the utility of another participant. The second efficiency criterion is allocative efficiency. A mechanism is called allocative efficient if it maximizes the sum of individual utilities.

3.2 Key Features

The motivational scenarios and the further requirements lead to four key features the of the EERM presented in this work.

Quality of Service. The first feature is quality of service. This can be broken down into two aspects. The first aspect is to assure adequate performance during normal operation of the resources. Overload situations can lead to reduced overall performance [17] and thereby can result in breaking QoS agreements between the provider and clients. Thus it is necessary to have a mechanism that ensures that jobs will not be accepted if they result in an overload situation. The second aspect of quality of service regards situations in which parts of the resources fail. To be able to fulfill all SLAs even in situations of partial resource failure it would be necessary to keep an adequate buffer of free resources. Where this is not feasible there should at least be a mechanism that ensures that those SLAs that can be kept with the available resources are fulfilled. This can be done by suspending or cancelling those jobs that can not be finished in time due to the reduced availability of resources.

Job Cancellation. Related to QoS is the feature automatic job suspension and cancellation. It is needed to ensure quality of service in situations where problems arise, i.e. parts of the resources fail or the estimations of the utilization were to optimistic. Cancellation of lesser important jobs to free capacity for incoming jobs with higher importance, i.e. jobs from a client with a higher classification or a jobs that deliver significantly more revenue is also possible.

Dynamic Pricing. Another enhancement is dynamic pricing based on various factors. [21] shows an approach for a pricing function depending on a base pricing rate and a utilization pricing rate. However the price can depend not only on current utilization but also on other factors such as projected utilization, client classification, and projected demand. Pricing should also be contingent on the demand on the market. This feature can be either implemented in the EERM or in a dedicated component responsible for trading. This agent would request a price based on factors including utilization of the resources and client classification. from the EERM and then calculate a new price taking the situation on the market into account.

Client Classification. Earlier giving clients different privileges was mentioned, e.g. by discriminating on factors like price and quality of service. This part describes the factors that can be used to differentiate various client classes.

Price Discrimination. Price discrimination or customer-dependent pricing is one way to differentiate between different classes of clients. One idea to achieve this is introducing Grid miles [16] in analogy to frequent flyer miles. Clients could be offered a certain amount of free usage of the resources or a 10% discount after spending a certain amount of money.

Reservation of Resources. For certain users it may be very important to always have access to the resources. This class of users could be offered a reservation of a certain amount of resources. One option is to reserve a fixed share of resources for a certain class of users another possibility is to vary this share depending on the usage of the system.

Priority on Job Acceptance. Another option is to use a client priority on job acceptance. When the utilization of the system is low jobs from all classes of clients are accepted but when the utilization of the resources rises and there is competition between the clients for the resources, jobs from certain clients are preferred. There are two types of priorities: strict priorities and soft priorities.

- *Strict priority* means that if a job from a standard client and a client with priority compete for acceptance, the job from the client with priority always wins. Jobs from clients with priority are always preferred, thus there is no real competition between the different classes of clients.
- *Soft priority* means jobs from clients with priority are generally preferred but standard clients have the chance to outbid clients with priority. Thus soft priority is essentially a discount on the reservation price or bid that may only apply in certain situation, i.e. when utilization exceeds a certain threshold.

Quality of Service. Another factor where differentiation for classes of clients is possible is quality of service. For some classes of clients quality of service is offered, for others not. Offering different levels of quality of service for different classes of clients is also possible. An example for this would be offering different risk levels [7].

4   Architecture of the EERM

This part describes the architecture of the EERM that was designed based on the requirements and key features. It includes a description of the components

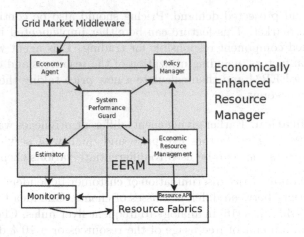

Fig. 1. EERM Architecture

with the aid of sequence diagrams. An overview of the architecture of the EERM can be seen in Fig. 1.

The EERM interacts with various other components, namely a Grid Market Middleware, a Monitoring component and the Resource Fabrics. The Grid Market Middleware represents the middleware responsible for querying prices and offering the services on the Grid market. The Monitoring is responsible for monitoring the state and the performance of the resources and notifying the System Performance Guard in case of problems. Additionally data collected by the Monitoring is used by the Estimator component to for its predictions. Resource Fabrics refers to Grid Middlewares such as Condor [15] or Globus [9].

Economy Agent. The Economy Agent is responsible for deciding whether incoming jobs are accepted and for calculating their prices. These calculations can be used both for negotiation as well as bids or reservation prices in auctions. Figure 2 shows the sequence of a price request. First the Economy Agent receives a request from a market agent. Then it checks whether the job is technically and

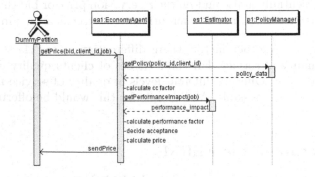

Fig. 2. Sequence Diagram Price Request

economically feasible and calculates a price for the job based on client category, resource status, economic policies and predictions of future job executions from the estimator component.

Estimator. The Estimator component calculates the expected impact on the utilization of the resources. This is important to prevent reduced performance due to overload [17], furthermore the performance impact can be used for the calculation of prices. This component is based on work by Kounev et al. using online performance models [13],[14].

System Performance Guard. The System Performance Guard is responsible for ensuring that the accepted SLAs can be kept. In case of performance problems with the resources it is notified by Monitoring. After checking the corresponding policies it determines if there is a danger that SLAs cannot be fulfilled. It then takes the decision to suspend or cancel jobs to ensure the fulfilment of the other SLAs and maximize overall revenue. Figure 3 shows a typical sequence for such a scenario. Jobs can also be cancelled when capacity is required to fulfil commitments to preferred clients.

Policy Manager. To keep the EERM adaptable the Policy manager stores and manages policies concerning client classification, job cancellation or suspension, etc. Policies are formulated in the Semantic Web Rule Language (SWRL) [11]. All features of the EERM require the respective components to be able to communicate with the Policy Manager and base their decisions on the corresponding policies. A simple example from a pricing policy in SWRL is the following rule which expresses that if the utilization is between 71% and 100% there is a surcharge of 50:

$$Utilization(?utilization) \land InsideUtilizationRange(?utilization, "71\%-100\%")$$

$$\Rightarrow SetSurcharge(?utilizationsurcharge, "50")$$

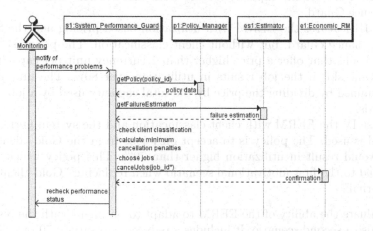

Fig. 3. Sequence Diagram System Performance Guard

Economic Resource Management. The Economic Resource Management is responsible for the communication with the local resource managers and influences the local resource management to achieve a more efficient global resource use.

5 Evaluation

The evaluation of this proposal consists of three parts. First an example scenario is described, then the results are presented and last the economic design criteria [4],[20] are considered.

5.1 Example Scenario

The first part of the evaluation considers an example scenario. For this evaluation there were no actual jobs executed. The policies were manually applied and revenue, utilization and utilities calculated. The job information given in Table 1 and the following assumptions were used.

The total capacity is 100 and the jobs given in Table 1 will be available during the run. The system receives information about jobs one timeslot before they become available. The Gold-client only uses the provider when he is assured access to a certain capacity. In this case he is guaranteed a total capacity of 60 units, while being able to launch new jobs of up to 30 units per period.

Using this scenario as a basis four policies are evaluated. The following policies represent the some of the basic options - no enhancements, client classification with fixed reservation, dynamic pricing with reservation prices, and client classification with strict priority:

- Case I is an example without any EERM. In this case any job is accepted if there is enough capacity left to fulfil it.
- Case II features a simple form of client classification, there is a fixed reservation of 60% of the capacity for the Gold-client. There is no System Performance Guard.
- Case III features the EERM with utilization-based pricing and the System Performance Guard, but without client classification. The policy is to only accept jobs that offer a price higher than 1 currency unit per capacity unit and time slot if the job results in utilization over 80%. The unit price is determined by dividing the price by the total capacity used by a job over all periods.
- In case IV the EERM with client classification and the system performance guard is used. The policy is to accept only jobs from the Gold-client if the job would result in utilization higher than 70%. This policy which can be applied to the first motivational scenario when replacing "Gold-client" with "Internal".

To evaluate the ability of the EERM to adapt to problems with the resources there is also a second scenario. It includes a reduced capacity of 70 in t=7 and a capacity of 60 in t=8 due to unpredicted failure of resources. In the cases without

Table 1. Example scenario

Job	Start Time	End Time	Capacity/t	Client Class	Price	Penalty
A	1	3	55	Standard	330	-82.5
B	1	5	24	Standard	180	-30
C	1	7	20	Standard	140	-17.5
D	2	4	20	Standard	120	-30
E	3	5	15	Standard	90	-22.5
F	3	8	20	Standard	120	-15
G	4	7	20	Standard	160	-40
H	4	9	15	Standard	135	-22.5
I	5	10	30	Standard	180	-22.5
K	5	8	30	Standard	240	-60
L	6	8	30	Standard	90	-11.25
M	6	9	12.5	Standard	50	-6.25
O	8	10	20	Standard	90	-15
P	9	10	21	Standard	84	-21
Q	9	10	30	Standard	90	-15
R	2	6	30	Gold	375	-187.5
S	5	8	30	Gold	300	-150
T	7	10	7.5	Gold	75	-37.5
U	7	9	20	Gold	150	-75
V	9	10	20	Gold	100	-50

the System Performance Guard it is assumed that the SLAs of jobs that end in periods that are overloaded due to the reduced capacity fail and that jobs that continue to run for more time can catch up and fulfil their SLAs. To assess the benefits of the EERM for providers and clients we compare their utility without and with EERM. Utility functions depend on the preferences of providers and clients. For the provider revenue is obviously a key criteria. Another factor that has to be considered is utilization. To allow for maintenance and reduce the effects of partial resource failure it is not in the interest of the provider to have a utilization that approaches 100%. In this example an average utilization of 80% is considered optimal. The following utility function for the provider considers the revenue per utilization while taking into account the optimal average utilization of the resources:

$$u_{provider} = \frac{revenue}{n * averageutilization} * \frac{1}{1 + |0.8 - \frac{averageutilization}{100}|}$$

For the client utility the idea is to weight the capacity needed by a job with its importance. The importance of a job for a client is expressed in the price per capacity unit the client is willing to pay for it. This results in the following client utility function, where P is the set of prices per unit (i.e. 2.5, 2, 1.5, 1):

$$u_{client} = \sum_{l \in P} unitprice_l * \frac{allocatedjobcapacity_l}{totaljobcapacity_l}$$

Table 2. Result of the four cases

Case	Completed	Revenue	Avg Load	$u_{provider}$	u_{client}	Revenue Proportion
I	A, B, C, G, H, L, M, O , P, Q	1349	89.7	1.37	3.10	76.65%
II	C, D, M, O, R, S, T, U, V	1400	71	1.81	3.33	79.55%
III	A, B, D, G, H, K, M, O, P, Q	1479	84.7	1.67	3.25	84.03%
IV	A, G, H, R, S, T, U, V	1625	73.5	2.08	3.87	92.33%
A*	A, E, G, O, Q, R, S, T, U, V	1760	81	2.15	3.53	100%

Table 3. Result with partial resource failure

Case	Completed	Failed	Revenue	Avg Load	$u_{provider}$	u_{client}	Revenue Proportion
I	A, B, H, L, M, P, Q	C, G	901.5	79.7	1.13	2.28	56.56%
II	D, P, R, M, T, U, V	C, S	786.5	66.2	1.04	2.23	49.34%
III	A, B, D, G, H, K, P, Q	M	1332.75	74.95	1.69	3.05	83.61%
IV	A, P, R, S, T, U, V	G, H	1351.5	71.2	1.74	3.31	84.79%
A*	A, E, P, Q, R, S, T, U, V		1594	71.2	2.06	3.11	100%

5.2 Results

Table 2 shows the results of applying these 4 cases to the scenario given in Table 1. A* is the allocation with maximum revenue that could be achieved if all job information was available before the first period. The table includes information about the completed jobs, the realized revenue, the average utilization of the providers resources, the provider utility, the client utility, and the proportion of the revenue in comparison with the A*. Cases II, III and IV perform significantly better than the standard case I without any economical enhancements. Although the average utilization is lower than in case I the yielded revenue as well the provider's and the client's utility is higher.

Table 3 shows the results of applying the policies to the scenario with failure of parts of the resources in t=7 and t=8. It can be seen that the improvement in cases III and IV is even more significant. In this situation case II is worse than case I regarding revenue, provider utility and client utility. This is caused by the lack of the System Performance Guard and thereby higher cancellation penalties of the jobs of the Gold-client. Case III delivers a significantly better utility for providers and clients. However since it doesn't include client classification it does not address the needs of the Gold-client. Case IV again delivers the best results regarding revenue, provider utility and client utility. Figure 4 shows the revenue proportion in the standard case (a) and with partial resource failure (b). In both cases III and IV the benefit of the System Performance Guard can be clearly seen, they deliver of 83.61% and 84.79% compared to the 56.56% of the standard case I.

5.3 Economic Design Criteria

In cases III and IV of the example both sides have a clear benefit from using the respective mechanisms, hence *individual rationality* is achieved in these

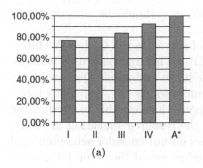 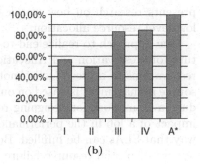

(a)   (b)

Fig. 4. Proportion of maximum revenue in the four cases

cases. The proposed mechanisms introduce some additional *complexity* but this is encapsulated in the EERM. They, however, do not introduce any NP-hard problems into the mechanism and the additional *computational cost* is limited. Another key characteristic for resource providers is *revenue maximization*. In the given example the policy of case IV delivers the highest revenue, provider utility and client utility among the options.

Whether the criterion of *incentive compatibility* is fulfilled depends on the market and classification mechanisms the EERM is embedded in. For the example a simple pay as you bid scheme is assumed. As this means that clients set their bids lower than their evaluation to achieve a benefit incentive compatibility is not given in the example. If, however, embedded in an allocation and pricing mechanism that is incentive compatible the EERM serves to give more precise valuations for the jobs.

Efficiency also depends on the mechanisms the EERM is embedded in. For the example we used a scenario where the provider decides whether jobs are accepted. Obviously the provider is mainly concerned with maximizing its own utility. Therefore it does not guarantee the maximum total utility. The EERM can also be embedded in efficient market mechanisms serving to give a more precise valuations of the jobs.

6   Related work

There is related work covering different aspects of our work. [8] discusses the application of economic theories to resource management. [1] presents an architecture for autonomic self-optimization based on business objectives. Elements of client classification such as price discrimination based on customer characteristics have been mentioned in other papers [16], [3]. They did however not consider other discrimination factors. [6] describes data-mining algorithms and tools for client classification in the electricity grids but concentrate on methods for finding groups of customers with similar behaviour. An architecture for admission control on e-commerce websites that prioritizes user sessions based on predictions about the user's intentions to buy a product is proposed in [18]. [2]

presents research on how workload class importance should be considered for low-level resource allocation.

One approach to realize end-to-end quality of service is the Globus Architecture for Reservation and Allocation (GARA) [10]. This approach uses advance reservations to achieve QoS. Another way to achieve autonomic QoS aware resource management is based on online performance models [13], [14]. They introduce a framework for designing resource managers that are able to predict the impact of a job in the performance and adapt the resource allocation in such a way that SLAs can be fulfilled. Both approaches do not consider achieving QoS in case of partial resource failure. Our work makes use of the second approach and adds the mechanism of Job Cancellation.

The introduction of risk management to the Grid [7] permits a more dynamic approach to the usage of SLAs. It allows modelling the risk that the SLA cannot be fulfilled within the service level agreement. A provider can then offer SLAs with different risk profiles. However such risk modelling can be very complex. It requires information about the causes of the failure and its respective probabilities. Clients need to have the possibility to validate the accuracy and correctness of the providers risk assessment and risks have to be modelled in the SLAs.

7   Conclusion and Future Work

In this work various economical enhancements for resource management were motivated and explained. We presented a mechanism for assuring Quality of Service and dealing with partial resource failure without introducing the complexity of risk modelling. Flexible Dynamic Pricing and Client Classification was introduced and it was shown how these mechanisms can benefit service providers. Various factors and technical parameters for these enhancements were presented and explained.

Furthermore the preliminary architecture for an Economically Enhanced Resource Manager integrating these enhancements was introduced. Due to the general architecture and the use of policies and a policy manager this approach can be adapted to a wide range of situations.

The approach was evaluated considering economic design criteria and using an example scenario. The evaluation shows that the proposed economic enhancements enable the provider to increase his benefit. In the standard scenario we managed to achieve a 92% of the maximum theoretically attainable revenue with the enhancements in contrast to 77% without enhancements. In the scenario with partial resource failure the revenue was increased from 57% to 85% of the theoretical maximum.

The next steps will include refinement of the architecture as well as the implementation of the EERM. During this process further evaluation of the system will be done, e.g. by testing the system and running simulations. Another issue that requires further consideration is the generation of business policies for the EERM. Further future work in this area includes the improvement of price calculations with historical data and demand forecasting; the design of mechanisms for using collected data to determine which offered services deliver the most revenue and concentrate on them; the introduction of a component for automatic evaluation and improvement of client classification.

Acknowledgments

This work is supported by the Ministry of Science and Technology of Spain and the European Union (FEDER funds) under contract TIN2004-07739-C02-01 and Commission of the European Communities under IST contract 034286 (SORMA).

References

1. Aiber, S., Gilat, D., Landau, A., Razinkov, N., Sela, A., Wasserkrug, S.: Autonomic self-optimization according to business objectives. In: ICAC 2004. Proceedings of the First International Conference on Autonomic Computing, pp. 206–213 (2004)
2. Boughton, H., Martin, P., Powley, W., Horman, R.: Workload class importance policy in autonomic database management systems. In: POLICY 2006. Proceedings of the Seventh IEEE International Workshop on Policies for Distributed Systems and Networks, pp. 13–22 (2006)
3. Buyya, R.: Economic-based Distributed Resource Management and Scheduling for Grid Computing. PhD thesis, Monash University (2002)
4. Campbell, D.E.: Resource Allocation Mechanisms. Cambridge University Press, London (1987)
5. Carr, N.G.: The end of corporate computing. MIT Sloan Management Review 46(3), 32–42 (2005)
6. Chicco, G., Napoli, R., Piglione, F.: Comparisons among clustering techniques for electricity customer classification. IEEE Transactions on Power Systems 21(2), 933–940 (2006)
7. Djemame, K., Gourlay, I., Padgett, J., Birkenheuer, G., Hovestadt, M., Kao, O., Voß, K.: Introducing risk management into the grid. In: eScience2006. The 2nd IEEE International Conference on e-Science and Grid Computing, Amsterdam, Netherlands, p. 28 (2006)
8. Ferguson, D.F., Nikolaou, C., Sairamesh, J., Yemini, Y.: Economic models for allocating resources in computer systems, 156–183 (1996)
9. Foster, I., Kesselman, C.: Globus: A metacomputing infrastructure toolkit. International Journal of Supercomputer Applications and High Performance Computing 11(2), 115–128 (1997)
10. Foster, I., Kesselman, C., Lee, C., Lindell, B., Nahrstedt, K., Roy, A.: A distributed resource management architecture that supports advance reservations and co-allocation. In: IWQoS 1999. Proceedings of the 7th International Workshop on Quality of Service, London, UK, pp. 62–80 (1999)
11. Horrocks, I., Patel-Schneider, P.F., Boley, H., Tabet, S., Grosof, B., Dean, M.: Swrl: A semantic web rule language combining owl and ruleml. Technical report, W3C Member submission (2004)
12. Kenyon, C., Cheliotis, G.: Grid resource commercialization: economic engineering and delivery scenarios. Grid resource management: state of the art and future trends, 465–478 (2004)

13. Kounev, S., Nou, R., Torres, J.: Autonomic qos-aware resource management in grid computing using online performance models. In: VALUETOOLS 2007. The 2nd International Conference on Performance Evaluation Methodologies and Tools, Nantes, France (2007)
14. Kounev, S., Nou, R., Torres, J.: Building online performance models of grid middleware with fine-grained load-balancing: A globus toolkit case study. In: EPEW 2007. The 4th European Engineering Performance Workshop, Berlin, Germany (2007)
15. Litzkow, M.J., Livny, M., Mutka, M.W.: Condor - A hunter of idle workstations. In: Proceedings of the 8th International Conference of Distributed Computing Systems (1988)
16. Newhouse, S., MacLaren, J., Keahey, K.: Trading grid services within the uk e-science grid. Grid resource management: state of the art and future trends, 479–490 (2004)
17. Nou, R., Julià, F., Torres, J.: Should the grid middleware look to self-managing capabilities? In: ISADS 2007. The 8th International Symposium on Autonomous Decentralized Systems, Sedona, Arizona, USA, pp. 113–122 (2007)
18. Poggi, N., Moreno, T., Berral, J.L., Gavaldà, R., Torres, J.: Web customer modeling for automated session prioritization on high traffic sites. In: Proceedings of the 11th International Conference on User Modeling, Corfu, Greece (2007)
19. Rappa, M.A.: The utility business model and the future of computing services. IBM Systems Journal 43(1), 32–42 (2004)
20. Wurman, P.R.: Market structure and multidimensional auction design for computational economies. PhD thesis, University of Michigan, Chair-Michael P. Wellman (1999)
21. Yeo, C.S., Buyya, R.: Pricing for Utility-driven Resource Management and Allocation in Clusters. In: ADCOM 2004. Proceedings of the 12th International Conference on Advanced Computing and Communications, Ahmedabad, India, pp. 32–41 (2004)

Enhancing Web Services Performance Using Adaptive Quality of Service Management

Abdelkarim Erradi[1] and Piyush Maheshwari[1,2]

[1] School of Computer Sc. and Eng., The University of New South Wales, Sydney, Australia
[2] IBM India Research Lab, New Delhi, India
aerradi@cse.unsw.edu.au, pimahesh@in.ibm.com

Abstract. The variation of contexts in which a Web service could be used and the resulting variation in Quality of Service (QoS) requirements motivates further research to extend Web services management platforms with automatic and adaptive management mechanisms in order to achieve differentiated service offerings and to improve quality of service in terms of availability, response time and throughput. However, most Web services platforms are based on a best-effort model, which treats all requests uniformly, without service differentiation. This paper presents WS-DiffServ, a service differentiation middleware based on adaptive scheduling of service requests that prioritizes requests depending on their associated class of service and the current degree of service level conformance. The goal of the proposed approach is to increase conformance to negotiated service levels particularly in case of overloads such that the incurred penalties for violations are minimized. The paper first explores the typical requirements of a differential QoS support for Web services. We then present the design of WS-DiffServ. The effectiveness of the proposed approach is analyzed using supply chain management scenarios.

Keywords: Adaptive Quality of Service Management, Differential QoS.

1 Introduction

Web services are increasingly used to ease interoperability between heterogeneous and autonomous systems both for internal and external integration. Web services promise business agility and cost savings through re-use and standards-based integration. However, Web services based integration builds a web of interdependencies between participating systems and introduces various challenging interoperability and management issues [1]. One of the major issues with Web services is availability, which refers to the capability of serving requests in timely manner. Typically at the service provider side, if the volume of service requests temporarily exceeds the capacity for serving them the service becomes overloaded. As result this might cause clients to turn to competitor services and lead to loss of revenue. This motivates the need to formally specify and manage QoS coupled with the ability to deliver differentiated services with varying functionality and QoS attributes. For example, a financial Web service could be offered in either premium or basic grades. The service offerings may vary in the depth of the financial analysis, the rate of notification to clients and the guaranteed response time. This calls for

B. Benatallah et al. (Eds.): WISE 2007, LNCS 4831, pp. 349–360, 2007.

development of architectural principles for building such systems, and devising specialized middleware to ease the development and delivery of differentiated services. For example, consider a finance portal offering different financial Web services, such as stock quotes and equity trading services. The portal could be accessed by diverse sets of users, such as investors, brokers and partners. These services could be accessed either programmatically by a business partner by directly plugging into the Web services or by using a Web client or even from a mobile device. In case of high traffic to the Web services, a paying user should expect a shorter waiting time than a guest visitor browsing the site. Hence, business policies and the context in which a service is used as well as the client device capabilities create the need for differentiation in the order in which the requests are served. Differentiated services are further motivated by the fact that Web services users (i.e., applications or individuals) can have varied profiles, different requirements, and different levels of willingness to pay for different service properties. However, in the best-effort service delivery models of today, all requests are treated equally based on their arrival time and differentiation cannot be easily provided. Our framework, named WS-DiffServ, uses admission control and adaptive request scheduling to enable a Web service to remain responsive even when the load exceeds the capacity of the service, particularly enabling higher levels of responsiveness for the higher priority Web service consumers. The reminder of this paper is organized as follows. Section 2 overviews the relevant background on Web services QoS. Section 3 describes the logical architecture and features of the WS-DiffServ. Section 4 presents WS-DiffServ evaluation using a supply chain management case study. Section 5 concludes the paper and outlines some directions for future work.

2 Background and Problem Area

Web services QoS is an important theme in service-oriented computing and it is being studied in many ways. In the context of Web services, the notion of QoS covers specifying, negotiating, providing, monitoring, and measuring the non-functional aspects of a service. Several ongoing research efforts are looking at addressing the following key dimensions as depicted in Figure 1:

- QoS modeling, negotiation and agreement to manage the service delivery [2]
- QoS measurement and monitoring to assess QoS compliance and to trigger the adjustment of resource allocation [3]
- QoS provisioning, differentiation and adaptation using different policies like admission control, content adaptation and automated resource management to meet QoS assurances [1].

Although there is a growing body of knowledge in QoS for Web services, QoS provisioning remains open for further research. The biggest challenge facing the provision of differentiated services is the automated mapping of service offerings or SLAs requirements into sufficient computing and communication resources across various abstraction layers in order to meet the promised QoS assurances. This requires accurate service performance modeling, capacity management and timely information on available resources. The second challenge is the integration of differential QoS mechanisms available at different resource managers along the message path:

network, application servers, database servers, Web servers and SOAP layer. These challenges are further complicated by changes in resources load and availability as well as the virtualization introduced by services and their composition.

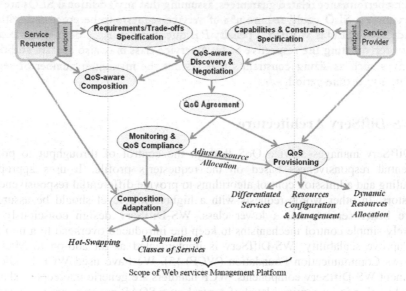

Fig. 1. Web services QoS dimensions

2.1 WS-DiffServ Quality of Service Model

SLAs, such the ones advocated by WS-Agreement [4], specify an agreement between a service consumer and a service provider in terms of one or more service level objectives (SLO) which state the requirements and capabilities of each party on service qualities. For example, an agreement may provide assurances on the service response time limits or service availability. Often an agreement contains scopes for which the guarantee is applicable, qualifying conditions (QC) which must be satisfied in order for the guarantee on the SLO to be valid, and business values (BV), such as rewards and penalties which incur if the SLO is violated. It may also contain multiple quality of service classes that consumers can choose from. An example for a service with multiple service classes is shown in the following table:

Table 1. Example of classes of service for GetStockPrice service operation

CS_1	CS_2	CS_3
SLO: ResponseTime less than 4 sec for 95% of Requests. QC: DayOfWeek = WeekDay and NumInvocations < 1000 Price per violation: $3 Penalty: 5$ per invocation with $P_{max} = \$2000 \ per \ week$	*SLO*: ResponseTime less than 6 sec for 90% of Requests. QC: DayOfWeek = WeekDay and NumInvocations < 2000 Price per violation: $2 Penalty: 3$ per violation with $P_{max} = \$1000 \ per \ week$	*SLO*: ResponseTime less than 10 secfor 80% of Requests. QC: NumInvocations < 4000 Price per invocation: $1 Penalty per violation: 0.5$ $P_{max} = \$500 \ per \ week$

In WS-DiffServ QoS model, a service may offer multiple Classes of Service $\{CS_1|\ CS_2\ ...\ :\ CS_n\}$ with alternative sets of Guarantee Terms G such that: $CS_i = \{G_1, G_2, ...\ G_n\}$ and $G_i = \{Obligated, Scope, SLO, QC, BV\}$. We concentrate on fulfilling performance related guarantees, assuming that any additional SLOs are met. For example a SLO could require n% of service requests to be processed within x seconds. If the SLO is violated, a penalty P is due. Further, P_{max} defines a maximum penalty for violating the respective SLO. A service class may also contain additional objectives such as sizing constraints that restrict the maximum number of service invocations per time period.

3   WS-DiffServ Architecture

WS-DiffServ manages service QoS through the control of throughput to provide differential responsiveness based on the requester's profile. It uses appropriate scheduling and admission control algorithms to provide differential responsiveness to requestors. In other words requests with a higher QoS level should be assured to receive better service that a lower class. WS-DiffServ design consciously uses relatively simple control mechanisms to keep the introduced overhead to a minimum and improve scalability. WS-DiffServ is implemented as an add-on to Microsoft Windows Communication Foundation (WCF) [5]. We have used WCF handlers to implement WS-DiffServ components. WCF handlers are generic interceptors that can be used to plug in customized level of control over SOAP message processing. WS-DiffServ can be placed between the client applications and the service implementations to transparently intercept service requests and manage the provided QoS. This is realized via intercepting arriving requests and providing the admission control and scheduling functionality. The admission control limits the number of simultaneously executed requests by queuing newly arriving requests when the service reaches its capacity. Queued requests are scheduled for execution based on their priority and execution deadlines as well as their impact on the overall profitability. Additionally, the service operations execution time is monitored and the average response time as well as the SLA conformance rate per service class is maintained. WS-DiffServ comprises the following components (Figure 2): Classifier, Scheduler and QoS Measurement Service (the latter is covered in details in [6]).

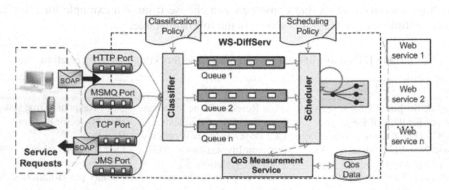

Fig. 2. WS-DiffServ Logical Architecture

3.1 Request Classifier

The Classifier is responsible for assigning a service class to incoming requests according to a predefined classification policy to enable prioritization. The request is then placed on the appropriate priority queue according to its assigned service class. This component also performs policing to make sure that the number of request per customer is within a pre-defined limit. Requests exceeding the maximum throughput limit are assigned a low priority class.

The classification schemas can be divided into two board categories: provider-based and client-based. The provider-based method classifies requests according to the requested service or the requested service operation. The client-based method characterizes requests according to the source of the request (e.g., user authentication/authorization credentials or Client IP address), the access channel or the client device used to reach the service. Using WS-DiffServ configuration the service provider can create a set of service classes and map each tuple of <User, Service-Operation, Service-Grade> into a specific service class. The service provider can assign QoS targets to each service class, such response time and throughput. WS-DiffServ classifier component uses the requester attributes metadata (e.g., the user's role) generated by the authentication handler, to map the request to a service class. Service classes (such as Gold, Silver and Bronze classification) managed by WS-DiffServ are encoded using WSPolicy4MASC [7].

The classifier mapping component also supports Web service calls to take the values of dynamic variables into account for the request classification. This allows WS-DiffServ to support complex requests classification schemas where a service level can be determined based on dynamic parameters, a simple example could be that a business partner providing higher business volumes be given higher QoS.

3.2 Admission Control

Admission control is used to control throughput and prevent overload by queuing requests that exceed the service capacity. Admission control requires two prerequisites: determining the load that a particular service invocation will generate on a system, and knowing the service capacity [8]. For the first requirement, WS-DiffServ uses online measurements to estimate the load that each service call imposes on the service. This is done by maintaining a moving average response time measured based on the QoS metrics gathered from recent invocations. The measured response time include the time from when the request is received until the response is sent back to the requester. Given a service provider exposes a finite number of operations, it is manageable to maintain per-operation estimates.

For the second requirement, we determine the capacity of the service offline by leveraging off the shelf tools, such as LoadRunner [9]. The service capacity limits the number of requests admitted to the system; excess requests are queued. The capacity of the service should be recomputed if there is a change in the system configuration, for example hardware upgrade.

WS-DiffServ maintains a running estimate of system load as the sum of the estimated service times of all executing requests. Before a service request is processed, the load estimate for that service operation is examined and the request is admitting only if it does not exceed the service capacity. If the request is allowed to proceed then the running load value is incremented accordingly. Otherwise, the request gets queued for later execution. As a service operation completes execution, WS-DiffServ decrements the current service load value accordingly and any pending requests are admitted in turn as long as they do not overload the service.

3.3 Scheduler

The scheduler manages dispatching queued requests to the service. When the processing of admitted requests in completed, pending requests are admitted from the waiting queues until the capacity threshold is reached. Various scheduling policies are supported by WS-DiffServ:

• Strict priority: This policy ensures that higher-class requests are always served first before servicing any of the lower-priority requests. The main limitation of this model is that low priority requests can starve. However, this can be addressed by periodically adapting priorities based on aging mechanism. The latter enforces an upper bound on the waiting time in the queue. The maximum waiting time is configurable, for example it can be defined as a multiple of the expected service time for the request. E.g., a request should not wait more than three times its expected service time otherwise it should auto-promoted to a higher priority.
• Weighted Round Robin (WRR): This policy schedules a service class based on its weighted importance. For example, one class will receive twice as many scheduled requests if its class weight is twice that of another.
• Earliest Deadline First (EDF): This schema schedules requests based on the deadline for completion of each request. This can be used to provide a guaranteed predicted response time.
• Shortest Job First (SJF): This policy allows sorting and placing arriving requests in the admission queue based on their expected processing times.

Beside the above static scheduling algorithms, WS-DiffServ adaptively prioritizes requests in a way that minimizes the overall penalty while achieving the highest compliance to established SLAs. The proposed adaptive model is based on the Weighted Round Robin (WRR) queuing discipline where an integer linear programming (ILP) task dynamically calculates optimal values of weights (i.e., the number of requests to process per service class each scheduling round), based on the QoS requirements, the current SLA compliance rate per service class and information about incurred penalties for SLO violations.

The central concept of our QoS management is an economic model that adaptively prioritizes individual requests based on their associated class of service while taking into account their impact on the overall profitability and the current degree of SLA compliance that the particular service class exhibits. The compliance rate is monitored per service class associated with the processed requests. We define C_{slc} as:

$$C_{slc} = \frac{Number\ of\ serviced\ requests\ within\ time\ limit}{Total\ number\ of\ serviced\ requests}.$$

The goal of the scheduling policy is to create a schedule of the pending requests, such that such that the sum of incurred penalties is minimized.

The overall $Penalty = \sum_{i=1}^{n}(P_i)$. With n is the number of pending requests. $P_i = (Number\ of\ serviced\ requests\ above\ time\ limit \times Penalty_{rate}) \times \propto$

With $\propto = (if\ C_{slc} > GoalConformanceRate_{slc}\ then\ 1\ else\ 0)$

and $GoalConformanceRate_{slc}$ is the minimum SLA Compliance Rate per

Service Class slc.

If $P_i > P_{max}$ then $P_i = P_{max}$. P_{max} is the Maximum Penalty per service class.

Let pf_r denote the penalty function attached to a pending request r. Furthermore, let Pen_r denote the penalty at the current time t, i.e., $Pen_r = pf_r(t)$, and let $NextPen_r$ denote the penalty at time t+1. The prioritization algorithm orders the queued requests by descending values using the following algorithm:

For each Request r in Queued Requests R do:

Compute $\Delta_r = \dfrac{NextPen_r - Pen_r}{Pt_r}$ with Pt_r the processing time of request r

End for

Sort the requests in descending order of Δ_r.

The algorithm above first schedules the requests that cause the largest penalty increase. The algorithm basically computes for all pending requests the increase of penalty between the current time t and the next scheduling time t+1. Then the requests are sorted in descending order according to the increase of penalty.

WS-DiffServ assigns a Queue Pool and a Listener Pool for each service class. The size of the queue/listener pools are reconfigured depending on runtime conditions. For example if high priority queues approach their threshold limit, the Queue Manager can assign more queues from low priority class Queue Pool, while doing so if there are pending requests in the 'borrowed' queues, they can be persisted for later processing.

4 WS-DiffServ Evaluation

This section describes the analysis and the results of the conducted performance testing to demonstrate the effectiveness of our approach. We compared our dynamic prioritization algorithm to three alternative approaches commonly used in scheduling. (1) FCFS where requests are submitted directly without any admission control and scheduling. (2) Static Weighted Round Robin (WRR) which statically defines the number of requests to process from each queue, for example 4 request from Gold queue, 2 requests from Silver queue and 1 request from Bronze queue (3) Earliest deadline first (EDF) where the requests that have approached or passed their deadline are scheduled with maximum priority. WS-DiffServ prototype is implemented using WCF) [5]. Exchanged messages between participating Web services are passed through series of handlers in the form of MessageContexts. In-Queues are used to categorize different types of incoming requests as each In-Queue is used for a different class of service. As the requests come in, they are placed appropriately in the In-Queue, to be processed in a prioritized manner. The Scheduler module is responsible for controlling throughput for each In-Queue based on configured priority values (as discussed in Section 3). Finally the Dispatcher module forwards the prioritized requests to the destination service.

We conducted a series of benchmarking tests to assess the effectiveness of WS-DiffServ in providing differentiated responsiveness of Web services. We used an extended .Net-based implementation of WS-I Supply Chain Management (SCM) application [10]. The SCM scenarios, as shown in Figure 3, are designed as Web services based interactions that simulate the business activity of an online supplier of electronic goods. We have extended the use cases so that each interaction is subject to a response time constraint depending on the customer profile. First a Web client calls the Retailer service's **GetCatalog** operation. When the user submits the order, the Web client calls the Retailer service's **SubmitOrder** operation. To fulfill orders, the Retailer Web service manages stock levels in three warehouses (WA, WB, and WC). If Warehouse A cannot fulfill an order, the Retailer checks Warehouse B; if Warehouse B cannot, the Retailer checks Warehouse C. When an item in stock falls below a certain threshold, the Warehouse must restock the item from the Manufacturer's inventory (MA, MB, and MC). Each use case includes a logging call to a logging facility in order to monitor the activities of the services. The customer can track orders by using the **GetOrderStatus** operation of the logging facility Web service. During the SCM process enactment, participating Web services can log event by calling the **LogEvent** operation of the logging facility Web service.

Optionally, there is a Configurator Web Service that lists all of the implementations registered in the UDDI registry for each of the Web Services in the sample application.

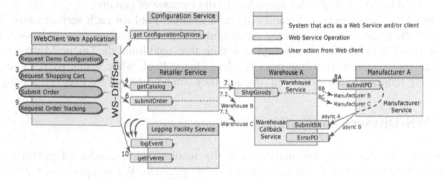

Fig. 3. SCM Application Control flow

Our goal is to study the impact of introducing WS-DiffServ on the throughput and the round-trip response time of Web services used in the above SCM application. Additionally, our aim is to discover areas of the platform that needs further improvement. To this end, we simulated multiple concurrent Web service clients, each of which invokes deployed services multiple times. We used a custom load generator to generate the workload and to measure the observed performance.

We deployed WS-DiffServ and the SCM backend Web services (The Retailer, the Manufacturer and Warehouses Web services) in a P4 2.8GHz, 1GB RAM PC running Windows Vista. The load generator was deployed in a Windows Vista laptop with 1 GB RAM. The machines were connected by 100MB LAN.

Table 2. Transaction Mix and Response Time Constraints

Transaction Type	% of Mix	Response Time SLO (in seconds)
GetCatalog	55	5
SubmitOrder	35	10
GetOrderStatus	10	4

To be able to compare the response time and throughput metrics, our experimental setup consists of two runtime configurations: one with WS-DiffServ placed in front of the backend Web services and other without WS-DiffServ mediation. Both experiments use the identical application logic implemented in C#.

Table 2 summarizes the transaction mix and the response time constraints configured for the conducted experiments. For each service invocation, we defined an SLA that requires 90% of the transaction invocations to be completed in less than the corresponding response time requirement. A violation of this constraint incurs a penalty P that depends on the service requester. In our test scenario, 15% of the requests incurred high penalty ($P = 5$), 35% incurred medium penalty ($P = 3$), and the remaining requesters incurred low penalty ($P = 1$). This mix constitutes a service with some important consumers and many regular customers that must be preferably served compared to normal users.

The graph in Figure 4 summarizes some of the performance test results. During the testing we focused our attention on two key metrics:

- The average response time per service class is calculated as the Total time taken for message processing / Number of messages processed. The measured response time includes the execution time on the Web server, the application and database servers as well as the queue waiting time (if any).
- Throughput represents the average number of successful requests processed in a sampling period. Throughput equals to Total number of messages processed / Total processing time

Each data point in Figure 4 represents the average response time for SubmitOrder requests over three independent runs of 200 requests each.

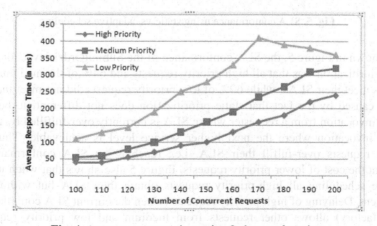

Fig. 4. Average response time using 3 classes of service

Figure 4 shows that the average response time for high class requests is always the lowest compared with ones for medium and low classes. This demonstrate that WS-DiffServ can guarantee that requests with higher priority classes will receive better services than classes with lower priorities and achieve the basic goals of service differentiation, especially when the system is under heavy load and not enough server resources to allocate.

Table 3. Throughput per service class type

Request Type	Number of Incoming requests	Observed Throughput
High Priority	80	65
Medium Priority	52	25
Low Priority	68	9

Table 3 presents a summary of the observed throughput that measures the number of processed requests per second. It shows how WS-DiffServ is able to maintain a higher throughput for high priority requests, even though the incoming requests are randomly arranged.

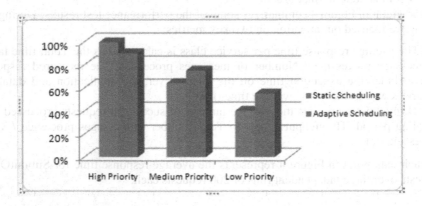

Fig. 5. SLA conformance using static vs. adaptive scheduling

As shown in Figure 5, the SLA conformance of the SubmitOrder invocations using static prioritization was that all high priority requests and 64% of the medium priority requests meet their SLAs, while 60% of low-priority requests did not meet their SLA. This is caused by the fact that static prioritization does not differentiate between a service invocation from a requester whose SLA is currently over-fulfilled by far and a service invocation where the next higher service level is achievable. Thus, high-priority requests over-fulfill their SLA, i.e., reaching an SLA conformance near 100%, at the cost of lower priority requests. Figure 5 also shows that when using the adaptive scheduling all high-priority requests obtain their SLA but without over-fulfillment. Delaying of high-priority requests (when the current SLA compliance rate is satisfactory) allows other requests from medium and low priority requests to achieve better SAL compliance. Basically using adaptive scheduling, if the current conformance ratio is above the target, high priority requests can be delayed.

The overhead of our adaptive scheduling algorithm varies between 0.65 milliseconds and 8.38 milliseconds with an average queue length of 200 in our benchmarks.

5 Related Work and Discussion

This section briefly highlights the research efforts closely related to our work. We also discuss how our work complements existing body of knowledge in the area of differentiated Web services. The state-of-practice regarding QoS differentiation is to provide different service interfaces for different service levels or to specify additional parameter in the service operation signature. Differentiation in terms of performance and throughput is usually achieved by using hardware-based techniques to either allocate dedicated resources for different QoS levels (i.e., physical partitioning) or over-provisioning of resources to absorb traffic fluctuations. The major drawbacks of these mechanisms are added complexity, higher cost and inefficient resource utilization. Regarding the state of the art for differentiated QoS, beside DiffServ architecture studied by the networking community [11], multiple approaches have been proposed for addressing differentiated QoS on Web servers and application servers [11]. Most of these approaches usually use some form of classification based on client IP address, requested URL or cookie to assign a service level to a request. To achieve differentiation, these approaches include admission control and request scheduling. Some have used observation based measurements for providing QoS guarantees while others have leveraged control theory. Our work complements and extends the above approaches by applying differentiation techniques for Web services rather than simple static or dynamic Web content. We focus on managing a multitude of service classes to achieve differential QoS using adaptive prioritization of requests.

A common limitation among these approaches is that they are not easily portable as they require extensive modifications to the web server or the operating system kernel in order to incorporate the service differentiation mechanisms. They also operate at lower protocol layers, such as HTTP or TCP. On the other hand our approach is portable, transparent and allows content-based request classification as it operates at the SOAP layer. Furthermore, our proposed solution in transparent and not invasive, it does not require any changes to the operating system, the application server, the service client or the service implementation. Hence, unmodified commodity Web services platforms can be used without costly changes while avoiding specific hardware or OS prerequisites.

6 Conclusions and Future Work

The notion of differentiated services is relevant for both business services and third-party Web and Grid services to meet the need for variety and flexibility. Web services capable of supporting different levels of service provide greater flexibility and achieve higher levels of reuse and adaptability. Advanced service level management techniques providing support for differentiated services are needed because the best effort delivery model is not enough to satisfy various categories of client applications accessing services in different contexts with varying service level expectations. This paper presented the architecture and features of WS-DiffServ, a practical middleware for differentiated Web services responsiveness from a service provider's perspective. Our architecture leverages the profiles of the service users to prioritize incoming requests coupled with static and adaptive scheduling strategies to regulate the service throughput at a desirable level and improve the response time for the preferred classes of requests.

Experimental results proved the feasibility and effectiveness of our approach and suggest the potential of WS-DiffServ in adding predictable differential QoS management for Web services in non-invasive way, particularly in terms of increased responsiveness and system throughput; the overhead incurred is limited. In future work we plan to complete the development of the framework, fine-tune it, and further evaluate it experimentally in realistic scenarios to assess its scalability. Furthermore, we plan to study the impact of using other scheduling policies on the end-to-end service response time and throughput.

Acknowledgment

This work is jointly funded by the Australian Research Council and Microsoft Australia (ARC Linkage Project LP0453880). We thank Prof. Boualem Benattallah for his insightful discussions and comments.

References

1. Dan, A., Davis, D., Kearney, R., King, R., Keller, A., Kuebler, D., Ludwig, H., Polan, M., Spreitzer, M., Youssef, A.: Web Services on demand: WSLA-driven Automated Management. IBM Systems Journal, Special Issue on Utility Computing 43, 136–158 (2004)
2. Zhou, C., Chia, L.-T., Lee, B.-S.: Web Services Discovery with DAML-QoS Ontology. International Journal of Web Services Research 2, 43–66 (2005)
3. Liu, Y., Ngu, A.H., Zeng, L.Z.: QoS computation and policing in dynamic web service selection. In: Proceedings of the 13th international World Wide Web conference on Alternate track papers & posters, pp. 66–73 (2004)
4. Web Services Agreement Specification (WS-Agreement) (2007)
5. MS: .NET Framework 3.5 (2007), http://www.netfx3.com
6. Erradi, A., Tosic, V., Maheshwari, P.: Policy-based Monitoring of Composite Web Services with the MASC Middleware. In: ECOWS 2007. 5th IEEE European Conference on Web Services, Halle (Saale), Germany (2007)
7. Tosic, V., Erradi, A., Maheshwari, P.: WS-Policy4MASC - A WS-Policy Extension Used in the Manageable and Adaptable Service Compositions (MASC) Middleware. In: SCC 2007. IEEE Int. Conf. on Services Computing, Utah, USA (2007)
8. Elnikety, S., Nahum, E., Tracey, J., Zwaenepoel, W.: A method for transparent admission control and request scheduling in e-commerce web sites. In: WWW 2004. 13th international conference on World Wide Web, New York, NY, USA, pp. 276–286 (2004)
9. Mercury LoadRunner (2007), http://www.mercury.com/us/products/performance-center/loadrunner/
10. WS-I: SCM Sample Application (2007), http://www.ws-i.org/deliverables/workinggroup.aspx?wg=sampleapps
11. Zhou, X., Wei, J., Xu, C.-Z.: Quality-of-service differentiation on the Internet: a taxonomy. Journal of Network and Computer Applications (2005)

Coverage and Timeliness Analysis of Search Engines with Webpage Monitoring Results

Yang Sok Kim and Byeong Ho Kang

School of Computing, University of Tasmania,
Private Bag 100 Hobart TAS 7001 Australia
{yangsokk, bhkang}@utas.edu.au

Abstract. Web monitoring systems and meta-search engines were designed to provide time and coverage critical services, where time critical means that new information should be provided as soon as it is publicized on the web and coverage critical means that any information should not be missed by the systems. We have analyzed coverage and timeliness of three commercial search engines with the web page monitoring results to investigate how rapidly and how efficiently web monitoring system and meta-search engines collect and provide newly published web information. We have also assessed how the meta-search engines might improve coverage and timeliness by providing collective services. Our experiment results show that commercial search engines still cover 65% ~ 75% of newly published information, taking from five to 13 days to retrieve the information. Theoretically, meta-search engines discover up to 86% of all published data and shorten delay time up to 8 days.

Keywords: Search Engines, Coverage of Search Engines, Freshness of Search Engines.

1 Introduction

The Web was developed to support information sharing by using hypertext. Tim Berners-Lee [1] proposed a distributed hypertext system to solve information management problems at CERN. In his proposal, he suggested the basic concept of information finding on the Web be as follows.

"To find information, one progressed via the links from one sheet to another..."

Even though there has been significant Web technology development, this basic information finding concept is still effective. As the Web grew exponentially over a very short time period, manual information finding mechanism revealed its limits. Search engines were heavily researched as an alternative from the early 1990s and manual information finding mechanisms were replaced by computer programs called crawlers or robots.

Though the web search engine is a principal system for finding information on the web, other alternative systems have been proposed to address their search limitations and to address different user requirements. Meta-search engines try to extend

B. Benatallah et al. (Eds.): WISE 2007, LNCS 4831, pp. 361–372, 2007.

coverage and relevancy performance by combining search results from different search engines. Until now, most search engines have focused on coverage and ranking differences, but temporal differences among search engines can also be used to improve coverage as well as freshness. Web monitoring systems focus on timeliness of information provision with complete coverage. The main problem in web monitoring systems is appropriate revisit scheduling for registered web pages and monitor site locating, which is considered a cost inefficient process.

This paper focuses on the coverage and timeliness of search engines to obtain a deeper understanding of how they work in the real world through the analysis of their coverage and timeliness, and furthermore to uncover their practical benefits. This study compares the coverage and timeliness of three commercial search engines - Google, Yahoo, and MSN. For this purpose, we collected web publications from 268 Australian government web pages and analyzed how completely and rapidly they are provided search results according to web page characteristics such as publication amounts and publication agent type (e.g. federal / local government or media release / home pages).

2 Related Work

2.1 Sampling Web Pages

Appropriate sampling techniques for the Web are necessary to discuss web evolution, measuring the size of the web, search engines coverage, and the timeliness. Until now, three significant sampling methods were proposed by different researchers:

(1) *IP based sampling*: O'Neil et al [2] and Lawrence and Giles [3] proposed an approach that uses IP addresses for random sampling. Random IP addresses were selected and checked as to whether they hosted web sites. If so, assessable web pages were harvested for further research such as measuring the size of the Web. However, this sampling method is not perfectly random, because it includes some sources of bias such as multiple hosts sharing the same IP address or hosts being spread over multiple IP address [2].

(2) *Random walk based sampling*: Bar-Yossef et al.[4] (Berkeley algorithm) and Henzinger et al [5, 6] (Compaq algorithm) independently proposed a method for sampling web pages by using random walks on the Web, Based on [4], Rusmevichientong et al.[7] (NEC algorithm) also proposed another random walk based sampling method, which in theory should lead to uniform random samples. All three algorithms consist of a *walking phase* and a *sampling phase*. In the walk phase, a memoryless random walk is performed on the web graph, and the nodes visited are sub-sampled in a possibly biased way. Baykan et al. [8] conducted a comparison study for these three sampling techniques. Their results showed that the Berkeley algorithm and the NEC algorisms more correctly reflect the out-degree distribution of the Web than the Compaq algorithm.

(3) *Random query based sampling*: This approach uses search engines for sampling the Web. Lawrence and Giles [9] queried major search engines to estimate the overlap in their databases. They inferred a bound on the size of the indexable web and estimate the search engines' relative coverage. Bharat and Broder [10]

attempted to generate random pages in a search engine index by constructing random queries. These pages were then used to estimate the relative size and overlap of various search engines. Their query words were selected from a dictionary of 400,000 words collected from about 300,000 documents in the Yahoo! directory. Both disjunctive (OR) and conjunctive (AND) queries were used in the sampling process. The first 100 results were considered and a random page selected. This introduced a *ranking bias* because the first 100 results depend on the search engine's ranking strategy. Furthermore, the resulting sample was therefore biased towards content rich pages, which is termed a *query bias* [10]

Even though these studies proposed some possible random sampling methods, the practicality of scaling to the entire Web is still not clear because no one knows the extent of the Web [11]. For this reason, many researchers who investigated web evolution and timeliness, intuitively select their sampling web sites, assuming that they represent some characteristics of complete web trends. For this reason we did not choose any of the random sampling techniques discussed above. Instead we considered two characteristics of Web pages, namely, reachability by crawlers and the frequency of their content updating.

2.2 Web Evolution

Many researchers [12-15] have focused on the changing characteristics of the Web and tried to find methods that determine the frequency at which web pages change.

(1) **Changing characteristics of the web.** Brewington and Cybenko [12, 13] leveraged their web clipping service, called Informant, to study the rate of change of web pages. The service collected 100,000 web pages daily and recorded the last-modified time stamp, the time of observation, and stylistic information (number of images, tables, links, and similar data) between March and November 1999. They reported 56% of the monitored Web pages were not changed during the duration of the study and only 4% changed every single time. Cho and Garcia-Molina [14] crawled a set of 720,000 pages from 270 'popular' web severs on a daily basis over four months, and counted web pages as having changed if their MD5 checksum changed. They founded that 40% of all web pages in their data set changed within a week, and 23% of those web pages that fell into the .com domain changed daily. Fettery et al.[15] investigated the evolution of over 150 million web pages downloaded weekly over a span of eleven weeks in 2002. They recorded variables including the length of each downloaded document, the number of non-markup words in each document, and a set of variables related to their syntactic similarity. Their observations showed that pages in the .com domain change more frequently than those in .gov and .edu domains, reflecting results similar to those obtained by Cho and Garcia-Molina. In addition, they found that document size was related to both the frequency and degree of change.

(2) **Determining the frequency of the web page changes.** Brewington and Cybenko [12, 13] developed an exponential probabilistic model for the time between individual web page changes and a model for the distribution of the

change rate defining those exponential distributions. They introduced (α, β)-currency which defines their notion of being up-to-date by using a probability, α, that a search engine is current, relative to a grace period, β, for randomly selected web page. A search engine for a collection is said to be of (α, β)-currency if a randomly chosen page in the collection has a search engine entry that is β-current with probability at least α. Cho and Garcia-Molina [16] and Matloff [17] independently proposed methods for estimating the frequency of change of individual web sites / pages based on Poisson-process model.

2.3 Timeliness Analysis of Search Engines

Lewandowski et al.[18] conducted timeliness analysis on three commercial search engines (Google, Yahoo, and MSN). They selected new articles from 38 web sites that were updated on a daily basis for six weeks, and analyzed freshness by submitting URLs to the search engines and comparing cached date with the actual date displayed on the Web. Their experiment shows that the three search engines collected new information relatively quickly. The search engine results that were not older than one day were Google (82.7%), MSN(48.0%), and Yahoo(41.9%). They also reported on the mean and median up-to-datedness with mean up-to-datedness being 3.1 days for Google, 3.5 days for MSN, and 9.8 days for Yahoo. Median up-to-datedness was 1 day for Google and MSN, and 4 days for Yahoo. These results for up-to-datedness may be an overestimate because they used news web sites. It is assumed that crawlers visit those kinds of web sites and update their index accordingly more frequently. However, even though this assumption is widely accepted, we could not find prior studies that supported this assumption. In our research, we used more moderately sampled web pages that have various publication frequencies to test this. In addition we analyzed the relationship between freshness and other web characteristics such as coverage, publication frequency, and source differences.

3 Method

3.1 Selection of Monitoring Web Pages

In our research, we did not choose any of the random sampling techniques described in Section 2.1. Instead, we selected 268 federal, state and local government web pages, consisting of 21 federal government homepages, 119 federal government media release pages, 78 Tasmanian government homepages, 16 Tasmanian government media release pages, and 34 Tasmanian council homepages. Obviously this sample set will not test the overall performance of Web search engines, but we believe that the chosen pages are not extreme cases with respect to reachability by crawlers, and with regard to the frequency of content updating.

3.2 Web Monitoring System

We used a web monitoring system, called WebMon, to collect newly uploaded publications from the selected web pages [19]. The terms used in our discussion are defined as follows:

- "Link text" is located between <a> and </a>tags and displayed in a Web browser.
- "URL" indicates the location of the content document.
- "Linked content" is the main content to be read by users.

Our process is as follows. The system sends an HTTP request message to the web server of the registered web page according to the revisit time ($T_{revisit}$), which is decided by various factors including publication frequency of the source pages and the user's need for information. For example, if the publication frequency is high, $T_{revisit}$ should be diminished to cover all new web information. Otherwise, some information may be omitted from the target web page, because some web pages such as online news pages usually have limited numbers of hyperlink slots.

As we had no prior information about the publication patterns of the selected domains and the focus of the research is on analyzing the timeliness performance of search engines using a web monitoring system rather than scheduling strategies for web monitoring, we employed a single fixed scheduling strategy. In this strategy, $T_{revisit}$ is set as a fixed interval such as every 30 minutes, every day at 9:00 am, or 9:00 am every Monday. In our study, we set the revisit time ($T_{revisit}$) as 2 hours for all the Web information source pages.

When the system received the HTTP response message from the server, it extracted URLs and their link texts from the HTTP response body (H_c) and compared them with those of the previous HTTP response body (H_p).

H_c and H_p are defined as follows

$$H_c = \{(U_i, T_i)\},$$

where H_c is a set of URLs and link texts of the current HTTP response message body. (U_i, T_i) is the ith pair of hyperlink and link text of a hyperlink.

$$H_p = \{(U_j, T_j)\},$$

where H_p is a set of URLs and link texts of the last HTTP response message body. (U_j, T_j) is the jth pair of URL and link text of a hyperlink.

Newly updated information is

$$I_n = (H_p \cup H_c) - H_p$$

We eliminated some URLs from the new URLs by registering filtering URLs. For example, advertising URLs are collected as new information, but our URL filter (H_f) eliminates those URLs before they are recorded into the database. We also excluded URLs that had already been recorded in the database. Some URLs may have been already harvested before the last session (H_e). New information is redefined as follows:

$$I_n = (H_p \cup H_c) - H_p - H_f - H_e$$

We did not use data that were collected from the first session, because they included the navigational information as well as old information published before the monitoring started.

3.3 Timeliness Measure

We defined total delay time (D_t) by measuring time difference between publication time (T_p) to servicing time (T_s).

$$D_t = T_s - T_p \qquad (1)$$

In fact, T_p is the time when a URL is actually added to the target Web pages, but we used the monitored time (T_m) as T_p because it is difficult to know the exact publication time.

$$d_t = T_s - T_m \qquad (2)$$

Therefore, there are gaps of less than two hour between actual creation time and the collected time.

The concepts of harvesting delay and internal processing delay are further addressed below. d_t is the sum of *harvesting delay (D_h)* and *internal processing delay (D_i)* (see Fig. 1).

$$d_t = D_h + D_i, \qquad (3)$$

where D_h equals $T_h - T_m$ and D_i equals $T_s - T_h$.

(1) **Harvesting delay (D_h):** Crawlers usually harvest new URLs by revisiting known URLs. D_h is calculated by subtracting harvesting time (T_h) from monitored time (T_m), which is decided by various factors among which the crawler's revisit strategy and the size of known URLs are main contributors. As some crawlers do not change their revisiting schedule, other crawlers adaptively change their revisiting schedule according to site importance, publication styles, which impacts on the crawler's revisit strategy. The actual strategy that is employed by each search engine is generally a business secret and not available to the public. If the size of known URLs is relatively small, it may not be a significant factor in deciding on D_h. However, nowadays it is already big enough not to be ignored. Even though many distributed systems are used to enhance crawling performance, it is still difficult to harvest all new information from known URLs.

(2) **Internal processing delay (d_i):** Information that is collected by crawlers is not directly provided. It requires further processing such as indexing and ranking, which takes time due to the vast amount of collected information. Detailed methods for indexing and ranking are not publicly known, such as the crawler's revisit strategy.

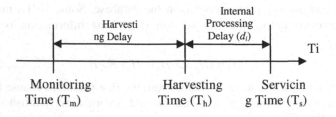

Fig. 1. Decomposition of Delay between Publishing and Servicing Time

One problem associated with this definition is that we do not know the exact harvest time (T_h) of each publication because it is an internal process of the search engine companies. The cached time (T_c) can be used as a substitute of T_h.

$$d_t = d_h + d_i \tag{4}$$

where d_h equals $T_c - T_m$ and d_i equals $T_s - T_m$.

However, this approach still has weaknesses because we do not know when T_c is recorded. Search engines may record it when the crawlers collect the information or when internal processing such as indexing and ranking is finished. For this reason, we only use the formula (2) to measure timeliness even though formula (3) and (4) are a more exact measure than formula (2). We developed an up-to-datedness checking system to measure the up-to-datedness of the search engines. Our system retrieves collected documents (URL + link text) from the document pool, and then submits each URL as a unique query word, because it is assumed that each URL is unique. If there is (are) the same URL(s) in the returned results, the system calculates the delay time by subtracting the current servicing time (T_s) from the monitoring time (T_m) and thus deletes the documents from the document pool.

4 Publication Analysis

Newly published documents were collected from the registered web pages for twelve weeks from December, 2006 to March, 2007. The numbers of 'active web pages' that published new documents during the monitoring period were 174 sites (64%). The active web pages ratio (number of active pages / number of registered pages) was the highest in the federal government domain, followed by the local government media release pages, the local government homepages, the municipal government homepage, and lastly the federal government media release pages (see Table 1). This ratio is higher than prior research results [12, 13], because government web pages are well managed and prolific compared to personal homepages.

Table 1. Active Web Pages by Domain

Site Type		Registered Pages	Active Pages	Ratio
Homepage	Federal	21	21	100%
	Local	78	48	62%
	Municipal	34	21	62%
Media page	Federal	119	70	59%
	Local	16	12	75%
Total		268	172	64%

Fig. 2 illustrates weekly and day-of-week publication trends. Publications were usually posted during working hours. For this reason, there were very small numbers of publications at the end of December and the beginning of January (week 2 ~ week 5) because of the Christmas and New Year holidays.

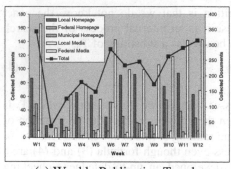
(a) Weekly Publication Trends

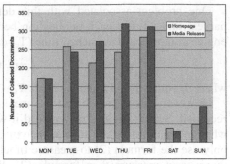
(b) Publications by Day of Week

Fig. 2. Publication trends over the experiments period

Table 2. Monitoring Results by Domain

Monitoring Page Types		Total Collected Docs	Active Pages	Average Docs per Pages
Homepage	Local	727	48	15.1
	Federal	290	21	13.8
	Municipal	181	21	8.6
Media Release	Local	214	12	17.8
	Federal	1,227	70	17.5
Total		2,639	172	15.3

Table 3. Monitoring Results by Publication Frequency

Monitoring Page Types	Total	Active Pages	Average Docs per Pages
Top 5 Webpages	727	5	145.4
Others	1,912	167	11.4

Monitoring results by domain and by publication frequency levels are summarized in Table 2 and 3. Table 2 shows publications by each of the five domains. The federal media release pages domain published the highest numbers of documents (1,277), followed by the local government homepage (727), the federal home pages (290), the local media release pages (214) and the municipal homepages (181). To compare the publication frequency of a single web page by domains, the mean average publication of each domain was compared. The comparison shows a different rank order because of the differences in the number of active pages. The highest domain is the local media release (17.8%), followed by the federal media page (17.5%), the local homepage (15.1%), the federal homepage (13.8%), and the municipal homepage (8.6%). We will later provide analytical results as to why timeliness is different among these domains. Monitoring results by publication frequency level is displayed in Table 3, where we compared the top 5 publication web pages with other web pages. Though the top 5 web pages comprise only 3% of all the monitoring pages, they cover

28% of all collected documents. In Section 5, we analyze how coverage and timeliness varies among different publication levels.

5 Coverage Analysis

The search engine's coverage (C) is defined as follows:

$$C = \frac{D_c}{D_m},$$

where D_c is the number of documents covered by a search engine and D_m is the number of documents collected by a monitoring system.

Table 4. Coverage Rates by Domains

Site Type		Google	Yahoo	MSN	Collective Coverage
Home page	Local	73.9%	60.0%	57.1%	83.1%
	Federal	89.7%	66.9%	79.3%	96.2%
	Municipal	84.5%	74.6%	80.7%	92.8%
Media page	Local	65.9%	60.3%	26.6%	72.0%
	Federal	76.7%	66.6%	72.7%	87.5%
Overall		77.0%	64.8%	65.9%	86.4%

Table 5. Coverage Rates by Publication Level

Site Type	Google	Yahoo	MSN	Collective Coverage
Top 5 Web pages	65.6%	61.8%	41.4%	77.0%
Others	81.3%	66.0%	75.3%	89.9%

Table 4 summarizes each search engine's coverage results by domains. Firstly, overall, as can be seen from Table 4, Google (77%) shows the best overall coverage, followed by MSN (65.9%) and Yahoo (64.8%). These results imply that search engines still miss 23% ~ 35% of published web information from known web pages. Coverage can be improved by employing web monitoring and/or meta-search technology. Whereas web monitoring systems try to cover all new information from the selected web pages, meta-search engines endeavor to cover more than only one search engine by combining search results. Our experiment results demonstrate that a meta-search engine can collect up to 86.4% of newly published web information by collectively considering each search engine's coverage. The fifth column in Table 4 shows the collective coverage of all search engines, and clearly illustrates that coverage performance can be improved from 9.4% to 21.6% using a meta-search engine. However, this improvement is the maximum possible, because when the user submits a query to the search engine, he/she probably does not know the exact URLs of the information they are trying to find. Secondly, there are significant patterns in coverage among domains. Even though the municipal government homepages are better covered by Yahoo and MSN, the three search engines provided relatively higher coverage in the federal government sites compared to those associated with local government. Collective coverage of all search engines also show that the federal government homepage and media release pages are better covered than those of local government. From these results, we can presume that search engines apply different crawling schemes to the web pages that have different importance. Thirdly, the results show that search engines miss more information when the web sites are more dynamically changed. Now Table 5 indicates that the coverage of the top 5 web pages is significantly lower than that of other web pages. This is caused by the mismatch of the revisiting schedule search engine and changing frequency of the web pages. Certain web pages can disappear before the crawlers revisit them, or where there may be too many new documents posted between two consecutive revisiting processes.

6   Timeliness Analysis

The average delay (D) in providing web information by search engine is defined as follows:

$$D = \frac{\sum d_t}{N},$$

where $\sum d_t$ is sum of T_m - T_s of each active web pages and N is total number of active web pages

Table 6. Average Delay by Domains (Day)

Site Type		Google	Yahoo	MSN	Collective Delay
Home page	Local	4.5	9.1	11.7	4.8
	Federal	3.9	7.9	14.0	2.8
	Municipal	4.8	10.5	7.8	4.2
Media page	Local	3.1	4.8	8.8	3.4
	Federal	6.1	7.8	14.0	4.0
Overall		5.1	8.1	12.8	4.3

Table 7. Average Delay by Publication Level

Site Type	Google	Yahoo	MSN	Collective Delay
Top 5 Web pages	4.6	5.6	11.5	3.6
Others	5.2	9.0	13.0	4.5

Average delay results are summarized in Table 6 and Table 7. In delay analysis, we do not include document that search engines do not provide. We found the following results from this experiment: Firstly, as can be seen from Table 6, the overall timeliness performance is inconsistent with the coverage performance results. The results illustrate that Google had the shortest average delay time (5.1 days) followed by Yahoo (8.1 days) and MSN (12.8 days). Though Yahoo and MSN achieved a very similar coverage rate, there are significant time gap differences between them. The collective delay of the three search engines significantly improves in prospect of timeliness, from 0.8 to 8.5 days overall. The collective delay was produced by taking the minimum average delay of each search engine, caused by the temporal differences of each search engine. Though Yahoo and MSN have a longer average delay, their revisiting schedule for specific web sites is less than that of Google. This implies that users can get fresh information by using a meta-search engine which exploits temporal differences among search engines. Secondly, as is evident in Table 6, the more frequently web documents are published, the more frequently the crawlers visit the web pages. However, there are some differences in the average delay among domains. In the homepage domain the federal government web pages are subject to less delay compared to the local government web pages, but in the media pages domain the local government pages have less delay when compared to that associated with the federal pages, which may be caused by the frequency distribution among the monitoring web page of each domain. However, such an assumption requires further investigation.

7 Conclusions

In this paper, we examined timeliness and coverage of three well known commercial search engines – Google, Yahoo, and MSN – with web monitoring results of Australian federal and local government homepage and media release pages. The results show that commercial search engines still cover 65% ~ 75 % of newly published information and they take considerable time to collect new information (5.2 ~ 13.0 days delay). Furthermore, the results also support that coverage and timeliness of search engines are significantly improved by combining each search engines' capabilities. These results are significant evidence of the need for web monitoring systems and meta-search engines.

Acknowledgement. This work is supported by the Asian Office of Aerospace Research and Development (AOARD) (Contract Number: FA5209-05-P-0253).

References

1. Berners-Lee, T.: Information Management: A Proposal, CERN (1989)
2. O'Neill, E.T., McClain, P.D., Lavoie, B.F.: A Methodology for Sampling the World Wide Web. Journal of Library Administration 34(3-4), 279–291 (2001)
3. Lawrence, S., Giles, C.L.: Accessibility of information on the Web. Intelligence 11(1), 32–39 (2000)
4. Bar-Yossef, Z., et al.: Approximating Aggregate Queries about Web Pages via Random Walks. In: VLDB 2000. 26th International Conference on Very Large Data Bases, pp. 535–544. Morgan Kaufmann, Cairo, Egypt (2000)

5. Henzinger, M.R., et al.: On near-uniform URL sampling. In: Ninth international World Wide Web conference on Computer networks, pp. 295–308. North-Holland Publishing Co., Amsterdam, Netherlands (2000)
6. Henzinger, M.R., Heydon, A., Mitzenmacher, M., Najork, M.: Measuring Index Quality Using Random Walks on the Web. In: Eighth International World Wide Web Conference, Toronto, Canada, pp. 1291–1303 (1999)
7. Rusmevichientong, P., Pennock, D.M., Lawrence, S., Giles, C.L.: Methods for Sampling Pages Uniformly from the World Wide Web. In: AAAI Fall Symposium on Using Uncertainty Within Computation, pp. 121–128. North Falmouth, Massachusetts (2001)
8. Baykan, E., Castelberg, S.d., Henzinger, M.: A Comparison of Techniques for Sampling Web Pages. In: IIWeb 2006. Workshop on Information Integration on the Web, Edinburgh Scotland (2006)
9. Lawrence, S., Giles, C.L.: Searching the World Wide Web. Science, 280 (1998)
10. Bharat, K., Broder, A.: A technique for measuring the relative size and overlap of public Web search engines. In: WWW7. The Seventh International World Wide Web Conference, Brisbane, Australia, pp. 379–388 (1998)
11. Henzinger, M.R.: Algorithmic Challenges in Web Search Engines. Internet Mathematics 1(1), 115–126 (2003)
12. Brewington, B.E., Cybenko, G.: How dynamic is the Web? Computer Networks 33, 257–276 (2000)
13. Brewington, B.E., Cybenko, G.: Keeping Up with the Changing Web. Computer 33(5), 52–58 (2000)
14. Cho, J., Garcia-Molina, H.: The Evolution of the Web and Implications for an Incremental Crawler. In: 26th International Conference on Very Large Data Bases, Cairo, Egypt, pp. 200–209 (2000)
15. Fetterly, D., et al.: A large-scale study of the evolution of Web pages. Software - Practice and Experience 34(2), 213–237 (2004)
16. Cho, J., Garcia-Molina, H.: Estimating frequency of change. ACM Transactions on Internet Technology (TOIT) 3(3), 256–290 (2003)
17. Matloff, N.: Estimation of internet file-access/modification rates from indirect data. ACM Transactions on Modeling and Computer Simulation (TOMACS) 15(3), 233–253 (2005)
18. Lewandowski, D., Wahlig, H., Meyer-Bautor, G.: The Freshness of Web search engine databases. Journal of Information Science 32(2), 131–148 (2006)
19. Park, S.S., Kim, S.K., Kang, B.H.: Web Information Management System: Personalization and Generalization. In: The IADIS International Conference WWW/Internet 2003, pp. 523–530 (2003)

Managing Process Customizability and Customization: Model, Language and Process

Alexander Lazovik[1] and Heiko Ludwig[2]

[1] University of Trento, Italy
lazovik@dit.unitn.it
[2] IBM TJ Watson Research Center, USA
hludwig@us.ibm.com

Abstract. One of the fundamental ideas of services and service oriented architecture is the possibility to develop new applications by composing existing services into business processes. However, only little effort has been devoted so far to the problem of maintenance and customization of already composed processes. In the context of global service delivery, where process is delivered to clients, it is critical to have a possibility to customize the standardized reference process for each particular customer. Having a standardized delivery process yields many benefits: interchangeable delivery teams, enabling 24/7 operations, labor cost arbitrage, specialization of delivery teams, making knowledge shared between all customers, optimization of standardized processes by re-engineering and automation. In the paper we propose an approach, where reference process models are used explicitly in the process lifecycle, where customer-specific process instantiations are obtained by a series of customization steps over reference processes. To show the feasibility of the approach, we developed a process-independent language to express different customization options for the reference business processes. We also provided an implementation that extends WebSphere BPEL4WS Editor to introduce process customizations to BPEL4WS processes.

1 Introduction and Motivation

Customizing processes from reference processes or templates is common practice to reduce the time and effort to design and deploy processes on all levels. It finds application in high-level business processes design and execution, Web services compositions, and also workflows on a technical level such as integration processes and scientific workflows. Customizing reference processes usually involves adding, removing or modifying process elements such as activities, control flow and data flow connectors. Oftentimes, however, there is a business need to limit the customizability of a process, the extent to which a process can be modified.

We look at the area of management of IT service delivery as an application scenario as managing process variations is a big challenge [10]. It comprises processes for provisioning servers, creating user IDs, managing storage and the like. In an outsourcing scenario, a service provider runs similar processes for different

B. Benatallah et al. (Eds.): WISE 2007, LNCS 4831, pp. 373–384, 2007.

clients but more often than not clients will require that their specific needs be addressed. While a service provider often maintains a set of best practices as reference processes, these processes will be customized when signing up a new client. However, there are limitations to these customizations based on specific technical properties and dependencies of the system used to implement the service, staff qualifications and availability, time zone constraints and many other reasons. For example, certain activities cannot be removed as they are mandatory for subsequent activities or a regulatory requirement. In server provisioning, an operating system image must be installed prior to installing an application. Other activities must be performed in specific sequence, no other activities added in between. This is particularly the case if subsequent activities are performed by the same person in a particular location in the data center. It might not be economical to let the person wait while something else happens.

Creation and customization of processes is a topic that has undergone significant research attention, e.g., by Rosemann et al. [14] and Becker et al. [3], and is a standard practice in business consulting and system integration of large standard application packages such as ERP products. Limiting and shaping customization of process reference models, however, is a nascent topic that has only received limited attention so far.

In this paper, we propose a model of reference process that includes a definition of the limitations of how it can be customized. In addition, we propose an XML-based language to express these annotations and associate them with a business process definition to express our extended notion of a reference process.

We proceed as follows: In the next section, we discuss in more detail the types of customization operations that might be required for process models. Subsequently, we introduce a formal model of a business process, its customizing operations and the constraints that can be applied. Section 5 briefly introduces an implementation of a process editor extension to the WebSphere Integration Developer BPEL editor. Finally, we discuss related work and conclude.

2 Managing the Customization Life-Cycle

The use of reference processes primarily improves and accelerates the definition of process models or templates to be instantiate at runtime. They are particularly useful in use cases where similar processes are defined for different organizational deployments, e.g., by a business process outsourcer or a consulting or system integration company. The use of reference processes extends the traditional process build time run time life-cycle model to a model of reference modeling customization run time. Reference process and customization are new conceptual entities that must be subjected to a government process in an organization. Staying with our scenario of service delivery management, we will use an example of a simple server provisioning process to illustrate what needs for customization and which limitations may exist.

In the above example process, the first step may be left out if the process is to be deployed in a re-provision scenario in which hardware is already in place.

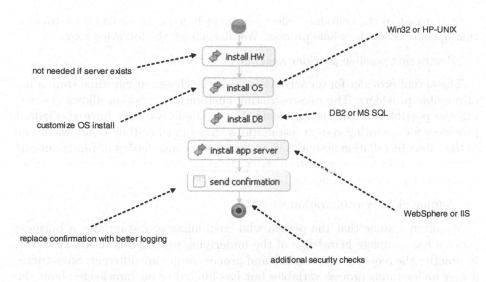

Fig. 1. Customization examples for a server build process

OS installation customization may be limited to the choice between a Windows or UNIX installation sub-process; likewise for applications running on the system. The send confirmation step may be replaced with any activity, e.g., writing a log entry. While we may want to maintain the specific order of these steps, additional steps might be added to the end of the process, e.g., for additional security checks. While this process offers many opportunities to customize it, the environment of process execution, e.g., the operations of a service provider data center limit the extent to which the service provider can accept modifications to this process.

2.1 Customization Operations

While customization refers to the process of adapting a copy of a reference process in general, a specific *customization operation* is an operation on a copy of a reference process that changes (customizes) it. The specific place where a customization operation is applied is called *customization point*. The potential customization operations at a customization point are called customization options. We consider two types of customization operations: *object-based* and *process-based*. Object-based ones target a single process element without affecting the rest of the process. Process elements are primarily the activities of a process but also control flow elements such as joins and forks. Object-based customization operations are the following:

- Adding a new process element;
- Removing a process element;
- Replacing one process element with another. This also comprises a change to the definition of an object, e.g., changing an activity description or a control flow condition.

In contrast to the contained effect of object-based Process-based customization options affect the whole process. We distinguish the following cases:

– Restricting possible provider assignment for a role.

The actual provider for an activity is typically chosen at run-time from a list of possible providers. The role-restricting customization option allows restricting the possible provider assignments. For example, we may have specialized providers for operating system installations that can run either Windows or one of the Linux installation activities. If we require the installation of Linux-specific software on a server, we want to restrict the possible operating system installation provider to the one offering Linux installations.

– Adding global customization variables.

We often assume that the person who customizes or instantiates a business process has complete knowledge of the underlying process semantics. However, in practice the requester, customizer and process owner are different. Sometimes, a user understands process variables but has limited or no knowledge about the process itself. Consider the following example: When buying a computer using an on-line configuration tool, a user can customize a computer by expressing his or her preference in terms of memory size, processor, mount unit format etc. However, the choices of process variables determine the customization of the assembly process implicitly. We want to be able to express this relationship and enable global customization variables. Based on explicitly defining the set of potential customization operations and their subjects, we can define means of expressing the limits of customization that must be observed.

3   Formal Model for Customizable Reference Process

Business process is a series of a linked set of activities that is designed to accomplish some goal, where activities represent some form of atomic operations. Business process can be seen as a generic process model that represents the best practices for some particular business domain. For web service domain, activities are web services or other processes. There are many process languages used for process specification, for example, BPEL for web services. However, ideas introduced in this paper are process-independent, that is, developed customization rules and mechanisms can be applied to any of the process specification, as far as it consists of activities, control flow constructs, etc. To be precise, let us introduce a formal definition of some abstract process specification that can be used as a reference point for our customizations. In practice, we assume that these formal definitions refer to some actual specifications language. For our example introduced later are developed for processed expressed in BPEL.

Definition 1 (Process). *A process is a tuple* $B = \langle name_B, A, M, F, T, D, R, r \rangle$, *where:*

– *$name_B$ is the name of the process;*
– *A is a set of activities with one activity marked as start activity a_0;*

- M *is a set of possible process messages, every message is defined as a set of variables;*
- F *is a set of connectors, all of type* $\{XOR, OR, AND\}$;
- $D_{in} : A \rightarrow M$ *is a data transition function,* $D(a)$ *defines the input parameters for the activity a;*
- $D_{out} : A \rightarrow 2^M$ *is a data transition function,* $D(a)$ *defines the output parameters for the activity a;*
- $T : U \rightarrow 2^U$ *is a transition function, where* U *is a set of process elements and is defined as* $U = A \cup F$. *Every transition* $T(a)$ *corresponds to one of the completion states of* a, *and, from this it follows that* $|D_{out}(a)| = T(a)|$;
- R *is a set of roles, with* $r : A \rightarrow R$ *being a function that associates the activity with its role.*

To make our formal model simple, the definition of D_{in} indicates that each activity has only one input message, which is not necessarily always the case, since an activity can have more than one input messages. However, since every message is represented as a set of variables (and, by that, it may actually consist of several other messages), this assumption does imply any serious restrictions on the formal model. In general, a process may be seen as an atomic operation itself. In this case we assume, that for such process B : $D_{in}(B) = D_{in}(a_0)$, and $D_{out}(B) = \{D_{out}(a_i)\}$, where a_0 is a starting activity for the process B, and a_i are terminating activities. We use service registries to map activities to possible providers. Providers are usually chosen at the begin of or during process execution. We define service registries as follows:

Definition 2 (Service Registry). *A service registry S for process B is defined by the following tuple:* $S = \langle name, P, s \rangle$, *where:*

- *name is a service registry reference name;*
- P *is a set of providers;*
- $s : R \rightarrow 2P$ *is a function that associates the roles R from process B with providers P.*

A process with customization points is called a reference process. Conceptually, a reference process defines a set of possible processes. It can be seen as a function $B(c_1, \ldots, c_n)$ over customization operations c_i. Assignments to c_i instantiate the reference process. Depending on the resolved customization options, different business processes can be potentially instantiated. This is *generating function* for the customizable reference process B_c. The outputs of the function are *instances* of the reference processes. However, this functional dependency is not function with its traditional meaning, since customization may have interrelations. That is, by resolving of some customization options other customization options may be affected. In our setting, reference process is defined on top of the business process, and seen as original process specification plus possible customization options. Formally, we define a reference process as follows:

Definition 3 (Customizable Reference Process). *A customizable reference process B_c is a reference process B and its service registry S enriched with customizable elements and is defined by the following tuple $B_c = \langle name, name_B, S, C, t \rangle$, where:*

- *name is a name of the customizable reference process;*
- *$name_B$ is a name of the process that is made customizable;*
- *C is a set of allowed customization operations for the business process;*
- *$t(C)$ is a function that defines a set of additional properties for each of the customization, as when the customization is resolved, by what role, etc.*

In practice we work with more specific classes of customization operations, and often we define customizations implicitly, as it is described in the following sections.

4 Customizations

In this section we define possible customization operations and their formal semantics. As a result of customization we have a customized process that is identical to the original one with changes applied in potential customization points according to customization operational semantics. Now we are ready to formally define a customization of the reference process:

Definition 4 (Customization of the Reference Process). *A customization of the reference process $B_1 = \langle \ldots \rangle$ is a process $B_2 = \langle \ldots \rangle$, that is derived from B_1 by resolving customization operations.*

A customization option can be seen as an operation of modification of the process. In practice, these possible changes are defined implicitly, since the actual process customization is performed in two steps: first, by describing of the possible customization options, and, as a second, resolving the customization option to some value. In the rest of the section we discuss semantic transition rules for the customization options introduced in Section 2.1.

Adding new activity a2 between activities a1 and a3

Requirements:

- $D_{in}(a_2) \subseteq D_{out}(a_1)$;
- $D_{in}(a_3) \subseteq D_{out}(a_2)$.

Semantic transition rules:

- $A_2 = A_1 + a_2$;
- $T_2(a_1) = T_1(a_1) \backslash a_3 + a_2$;
- $T_2(a_2) = a_3$.

Table 1. Definition of customization operations

Customization operation	Requirements	Semantic transition rule
Adding a_2 between a_1 and a_3	$D_{in}(a_2) \subseteq D_{out}(a_1)$ $D_{in}(a_3) \subseteq D_{out}(a_2)$	$A_2 = A_1 + a_2$ $T_2(a_1) = T_1(a_1)\backslash a_3 + a_2$ $T_2(a_2) = a_3$
Removing a between a_1 and a_2	$D_{in}(a_2) \subseteq D_{out}(a_1)$	$A_2 = A_1\backslash a$ $T_2(a_1) = T_1(a_1)\backslash a + a_2$
Replacing $a1$ with $a2$ between $a3$ and $a4$	$D_{in}(a_2) \subseteq D_{out}(a_3)$ $D_{in}(a_4) \subseteq D_{out}(a_2)$	$A_2 = A_1\ a_1 + a_2$ $T_2(a) = T_1(a)\ a_1 + a_2$ $T_2(a_2) = $ one of $T(a_1)$
Restricting provider assignment for a role r to a set of providers I	$I \subseteq s_1(r)$	$s_2(r) = I$
Adding global customization variables	-	-

Removing activity a between activities a1 and a2

Requirements:

- $D_{in}(a_2) \subseteq D_{out}(a_1)$.

Semantic transition rules:

- $A_2 = A_1\backslash a$;
- $T_2(a_1) = T_1(a_1)\backslash a + a_2$.

Replacing activity a1 with a2 after activity a

Requirements:

- $D_{in}(a_2) \subseteq D_{out}(a_3)$;
- $D_{in}(a_4) \subseteq D_{out}(a_2)$.

Semantic transition rules:

- $A_2 = A_1\ a_1 + a_2$;
- $T_2(a) = T_1(a)\ a_1 + a_2$;
- $T_2(a_2) = $ one of $T(a_1)$.

Restricting a provider assignment for a role r to a set of providers

Requirements:

- $I \subseteq s_1(r)$.

Semantic transition rules:

- $s_2(r) = I$.

Adding global customization variables

Requirements:

– A variable may be taken from the process variables and named as customization variable.

Semantic transition rules:

– No direct impact. The actual impact of the global customization variables is performed through guidelines. This type of customizations may be seen as an abstraction that allows implicitly orchestrate the customization process without getting too much into details of actual process semantics.

A summary of atomic customization operations is provided in Table 1. Consider, for example, adding an activity a_1 between activities a_2 and a_3. When described, a customization operation usually explicitly defines the activities a_2 and a_3, that is, the place where some (not yet defined) activity may be inserted. This type of customizations may be seen as an abstraction that allows implicitly orchestrate the customization process without getting too much into details of actual process semantics. A variable is taken from the process variables and named as customization variable.

From now, when we refer to customization option, we assume that it is defined in some parametric way as described above. Within this setting, resolving the customization option is the actual customization choice (activity to be inserted in the above example).

5   Implementation

We implemented an extension to the BPEL editor of the WebSphere Integration Developer (WID) application. It is realized in the form of an Eclipse plug-in. This allows us to use BPEL Editor also as an editor for reference processes, with BPEL process enriched with customization elements. Customizations are stored in a separate file in the same folder as the customized BPEL process.

The reference implementation architecture is shown in Figure 2. WID uses the same scheme of extensions based on plug-ins as the original Eclipse platform (www.eclipse.org). To allow customizations of BPEL processes, we extended the BPEL Editor with a set of plug-ins:

– EMF customization model extends the BPEL EMF model, and is recognized by the BPEL editor as BPEL extension elements;
– To save/load BPEL processes, if the process contains any customization extensions, they are saved/loaded using in/from a separate file, making the on-the-fly validation against customization XML schema. If the validation fails, the problems tab points to a set of validation errors;
– To fully support visual editing of customizations, for each element we added a separate visual extension element to BPEL Editor, as shown in Figure 3.

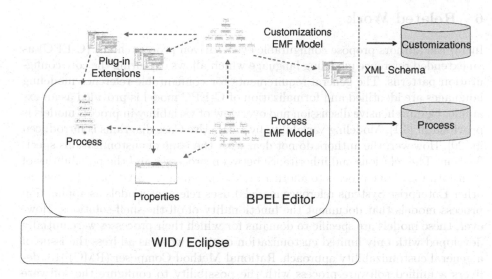

Fig. 2. Implementation of a customizability annotator as extension to a BPEL editor

Context-aware property tabs are activated, when the customized element is chosen. To define relations between customizations in the form of guidelines and requirements the separate view is used as shown in Figure 3.

The reference process XML representation generated by the tool is used as input to a customization tool.

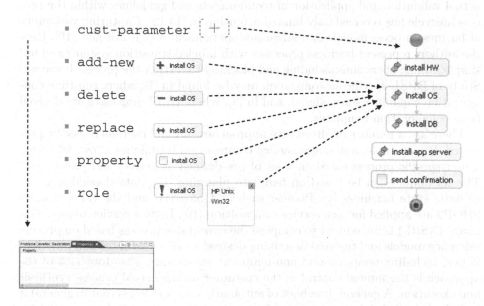

Fig. 3. Customization elements attached to a process activity

6   Related Work

In[14] the authors propose configurable event-driven process chains (C-EPC) as an extended reference modeling language which allows capturing the core configuration patterns. The general requirements for configurable reference modeling languages are identified and formalization of C-EPC model is provided as an example. Comprehensive discussion and overview of variability in process models is provided in [6,1]. Modeling variability by UML Use Case diagrams is introduced in [20]. However, the authors do not deal with the issue of customizations specification. The relations and inheritance between processes and the possibilities of transforming of one process to another are discussed in [2]. SAP [5,15] (as well as other Enterprise Systems reference models) uses reference models as application process models that document the functionality of off-the-shelf-solutions. However, these models are specific to domains for which their processes were initially developed with only limited customization options whereas address the issue of a general customizability approach. Rational Method Composer (RMC) [13] delivers a unified software process with the possibility to configure the software process in a customer-specific way on a base of initial configuration. While the customization possibilities of RMC are extensive, there is no or little control over how the customization process happens. In contrast, this work not only provides the language for customization options, but also the mechanism to control and validate the customization process by means of requirements and guidelines. The concept of guidelines and requirements is very similar to those defined in [12]. Requirements allow designers to limit the configuration space, while guidelines allow them to shape the configuration process. However, the actual validation and application of requirements and guidelines within the process lifecycle has received only limited attention in [14,12]. Capturing variability of business process models in a language-independent way is not new [19]. Here the authors represent business processes with labeled transition systems and use simple atomic operations as hiding and blocking to restrict the process behavior. Study of BPEL-specific customization may be found in [7], where run-time configuration capabilities are offered, and in [8], where BPEL processes are derived from collaboration templates.

There are a number of alternative approaches to build customer-specific processes. For example, automatic service composition techniques allow delivering a user-specific process based on a set of pre-defined objectives and preferences. The techniques can be based on template planning [17], data dependency [18] or finite state machines [4]. Planner such as Golog [11] and the HTN planner SHOP2 are applied for web service composition [16]. In [9] a service request language (XSRL) is introduced to compose the executable process based on process reference models and the goal describing desired service attributes and functionalities, including temporal and non-temporal constraints. The drawback of the approach is the limited control of the customer on the actual process synthesis and execution. A general drawback of automatic composition is that in general it is difficult to maintain the semantic descriptions of process activities up-to-date, making altering the service implementations difficult. Hence, while automated generation approaches implicitly contain a constraint on the set of process that can be created it is usually not in a form in which it can be usefully managed by a person who wants to restrict reference process use.

7 Conclusion

This paper outlines the issue of constraining the customizability of reference processes and introduces a model and representation for defining customization options on reference processes. The proposed approach allows a reference process editor to explicit control the customizability of the reference process. We defined a formal model for customizations with detailed operation semantics rules for each customization type. The formal model is built on top of existing process definitions. The XML-based language for customization definitions can be used in the context of existing process definition language. It is designed with an idea of being independent from particular process notation. To demonstrate the feasibility of the approach, we developed a reference prototype as an extension to WebSphere Integration Developer BPEL Editor.

This is an important contribution to the management of process variability that is in particular relevant for service providers that deal with numerous instances of a similar process entailing high life-cycle costs.

In the next step we plan to further investigate the implication of customization operations. For example, application of object-based customization operations (like removing or modifying an activity) does not always have only local implications. What if there are connectors in-between the activities where a new activity is inserted? It would be also interesting to understand the implications on the behavioral correctness (e.g. removing an activity may lead to a deadlock). As a separate line of research, we also plan to address different ways of executing customization options, which depend to a large extent on the richness of the choices and the level of skills of the customizer. We plan to extend the provided implementation to validate and simulate customization options. Extensions for other process notations, e.g., BPMN, are also planned.

References

1. Bachmann, F., Bass, L.: Managing Variability in Software Architecture. ACM Press, New York (2001)
2. Basten, T., van der Aalst, W.M.P.: Inheritance of behavior. J. Log. Algebr. Program. 47(2), 47–145 (2001)
3. Becker, J., Delfmann, P., Knackstedt, R.: Adaptative reference modeling: Integrating configurative and generic adaptation techniques for information models. In: Reference Modeling Conference (2006)
4. Berardi, D., Calvanese, D., De Giacomo, G., Lenzerini, M., Mecella, M.: Automatic composition of e-services that export their behavior. In: Orlowska, M.E., Weerawarana, S., Papazoglou, M.M.P., Yang, J. (eds.) ICSOC 2003. LNCS, vol. 2910, Springer, Heidelberg (2003)

5. Curran, T., Keller, G.: SAP R/3 Business Blueprint: Understanding the Business Process Reference Model. Upper Saddle River (1997)
6. Halmans, G., Pohl, K.: Communicating the variability of a software-product family to customers. Software and System Modeling 2(1), 15–36 (2003)
7. Karastoyanova, D., Leymann, F., Nitzsche, J., Wetzstein, B., Wutke, D.: Parameterized bpel processes: Concepts and implementation. Business Process Management, 471–476 (2006)
8. Khalaf, R.: From rosettanet pips to bpel processes: A three level approach for business protocols. Business Process Management, 364–373 (2005)
9. Lazovik, A., Aiello, M., Papazoglou, M.: Planning and monitoring the execution of web service requests. Journal on Digital Libraries (2005)
10. Ludwig, H., Hogan, J., Jaluka, R., Loewenstern, D., Kumaran, S., Gilbert, A., Roy, A., Surendra, M., Nellutla, T.: Catalog-based service request management in the bluecat environment. IBM Systems Journal 46(3) (2007)
11. McIlraith, S., Son, T.C.: Adapting Golog for composition of semantic web-services. In: KR. Conf. on principles of Knowledge Representation (2002)
12. Mendling, J., Recker, J., Rosemann, M., van der Aalst, W.M.P.: Towards the interchange of configurable EPCs. In: EMISA, pp. 8–21 (2005)
13. RMC. Rational method composer, http://www-128.ibm.com/developerworks/rational/products/rup
14. Rosemann, M., van der Aalst, W.: A configurable reference modelling language. Inf. Syst. 31(1), 1–32 (2007)
15. SAP. Online documentation mySAP ERP (2005), http://help.sap.com
16. Sirin, E., Parsia, B., Hendler, J.: Filtering and selecting semantic web services with interactive composition techniques. IEEE Intelligent Systems 19(4) (2004)
17. Thakkar, S., Ambite, J., Knoblock, C., Shahabi, C.: Dynamically composing web services from on-line sources. In: AAAI Workshop Intelligent Service Integr. (2002)
18. Thakkar, S., Ambite, J., Knoblock, C., Shahabi, C.: Dynamically composing web services from on-line sources. In: AAAI Workshop Intelligent Service Integr. (2002)
19. van der Aalst, W., Dreiling, A., Gottschalk, F., Rosemann, M., Jansen-Vullers, M.: Configurable process models as a basis for reference modeling. In: BPM Workshops, pp. 512–518 (2005)
20. von der Massen, T., Lichter, H.: Modeling variability by uml use case diagrams. In: REPL, Technical Report: ALR-2002-033, AVAYA labs (2002)

A WebML-Based Approach for the Development of Web GIS Applications

S. Di Martino, F. Ferrucci, L. Paolino, M. Sebillo, G. Vitiello, and G. Avagliano

University of Salerno, Via Ponte Don Melillo, I-84084 Fisciano (SA), Italy
{sdimartino, fferrucci, lpaolino, msebillo, gvitiello}@unisa.it

Abstract. The goal of disseminating and manipulating spatial knowledge over the Internet, has led to the development of Web applications, known as *Web Geographic Information Systems (Web GIS)*. Web GIS can be considered as a particular class of data-intensive Web applications, since they are mainly devoted to handle (spatial) information to and from the user. The success of *WebML (Web Modeling Language)* for designing traditional data-intensive web applications suggested use to extend this visual formalism to model relevant interaction and navigation operations typical of Web GIS. In the paper, we describe the proposed extension and provide an example of its application.

1 Introduction

Geographic Information Systems (GIS) are a kind of software system specifically suited to deal with data and attributes spatially referenced [9][10]. They turn out to be fundamental in a great variety of contexts, ranging from urban planning to marketing, route planning, natural disaster prevention, archaeology, and so on.

Recently, the growing need of sharing and exchanging geodata via Web has affected researchers and professionals work, leading to the development of Web-based GIS applications, known as *Web GIS*, intended for dissemination and manipulation of spatial knowledge in specific domains. A Web GIS holds the potential to make geographic information available to a worldwide audience, allowing the Internet users to access GIS applications from their browsers without purchasing proprietary GIS software. The requirement of GIS functionality via Web is also proven by the success of Google Earth[TM] and Google Maps[TM], which represent examples of free solutions for navigating and querying thematic maps. The development of Web GIS applications will be further boosted by the standardization efforts made by the Open Geospatial Consortium, Inc.[®] (OGC) [11].

With the increased time-to-market pressure, it is no longer possible to deal with low-level issues, and create Web GIS applications from scratch. Thus, there is a growing need for tools and methodologies that allow us to rapidly develop this kind of Web applications and to rapidly modify them to meet the ever-changing business needs. For that reason, leading companies, such as ESRI and Intergraph, have provided users with proper extensions of their GIS products for developing Web GIS applications [16][17][18][19]. GRASS and MapServer are well-established open source GIS software which embed components for Web GIS implementation.

B. Benatallah et al. (Eds.): WISE 2007, LNCS 4831, pp. 385–397, 2007.

However, such environments supporting the development of Web applications can be very hard to use and require specialized skills [12].

For high-level development of traditional Web applications, many solutions are currently available [1][3][7][8][15]. Among them, *WebML* (*Web Modeling Language*) is a formal visual language specifically conceived to design data-intensive Web applications [3][4][20]. It provides suitable models, a development process and a CASE tool (*WebRATIO*) [5] to build applications that rely mainly on data management and movements. WebML has been successfully applied in many different contexts, from industry to academia, with positive results.

Web GIS can be considered as a particular class of data-intensive Web applications, since they are mainly devoted to handle (spatial) information to and from the user. Thus, in this paper we present an adaptation of the WebML approach, to deal with the Web GIS context. The resulting methodology consists of two orthogonal perspectives: *Geodata and Metadata Conceptual Model* and *Hypertext and Web Mapping Model*. In particular, the Spatial E-R model by Calkins [2] has been exploited to design the conceptual schema of (spatial) data. While, the WebML Hypertext Model has been extended to include new notations specifically tailored for GIS concepts and tasks. Moreover, the extension of the CASE tool WebRATIO will provide potential Web GIS designers with a Rapid Application Development environment, where they will be able to design in a visual fashion the application that will be automatically generated by the tool.

The remainder of the paper is organized as follows: in Section 2 we recall the main concepts of WebML, while in Section 3 we describe the proposed modeling concepts and notations which extend WebML for Web GIS. In Section 4 we present an example of use of the proposed notations, while the architecture of the WebGIS applications generated by the WebRatio tool is described in Section 5. Some remarks conclude the paper.

2   The *WebML* Methodology for Data-Intensive Web Applications

Graphical and diagrammatic representations play a central role in the field of software and Web engineering [6]. They are widely employed to support many activities of the software development process, such as specification, analysis and design, since they help to describe and understand complex systems, by means of abstractions and different views. *WebML* is a modeling language suited to support users in designing data-intensive Web applications [4], by providing a set of visual notations to model the content structure, data and navigational aspects of a Web application. Indeed, the design of a Web application is based on two orthogonal perspectives: data and navigation. The former is described by means of the well-known E-R data model, to represent all the relevant entities and relationships. The latter is modeled by the *WebML Hypertext Model*, which describes how the previously defined contents will be arranged and rendered within the Web application. In particular, the application structure and appearance are defined through four main levels of abstraction: *site views*, *areas*, *pages* and *content units*.

The site view is the highest level and is used to model the hypertextual structure of the whole web site. A site view is composed of many areas that model the sections of the application. Each area can contain both other areas (in a recursive way) and pages, that are the containers of the information. Each page is composed of some content units, the elementary pieces of information taken from the data schema. To bind a unit with the corresponding data, the *WebML Hypertext Model* requires to specify a source and a selector. The former is an entity of the database, the latter is a condition over this entity, to filter the rendered data. These objects can be interconnected by links, to model the resulting navigational aspects. Furthermore, links can be also used to represent data movements, or parameter passing, between units. As an example, an information gathered from the user via a web form, can be used in the selector condition for extracting the data instances to be displayed. Several different types of content units are defined in the *WebML Hypertext Model*. The well-defined semantics and syntax of the *WebML* notation allowed for the construction of a specifically conceived CASE tool, *WebRatio*, which automatically generates the Web application starting from the corresponding design models, thus effectively reducing time and costs of software development. Within this tool, each stereotype has both a visual and an XML-based textual representation, and is used to specify additional detailed properties, such as the grammar for the combination rules. The availability of an XML specification has two main advantages: it enables the deployment of the same design into multiple rendering formats, and it supports the definition of customized extensions of the language.

3   An Extension of WebML for Web GIS

Web GIS are a special case of data-intensive Web applications, meant to deal with the geographic data and share them across several users for different business goals. From a design point of view, Web GIS present many specific characteristics, which make them different from traditional data-intensive Web applications. As for the data model, the complex nature of geographic data, where the two components, descriptive and spatial, should be analyzed and managed in a joint manner, requires the use of a proper modeling approach, named *Geodata and Metadata Conceptual Model*. In our approach, we suggest the adoption of the Spatial E-R model [2], where each set of geodata is described as a *spatial entity* characterized by a set of attributes, a geometry and a couple of coordinates, whereas spatial associations among entities are expressed in terms of topological relationships. In addition, since a Web GIS may also be aimed at supporting data sharing, metadata management is a crucial issue to be accomplished. Making interoperable data, which are heterogeneous in terms of formats and/or sources, may guarantee better performances during data exchanging and retrieving.

As for the navigational model, specific functionality meant to dynamically navigate a geographic map, commonly known as *Web mapping*, have to be taken into account, giving rise to the *Hypertext and Web Mapping Model* as an extension of the *WebML Hypertext Model*. In particular, to exploit *WebML* for modeling Web GIS, we have considered new interaction metaphors able to capture the characteristics of this kind of applications. Besides the common interactions a user performs when navigating and executing a web application, we have taken into account the actions directly performed on a map, when carrying out web mapping or spatial query tasks. Such actions usually manipulate and produce data, which are not presented in a new web page but are rather rendered onto the same map. Thus, new *WebML* units are needed to model specific Web GIS interaction tasks, such as map visualization and navigation. In the following we will describe some of them.

The MultiMap Unit

Basically, the *MultiMap* Unit is a visual metaphor for a map viewer, i.e. a graphic component within a Web browser, able to render both vector and raster data arranged in layers. It is characterized by the following properties:

- Name: the name chosen by the Application Architect for the MultiMap Unit.
- Raster Sources: the reference to raster entities. This source is optional, because a Web GIS application may not require raster layers.
- Raster Selector [optional]: predicate useful to determine which objects belonging to the raster entities share in the unit content.
- Vector Sources: the reference to vector entities. This source is optional, because a Web GIS application may not require vector layers.
- Vector Selector [optional]: predicate useful to determine which objects from the vector entities share the unit content.

Moreover, to correctly render source data, the *MultiMap* Unit requires a global parameter, named *Extension*, which contains coordinates of the upper-left and lower-right corners of the presented area. Each time this parameter is modified by an external operation, the area rendered in the *MultiMap* Unit is updated accordingly.

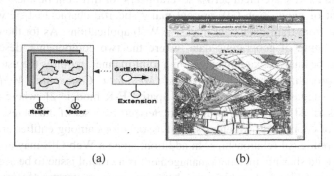

(a)   (b)

Fig. 1. Graphic notation of an instance of a *MultiMap* Unit (a) and its representation in a Web browser (b)

Figure 1(a) depicts the visual notation of a *MultiMap* Unit instance, namely *TheMap*, as well as the associated global parameter. The resulting representation in a Web browser is shown in Figure 1(b). Here, both raster and vector layers are displayed, as determined by invoking both source entities, while no selectors are set, thus loading all the instances of the underlying data. Differently, when a Web GIS application is expected to allow the activation and display of layers from a Table of Contents on users' demand, a different modeling is required as shown in Figure 2(a). It contains a *MultiMap* Unit to render the layers, and two *Multi-choice Index* Units, to represent the Table of Contents which is made up of two different legends. This scenario also requires selectors associated with the *MultiMap* Unit sources, which are used to select data in agreement with parameters set through the two instances of the *Multi-choice Index* Units of WebML, namely *RasterLegend* and *VectorLegend*. The resulting application on the Web browser has a checkbox, as shown in Figure 2(b), by which users may activate a layer and visualize the corresponding data on the map. Depending on aesthetic/usability factors, design layouts can also be provided where legends are integrated into a homogeneous format.

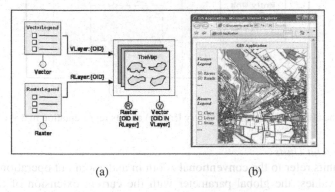

(a)   (b)

Fig. 2. Modeling the layer selection associated with a map

The Geometry Entry Unit

In a Web GIS, the spatial selection is one of the most common modalities of interaction between the user and the application. It can be used either to select an area of interest to magnify/reduce, or to select a set of geodata on which a basic functionality may be applied. In order to model the associated geometry acquisition, we have introduced the *Geometry Entry* Unit, which may be used to represent interaction metaphors, typical of a Web GIS environment. This unit is characterized by a name and the geometry type on which the interaction is based, which can be a:

- **Point** - it is a single point expressed as a coordinate (x,y). It corresponds to a mouse click on the viewer map.
- **Multipoint** - it is a sequence of n points, corresponding to n mouse clicks on the map.
- **Line** - it is constituted by two points, corresponding to 2 mouse clicks on the map. The two points are then connected by a segment.
- **Polyline** – it is constituted by n points, corresponding to n mouse clicks on the map. The points are connected by a sequence of consecutive lines, forming a single polyline.

- **Rectangle** - it is made up of two points, corresponding to 2 mouse clicks on the map, which are associated with the opposite vertexes of the rectangle diagonal.
- **Polygon** - it is constituted by *n* points, corresponding to *n*+1 mouse click on the map (the first and last point are the same). It determines a polygon with *n* sides.
- **Drag** - it is constituted by 2 points, the former is captured when the user presses the mouse button, the latter corresponds to its release.

Figure 3(a) shows the graphic notation of an instance of the *Geometry Entry* Unit, which refers to a rectangle selection, namely the *Rectangle Entry* unit. The corresponding Web representation is shown in Figure 3(b).

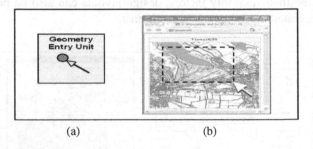

(a)   (b)

Fig. 3. Graphic notation of a *Geometry Entry* Unit (a) and its representation in a Web browser (b)

The Zoom In/Out Units
These two units refer to the conventional zoom in and zoom out operations. Both take three input values: the global parameter with the current extension of the map, the global parameter with a magnification/diminishing factor and the instance *Point* of *Geometry Entry* Unit, which is used to identify the area on which the operation should be applied. As a result, if the operation succeeds, it sets a new value for the *Extension* global parameter, according to the zoom factor. This arrangement is graphically depicted in Figure 4 where an OK link leads to a new value for the *Extension*, while a KO redirects towards a generic error page.

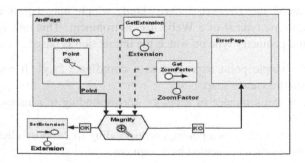

Fig. 4. An example of *ZoomIn* Unit

The Pan Unit
This unit represents the pan operation that shifts the focus of a map toward a specified direction. This unit takes three input values, namely the two points of the map where the user presses and releases the mouse button, and the global parameter with the current extension of the map. If the operation succeeds, a new value for the *Extension* global parameter is set, i.e. the new area of the map which is visualized. In Figure 5, visual symbols involved in a pan operation are shown.

The CreateOverlay Unit
In a Web GIS application, a query processing usually outputs a set of geodata which satisfy user's criteria. To display such geodata onto a map, they are organized in a new temporary layer, which may be overlaid onto the existing ones. Users may use them to perform other basic operations and derive new geographic information. To describe such a functionality through a WebML-like approach, we introduced the *CreateOverlay* Unit, which can be exploited to model scenarios where users pose queries and apply further spatial operations to the corresponding output in a cascade-like fashion. The Unit requires two parameters, namely a spatial filter and a layer on which the filter should be applied. In Fig. 9, an instance of the *CreateOverlay* unit is depicted, namely *NewTheme*, where the spatial filter is obtained by a *Geometry Entry* Unit which returns a rectangle, and the layer is the active one. In case the operation succeeds, the OK link indicates that the OID of the newly created layer will be forwarded to the *MultiMap* unit, which will eventually render it.

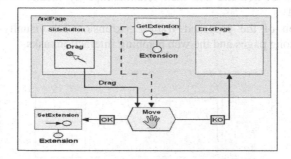

Fig. 5. An example of the Pan operation

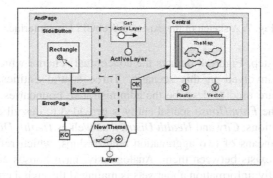

Fig. 6. An example of the *CreateOverlay* operation

4 Modeling a Web GIS Application for Farm Houses Monitoring

The described web mapping units have been implemented in *WebRatio*, and adopted in a preliminary prototype of the tool, to prove the effectiveness of the proposed methodology. The prototype exploits a visual environment previously developed by authors [13], which allows to design Spatial ER schemas and to generate a corresponding relational database and an XML specification according to the *WebRatio* format so that it can be used as underlying layer for the various units (both new and old ones). In this section we describe the use of the tool in the design of an application meant to monitor farm houses located in Campania, a region of Southern Italy. Here, Health Districts may control phenomena related to distribution of livestock and disease diffusion, by navigating a map, displaying geodata and posing queries about areas of interest. Moreover, the resulting Web GIS application may support vets and administrators in making decisions about policy and preventative plans.

Figure 7 depicts the framework for an interface layout for this kind of Web application. It consists of a Map Working Area where geodata are displayed and users may navigate, a Table of Contents, where a legend allows to select and activate layers, a palette of GIS Tools to invoke GIS functionalities, a Query Frame where a query may be posed, and a KeyMap which supports users in orienting themselves.

As for the design phase, in this section we report on two main aspects of the entire process, namely:

- the description of geodata and their relationships by means of the Spatial E-R extension, and
- the application of the proposed *WebML* extension for visually describing the transitions among pages and the web mapping interaction tasks.

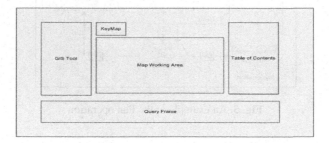

Fig. 7. A sketch of the expected WebGIS User Interface

As for the first step, Figure 8 depicts the conceptual schema which models entities and relationships involved in this scenario. Three spatial entities, *Region, Health District* and *City,* are used to describe the administrative boundaries of land boards in Campania, while the *FarmHouse* spatial entity is provided to describe both farm house properties and positions. *City* and *Health District*, as well as *Health District* and *Region*, are connected by means of two aggregation relationships, which reflect the "part of" association which exists between them. Analogously, farm houses are contained in at least one city. Finally, information about vets is managed through a conventional entity.

In this case, a relationship between the *Veterinary* and *Health District* has been defined in order to establish that a vet works just in one health district.

As for the *Hypertext & Web Mapping* design, the Web GIS site is composed by three ANDed pages (see Fig. 12):

- a *SideButton* page that contains buttons to activate the GIS functionality, such as *Zoom In*, *Pan* and *SelectByLayer*;
- a *Central* page, containing the *ViewerGIS* instance of the *MultiMap* Unit, the *Multi-choice* units to de/select raster and vector layers within the map, an *Index Unit* to select the active layer where the user computes an operation, and finally some global parameters to store the active layer name, the map *Extension* and the *ZoomFactor*;
- an *Error Page*.

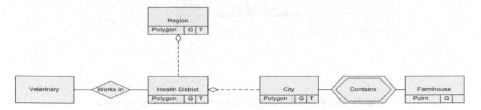

Fig. 8. The Spatial E-R schema of the running example

In order to describe the zoom in functionality, we exploited the *Zoom In* Unit, namely *Magnify*, which is fed through the *ZoomFactor* and the *Extension* global parameters and an instance of a *Geometry Entry* Unit, which returns a point. The result is a new map extension which will be stored into the *Extension* parameter. The *ViewerGIS* Unit will automatically be refreshed.

For the *Pan* operation, the *Move* instance of the *Pan* Unit handles three parameters, namely the present extension, the direction and the distance to move along that direction. Also, the *Drag* instance of the *Geometry Entry* unit is modeled which allows users to visually specify direction and distance by means of a rectangle drawn by a drag operation on the map. Such values, together with coordinates received from the *Extension* global parameter, are fed to the *Move* unit in order to compute the new map extension. Finally, to visualize the new map boundaries we put the *Move* result into the *Extension* parameter.

The last operation to model corresponds to a spatial query aimed to identify all the farm houses within a user drawn rectangle. Then, in order to get graphical results, the following steps are necessary:

- to add a new entry into the *ViewerGIS* source,
- an instance of the *GeometricEntry* Unit to draw the rectangle of the area of interest on the map;
- the *NewTheme* instance of the *CreateOverlay* Unit in order to manage output rendering.

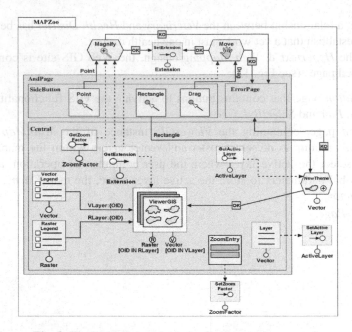

Fig. 9. The WebML-based schema for the running example

Figure 10 shows the resulting Web GIS application, where, by following the same approach, also other functions have been considered.

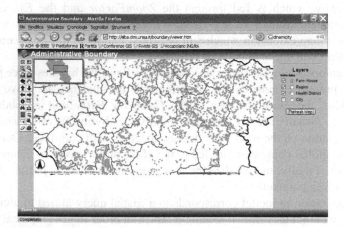

Fig. 10. The UI of the modeled Web GIS application

5 Model-Driven Generation of Web GIS Applications

Extendibility and flexibility are some of the key features characterizing *WebRatio*, the CASE tool supporting the *WebML* methodology. Indeed, it provides a means to define new symbols and links, and, for each of them, to specify the rules to generate the corresponding J2EE-based code for the web applications. By exploiting this mechanism, we have extended *WebRatio* to support the design of Web GIS applications and their automatic generation starting from the specified Geodata and Metadata Conceptual Model, and the Hypertext and Web Mapping Model. In particular, the tool generates the server side code of the Web GIS application and the corresponding database definition. The architecture of a generated Web GIS application together with the employed technologies is depicted in Fig. 14.

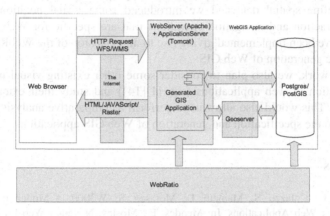

Fig. 11. The client-server architecture of a generated Web GIS application

The application is based on a multi-tiered architecture and on a set of standard and open technologies such as Apache/Tomcat, Postgres/Postgis, Geoserver, Web Feature Service (WFS) and Web Map Service (WMS) protocols, which guarantee the interoperability among server modules and the client browser. Basically, the *WebML* units are translated into a JSP/Servlet program and successively mounted as a Tomcat/Apache web application. This application will be able to manage requests coming from the Internet and switch them to either the Geoserver or the PostGIS as required. On the other side, the conceptual schema described through the Spatial ER diagram is translated in an XML specification and input to *WebRatio* in order to build the underlying database.

At run-time actions performed by the user on the client side are translated into HTTP request by JavaScript code and sent to the server where data satisfying the request are selected and sent back to the client as HTML, JavaScript and raster data. In detail, a GIS query is transmitted as a GET or POST request written according to either the WFS or WMS specifications. The request is therefore captured by the generated GIS application and forwarded to Geoserver which, in turn, interprets the query, composes the SQL statement according to the PostGIS DML and send it to the DBMS. Once the DBMS computes the query, results are gathered by Geoserver to create the answer. In particular, in case of a WMS request, Geoserver computes a raster map containing the results encoded as a standard picture format (GIF, PNG, SVG, etc.). In case of a WFS request, Geoserver collects data from the DBMS and returns a GML (Geography Markup Language) encoded data to the generated GIS server application. The latter further processes the resulting GML data, by sending it back to the client side in HTML format

6 Conclusions

In the present paper, we described an approach for the design of Web GIS applications based on WebML. In particular, we proposed to exploit the Spatial E-R model to define spatial data and we introduced some visual notations to model relevant interaction and navigation operations that are specific for Web GIS. These new units have been implemented giving rise to an extension of the WebRatio tool for the automatic generation of Web GIS.

As future work, we also plan to consider some other existing visual notations for the specification of web applications (e.g. [1][14]) and study their extension to the GIS domain. This would also allow us to perform a comparative analysis of different approaches to the specification and generation of Web GIS applications.

References

[1] Baresi, L., Colazzo, S., Mainetti, L., Morasca, S.: W2000: A Modeling Notation for Complex Web Applications. In: Mendes, E., Mosley, N. (eds.) Web Engineering, pp. 335–408. Springer, Heidelberg (2006)
[2] Calkins, H.W.: Entity Relationship Modeling of Spatial Data for Geographic Information Systems. International Journal of Geographical Information Systems (January 1996)
[3] Ceri, S., Fraternali, P., Bongio, A.: Web Modeling Language (WebML): a Modeling Language for Designing Web Sites. Computer Networks: The International Journal of Computer and Telecommunications Networking 33(1-6), 137–157 (2000)
[4] Ceri, S., Fraternali, P., Bongio, A., Brambilla, M., Comai, S., Matera, M.: Designing Data-Intensive Web Applications. Morgan-Kaufmann Publishers, San Francisco (2002)
[5] Ceri, S., Fraternali, P., Matera, M.: WebML Application Frameworks, a Conceptual Tool for Enhancing Design Reuse. In: Procs. "Web Engineering", Hong Kong (May 2001)
[6] Ferrucci, F., Tortora, G., Vitiello, G.: Exploiting Visual Languages in Software Engineering - Handbook of Software Engineering and Knowledge Engineering (2001), ISBN: 981-02-4973-X
[7] Garzotto, F., Paolini, P., Schwabe, D.: HDM- A Model-Based Approach to Hypertext Application Design. ACM Transactions on Information Systems 11(1), 1–26 (1993)
[8] Isakowitz, T., Stohr, E.A., Balasubramanian, P.: RMM: A Design Methodology for Structured Hypermedia Design. Communications of the ACM 38(8), 34–44 (1995)
[9] Longley, P.A., Goodchild, M.F., Maguire, D.J., Rhind, D.W.: Geographic Information Systems and Science, p. 472. Wiley, Chichester (2001)
[10] Longley, P.A., Goodchild, M.F., Maguire, D.J., Rhind, D.W.: Geographical Information System Principles, 2nd edn. Wiley, Chichester (2001)
[11] Open Geospatial Consortium, Inc. ® (OGC), http://www.opengeospatial.org

[12] Paolino, L., Sebillo, M., Tortora, G., Vitiello, G.: VIEW-GIS - An Environment Supporting Web GIS Developers. In: VLC. Procs 12th International Conference on Distributed Multimedia Systems - Workshop on Visual Languages and Computing, Grand Canyon USA, pp. 141–146 (August 2006)

[13] Paolino, L., Sebillo, M., Tortora, G., Vitiello, G.: An OpenGIS® based approach to define continuous field data within a visual environment. In: Bres, S., Laurini, R. (eds.) VISUAL 2005. LNCS, vol. 3736, pp. 83–93. Springer, Heidelberg (2006)

[14] Pinet, F., Lbath, A.: A visual modelling language for distributed geographic information systems. In: Proceedings IEEE International Symposium on Visual Languages (September 10-13, 2000)

[15] Schwabe, D., Rossi, G.: An Object Oriented Approach to Web based Applications Design. Theory and Practice of Object Systems 4(4), 207–225 (1998)

[16] AutoDesk MapGuide, web site: http://usa.autodesk.com/adsk/servlet/index

[17] ESRI. ArcIms, web site: http://www.esri.com/software/arcgis/arcims/index.html

[18] Intergraph. WebMap, web site: http://imgs.intergraph.com/gmwm/

[19] Mapcell Web Site: http://www.territoriumonline.com/it/gis/mapaccel/

[20] WebML site. Available at http://www.Webml.org

An Object-Oriented Version Model for Context-Aware Data Management

Michael Grossniklaus and Moira C. Norrie

Institute for Information Systems, ETH Zurich
8092 Zurich, Switzerland
{grossniklaus,norrie}@inf.ethz.ch

Abstract. Context-aware computing is a major trend in mobile computing, pervasive computing and web engineering. Several models, frameworks and infrastructures have been developed to represent, process and manage context. While most of these approaches support the adaptation of application logic based on context, the requirements of context-aware systems in terms of data management have received little attention. This is most apparent in the field of web engineering as many web sites are data-intensive and require context-dependent content adaptation to support internationalisation, personalisation and multiple channels. We present a version model featuring alternative versions for context-aware data management and query processing that has been integrated in an object-oriented database system. Finally, we also describe the implementation of a mobile tourist information system based on this system.

1 Introduction

Context-aware computing has become an important issue in several application domains. Although the requirements that motivate the need for context-awareness vary from one field of computing to the next, context is widely regarded as a valuable resource that can help to improve applications. In mobile computing context information is used to address the challenges that arise from the reduced interaction bandwidth that characterises most portable devices. Working with a device that has a small screen and limited input facilities can be made less awkward by adapting the application to the current environment of the users. In pervasive computing, the role of context is even more prominent as applications are often not built for standard devices with traditional user interfaces. Rather these applications are completely embedded in the user's environment and have to rely on sensed context information for interaction.

In this paper, we focus on context-awareness in web engineering. As the Web was originally developed to exchange information, the management and delivery of information has always been a primary concern. Also, due to requirements resulting from globalisation, personalisation and multi-channel access, many different forms of context information, both technical and cultural in nature, can be witnessed in web engineering. Over time, the Web has evolved into an omnipresent platform that nowadays also supports complex applications. As a consequence, well-established web standards are no longer only used to build web

B. Benatallah et al. (Eds.): WISE 2007, LNCS 4831, pp. 398–409, 2007.

sites, but have also been adopted in the realisation of both mobile and pervasive systems. In a sense, web engineering has come full circle as its technologies are now being used to build context-aware systems that were originally motivated by ubiquitous computing.

Efforts towards context-awareness in web engineering are well documented by several model-based approaches that have been extended or designed to provide such features. The personalisation of web sites constitutes one form of adaptation to the user's context and is supported by several methodologies such as OOHDM [1], OO-H [2] and SiteLang [3]. In contrast to these systems that adapt to a single user, WSDM [4] proposes audience-driven personalisation to adapt to a group of users. The Hera [5] methodology combines concepts from adaptive hypermedia with semantic web technologies and therefore supported adaptivity from the start. In Hera, adaptability denotes the ability to adapt a web site to users or device capabilities at design-time, while adaptivity is the adaptation of an application at run-time. Finally, WebML [6] which maybe marks the most comprehensive approach to model-driven web site development has been extended to model context-aware and adaptive web applications [7].

While all of these model-based approaches provide possibilities to specify context-aware adaptation, general platforms that support the implementation of this functionality have not yet emerged. Some design methodologies such as Hera and WebML have introduced their own run-time systems. However, these platforms are often tailored to proprietary features of a methodology or do not even support all capabilities of the model. To adapt the application logic of a web application, the use of aspect-oriented programming has been proposed [8] in the setting of UML-based Web Engineering (UWE) [9]. Unfortunately, similar concepts that would allow the adaptation of the application content at the level of database systems have not yet emerged. In this paper, we present a solution for context-aware data management that consists of a version model supporting context-dependent alternatives and a matching algorithm as the basis for context-aware query processing.

We begin in Sect. 2 with a discussion of related work and background information documenting the origins of our approach. The version model for context-aware data management is presented in Sect. 3 together with the definition of the notion of context used in our work. Section 4 discusses context-aware query processing based on this version model. In Sect. 5, we present a mobile information system that was implemented using a database system extended with the presented version model. Finally, in Sect. 6 we give concluding remarks.

2   Related Work and Background

An early approach to deliver content according to the device context of the client is generally known as distilling or transcoding [10,11,12] on a proxy server. Common forms of adaptation include splitting large documents into smaller fragments that can be handled by the client as well as changing the format, colour depth or size of an image. Usually, these transformations are effected online and thus result in reduced performance. Several solutions to this problem have been proposed, among them the caching of previous results [13] of the distillation process and the pre-computation of content variants [14].

Another solution is proposed by the web authoring language Intensional HTML (IHTML) [15]. Based on version control mechanisms, IHTML supports web pages that have different variants for different user-defined contexts. The concepts proposed by IHTML were later generalised to form the basis for Multidimensional XML [16] which in turn provided the foundation for Multidimensional Semistructured Data (MSSD) [17]. MSSD is represented using a graph model with multidimensional nodes and context edges. Multidimensional nodes capture entities that have multiple variants by grouping the nodes representing the facets. These variants are connected to the multidimensional node using context edges. The label of a context edge specifies in which context the variant pointed to is appropriate. Based on this graph representation, a Multidimensional Query Language [18] has been defined that allows the specification of context conditions at the level of the language. Thus, it can be used to formulate queries that process data across different contexts.

As mentioned previously, some model-based approaches provide a proprietary implementation platform designed to support some or all context-aware features of the methodology. For example, the WebML run-time which is part of the WebRatio Site Development Studio [19] has been extended to allow the execution of context-aware and adaptive web applications. WebRatio is implemented using industry standards such as relational database management systems, Java-based application servers, XML and XSLT. Both the data and navigation models are rendered in XML and passed together with an XSLT representation of the presentation design to an automatic code generator. Apart from this data, the code generator also uses metadata describing data mappings and WebML-specific tag libraries to produce Java Server Pages that are deployed to an application server.

Similar to the approach taken by WebML, web sites designed with Hera can be implemented using the Hera Presentation Generator [20]. It combines the data stored using the Resource Description Framework (RDF) with the models represented in RDF Schema to generate an adapted presentation. This presentation is rendered as a set of static documents that contain the mark-up and content for one particular class of clients. A more dynamic implementation platform for Hera has been proposed based on the AMACONT project [21]. Based on a layered component-based XML document format, reusable elements of a web site can be defined at different levels of granularity. Adaptation is realised by allowing components of all granularities to have variants. A variant of a component specifies an arbitrarily complex selection condition as part of the metadata in its header. AMACONT's publishing process is based on a pipeline that iteratively applies transformations to a set of input documents to obtain the fully rendered output documents.

A more general and extensible architecture to support context-aware data access has been proposed based on profiles and configurations. Context is represented as a collection of profiles that each specify one aspect of context such as the user or device. Profiles are expressed according to the General Profile Model [22] that provides a graphical notation and is general enough to capture a wide variety of formats currently in use to transmit context information as well as transforming from one format to another. Configurations have three parts matching the general architecture of web information systems in terms of content, structure and presentation. Configurations are stored in a repository on the server side and matched to the profiles submitted by the client as part of its request. The matching is based on adaptation rules consisting of a parametrised profile, a condition and a parametrised configuration [23]. During the matching process, the client profile is compared to adaptation rules. If the client profile matches the parametrised profile of the rule and the specified values fulfil the condition, the parametrised configuration is instantiated and applied.

Although all of these approaches provide some functionality for creating context-aware systems, none of them offers support for the management of context-aware data. As a consequence, this neglect requires that context information about application data is modelled explicitly in the data model. For example, to implement a simple multi-lingual web site otherwise elegant and intuitive data models have to be extended with additional concepts to capture the language context and thus become overcrowded and cumbersome. Another drawback of such stratum approaches is the fact that they scale very poorly when multiple context dimensions are required. Therefore, instead of building frameworks for context-aware data access on top of existing database management systems, we believe that this issue needs to be addressed at the level of data storage and querying inside the database itself. Inspired by version models proposed in the domain of engineering databases [24] and software configuration systems [25], we propose a two-dimensional version model that, in addition to revisional versions, supports alternative versions to represent a data object in different contexts. While the basic organisation of our version model is closely related to previous approaches, the process of evaluating context-aware queries is very different to that used in existing version models to compute configurations. In engineering databases and software configuration systems, "context" information supplied with the query constitutes an integral part of the specification of the query results. Context as seen in mobile and ubiquitous computing as well as web engineering is instead information that can be used to refine the result of a query that otherwise has well-defined default behaviour. As a consequence, our query processor uses an algorithm based on a best match rather than an exact match to select the variant that is most appropriate in a given context. In the next sections, we will present our version model and give further detail how this notion of context has affected its design.

3   Version Model

In the setting of our context-aware data management system, context information is regarded as optional information used by the system to augment the result of a query rather than specifying it. As a consequence, such a system also needs a well-defined default behaviour that can serve as a fall-back in the absence of context information. In our approach, context information is gathered outside the database system by the client application. Therefore, it is necessary that client applications can influence the context information used during query processing. To support this, a common context representation that is shared by both components is required. Since several frameworks for context gathering, management and augmentation already exist, our intention was to provide a representation that is as general as possible. The following definitions specify the notion and representation of context as used by our system. We will use the given sets NAMES and VALUES to denote the sets of legal context dimension names and context values, respectively.

Definition 1. *A context space represented by S denotes which context dimensions are relevant to an application of the version model for context-aware data management. It is defined as $S = \{name_1, name_2, \ldots name_n\}$ such that $\forall i : 1 \leq i \leq n \Rightarrow name_i \in$ NAMES and therefore $S \subseteq$ NAMES.*

Definition 2. *A context value c is defined as a tuple $c = \langle name, value \rangle$ where $name \in$ NAMES and $value \in$ VALUES.*

Definition 3. *$C(S)$ denotes a context for a context space S and is represented as an unordered set of context values*

$$C(S) = \{\langle name_1, value_1 \rangle, \langle name_2, value_2 \rangle, \ldots, \langle name_m, value_m \rangle\}$$
$$= \{c_1, c_2, \ldots, c_m\}$$

such that $\forall i : 1 \leq i \leq m \Rightarrow name_i \in S$ and $\forall c_i, c_j \in C : i \neq j \Rightarrow name_i \neq name_j$.

Definition 4. *A context state denoted by $C_\star(S)$ is a special context, where $\forall name \in S : \exists \langle name, value \rangle \in C_\star(S)$.*

Our version model for context-aware data management has been specified within the framework of an object-oriented database management system developed at our institute. As this database management system is built on the concepts defined by the OM data model [26], we defined our model as an extension of OM. OM is a rich and flexible object-oriented data model that features multiple instantiation, multiple inheritance and a bidirectional association concept. This model was chosen as, due to its generality, it can be used to represent other conceptual models such as the Entity-Relationship model or RDF.

As the OM model supports multiple instantiation, an object is represented by a number of instances—one for every type of which the object is an instance. All instances of an object share the same object identifier but are distinguishable based on their instance type. For the purpose of our version model, we have broken this relationship between the object and its instances and introduced the additional concepts of alternative and revisional versions. As shown in Fig. 1, in the extended OM model, an object is associated with a number of alternative

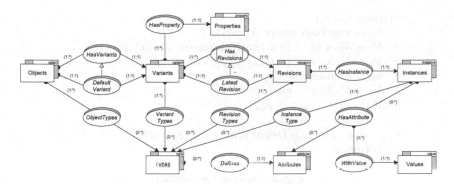

Fig. 1. Conceptual data model of an object

versions (Variants) which in turn are each linked to a set of revisional versions (Revisions). Finally, each revision is connected to the set of instances containing the actual data. As can be seen from the figure, our model supports two versioning dimensions. Variants are intended to enable context-aware query processing while revisions support the tracking of the development process. However, for the scope of this paper we will focus on variants exclusively and neglect the presence of revisional versions in the model. Note that all versions of an object still share the same object identifier tying them together as a single conceptual unit. A variant of an object is identified by a *variant context* $C_{var}(S)$ that is stored as a set of properties represented as $\langle name, value \rangle$ context values.

4   Context-Aware Query Processing

Context-aware queries over multi-variant objects are evaluated using the matching algorithm shown in Fig. 2. Whenever an object is accessed by the query processor, the algorithm selects the object variant that best matches the current context state of the system $C_*(S)$. First it retrieves all variants of the input object o linked to it through the *HasVariants* association. After building the context state of each variant from the properties associated to it through *HasProperty*, the algorithm applies a scoring function f_s to this variant context state that returns a value measuring how appropriate the variant is in the current context. It then returns the variant of o with the highest score s_{max}. If the highest score is below a certain threshold s_{min} or if there are multiple variants with that score, the default variant is returned. This fall-back mechanism has been introduced into the algorithm to guarantee a well-defined default behaviour.

Similar to context dimensions, the concrete definition of the scoring function depends on the requirements of a context-aware application. Our system therefore allows the default scoring function to be substituted with an application-specific function. In order to give an indication of what is involved in designing such a scoring function, we will describe the *general scoring function*, as specified in Def. 5, that is used in our system as a default. To allow greater flexibility in specifying

MATCH$(o, C_\star(S))$
1   $V_0 \leftarrow rng(\textit{HasVariants } dr(\{o\}))$
2   $V_1 \leftarrow V_0 \propto (x \rightarrow (x \times rng(\textit{HasProperty } dr(\{x\}))))$
3   $V_2 \leftarrow V_1 \propto (x \rightarrow (dom(x) \times f_s(C_\star(S), rng(x))))$
4   $s_{max} \leftarrow max(rng(V_2))$
5   $V_3 \leftarrow V_2 \% (x \rightarrow rng(x) = s_{max})$
6   **if** $|V_3| = 1 \wedge s_{max} \geq s_{min}$
7   **then** $v \leftarrow V_3 \; nth \; 1$
8   **else** $v \leftarrow rng(\textit{DefaultVariant } dr(\{o\})) \; nth \; 1$
9   **return** v

Fig. 2. Matching algorithm

Table 1. Value syntax

Set	Syntax	Description	Examples
ATOM	x	Atomic value	en, 27
SET	$x_1\{: x_i\}$	Set of atomic values $S = \{x_1, \ldots, x_n\}$	at:ch:de, red:blue
RANGE	$x_{min}..x_{max}$	Range of atomic values $I = [x_{min}, x_{max}]$	5.5..7.0, a..f
STAR	$*$	Wildcard	*

the current system context as well as to describe object variants, we have partitioned the given set VALUES into four subsets as shown in Tab. 1. Based on these sets, the *value* field of a context value c can be used to specify more than one value. The behaviour resulting from using identity to define equality can still be obtained by using values of set ATOM. In addition, collections of atomic values can be expressed using a value of set SET. Values defined by RANGE have been introduced to allow intervals to be defined based on a lower and upper bound. Finally, if all possible values of a context dimension are permissible, the wildcard value from set STAR can be used. Providing a wildcard value in this setting might seem paradoxical at first as one might argue that the corresponding context value could simply be omitted. Its right to exist will become clear when we present how the scoring function computes the score value based on the matching values.

A matching condition ($\cong$) for context values has been defined. As context values are used in our approach to describe both the current context state of the system and the context of a variant, any combination of these four given sets has to be considered. As the discussion of all of the resulting sixteen cases is clearly out of the scope of this paper, we refrain from going into more details here. Intuitively, two context values match if they are either the same, contained in one another or if a wildcard value is involved. Our general scoring function also supports two mechanisms that have been created to give more control over the matching process to the client of our system. On the one hand, an application can assign weights to each of its context dimensions to be used when computing the score of an object variant. Required and illegal matches can also be specified by prefixing the corresponding context value with $+$ or $-$, respectively. If, for example, a client requires that all data is delivered in English, it would specify the context value $\langle lang, +en \rangle$ in the current context state $C_\star(S)$. Analogously, an object variant that is not intended for a novice user could be described by the context value $\langle level, -novice \rangle$. To compare such prefixed context values, a second matching condition ($\cong_\pm$) has been introduced that also takes required and illegal values into consideration.

Definition 5. *The* general scoring function f_s *takes two contexts* C_1 *and* C_2 *as arguments and returns a scoring value representing the number of matching context dimensions of the two contexts normalised by* $|N|$. *It is defined as*

$$f_s(C_1, C_2) = \frac{1}{|N|} \sum_{n \in N} (w(n) \times f_i(n, C_1, C_2)) \times \prod_{n \in N} f_\pm(n, C_1, C_2)$$

where N *denotes the union* $N_1 \cup N_2$ *of the sets of all names of context values specified either by* C_1 *or* C_2, *respectively. The indicator function* f_i *is defined as*

$$f_i(n, C_1, C_2) = \begin{cases} 1 & \exists\, c_1 \in C_1, c_2 \in C_2 : \\ & name_1 = name_2 = n \wedge value_1 \cong value_2 \\ 0 & otherwise. \end{cases}$$

The context dimension weight function w *returns a weight* $w(n) \in \mathbb{R}^+$ *for every name* $n \in N$, *where* N *represents the set of the names of all context dimensions. Finally, the matching function for prefixed values* $f_\pm$ *is defined as*

$$f_\pm(n, C_1, C_2) = \begin{cases} 1 & \exists\, c_1 \in C_1, c_2 \in C_2 : \\ & name_1 = name_2 = n \wedge value_1 \cong_\pm value_2 \\ 0 & otherwise. \end{cases}$$

5   Mobile Information System

The version model and query processing presented has been implemented in the OMS database system which provides native support for concepts of the OM data model. Using this extended OMS system, we have also built the Extensible Content Management System (XCM) [27] based on the separation of the four concepts of content, structure, view and layout [28]. To corroborate our claim that it is vital to place context-awareness also at the level of the database system, we will describe how XCM served as a content server in a multi-channel mobile tourist information system designed to assist visitors to the Edinburgh festivals held each year during the month of August. An overview of the architecture of the so-called EdFest system [29] is shown in Fig. 3.

The clients supported by EdFest are shown on the left of the figure. Apart from traditional clients such as desktop PCs and PDAs, EdFest introduced a novel interaction channel based on interactive paper [30]. Our context-aware

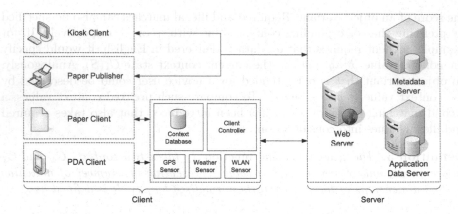

Fig. 3. Overview of the EdFest architecture

content management system is shown on the right of the figure. It consists of a web server handling the communication with the clients, a server managing publishing metadata and an application database that stores the content of the EdFest system. While the kiosk and PDA clients are implemented using standard HTML, the paper client actually consists of two publishing channels. The paper publisher [31] channel is used at design time to author and print the interactive documents from the content stored in the application database. The paper client channel is then active at run-time when the system is used and is responsible for delivering additional information about festival venues and events through voice feedback when the tourists interact with the paper documents.

As the paper and PDA client are mobile, they have been integrated with the platform shown at the centre of the figure that manages various aspects of context. Sensors gather location, weather and network availability information that is then managed in a dedicated context database. Context information is sent from the client to the server by encoding it in the requests sent to the content management server. This is done by the client controller component that acts as a transparent proxy that enriches requests with the current context state stored in the context database. For instance, to access the `getEventInfo` information service, the paper client would send the following request to the server.

`http://.../xcm?anchor=getEventInfo&event=o2451`

After intercepting and augmenting the request, the extended URI given below is issued to the XCM server by the client controller proxy.

`http://.../xcm?anchor=getEventInfo&event=o2451&format=+vxml&`
`        lang=en&user=guest`

By inserting these parameters into the request, the client provides information about the format it requires, the preferred language and the current system user.

To discuss how context-aware data management is used in XCM, it is necessary to understand its four-step publishing process shown in Fig. 4. As XCM uses

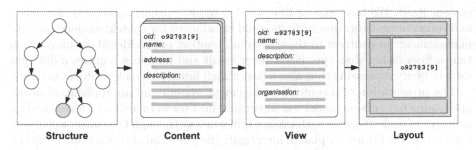

| Structure | Content | View | Layout |

Fig. 4. Publishing process

a composite pattern to capture structure, the first step depends on whether the object to publish is a container or content object. If the client requests a container object, the publishing process has to be applied recursively to all components within the container. If a content object is requested, the publishing process is executed only once. Assume that the grey circle in the figure represents the component object requested by the client. Since it is one of the leaf nodes of the structure tree, it represents a content object and can be published directly. As shown in the figure, the content object could represent a person object with attributes *name*, *address* and *description*. The content publisher then retrieves the view object associated with the content object and applies it. In our example, the application of the view object results in the hiding of attribute *address* and the appearance of aggregated content about the *organisation* this person works for. Finally, the publishing component retrieves the presentation objects for the component object. By applying the template to the object, the final presentation is generated.

XCM uses dedicated databases to manage application data and publishing metadata. Both of these databases have been extended with the presented version model. Thus, in each step of the publishing process, a context-aware query is issued to access both data and metadata. As a consequence, XCM supports context-dependent adaptation for content, structure, view and layout. In the EdFest system, this built-in capability of the database system was used to provide location-based services as well as multi-channel and multi-modal access to the system in a uniform way. Further, since in OM everything—even concepts of its metamodel—is represented as objects, it has been possible to address the need for context-aware interactions [32] for different publishing channels with the concept of object variants for method objects.

6 Conclusions

We have motivated the need for implementation platforms that support context-aware data management by collecting requirements from mobile and pervasive computing as well as from model-based design methods for web engineering. We believe the issues related to context-aware data management can only be addressed by providing adequate concepts at the level of the data management system itself. We have shown how version models from engineering databases and software configuration systems could be adapted for context-dependent data management. In contrast to these systems, context is considered as information that refines rather than specifies a query result and therefore requires a different form of query processing that uses a best match function to select object variants.

The presentation of a content management system and an application developed using our version model has illustrated how we were able to address several concerns of context-aware systems in a uniform way through the concept of object variants. In the future, we plan to investigate the potential of this version model by integrating our platform with an existing model-based web design methodology.

References

1. Rossi, G., Schwabe, D., Guimarães, R.: Designing Personalized Web Applications. In: Proc. Intl. Conf. on World Wide Web, Hong Kong, China (2001)
2. Garrigós, I., Casteleyn, S., Gómez, J.: A Structured Approach to Personalize Websites Using the OO-H Personalization Framework. In: Proc. Asia Pacific Web Conf., Shanghai, China (2005)
3. Schewe, K.D., Thalheim, B.: Reasoning About Web Information Systems Using Story Algebras. In: Benczúr, A.A., Demetrovics, J., Gottlob, G. (eds.) ADBIS 2004. LNCS, vol. 3255, Springer, Heidelberg (2004)
4. Casteleyn, S., De Troyer, O., Brockmans, S.: Design Time Support for Adaptive Behavior in Web Sites. In: Proc. ACM Symp. on Applied Computing, Melbourne, FL, USA (2003)
5. Frăsincar, F., Barna, P., Houben, G.J., Fiala, Z.: Adaptation and Reuse in Designing Web Information Systems. In: Proc. Intl. Conf. on Information Technology: Coding and Computing, Las Vegas, NV, USA (2004)
6. Ceri, S., Fraternali, P., Bongio, A., Brambilla, M., Comai, S., Matera, M.: Designing Data-Intensive Web Applications. The Morgan Kaufmann Series in Data Management Systems. Morgan Kaufmann Publishers Inc., San Francisco (2002)
7. Ceri, S., Daniel, F., Matera, M., Facca, F.M.: Model-driven Development of Context-Aware Web Applications. ACM Trans. on Internet Technology 7(2) (2007)
8. Baumeister, H., Knapp, A., Koch, N., Zhang, G.: Modelling Adaptivity with Aspects. In: Lowe, D.G., Gaedke, M. (eds.) ICWE 2005. LNCS, vol. 3579, Springer, Heidelberg (2005)
9. Koch, N.: Software Engineering for Adaptive Hypermedia System. PhD thesis, Ludwig-Maximilians-University Munich, Munich, Germany (2000)
10. Fox, A., Brewer, E.A.: Reducing WWW Latency and Bandwidth Requirements by Real-time Distillation. Computer Networks and ISDN Systems 28(7-11) (1996)
11. Fox, A., Gribbe, S.D., Brewer, E.A., Amir, E.: Adapting to Network and Client Variability via On-Demand Dynamic Distillation. Computer Architecture News 24(Special Issue) (1996)
12. Smith, J.R., Mohan, R., Li, C.S.: Transcoding Internet Content for Heterogeneous Client Devices. In: Proc. Intl. Symp. on Circuits and Systems, Monterey, CA, USA (1998)
13. Singh, A., Trivedi, A., Ramamritham, K., Shenoy, P.: PTC: Proxies that Transcode and Cache in Heterogeneous Web Client Environments. World Wide Web 7(1) (2004)

14. Mohan, R., Smith, J.R., Li, C.S.: Adapting Multimedia Internet Content for Universal Access. IEEE Transactions on Multimedia 1(1) (1999)
15. Wadge, W.W., Brown, G., Schraefel, M.C., Yildirim, T.: Intensional HTML. In: Munson, E.V., Nicholas, C., Wood, D. (eds.) PODDP 1998 and PODP 1998. LNCS, vol. 1481, Springer, Heidelberg (1998)
16. Stavrakas, Y., Gergatsoulis, M., Rondogiannis, P.: Multidimensional XML. In: Proc. Intl. Workshop on Distributed Communities on the Web, Quebec City, Canada (2000)
17. Stavrakas, Y., Gergatsoulis, M.: Multidimensional Semistructured Data: Representing Context-Dependent Information on the Web. In: Pidduck, A.B., Mylopoulos, J., Woo, C.C., Ozsu, M.T. (eds.) CAiSE 2002. LNCS, vol. 2348, Springer, Heidelberg (2002)
18. Stavrakas, Y., Pristouris, K., Efandis, A., Sellis, T.: Implementing a Query Language for Context-Dependent Semistructured Data. In: Benczúr, A.A., Demetrovics, J., Gottlob, G. (eds.) ADBIS 2004. LNCS, vol. 3255, Springer, Heidelberg (2004)
19. WebRatio: Site Development Studio (2001), http://www.webratio.com
20. Frăsincar, F.: Hypermedia Presentation Generation for Semantic Web Information Systems. PhD thesis, Technische Universiteit Eindhoven, Eindhoven, The Netherlands (2005)
21. Fiala, Z., Frăsincar, F., Hinz, M., Houben, G.J., Barna, P., Meissner, K.: Engineering the Presentation Layer of Adaptable Web Information Systems. In: Proc. Intl. Conf. on Web Engineering, Munich, Germany (2004)
22. De Virgilio, R., Torlone, R.: Management of Heterogeneous Profiles in Context-Aware Adaptive Information System. In: Proc. OTM Workshop on the Move to Meaningful Internet Systems, Agia Napa, Cyprus (2005)
23. De Virgilio, R., Torlone, R., Houben, G.J.: A Rule-based Approach to Content Delivery Adaptation in Web Information Systems. In: Proc. Intl. Conf. on Mobile Data Management, Nara, Japan (2006)
24. Katz, R.H.: Toward a Unified Framework for Version Modeling in Engineering Databases. ACM Computing Surveys 22(4) (1990)
25. Conradi, R., Westfechtel, B.: Version Models for Software Configuration Management. ACM Computing Surveys 30(2) (1998)
26. Norrie, M.C.: An Extended Entity-Relationship Approach to Data Management in Object-Oriented Systems. In: Proc. Intl. Conf. on the Entity-Relationship Approach, Arlington, TX, USA (1994)
27. Belotti, R., Decurtins, C., Grossniklaus, M., Norrie, M.C., Palinginis, A.: Interplay of Content and Context. Journal of Web Engineering 4(1) (2005)
28. Grossniklaus, M., Norrie, M.C.: Information Concepts for Content Management. In: Proc. Intl. Workshop on Data Semantics and Web Information Systems, Singapore (2002)
29. Norrie, M.C., Signer, B., Grossniklaus, M., Belotti, R., Decurtins, C., Weibel, N.: Context-Aware Platform for Mobile Data Management. Wireless Networks (2007), http://dx.doi.org/10.1007/s11276-006-9858-y
30. Signer, B.: Fundamental Concepts for Interactive Paper and Cross-Media Information Spaces. PhD thesis, ETH Zurich, Switzerland (2006)
31. Norrie, M.C., Signer, B., Weibel, N.: Print-n-Link: Weaving the Paper Web. In: Proc. ACM Symp. on Document Engineering, Amsterdam, The Netherlands (2006)
32. Grossniklaus, M., Norrie, M.C.: Using Object Variants to Support Context-Aware Interactions. In: Proc. Intl. Workshop on Adaptation and Evolution in Web Systems Engineering, Como, Italy (2007)

Extending XML Triggers with Path-Granularity

Anders H. Landberg, J. Wenny Rahayu, and Eric Pardede

Department of Computer Science & Computer Engineering
La Trobe University, Melbourne, Australia
{ahlandberg@students.,w.rahayu@,e.pardede@}latrobe.edu.au

Abstract. Triggers are a well-founded concept in relational databases that provide reactive behaviour in response to database modifications [1] [2]. For XML databases, however, there still does not exist a standardised trigger mechanism. In order to make trigger functionality for XML databases as practical and applicable as it is in the relational context, the hierarchical nature of the XML data model must be considered. Trigger granularity is a fundamental concept that is closely related to the structure of the data that the trigger operates on. We must explore how update operations on particular XML document nodes impact the structure of the data. This paper addresses this issue and proposes path-level granularity, a novel extension to XML trigger granularity. It introduces definitions and methodologies for performing path-level granularity in XML triggers. An implementation of an XML trigger engine and case study are used as proof of concept. Further, a cost-based evaluation of the proposed concepts is also included.

1 Introduction

Trigger granularity is a fundamental and low-level underlying concept that discusses the level of detail in which modified data is to be treated by a trigger. This involves determining the chunks of information that make up a particular grain. As triggers provide reactive functionality in response to XML data updates, it is crucial to identify which parts within an XML document are impacted during the update [3]. This is so important because the structure of XML is hierarchical instead of flat. Therefore, in the XML context, one type of granularity may be the set including a particular target node together with its parent and child nodes [4]. The relationship within this set of nodes can be seen as a hierarchical relationship, as each node in the set is either hierarchically above or below others, or both.

When examining data with this granularity, each target node is not being considered by itself alone, but as part of this hierarchical relationship. Similar to how sibling nodes can be put into a horizontal relationship, namely to be on the same hierarchical level, nodes can also be put into a vertical relationship which considers neighbouring nodes on different hierarchical levels. The consideration of such a vertical, hierarchical node relationship in regards to trigger granularity is vital because it strongly draws focus to the concept of data hierarchy that is one of the major discrepancies between relational and XML data models.

B. Benatallah et al. (Eds.): WISE 2007, LNCS 4831, pp. 410–422, 2007.
© Springer-Verlag Berlin Heidelberg 2007

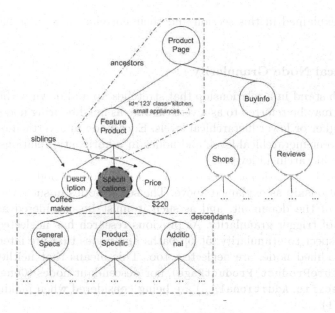

Fig. 1. Ancestor and descendant nodes

Consider Figure 1. The highlighted node Specifications has been modified by a database update operation. When making changes to Specifications, it is possibly necessary to make changes to attributes in nodes GeneralSpecs, ProductSpecific, ... Also, it is possibly necessary to even consider checking FeatureProduct and ProductPage. Also, it would be useful to have access to all of the modified node's leaf nodes in a quick and efficient manner.

This paper proposes *path-level granularity*, a novel concept for XML database triggers that considers vertical node relationships during updates on XML data. XML trigger with this new functionality will overcome current approaches' limitation in respect to node-relationships and granularity and make the trigger more applicable and useful for more complex constraint checking.

Organisation. Section 2 gives background information about trigger granularity. Section 3 proposes definitions and methodologies to specify the new level of granularity and apply it to the XML context. Sections 4 and 5 describe the implementation that was done in order to prove and evaluate the proposed concept. Related work is summarised in section 6 Finally, Section 7 concludes the paper and offers ideas for future work.

2 Preliminaries

Vertical node granularity is based on the ancestor and descendant paths of a particular node. It regards each node as belonging to a vertical relationship that includes ancestor and descendant nodes. Similarly, horizontal node granularity addresses a particular node and its siblings. These types of granularity will further be explained in this section and their correlation with triggers will be explored.

2.1   Vertical Node Granularity

Nodes which stand in a relationship that stretches up and down within the document tree, may be referred to as a vertical one, as neighbouring nodes are to be found on higher or lower hierarchical levels. Each node in a vertical relationship is on a different hierarchical level, and nodes in a horizontal relationship are all on the same hierarchical level.

A path can be regarded as a set of related nodes, namely nodes that belong to vertical relationship as explained above. The set of nodes in such a relationship is a subset of the document, and as such it must be considered as an additional level of trigger granularity. As previous research has neglected the path notion in respect to granularity [5] [6], paths and nodes that are interconnected with the modified node, are neglected, too. This means that neither ancestor nodes (FeatureProduct, ProductPage), nor descendant nodes (GeneralSpecs, ProductSpecific, Additional, ...) are being considered when making updates (see Figure 1).

2.2   Horizontal Node Granularity

Horizontal node granularity considers nodes which are on one and the same hierarchical level, and as such stand in a horizontal relationship towards one another. As for the granularity, current approaches introduce two different levels, namely node-level and document-level granularities [7]. They represent the XML equivalent of row and statement granularities in the relational context. Node-level granularity considers each trigger event separately, and the XPath expression in the trigger's event part is associated with a single event. Document-level granularity assumes that each trigger considers at once all relevant event instances.

Hence, a document-level trigger is fired once for all modified nodes. Although this mapping from the relational to the XML context works for some cases, it neglects the basic discrepancies between the different data models. The next paragraphs give insight in resulting issues that arise from these differences between relational and XML data in respect to triggers.

In Figure 2, nodes 10 and 11 have been modified by an update statement and have been matched by a trigger event path. In the left example, the (node-level) trigger will be fired twice for node 10 and 11. In the right example, the (document-level) trigger will be fired once for both nodes.

As can clearly be seen, neither of these existing approaches to granularity solve the above described problem, as they only consider horizontal node relationships and do not focus on vertical node relationships. The next section provides definitions that can be used to describe path-level granularity, and equip XML triggers with this novel concept which makes it aware of the previously explained vertical node relationships.

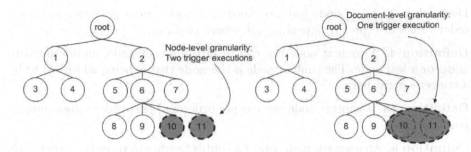

Fig. 2. Node and Document level granularity

3   Proposed Method

This section formalises and defines the notion of *context path*, explains how context paths are identified, and introduces a methodology for incorporating path-level granularity into XML triggers. The background of the proposed method is based on the fact that an XML document can be represented as a tree of interconnected nodes which are hierarchically ordered.

When a node in this tree is modified, it is necessary to consider nodes that are hierarchically above and below it, as these nodes are directly or indirectly related to it. To realise this, the method considers all parent and child nodes of the modified node. Such a set of parent and child nodes together with the modified node will be referred to as a path and will be defined in the next paragraphs.

Choosing path-level as an additional level of granularity is necessary to make the trigger aware of interconnected nodes throughout the document tree and to conform to the XML tree structure. Further, it improves the trigger functionality by adjusting it to the XML data model's peculiarities.

3.1   Definitions

Following definitions are necessary to describe a context path. First, it is important that the nodes in a document tree are be labelled in a way that uniquely identifies them. Also, the labelling must allow further definitions being dependent on it. In our approach, we use the pre-order and post-order scheme [8] [9].

Notation. A '/' is a single separator between two nodes A and B, where B is a child node of A.

Definition 1. A root node *root* is defined as a node whose pre-order and post-order values are $(1, 2n)$, where n is the number of nodes in the tree.

Definition 2. An intermediate node *interm* is defined as a node whose pre-order and post-order values are $(\text{pre}_{interm}, \text{post}_{interm})$, where $\text{pre}_{interm} > 1$ and $\text{post}_{interm} < 2n$ and $\text{post}_{interm} - \text{pre}_{interm} > 1$.

Definition 3. A leaf node *leaf* is defined as a node whose pre-order and post-order values are (pre_{interm},$post_{interm}$), where $post_{interm}$ - pre_{interm} = 1.

Definition 4. A context node *con* can be either a root node, an intermediate node, or a leaf node. The context node is the node that is being addressed to by a trigger's event path.

Definition 4a. A context node *con* has pre-order and post-order values (pre_{con}, $post_{con}$).

Definition 5. An ancestor node *anc* of a context node can be either a root node or an intermediate node.

Definition 5a. An ancestor node *anc* of a context node has pre-order and post-order values (pre_{anc},$post_{anc}$) where pre_{anc} < pre_{con} and $post_{anc}$ > $post_{con}$.

Definition 6. A descendant node *desc* of a context node can be either an intermediate node or a leaf node.

Definition 6a. A descendant node *desc* of a context node has pre-order and post-order values (pre_{desc},$post_{desc}$) where pre_{desc} > pre_{con} and $post_{desc}$ < $post_{con}$.

Definition 7. A *Valid Path* $VP = N_1/N_2/N_3/.../N_m/$, with length m, where N_1 is a root node, and node N_{i+1} is a descendant node of N_i, and N_m is a leaf node ($1 \leq i \leq m\text{-}1$).

Definition 8. A *Context Path CP* is a valid path that contains a context node.

Definition 8a. A *Context Path* $CP = \{A,c,D\}$, where the *set of ancestor nodes* $A = \{a_1,a_2,a_3,...\}$ and the context node c and the *set of descendant nodes* $D = \{d_1,d_2,d_3,...\}$.

Definition 9. The number of context paths of a context node is equal to the number of leaf nodes that are descendants of that context node.

Definition 9a. A root context node has k context paths, where k = *all valid paths in the tree*. Hence, context paths = valid paths.

Definition 9b. A leaf context node has 1 context path.

Using the above listed definitions, it is now possible to label an XML document tree, and to identify context paths relative to a particular context node. Further, it is also possible to determine the number of context paths. With these definitions, we can now describe an XML document by listing its nodes, and describe context nodes and context paths within it.

To illustrate the above defined node-labelling scheme and context node and context path definitions, consider Figure 3. In this example, all nodes have a pre-order and post-order value associated to them, as well as some additional labels that specify their relative position in the document. The context node *con* is marked with a bold circle, and all nodes that stand in a vertical node relationship to it are shaded in grey colour. All the nodes that are coloured grey

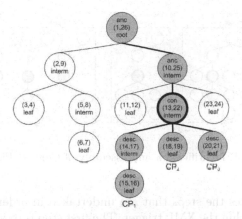

Fig. 3. Labelled document tree

will be considered by the path-level granularity. The remaining nodes are of no importance. The context node with labelling (13,22) has the following context paths related to it:

$CP_1 = /(1,26)/(10,25)/(13,22)/(14,17)/(15,16)$
$CP_2 = /(1,26)/(10,25)/(13,22)/(18,19)$
$CP_3 = /(1,26)/(10,25)/(13,22)/(20,21)$

Note that the total number of context paths within a document is equal to the sum of context path of all context nodes. So given that there are n context nodes with p_n context paths each, then the total number of context paths is $p_1 + p_2 + ... + p_{n-1} + p_n$.

The following part of the section will now apply these definitions in order to realise the functionality of path-level granularity. Also, it explains how the notion of a context path is used in correlation with XML triggers.

3.2 Trigger Methodology

Based on the proposed concepts presented above, this section outlines how we use the above concepts to form a path-level trigger methodology. Our main contributions aim to extend XML trigger methodologies, and are as follows:

1. Applying the concept of a context path into trigger execution based on the context path definition
2. Expressing the new path-granularity using a proposed syntax

First, we will discuss the concept of context paths. To do this, it is necessary to analyse which nodes in a document tree will possibly be affected by an update. As defined previously, the nodes in an XML document that are modified during an insert, delete, or update operation and match a trigger's event path, are addressed to as context nodes. In existing research papers, context nodes are often referred to as affected nodes [6] [10]. Similarly, a context document is a document that is modified, i.e. that contains context nodes.

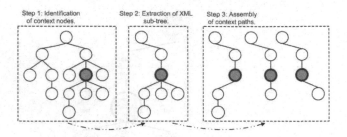

Fig. 4. Identification and assembly of Context Paths

Figure 4 illustrates the steps that are undertaken in order to perform path-level granularity within the XML trigger. The first step is to identify the context node. This is the node, or set of nodes, that is affected by an update statement, such as an insertion, deletion, or modification of nodes. All ancestor and descendant nodes of the context node are extracted, they represent a sub-tree of the entire document tree. The second step traverses this sub-tree and assembles the context paths. These context paths are then available to the trigger in the form of transition variables during execution.

The advantage of this level of granularity is that content of various node types in an entire document sub-tree can be considered by a single trigger. Also, it has the advantage that all relevant paths that traverse the context node are considered. A path-level trigger will therefore create one event instance for every path that traverses the context node. Each of these events is considered separately by the trigger.

Next, we will focus on the syntax that is necessary to describe a trigger that incorporates path-level granularity. The below listing shows the general syntax for creating an XML trigger. The expression DO FOR EACH [doc|path|node] on line 5 in the listing specifies the level of granularity. *doc* and *node* denote document-level and node-level granularities respectively, and *path* denotes path-level granularity.

When a trigger is fired and after its condition (refer to line 4 in the below listing) has evaluated to true, it will be executed according to its level of granularity (see line 5). In accordance with our definitions, a trigger with path-level granularity will therefore be fired once for each context path that traverses the context node. If the number of context nodes in a document is n, then the trigger will be fired for each context path of each of the n context nodes. When a trigger is fired, its action body, denoted by <XQuery-expr> in the listing, is executed according to the level of granulariy.

XML Trigger Syntax

```
1. CREATE OR REPLACE XMLTRIGGER <trigger-name>
2. ON [insert|modify|delete]
3. OF doc(<document-name>)/<update-path>
4. IF <XPath-qualifier>
```

```
5. DO FOR EACH [doc|path|node] [<XQuery-expr>]
6. END XMLTRIGGER;
```

This section has explained the methodologies for performing path-level granularity in XML triggers. The notion of a context path is hereby used to specify the vertical node relationships between a context node and its related nodes. In order to examine these context paths, a trigger with this new level of granularity will be executed once for each context path that has been identified in a document during an update operation. The next section introduces a case study that was conducted to prove the proposed concept. As there is no trigger functionality available in existing XML databases, a wrapper program was developed, which represents a trigger engine and incorporates path-level granularity as well as the above suggested XML trigger syntax.

4   Implementation

Figure 5 gives a rough overview over the components that were necessary to realise the XML trigger support prototype. The user interface component was added in to make interaction with the database and the trigger functionalities easier and better visible to the user. The BDB XML component represents the Oracle Berkeley XML database [11] which can be accessed by the XML trigger support system using the BDB XML Connection component. It was necessary to develop two parsers in order to support (i) the XML trigger syntax (to create a new XML trigger) and (ii) the update syntax which had to be adjusted to be more suitable for the system.

Although the main functionalities all are carried out by the XML trigger engine component, several additional classes with helper- and utility methods were developed to support the XML trigger functionality. Figure 6 shows a screen shot of the interaction with the command line prompt when creating a new XML trigger.

The communication between the components works as follows. After the user enters a command which is recognised by the system, it is sent to the parser component for further processing. If the user has entered a trigger create statement or a database insert- or delete statement, the parser will return the respective trigger- or query object. This parsed input is then passed to the XML trigger engine which determines the type of database operation and initiates associated trigger processing. When data needs to be analysed or evaluated which resides in the database, then the BDB XML Connection component is invoked. It has methods that enable access to the XML database (represented by the BDB XML component) in various ways.

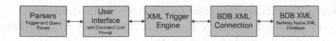

Fig. 5. Trigger support system architecture

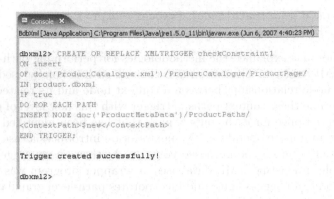

Fig. 6. Create a path-level XML trigger

The XML trigger engine is the core component of the XML trigger support system. Its main functionalities include the detection of triggers that can be fired in response to an update statement, the evaluation of these possible triggers, and the execution of triggers whose evaluation was successful.

Implementation tools and system. The prototype is implemented in JAVA SE 5.0 and utilises an existing API to interact with the Berkeley XML database. The Eclipse IDE 3.2.2 was used to develop the system.

5   Case Study and Evaluation

This section demonstrates how XML triggers with the new path-level granularity can be applied in a database system. The proposed syntax is used to implement the XML triggers. Then, tests on real XML data will be conducted. As the database system that will be used does not support the new trigger functionality yet, this case study attempts to simulate and work-around the limitations.

The tests described in this section were conducted using the above described implementation of a trigger engine on a machine equipped with a 2.0GHz Intel Pentium M 760 processor and 1GB DDR2 memory, running Windows XP professional as operating system.

5.1   Case Study

In the experiments (illustrated in Figure 7), we used an XML document that represents a product catalogue. For each product, there is a `ProductPage` record in the document. Whenever a new product is added or taken out of the catalogue, the respective element within the document is inserted or deleted. Hence, there are two different operations that can take place in this database.

Two triggers are used to check insertion and deletion of product elements inside the catalogue document, and to take appropriate action under certain conditions. Both triggers use path-level granularity and are applied to the product catalogue XML document.

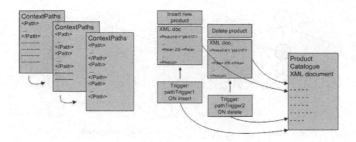

Fig. 7. Product catalogue scenario

Updates. The update operations that modify the database are as described above, inserting and deleting a product element.

Triggers. The first trigger will be applied to a `ProductPage` element and is fired every time a new product is being inserted into the database. Its condition is set to `true`, which means that we do not perform any further condition checking, but execute the trigger immediately. As the event path of both these triggers is set to be the `ProductPage` element, it will be executed once for each context path which runs through the context node. The context node in this case is the `ProductPage` node. Likewise, when deleting a document, the trigger that checks for product deletions will also be applied to the `ProductPage` node. The context paths that are identified by the trigger are XML fragments, which contain all nodes beginning from the root node, traversing through the context node, and ending in a leaf node below the `ProductPage` node.

For this experiment, we have chosen both triggers' action parts to extract the leaf node of the respective context path, record its value, and then check if there are `idref` attributes which can possibly refer to nodes in the same context path. For this example, the trigger will not carry out any further actions, as we are primarily interested in the correct identification of context paths.

5.2 Evaluation

In this section, we evaluate the effectiveness of the proposed trigger methodology. We compare it to a more traditional approach, whereby the trigger is replaced by a sequence of manual XQuery updates. Figure 8 shows how the comparison is made against manual XQuery updates which perform similar or the same functionality. As can be seen, the traditional approach requires context path traversals of all possible context nodes. In the new method, the context node is known, and using our proposed trigger methodology, the trigger will only be executed on the relevant context paths.

We are experimenting with three different types of XML trees, as shown in Figure 9. The first dimension of the experiments' parameters is represented by different types of sample tree structures: deep and narrow, shallow and wide, deep and wide. The second dimension specifies whether: number of possible context nodes is equal to number of context nodes, number of possible context nodes

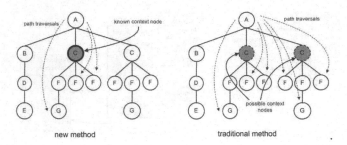

Fig. 8. Tree traversal for identifying context paths

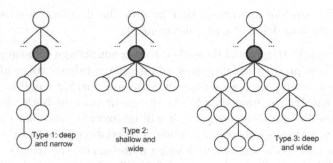

Fig. 9. Tree structures used in the experiments

is greater than number of context nodes. The sample tree structures are auto-generated and illustrated in Figure 9. Figure 8 illustrates the possible context nodes versus actual, identified context nodes.

Figure 10 shows the results of our experiments. The efficiency is evaluated in terms of total cost units. The costs are represented in four major operations, namely: path assemblies, parent traversal, child traversal and node checks.

Referring to graph (a) in Figure 10, we observe that the new method requires only slightly less operations than the traditional approach. The reason for this is that there are no redundant path traversals to be performed. Graphs (b) and (c) show that with increasing number of possible context nodes, the traditional method is clearly outperformed by the new approach, particularly for deep and wide trees. As this kind of tree represents the most common XML document structure, the path-granularity trigger will be of great benefit when a large number of context nodes are evaluated.

Summarising the above graphs, it can be said that if the number of context nodes for which the trigger executes is equal to the number of possible context nodes, the traditional method performs nearly at the same efficiency, as nearly no redundant traversals and node checks are necessary. However, with increasing difference between possible context nodes and identified context nodes, the new method clearly outperforms the traditional approach. The graphs also show

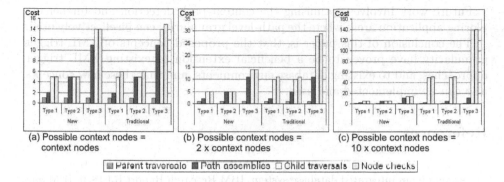

(a) Possible context nodes = context nodes

(b) Possible context nodes = 2 x context nodes

(c) Possible context nodes = 10 x context nodes

Parent traversals   Path assemblies   Child traversals   Node checks

Fig. 10. Evaluation results

which costs each type of tree incurs proportionally. Given that the cost of assembling the context paths is the most expensive operation, deep and narrow trees will benefit most from the new method.

6   Related Work

Despite the big differences between relational and XML data, the levels of granularity were almost directly adopted from the relational model by all existing literature. What is in the relational context row-level and statement-level has now become node-level and document-level granularity, as proposed by James Bailey et al [5]. Some approaches by Bonifati refer to document-granularity as set-granularity [10] or statement-granularity [7]. Rekouts only proposes the node-level granularity for the reason of simplicity in their approach [6]. Although different terms are used, they describe one and the same thing. So is the node-level trigger equivalent to the row-level trigger and the document- or set-level trigger equivalent to the statement- level trigger, for XML and relational contexts respectively. Other levels of granularity have not been proposed.

As a summary, it can be said that the concept of trigger granularity has not been explored into enough detail. Yet it is one of the most fundamental aspects that must be considered for XML triggers.

7   Summary and Future Work

This paper has proposed path-level granularity, a new concept for XML triggers. The definitions and methodologies that are introduced in this paper were used to incorporate path-level granularity into XML trigger execution. Hence, the triggers' functionality has been extended, so that context nodes can be identified, and related context-paths can be extracted during trigger execution. An implementation of an XML trigger engine and a case study were used to demonstrate the proposed concepts. The cost-based evaluation shows the new approach in comparison to an alternative, traditional approach.

Future work involves optimising the context-path detection and further exploring application areas where the path-level granularity can be effectively used.

The conclusion of this paper is, that the proposed concepts of XML trigger granularity have proven to be a beneficial extension to existing XML trigger approaches, and to existing database environments, and that these new concepts also can be implemented and used.

References

1. Eswaran, K.P.: Specifications, implementations and interactions of a trigger subsystem in an integrated database system. IBM Research Report RJ 1820, IBM San Jose Research Laboratory, San Jose, California (1976)
2. Ceri, S., Cochrane, R., Widom, J.: Practical Applications of Triggers and Constraints: Success and Lingering Issues (10-Year Award). In: VLDB Conference, pp. 254–262 (2000)
3. Tatarinov, I., Ives, Z.G., Halevy, A.Y., Weld, D.S.: Updating XML. In: SIGMOD Conference, pp. 413–424 (2001)
4. Barbosa, D., Mendelzon, A.O., Libkin, L., Mignet, L., Arenas, M.: Efficient Incremental Validation of XML Documents. In: ICDE, pp. 671–682 (2004)
5. Bailey, J., Poulovassilis, A., Wood, P.T.: An event-condition-action language for XML. In: WWW Conference, pp. 486–495 (2002)
6. Rekouts, M.: Incorporating Active Rules Processing into Update Execution in XML Database Systems. In: DEXA Workshops, pp. 831–836 (2005)
7. Bonifati, A., Braga, D., Campi, A., Ceri, S.: Active XQuery. In: ICDE, pp. 403–412 (2002)
8. Grust, T.: Accelerating XPath location steps. In: SIGMOD Conference, pp. 109–120 (2002)
9. Jagadish, H.V., Al-Khalifa, S., Chapman, A., Lakshmanan, L.V.S., Nierman, A., Paparizos, S., Patel, J.M., Srivastava, D., Wiwatwattana, N., Wu, Y., Yu, C.: TIMBER: A native XML database. VLDB Journal 11(4), 274–291 (2002)
10. Bonifati, A., Ceri, S., Paraboschi, S.: Active rules for XML: A new paradigm for E-services. VLDB Journal 10(1), 39–47 (2001)
11. Brian, D.: The Definitive Guide to Berkeley DB XML (Definitive Guide). Published by Apress, Berkely, CA, USA (2006), ISBN = 1590596668

A Replicated Study Comparing Web Effort Estimation Techniques

Emilia Mendes[1], Sergio Di Martino[2], Filomena Ferrucci[2], and Carmine Gravino[2]

[1] The University of Auckland, Private Bag 92019
Auckland, New Zealand, 0064 9 3737599 ext. 86137
emilia@cs.auckland.ac.nz
[2] The University of Salerno, Via Ponte Don Melillo, I 84084 Fisciano (SA)
Italy, 0039 089963374
{sdimartino,fferrucci,gravino}@unisa.it

Abstract. The objective of this paper is to replicate two previous studies that compared at least three techniques for Web effort estimation in order to identify the one that provides best prediction accuracy. We employed the three effort estimation techniques that were mutual to the two studies being replicated, namely Forward Stepwise Regression (SWR), Case-Based Reasoning (CBR) and Classification & Regression Trees (CART). We used a cross-company data set of 150 Web projects from the Tukutuku data set. This is the first time such large number of Web projects is used to compare effort estimation techniques. Results showed that all techniques presented similar predictions, and these predictions were significantly better than those using the mean effort. Thus, all the techniques can be exploited for effort estimation in the Web domain, also using a cross-company data set that is specially useful when companies do not have their own data on past projects from which to obtain their estimates, or that have data on projects developed in different application domains and/or technologies.

Keywords: Cost estimation, effort estimation, stepwise regression, case-based reasoning, regression tree, Web projects, Web applications.

1 Introduction

Web development, despite being a relatively young industry, initiated just 13 years ago, currently represents a market that increases at an average rate of 20% per year, with Web e-commerce sales alone surpassing 95 billion USD in 2004 (three times the revenue from the world's aerospace industry)[1][19]. Unfortunately, in contrast, most Web development projects suffer from unrealistic project schedules, leading to applications that are rarely developed on time and within budget [19]. Therefore Web effort estimation is a very important area within the field of Web engineering and evidence from empirical studies is fundamental in order to build a body of knowledge in Web engineering, which can be used to inform practitioners and researchers alike.

---

[1] http://www.aia-erospace.org/stats/aero_stats/stat08.pdf
http://www.tchidagraphics.com/website_ecommerce.htm

B. Benatallah et al. (Eds.): WISE 2007, LNCS 4831, pp. 423–435, 2007.
© Springer-Verlag Berlin Heidelberg 2007

To date numerous studies have investigated techniques and/or measures for Web effort estimation [1][5][6][9][10][11][12][13][14][15][16][18][19][21], however of these only two studies [5][18] have compared Web effort estimation techniques using three or more effort estimation techniques. Their data sets, size measures, choice of cross-validation, and some of the techniques differed slightly, however there were also similarities: i) neither used costs drivers, in addition to size measures, to estimate effort, i.e., the only independent variables employed were size measures; ii) Length was a class of size measures [12] common to both studies (examples of Length measures are number of Web pages, number of images/medias, number of internal link); iii) the effort estimation techniques Stepwise Regression (SWR), Case-based Reasoning (CBR) and Classification and Regression Trees (CART) were also used in both studies; iv) both studies looked at a common research question: Which of the techniques employed gives the most accurate predictions for the data set?; v) Prediction accuracy was measured in both case studies using the three common accuracy measures, namely the Mean Magnitude of Relative Error (MMRE), Prediction at level 25 (Pred(25)), and Median Magnitude of Relative Error (MdMRE) [22].

Costagliola et al. [5] found that none of the effort estimation techniques used in their study was statistically significantly superior; however, the three accuracy measures used suggested that Stepwise Regression presented overall the best and the worst accuracy using elements of Functionality and Length size measures, respectively. The class of Functionality measures included the nine components that are part of Web Objects [19]. Mendes et al. [18] found that Stepwise Regression showed statistically significantly superior predictions than the other techniques when using Length size measures. These results seem contradictory since Length size measures used in combination with SWR provided the worst results in [5], based on accuracy measures, and the best results in [18]. However, the data sets used in these two studies differed largely: Mendes et al. [18] used a data set containing data on 37 Web hypermedia applications developed by postgraduate and MSc students attending a Hypermedia and Multimedia Systems course at the University of Auckland (NZ), whereas Costagliola et al. [5] used a data set containing data on 15 real Web software applications developed by a single Web company.

This paper replicates these two previous studies, using the three common effort estimation techniques they employed, and a data set containing a mix of data on 150 Web hypermedia and Web software applications from the Tukutuku database [11]. Each Web project in the Tukutuku database (see Section 2) is characterised using Length & Functionality size measures, and also some cost drivers. These size measures and cost drivers were obtained from the results of a survey investigation [11], using data from 133 on-line Web forms aimed at giving quotes on Web development projects, and were also confirmed by an established Web company and a second survey involving 33 Web companies in New Zealand. Therefore it is our belief that these size measures and cost drivers are both meaningful to Web companies and as a consequence both should be included in our analysis.

The remainder of the paper is organised as follows: Section 2 describes the data set used for the case study. The results of the empirical analysis obtained using SWR, CBR, and CART are presented in Section 3. A discussion of the results is provided in Section 4, and conclusions are given in Section 5.

2 Data Set Description

The analysis presented in this paper was based on data from the Tukutuku database [11], which aims to collect data from completed Web projects to develop Web cost estimation models and to benchmark productivity across and within Web Companies. The Tukutuku includes Web hypermedia systems and Web applications [3]. The former represents static applications using a very rich navigational structure. Conversely, the latter represents software applications that depend on the Web or use the Web's infrastructure for execution and are characterized by functionality affecting the state of the underlying business logic.

The Tukutuku database has data on 150 projects where:

- Projects come from 10 different countries, mainly New Zealand (56%), Brazil (12.7%), Italy (10%), Spain (8%), United States (4.7%), and England (2.7%).
- Project types are new developments (56%) or enhancement projects (44%).
- Applications are mainly Legacy integration (27%), Intranet and eCommerce (15%).
- Languages used are mainly HTML (88%), Javascript (DHTML/DOM) (76%), PHP (50%), Graphics Tools (39%), ASP (VBScript, .Net) (18%), and Perl (15%).

Each Web project in the database was characterized by 25 variables, related to the application and its development process (see Table 1).

Table 1. Variables for the Tukutuku database

Variable Name	Scale	Description
COMPANY DATA		
Country	Categorical	Country company belongs to.
Established	Ordinal	Year when company was established.
nPeople	Ratio	Number of people who work on Web design and development.
PROJECT DATA		
TypeProj	Categorical	Type of project (new or enhancement).
nLang	Ratio	Number of different development languages used
DocProc	Categorical	If project followed defined and documented process.
ProImpr	Categorical	If project team involved in a process improvement programme.
Metrics	Categorical	If project team part of a software metrics programme.
DevTeam	Ratio	Size of a project's development team.
TeamExp	Ratio	Average team experience with the dev. language(s) employed.
TotEff	Ratio	Actual effort in person hours used to develop a Web application.
EstEff	Ratio	Estimated effort in person hours to develop a Web application.
Accuracy	Categorical	Procedure used to record effort data.
WEB APPLICATION		
TypeApp	Categorical	Type of Web application developed.
TotWP	Ratio	Total number of Web pages (new and reused).
NewWP	Ratio	Total number of new Web pages.
TotImg	Ratio	Total number of images (new and reused).
NewImg	Ratio	Total number of new images created.
Fots	Ratio	Number of features reused without any adaptation.
HFotsA	Ratio	Number of reused high-effort features/functions adapted.
Hnew	Ratio	Number of new high-effort features/functions.
TotHigh	Ratio	Total number of high-effort features/functions
FotsA	Ratio	Number of reused low-effort features adapted.
New	Ratio	Number of new low-effort features/functions.
TotNHigh	Ratio	Total number of low-effort features/functions

Within the context of the Tukutuku project, a new high-effort feature/function requires at least 15 hours to be developed by one experienced developer, and a high-effort adapted feature/function requires at least 4 hours to be adapted by one experienced developer. These values are based on collected data.

Summary statistics for the numerical variables from the Tukutuku database are given in Table 2, and Table 3 summarises the number and percentages of projects for the categorical variables.

Table 2. Summary Statistics for numerical variables

	Mean	Median	Std. Dev.	Min.	Max.
nlang	3.75	3.00	1.58	1	8
DevTeam	2.97	2.00	2.57	1	23
TeamExp	3.57	3.00	2.16	1	10
TotEff	564.22	78.00	1048.94	1	5000
TotWP	81.53	30.00	209.82	01	2000
NewWP	61.43	14.00	202.78	0	1980
TotImg	117.58	43.50	244.71	0	1820
NewImg	47.62	3.00	141.67	0	1000
Fots	2.05	0.00	3.64	0	19
HFotsA	12.11	0.00	66.84	0	611
Hnew	2.53	0.00	5.21	0	27
totHigh	14.64	1.00	66.59	0	611
FotsA	1.91	1.00	3.07	0	20
New	2.91	1.00	4.07	0	19
totNHigh	4.82	4.00	4.98	0	35

Table 3. Summary of number of projects and percentages for categorical variables

Variable	Level	Num. Projects	% Projects
TypeProj	Enhancement	66	44
	New	84	56
DocProc	No	53	35.3
	Yes	97	64.7
ProImpr	No	77	51.3
	Yes	73	48.7
Metrics	No	85	56.7
	Yes	65	43.3

As for data quality, Web companies that volunteered data for the Tukutuku database did not use any automated measurement tools for effort data collection. Therefore in order to identify guesstimates from more accurate effort data, we asked companies how their effort data was collected (see Table 4).

Table 4. How effort data was collected

Data Collection Method	# of Projects	% of Projects
Hours worked per project task per day	93	62
Hours worked per project per day/week	32	21.3
Total hours worked each day or week	13	8.7
No timesheets (guesstimates)	12	8

At least for 83% of Web projects in the Tukutuku database effort values were based on more than guesstimates.

3 Effort Estimation Techniques

The three effort estimation techniques were applied to the same subset of 120 randomly selected projects from the Tukutuku database (*training set*). The remaining 30 Web projects from the database (*validation set*) were used later to measure the techniques' prediction accuracy. There is no agreement as to what should be the ideal size for the training and validation sets. Therefore our choice was subjective.

3.1 Regression-Based Web Effort Model

Stepwise Regression (SWR) [8] is a statistical technique whereby a prediction model (Equation) is built to represent the relationship between independent and dependent variables. This technique aims to find the set of independent variables (predictors) that best explains the variation in the dependent variable (response) by adding, at each stage, the independent variable with the highest association to the dependent variable, taking into account all variables currently in the model. Before building a model it is important to ensure assumptions related to using such technique are not violated [8]. The One-Sample Kolmogorov-Smirnov Test (K-S test) confirmed that none of the numerical variables were normally distributed, and so were transformed into a natural logarithmic scale to approximate a normal distribution. Several variables were clearly related to each other (e.g. TotWP and NewWP; TotImg and NewImg; totNHigh, FotsA and New) thus we did not use the following variables in the stepwise regression procedure:

- *TotWP* – associated to *NewWP*.
- *TotImg* – associated to *NewImg*.
- *Hnew* and *HFotsA* – associated to *TotHigh*; both present a large number of zero values, which leads to residuals that are heteroscedastic.
- *New* and *FotsA* – associated to *TotNHigh*; both also present a large number of zero values, which leads to residuals that are heteroscedastic.

We created four dummy variables, one for each of the categorical variables *TypeProj*, *DocProc*, *ProImpr*, and *Metrics*.

To verify the **stability** of the effort model the following steps were used [8]:

- Residual plot showing residuals vs. fitted values to check if residuals are random and normal.
- Cook's distance values to identify influential projects. Those with distances higher than 4/n are removed to test the model stability. If the model coefficients remain stable and the goodness of fit improves, the influential projects are retained in the data analysis.

The prediction **accuracy** of the regression model was checked using the 30 projects from the validation set, based on the raw data (not log-transformed data).

The regression model selected six significant independent variables: *LTotHigh*, *LNewWP*, *LDevTeam*, *ProImprY*, *LNewImg*, and *Lnlang*. Its adjusted R^2 was 0.80. The residual plot showed several projects that seemed to have very large residuals, also confirmed using Cook's distance. Nine projects had their Cook's distance above the cut-off point (4/120). To check the model's stability, a new model was generated without the nine projects that presented high Cook's distance, giving an adjusted R^2 of 0.857. In the new model the independent variables remained significant and the coefficients had similar values to those in the previous model. Therefore, the nine high influence data points were not permanently removed. The final equation for the regression model is described in Table 5.

Table 5. Best Model to calculate LTotEff

Independent Variables	Coeff.	Std. Error	t	p>|t|
(Constant)	1.636	.293	5.575	.00
LTotHigh	.731	.119	6.134	.00
LNewWP	.259	.068	3.784	.00
LDevTeam	.859	.162	5.294	.00
ProImprY	-.942	.208	-4.530	.00
LNewImg	.193	.052	3.723	.00
Lnlang	.612	.192	3.187	.002

When transformed back to the raw data scale, we get the Equation:

$$TotEfft = 5.1345\ TotHigh^{0.731} NewWP^{0.259} DevTeam^{0.859} e^{-0.942 ProImpr} NewImg^{0.193} nlang^{0.612} \qquad (1)$$

The residual plot and the P-P plot for the final model suggested that the residuals were normally distributed.

3.2 CBR-Based Web Effort Estimation

Case-based Reasoning (CBR) is a branch of Artificial Intelligence where knowledge of similar past cases is used to solve new cases [22]. Herein completed projects are characterized in terms of a set of p features (e.g. *TotWP*) and form the case base. The new project is also characterized in terms of the same p attributes and it is referred as the target case. Next, the similarity between the target case and the other cases in the p-dimensional feature space is measured, and the most similar cases are used, possibly with adaptations to obtain a prediction for the target case. To apply the method, we have to select: the relevant project features, the appropriate similarity function, the number of analogies to select the similar projects to consider for estimation, and the analogy adaptation strategy for generating the estimation. The similarity measure used in this study is the one used in both [5][18], namely the Euclidean distance. Effort estimates were obtained using the five adaptations used in both [5][18]: effort for the most similar project in the case base (A1); the average of the two (A2) and three (A3) most similar projects; the inverse rank weighted mean of the two (IRWM2) and the three (IRWM3) most similar cases. In addition, common to [5][18], all project attributes considered by the similarity function had equal influence upon the selection of the most similar project(s). We used the commercial CBR tool CBR-Works version 4 to obtain effort estimates.

3.3 CART-Based Web Effort Estimation

The objective of CART [2] models is to build a binary tree by recursively partitioning the predictor space into subsets where the distribution of the response variable is successively more homogeneous. The partition is determined by splitting rules associated with each of the internal nodes. Each observation is assigned to a unique leaf node, where the conditional distribution of the response variable is determined. The best splitting for each node is searched based on a "purity" function calculated from the data. The data is considered to be pure when it contains data samples from only one class. The least squared deviation (LSD) measure of impurity was applied to our data set. This index is computed as the within-node variance, adjusted for frequency or case weights (if any). In this study we used all the variables described in Tables 2 and 3. We set the maximum tree depth to 10, the minimum number of cases in a parent node to 5 and the minimum number of cases in child nodes to 2. We looked to trees that gave the small risk estimates (SRE), which were set at a minimum of 90%, and calculated as:

$$SRE = 100 * \left(1 - \frac{node-error}{explained-variance}\right) \quad (2)$$

where *node-error* is calculated as the within-node variance about the mean of the node. *Explained-variance* is calculated as the within-node (error) variance plus the between-node (explained) variance. By setting the SRE to a minimum of 90% we believe that we have captured the most important variables.

Our regression trees were generated using SPSS Answer Tree version 2.1.1, and our final regression tree is shown in Fig. 1.

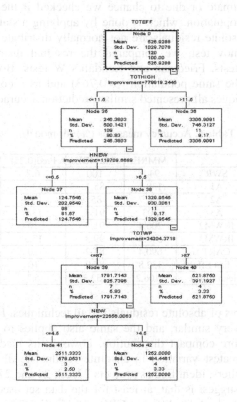

Fig. 1. CART Tree used to obtain effort estimates

4 Comparing Prediction Techniques

The 30 Web projects in the validation set were used to measure the prediction accuracy of the three different techniques presented in this paper. As mentioned above, to assess the accuracy of the derived effort prediction models, we have employed de facto standard accuracy measures, such as the mean Magnitude of Relative Error (MMRE), median MRE (MdMRE), and Prediction at 25% (Pred(25)) [4].

Here we compare the prediction between effort techniques taking into account first, the accuracy statistics and second, the boxplots of absolute residuals, where residuals are calculated as (actual effort – estimated effort). Moreover, as suggested by Mendes and Kitchenham [6][9] we also analyzed MMRE, MdMRE, and Pred(0.25) obtained by considering a model based on the mean effort (i.e., MeanEffort) and the median effort (i.e., MedianEffort) as predicted values. The aim is to have a benchmark to assess whether the estimates obtained with SWR, CBR, and CART are significantly better than estimates based on the mean or median effort.

Table 6 shows their MMRE, MdMRE and Pred(25). MMRE seems to indicate large differences in accuracy between SWR and CART/IRWM3; however, the median MRE (MdMRE) accuracy seems similar throughout all techniques. Pred(25) also shows some difference between SWR and the other techniques, however the difference is not large. To verify if the differences observed using MMRE, MdMRE and Pred(25) are legitimate or due to chance we checked if the absolute residuals came from the same population, which is done by applying a statistical significance test. Given that the absolute residuals are not normally distributed, confirmed using the Kolmogorov-Smirnov test, we compared the residual distributions with two nonparametric paired tests, Friedman and Kendall's W tests. Both showed that all residuals came from the same population (p=0.763) and as a consequence that the effort estimation techniques all presented similar prediction accuracy.

Table 6. Accuracy measures (%) for models

	MMRE	MdMRE	Pred(0.25)
SWR	94.8	100	6.7
A1	138.1	85.1	13.3
A2	134.7	85.	13.3
A3	203	91.7	13.3
IRWM2	206.4	91.5	10
IRWM3	546.3	99.3	13.3
CART	690.4	83.2	20
Median Effort	132.8	85.9	10
Mean Effort	1106.2	252.3	6.7

Fig. 2 shows boxplots of absolute residuals for all techniques. Except for IRWM3, all the medians look very similar, and the same also applies to their distributions. IRWM3 presents a more compact distribution, however its median is the highest; CART presents the greatest variation in absolute residuals. All these distributions share the same four outliers, identified as projects 22, 23, 24 and 27.

What these results suggest is that, at least for the data set used, any of the effort techniques compared (and all the different CBR combinations) would lead to similar effort predictions. Our results corroborate those of Costagliola *et al.* [5], who used a much smaller data set of industrial Web software applications than ours. As for the prediction accuracy of the effort techniques, if we use the minimum threshold proposed by Conte *et al.* [4] as means to indicate if a model is accurate (MMRE & MdMRE <= 25%, and Pred(25) >= 75%) our conclusion would be that all models were poor; however, we also must take into account another benchmark, which is obtained using the median (59.1) and the mean (526.9) effort models (see Table 6). The statistical significance tests confirm that all the effort models are significantly superior to the mean effort model, however similar to the median effort model. This means that if a Web company were to use the mean effort of past projects as the estimate for a new project, this estimate would be significantly worse than using either the median effort or any of the effort techniques in this paper.

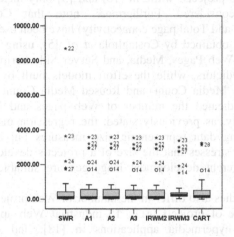

Fig. 2. Absolute residuals for SWR (SWR), CBR (A1, A2, A3, IRWM2, IRWM3) and CART (CART) models

The Tukutuku database does not contain a random sample of Web projects, therefore our results only apply to the Web companies which projects were used in this paper and Web companies that develop similar projects to those from the Tukutuku database.

5 Comparing Studies

In this section we compare our case study results with those reported in [18] and [5], taking into account descriptive statistics characterizing the three data sets employed and the accuracy of the estimations obtained.

First, we compare the size measures used in the three case studies. Mendes *et al.* employed 5 size measures (i.e., Page Count, Media Count, Program Count, Reused Media Count, and Reused Program Count), and 2 Complexity measures (i.e., Connectivity Density and Total Page Complexity) [18]. Costagliola *et al* employed 8 Length measures (i.e., Web Pages, New Web Pages, Media, New media, Client Side

Scripts and Application, Server Side Scripts and Applications, Internal Link, External References), which in some cases coincide with some of the size measures used in Mendes *et al.*, and 9 Functional measures that are all part of the *Web Objects* size measure [19]. Each Web project used in our case study was characterized by eleven size measures (denoting Length and Functional measures), and seven cost drivers. Note that despite our case study and the case study in [18] both use Length and Functional measures, there are differences between these measures. In particular, Costagliola *et al.* carried out the analysis separately on the distinct sets of Length and Functional measures. Conversely, in this paper we considered Length and Functional measures as a single set on which SWR, CART, and CBR were applied to. Another difference in the three case studies relates to the information the variables represent. In this paper we used size variables that characterise Web applications but cost drivers that characterise Web projects; while in [18] and [5] only measures related to Web applications were considered. Furthermore, note that Complexity measures (Connectivity density and Total page connectivity) have been used in [18] only.

The effort models obtained by Costagliola *et al.* [5], using Linear and Stepwise Regression, selected Web Pages, Media, and Server Side Scripts & Applications as significant effort predictors, while the effort models built by Mendes *et al.* [18] selected Page Count, Media Count, and Reused Media Count as predictors. Thus, both case studies indicated the number of Web pages and Media as significant predictors. Conversely, as previously stated, the regression model obtained in this paper was built using data representing size measures but also cost drivers. In particular, the model stresses that the size of a project's development team and the number of used different development languages are suitable predictors of Web development effort.

The three case studies also differ in the number of Web projects contained in each data set and the type of projects used: 15 industrial Web applications in [5], 37 student-based Web hypermedia applications in [18], and 150 industrial Web applications in our case study (see Table 8). Moreover, the descriptive statistics presented in Table 8 show that the projects used in [5] have much larger median and mean effort and number of Web pages than those used in [18] and in this paper. In particular, the mean effort of projects in [5] is 23.9 times and 4.7 times the mean effort of projects in [18] and in this case study respectively.

Table 7 also shows that the mean and standard deviation for the number of (static) Web pages are quite similar for the projects used in case studies [5] and [18], and much smaller than the mean and standard deviation for the projects used in our case study. The median, minimum and maximum values differ across the three case studies, and suggest that the applications used in [18] were overall larger than the applications used in the other two case studies (median = 53).

Finally, studies [5] and [18] reported effort accuracy that was overall above the minimum thresholds suggested by Conte *et al.* [4]. These thresholds advocate that good prediction is only achieved when MMRE & MdMRE <= 25%, and Pred(25) >= 75%. Conversely, the prediction accuracy for all the effort techniques used in this case study showed MMRE & MdMRE values > 25% and Pred(25) < 75%, thus suggesting, if we apply Conte *et al.*'s thresholds, that predictions were poor.

These results are not surprising since the estimations obtained in [5] and [18] were based on single-company data sets (i.e., data coming from 15 Web projects provided

Table 7. Summary statistics for effort and Web pages

	EFFORT				
	Mean	Median	Std. Dev.	Min.	Max.
Costagliola *et al.* - ICWE 2006	2,677.87	2,792	827.12	1,176	3,712
Mendes *et al.* - ESE 2003	111.89	114.65	26.43	58.36	153.78
Current case study	564.22	78	1048.94	1	5000
	NUMBER OF PROJECTS				
Costagliola *et al.* -ICWE 2006	15				
Mendes *et al.* - ESE 2003	37				
Current case study	150				
	NUMBER OF (STATIC) WEB PAGES				
	Mean	Median	Std. Dev.	Min.	Max.
Costagliola *et al.* - ICWE 2006	17..2	11	12.167	2	46
Mendes *et al.* - ESE 2003	24.82	53	11.26	33	100
Current case study	81.53	30	209.82	1	2000

by a single company [5] and from 37 Web hypermedia applications developed by postgraduate students of similar background and competences [18]) while the predictions obtained in this paper have been obtained taking into account a cross-company data set (i.e., a data set that contains project data volunteered by several companies [11]). Indeed, several empirical studies have shown that estimations obtained using a single-company dataset are superior to predictions obtained from a cross-company dataset [6][9][17]. However, the effort estimations obtained in this case study using effort techniques are significantly superior to the estimations obtained taking into account the mean effort. Thus, the use of a cross-company data set seems particularly useful for companies that do not have their own data on past projects from which to obtain their estimates, or that have data on projects developed in different application domains and/or technologies.

One possible reason for the better performance of the single-company data set, compared to the cross-company one, may be related to the size of the single-company data sets. Indeed, recently, Kitchenham *et al.* [7], by means of a systematic review, found that all studies where single-company predictions were significantly better than cross-company predictions employed smaller number of projects than in the cross-company models, and such data sets were characterized by a smaller maximum effort. They speculated that as single-company data sets grow, they incorporate less similar projects so that differences between single- and cross-company data sets cease to be significant.

6 Conclusions

In this paper we have replicated two previous studies that compared several effort estimation techniques to identify which technique provided the best prediction accuracy to estimate effort for Web applications [5][18]. We employed the three effort estimation techniques that were mutual to the two studies being replicated, namely Forward Stepwise Regression, Case-Based Reasoning, and Classification and Regression Trees. Our case study employed data obtained from 150 Web projects from the Tukutuku database [11]. This is the first time such large number of Web projects is used to compare Web effort estimation techniques. The empirical results showed that all techniques presented similar predictions, and these predictions were significantly better than those using the mean effort. Thus, we have corroborated the results presented by Costagliola *et al.* in [5], who found that none of the effort estimation techniques used in their study was statistically significantly superior.

We also showed that the two previous case studies presented effort estimates with superior accuracy to the estimates obtained in our case study, and that these results may have been caused by the use of a cross-company dataset in our study, compared to single-company datasets in [5] and [18].

Finally, the Web projects used in [5] presented the largest median effort and smallest median number of static Web pages.

Acknowledgments. We thank all companies that volunteered data to the Tukutuku database.

References

1. Baresi, L., Morasca, S., Paolini, P.: Estimating the design effort for Web applications. In: Proc. Metrics, pp. 62–72 (2003)
2. Brieman, L., Friedman, J., Olshen, R., Stone, C.: Classification and Regression Trees. Wadsworth Inc., Belmont (1984)
3. Christodoulou, S.P., Zafiris, P.A., Papatheodorou, T.S.: WWW2000: The Developer's view and a practitioner's approach to Web Engineering. In: Proc. ICSE Workshop on Web Engineering, Limerick, Ireland, pp. 75–92 (2000)
4. Conte, S.D., Dunsmore, H.E., Shen, V.Y.: Software Engineering Metrics and Models, Benjamin-Cummins (1986)
5. Costagliola, G., Di Martino, S., Ferrucci, F., Gravino, C., Tortora, G., Vitiello, G.: Effort estimation modeling techniques: a case study for web applications. In: ICWE 2006. Procs. Intl. Conference on Web Engineering, pp. 9–16 (2006)
6. Kitchenham, B.A., Mendes, E.: A Comparison of Cross-company and Single-company Effort Estimation Models for Web Applications. In: Procs. EASE 2004, pp. 47–55 (2004)
7. Kitchenham, B.A., Mendes, E., Travassos, G.: A Systematic Review of Cross- and Within-company Cost Estimation Studies. In: Proceedings of Empirical Assessment in Software Engineering, pp. 89–98 (2006)
8. Maxwell, K.: Applied Statistics for Software Managers. Software Quality Institute Series. Prentice Hall, Englewood Cliffs (2002)
9. Mendes, E., Kitchenham, B.A.: Further Comparison of Cross-company and Within-company Effort Estimation Models for Web Applications. In: Proc. IEEE Metrics, pp. 348–357 (2004)
10. Mendes, E., Counsell, S.: Web Development Effort Estimation using Analogy. In: Proc. 2000 Australian Software Engineering Conference, pp. 203–212 (2000)
11. Mendes, E., Mosley, N., Counsell, S.: Investigating Web Size Metrics for Early Web Cost Estimation. Journal of Systems and Software 77(2), 157–172 (2005)
12. Mendes, E., Counsell, S., Mosley, N.: Towards a Taxonomy of Hypermedia and Web Application Size Metrics. In: Lowe, D.G., Gaedke, M. (eds.) ICWE 2005. LNCS, vol. 3579, pp. 110–123. Springer, Heidelberg (2005)
13. Mendes, E., Mosley, N., Counsell, S.: Web Effort Estimation. In: Mendes, E., Mosley, N. (eds.) Web Engineering, pp. 29–73. Springer, Heidelberg (2005)

14. Mendes, E., Mosley, N., Counsell, S.: Early Web Size Measures and Effort Prediction for Web Costimation. In: Proceedings of the IEEE Metrics Symposium, pp. 18–29 (2003)
15. Mendes, E., Mosley, N., Counsell, S.: Comparison of Length, complexity and functionality as size measures for predicting Web design and authoring effort. IEE Proc. Software 149(3), 86–92 (2002)
16. Mendes, E., Mosley, N., Counsell, S.: Web metrics - Metrics for estimating effort to design and author Web applications. IEEE MultiMedia, 50–57 (January-March 2001)
17. Mendes, E., Di Martino, S., Ferrucci, F., Gravino, C.: Effort Estimation: How Valuable is it for a Web Company to use a Cross-company Data Set, Compared to Using Its Own SingleCompany Data Set? In: Proceedings of WWW 2007 (accepted for publication, 2007)
18. Mendes, E., Watson, I., Triggs, C., Mosley, N., Counsell, S.: A Comparative Study of Cost Estimation Models for Web Hypermedia Applications. ESE 8(2), 163–196 (2003)
19. Reifer, D.J.: Web Development: Estimating Quick-to-Market Software. IEEE Software, 57–64 (November-December 2000)
20. Reifer, D.J.: Ten deadly risks in Internet and intranet software development. IEEE Software (March-April 12-14, 2002)
21. Ruhe, M., Jeffery, R., Wieczorek, I.: Cost estimation for Web applications. In: Proc. ICSE 2003, pp. 285–294 (2003)
22. Shepperd, M.J., Kadoda, G.: Using Simulation to Evaluate Prediction Techniques. In: Proceedings IEEE Metrics 2001, London, UK, pp. 349–358 (2001)

Development Process of the Operational Version of PDQM

Angélica Caro[1], Coral Calero[2], and Mario Piattini[2]

[1] Department of Computer Science and Information Technologies, University of Bio Bio
Chillán, Chile
mcaro@ubiobio.cl
[2] Alarcos Research Group. Information Systems and Technologies Department
UCLM-INDRA Research and Development Institute.
University of Castilla-La Mancha
{Coral.Calero, Mario.Piattini}@uclm.es

Abstract. PDQM is a web portal data quality model. This model is centered on the data consumer perspective and for its construction we have developed a process which is divided into two parts. In the first part we defined the theoretical version of PDQM and as a result a set of 33 data quality attributes that can be used to evaluate the data quality in portals were identified. The second part consisted of the conversion of PDQM into an operational model. For this, we adopted a probabilistic approach by using Bayesian networks. In this paper, we show the development of this second part, which was divided into four phases: (1) Definition of a criterion to organize the PDQM's attributes, (2) Generation of a Bayesian network to represent PDQM, (3) Definition of measures and the node probability tables for the Bayesian network and (4) The validation of PDQM.

Keywords: Data Quality, Information Quality, Web Portal, Data Quality Evaluation, Bayesian Network, Measures.

1 Introduction

In literature, the concept of Data or Information Quality (hereafter referred to as DQ) is often defined as "fitness for use", i.e., the ability of a collection of data to meet a user's requirements [3, 14]. This definition and the current view of assessing DQ, involve understanding DQ from the users' point of view [8].

Advances in technology and the use of the Internet have favoured the emergence of a large number of web applications, including web portals. A web portal (WP) is a site that aggregates information from multiple sources on the web and organizes this material in an easy user-friendly manner [16]. In the last years the number of organizations which own WPs has grown dramatically. They have established WPs with which to complement, substitute or widen existing services to their clients [17]. Many people use data obtained from WPs to develop their work and to make decisions. These users need to be sure that the data obtained from the WPs are appropriate for the use they need to make of it. Likewise, the WP owners need to deliver data that meet the user's requirements in order to achieve the user's preference. Therefore, DQ represents a common interest between data consumers and WP providers.

B. Benatallah et al. (Eds.): WISE 2007, LNCS 4831, pp. 436–448, 2007.

In recent years, the research community has started to look into the area of DQ on the web [7]. However, although some studies suggest that DQ is one of the relevant factors when measuring the quality of a WP [10, 17], few address the DQ in WPs. Along with this, another important factor to consider is the relevance of users (or data consumers) in DQ evaluation and the necessity of proposals dealing with this topic [2, 3, 7].

Consequently, our research aims is to create a DQ model for web portals, named PDQM, which focuses upon the data consumer's perspective. To this end, we have divided our work into two parts. The first consisted of the theoretical definition of PDQM [4], which resulted in the identification of 33 DQ attributes that can be used to assess a portal's DQ. The second, presented in this paper, is concerned with converting the theoretical model into an operational one. This conversion consists of specifying the DQ attributes of PDQM in an operational way. This means defining a structure with which to organize the DQ attributes, and to associate measures and criteria for them.

Considering the subjectivity of the data consumer's perspective and the uncertainty inherent in quality perception [6], we chose to use a probabilistic approach by means of Bayesian networks, such as that proposed in [9], to transform PDQM into an operational model. A Bayesian network (BN) is a directed acyclic graph where nodes represent variables (factors) and arcs represent dependence relationships between variables. Arcs in a BN connect parent to child nodes, where a child node's probability distribution is conditional to its parent node's distribution. Arcs, nodes and probabilities can be elicited from experts and/or empirical data, and probabilities are conveyed by using Node probability tables (NPTs) which are associated to nodes [13]. In our context, BNs offer an interesting framework with which it is possible to: (1) Represent the interrelations between DQ attributes in an intuitive and explicit way by connecting influencing factors to influenced ones, (2) Deal with subjectivity and uncertainty by using probabilities (3) Use the obtained network to predict/estimate the DQ of a portal and (4) Isolate responsible factors in the case of low data quality.

This paper focuses on the process of converting PDQM into an operational model. That is, the creation of the Bayesian network that represents PDQM, its preparation in order to use it in the DQ assessment and its validation.

The rest of the paper is organized as follows. Section 2 presents a summary of the theoretical definition of PDQM and describes the process used to generate the operational model. Section 3 presents the criterion used to organize the PDQM's attributes. Section 4 shows the generation of the structure for the BN that will support PDQM. The definition of measures and NPTs for the BN is described in Section 5. The validation of PDQM is explained in Section 6. Finally, conclusions are given in Section 7.

2 Defining a Data Quality Model

To produce the portal data quality model (PDQM), we have defined a process which we have divided into two parts. The first part corresponded to the theoretical definition of PDQM and was based on the key aspects that represent the data consumer's perspective and the main characteristics of WPs. As a result of this first part we obtained a set 33 of DQ attributes that can be used to assess DQ in WPs (see Table 1). All the details of the development of the theoretical version of PDQM can be found in [4].

Table 1. Data Quality Attributes of PDQM

Accessibility	Consistent Representation	Novelty	Timeliness
Accuracy	Customer Support	Objectivity	Traceability
Amount of Data	Documentation	Organization	Understandability
Applicability	Duplicates	Relevancy	Currency
Attractiveness	Ease of Operation	Reliability	Validity
Availability	Expiration	Reputation	Value added
Believability	Flexibility	Response Time	
Completeness	Interactivity	Security	
Concise Representation	Interpretability	Specialization	

The second part consists of the transformation of the theoretical model into an operational one. To do this, we decided to use a probabilistic approach by using BN. The second part is composed of four phases. During the first phase, we have defined the criteria with which to organize the DQ attributes of PDQM. In the second phase, we have generated the graphical structure of PDQM (BN graph). In the third phase, we have prepared PDQM to be used in an evaluation process. Finally, the fourth phase corresponds to the model validation (see, Fig. 1).

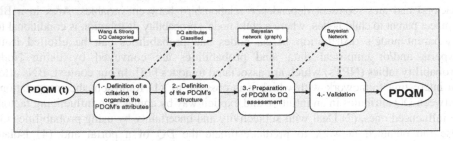

Fig. 1. The development process of the operational version of PDQM

The following sections describe the conversion of PDQM into an operational model.

3 Phase 1: Definition of a Criterion to Organize the PDQM's Attributes

As explained in [11], a BN can be built by starting from semantically meaningful units called network fragments. A fragment is a set of related random variables that can be constructed and reasoned about separately from other fragments. Thus an initial phase when building the BN for PDQM, was to define a criterion that allowed us to organize the DQ attributes into a hierarchical structure, with the possibility of creating network fragments.

We used the conceptual DQ framework developed in [14] as a criterion for organizing the DQ attributes. However, in our work we have renamed and redefined the Accessibility category, calling it the Operational category. The idea was to emphasize the importance of the role of systems, not only with respect to accessibility and security, but also with respect to aspects such as personalization, collaboration, etc. Having done all this, and taking the definition of each DQ category into account, we have classified all the DQ attributes of PDQM into the categories seen below in Table 2. Thus, we have identified 4 network fragments based on this classification, one per category.

Table 2. Classification of DQ Attributes of PDQM into DQ Categories

DQ Category	DQ Attributes
Intrinsic: This denotes that data have quality in their own right.	Accuracy, Objectivity, Believability, Reputation, Currency, Duplicates, Expiration, Traceability
Operational: This emphasizes the importance of the role of systems; that is, the system must be accessible but secure in order to allow personalization and collaboration, amongst other aspects.	Accessibility, Security, Interactivity, Availability, Customer support, Ease of operation, Response time
Contextual: This highlights the requirement which states that DQ must be considered in the context of the task in hand.	Applicability, Completeness, Flexibility, Novelty, Reliability, Relevancy, Specialization, Timeliness, Validity, Value-Added
Representational. This denotes that the system must present data in such a way as to be interpretable and easy to understand, as well as concisely and consistently represented.	Interpretability, Understandability, Concise Representation, Consistent Representation, Amount of Data, Attractiveness, Documentation, Organization

4   Phase 2: Definition of the PDQM's Structure

In order to generate new levels in the BN, we established relationships of direct influences between the attributes in each category. These relationships were established by using the DQ categories and the DQ attributes definitions, together with our perceptions and experience. Thus, each relationship is supported by a premise that represents the direct influence between an attribute and its parent attribute. As an example of how this works, Table 3 shows the relationships established in the DQ Representational category.

Table 3. Relationships between DQ attributes in the DQ Representational category

	Relation of Direct Influence		Premise that supports the direct influence relationships
	Level 2	Level 3	
DQ Representational (Level 1)	Concise Representation	-	If data are compactly represented without superfluous elements then they will be better represented.
	Consistent Representation	-	If data are always presented in the same format, are compatible with previous data and consistent with other sources, then they will be better represented.
	Understandability	Interpretability	If data are appropriately presented in language and units for users' capability then they will be understood better.
		Amount of data	If the quantity or volume of data delivered by the WP is appropriate then they will be understood better.
		Documentation	If data have useful documents with meta information then they will be understood better.
		Organization	If data are organized with a consistent combination of visual settings then they will be understood better.
	Attractiveness	Organization	If data are organized with a consistent combination of visual settings then they will be more attractive to data consumers.

Taking these relationships as basis, we built the BN graph which represents PDQM, see Fig. 2.

In the BN which was created, four levels can be distinguished. Level 0, where PDQ is the node that represents DQ in the whole WP. Level 1, where nodes represent DQ in each DQ category in a WP (obviously, the PDQ node is defined in terms of the others 4). Level 2, where nodes represent the DQ attributes with a direct influence upon each of the DQ categories, and Level 3, where nodes represent the DQ attributes with a direct influence upon each of the DQ attributes in Level 2.

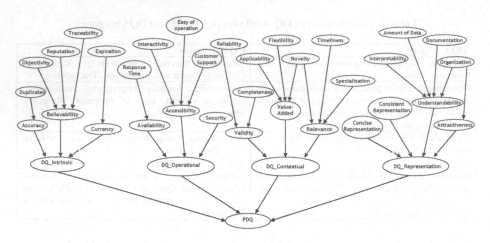

Fig. 2. BN graph to represent PDQM

In the following phase the BN must be prepared to be used in an evaluation process.

5 Phase 3: Preparation of PDQM for DQ Assessment

Having taken into consideration the size of the BN generated in the previous phase, and although our final objective is to create a comprehensive BN model for PDQM, we decided to develop this phase separately for each fragment network (DQ_Intrinsic, DQ_Operational, DQ_Contextual and DQ_Representational). In this paper we shall in particular work with the DQ_Representational fragment. To prepare the fragment network to be used in the DQ assessment, the following sub-phases will be developed:

a. If necessary, artificial nodes will be created to simplify the fragment network, i.e., to reduce the number of parents for each node.
b. Measures for the quantifiable variables (entry nodes) in the fragment network will be defined.
c. The NPTs for each intermediate node in the fragment network will be defined.

Phase a: Simplifying the fragment network. The original sub-network had two nodes with four parents (Understandability and DQ_Representational) so we decided to create two synthetic nodes (Representation and Volume of Data) in order to reduce the combinatory explosion in the following step during the preparation of the NPTs. In Fig. 3 we will show the original sub-network (graph 1) and the sub-network with the synthetic nodes created (graph 2).

Phase b: Defining qantifiable variables for fragment network. This sub-phase consists of the definition of measures for the quantifiable variables in the DQ_Representational fragment. We therefore defined an indicator for each entry node in the fragment (see Fig. 3, graph 3). In general, we selected and defined measures according to each attribute's (entry node) definition. To calculate each indicator we followed two methods: (1) base and derived measures are used when objective measures can be obtained (a measure is derived from another base or from derived measures [1]) or (2) data consumer valuations are used when the attribute is subjective. In both cases the indicators will take a numerical value of between 0 and 1. For each indicator, the labels that represent the fuzzy sets associated with that indicator were defined by using a fuzzy approach, and by considering the possible values that the indicator may take. Finally, a membership function was defined to determine the degree of membership of each indicator with respect to the fuzzy labels. In Table 4 we show a summary of the indicators defined for this fragment.

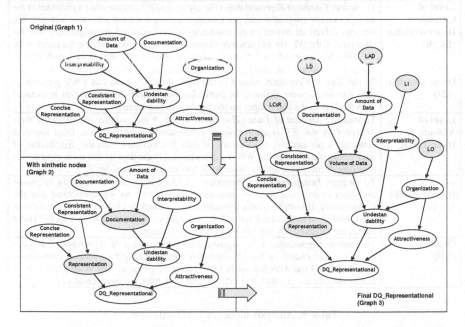

Fig. 3. Preparation of Sub-network DQ_Representational for assessment

As an example, we will explain the definition of the LAD indicator. To calculate the LAD indicator we have established an *Analysis Model* that includes a *formula* which gives us a numerical value and a *Decision Criteria* in the form of a membership function (see Table 5). This will later allow us to determine the degree of membership of each measure with respect to fuzzy labels. Note, as is explained by Thomas in [15], that the membership function degrees can be used as probabilities with the condition that both the fuzzy clustering algorithm and the approximation method preserve the condition that the sum of the membership degrees is always equal to 1.

The analysis model (formula and decision criteria) attached to each of these indicators was determined from an analysis of literature, or from common-sense assumptions about the preferences of data consumers in WPs.

In the case of the LAD indicator, our intention is to use the formula to represent the fact that data consumers estimate the amount of data that exists, by assessing the amount and distribution of images, links and words that a WP delivers on each page. We assign more importance (0.4) to the amount of words because it has more impact on users: they do not feel comfortable if they have to read too much [12].

Table 4. Indicators defined for the fragment network

Name	Description
Level of Concise Representation (LCcR)	To measure *Concise Representation* (*The extent to which data are compactly represented without superfluous or not related elements*). To calculate *LCcR*, measures associated with the amount and size of paragraphs and the use of tables to represent data in a compact form were considered.
Level of Consistent Representation (LCsR)	To measure *Consistent Representation* (*The extent to which data are always presented in the same format, are compatible with previous data and are consistent with other sources*). The measures defined are centred on the consistency of the format and on compatibility with the other pages in the WP. For this indicator measures based on the use of Style in the pages of the WP and in the correspondence between a source page and the destination pages were defined.
Level of Documentation (LD)	To measure *Documentation* (*Quantity and utility of the documents with metadata*). To calculate LD measures related to the basic documentation that a WP presents to data consumers were defined. In particular, the simple documentation associated with the hyperlinks and images on the pages was considered.
Level of Amount of Data (LAD)	To measure *Amount of Data* (*The extent to which the quantity or volume of data delivered by the WP is appropriate*). Understanding the fact that, from the data consumer's perspective, the amount of data is concerned with the distribution of data within the pages in the WP. To calculate LAD the data in text form (words), in hyperlink form (links) and in visual form (images) were considered.
Level of Interpretability (LI)	To measure *Interpretability* (*The extent to which data are expressed in language and units appropriate for the consumer's capability*). As we considered that the evaluation of this attribute was too subjective, a check list for its measurement was used. Each item in the check list will be valuated with a number from 1 to 10. These values are subsequently transformed into a [0, 1] range.
Organization (LO)	To measure *Organization* (*The organization, visual settings or typographical features (color, text, font, images, etc.) and the consistent combinations of these various components*). To calculate LO measures that verify the existence of groups of data in the pages (tables, frames, etc.), the use of colors, text with different fonts, titles, etc. were considered.

Table 5. Analysis model of LAD indicator

LAD (Level of Amount of Data)	
Formula to Calculate LAD	Decision Criteria
LAD = DWP * 0.4 + DLP * 0.3 + DIP * 0.3	
Derived Measures	
DWP: Distribution of Words per page DLP: Distribution of Links per page DIP: Distribution of Images per page	

For the decision criteria (taking into consideration that several studies show that users prefer data presented in a concise form [12]), the membership function transforms the numerical value of the indicator into one of the following labels: Good, Medium and Bad.

Phase c: Defining the NPTs for fragment network. The intermediate nodes are nodes which are defined by their parents and are not directly measurable. Thus, their NPTs were made by expert judgment.

On the other hand, after having taken into account the importance of considering the task context of users and the processes by which users access and manipulate data to meet their task requirements [14] in a DQ evaluation process, we considered that the probability distribution may differ according to the WP context.

This implies that sub-phase c must be developed by considering a specific domain of WPs. In this work we have started to consider the educational context and we have defined the NPTs by considering that this fragment will be applied to university WPs. Table 6 shows the NPTs for the nodes in the third level of the fragment.

Thus, the DQ_Representational fragment is prepared to evaluate the DQ in university WPs. The last phase in generating the operational model for this fragment is that of its validation. In the next section we shall describe the validation experiment which was developed and the results which were obtained.

Table 6. Node probability tables for Level 2 in fragment

Level 2													
Consistent Representation				**Volume of Data**									
LCsR	Low	Medium	High	Documentation	Bad			Medium			Good		
Bad	0.9	0.05	0.01	Amount of Data	Bad	Medium	Good	Bad	Medium	Good	Bad	Medium	Good
Medium	0.09	0.9	0.09	Bad	0.9	0.8	0.5	0.8	0.3	0.15	0.5	0.15	0.01
Good	0.01	0.05	0.9	Medium	0.09	0.15	0.3	0.15	0.4	0.25	0.3	0.25	0.09
				Good	0.01	0.05	0.2	0.05	0.3	0.6	0.2	0.6	0.9
Concise Representation				**Interpretability**				**Organization**					
LCcR	Low	Medium	High	LI	Low	Medium	High	LO	Low	Medium	High		
Bad	0.9	0.05	0.01	Low	0.9	0.05	0.01	Bad	0.9	0.05	0.01		
Medium	0.09	0.9	0.09	Medium	0.09	0.9	0.09	Medium	0.09	0.9	0.09		
Good	0.01	0.05	0.9	High	0.01	0.05	0.9	Good	0.01	0.05	0.9		

6   Phase 4: Validation of PDQM

The method defined to validate PDQM consisted of using two different strategies to evaluate the representational DQ in a given WP. One of them evaluated the DQ with a group of subjects and the other evaluated it with PDQM. We next compared the results obtained to determine whether the evaluation made with PDQM was similar to that made with the subjects. That is, whether the model represented the data consumer's perspective.

Therefore, for the first strategy we developed an experiment by which to obtain the judgments of a group of subjects about a DQ representational in a university WP. In this experiment, the subjects were asked for their partial valuations of each DQ attribute in the fragment and for their valuation of the global representational DQ in the WP.

For the second assessment strategy, we built a tool that implements the fragment of the DQ_Representational. This tool allows us to automatically measure the quantifiable variables and, from the values obtained, to obtain the entry data for the BN that will give us the evaluation of PDQM. In the following subsections we will describe the experiment, the automatic evaluation and the comparison of both results in greater detail.

6.1   The Experiment

The subjects who took part in the experiment were a group of students from the University of Castilla-La Mancha in Spain. The group was composed of 79 students enrolled in the final year (third) of Computer Science (MSc). All of the subjects had previous experience in the use of WPs as data consumers. The experimental material was composed of one document including: the instructions and motivations, the URL of a university WP, three activities to be developed in the WP, and a set of 9 questions in which we requested their valuations for the DQ representational in the WP. The first 8 valuations were requested for each of the DQ attributes in the DQ_Representational fragment and the last question attempted to gauge the global DQ Representational in the WP. As a result of this experiment we obtained the valuations shown in Table 6.

Table 6. Valuations given by the subjects for the DQ Representational

Attribute Evaluated	Valuations		
	Low/Bad	Medium	High/Good
Attractiveness	30%	61%	9%
Organization	37%	44%	19%
Amount of Data	18%	49%	33%
Understandability	32%	47%	21%
Interpretability	6%	45%	48%
Documentation	16%	49%	34%
Consistent Representation	18%	53%	29%
Concise Representation	16%	52%	32%
Portal	**17%**	**68%**	**16%**

6.2 The Automatic Evaluation

The PoDQA tool [5] is the application that will support the PDQM model. Its aim is to give the user information about the DQ level in a given WP (at present it is just a prototype and only the Representational DQ is supported). The tool downloads and analyzes the pages of the WP, in order to calculate the defined measures using the public information in WPs.

Thus, for a given WP PoDQA will calculate the measures associated with the indicators: LCsR, LCcR, LD, LAD, LI, LO. Each indicator will take a value of between 0 and 1. This value will be transformed into a set of probabilities for the corresponding labels. Each of these values will be the input for the corresponding input node. With this value, and by using its probability table, each node generates a result that is propagated, via a causal link, to the child nodes for the whole network until the level of the DQ Representational is obtained.

We used PoDQA to evaluate the same university WP that was used in the experiment. As a result we obtained the values for each indicator (see Table 7) which were transformed into a valid entry for the BN. These values were entered in the BN which, finally, generated the level of DQ representational in the WP (like as in Fig. 4).

Table 7. Values obtained for the indicators of the DQ_Representational fragment

LCsR	LCcR	LD	LAD	LI	LO
0.12	0.99	0.46	0.99	0.5	0.44

6.3 Comparing the Results Obtained

When comparing the results obtained with the two evaluation strategies, we can observe that, in general, they are very different, see Table 8.

In effect, with regard to the final evaluation, Table 8 (last row) shows that while in the experiment the subjects evaluated the DQ at a *Medium* level (68%), with the automatic evaluation the DQ was evaluated at the same value for the *Medium* and *High* levels (in both cases 40%). With regard to the partial values, that is, for each DQ

Table 8. Valuations obtained from the experiment and valuations calculated automatically

Attribute Evaluated	Low/Bad		Medium		High/Good	
	Subj.	PDQA	Subj.	PDQA	Subj.	PDQA
Attractiveness	30%	34%	61%	44%	9%	22%
Organization	37%	26%	44%	66%	19%	8%
Amount of Data	18%	6%	49%	13%	33%	81%
Understandability	32%	52%	47%	23%	21%	25%
Interpretability	6%	43%	45%	49%	48%	7%
Documentation	16%	9%	49%	82%	34%	9%
Consistent Representation	18%	81%	53%	13%	29%	6%
Concise Representation	16%	6%	52%	13%	32%	81%
Portal	**17%**	**20%**	**68%**	**40%**	**16%**	**40%**

attribute, the results are also very different. The reason for this is, in our opinion, that the results given for the indicators are, in some cases, very extreme (see for example the values for LCcR and LCsR). Consequently, the nodes with most differences are the child nodes of the nodes that represent the indicators that take these extreme values.

A preliminary interpretation of these results is that PDQM is more demanding than the subjects and needs to be adjusted. Thus, we attempted to reduce these differences by adjusting the NPTs and recalculating the representational DQ. The results obtained can be observed in Fig. 4, which shows the BN and the values calculated for it to each DQ attribute and the representational DQ level, and Table 9, which allow us to compare the values obtained.

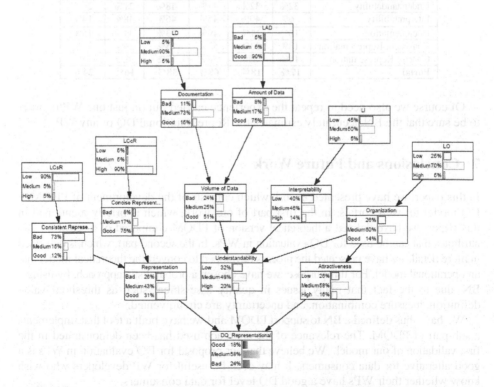

Fig. 4. New results adjusting the node probability tables in the BN of DQ Representational

As a result of this new configuration the general result of the automatic evaluation is closer to the subjects' evaluations. However, in spite of the fact that both evaluations gave their result as *Medium*, total coincidence between the values calculated does not exist (see last row in Table 9). Moreover, the partial values also have a better fit than in the first calculation, but do not totally coincide. See for example the differences between the Interpretability and Consistent Representation attributes for the valuations Low/Bad. We again believe that the main reason for this is the extreme values of the indicators. But this is not the only reason. Together with the former problem, we believe that the design of the WP evaluated may also influence this result. For example, to calculate the *Level of Amount of Data*, it is necessary to know the *distribution of words per page*. The measured WP presents values for this measure which can be considered as outliers (they take extreme values that do not follow a uniform distribution). Obviously, these values need to be removed from the calculation of the measure. Because of this we are now refining the calculations made by the tool by detecting and eliminating the outliers in our measures.

Table 9. New valuations obtained from PDQA with the new configuration of PDQM

Attribute Evaluated	Low/Bad		Medium		High/Good	
	Subj.	PDQA	Subj.	PDQA	Subj.	PDQA
Attractiveness	30%	26%	61%	58%	9%	16%
Organization	37%	26%	44%	60%	19%	14%
Amount of Data	18%	8%	49%	17%	33%	75%
Understandability	32%	32%	47%	48%	21%	20%
Interpretability	6%	40%	45%	46%	48%	14%
Documentation	16%	11%	49%	73%	34%	15%
Consistent Representation	18%	73%	53%	15%	29%	12%
Concise Representation	16%	8%	52%	17%	32%	75%
Portal	**17%**	**18%**	**68%**	**58%**	**16%**	**24%**

Of course we also need to repeat the experience carried out on just one WP in order to be sure that the BN accurately estimates the Representational DQ of any WP.

7   Conclusions and Future Work

In this paper, we have presented a work which consists of the development of PDQM, a DQ model for web portals. In the first part of our work, which is briefly mentioned in this paper, we have defined a theoretical version of PDQM composed of a set of 33 DQ attributes that can be used for DQ evaluation in WPs. In the second part, which is described in more detail, we have presented the process developed to convert the theoretical model into an operational model. For this purpose, we have chosen a probabilistic approach, by using a BN, due to the fact that many issues in quality assessment such as threshold value definition, measure combination, and uncertainty are circumvented.

We have thus defined a BN to support PDQM and we have built a tool that implements a sub-part of PDQM. The relevance of the approach used has been demonstrated in the first validation of our model. We believe that our proposal for DQ evaluation in WPs is a good alterative for data consumers. It may even be useful for WP developers who wish know whether their WPs have a good DQ level for data consumers.

We believe that one of the advantages of our model will be its flexibility. Indeed, the idea is to develop a model that can be adapted to both the goal and the context of evaluation. From the goal perspective, the user can choose the fragment that evaluates the characteristics he/she is interested in. From the context point of view, the parameters (NPTs) can be changed to consider the specific context of the WP evaluated.

As future work, we first plan to develop new validations which consider a greater number of WPs and which will allow us to refine PDQM. Another aspect to be considered is that of extending the definition of PDQM to other WP contexts. Lastly, we plan to extend the model to the other fragments and to include them in the PoDQA tool.

Acknowledgments. This research is part of the following projects: ESFINGE (TIC2006-15175-C05-05) granted by the Dirección General de Investigación del Ministerio de Ciencia y Tecnología (Spain), CALIPSO (TIN20005-24055-E) supported by the Minis-terio de Educación y Ciencia (Spain), DIMENSIONS (PBC-05-012-1) supported by FEDER and by the "Consejería de Educación y Ciencia, Junta de Comunidades de Castilla-La Mancha" (Spain) and COMPETISOFT (506AC0287) financed by CYTED.

References

1. Bertoa, M., García, F., Vallecillo, A.: An Ontology for Software Measument. In: Calero, C., Ruiz, F., Piattini, M. (eds.) Ontologies for Software Engineering and Software Technology (2006)
2. Burgess, M., Fiddian, N., Gray, W.: Quality Measures and The Information Consumer. In: Proceeding of the 9th International Conference on Information Quality (2004)
3. Cappiello, C., Francalanci, C., Pernici, B.: Data quality assessment from the user's perspective. In: International Workshop on Information Quality in Information Systems, Paris, Francia, ACM, New York (2004)
4. Caro, A., Calero, C., Caballero, I., Piattini, M.: Defining a Data Quality Model for Web Portals. In: Aberer, K., Peng, Z., Rundensteiner, E.A., Zhang, Y., Li, X. (eds.) WISE 2006. LNCS, vol. 4255, Springer, Heidelberg (2006)
5. Caro, A., Calero, C., de Salamanca, J.E., Piattini, M.: A prototype tool to measure the data quality in Web portals. In: The 4th Software Measurement European Forum, Roma, Italia (2007)
6. Eppler, M.: Managing Information Quality: Increasing the Value of Information in Knowledge-intensive Products and Processes. Springer, Heidelberg (2003)
7. Gertz, M., Ozsu, T., Saake, G., Sattler, K.-U.: Report on the Dagstuhl Seminar "Data Quality on the Web". SIGMOD Record 33(1), 127–132 (2004)
8. Knight, S.A., Burn, J.M.: Developing a Framework for Assessing Information Quality on the World Wide Web. Informing Science Journal 8, 159–172 (2005)
9. Malak, G., Sahraoui, H., Badri, L., Badri, M.: Modeling Web-Based Applications Quality: A Probabilistic Approach. In: 7th International Conference on Web Information Systems Engineering, Wuhan, China (2006)
10. Moraga, M.Á., Calero, C., Piattini, M.: Comparing different quality models for portals. Online Information Review. 30(5), 555–568 (2006)
11. Neil, M., Fenton, N.E., Nielsen, L.: Building large-scale Bayesian Networks. The Knowledge Engineering Review 15(3), 257–284 (2000)

12. Nielsen, J.: Designing Web Usability: The Practice of Simplicity. New Riders Publishing, Indianapolis (2000)
13. Pearl, J.: Probabilistic Reasoning in Intelligent Systems: Networks of Plausible Inference. Morgan Kaufmann, San Francisco (1988)
14. Strong, D., Lee, Y., Wang, R.: Data Quality in Context. Communications of the ACM 40(5), 103–110 (1997)
15. Thomas, S.F.: Possibilistic uncertainty and statistical inference. In: ORSA/TIMS Meeting, Houston, Texas (1981)
16. Xiao, L., Dasgupta, S.: User Satisfaction with Web Portals: An empirical Study. In: Gao, Y. (ed.) Web Systems Design and Online Consumer Behavior, pp. 193–205. Idea Group Publishing, USA (2005)
17. Yang, Z., Cai, S., Zhou, Z., Zhou, N.: Development and validation of an instrument to measure user perceived service quality of information presenting Web portals. Information and Management, vol. 42, pp. 575–589. Elsevier, Amsterdam (2004)

A New Reputation Mechanism Against Dishonest Recommendations in P2P Systems*

Junsheng Chang, Huaimin Wang, Gang Yin, and Yangbin Tang

School of Computer, National University of Defense Technology,
HuNan Changsha 410073, China
cjs7908@163.com

Abstract. In peer-to-peer (P2P) systems, peers often must interact with unknown or unfamiliar peers without the benefit of trusted third parties or authorities to mediate the interactions. Trust management through reputation mechanism to facilitate such interactions is recognized as an important element of P2P systems. However current P2P reputation mechanism can not process such strategic recommendations as correlative and collusive ratings. Furthermore in them there exists unfairness to blameless peers. This paper presents a new reputation mechanism for P2P systems. It has a unique feature: a recommender's credibility and level of confidence about the recommendation is considered in order to achieve a more accurate calculation of reputations and fair evaluation of recommendations. Theoretic analysis and simulation show that the reputation mechanism we proposed can help peers effectively detect dishonest recommendations in a variety of scenarios where more complex malicious strategies are introduced.

1 Introduction

P2P (Peer-to-Peer) technology has been widely used in file-sharing applications, distributed computing, e-market and information management [1]. The open and dynamic nature of the peer-to-peer networks is both beneficial and harmful to the working of the system. Problems such as free-riders and malicious users could lead to serious problems in the correct and useful functioning of the system. As shown by existing work, such as [2, 3, 4, 5, 6, 7], reputation-based trust management systems can successfully minimize the potential damages to a system by computing the trustworthiness of a certain peer from that peer's behavior history. However, there are some vulnerabilities of a reputation-based trust model. One of the detrimental vulnerabilities is that malicious peers submit dishonest recommendations and collude with each other to boost their own ratings or bad-mouth non-malicious peers [7]. The situation is made much worse when a group of malicious peers make collusive attempts to manipulate the ratings [8, 9].

---

* Supported by National Basic Research Program of china under Grand (No.2005CB321800), National Natural Science Foundation of China under Grant (No.90412011, 60625203) and National High-Tech Research and Development Plan of China under Grant (2005AA112030).

B. Benatallah et al. (Eds.): WISE 2007, LNCS 4831, pp. 449–460, 2007.

In this paper, we present a reputation mechanism for P2P systems. Within this system a peer can reason about trustworthiness of other peers based on the available local information which includes past interactions and recommendations received from others. Our contributions include: (1) Robustness against strategic recommendations, such as correlative and collusive ratings, is improved by decoupling service trust and recommendation trust. (2) A recommender's credibility and level of confidence about the recommendation is considered in order to achieve a more accurate calculation of reputations and fair evaluation of recommendations. (3) To assess the effectiveness of our approach, we have conducted extensive analytical and experimental evaluations. As a result, the reputation mechanism we proposed can help peers effectively detect dishonest recommendations in a variety of scenarios where more complex malicious strategies are introduced.

The reminder of this paper is structured as follows. In the second section, the related works are introduced; the new reputation mechanism we propose will be illuminated in the third section; and in the fourth section, a simulation about this reputation mechanism is laid out. Conclusions and future works are in the end.

2   Related Work

In this section, we survey existing reputation mechanisms for P2P system, especially focusing on their handing of recommendations.

Effective protection against unfair ratings is a basic requirement in order for a reputation system to be robust. The methods of avoiding bias from unfair ratings can broadly be grouped into two categories, endogenous and exogenous [10], as described below.

1. Endogenous Discounting of Unfair Ratings. This category covers methods that exclude or give low weight to presumed unfair ratings based on analysing and comparing the rating values themselves. The assumption is that unfair ratings can be recognized by their statistical properties only. Proposals in this category include Dellarocas [11] and Chen & Singh [12]. The implicit assumption underlying endogenous approaches is that the majority of recommendations are honest such that they dominate the lies. Therefore, a recommendation that deviates from the majority is considered a lie. This assumption is not solid in open environments where recommendations can be very few in number, most of which can be untruthful.

2. Exogenous Discounting of Unfair Ratings. This category covers methods where external factors, such as recommendation credibility, are used to determine the weight given to ratings. In order to calculate the recommendation credibility, PeerTrust [5] proposes to use a personalized similarity measure (PSM for short) to rate the recommendation credibility of another node x through node n's personalized experience, the evaluation of recommendation credibility is depending on the common set of peers that have interacted with requestor and the recommendatory peers. As the increase of peers' quantity, the common set is always very small [13]. This evaluation algorithm of recommendation credibility is not reliable. Eigentrust [3] considers the recommendation trust as being equal to the service trust. This metric is not suitable in circumstances where a peer may maintain a good reputation by providing high quality services but send malicious feedbacks to its competitors.

Research [14] presents a new recommendation credibility calculation model, but there exists unfairness to blameless peers. Research [6] proposes the weighted majority algorithm (WMA), the main idea is to assign and tune the weights so that the relative weight assigned to the successful advisors is increased and the relative weight assigned to the unsuccessful advisors is decreased. But, the approaches mentioned above don't consider more complex malicious strategies, for example, peers could try to gain trust from others by telling the truth over a sustained period of time and only then start lying, colluding peers could inflate reputation using unfair ratings flooding. Moreover, a peer may be penalized for an incorrect opinion that was based on a small number of interactions and/or a large variation in experience. Then, honest peers will be falsely classified as lying.

3 Reputation Management Mechanism

We adopt the terminology witness to denote a peer solicited for providing its recommendation. Finding the right set of witnesses is a challenging problem since the reputation value depends on their recommendations. Our approach for collecting recommendations follows the solution proposed by Yu et al [6], in which recommendations are collected by constructing chains of referrals through which peers help one another to find witnesses.

Providing services and giving recommendations are different tasks. A peer may be a good service provider and a bad recommender at the same time. Our reputation mechanism separates trust for providing good service and trust for providing fair recommendation, respectively denoted as service trust and recommendation credibility. Moreover, a recommender's credibility and level of confidence about the recommendation is considered in order to achieve a more accurate calculation of reputations and fair evaluation of recommendations.

In this section, we first introduce our service trust valuation algorithm used by a peer to reason about trustworthiness of other peers based on the available local information which includes past interactions and recommendations received from others. Then, we present our recommendation credibility calculation model, which can effectively detect dishonest recommendations in a variety of scenarios where more complex malicious strategies are introduced.

3.1 Trust Evaluation Algorithm

In respect that peers may change their behaviors over time, and the earlier ratings may have little impact on the calculation result, it is desirable to consider more recent ratings, that is, to consider only the most recent ratings, and discard those previous ones. Such a restriction is motivated by game theoretic results and empirical studies on ebay that show that only recent ratings are meaningful [9]. Thus, in the following, only interactions that occur within a sliding window of width D are considered. Moreover, by doing so, the storage costs of our reputation system are reasonable and justified given its significant benefits.

There are two kinds of trust relationships among peers, namely, direct trust and indirect trust [15]. The direct trust of peer i to peer j can be evaluated from the direct transaction feedback information between i and j. The indirect trust of i to j can be computed according to the transaction feedback information of peers who have interacted with j. The overall trust of peer i to peer j is produced by combining these two trust value.

Direct Trust. Any peer in our system will maintain a direct trust table for all the other peers it has interactions with directly. Suppose peer i has some interactions with peer j during the last D time units, the entry for peer j is denoted as $Exp(i, j) = \langle n, EXP\_LIST \rangle$, where n is the number of ratings, and EXP_LIST is an index in which these ratings are kept. The rating is in the form of r = (i, j, t, v). Here i and j are the peers that participated in interaction, and v is the rating peer i gave peer j. The range of v is [0, 1], where 0 and 1 means absolutely negative, absolutely positive respectively. t is the time stamp of the interaction. A rating is deleted from the direct trust table after an expiry period D.

From the direct trust table, the direct trust valuation of peer i to peer j at time t is represented as $< D_{ij}^{t}, \rho_{ij}^{t} >$, where D_{ij}^{t} is the direct trust value and ρ_{ij}^{t} is introduced to express the level of confidence about this direct trust value. Although there are a lot of elements that can be taken into account to calculate the level of confidence, we will focus on two of them: the number of experiences used to calculate the direct trust value and the variability of its rating values. D_{ij}^{t} is calculated by the following formula:

$$D_{ij}^{t} = \frac{\sum_{e \in Exp(i,j).EXP\_LIST} e.v * \alpha^{(t-e.t)}}{\sum_{e \in Exp(i,j).EXP\_LIST} \alpha^{(t-e.t)}} \tag{1}$$

Where e.v is the value of the rating e, and α is the decay factor in the range of (0, 1). A malicious node may strategically alter its behavior in a way that benefits itself such as starting to behave maliciously after it attains a high reputation. In order to cope with strategic altering behavior, the effect of an interaction on trust calculation must fade as new interactions happen [9]. This makes a peer to behave consistently. So, a peer with large number of good interactions can not disguise failures in future interactions for a long period of time.

Let CIN_{ij}^{t} be the level of confidence based on the number of ratings that have been taken into account in computing D_{ij}^{t} . As the number of these ratings ($Exp(i, j).n$) grows, the level of confidence increases until it reaches a defined threshold (denoted by m).

$$CIN_{ij}^{t} = \begin{cases} \dfrac{Exp(i, j).n}{m} & if \ Exp(i, j).n \le m \\ 1 & otherwise \end{cases} \tag{2}$$

Hence, the level of confidence CIN_{ij}^{t} increases from 0 to 1 when the number of ratings $Exp(i, j).n$ increases from 0 to m, and stays at 1 when $Exp(i, j).n$ exceeds m.

Let CID_{ij}^t be the level of confidence based on the variability of the rating values. CID_{ij}^t is calculated as the deviation in the ratings' values:

$$CID_{ij}^t = 1 - \frac{1}{2} * \frac{\sum_{e \in Exp(i,j).EXP\_LIST} \alpha^{(t-e.t)} * |e.v - D_{ij}^t|}{\sum_{e \in Exp(i,j).EXP\_LIST} \alpha^{(t-e.t)}} \tag{3}$$

This value goes from 0 to 1. A deviation value near 0 indicates a high variability in the rating values (this is, a low confidence on the direct trust value) while a value close to 1 indicates a low variability (this is, a high confidence on the direct trust value).

Finally, the level of confidence ρ_{ij}^t about the direct trust value D_{ij}^t combines the two reliability measures above:

$$\rho_{ij}^t = CIN_{ij}^t * CID_{ij}^t \tag{4}$$

Indirect trust. After collecting all recommendations about peer j using the rating discovery algorithm proposed by Yu et al [6], peer i can compute the indirect trust about peer j. Let $T_i = \{p_1, p_2, \ldots, p_{ti}\}$ be the set of trustworthy peers which reply the request. If $p_k \in T_i$ had at least one service interaction with p_j, it replies recommendation $< \operatorname{Re} c_{kj}^t, CI_{kj}^t >$ based on the local rating records with peer j. For an honest peer p, we have $\operatorname{Re} c_{kj}^t = D_{kj}^t$ and $CI_{kj}^t = \rho_{kj}^t$. The inclusion of level of confidence in the recommendation sent to the recommendation requestor allows the recommendation requestor to gauge the how much confidence the peer itself places in the recommendation it has sent. To minimize the negative influence of unreliable information, the recipient of these recommendations weighs them using this attached level of confidence and the credibility of the sender. If the level of confidence has a small value, the recommendation is considered weak and has less effect on the reputation calculation. Credibility is evaluated according to the past behavior of peers and reflects the confidence a peer has in the received recommendation. Credibility computation is presented in the next subsection.

The indirect trust value of peer j according peer i, denoted by R_{ij}^t, is given by the following formula:

$$R_{ij}^t = \frac{\sum_{p_k \in T_i} Cr_{ik}^t * CI_{kj}^t * \operatorname{Re} c_{kj}^t}{\sum_{p_k \in T_i} Cr_{ik}^t * CI_{kj}^t} \tag{5}$$

Where Cr_{ik}^t is the credibility of peer k according to peer i, CI_{kj}^t denotes the level of confidence about the recommendation value $\operatorname{Re} c_{ij}^t$. So peer i gives more weight to recommendations that are considered to be of a high confidence and that come from peers who are more credible.

Overall Trust. Base on the direct trust value and indirect trust value calculated above, the overall trust value of peer i to peer j's service (denoted by O_{ij}^t) is defined in formula (6).

$$O_{ij}^t = \lambda * D_{ij}^t + (1-\lambda) * R_{ij}^t \qquad (6)$$

Where D_{ij}^t denotes the direct trust value of i to j, R_{ij}^t is the indirect trust of peer j according to peer i, the "self-confidence factor" is denoted by λ, which means that how a peer is confident to its evaluation of direct trust value. $\lambda = Exp(i, j).n / m$, $Exp(i, j).n$ is the number of the direct interactions considered, and m is the maximum number to be considered for a peer, and the upper limit for λ is 1.

3.2 Recommendation Credibility

Any peer in our system will also maintain a recommendation credibility table for all the other peers it has got recommendations from. Suppose peer i has got some recommendations with peer k during the last D time units, the entry for peer k is denoted as $REC(i, k) = \langle n, REC\_LIST \rangle$, where n is the number of credibility ratings, and REC_LIST is an index in which these credibility ratings are kept.

In more detail, after having an interaction with peer j, peer i gives its rating about j's service performance as V_{ij}. Now, if peer i received recommendation $<V_{kj}, CI_{kj}>$ from peer k, then the credibility rating value V_w for peer k about this recommendation is given in the following formula:

$$V_w = 1 - |V_{kj} - V_{ij}| \qquad (7)$$

The credibility rating value V_w is set to be inversely proportional to the difference between a witness recommendation value and the actual performance (e.g. higher difference, lower credibility).

The rating about peer k's credibility — $r = (i, k, t, V_w, CI_w)$ — is then appended to peer i's recommendation credibility table. t is the time of peer k providing peer i the recommendations about peer j, $CI_w = CI_{kj}$.

The recommendation credibility of peer i to peer j at time t is denoted by Cr_{ij}^t, it is a [0, 1]-valued function which represents the confidence formed by peer i about the truthfulness of j's recommendations. This function is local and is evaluated on the recent past behavior of both peer i and j. It is locally used to prevent a false credibility from being propagated within the network. The credibility trust value Cr_{ik}^t is calculated as follows:

$$Cr_{ik}^t = \begin{cases} \dfrac{\sum_{b \in Rec(i,k).REC\_LIST} e.V_w * \alpha^{-e.t} * e.CI_w}{\sum_{b \in Rec(i,k).REC\_LIST} \alpha^{-e.ts} * e.CI_w} & if \ Rec(i,k).REC\_LIST \neq \varnothing \\ c_0 & otherwise \end{cases} \qquad (8)$$

Where α is the decay factor in the range of (0, 1) using equation (1), So, it can cope with more complex malicious strategies, for example, peers could try to gain trust from others by telling the truth over a sustained period of time and only then start lying. $e.CI_w$ is the level of confidence about the recommendation value, and $e.V_w$ is the value of the credibility rating e. If no such ratings has been recorded, we will assign the default credibility trust value, denoted by c_0, to peer j.

Our credibility model considers the level of confidence about the recommendation value. Giving incorrect recommendation can decrease the recommendation credibility of a peer. So, a peer can lower the level of confidence for opinions about which it is not very sure, therefore risking less loss of credibility in case its judgment is incorrect. If a weak recommendation is inaccurate, the recommendation credibility does not diminish quickly. A peer can not be penalized as much for an incorrect opinion that was based on a small number of interactions and/or a large variation in experience. Then, honest peers will not be falsely classified as lying.

4 Experimental Evaluation

We will now evaluate the effectiveness of the reputation mechanism we proposed by means of experiments. Our intention with this section is to confirm that it is robust against the collusion and badmouthing attacks, that it can effectively detect dishonest recommendations in a variety of scenarios where more complex malicious strategies are introduced.

4.1 Simulation Setup

In our simulation, we use the topology of the system and the deception models as Bin Yu's reputation mechanism. In order to empirically evaluate our new reputation mechanism against more complex strategies, we make some changes. In our simulation experiment, the quality for a peer to be a SP (service provider) is independent of the quality for a peer to be a rater which provides recommendation. We first define the types of qualities of both SPs and raters used in our evaluation. Three types of behavior patterns of SPs are studied: good peers, fixed malicious peers and dynamic malicious peers. Good peers and fixed malicious peers provide good services and bad services without changing their qualities once the simulation starts respectively. Dynamic malicious peers alter their behavior strategically. The behaviors of peers as raters can be one of the three types: honest peers, fixed dishonest peers and dynamic dishonest peers. Honest and fixed dishonest peers provide correct and incorrect feedback without changing their patterns respectively. Dynamic dishonest peers provide correct feedback strategically, for example, the dynamic dishonest peers which tell the truth over a sustained period of time and only then start lying.

Our initial simulated community consists of N peers, N is set to be 128. The percentage of the bad SPs is denoted by pb, the percentage of the bad raters is denoted by pf. Table 1 summarizes the main parameters related to the community setting and trust computation. The default values for most experiments are listed. In the default setting, 50% malicious peers are fixed malicious service providers, 50% malicious peers are dynamic malicious service providers, with 50% probability giving service maliciously. The dishonest raters are fixed dishonest peers which give complementary rating and the level of confidence is set to 1. We divide the total simulation time into multiple simulation time units. In every time unit, each peer initiates a single request that can be satisfied by any of the potential service providers. Every peer issues one service request per simulation round. When a peer receives a query, it answers it with res probability, or refers to other peers. res is set to 0.1 in the experiments. Two transaction settings are simulated, namely random setting and trusted setting. In random setting, peers randomly pick a peer from candidate peers who answer the service request to perform transactions. In trusted setting, peers select the reputable peer who has the maximal reputation value. The simulation program has been implemented in Java programming language.

Table 1. Simulation Parameters

Parameter	Description	Default
N	number of peers in the community	128
pb	percentage of malicious peers in the community	80%
pf	percentage of dishonest raters in the community	80%
res	probability peer responds to a service request	0.1
λ	Self-confidence factor	dynamic
α	the decay factor	0.9
c_0	initial credibility value	0.5
TTL	bound of the referral chain's length	4
B	branching factor	2
ρ	exaggeration factor	1
D	sliding time window	10
M	the threshold number of interactions in formula (2)	5

4.2 Effectiveness of the Reputation Mechanism

This simulation evaluates the immunity of our reputation mechanism to the collusion and badmouthing attacks. This set of experiments demonstrates the benefit of reputation mechanism we proposed, peers compare the trustworthiness of peers and choose the peer with the highest trust value to interact with. A transaction is considered successful if both of the participating peers are good peers, otherwise is a failure transaction. We define successful transaction rate as the ratio of the number of successful transactions over the total number of transactions in the community up to a certain time. A community with a higher transactions success rate has a higher productivity and a stronger level of security. The experiment is performed in both non-collusive setting and collusive setting. We show the benefit of our reputation mechanism compared to a community without any trust scheme. We also compare the performance of our scheme against the trust management scheme proposed by Bin Yu.

Figure 1 shows the rate of success transactions with respect to the number of time units in collusive and non-collusive setting. We can see an obvious gain of the transaction success rate in communities equipped with a trust mechanism either in non-collusive setting or in collusive setting. Both our reputation mechanism and Bin Yu's scheme can help peers avoid having interactions with malicious service providers in both setting, malicious peers are effectively identified even when they launch a collusion attack. This confirms that supporting trust is an important feature in a P2P community as peers are able to avoid untrustworthy peers. While in the collusive setting, dishonest peers' collusive behaviors hardly disturb honest peers' judgment. It needs more interactions to differentiate good peers from bad peers. Moreover, it is observed that Bin Yu's scheme needs more interactions to differentiate good peers from bad peers in both setting, so our reputation mechanism outperforms Bin Yu's reputation mechanism.

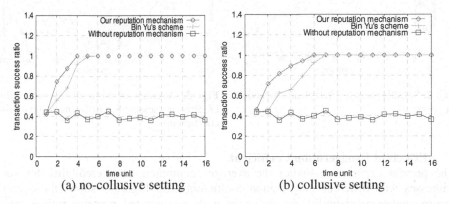

(a) no-collusive setting   (b) collusive setting

Fig. 1. Effectiveness against the collusion and badmouthing attacks

4.3 Predicting Honesty

We now define a useful metric to evaluate the performance of our proposed recommendation credibility model.

Definition 1. The average recommendation credibility of a witness W_j is

$$C r e_j = \frac{1}{N} \sum_{i=1}^{N} C r_{ij} \qquad (9)$$

Where Cr_{ij} is the credibility value of witness W_j from peer Pi's acquaintance model [6], and N is the number of peers in whose acquaintance model W_j occurs.

- **Sensitiveness to Strategically Alter Behaviors of Peers**

The goal of this experiment is to show how credibility model we proposed works against strategic dynamic personality of peers. We simulated a community with all good peers but a dynamic malicious rater with dynamic personality. We simulated two changing patterns. First, peer could try to gain trust from others by telling the truth over a sustained period of time and only then start lying. Second, the peer is trying to improve its recommendation trust by telling the truth.

Figure 2 illustrates the changes of recommendation credibility of both the peer who is milking its recommendation trust and the peer who is building its recommendation trust in the whole process. The results indicate that our reputation mechanisms is very sensitive to peers' strategically alter behaviors.

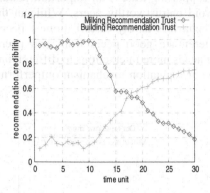 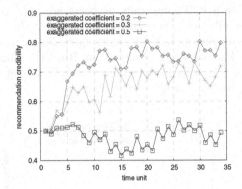

Fig. 2. Sensitiveness to strategically behaviors of peers

Fig. 3. Average recommendation credibility of alter witnesses for different exaggeration coefficients

- **Effects of Exaggeration Coefficient**

The present experiment studies the average recommendation credibility for such witnesses with different exaggeration coefficients [6]. Figure 3 shows the average recommendation credibility for witnesses with exaggerated negative ratings when exaggeration coefficient ρ is set to 0.2, 0.3, and 0.5, respectively. The results indicate that our approach can effectively detect witnesses lying to different degrees. For the witnesses with exaggerated negative ratings, their average recommendation credibility reaches to about 0.8, 0.7, and 0.5, respectively, after 10 time unit. So, the marginal lying cases can be detected.

- **Impact of Level of Confidence**

In the above two experiments, we only considered peers providing service with fixed personality. This experiment considers dynamic attack. An attacker, with x% probability, behaves maliciously by giving malicious service. In the other times, it behaves as a good peer. In this experiment, 80% peers are dynamic attackers with 50% probability giving service maliciously, other peers are good peer, and all the peers provide honest recommendations. The recommendation trust metrics has been observed to understand if honest peers assign fair recommendation trust values to each other.

Figure 4 shows the recommendation credibility of honest peers in this setting, we can conclude that: without level of confidence, a peer may be penalized for an incorrect opinion that was based on a large variation in experience. Our approach allows a peer to determine the level of confidence about its recommendation. Giving incorrect recommendation can decrease the credibility of a peer. So, a peer can lower the level of confidence for opinions about which it is not very sure, therefore risking less loss of credibility in case its judgment is incorrect. If a weak recommendation is inaccurate, the recommendation credibility does not diminish quickly. A peer can not be penalized as much for an incorrect opinion that was based on a small number of interactions and/or a large variation in experience. Then, honest peers will not be falsely classified as lying.

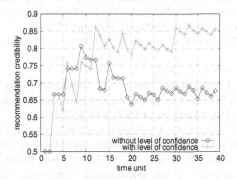

Fig. 4. Impact of level of confidence

5 Conclusions and Future Work

We present a new P2P reputation management system. Within this system a peer can reason about trustworthiness of other peers based on the available local information which includes past interactions and recommendations received from others. Recommender's credibility and confidence about the information provided are considered for a more accurate calculation of reputations and fair evaluation of recommendations. Theoretic analysis and simulation show that the reputation mechanism we proposed can help peers effectively detect dishonest recommendations in a variety of scenarios where more complex malicious strategies are introduced. The precision of inaccuracy detection is improved, more marginal lying cases can be detected, and honest witnesses will not be falsely classified as lying because of an increased fluctuation in a provider's performance.

Interactions (service providing and recommendation providing) that occur within a sliding window of width D are considered, therefore, storage requirements for storing trust information are tolerable. The main overhead of our reputation mechanism comes from the reputation queries. In a service session, one provider is selected but reputation values about other providers are deleted. Thus, reputation values about unselected providers can be cached. Since a peer obtains more acquaintances with time, number of cache entries and cache hit ratio increase with time. By this way, we can reduce the overhead of our reputation mechanism comes from the reputation queries.

As a next step, we will be evaluating our reputation mechanism as applied to a peer-to-peer network.

References

1. Oram, A.: Peer to Peer: Harnessing the power of disruptive technologies (2001), ISBN 0-596-00110-X
2. Aberer, K., Despotovic, Z.: Managing Trust in a Peer-2-Peer Information System. In: The Proceedings of Intl. Conf. on Information and Knowledge Management (2001)
3. Kamwar, S.D., Schlosser, M.T., Garcia-Molina, H.: The Eigentrust Algorithm for Reputation Management in P2P Networks. In: The Proceedings of the twelfth international conference on World Wide Web, Budapest, Hungary (2003)
4. Damiani, E., De Capitani di Vimercati, S., Paraboschi, S., Samarati, P.: Managing and sharing servents' reputations in p2p systems. IEEE Transactions on Data and Knowledge Engineering 15(4), 840–854 (2003)
5. Xiong, L., Liu, L.: PeerTrust: Supporting reputation-based trust in peer-to-peer communities. IEEE Transactions on Data and Knowledge Engineering, Special Issue on Peer-to-Peer Based Data Management 16(7), 843–857 (2004)
6. Yu, B., Singh, M.P., Sycara, K.: Developing trust in large-scale peer-to-peer systems. In: Proceedings of First IEEE Symposium on Multi-Agent Security and Survivability (2004)
7. Junsheng, C., Huaimin, W., Gang, Y.: A Time-Frame Based Trust Model for P2P Systems. In: Proceedings of 9th International Conference on Information Security Cryptology, Seoul, Korea (2006)
8. Lam, S.K., Riedl, J.: Shilling recommender systems for fun and profit. In: Proceedings of the 13th World Wide Web Conference (2004)
9. Srivatsa, M., Xiong, L., Liu, L.: TrustGuard: countering vulnerabilities in reputation management for decentralized overlay networks. In: WWW 2005, pp. 422–431 (2005)
10. Withby, A., Jøsang, A., Indulska, J.: Filtering Out Unfair Ratings in Bayesian Reputation Systems. In: AAMAS 2004. Proceedings of the 7th Int. Workshop on Trust in Agent Societies, ACM, New York (2004)
11. Dellarocas, C.: Immunizing Online Reputation Reporting Systems Against Unfair Ratings and Discriminatory Behavior. In: ACM Conference on Electronic Commerce, pp. 150–157 (2000)
12. Chen, M., Singh, J.: Computing and Using Reputations for Internet Ratings. In: EC 2001. Proceedings of the Third ACM Conference on Electronic Commerce, ACM, New York (2001)
13. Feng, Z.J., Xian, T., Feng, G.J.: An Optimized Collaborative Filtering Recommendation Algorithm. Journal of Computer Research and Development 41(10) (2004) (in Chinese)
14. Huynh, T.D., Jennings, N.R., Shadbolt, N.: On handling inaccurate witness reports. In: Proc. 8th International Workshop on Trust in Agent Societies, Utrecht, The Netherlands, pp. 63–77 (2005)
15. Beth, T., Borcherding, M., Klein, B.: Valuation of Trust in Open Networks. In: Gollmann, D. (ed.) Computer Security - ESORICS 1994. LNCS, vol. 875, Springer, Heidelberg (1994)

A Framework for Business Operations Management Systems

Tao Lin, Chuan Li, Ming-Chien Shan, and Suresh Babu

SAP Labs, LLC
3475 Deer Creek Road
Palo Alto, CA 94304, USA
{tao.lin, chuan.li, ming-chien.shan, suresh.babu}@sap.com

Abstract. Today's common business phenomena, such as globalization, outsourcing of operations, mergers and acquisition, and business renovation require rapid changes of business operations. In order to support such changes, it is critical to gain the total visibility of the entire business operation environment and to understand the impact of the potential changes so that people can react properly in time. Business OPerations Management environment (**BOPM**) is a system that is on the top of existing business application systems and other IT systems and is responsible for users to understand, manage and change business operations. The development of an effective BOPM system is necessarily complex since it needs to deal with multiple dimensions of artifacts – people, information, processes, and systems – across the extent of a company's business operations, both within and outside the company boundaries, to provide the visibility and control, to discover where to make changes, to evaluate new business processes or changes, and to govern all business operations, particularly new or changed ones. This paper presents a conceptual architecture for BOPM framework and discusses some key technical features.

Index Terms: Business Operations Management, Interactive Data Visualization, Dynamic Data Modeling, Model Structure.

1 Introduction

Globalization, business operation outsourcing, contract manufacturing, mergers and acquisitions, and business renovation require a company to shift its business strategy more globally and dynamically. The shifting business environment makes the management of business operations extremely challenging to an enterprise. In order to face the challenges posed by constant business shifts, enterprises need unprecedented capabilities to gain the entire and holistic picture of their business operations across the value chain in order to gain the total visibility of their business operations; to analyze the impact of these changes to business operations; to understand potential responses to the shifting business environment and underlying constraints through analysis and simulation; and to implement and monitor the targeted operational changes and the total cost of ownerships with the changes of business operations or new ones.

Nowadays, enterprises are operated in an environment that supports different business roles employed in various functions across the enterprise, as well as beyond

B. Benatallah et al. (Eds.): WISE 2007, LNCS 4831, pp. 461–471, 2007.
© Springer-Verlag Berlin Heidelberg 2007

the company boundaries; business processes that span the company operations and go across boundaries to include the operations of business partners, such as customers and suppliers; business information that is extensive and distributed; and business applications and systems, such as Enterprise Resource Planning (ERP), Customer Relationship Management (CRM), Supply Chain Management (SCM), Supply Relationship Management (SRM), Product Lifecycle Management (PLM), and a multitude of other systems that are required to operate the enterprise. Any change in business strategy, whether it is a merger, an outsourcing decision, or a competitive challenge, requires the enterprise to rationalize its operations across the value chain, which involves managing the key dimensions of enterprise artifacts: people, information, processes, and systems, against targeted operational goals.

Today, new trends in technology, such as enterprise service-oriented architecture, software as a service, Enterprise 2.0 technologies, event-driven architecture, embedded analytics, RFID/Auto-ID technology, and Web 3.0 provide new opportunities to enrich the existing business applications. Evaluating the potential impact of these new technologies to the existing IT infrastructure and raising the requirements for changes are critical to manage the success of the business for a company.

A new system on the top of the expanded IT infrastructure including all business application systems and other IT systems from related enterprise organizations is required. This new system can support users to explore the artifacts and the relationships between the artifacts (understand/analyze), to find where to make changes for a new business need (discover), to create a new business process or modify an existing business process to support the new business need (enhance), to evaluate the impact of the change, and to monitor the entire environment, in particular the impact of the changes (monitor). This system will help business users to untangle the current complex web operations that characterizes many enterprises and show how the key dimensions (people, information, processes, and systems) are interrelated to that enterprise, how business shocks and shifts impact these associations and what constraints (such as unmatched supply, extent of product overlap in a merger, gaps in service coverage, etc) lie ahead for the operational changes contemplated. A Business OPeration Management (BOPM) is such a new system that sites on the top of existing business application and other IT systems for supporting users to understand, change, manage, and monitor these artifacts. BOPM can also be used to evaluate the impact of new technology trends to existing IT infrastructure. For convenience, we define a BOPM environment that consists of a BOPM system, all business application systems and other IT systems to be dealt by the BOPM system.

The rest of the paper is organized as follows. Section 2 describes two use cases: one with business process renovation driven by a new tax policy, and the other one with transportation business process outsourcing. Based on these two use cases, both business and technical requirements of a BOPM are discussed in Section 3. Section 4 presents a conceptual architecture and some key technical features for the framework of the BOPM environments. Section 5 concludes this paper, shows some related works, and discusses future works.

2 Use Cases

This section will describe two use cases: a business operation renovation use case (modifying business operations due to a new tax policy), and a business process outsourcing use case (outsourcing transportation services). These use cases highlight the challenges in business operations management, and provide insight into the key requirements for a BOPM business application framework, which is presented in Section IV.

A Business Process Renovation Use Case

Situation: A country, a major trading area and manufacturing base of an enterprise, issues a new tax policy that will take effect within a year. This enterprise needs to change all its business operations to be compliant to the new tax policy.

Implications: The new tax policy has numerous operational implications for the company's operations. The enterprise needs to identify the key artifacts impacted – the roles, the information changes (master data, transactions, etc), the business processes, and the systems – to understand how operational changes can be implemented and tracked. A key initial phase is to understand the business areas, business operations, and activities affected.

A tax policy change has wide ramifications for operations: sales orders have to include tax provisions; purchasing and vendor approved processes have to be revised to accommodate the new regulations such as governmental declaration online or value-added determination or withholding; roles that handle taxes have to be retrained to recognize new needs and to monitor compliance; pricing may be affected by the new provisions; reporting, governance, and declaration needs will change; the underlying systems may need comprehensive integration with the country's government systems for verification, tax processing, and compliance; and so on. The company needs to plan for the tax policy transition that will take effect within a year.

Implementation: We start with analyzing of what business processes is affected by tax policies. Then we select a specific business process to understand how this specific business process should be modified. Once the business process is changed, we will monitor the execution of the modified business process with simulated data to check whether the modified process can meet the requirements. Further modifications will be made till the business process meets the requirements.

Ideally, the associations between all the relevant artifacts have been established. Then we can check which business process needs to deal with the functions or operations that the selected business process have modified. Here we adopt another approach. For the functions or operations need to be modified for this business process, we trigger message events to report which business processes are using these functions or operations. Through this exercise, we can identify the business processes that need to be modified.

A Business Process Outsourcing Use Case

Situation: In order to reduce the cost and to improve the efficiency for handling the transportation operations in an organization, outsourcing the transportation services to a 3rd party organization has become a common practice. This use case is about the development of a 3rd party transportation service system.

Implications: Traditionally, each individual organization is used to manage the transportations of the shipments to its customers. It turns out that each individual organization needs to deal with all carriers in order to find the best deal. Also, each individual organization has to monitor the entire transportation process of a delivery. The engagement and negotiation with carriers, and the monitoring of the transportation processes have become common tasks to all individual organizations.

With a 3rd party transportation service system, the organization will inform the transportation service company (customer) when an organization confirms a purchase order from a customer. The transportation service company will engage with all available carrier companies to find the best deal for this delivery. The transportation service company can also monitor the delivery processes and record the billing and paying operations to provide further services.

The 3rd party transportation service company can provide better services as it can access more carriers and have more information on weather, transportation, and carrier performance. What is more, as this transportation service company can also monitor the paying and the engagement behaviors of the customers, it can also provide better services to carriers as well. Some additional services, such as detecting wrong delivery, delivery reconciliation, fraud detection, and anti-counterfeit, can be supported by using this new transportation mechanism.

To switch from current transportation processing mechanism to the outsourcing transportation mechanism, we have to provide the same security and authentication capability to ensure the data privacy for each individual organization. Furthermore, as the communication with business applications in each individual organization is the key, communication will be through web services. As there are strong dependency of the services while composing services for this transportation service system, maintaining the dependency of the services is also a critical issue.

Implementation: In this new 3rd party transportation service system, there are two main business process flows: shipment arrangement; and shipment delivery monitoring. The shipment arrangement process is triggered by purchase order confirmation from the customer organization. Based on the purchase order, the transportation system will request the offers from all carriers and select the best offer from the qualified ones. The shipment arrangement process will remind the carrier to pick the delivery and also inform the customer organization to pay the carrier once the shipment has been picked or delivered. The shipment delivery monitoring process is to monitor the delivery process of a delivery. The customer organization and carrier will be informed whether traffic situation can cause the delay to any deliveries. Moreover this process will develop credit records for all customer organizations and carriers. The 3rd party transportation service system can find better deals for customer organizations and carriers with better credit history.

We start with the analysis of the transportation related business processes for an individual organization. Based on these business processes, we add the 3rd party transportation service system as an external system based on enterprise service composition.

3 Requirements

The requirements can be divided into two categories: business requirements and technical requirements.

Business Requirements

From business perspective, it requires a BOPM to provide the following features:

- *To gain the total visibility of the business operations.* A BOPM needs to handle different artifacts: people, information, processes, and systems. To understand, modify and monitor the business operations, a mechanism is required to help users to gain the total visibility of the artifacts for a BOPM environment.

- *To extract static relationships between artifacts.* Examples of static relationships are the relationships between processes and systems, between people's roles and processes, between database tables and business processes, and between object classes. Static relationships between artifacts are important for users to understand a BOPM environment.

- *To dynamically extract relationship between artifacts.* Run-time relationships between artifacts are also very important in order to understand system performances and execution behaviors. Examples of dynamic relationships are the relationships between users' interactions and business process instances, and between the triggered events and business process instances.

- *To enable dynamic modeling.* There is a large volume of data related to the artifacts and potential relationships. It requires a user-friendly environment that enables the user to focus on what she/he is interested. To model what the user is interested and to provide tools to extract relationships based on what the user is interested are critical, for example, the creation of a model with a business process and the events associated to the execution of this business process. From the report of the association of the events with the business process, one can monitor the execution of the business process.

- *To dynamically extract interested information.* As the data can be retrieved from a BOPM environment is huge and a large amount of the data is dynamic. Dynamic retrieving of the interested data can greatly reduce the cost and provide users with updated information.

- *To be able to add and enhance a business process.* As business operations keep changing, new business operation processes need to be added and existing ones need to be modified. This requires a business process to be easily added or modified in a BOPM environment.

- *To validate new or changed business processes.* A BOPM is always a mission-critical environment. It is important to have a stable and reliable environment for a BOPM. Therefore, it is required to validate new or modified business processes before deployment. To monitor the execution of business processes and report their behaviors is the basics for a BOPM system.

- *To model and extract relationships cross organizations.* A BOPM is a cross organization environment. To gain the understanding of a BOPM, it is important to be able to model and extract relationships cross organizations. However, the data privacy, security and authentication must be managed properly.

- *To model and validate with constraints.* There are various constraints within a BOMP, for examples, organization boundary constraints, performance constraints, and accessing constraints. While modeling and validate new artifacts, those constraints need to be retained.

Technical Requirements

A BOPM needs to support the common functionalities for business management systems. In additional, in order to support the business requirements discussed above, a BOPM needs the following functionalities:

- *Modeling complex relationships.* In order to understand different artifacts: people, information, processes and systems, the static and dynamical relationships between those artifacts, an effective modeling mechanism is required to model the complex relationship. The modeling mechanism needs to support the modeling of any categories of the artifacts, the relationships between the entities of one category and the entities in another category, and also the operations for extracting those relationships.
- *Dynamic and interactive modeling.* Due to the volume of artifacts and the dynamic nature of these data, a user needs to explore the data based on the needs. Through dynamic and interactive modeling mechanism, a user can interactively build the model at run-time.
- *Interactive visualization and data extraction.* A picture is worth a thousand words. In particular visualization is very powerful to help people to understand complicated relationships. However, due to the potential volume of data in a BOPM system, a user can only browse a portion of data based on his/her current interests. It is required to support interactive visualization mechanism that allows the user to control the visualization to display the interested data. Data extraction needs to be integrated with visualization coherently. Through the visualization mechanism, a user can extract more data based on his/her interests.
- *Simulation.* To validate a new or a modified business process before deployment, some thorough testing is needed as a BOPM is always a mission-critical environment. A simulation is needed to generate relevant data and data flow for validating proposed business processes.
- *Embedded analysis.* As the large scope of systems that a BOPM needs to support, it is impossible to retrieve all related data into centralized data storage to analyze. The environment needs to support embedded analysis capability. Thus BOPM can manage and distribute business logics for embedded analysis with the business application and other IT systems.
- *Flexible event triggering and monitoring.* Through triggered events, a user can monitor the execution of a business process. Due to the potential number of business process instances in a BOPM environment, a user can only deal with limited number of business process instances. A flexible event triggering and monitoring mechanism enables the trigger and monitor of events to base on the interested business processes.

4   Basic Services

The basic services for a BOPM can be divided into four categories illustrated in Figure 1:

- *Analyze and understand*. To enable the users to analyze and browse the artifacts and their relationships in a BOPM environment statically and dynamically.
- *Discover*. To identify where to make changes in order to meet a new business need. In the use cases presented in Section 2, two business needs are presented: business renovation based on a new policy; and business process outsourcing. In both use cases, it requires to find the relevant artifacts to be affected by a new business need.
- *Enhance*. To change or create a business process or to create new systems to support the business needs. In the use case discussed in Section 2, first use case requires the change of a business process and second one needs to develop a new 3rd party transportation service system.
- *Monitor*. To monitor the behavior or performance of the changed BOPM environment. Through the monitoring service, it aims to tell whether the changed environment can meet the new business needs. The monitoring services are needed by both simulated and deployed environments.

The basic services in these four categories form a cycle. Through monitoring services, a new business change cycle will start if issues with business processes been identified then the new iteration of business renovation starts.

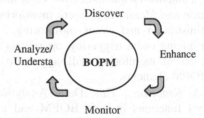

Fig. 1. Basic services with a BOPM system

Conceptual Architecture

Figure 2 presents a conceptual architecture for a BOPM environment that consists of BOMP elements and non-BOPM elements. BOPM elements are the modules of a BOPM system. Non-BOPM elements are the business application and other IT systems including existing or new systems. If a new business services is added to an existing business application, it is considered as part of a non-BOPM element.

A BOPM system consists of the following major elements:

- *Data Modeling & Retrieving*. This module is for modeling and retrieving data elements in particular relationships between data elements. Modeling defines the categories of artifacts, the relationships between the artifacts within or cross

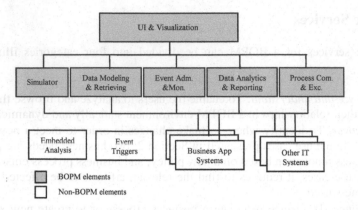

Fig. 2. Conceptual Architecture for a BOPM environment

categories, and the operations for retrieving information of the artifacts and their relationships. The data retrieving can be divided into two categories: static and dynamic. Dynamic data are the data been generated during the executions. There are two channels to obtain dynamic data: data retrieving (pulling); and updating (pushing). Data modeling needs to support the modeling of static and dynamic data including both data retrieving and data updating.

- *Process Composition & Execution.* In a BOPM environment, a business process consists of business-steps which are conducted by a business application system or one of other IT systems. Business Composition & Execution module is responsible for the composition and execution of business processes.

- *Event Administration and Monitoring.* Event monitoring module consists of two parts: event administration and event monitoring. Event administration is responsible for arranging event triggering operations in both BOPM and non-BOPM elements. Event monitoring will aggregate the events triggered by BOPM and non-BOPM elements.

- *Data Analytics & Reporting.* This Data Analytics module manages the embedded analytics functions in both BOPM and non-BOPM elements and aggregates the data collected by the embedded analytical functions from BOPM and non-BOPM elements. The Data Reporting module generates reports from the aggregated data. Visualization is for users to browse the data generated by the Data Reporting module. However the data generated by Reporting Module can also be stored for later analysis or browsing.

- *Simulator.* The simulator will generate data and data flow based on the business processes that need to be executed. The rest of BOPM environment will behave the same as dealing with normal process executions. Simulation is used to validate either new or modified business processes. Once the validation is finished, the simulated data will be removed from the environment.

- *UI & Visualization.* UI and visualization module is the operation windows for users to interact with a BOPM environment. Through this module, a user can create data model, interactively retrieve data, compose new business processes, validate new or changed business processes and monitor the behavior of the entire BOPM environment. The visualization module can also report the data pushed by event triggering and embedded analytics during the execution of business processes.

To support embedded analysis, libraries for embedded analytics can be added into the business applications and other IT systems. Based on the needs, the business logic will be pushed into the embedded analysis functions within business application or other IT systems from Data Analytics &Reporting module in a BOPM system. The embedded analysis functions will conduct analytical processing locally in those systems and aggregate information to report to the Data Analytics &Reporting module in a BOPM system. Similarly, event trigger is also a library that is added with the non-BOPM elements. Based on the needs, the logic for triggering events will be pushed down by Event Monitoring module in a BOPM system.

Some Key Features

Here we intend to discuss some key features for a BOMP framework.

- Complex Relationships Modeling

$M = \{N, R\}$;

$N = \{N_1, N_2, ..., N_i, ..., N_n\}$;

$R = \{R_1, R_2, ..., R_j, ..., R_m\}$;

$N_i = \{d_i^a, \{n_1^i, n_2^i, ..., n_s^i\}, \{op_1^i, op_2^i, ..., op_u^i\}, \{c_1^i, c_2^i, ..., c_v^i\}\}$;

$R_j = \{d_j^r, \{r_1^j, r_2^j, ..., r_t^j\}, \{op_1^j, op_2^j, ..., op_x^j\}, \{c_1^j, c_2^j, ..., c_y^j\}\}$;

$d_i^a = \{s_i^a, \{S_1^i, S_2^i, ..., S_k^i\}\}$;

$d_j^r = \{s_j^r, \{N_1^j, N_2^j, ..., N_l^j\}\}$.

This framework supports an extensible and flexible modeling mechanism to model complex relationships with multiple sets of artifacts. Model M consists of a set of nodes $N = \{N_1, N_2, ..., N_i, ..., N_n\}$, and a set of relationships $\{R_1, R_2, ..., R_j, ..., R_m\}$ which are extracted between the elements in the same node set.

A Node model consists of the definition d_i^a, the Node elements $\{n_1^i, n_2^i, ..., n_s^i\}$, operations $\{op_1^i, op_2^i, ..., op_u^i\}$, and constraints $\{c_1^i, c_2^i, ..., c_v^i\}$. The definition consists of the data schema definition s_i^a and the sources $\{S_1^i, S_2^i, ..., S_k^i\}$ from which the Node elements of the Node set are extracted. The examples of a source can be a system, a database table, and a process. The operations include the basic services for handling the Node elements and the constraints define conditions and restrictions for these operations. A Relation model is similar to the one for a Node model. In a Relation Model, the relationships are extracted from the Node sets rather than the sources directly.

A model structure is defined as the model mechanism described above without the Node elements and Relation elements.

- **Dynamic Modeling and Interactive Exploration**
In the modeling mechanism described above, model structure is part of the model. A user can define the elements for the model structure at run-time. The data elements in both Node and Relation data sets can be retrieved or generated by users' interests. This provides a flexible and extensible mechanism for a user to explore the data in a BOPM environment dynamically. By providing a navigation tool based on the model structure, a user can interactively define the model structure and control the data retrieving and updating.

- **Model-Based Process Correlation**
In the two use cases described in Section 2, both business renovation and business process outsourcing are required to associate business processes with other artifacts in a BOPM environment, such as system, database tables, different user roles, and other business processes.

A business process is defined with BPMN [1]. The relationships between a business process and the elements of the other artifacts can be easily defined and retrieved. Data retrieving and data updating can be arranged based on the interested spots.

- **Event Triggering and Embedded Analytics**
Event triggering and embedded analytics are plug-ins to both BOPM and non-BOMP modules. Through the plug-ins, data can be pre-processed, aggregated and updated to the BOPM system. The Event Monitoring and Data Analytics modules in BOPM will further process the data and generate corresponding reports.

Taking the use cases described in Section 2 again. Through the modeling mechanism, one can easily get the information of the related systems, business processes, and other artifacts related to the interested business processes or database tables, and the Event Administration and Data Analytics Administration modules can manage event triggering and embedded analytics in the corresponding systems, business processes and operations.

- **Visualization and Monitoring**
Visualization module in this system has three major functions:

- Interactively exploring data in the model. The visualization module can display data of the model through graph-based display. Multiple Node sets and Relation sets can be displayed in the same view. The user can further include related Node sets or Relation sets in the display and also remove Node sets and Relation sets.
- Interactively generating data in the model. As the size of the potential data in a BOPM can be huge, it can be very inefficient if updating the entire data model all the time. Through a navigation tool, a user can select the data sets including both Node and Relation according to his/her focuses.
- Interactively developing model structure. A BOPM is an evolving environment. To support new business requirements, a user needs to further extend the model. Through visualization, the user can interactively extend and modify the existing model at run-time.

Based on users' focus, the profile of event triggering and embedded analytics will be dynamically updated with a BOPM environment. The updated data will be combined with the existing displays and also be highlighted.

5 Summary

Currently, common business phenomena, such as globalization, business operation outsourcing, mergers and acquisitions, and business renovation, require rapid changes of business strategy and operations. A BOPM is a system that is on the top of existing business application systems and other IT systems for users to understand, discover, enhance, and monitor business operations for new business needs.

This paper presents a framework for a BOMP system. To support the basic services in a BOPM, an extensible and flexible modeling mechanism is critical. This paper presents such a modeling mechanism that is the core of the framework. Based on this modeling mechanism, the users can easily explore the data through visualization, extend and modify an existing model, request data update from the event triggering and embedded analytics operations by using relevant systems and processes only. This mechanism can help the users to gain the visibility of the entire BOPM environment. Through the interactive visual tool, the users can interactively model and navigate the entire artifact space in a BOMP environment.

The success of the modeling of Entity-Relationship [2] has greatly changed the state-of-art of the database development. UML [3] has been a long-time effort for the modeling of elements associated to software development including both static and dynamic aspects. We have applied multiple-layer and multiple relationship modeling and visualization technology for both the understanding of complex software systems and also complicated geological models [5, 6, 7].

The framework presented in this paper is still under development. Most modules discussed are based on our previous works and 3rd party contributions. More coherent integrations of the concepts and systems are also required for our future work.

References

[1] White, S.A.: Business Process Modeling Notation (BPMN) Version 1.0 (May 2004), At the BPMN website http://www.bpmn.org
[2] Chen, P.P.: The Entity-Relationship Model - Toward a Unified View of Data. ACM Transactions on Database Systems 1(1), 9–36 (1976)
[3] Ambler, S.W.: The Object Primer: Agile Model Driven Development with UML 2. Cambridge University Press, Cambridge (2004)
[4] Booch, G., Rumbaugh, J., Jacobson, I.: The Unified Modeling Language User Guide. Addison Wesley Longman Publishing Co., Inc., Redwood City, CA (1999)
[5] Lin, T.: Visualizing relationships for real world applications. Workshop on New Paradigms. In: The Proceedings of ACM Information Visualization and Manipulation Workshop, pp. 78–81 (1998)
[6] Lin, T., O'Brien, W.: FEPSS: A Flexible and Extensible Program Comprehension Support System. In: The Proceeding of IEEE 5th Working Conference on Reverse Engineering, pp. 40–49 (1998)
[7] Lin, T., Cheung, R., He, Z., Smith, K.: Exploration of Data from Modeling and Simulation Through Visualization. In: The Proceedings of International SimTecT Conference, pp. 303–308 (1998)

A Practical Method and Tool for Systems Engineering of Service-Oriented Applications

Lisa Bahler, Francesco Caruso, and Josephine Micallef

Telcordia Technologies, One Telcordia Drive Piscataway, NJ 08854-4157 US
{bahler, caruso, micallef}@research.telcordia.com

Abstract. As software organizations develop systems based on service-oriented architectures (SOAs), the role of systems engineers (SEs) is crucial. They drive the process, in a top-down fashion, from the vantage point of the business domain. The SE, utilizing tools that allow work at a suitably high level of abstraction, creates service description artifacts that document service contracts, the obligations that govern the integration of services into useful applications. This paper describes a practical systems engineering methodology and supporting toolset for SOA that has been successfully used within the telecommunications domain. The methodology and toolset, named the STructured Requirements and Interface Design Environment (STRIDE), are based upon a high-level service description meta-model, and as such, encourage top-down service design. STRIDE promotes reuse of service models, as well as of the artifacts generated from those models, across the enterprise. STRIDE also embodies an effective service evolution and versioning strategy.

Keywords: SOA, Service Design Methodology, Model-Driven Development, Systems Engineering, SOA Tools, Enterprise Application Integration.

1 Introduction

Enterprises are increasingly adopting service-oriented architectures (SOAs) to achieve benefits of flexibility, agility, reuse, and improved ease of integration. In large-scale enterprises, it is typically the systems engineer who specifies what the services will do and how the services interact, in a top-down, service-description-centric fashion. This paper presents a practical approach for systems engineering of SOAs that has been successfully used in commercial and military telecommunications applications.

To be able to play this role effectively, the SE must be equipped with a service design methodology and tools that allow him to focus as much as possible on his area of expertise – the business domain in which the enterprise operates. The methodology and supporting tools described in this paper employ a service model abstraction that is simple, but sufficiently expressive for the needs of the SE. A service description meta-model that clearly defines the semantics of the design elements, as well as their composition into services, enables both design-time validation and model-driven development, in which the service model is later transformed into code artifacts for a target technology platform.

B. Benatallah et al. (Eds.): WISE 2007, LNCS 4831, pp. 472–483, 2007.

A service design abstraction for systems engineering has the added benefit of insulating designs from technology changes. As SOA-related standards evolve, the service designs, the core assets of the organization, are not impacted, and only the "generators" of downstream development artifacts need to be modified. The updates to the "generators" to accommodate changing standards can be managed consistently across the enterprise, rather than on an ad hoc basis for each service.

The rest of the paper is organized as follows: Section 2 presents the requirements for a SOA methodology that is robust and appropriate for enterprise-scale use. Section 3 describes a comprehensive service description model and its role in the methodology. The STRIDE toolset supporting the methodology is introduced in Section 4. Section 5 compares our approach with other work. We conclude in Section 6 with areas for enhancement based on our experiences with the methodology and tool in developing SOA applications.

2 Requirements for an Enterprise-Scale SOA Methodology

Within a software enterprise, good software engineering practice dictates design for reuse and management of change. The former is facilitated, to the extent that is practical, by enterprise-wide information models that are stored in a common service repository and the second by a coherent change management and service versioning strategy. These considerations lay the groundwork for enhanced interoperability of services that may be developed by separate development teams, but deployed together into a more comprehensive solution. These considerations should also be core to the service modeling abstractions provided for use by SEs.

Further, an enterprise-wide view of service assets modeled at the business level should translate into a uniform set of implementation artifacts that are generated directly from the service description models for use throughout the enterprise. The hand-off from the SE to the developers who implement service details must be handled in a uniform way. The SE will need to be guided by design governance rules, optimally built into tools.

2.1 Shared Information Model and Service Repository

A shared information model facilitates semantic interoperability across service interactions. In an enterprise with legacy software assets that have evolved independently, each legacy system commonly has an information model that is optimized for its own functionality. Often, the same semantic concepts are represented in different ways in the various models. By developing a common information model façade and mapping the individual legacy information models to this façade, the cost of service integrations is reduced.

Storage of the service models (as well as artifacts derived from them) in a common enterprise-wide repository promotes reuse of the models by both legacy and new applications. An enterprise-wide service naming and classification strategy aids in the organization of this repository.

2.2 Versioning Strategy

Deployed services evolve at different rates, and it is not practical to synchronize the upgrade of all service consumers simultaneously. A versioning strategy that allows incremental service upgrade is essential. As is common practice, semantic or syntactic changes to a service interface which make the new version incompatible with the previous one lead to an incremented major version number. Changes of a more minor nature, which can be handled by run-time message transformations for consumers and services at differing upgrade levels, result in an incremented minor version number.

The versioning strategy should be consistently employed throughout the enterprise, with version information being attached to service namespaces and generated implementation and documentation artifacts. Version-aware transformation code that handles run-time minor version incompatibilities should also be generated from the service descriptions.

2.3 Governance of Service Design and Implementation

Enterprise architects commonly devise a set of rules and guidelines for SEs and developers to follow in producing services. Such rules are intended to enhance service asset reuse and service interoperability, usability, maintainability, and ease of deployment. Governance rules may mandate adherence to the corporate naming and versioning strategies, the use and organization of a service repository, and the types of documentation and other service artifacts that are produced to describe services.

Governance rules can dictate the semantics and syntax of the higher level service modeling language employed by the SEs and how these models are transformed into lower level service assets like XML schemas and WSDL files. For instance, rules can constrain the style of schemas generated [7]. They can also restrict the use of constructs used in schema and WSDL files to subsets that are known to work well with the majority of deployment platforms or are commonly used in practice. For example, the rules can constrain the WSDLs to be WS-I (Web Service Interoperability organization) [3] compliant, as adherence to the WS-I guidelines is common best practice. (Similar guidelines to govern the use of XML schema are just evolving [4], necessitating reliance upon the experience of enterprise architects with target deployment platforms to determine what schema constructs may be used.)

There will always be a human component to governance. For example, a corporate governance board may decide what the level of granularity of services should be and what operations they should expose. The board may also oversee use of the service repository and ensure that service assets are reused when possible. The board needs to certify that written SOA guidelines have been followed and the proper tools used to produce the service artifacts. Indeed, without tools based upon an enterprise-wide meta-model, such as described in this paper, the manual governance would be very expensive and time consuming, ultimately involving the manual inspection of service description artifacts that have been hand-crafted, to ensure that they comply with the enterprise best practice guidelines.

Ideally, tools should embed the corporate governance rules, and automatically generate compliant lower-level service description artifacts, such as WSDLs and XSDs, from the higher-level models. Not only are these artifacts generated according to the enterprise-wide best practice guidelines, but the tools make it unnecessary for most people in an enterprise to master all the intricate details of technologies such as WSDL and XSD.

3 The Central Role of the Service Description

The service description (SD), which is a formal specification of the service, is a pivotal element of a SOA methodology, as it decouples the service implementation from its consumers by clearly defining the contract the service provider needs to fulfill to meet the service consumers' expectations and usage patterns. The SD comprehensively describes a service, including both functional and non-functional aspects. Importantly, the SD is independent of the implementation technology, and enables multiple conforming service implementations to be deployed within different environments. These key points distinguish the SD that is central to the SOA methodology presented in this paper from service descriptions authored in WSDL.

The service description plays a role in the entire service lifecycle, as illustrated in Figure 1. The lifecycle includes the analysis and design of the service, its implementation and assembly into a solution, the deployment and operation of that solution, and the folding of changes into a new revision of the service requirements. As the service progresses through its lifecycle, the technology-neutral specification is transformed into a concrete service implementation targeted toward a particular environment.

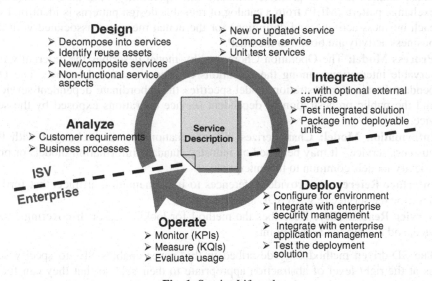

Fig. 1. Service Lifecycle

The SOA methodology presented in this paper exploits the rich service description with a model-driven approach [10] to support the systems engineering process through service design, validation of the design, and generation of downstream developmental artifacts. To be able to do this, the service description must be structured (i.e., machine processable) and have well-defined semantics. An overview of the SD meta-model is shown in Figure 2. Our SD meta-model has been inspired by the OASIS SOA Reference Architecture [11], which we have enhanced with message exchange patterns and service choreography and dependencies.

The service description contains or refers to the following key elements of the meta-model:

- **Identity & Provenance:** Identifies the service by name, provides various classification dimensions (such as by domain, functional area, or business process area), identifies the owner/provider of the service.
- **Requirements:** Includes both the observable and measurable real world effects of invoking the service (functional requirements), as well as non-functional aspects of the service (e.g., availability, reliability, throughput). Typically managed by a requirements management system.
- **Service Interface:** Specifies the business capability provided by the service, which may involve multiple business activities (or operations), each of which may have real world effects observable after it completes.
- **Behavior Model:** Describes the essential properties and constraints to correctly interact with the service through the action and process models.
- **Action Model:** Specifies the business activities (or operations) that constitute the service by describing the service message exchanges. The message definitions are based on a common information model (at the enterprise or departmental or functional domain level) to promote semantic interoperability. An abstract message exchange pattern (MEP) from a catalog of reusable design patterns is identified for each business activity, and definitions for the actual messages associated with the business activity are bound to the MEP.
- **Process Model:** The Operation Choreography model describes the externally observable interactions among the operations of a conversational service. The Dependent Service Orchestration model specifies the subordinate dependent services and the public orchestration of dependent service operations exposed by the service.
- **Information Model:** Characterizes the information that is exchanged with the business service. It may be based on industry standard information models or proprietary models common to the enterprise.
- **Interface References:** Provides references to implementation artifacts that further describe the service (e.g., WSDL and XSD files).
- **Service Reachability:** Describes the method for looking up or discovering a service, and the service access point.

The SD-driven methodology described in this paper enables SEs to specify services at the right level of abstraction appropriate to their role so that they can focus their efforts on the business domain. It also facilitates the bootstrapping of the service implementation via code generation based directly upon the comprehensive and structured systems engineering models produced by the SEs. We have contributed the SD meta-model to the TeleManagement Forum for consideration within the New Generation Operations Support Systems (NGOSS) Technology Neutral Architecture (TNA) working group.

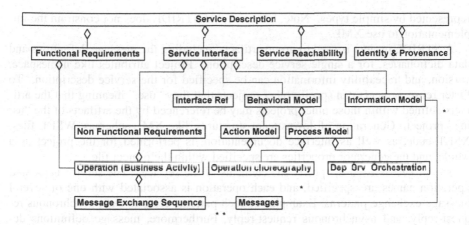

Fig. 2. Service Description Meta-model

4 STRIDE: STructured Requirements and Interface Design Environment

STRIDE is a service description authoring and generation tool used by SEs within Telcordia to develop telecommunications applications. Before STRIDE was developed, a documentation-centric SE methodology was used to design and describe message-oriented application interfaces. SEs produced tables showing both the structure of the messages exchanged over an interface and the data dictionaries of basic types upon which those structures depended. The decision was made at the outset of STRIDE development to create a methodology and tool that would allow the SEs to work with higher-level interface abstractions very similar to those they had been using. The tool would improve the process by enabling SEs to create models conforming to a well-defined interface description meta-model, from which would be generated documentation and other artifacts. The objectives for this approach were: the elimination of error-prone, manual work needed to translate SE outputs to developmental artifacts; the guarantee of conformance of those artifacts to governance guidelines; the ease of maintenance as system interfaces evolved; and support for reuse of interface components across the enterprise.

4.1 STRIDE Support of the Service Description

STRIDE supports the service description meta-model outlined in Section 3. The service description artifacts are stored as XML files that are described by XML schemas. These SD artifacts can be put under configuration control with other system source files.

The most basic STRIDE artifacts are message definitions, which are used to specify message structures, and data dictionaries, which are used to specify the primitive types used in message structures. Service input and output messages are described by a composition of message definition structures. When translated into XML schemas, message definitions are represented by complex types, while data dictionaries are represented by simple types. Note, however, that STRIDE does not constrain the implementation to use XML.

A STRIDE project is used to group artifacts, including the message definitions and data dictionaries, for a single service description. Project attributes like namespace, version, and traceability information can be specified for the service description. To foster reuse, projects can specify other projects that they "use," meaning that the artifacts defined within those other projects may be referenced by the artifacts of the "using" project. Generation of downstream structures, like XML schemas, WSDL files, XSLT code, as well as interface documentation, is performed for the project as a whole, and the generator properties are specified within the project file.

The project is used to specify the elements of the service. The service name and operation names are specified, and each operation is associated with one of several message exchange patterns. Examples of such patterns are one-way, synchronous request-reply, and asynchronous request-reply. Furthermore, message definitions defined within the project are bound to each abstract message associated with the selected message exchange pattern.

4.2 Message Definitions

The message definition meta-model is semantically rich, but still easy to understand and use. A message definition contains a sequence of objects, most of which correspond directly to elements within a hierarchical message. Such message definition objects contain information specifying the names of message elements, their nesting levels within the hierarchical message, their types (if these are defined in other message definitions or in data dictionaries), their occurrence counts, and textual information describing the elements and their usage.

The message definition meta-model supports type reuse. As a message definition is created, the SE is presented with all the message definition structures and data dictionary entries that have already been created for the current project and the projects that it uses, so that he can assign those types to the elements he creates. The message definition author can also use types from external schemas, such as standards body schemas, that have been imported into the project.

4.3 Data Dictionaries

The simple types to which message definitions may refer are defined within data dictionaries. Multiple data dictionaries may be defined or reused within a project. The dictionary types are based upon a set of basic types like "Unsigned Integer" and "Alphanumeric."

Data dictionaries are best manipulated in a tabular format. In accordance with the meta-model, the columns of the data dictionary table are used to specify information such as the name of the dictionary type being defined, its base type, documentation describing the type and its intended usage, lexical lengths or numerical ranges (as appropriate), any enumeration of its allowed values, and any regular expression further constraining the basic type. The data dictionary model also allows SEs to specify how some of the basic types are to be interpreted; for example, an "Unsigned Integer" can be interpreted lexically (e.g., a telephone number), or numerically (e.g., a cost).

Only the name of the dictionary type and its base type are required fields, but the more complete the information entered into the data dictionary, the more complete will be the description of the service. Clients of the service will know what is expected of them and will be able to rigorously unit test the messages they send to the service. The XML schemas generated from the dictionary model can be used at runtime to validate the messages, so the more detailed they are, the less such validations need be done by application code.

4.4 Message Validation Rules

While message definitions and data dictionaries are the basic STRIDE inputs, they do not enable the formal specification of co-occurrence (cross-field) constraints. An example of such a constraint on a message is one specifying that the *ShipMethod* element cannot be "ground" if the *Destination* element is "Hawaii." STRIDE provides a rich rule language which allows users to annotate an interface's messages with formal validation rules. This language takes into account that a single message is usually composed of multiple message definitions, and it enables the expression of constraints that span them.

Users create and edit message validation rules that are implementation technology neutral. From these rules, STRIDE generates XSLT code that can be invoked at runtime to validate the service messages against the cross-field constraints; such validation can be done upstream as messages are processed. Based as it is directly upon systems engineering input, the validation code is kept up to date with the SE's definition of the message constraints. The inclusion of these cross-field message constraints in the STRIDE generated documentation also results in better-documented interfaces.

4.5 Generated Service Description and Code Artifacts

STRIDE is used to generate lower level service description artifacts like XML schemas and WSDL files, documentation, and even code from the SD model, as shown in Figure 3.

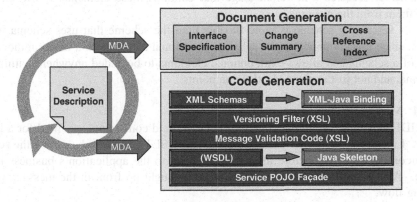

Fig. 3. Artifacts generated from the SD

4.5.1 WSDL Generation

Web Service Description Language files are used in most service-oriented applications today. Infrastructure platforms process WSDL files to produce server side skeletons, to which the service implementation is attached, and client side stubs, which embed the communication with the server. To improve service interoperability, STRIDE generates WS-I compliant WSDLs that adhere to the document/literal/wrapped style. Further, when a service contains asynchronous request/reply operations, STRIDE generates two complementary WSDL files, one for the request and one for the callback.

4.5.2 XML Schema Generation

The types section of a generated WSDL file refers to XML schema structures that are generated by STRIDE from the message definition and data dictionary tables. STRIDE produces a very basic style of schema that has been found by experience to work well with a multitude of tools. Similar work is ongoing in the W3C to produce usage guidelines for XML schema to promote the use of constructs that are uniformly supported by tools [4]. Having a central generator produce schemas for an enterprise leads to uniformity of style and format. As developers who write clients to invoke services may need to work with the schemas directly, such uniformity can make them much easier to understand.

4.5.3 Version Filter Generation

STRIDE supports the enterprise versioning strategy described in Section 2 through its Filter Generator tool. For a top-level message definition, usually composed of other message definitions, the Filter Generator produces an XSLT file that specifies the overall structure of the message, in terms of what elements appear, in what order, and with what cardinality. This generated filter is to be run, in a "receiver make right" fashion, whenever a client and the server agree in their view of the service's major version, but disagree in their view of the minor version. For example, a newer version of a service may define new optional elements in a message definition; if an older server receives a newer message that contains these elements, it will simply strip them from the message.

This versioning scheme is more flexible than the scheme that uses schema wildcards, since it accommodates the deletion of optional elements in a newer minor version of a schema and allows extra optional elements to be added anywhere within the schema, and not just at planned extension points.

4.5.4 Message Validation Code Generation

STRIDE generates validation code for the cross-field constraints specified for a message. Like the version filter code, the generated XSLT code is to be run by the server on incoming messages before they are passed on to the application's business logic code. Any cross-field validation errors will be caught up front in the message processing flow.

4.5.5 Service Java Code Generation
STRIDE allows the SE to jump-start the service development process. In addition to the XSLT code, STRIDE provides scripts to invoke JAXB (Java API for XML Binding) [12] generation of Java data binding code directly from the XML schemas.

For each service, the STRIDE Service Generator also generates a POJO (plain old Java object) façade, to be called by the service's platform-generated skeleton. This POJO code serves a number of purposes. It invokes JAXB unmarshaling (with optional validation) on each incoming request message to bind it to the Java data binding classes. The POJO façade similarly invokes the marshal command for the reply message. The POJO façade also runs the version filter if a minor version mismatch is detected between the server's view of its own minor version and that of a request message to it. Similarly, any message validation XSLT code is also run by the POJO façade.

Back end application code that implements the service operations is then invoked by the POJO façade. This code may have existed prior to the exposure of the application as a service. The intention behind the POJO façade is to shield this code, as much as possible, from needing to change in support of the new service.

4.5.6 Generated Documentation
The STRIDE models created by the SEs can be rendered in a compact tabular form that is well suited for insertion into the service documentation. The message definitions, data dictionaries, message validation rules, and the basic service description information (service name, operations, etc.) are rendered by STRIDE for use in documentation.

Another utility, called Project Compare, is crucial during the service design lifecycle. It generates enhanced versions of the tables, with mark-up showing information that has been added, deleted, or modified from one version of a service to the next.

The STRIDE Indexer utility generates a table that shows which dictionary types and message definitions are used by which other message definitions, and under what names, which is especially useful for large projects. Last, a utility called the Pruner is used to prune unreferenced dictionary types from the dictionary documentation, which is useful when only a portion of a shared dictionary is used by a service's message definitions.

5 Related Work

Model-driven development approaches often assume UML as the model framework. In our experience, we found UML to be useful for high-level system modeling but not practical for detailed service design. UML does not directly support message layout structures or data dictionaries, and it was deemed unusable to specify complex and large message structures by our SE population. We also found that we needed a simpler language targeted specifically for message co-occurrence constraint specification, rather than OCL. An interesting aspect that we are exploring is how to maintain traceability between high-level UML models and the more detailed service descriptions.

On the opposite end of the spectrum from UML modeling are low-level tools, such as XML schema and WSDL authoring tools, which allow the user to create lower-level service description artifacts directly. Such tools, however, tend to force the user to dwell on the piece parts of a service interface, obscuring the way they fit into the overall SOA enterprise architecture. Further, these tools are technology specific and do not support a model-driven approach to service design and development. The methodology and toolset described in this paper fills the gap between these two extremes by providing a rich enough meta-model that allows the details of a service to be fully specified, without the need to expose the SE to lower-level service implementation languages.

The decision to render the service models as tables was initially based upon the desire to allow the SEs to produce documents similar to the semi-structured ones originally produced. It has since become apparent that tables provide a nice, compact view of service interfaces. This is especially true given the STRIDE tool support for inlining, indexing, and pruning capabilities. Such a table-based approach has also been adopted for the Universal Business Language (UBL) [2].

The automatic generation of lower-level artifacts, like XML schema and WSDLs, from the STRIDE models guarantees that those artifacts will be generated according to the styles established by enterprise architects. Aside from the enhanced interoperability of the artifacts and ease of deployment on a variety of platforms, the issue of increased ease of maintenance cannot be stressed enough. It has been the authors' experience that it is difficult to directly maintain lower-level service description artifacts, such as schemas, especially when they have been hand-written with no enterprise-wide consistency in the style of authorship. Indeed, a study of 6,000 schemas of varying complexity across 60+ projects found an amazing lack of project-wide consistency of style [1].

The Service Component Architecture (SCA) is a programming language- and implementation-neutral model that focuses on the composition and deployment of services [13]. Our methodology is complementary to SCA, providing a rich service description for SEs to specify the assembly of components using SCA. Analogously, the STRIDE toolset is complementary to the Eclipse SOA Tools Platform [14] tools, which implement SCA.

6 Conclusions

The methodology and toolset described in this paper serve as the cornerstone for service design governance within our company, and have been used successfully to develop a number of service-oriented solutions in the telecommunications and military domains. We have seen dramatic cost reduction and quality improvements due to:

- Tool-enforced interface design best practice to minimize integration costs;
- Change management of interface specifications to handle requirements churn;
- Automated generation of XML schemas, WSDL, and code to improve productivity and minimize errors; and
- Model-driven interface design that can be adapted to or realized on different software platforms.

We are exploring a number of enhancements to our SOA systems engineering methodology. One is the ability to specify the non-functional aspects of a service, such as reliability and security, and then generate the appropriate middleware infrastructure configuration. Additionally, we plan to provide support for the generation of run-time validation code that is based directly upon the description of the service choreography and the service composition description. Since STRIDE is complementary to the SCA/STP, we are investigating the possibility of providing the STRIDE capabilities as an Eclipse plug-in to tie it to STP; SEs will then be able to specify in a top-down fashion the interfaces for the SCA components, as the basic service building blocks are called in that architecture.

References

1. Lämmel, R., Kitsis, S.D., Remy, D.: Analysis of XML Schema Usage. In: Proceedings of XML 2005, International Digital Enterprise Alliance, Atlanta (November 2005)
2. Universal Business Language v2.0 Standard, Organization for the Advancement of Structured Information Standards (OASIS) (December 2006)
3. Basic Profile Version 1.1, Web Services Interoperability Organization (WS-I) (April 2006)
4. Basic XML Schema Patterns for Databinding Version 1.0, W3C Working Draft (November 2006)
5. Web Services Description Language (WSDL) 1.1, W3C Note (March 2001)
6. XML Schema Part 0: Primer Second Edition, W3C Recommendation (October 2004)
7. Maler, E.: Schema Design Rules for UBL..and Maybe for You. In: Proceedings of XML Conference and Exposition 2002, International Digital Enterprise Alliance, Baltimore (December 2002)
8. Unified Modeling Language: Superstructure and Unified Modeling Language: Infrastructure, version 2.1.1, Object Management Group (February 2007)
9. Frankel, D.: MDA Journal: A Response to Forrester. Online publication Business Process Trends (April 2006)
10. Miller, J., Mukerji, J. (eds.): MDA Guide Version 1.0.1. Object Management Group (June 2003)
11. MacKenzie, C.M., et al. (eds.): Reference Model for Service Oriented Architecture, Organization for the Advancement of Structured Information Standards (OASIS) (February 2006)
12. Kawaguchi, K., Vajjhala, S., Fialli, J. (eds.): The Java™ Architecture for XML Binding (JAXB) 2.1., Sun Microsystems (December 2006)
13. Service Component Architecture Specification version 1.0, Open Service Oriented Architecture (OSOA) (March 2007)
14. Eclipse SOA Tools Platform Project, http://www.eclipse.org/stp

A Layered Service Process Model for Managing Variation and Change in Service Provider Operations

Heiko Ludwig[1], Kamal Bhattacharya[1], and Thomas Setzer[2]

[1] IBM T.J. Watson Research Center, 19 Skyline Dr., Hawthorne, NY-10532, USA
[2] Technische Universität München, Boltzmannstr. 3, 85748 Garching, Germany
{kamalb, hludwig}@us.ibm.com, thomas.setzer@in.tum.de

Abstract. In IT Service Delivery, the change of service process definitions and the management of service process variants are primary cost drivers, both for enterprise IT departments but also in particular for IT service providers. Consciously managing the evolution and variation of service specifications and understanding the impact on IT infrastructure in production or planning is necessary to manage – and minimize – the associated costs. Change in service definition management is typically driven from two sides: (1) high-level business processes, often based on a best practice, change and entail changes to the implementation of the business process; (2) an IT service on which a business process implementation is based is being revised. Variation is mostly common in large enterprise and service providers, offering the same kind of service to different customers or internal business units. This paper proposes a model of IT service processes that localizes and isolates change-prone process parts by introducing a layer of indirection. The model is borne out of experience gained in the large-scale service delivery organization of IBM. Enterprises and providers can use this model to manage their transition costs or to start offering variations of services as they now can assess better the impact.

Keywords: Change Management, Variation Management, Service Transition, Service Workflow Management, Service Process Evolution.

1 Introduction

Service Request Management (SRM) deals with the strategic, tactical and operational tasks for providing services to customers at a defined – often contracted – level of service [4]. Service requests are issued by customers for particular service request types against an endpoint where it is received by a service provider, which in turn executes a service process to fulfill the service request. SRM can be applied for services implemented in a multitude of technologies and offered at different types of endpoints, e.g., Web services, software-as-a-service using a Web interface as a front end, or phone services such as phone banking. SRM originates in the field of IT Service Delivery but can be equally applied in other business domains. In this paper, we focus on the example of IT services such as server provisioning or issuing a new laptop, which often combine automated and manual execution of a task.

B. Benatallah et al. (Eds.): WISE 2007, LNCS 4831, pp. 484–492, 2007.

Beyond the service implementation, SRM addresses the issues of service life-cycle management, service portfolio management and service contract management, i.e. which customer can request which service under which conditions, as well as operational issues such as entitlement check and performance management. While the various management topics mentioned above have obviously been addressed in research and by service providers in the past, the holistic view of SRM has been focus of discussion more recently, e.g., in the context of the Information Technology Infrastructure Library (ITIL) V3 [2] and research such as [4].

Service requests are fulfilled using service processes, which represent a high-level view on the course of action to take place. On this high level, the service process is often agreed on with the customer or follows a best practice for its kind of service. In IT service management, a number of standards have evolved such as the ITIL [1,2] and Control Objectives for Information and related Technology (COBIT) specifications. Other industries such as healthcare and banking are subject to government regulations in many legislations prescribing all or parts of service processes for particular types of services.

Services are rendered using a composition of atomic services. Atomic services are provided by the underlying IT systems of the service provider. In the case of IT services, atomic services may include a provisioning service, an asset management service, an ID management service, etc. Also, employees can be involved in service fulfillment. For example, in a *New Server Build* service of an IT organization, an employee may take on the rack and stack work, putting the server physically into its place and cabling it, while a provisioning service, e.g., Tivoli Provisioning Manager® or other products install an operating system image onto the server. The implementation of services requires a service designer to map a service process onto existing infrastructure/atomic services; or, in a dual, bottom-up viewpoint, to compose atomic services into aggregates that correspond to the agreed upon service processes. If a service process is complex, this is often done in multiple steps of composition or decomposition, respectively.

In current practice, service fulfillment processes are often very carefully designed and decomposed to very fine steps, prescribing the service fulfillment process in great detail. A server build process of an enterprise is often defined in a hundred steps or more. While this level of detail is necessary to run an operational service process, it makes reuse of process specifications very difficult. Hence, service providers are challenged to manage variations of their services processes for different account. Customers demand slight variations regarding the implementation of service processes for the same kind of service. In the context of IT service management, customers may want to use their own asset management system that needs to be invoked in a server procurement process or they mandate an additional security check ("server lockdown") in the course of a server's network configuration. In addition, service providers may have different ways to deliver services out of different service delivery centers, using different degrees of automation and different systems management software. Dealing with many variations of the same 100 step fulfillment process is costly and one of the reasons why process reuse by service providers is not as common as one would expect and, hence, so is reuse of service delivery infrastructure and personnel.

In another industry trend, IT service management has received much attention as organizations understand that operating their IT infrastructure is a large part of their operating costs. Change and release management are among the standard disciplines of IT service management and as such widely discussed in research and practice and subject to standardization and best practice, e.g., in the ITIL and COBIT specifications.

Service composition and implementation and – to some extent – production has been mostly viewed as an issue of software and service engineering, not so much from an SRM point of view. However, the operational variation or change of a service fulfillment process may impact the currently running service operations and entail significant costs. Applying change and transition management to service specifications and implementation, not just to the IT infrastructure on which it runs, has been a recent concern and has been subject of the ITIL V3 specification, which specifically addresses service transition.

While the standards give us high-level best practices processes to perform changes we also need to address how to structure our service implementation architecture to limit the impact of change and, hence, facilitate cost-effective changes to services.

This paper proposes a layered model of service implementation that decouples the service specification level from the atomic services level and thereby makes it more resilient to change and more amenable to dealing with service process variation. We will proceed as follows: In the next section, we introduce the layered service process model. Subsequently, we illustrate how variations of services can be dealt with in a localized manner. In the following section, we discuss how changes of services and service implementation can be managed minimally affecting the service system as a whole. Finally, we summarize and conclude.

2   Layered Model of Service Processes

The core contribution of this paper is a layered model of service processes. It enables us to manage the variation and change of service processes effectively. Figure 1 outlines and illustrates the layers of the model.

Service Fulfillment Layer: A Request for Service can be viewed as an approved, contractually agreed upon service request to be invoked by the customer. Figure 1 shows a server request that had been agreed upon as being one of the service requests to be handled by the service provider. The flow shows the different states of the Server Request object and follows a best practice as prescribed in an industry standard framework such as ITIL or IBM Tivoli Unified Process (ITUP) [3]. This process is initiated by the customer by sending in a Server Request with the appropriate level of requirements. From a service provider perspective this Server Request process may be a widely used service across various accounts. A service provider following a best practice for a request fulfillment should deploy the same process across all accounts that offer the service. This approach ensures appropriate measurability with consistent metrics for all accounts, thus facilitating comparison with respect to quality of service to the customer.

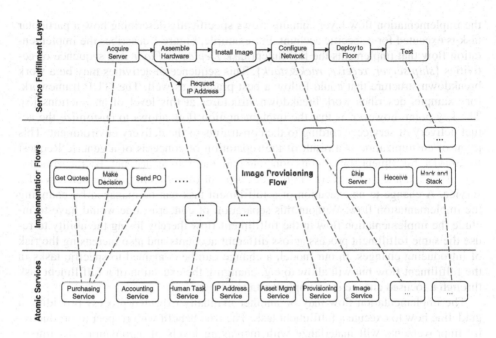

Fig. 1. Layered service process model

The simplest implementation of the Server Request process could be by simply managing the state of the Server Request object, which is updated after execution of each task. This would be a scenario where each account that provides this service will follow the best practice but manually (or tool assisted) update the Server Request object and measure the execution time for each task.

The fulfillment process describes what needs to be done but not how it needs to be done. The actual execution of the fulfillment process is separated from the fulfillment flow, which is the core value proposition of our approach. A best practices based approach should always focus on providing the basic context for execution but not prescribe the actual execution. This is especially important in a service provider scenario as different accounts, even though providing the same service to their customer may have different means to deliver the service. For example, one account is delivering all services based on inherited legacy applications that are mostly home grown versus another account that over the years has moved to a more industry platform based enterprise application model.

In a realistic scenario, not all of the functionality required to complete a given task of the fulfillment process can be provided by a given application. The implementation of a task may require a composition of different application functionalities or invocation of a long-running process. We introduce the notion of an *Implementation Flow Layer* to enable specific implementation flows that describe how work in the context of fulfillment tasks is executed.

Implementation Flow Layer: Whereas the fulfillment layer manages flow descriptions that manage fulfillment according to best practices applicable to any account, the implementation flow layer contains flows specifically describing how a particular task is executed for a specific account. For example, Figure 1 describes the implementation flow that implements the fulfillment task *Deploy to Floor* as a sequence of activities {*ship server, receive, rack&stack*}. This sequence of activities may be a work breakdown structure that could follow a best practice as well. The ITUP framework, for example, describes work breakdown structures at this level of abstractions [3]. The key point, however, is that the implementation flow allows to customize the actual delivery of service according to the constraints of the delivery environment. This point of customization is a result of the separation of concerns of a generic Request for Service fulfillment flow and its concrete realization in an account. By the same token, this separation of concerns introduces a point of variability in the delivery of a service. A change to an execution of a fulfillment task can be managed by changing the implementation flow. Without this separation of concerns, one would have to include the implementation flow in the fulfillment flow thereby losing the ability to reuse the same fulfillment process across different accounts and also increasing the risk of introducing changes. In our model, a change can be contained to specific tasks in the fulfillment flow but will allow to e.g. changing the execution of a fulfillment task through a focused change in an implementation flow.

The implementation flow may be another manually executed process providing a guideline how to execute a fulfillment task. The true benefit with respect to productivity improvements will materialize with increasing levels of automation. An implementation flow would invoke applications to execute each activity. We model a application functionality as so called atomic services.

Atomic Service Layer: An atomic service can be considered as black box application with a well-defined transaction boundary. The service delivers a specific functionality such as *generateIP* or *assignServername*. The internals of an atomic service are opaque to its invoking entity. An atomic service can be described by an interface such as WSDL and thus exposed as web service and the services descriptions managed and stored in a service registry for discovery and invocation. For a service provider the notion of atomic services provides the same benefits as typically considered for Service-oriented Architecture based efforts. In a given account, atomic services are ideally identified by decomposing the fulfillment processes to implementation flows down to application functionalities that enable specific activities. Performing process decomposition for various accounts will lead to a large set of atomic services and typically one can expect opportunities for consolidation by reusing atomic service across accounts. Finally, we want to note that an atomic service can be invoked from either an implementation flow or a fulfillment flow or both as shown in Figure 1.

Summarizing, we have introduced a layered approach to facilitate the following aspects: First, a fulfillment layer leverages industry best practices that describe how a service can be delivered depending on the type of request. Second, an implementation layer describes the actual delivery for each task of a fulfillment process. The implementation process can be based on industry standards as well but can also be customized to the need of the delivery organization. Third, atomic services are specific, black box functionalities provided by applications and invoked in the course of execution from either a fulfillment flow or an implementation flow or both.

3 Managing Variability

Managing variability is a primary concern of a service provider. On the one hand, customers require specific functionality to address their unique situation. On the other hand, best practices, be it corporate or drawn from public frameworks such as ITIL or COBIT are often under-specified, if not in the granularity of the steps but in any case in the level of technical details required to compose real, running atomic services into an actionable process. The implementation flow layer enables variability management by allowing the definition of variants of steps in the service fulfillment flow as alternative implementation flows, without impacting neither the service fulfillment flow nor atomic services layer.

As an example, consider the following 'new server request' example: a project member requests a new project server and submits a request for change (RFC) including detailed server requirement to the change management system. The authorization process solicits approval from a central coordinator as well as the project manager. Once the request is approved, the request is fulfilled in a number of tasks. Although one team member takes the lead, other roles like storage and security teams are included as needed. There is a multitude of tasks to proceed when managing a request for a new server, or a change request in general, like change filtering, priority allocation, categorization, planning and scheduling the change, testing, review, as defined for example in the ITIL change management reference process.

As described in Section 2, best practices framework just define a task 'Build Change' without detailed information how to do the actual implementation, as this highly depends on the type of change and the business one looks at. In Fig. 2, a reference process for a *server (re) build service process* is presented. The figure shows a

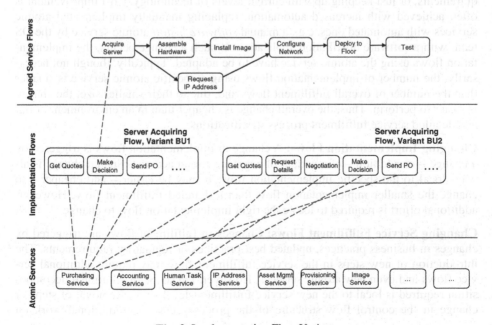

Fig. 2. Implementation Flow Variants

standardized, aggregated service flow adequate for all variants of server rebuild service requests. Consider two business units (BU) of a company with almost identical server rebuild service process implementations, only differing in the way the new server hardware is acquired. BU1 usually requires standard servers and simply submits server requirements and preferences to the company's central procurement department where all further activities are fulfilled by atomic services offered by the department (purchasing service) until the final server purchase order is placed.

BU2 instead requires high-price servers with special characteristics and individual service level agreements which need to be negotiated with a service vendor. Therefore, experts need to analyze the quotes, maybe request further information about the server and negotiate (perhaps with assistance of the financial department) the price and other attributes and conditions with a server vendor before making a purchase decision. If at some point in time, BU2's special requirements get standard server features in an acceptable price range, BU2 might switch their server acquiring implementation flow to the flow used by BU1, and vice versa.

4 Managing Change

As discussed above, the implementation flow layer of the model provides us with a level of indirection separating the Service Fulfillment Layer from the atomic services. By employing this level of indirection, we can deal with change in the service specification more efficiently, without redesigning many large high-level flows.

Changing Atomic Services: Organizations change implementations for specific atomics services continually, usually aiming at lower cost, satisfying regulatory requirements, or just keeping up with current levels of technology. Cost improvement is often achieved with increased automation, replacing manually implemented atomic services with automated ones, e.g., a manual *software install* atomic service by the OS team with a software distribution application. If such a change occurs, the implementation flows using the atomic service have to be adapted. Typically, though not necessarily, the number of implementation flows using a specific atomic service is smaller than the number of overall fulfillment flows and, due to their smaller size, the change is easier to perform. Thus, the overall change is cheaper than in an environment of flat and detailed service fulfillment process specifications.

Changing Implementation Flows: A change in implementation flows is triggered by a change – or addition – of a variant of a service for an account. Such a change is always local to the specific implementation flow. It may be easier – and cheaper – to change the smaller implementation flow than a detailed fulfillment flow. However, additional effort is required to locate the right implementation flow to change.

Changing Service Fulfillment Flows: Changes in fulfillment flows are triggered by changes in business practices, updated best practices or regulatory requirements. The introduction of new steps in the service fulfillment flow may require additional service flows and even new atomic services. However, the scope of additional decomposition required is local to the new service fulfillment-level steps. Removal of steps or change in the control flow structure of the process entails no additional work on implementation flows or atomic services, save some potential clean-up of obsolete implementation flows or unused atomic services.

5 Discussion and Conclusion

In this paper, we have introduced a layered approach to managing service processes: First, a fulfillment layer leverages industry best practices that describe how a service can be delivered depending on the type of request. Second, an implementation layer describes the actual delivery for each task of a fulfillment process. Third, atomic services are specific, black box functionalities provided by applications and invoked in the course of execution from either a fulfillment flow or an implementation flow or both.

The additional level of indirection enables a more efficient management of service fulfillment process variation and change than it would be the case of managing large flat fulfillment processes or a top-down, case-by-case decomposition approach as facilitated, for example, by the use of a model-driven architecture [7.9.10]. The proposed approach also builds on related approaches of dealing with variation, such as the use of reference models to derive variations faster [6] and in a codified process [5], which are a good complement to the proposed approach but do not address the structure of the process to be customized. Based on our experience using this level of indirection in the process of consolidating operations from multiple accounts, significant cost savings can be realized at the initial expense of normalizing and structuring service processes in these layers.

Future work will extend in multiple directions: We will further investigate the business impact of variation creation and change, in particular considering change scheduling, extending current work such as [11] based on the proposed multi-layer approach. Furthermore, on the basis of more data yielded from practical deployment, we will work on a better quantification of the benefits of the proposed approach.

References

1. Office of Government Commerce: ITIL Service Support. The Stationery Office, London (2000), OGC's Official ITIL Website: http://www.best-management-practice.com/IT-Service-Management-ITIL
2. Office of Government Commerce: Service Transition: Guidance for Practitioners. The Stationery Office, London (2007)
3. IBM Tivoli Unified Process Homepage, http://www-306.ibm.com/software/tivoli/governance/ servicemanagement/itup/tool.html
4. Ludwig, H., Hogan, J., Jaluka, R., Loewenstern, D., Kumaran, S., Gilbert, A., Roy, A., Nellutla, T.R., Surendra, M.: Catalog-Based Service Request Management. IBM Systems Journal 46(3) (2007)
5. Lazovik, A., Ludwig, H.: Managing Process Customizability and Customization: Model, Language and Process. In: Proceedings of WISE (2007)
6. Becker, J., Delfmann, P., Knackstedt, R.: Adaptative reference modeling: Integrating configurative and generic adaptation techniques for information models. In: Reference Modeling Conference (2006)

7. Kumaran, S., Bishop, P., Chao, T., Dhoolia, P., Jain, P., Jaluka, R., Ludwig, H., Moyer, A., Nigam, A.: Using a model-driven transformational approach and service-oriented architecture for service delivery management. IBM Systems Journal 46(3) (2007)
8. Information Systems Audit and Control Association (ISACA), IT Governance Institute (ITGI): Control Objectives for Information and related Technology (COBIT), Version 4.1 (2007), http://www.icasa.org/cobit
9. Kleppe, A., Warmer, J., Bast, W.: MDA Explained - The Model Driven Architecture: Practice and Promise. Addison-Wesley, Boston, USA (2003)
10. Mukerji, J., Miller, J.: MDA Guide Version 1.0.1. Technical report, OMG (2003) , http://www.omg.org/cgibin/ doc?omg/03-06-01
11. Rebouças, R., Sauvé, J., Moura, A., Bartolini, C., Trastour, D.: A Decision Support Tool for Optimizing Scheduling of IT Changes. In: 10th IFIP/IEEE Symposium on Integrated Management (2007)

Providing Personalized Mashups Within the Context of Existing Web Applications

Oscar Díaz, Sandy Pérez, and Iñaki Paz

ONEKIN Research Group, University of the Basque Country,
San Sebastián, Spain
sandy-perez@ikasle.ehu.es, {oscar.diaz,inaki.paz}@ehu.es
http://www.onekin.org

Abstract. There is an increasing tendency for Web applications to open their data silos and make them available through APIs and RSS-based mechanisms. This permits third parties to tap on those resources, combining them in innovative ways to conform the so-called mashup applications. So far, most of the approaches strive to facilitate the user to create *bright new* mashup applications which are regarded as stand-alone applications. However, the fact that these applications are data driven suggests that the mashup data is frequently used to achieve higher-order goals. Frequently, you are gathering data not just for the sake of the data itself but to help taking some decisions. Some of these decisions are conducted through Web applications. In this scenario, it would be most convenient to post the mashup data by the application where the decision is taken. To this end, the term *"mashup personalization"* is coined to describe the approach of using mashup techniques *for the end user* to enrich the content of *existing* Web applications. A proof-of-concept framework is introduced, MARGMASH, whose operation is illustrated through a running example.

Keywords: personalisation, mashup, wrapping.

1 Introduction

Personalization is the process of tailoring pages to individual users' characteristics or preferences *that will be meaningful to their goals*. However, it is not always easy for the designer to foresee the distinct utilization contexts and goals from where the application is accessed. *"No design can provide information for every situation, and no designer can include personalized information for every user"* [15]. Hence, traditional approaches can be complemented by mechanisms that allow end users to add their own content once the application is already deployed. This is akin to the *"do-it-yourself"* principle brought by mashups by allowing the layman to combine existing data from disparate sources in innovative ways. Likewise, the term *"mashup personalization"* is coined to describe the process whereby *recurrent* users can enrich existing applications with additional data using a mashup approach.

As an example, consider an online travel agency such as *Expedia.com*. Based on some trip criteria, this site provides information about flights, hotels, vacation packages and the like. Very often, making the right selection requires additional data not available at the *Expedia* site itself. For instance, if a user is looking for a hotel

B. Benatallah et al. (Eds.): WISE 2007, LNCS 4831, pp. 493–502, 2007.

for a conference, it is most convenient to see how the *Expedia* hotels are located with respect to the conference venue, and all the data being rendered through *Yahoo! Local.* By contrast, while searching for a "last-minute" trip, hotel location might not be so important comparing to the weather forecast for the available places. In this case, the weather forecast becomes crucial data. Yet, another user is looking for exotic places but where appropriate travel guides are available. Here, the vacation package data can be supplemented with a list of guides taken from *Amazon.*

In this scenario, current personalization approaches fall short as it can not be expected *Expedia.com* to foresee all possible data requirements, and hence, the user is forced to cumbersome, "daisy-like" roundtrips between *Expedia.com* and the other sites where the data is to be found. The point to stress is that this additional information largely depends on the user at hand and his hidden motivation. The user is the one that knows best why s/he is interacting with the site, and what additional data is required to accomplish the task at hand.

To this end, "mashup personalization" uses mashup techniques *for the end user* to enrich the content of *existing* Web applications. This additional content comes from external sources through APIs or RSS feeds so that the user can tap on this content in new, unexpected ways. Distinct tools are currently available (e.g. *JackBe, Kapow, Dapper, IBM (QEDWiki), Yahoo (Pipes)*) that strive to facilitate almost anyone to create the mashup they need, whenever they need it. Here, the resource to be capitalized on is the data, and the user comes up with a bright new application: the mashup. This makes sense when the mashup is an end in itself but frequently, this data is collected to serve a higher-order goal. Frequently, you are gathering data not just for the sake of the data itself but to help you to make some decisions (e.g. which hotel to book, which location to travel). And an increasing number of these decisions are conducted through Web applications (e.g. on-line travel agencies, on-line brokers, on-line course enrollment, on-line shopping, on-line auctions and so on). In this scenario, it would be most convenient to post the mashup data by the application where the decision is taken, which also provides the first feeds or inputs to enact the mashup pipe.

These ideas are realized through MARGMASH (from "MARGinal MASHup"), a tool for end users to add mashup fragments to their favorite web sites. The tool behaves as a lightweight wrapper that incorporates *"marginal mashups"* along the pages of the original application. Being targeted to end users, the selection of an application metaphor is most important for user adoption. To this end, pages of the original application are regarded as physical sheets where marginal mashups can be posted at specific places, highlighted by MARGMASH.

To facilitate user adoption and the development of the system, we opt for integrating Yahoo's pipes for mashup definition into MARGMASH. Yahoo's pipe is a very intuitive and powerful tool which is catching on very quickly[1]. Otherwise, the approach exhibits main mashup characteristics, namely, (1) mashups only affect the content, not the navigation or rendering of the *"mashuped"* application; (2) mashups are set by the user (the do-it-yourself (DIY) principle); and (3) the owner of the *"mashuped"* application is unaware of who and how is extending the application.

The paper starts introducing MARGMASH through an example.

---

[1] *http://pipes.yahoo.com/pipes/*

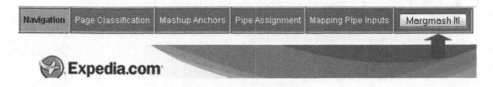

Fig. 1. Breadcrumb panel

2   MARGMASH by Example

Once logged into the MARGMASH framework, a URL of the existing Web application is provided by the user (e.g. *www.expedia.com*). Framed by MARGMASH, the user browses along *Expedia* till a page is reached where a mashup needs to be posted. This is indicated by clicking on the *"Margmash It!"* button (see figure 1). The process is supported through a breadcrumb panel.

First, the user needs to provide the visual clues (technically called *annotations*) on the selected page that will help to identify and classify that page. This identification should be general enough to select the right page regardless of the specific session or navigation that leads to the page. Notice that in this process, we are abstracting from a page-instance-based input selection to a page-class-level identification of the distinct elements on the page: *the page class*. A page class is defined as the set of pages that describes the same type of information and have a similar page structure. Thus, the user provides the visual clues by annotating distinct page instances which belong to the same class.

Figure 2 shows how this process is accomplished for a single page instance. The page instance can be the first of its class (thus, the system prompt for a class name, figure 2(a)) or belonging to an already existing class. Next, the user identifies the visual clues by selecting the markup chunks that univocally singularize the pages on this class. Figure 2(b) shows how the user selects these visual clues. The eligible markups are surrounded with a border line whenever the mouse moves over them. In this case, the page is singularized by clicking on the *"Hotel List View"* markup. The selection is internally identified through an absolute XPath over the DOM tree of the page[2].

However, *absolute* XPath expressions are very exposed to page structure changes. Improving the resilience of XPath expressions to structural changes involves the number of page instance be representative enough to come with a reasonable set of page instance. This set is then the input of an induction process which results in a more change-resilience XPath (see [11,14] for details).

In the next step the user identifies which markup fragments (i.e. nodes of the underlying DOM tree) play the role of the *mashup anchors*. The identification and location of these fragments go along a similar process as the one conducted to obtain page class but now at the level of fragments.

---

[2] Pages considered here are well-formed HTML documents. Documents that are not well-formed are converted to well-formed documents through document normalization using Mozilla GECKO (accessed through JRex APIs).

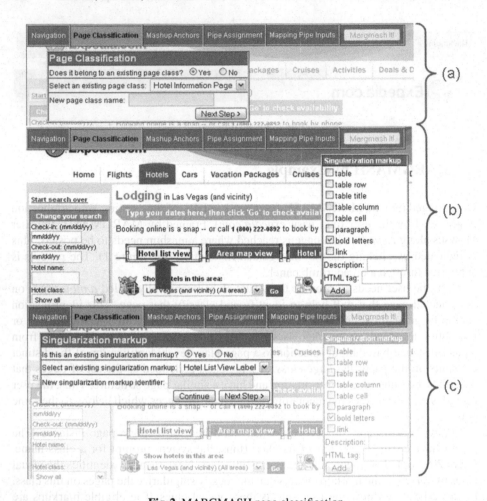

Fig. 2. MARGMASH page classification

Technically, mashup anchors play a double role. First, they hint for placing the mashup output. Second, they provide some feed data for the associated mashup. In so doing, it behaves as a scrapper for the contained markup from where it extracts the data. Next paragraphs describe the definition of a mashup anchor.

The definition of a mashup anchor includes:

1. the pipe's identifier, which the anchor holds as a reference. So far, this identifier corresponds to a Yahoo's pipe. By using Yahoo pipes, we externalize pipe definition to a popular tool, hence simplifying MARGMASH and facilitating user adoption. Hence, all the definition, debugging and publication of pipes take places within Yahoo's tool. MARGMASH acts as a proxy for Yahoo's pipes; therefore, collecting the data to feed the pipe and processing the output (see next paragraphs)
2. the pipe's starting feeds. Yahoo's tool uses a pipe metaphor for pulling and merging data from different applications from some initial parameters which are

prompted to the user. Now these parameters are automatically obtained from *Expedia*. To this end, "content feeds" are defined in the page classes using similar techniques as those used for page classification and anchor identification: a singularization-markup entry form pops up for MARGMASH to highlight specific markup chunks which are then selected by the user and associated with a given entry parameter of the pipe at hand. Thus, each pipe parameter is going to be fed by a tiny scrapper program that obtains the data from the containment page.

3. pipe's output layout, which addresses how to integrate pipe's output into the Web application. So far, the result of the pipe can be displayed as a separate layer or inlaid in the application page at either the left, right, top or bottom position with respect to the anchor situation.

4. anchor navigation mode. So far, two modes are considered: (1) *automatic* i.e. as soon as the page is loaded, the mashup is enacted, and its output rendered; (2) *manual* i.e. the mashup is explicitly enacted by the user by clicking the anchor.

Once the user has provided enough data, this process builds up a *margmash* wrapper for *Expedia*. Browsing along this *margmash* application will go along the *Expedia* site and simultaneously delivering the mashup content as a single experience. Figure 3 compares a row page from *expedia* website, and its "margmashed" counterpart.

3   The Architecture of a MARGMASH Application

This section looks inside a MARGMASH application. Figure 4 depicts an interaction diagram showing the main actors. On requesting a page load, the MARGMASH application behaves as a proxy that redirects the call to the "margmashed application" (e.g. *www.expedia.com*) through JRex. JRex is a Java Browser Component with set of API's for embedding Mozilla GECKO within a Java application. The aim is to facilitate java wrapping around the required Mozilla embedding interfaces. On receiving the petition, JRex just delegates the request to the *margmashed* application. However, when the page is returned, JRex APIs can be used to get the DOM structure counterpart of this page.

Once the HTML page has been converted into a DOM structure, embedded URLs are rewritten to point to MARGMASH so that latter interactions are always conducted through MARGMASH. Next, the current page is classified along the types identified at definition time (see previous section). If the current page belongs to a page class then, the content of the page must be enlarged with some mashups. To this end, the application first locates those markup fragments playing the role of mashup anchors.

The navigation mode of the mashup anchors can be *"manual"* and *"automatic"*. In the first case, MARGMASH dynamically inlays an AJAX script into the page that is returned to the user. In this case, the appearance of the page is just the same that the one originally returned by the existing application. The only difference stems from some markups been turned into anchors. On clicking one of these anchors, the corresponding AJAX script asynchronously enacts the associated pipe (see figure 4(b)). The pipe is executed at Yahoo's place and the output returned back to the AJAX engine. This time the AJAX script takes the Yahoo's page, extracts the pipe's output, and collage it as part of the rendered page.

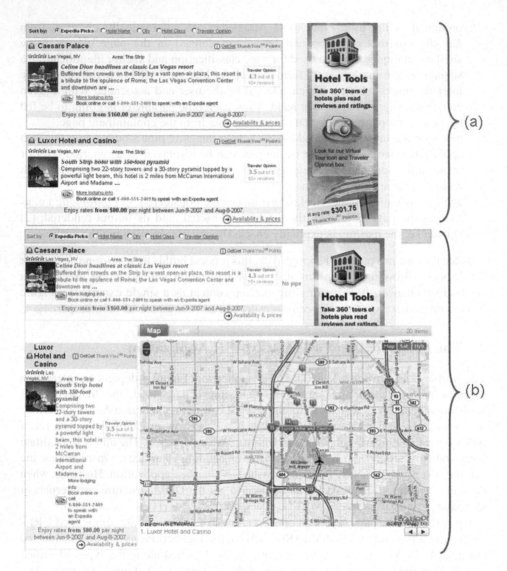

Fig. 3. Comparing original application with it "margmashed" counterpart

If the anchor's mode is automatic then, pipe enactment occurs right away without waiting for the user interaction. In this case, the pipe's output is inlaid immediately and the user gets the mashup data without additional interactions. Notice however that this navigation mode does not benefit from the advantages brought by the asynchronicity of Ajax. Hence, the user experience can suffer if this mode is heavily used due to the delay in building up the page out of the pipe's output.

One limitation of our approach, shared with most other mashup creation tools, is that the MARGMASH application can break if the underlying existing application changes. If a website changes its HTML such that the current wrapper (i.e. XPath expressions)

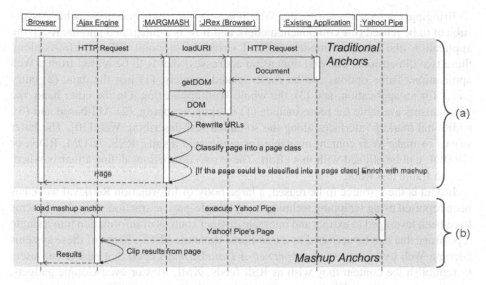

Fig. 4. Interaction diagram

no longer locates/classifies the right page or fragment, the MARGMASH application will not work correctly. However, advances in semantic enhancements will hopefully alleviate this problem.

Another problem arises from AJAX and JavaScript code. The existence of AJAX and JavaScript code on pages is supported by matching text patterns against standard navigation mechanisms (e.g. document.location, etc.) and introducing special functions that convert the URL parameter into a correct one. Thus AJAX requests are also captured and processed. However, whenever the AJAX response is processed at the client side (e.g. XML Document), MARGMASH might run into problems.

4 Related Work

This work can be framed within the distinct approaches that bring reuse to the Web. Besides the advantages brought by componentware, component aware technologies are being fueled by the increasing difficulty of processing the disparate, web-based sources of information that we have to consult on a daily basis. So as important sources of information continue to proliferate, people is craving for simplifying and centralizing their consumption [6].

Table 1. Dimensions for bringing reuse to the Web

What vs. How	Markup wrapping	API	Self-describing Web
data	Information Retrieval	Google, Amazon	RSS feed readers, Yahoo's pipe
entry page	Web clipping	Advert Engines	
whole application	Portletization	Portlet, Widgets	

Bringing componentware to the Web poses in turn two questions: which is the subject to be reused (the component model), and how is it integrated into the receiving application (the container model). Table 1 strives to pinpoint different efforts along these two dimensions. The first dimension addresses what is to be reused from a Web application. Three options can be contemplated, namely, (1) just the data, (2) entry pages for an application, and (3), the whole Web application. On the other hand, the mechanisms available for reuse include (1) markup wrapping, (2) API-based and (3), additional markup interfaces along the so-called self-describing Web [10]. The latter strives to make Web content available for automated agents. RSS, ATOM, RDFa or GRDDL can be aligned with this effort. These two dimensions define a matrix where we can frame this work.

If data is the resource to be reused, a large body of Information Retrieval work has been reported using wrapping techniques [8]. Single-page extraction wrapping systems have been leveraged to extract and integrate distinct sources of information into a single repository that can be accessed as a traditional database [2,13]. Some of these systems become Web tools (e.g. *www.dapper.net* or *DataMashups.com*) that enable end users to republish the content they wish as RSS feeds, XML files or even Google gadgets, enabling the creation of Web pages that syndicate the content of different data sources.

Another complementary tendency is that of application owners to permit access to their data silos through the use of Application Programming Interfaces (API). Amazon and Google are the best known examples. This removes the need for information extraction tools. Even more popular is providing the Web content through RSS or Atom channels where feed-reader agents can tap on to get the content out. The new crop of mashup frameworks such as Yahoo pipes can be framed here since they mainly integrate data coming from diverse sources. In a recent revision of commercial tools [7], this situation is highlighted when stated that *"there is a clear dividing line emerging between products that only provide data integration and recombination from remote Web services and tools which provide a way in which to build a real user interface for a mashup, with a two-way flow of data from the underlying services"*.

MashMaker [5] also presents an interesting approach to mashup using a spread-sheet-like approach. Similar to our proposal, data is integrated around an existing application. When the page is loaded, data is arranged in a tree. When a user is looking at a particular node, *MashMaker* will automatically suggest additional data functions that they might want to apply. Clicking on one of these buttons will insert a new node whose defining expression extracts data from a third-party application. These functions can be defined by the user (the so-called user-defined widgets) or suggested by the system based on previous interactions of other users with the same application.

However, data is a main but not the only asset provided by Web applications. Another important piece is the GUI itself. You can be interested in reusing not only the data but the presentation of this data as well .For instance, Web portals act as a front-end doorway to distinct applications (including Web applications) so that the user can benefit from portal commodities such as single sign-on. Hence, portal vendors offer mechanisms to include fragments of the entry page of existing Web applications [16,12]. This is referred to as *Web clipping* and interacting with the clipped entry page involves a navigation to the original Web application. *Web clipping* allows for a side-by-side visual composition of distinct Web applications in a single Web page [1,4,17].

Another related work is that of [9] where Ajax-based clips are created to be included on an authored Web page. Unlike our approach, this approach is intrusive in the sense that the code of the existing application is modified by the owner of the application. By contrast, our scenario is that of end users "extending" existing applications without this application being aware of who and how is extending it.

Finally, recently new approaches have emerged that permit to reuse a whole application (i.e. data, functionality and presentation): *portlets* and *widgets*. Portlets are interactive Web mini-applications to be delivered through a third party Web application (usually a Portal) [3]. Portlet technology permits to build portal pages by combining fragments from diverse portlets in a standalone server. In this line, portletization efforts [11] enable the construction of portlets by wrapping an existing Web application integrating it into a portal context. A related technology is that of gadgets. Promoted by Google, gadgets build pages by combining fragments with JavaScript on the client.

As for MARGMASH, the subject to be reused includes both a complete Web application and data coming from other places whereas this reuse is mainly achieved through wrapping techniques.

5   Conclusions

We have presented MARGMASH, a tool that permits end-users to do *mashuping* in the context of existing Web applications. Unlike previous approaches, the mashup is not perceived as an end in itself. Rather, the mashup is contextualized by delivering it within an existing website. In this way, the website acts as the initial data provider for the mashup input parameters as well as the container for its output. Additionally, MARGMASH capitalizes on the recent crop of mashup applications to deliver the content, taking Yahoo pipes as an example. Although currently only Yahoo's pipe is supported, future work includes defining the bindings with other frameworks so that the user can select the one that facilitates the task at hand.

However, it has yet to be demonstrated whether MARGMASH is usable by virtually anyone with any skill level using any browser in any language and without any training. This will be essential for mashup tools like ours to succeed with the general public.

Acknowledgments. This work was co-supported by the Spanish Ministry of Science & Education, and the European Social Fund under contract TIC2005-05610. Perez enjoys a doctoral grant from the Basque Government under the "Researchers Training Program".

References

1. Bauer, M., Dengler, D.: Infobeans - configuration of personalized information assistants. In: International Conference on Intelligent User Interfaces (1999)
2. Baumgartner, R., Gottlob, G., Herzog, M.: Interactively adding web service interfaces to existing web applications. In: SAINT 2004. International Symposium on Applications and the Internet (2004)

3. Bellas, F.: Standards for second-generation portals. IEEE Internet Computing 8(2), 54–60 (2004)
4. Bouras, C., Kounenis, G., Misedakis, I.: A web content manipulation technique based on page fragmentation. Journal of Network and Computer Applications 30(2), 563–585 (2007)
5. Ennals, R., Garofalakis, M.: Mashmaker: Mashups for the masses. In: ACM SIGMOD International Conference on Management of Data (2007)
6. Hinchcliffe, D.: Online ajax "desktops" try to change the rules of the game (2006), Published at http://blogs.zdnet.com/Hinchcliffe/?p=8
7. Hinchcliffe, D.: A bumper crop of new mashup platforms (2007), Published at http://blogs.zdnet.com/Hinchcliffe/?p=111
8. Laender, A.H.F., Ribeiro-Neto, B.A., da Silva, A.S., Teixeira, J.S.: A brief survey of web data extraction tools. ACM SIGMOD Record 31(2), 84–93 (2002)
9. Lingam, S., Elbaumda, S.: Supporting end-users in the creation of dependable web clips. In: International World Wide Web Conference (WWW 2007) (2007)
10. Mendelsohn, N.: The self-describing web. W3C (2007), at http://www.w3.org/2001/tag/doc/selfDescribingDocuments.html
11. Paz, I., Díaz, O.: On portletizing web applications.Submitted for publication to ACM Transactions on the Web, ACM TWeb (2007), Summary at http://www.onekin.org/margmash/tweb07.pdf
12. Oracle. Oracle9iAS Portal Web Clipping Portlet (2003), Published at http://portalcenter.oracle.com/
13. Pan, A., Viña, Á.: An alternative architecture for financial data integration. Communications of the ACM 47(5), 37–40 (2004)
14. Raposo, J., Álvarez, M., Losada, J., Pan, A.: Maintaining web navigation flows for wrappers. In: Lee, J., Shim, J., Lee, S.-g., Bussler, C., Shim, S. (eds.) DEECS 2006. LNCS, vol. 4055, Springer, Heidelberg (2006)
15. Rhodes, B.J.: Margin Notes: Building a Contextually Aware Associative Memory. In: International Conference on Intelligent User Interfaces (2000)
16. Smith, I.: Doing Web Clippings in under ten minutes. Technical report, Intranet Journal (March 2001), at http://www.intranetjournal.com/articles/200103/pic_03_28_01a.html
17. Tanaka, Y., Ito, K., Fujima, J.: Meme media for clipping and combining web resources. World Wide Web 9(2), 117–142 (2006)

Wooki: A P2P Wiki-Based Collaborative Writing Tool

Stéphane Weiss, Pascal Urso, and Pascal Molli

Nancy-Université, LORIA, INRIA-Lorraine
{weiss,urso,molli}@loria.fr

Abstract. Wiki systems are becoming an important part of the information system of many organisations and communities. This introduce the issue of the data availability in case of failure, heavy load or off-line access. We propose to replicate wiki pages across a P2P network of wiki engines. We address the problem of consistency of replicated wiki pages in the context of a P2P wiki system. In this paper, we present the architecture and the underlying algorithms of the wooki system. Compared to traditional wikis, Wooki is P2P wiki which scales, delivers better performances and allows off-line access.

1 Introduction

Currently, wikis are the most popular collaborative editing systems. They allow people to easily create and modify content on the web. This ease of interaction and operation makes a wiki an effective tool for collaborative writing. Collaborative writing is becoming increasingly common; often compulsory in academic and corporate work. Writing scientific articles, technical manuals and planning presentations are a few examples of common collaborative writing activities.

A lot of critical data are now under the control of wiki systems. Wikis are now used within enterprises or organizations. For example, United States intelligence community uses Intellipedia for managing national security informations. Wikis are now an important piece in the information system of many large organizations and communities. This introduces the issue of data availability in case of failure, heavy load or off-line access.

Current wiki systems are intrinsically centralized. Consequently, in case of failure or off-line work, data are unavailable. In case of heavy load, the system scales poorly and the cost linked to underlying hardware cannot be shared. Our objective is to replace the centralized architecture of a wiki server by a P2P network of wiki servers. This makes the whole wiki system fault tolerant, this allows to balance the load on the network and finally costs of the underlying hardware can be shared between different organizations.

This approach supposes that wiki data are replicated on a P2P network of wiki sites. Consequently, the main problem is how to manage replicate consistency between wiki sites. Traditional pessimistic replication approaches (Distributed Database, ...) ensure consistency but are not adapted to this context. They

B. Benatallah et al. (Eds.): WISE 2007, LNCS 4831, pp. 503–512, 2007.

scale poorly and do not support off-line work. Optimistic replication approaches suppose to know how to safely merge concurrent updates. Some previous work tried to build a P2P wiki [1,2] relying on distributed version control system (DVCS) approaches. The main problem with DVCS approach is correctness. An optimistic replicated system is considered as correct if it eventually converges i.e., when the system is idle all sites contain identical data. This is called *eventual consistency* [3]. DVCS have never ensured this property. Consequently, building a P2P system with a DVCS will not ensure eventual consistency. Other approaches ensure convergence [4,5] but are not compatible with P2P networks constraints. Finally, other approaches converge and are adequate with P2P constraints but do not support collaborative editing constraints.

We developed the woot [6] algorithm to manage consistency of a replicated linear structure in a P2P environment. This algorithm ensures convergence without managing versions. In this paper, we describe how we built Wooki. Wooki is a fully functional P2P wiki system based on this algorithm. We refined the original algorithm to achieve a better time complexity. We combined this new algorithm with a probabilistic dissemination algorithm for managing updates propagation on the overlay network and with an anti-entropy algorithm for managing failures and disconnected sites. Wooki is currently available under GPL license.

2   The Wooki Approach

A wooki network is a dynamic p2p network where any site can join or leave at any time. Each site has a unique identifier named *siteid*. Site identifiers are totally ordered. Each site replicates wiki pages of other sites. Each site only requires a partial knowledge of the whole network.

There is three main components in a wooki site (figure 1). The core component wooto which is in charge of generating and integrating operations affecting the documents. Another component is in charge of user interface. The last component is in charge of disseminating local operations and retrieving remote operations.

2.1   Wooto Approach

Wooto is an optimized version of woot [6]. A wooki page is identified by a unique identifier *pageid*. This identifier is set when the page is created. This identifier is the name of the created wiki page. If some sites create concurrently pages with the same name, their content will be directly merged by the wooto algorithm. A wooki page contains a sequence of four-tuples $< idl, content, degree, visibility >$, where each tuple represents a line of the wooki page.

idl is the unique identifier of the line. This id is a pair $(siteid, logicalclock)$. Each site maintains a logical clock [7]. Each time an operation is generated, the logical clock is incremented. The line identifiers are strictly totally ordered. *Let idl1 and idl2 be two line identifiers with their respective values $(s1, l1)$ and $(s1, l2)$. We get $idl1 <_{id} idl2$ if and only if (1) $s1 < s2$ or (2) $s1 = s2$ and $l1 < l2$.*

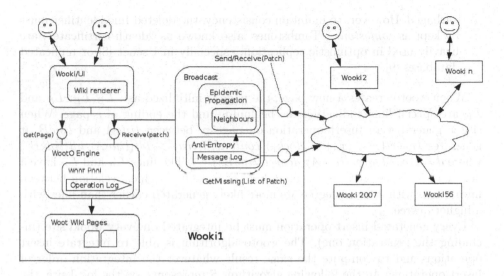

Fig. 1. Wooki architecture

content is a string representing the content of the wiki-readable line.

degree is an integer used by the wooto algorithm. The degree of a line is fixed when the line is generated. We will describe how the degree is computed when we will describe the generation of an operation.

visibility is a boolean representing if a line is visible or not. It means that in the wooki approach, we do not delete lines, we mark them as invisible.

By applying this storage model, if we have the following wiki page:

```
=== Three pigs ===
* [[Image:pig.png|thumb|left|100px|riri]]
fifi and loulou
```

Assuming these three lines were generated on site number 1 in this order and assuming there is no invisible lines, the wiki page will be internally stored as:

```
((1,1),=== Three pigs ===,0,true)
((1,2),* [[Image:pig.png|thumb|left|100px|riri]],1,true)
((1,3),fifi and loulou,2,true)
```

To manage the storage model, the Wooki handles two operations: the insertion and the deletion of a line. As in traditional version control systems, there is no operation of line update.

1. *Insert(pageid, line, l_P, l_N)* where *pageid* is the page where to insert the line *line* =< *idl, content, degree, visibility* >. l_P and l_N are the id of the previous and the next lines.

2. *Delete(pageid, idl)* sets visibility of the line identified by *idl* as false in the page identified by *pageid*. Optionally, the content of the deleted line can be

garbaged. However, to maintain consistency, the deleted line identifier must be kept as *tombstones*. Tombstones, also known as "death certificate", are heavily used in optimistic replication, especially in Usenet [8] or replicated databases [9].

When wooto creates a new page, the page is initialized as : L_B, L_E. L_B and L_E are special lines indicating the beginning and the ending of a page. When site x generates an insert operation on page p, between $lineA$ and $lineB$, it generates $Insert(p, < (x, + + clock_x), content, d, true >, idl(lineA), idl(lineB))$ where $d = max(degree(lineA), degree(lineB)) + 1$. The lines L_B and L_E have a degree of 0. The degree represents a kind of loose hierarchical relation between lines. Lines with a lower degree are more likely generated earlier than lines with a higher degree.

Every generated insert operation must be integrated on every wooki site (including the generation one). The wooto algorithm is able to integrate insert operations and to compute the same result whatever the integration order of insert operations. In the following algorithm, S represents, on the local site, the lines sequence of the page where a line have to be inserted.

```
IntegrateIns (l, lₚ, lₙ) :−
   let  S' := subseq(S, lₚ, lₙ);
   if  S' := ∅  then
      insert(S, l, position(lₙ));
   else
      let  i := 0,  d_min := min(degree, S');
      let  F := filter(S', λlᵢ.  degree(lᵢ) = d_min);
      while  (i < |F| − 1)  and  (F|i| <_id l)  do  i := i + 1;
      IntegrateIns (l,  F[i − 1],  F[i]);
   endif;
```

This algorithm selects the sub-sequence of lines present between the previous line and the next line. If this sequence is empty, the line l is inserted in the model just before the next line. Elsewhere, wooto filters the sub-sequence keeping only lines with the minimum degree. The remaining lines are sorted according the $<_{id}$ order [6]. Thus, l must be integrated at its place according $<_{id}$ between remaining lines. We then make a recursive call to place l among lines with higher degree.

However, since wooki sites can receive operations in any order, the operations have pre-conditions. When a site receives an operation, if its pre-conditions are verified, the operation can be integrated immediately by executing the wooto algorithm. If pre-conditions are false, the operation is placed on a waiting queue until its pre-conditions are verified. Pre-conditions are: a line can only be inserted on a site if its previous and next lines are already present in the model of this site. Similarly, only an existing line can be deleted.

Let's now illustrate wooto algorithm through an example (see figure 2). On this scenario, line L_1 and line L_2 were concurrently inserted when the page was empty, and line L_3 was inserted before line L_1. Without woot or wooto, such a scenario can lead to three different line orders L_B, L_2, L_3, L_1, L_E, L_B, L_3, L_2,

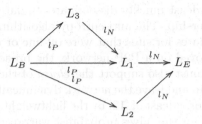

Fig. 2. Example

L_1, L_E or L_B, L_3, L_1, L_2, L_E each of them respecting the previous and next relationship. Wooto computes the unique order L_B, L_3, L_1, L_2, L_E.

According to definition, we have $degree(L_1) = degree(L_2) = 1$ and $degree(L_3) = 2$. Due to pre-conditions, there is no site where line L_3 is present and not line L_1. Now, let's assume that $L_1 <_{id} L_2 <_{id} L_3$.

Imagine a site, where line L_2 arrives when line L_1 and line L_3 are present. The wooto algorithm first selects the lines present between L_B and L_E and filters them to keep only line L_1. The reason of such a filtering is above. Indeed, since line L_3 is dependent to another line – here line L_1 –, it has a higher degree. Moreover, there could exist a site where line L_1 is present and not line L_3. Thus, line L_2 must be placed according to line L_1 earlier to line L_3. We obtain L_B, L_3, L_1, L_2, L_E. The wooto algorithm computes the same order L_B, L_3, L_1, L_2, L_E whichever the integration order respecting pre-conditions.

Finally, compared to woot original integration algorithm, the filtering is done on degree instead of the previous and next relationship of each line. Thus, wooto has a better time complexity : $O(n^2)$ instead of $O(n^3)$. This also allows to reduce the space required to store lines: an integer replacing two line identifiers. As woot, we have formally verified the eventual consistency of wooto with the Lamport's model-checker TLC [10].

2.2 User's Operation

Users do not directly edit the model. When a user opens a page for editing, he sees a view of the model which presents only the content of the visible lines. As in traditional wiki, the user makes all the modifications he wants and saves.

To detect operations, we use a diff algorithm [11] between the page the user requested at edition time and the page the user saves. We translate the operations given by the diff algorithm in terms of wooto operations. A delete of the line number n is translated into a delete of the nth visible line. An insert at position n is translated into an insert between the $(n-1)^{th}$ and the nth visible lines. These operations are integrated locally and broadcasted to the other wooki sites.

2.3 Wooki Broadcast

For operation dissemination, we use a broadcast protocol in the spirit of [12]. We combine a probabilistic broadcast algorithm with an anti-entropy algorithm [9].

The probabilistic broadcast quickly disseminates updates on the P2P network and to manage membership. The anti-entropy algorithm has the responsibility to recover missing updates for sites that were off-line or crashed.

In order to be deployed on a P2P network, the broadcast protocol must be scalable, reliable and must also support the *churn* of the network. Indeed, in a P2P network, nodes join and leave the network dynamically. Thus, we replace the standard probabilistic broadcast of [12] by the lightweight probabilistic broadcast (lpbcast) [13]. This algorithm gives probabilist warranties that sent operations will be delivered to all connected nodes.

In order to ensure a reliable dissemination of messages, each site must manage a table of neighbors. This table has a fixed size and contains only a partial nodes list of the entire P2P network. Lpbcast updates this table during messages propagation. Lpbcast gives probabilistic warranties that there is no clusters within P2P network. Moreover, since wooto does not require an ordering on message reception, the lpbcast can be unordered for higher efficiency.

The lpbcast algorithm ensures a reliable and scalable dissemination of operations on connected nodes. For managing off-line work, we combine it with an anti-entropy algorithm. We use the original anti-entropy algorithm of [9]. When a site starts an anti-entropy, it selects a neighbor at random in the local table of neighbors and send to him a digest of his own received messages. The selected site returns missing messages to caller. Using anti-entropy implies that each site keeps received messages in a log. As this log should not grow infinitely, we purge this log from time to time. Purging the log is an intrinsic problem of anti-entropy approach. If a site purges its log and then starts an anti-entropy with another site, it can receive previously purged messages. In the traditional approach, tombstones or death certificates are used to avoid this problem. Fortunately, Wooki supports re-integration of already integrated operations. Indeed, we drop an insert operation of a line which identifier is already present in the model. Also, reapplying deletion of a line causes no modification. Thus, we can purge the anti-entropy log without using operation tombstones. This is an interesting property of combining wooto and anti-entropy approach.

Finally, if a site is off-line for a long period of time, anti-entropy may not find missing messages i.e., missing messages which have been purged on all sites of the P2P network. In this case, this site cannot be synchronized. The only way to recover this site is to make a state transfer with a site. It is important to notice that, even in this case, off-line work is not lost since local operations are still applicable. The site can send its local operations and then make the state transfer.

3   Implementation

The wooki prototype has been implemented in Java as servlets in a Tomcat Server. Wooki pages are just stored in regular files. All network messages are transported by the http protocol.

Fig. 3. Wooki Interface

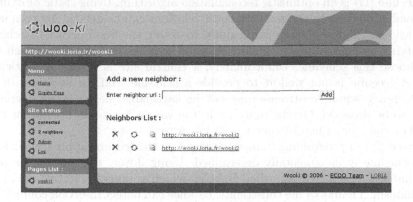

Fig. 4. Wooki Neighbors

In figure 3, we can see the same page entitled "wooki1" loaded from 2 different wooki sites: `http://wooki.loria.fr/wooki1` and `http://wooki.loria.fr/wooki2`. In this example, the same server "wooki.loria.fr" hosts 3 wooki sites. We can also see on figure 3 that both sites are connected and each site has two neighbors. We can observe the table of neighbors in figure 4. This interface lets the administrator manage manually this table. In the normal case, no administration is required except when starting wooki for the first time. The administrator has to connect the new wooki site to an existing wooki network by typing the http address of a wooki node. Once bootstrapped, the routing table is updated when messages from other sites are received. This interface also allows the administrator to start an anti-entropy mechanism or a state transfer. In the normal case, anti-entropy is activated at a regular interval of time. Administrator can force anti-entropy when he restarts the wooki system or when an off-line session is ended.

The wooki prototype is available with a GPL license at `http://p2pwiki.loria.fr`.

4   Related Work

Pessimistic replication (database [14] or consensus [15]) ensures that, at anytime, all replicas host the same content. Pessimistic replication is widely used and recognized for its safety. Unfortunately, pessimistic replication is also well-known for its poor scalability [3] and its incapacity to provide an off-line work.

Virtual Synchrony [16] is a middleware which allows sharing data among programs running on multiple machines. Virtual Synchrony provides a pessimistic replication with some optimizations to improve global performance. Consistency is ensured by enforcing a total ordered reception of modifications. Unfortunately, this ordering is costly and do not scale to a huge number of sites. In addition, Virtual Synchrony does not allow disconnected work.

Icecube [17] is an optimistic reconciliation algorithm. Using static or dynamic reconciliation constraints, some semantic relations can be defined between users' operations. The Icecube algorithm aimed to obtain the best operations' schedule satisfying all constraints. Therefore, Icecube requires a central site which have to choose this schedule. Unfortunately, a central site is a serious bottleneck, hence, Icecube is not well-fit to provide a scalable wiki. In addition, due to concurrency, some constraints may not be satisfiable, hence, some operations have to be dropped. On the contrary, in the wooto approach, operations never conflict and, consequently, they are all integrated.

Joyce [5] is a programing framework based on the Icecube approach. In Joyce, the schedule is incrementally determined. Using Joyce, the authors proposed a collaborative text editor called Babbles which supports selective undo/redo mechanism. Thanks to the constraints, babbles can detect insertion conflicts. On the contrary, we consider that two insertions never conflict. The main drawback of Babbles is the requirement of a primary site which have to resolve conflicts. This primary limits the scalability, and is a single point of failure.

The operational transformation (OT) approach [4] is composed by operations which express modifications, and the transformation functions used to modify concurrent operations toward local operations. A state vector is associated to each operations. The state vector size grows linearly with the number of sites and consequently, limits the scalability of this approach. The wooto approach depends only of the number of operations and not of the number of sites.

Distributed version control systems (DVCS) allow many users to edit the same documents concurrently. They provide the same features as CVS [18] or Subversion [19] without requiring a central site. DVCS do not express the notion of consistency. In addition, a well-known scenario from the OT approach [20] leads most of DVCS (Darcs, Mercurial, Git, Bazaar, ...) to the document inconsistency. Code Co-op [1] is a DVCS which allows to add a wiki to a replicated folder. Hence, we obtain a p2p wiki. As many DVCS, Code Co-op cannot ensure the wiki's pages consistency. Repliwiki is a P2P wiki based on the SHH-sync which is a DVCS. Unfortunately, as claimed before, DVCS algorithms failed to ensure consistency.

5   Conclusion

Wooki is a P2P wiki system. Compared to traditional centralized wikis, a P2P wiki system is fault-tolerant, improves global performances and allows user to work off-line. Wooki uses the wooto algorithm to manage consistency of copies. Wooto is an optimized version of the woot algorithm and ensures eventual consistency. Compared to other optimistic replication approaches, wooto is designed for P2P networks and is fully compatible with P2P network constraints.

In this paper, we described how to combine the wooto algorithm with an epidemic dissemination algorithm. This combination provides a reliable P2P wiki system that scales and tolerates the "churn" of P2P networks. It supports long-term disconnections without ever loosing off-line work. It also does not require any permanent sites, and the P2P network can be unstructured.

The wooki system has several open issues:

– We clearly make the choice to keep all tombstones in the wooto page model. It implies that wooki pages will grow infinitely. Nevertheless, keeping a tombstone has a very small space overhead. In the context of a wiki system, this choice is reasonable. In another context, it is interesting to design a distributed tombstone garbage collector that is compatible with P2P constraints.
– A P2P wiki system changes interactions between humans and the system. In a central wiki system, a user is aware about concurrent changes when he saves its page. If the wiki page has changed during the edition, the user have to resolve conflicts before committing his changes. In a P2P wiki system, a user can save its changes before concurrent ones arrive for integration. Wooto will solve the conflicts but the final page available on the site is a page computed by the system but not reviewed by a human. We need to add to Wooki an awareness engine that is responsible to warn about concurrent changes.
– Replicating a wiki site is not completely transparent. If a wiki page contains some macros that computes some statistics, the visible result will not be the same on all sites. Suppose a counter that counts the read number of a page. If the page is replicated, this counter can have different values on different sites. This does not violate eventual consistency. The source page is still the same on all sites, but the rendering of the page can be different.
– In collaborative editing, it is important to undo some operations. In traditional wikis, it is possible to revert to a previous version of the wiki. In the general case, it must be possible to undo any operations (not always the last one), anytime [21]. We have not yet provided such features in wooto. However, as we keep all informations within pages, it should be possible to undo an operation. We still need to design such an algorithm and prove eventual consistency in case of undo.
– In Wooki, we built a new P2P wiki system. We are working on integration of the wooki engine with existing wiki systems. The general approach is to manage optimistic replication by only using the web service interface of existing wiki systems.

References

1. Reliable Software: Code Co-op. (2006), http://www.relisoft.com/co_op
2. Kang, B.B., Black, C.R., Aangi-Reddy, S., Masri, A.E.: Repliwiki: A next generation architecture for wikipedia (unpublished)
   http://isr.uncc.edu/repliwiki/repliwiki-conference.pdf
3. Saito, Y., Shapiro, M.: Optimistic replication. ACM Computing Surveys 37(1), 42–81 (2005)
4. Ellis, C.A., Gibbs, S.J.: Concurrency control in groupware systems. In: SIGMOD Conference, vol. 18, pp. 399–407 (1989)
5. O'Brien, J., Shapiro, M.: An application framework for nomadic, collaborative applications. In: Eliassen, F., Montresor, A. (eds.) DAIS 2006. LNCS, vol. 4025, pp. 48–63. Springer, Heidelberg (2006)
6. Oster, G., Urso, P., Molli, P., Imine, A.: Data consistency for P2P collaborative editing. In: CSCW. Proceedings of the ACM Conference on Computer Supported Cooperative Work, Banff, Alberta, Canada (November 2006)
7. Lamport, L.: Time, clocks, and the ordering of events in a distributed system. Commun. ACM 21(7), 558–565 (1978)
8. Spencer, H., Lawrence, D.: Managing Usenet. O'Reilly (January 1998)
9. Demers, A., Greene, D., Hauser, C., Irish, W., Larson, J., Shenker, S., Sturgis, H., Swinehart, D., Terry, D.: Epidemic algorithms for replicated database maintenance. In: PODC. Proceedings of the ACM Symposium on Principles of Distributed Computing, pp. 1–12 (1987)
10. Yu, Y., Manolios, P., Lamport, L.: Model checking TLA+ specifications. In: CHARME 1999. Proceedings of Correct Hardware Design and Verification Methods, pp. 54–66 (1999)
11. Myers, E.W.: An o(nd) difference algorithm and its variations. Algorithmica 1(2), 251–266 (1986)
12. Birman, K.P., Hayden, M., Ozkasap, O., Xiao, Z., Budiu, M., Minsky, Y.: Bimodal multicast. ACM Trans. Comput. Syst. 17(2), 41–88 (1999)
13. Eugster, P.T., Guerraoui, R., Handurukande, S.B., Kouznetsov, P., Kermarrec, A.M.: Lightweight probabilistic broadcast. ACM Trans. Comput. Syst. 21(4), 341–374 (2003)
14. Bernstein, P.A., Hadzilacos, V., Goodman, N.: Concurrency Control and Recovery in Database Systems. Addison-Wesley, Reading (1987)
15. Lynch, N.A.: Distributed Algorithms. Morgan Kaufmann Publishers Inc., San Francisco, CA, USA (1996)
16. Birman, K.P., Joseph, T.A.: Exploiting virtual synchrony in distributed systems. In: Symposium on Operating Systems Principles (SOSP), pp. 123–138 (1987)
17. Kermarrec, A.M., Rowstron, A.I.T., Shapiro, M., Druschel, P.: The IceCube approach to the reconciliation of divergent replicas. In: PODC. Proceedings of the ACM symposium on Principles of distributed computing, pp. 210–218 (2001)
18. Berliner, B.: CVS II: Parallelizing software development. In: Proceedings of the USENIX Winter Technical Conference, Berkeley, California, USA, pp. 341–352 (1990)
19. CollabNet, Inc.: Subversion (2005), http://subversion.tigris.org/
20. Oster, G., Urso, P., Molli, P., Imine, A.: Tombstone transformation functions for ensuring consistency in collaborative editing systems. In: The International Conference on Collaborative Computing, CollaborateCom, IEEE Press, Atlanta, Georgia, USA (2006)
21. Sun, C.: Undo as concurrent inverse in group editors. ACM Transactions on Computer-Human Interaction (TOCHI) 9(4), 309–361 (2002)

Creating and Managing Ontology Data on the Web: A Semantic Wiki Approach

Chao Wang[1], Jie Lu[1], Guangquan Zhang[1], and Xianyi Zeng[2]

[1] Faculty of Information Technology, University of Technology, Sydney
PO Box 123, Broadway, NSW 2007, Australia
{cwang, jielu, zhangg}@it.uts.edu.au
[2] Ecole Nationale Supérieure des Arts et Industries Textiles
9 rue de l'Ermitage 59100 Roubaix, France
xianyi.zeng@ensait.fr

Abstract. The creation of ontology data on web sites and proper management of them would help the growth of the semantic web. This paper proposes a semantic wiki approach to tackle this issue. Desirable functions that a semantic wiki approach should implement to offer a better solution to this issue are discussed. Along with that, some key problems such as usability, data reliability and data quality are identified and analyzed. Based on that, a system framework is presented to show how such functions are designed. These functions are further explained along with the description of our implemented prototype system. By addressing the identified key problems, our semantic wiki approach is expected to be able to create and manage web ontology data more effectively.

1 Introduction

Ontology has been realized to be an essential layer of the emerging semantic web [1,2]. So the abundance of ontology related information is critical to the maturity of the semantic web. However, through the semantic web search engine Swoogle [3], while we've found that there are plenty of ontology schemas (which refer to classes, properties and their relations, as what "TBox" contains in description logics [4]) available over the web, ontology data (which refer to instances of classes or individuals as what "ABox" contains) do not seem to be very abundant in contrast.

One barrier for the problem is that it is not an easy job for ordinary users to create ontology data due to the complexity of the ontology languages, compared to the creation of normal web pages. Tools have been developed to assist the generation of ontologies with ease (e.g., Protege [5], OntoEdit [6], SWOOP [7]), but mostly they are not web-based tools and often used by ontology experts. In addition, while the ontology schemas, which usually reflect the domain knowledge, are relatively stable once developed, the ontology data are often prone to changes in a dynamic environment. Therefore, it may not be convenient to use offline tools to maintain such ontology data.

B. Benatallah et al. (Eds.): WISE 2007, LNCS 4831, pp. 513–522, 2007.

Recently, wikis [1] have been studied as an alternative way to create and maintain ontology data on the web. Several semantic wiki systems have been proposed. In contract to these studies, which mostly focus on some specific aspects or topics, this paper will first discusses the functions that a semantic wiki approach should implement to offer a better solution to web ontology data creation and management. Along with that, some key problems such as usability, data reliability and data quality will be identified and discussed. A system framework will be proposed to show how such functions are designed. These functions are further explained along with the description of our prototype system.

The rest of the paper is organized as follows. Section 2 discusses related work. Section 3 discusses the desirable functions for managing ontology data and the framework design. Section 4 and Section 5 present in detail the fundamental functions of browsing, editing, search and query along with the description of our prototype system. Section 6 outlines the design and/or implementation of other functions. Section 7 concludes the paper and discusses the future work.

2 Related Work

This section first discusses the general ideas of wiki systems. Then related work on typical semantic wiki systems or frameworks is analyzed.

The main idea of most wiki systems is to let web users create and maintain web contents in a collaborative way. These systems provide convenient functions so that an ordinary user with little knowledge of web page making can contribute contents to the web. Wikipedia[2], the most well-known online collaborative encyclopedia on the Web, is a successful application of wiki systems. However, most contents maintained in wiki systems are semi-structured web pages only for web users to search and read. Facts and knowledge expressed in those web pages lack proper syntax and semantics for computers to process.

Semantic wiki systems then emerge to introduce semantic web standards (e.g., RDF/RDFS , OWL) and techniques to the common wiki systems, trying to make the collaborated contents have more semantics for computers to understand and process. We will discuss these systems as follows.

Platypus Wiki [8] is an extended wiki system that allows users to input RDF and OWL statements in addition to plain text. However, how the system helps user to create such semantic statements has not been discussed in detail in [8].

Powl [9] is a web based platform that provides a collaborative environment for semantic web development. OntoWiki [10], which is built upon Powl, employs web 2.0 techniques to bring more features in authoring, editing, and searching semantic contents. Unlike Powl and OntoWiki which concentrate more on ontology editing, Semantic Wikipedia [11], as an extension to the popular MediaWiki software, brings semantics to texts and links in its wiki system. These systems usually have better user interfaces than those discussed before.

---

[1] http://en.wikipedia.org/wiki/Wiki
[2] http://www.wikipedia.org

There exist more other semantic wiki systems (e.g., WikSAR [12], COW [13], WikiFactory [14] and etc), which provide more features such as integration with the desktop [12], typical query mechanisms [13], domain-oriented design [14], and etc.

3   Design Considerations

3.1   Desirable Functions for Managing Web Ontology Data

The previous section shows that the existing work on semantic wikis is mostly focused on some specific topics. In contrast, we first discuss some desirable functions that are commonly required for managing web ontology data. By doing so, we try to provide some hints towards the design and development of better and more extendable semantic wiki systems.

Browsing, creating and editing. These functions involve the provision of facilities and interfaces for browsing the ontology schema and data, creating and updating the data as required. These functions are actually the most fundamental functions for any types of semantic wikis. However, how such functions are implemented matters. Since the formats of ontology data (e.g., OWL) are not intuitive to ordinary users, it is desirable to design these functions with more intuitive and interactive features.

Search and Query. Obviously, these functions help to retrieve created ontology data from the web site in response to the information need of the users. Generally, the search function may refer to some form of query upon free/unstructured texts such as the search provided by web search engines. In contrast, the query function is usually associated with the structured data (i.e., SQL for the relational databases). Since web ontology data is kind of semi-structured data, it is desirable to have both of these functions in semantic wikis.

Authentication and access control. These functions deal with the identification of the user and the control of their operations on the web ontology data. Existing semantic wiki systems seem to lack a sophisticated mechanism to implement this type of functions. The lack of this mechanism affects the trust on the contents from people, which hinders their use in some serious situations.

Data quality control. Functions of this type ensure the overall web ontology data contributed by users are of accepted quality. One issue of quality is that the data are not reliable. The deployment of proper functions on "authentication and access control" could help handle this issue as malicious editing or spam could be screened off by it. Another common issue is that the duplicated data may be generated in a distributed Web environment, which also decreases the overall data quality. In addition, incompleteness and inaccuracy are threats to data quality as well.

3.2   The System Framework

This subsection overviews the framework based on which the prototype system "robinet" is developed. It shows the general structure of the system and the relations among the functions. The system is implemented in Java.

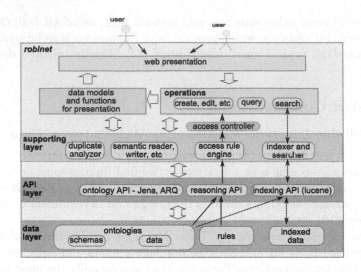

Fig. 1. The framework for the prototype system "robinet"

Fig. 1 illustrates the general structure of the robinet system and the inter-action between it and its managed web ontology data and users. At its lowest layer, the data layer, access rules and indexed data are stored in addition to the managed ontology data. Above the data layer is the API layer, which is made up of low level packages and the APIs they offer. Based on these APIs, a supporting layer provides customized tools which are used to build various operations, data models and functions for presentation. The propose of this layered design is to make the implemented system easy to be maintained and extended.

The framework is designed with principles such as modularity, friendly URLs, and ease of use. Along with these design principles, we have implemented es-sential functions to help web ontology data creation and management. Such functions allow users to browse ontology schemas, create and edit instances and etc. The following sections will discuss them in detail.

4   Browsing, Creating and Editing

4.1   Ontology Schema Browsing

Browsing ontology schemas helps ordinary users get familiar with the domain that the ontology schemas are intended to describe. It is clear that a user with-out an adequate understanding of the domain ontology schema may not be able to proceed and create the correct information for the domain. Some ontology specific editing tools (e.g., Protege [5], and etc) can provide graphical user in-terfaces that allow users to view the domain ontology schemas intuitively. But this requires users to install additional software. In addition, since such editing tools are mostly designed for domain experts or certain professionals to develop ontologies in a desktop environment, they are not web-based solutions preferred

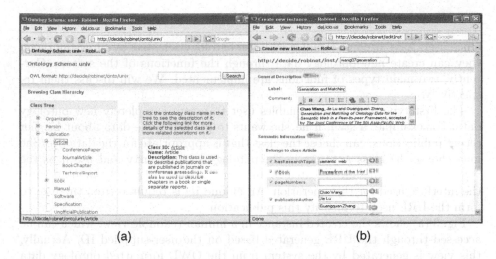

Fig. 2. The interfaces for browsing domain ontology schema and editing ontology data

by ordinary users. Therefore, it is desirable to have a web-based browsing mechanism so that users can just use any types of browsers to learn the domain ontology schemas before they are ready to contribute information.

Initially, the "robinet" system requires a setup process such as loading the ontology schemas for browsing. Once the initial setup is done, users can view the ontology schemas using common browsers. Fig. 2 (a) shows the web interface which renders the classes (concepts) in an ontology schema about the academic domain in a tree structure according to their is-a relationship. Users can get familiar with these classes and their relations by exploring this tree. In addition, if they are interested in a particular class, they can click it in the tree to see detailed comments (if supplied by the schema makers) on this class in the right-hand side panel. If further information is needed to know about this class, users can follow the link in the right-hand side panel to view such information.

In the system, the URLs used to identify their corresponding resources (classes or properties) is quite friendly to both end users and other semantic web applications which may run at other computers in the network. For example, while the URLs http://decide/robinet/onto/univ/ and http://decide/robinet/onto/univ/Article lead to human-readable web pages about the ontology schema univ and the its class Article respectively, the URLs http://decide/robinet/onto/univ and http://decide/robinet/onto/univ#Article identify the original OWL formatted versions which are more machine-understandable. Therefore, other semantic web applications can use such URLs to import the interested ontology schemas into their application space remotely. This mechanism could help the development of semantic web applications and intelligent software agents over the web.

4.2 Ontology Data Creation and Editing

Once users have become familiar with the ontology schemas by browsing them, they can create new ontology data through the functions of the system. Currently, two main types of functions are implemented. We will discuss them respectively.

The first type of functions enables user to create completely new ontology data. For example, a staff, if he/she wants to publish some data about his/her recent publications, can choose the class that is appropriate for the data. He/She may choose the class `Article` in the ontology schema `univ` and use it as the type for the new instance created to annotate a paper. Fig. 2 (b) demonstrates the interface used for this operation. An unique ID is required from the user to form the URL used to identify this publication.

Fig. 3 (a) shows the created instance in a human-readable view, which can be accessed through the URL generated based on the user-supplied ID. Actually, this view is generated by the system from the OWL formatted ontology data which can also be accessed through a browser, as shown in Fig. 3 (b). Obviously, the actual ontology data is not comfortable for ordinary users to read. However, computers on the other hand is good at processing such data. Since they are aligned with the given ontology schema, well designed semantic web applications that are aware of this domain ontology can use them to perform further complicated tasks.

After instances are created, they can be further edited by clicking the "Edit" button as shown in Fig. 3 (a).

The second type of functions allows users to create new ontology data which are related to the existing data.

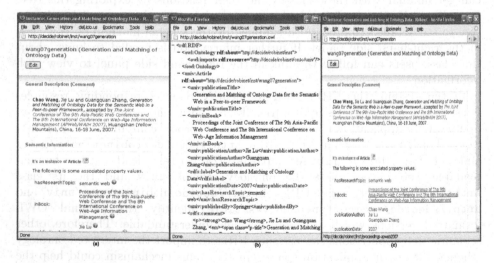

Fig. 3. The created instance of the class `Article`. (a) The instance shown in a human readable view; (b) The actual ontology data created by the user. (c) The instance after related instances are created and automatically linked with it.

First we discuss why we need this type of functions. This type of functions rely on the existence of a type of properties called "object properties" which link one instance to another with a particular semantic relation [15]. They offer the functionality similar to hyperlinks which are used over the current web to link different web pages together. The importance of hyperlinks has ready been obvious. Without them, The web could not form and grow into such a huge connected information repository as currently is; nor could modern search engines build a large information base and rank the related information with certain measurements such as authority or popularity [16][17]. Similarly, it is also very important to use object properties to link instances together. The resulting relational feature in ontology data could be even more essential than using normal hyperlinks in web pages as it delivers more specific semantic meanings than normal hyperlink mechanism. Therefore, it is very desirable to enable users to create ontology data which are linked to other ontology data via certain object properties.

Then we show how easy it is to create related new ontology data based on existing ontology data in our system. As explained before, when a completely new instance is created, initially all the property values are supplied as strings by users, regardless of the types of properties. However, after the instance is created, the system will detect according to the ontology schema whether a property value should be a certain data type or an instance. For those who should be instances via given object properties, it provide the function to create them. As shown in Fig. 3 (a), those property values that should be instances have question marks associated. By clicking the question mark, users will be presented an instance creation interface which is similar to the one shown in Fig. 2 (b). After creating these new instances, they are related to the existing instance with the specified object properties. Fig. 3 (c) shows the web pages of the existing instance after its related instances are created. The question marks has disappeared. Instead, those object property values are linked to the web pages of it's related instances.

5   Search and Query

As mentioned earlier, we distinguish search and query as two types of retrieval functions with their different focuses. The following discusses them respectively.

5.1   Free Text Search

Free text search doesn't require users to know much about query languages. One or more keywords are enough as input to invoke a search, just like the experience of using a web search engine. This seemingly basic function has become pervasive. For a conventional web site especially a web portal, having this type of search function offers users a simple and straightforward way to find and locate information. The same is applied to a semantic web site (or portal) that hosts ontology data.

Currently, we implement the free text search function using Lucene [3], a convenient java package for indexing and searching. In our implementation, every instance is treated as a virtual document. Its properties (including annotation information such as labels and comments) are concatenated together to form the contents of the document. This virtual document is then tokenized and indexed, therefore, able to be searched.

5.2 Structured Query

Structured queries are very useful to retrieve specific information out of the data source. This usually requires the knowledge of the data structure or the schema. Users are also required to know certain query languages that are used to issue queries. SPARQL [18] is one of the languages designed for the query of semi-structured RDF data like ontology data. We use ARQ [4], a SPARQL engine to process queries upon the created ontology data. Users are assumed to know about the query language. By browsing the ontology schema through the system, they could get familiar enough with it to compose their queries to get desired data.

Since most users are reluctant to learn specific query languages, it would be better to use structured query in some indirect ways other than this straight way. Similar to the idea in [13], queries can be specified by some competent users and they can be embedded in certain web pages. This mechanism allows the creation of typical and useful views on the ontology data. Furthermore, external web related applications or web agents can use the query mechanism to exploit the repository of the web ontology data effectively.

6   Discussion on Other Functions

Due to limited space, this section briefly outlines the design and/or implementation of the rest of the two functions as follows.

Authentication and Access Control. Some serious situations require authentication and access control for web ontology data management. We exploit the semantics of the ontology and use a set of access rules to provide a flexible mechanism for the implementation of this type of functions. To do so, we treat the system as a model with tuple $\Lambda(C, O, P, R)$ where C denotes *Semantic Contents*, O denotes *Operations*, P denotes *Participants* and R denotes *Access Rules*. In the system, C, O and P are all modeled using ontology. R is then composed using the elements from these three sets. When an access to the system happens, the access pattern is identified and reasoned against the rules together with the ontology to determine whether the access is accepted or rejected. Fig. 1 also shows the general components that are designed to work together for this type of functions.

---

[3] http://lucene.apache.org/
[4] http://jena.sourceforge.net/ARQ/

Data Quality Control. One major issue related to data quality is data dupli-cation and overlapping, which often happens when many users contribute their data simultaneously. To tackle this issue, first the duplicated data should be de-tected. We have developed particular methods [19][20] to accomplish this task. our methods explore the features of ontology data, therefore, achieve better re-sults compared to other conventional methods according to our experiments. We are currently in the process of integrating these methods into the system. So the system will be able to detect potential duplicated instances created by different users. They will be labeled (for example using the annotation property rdfs:seeAlso) for further review from users. Once the positive feedbacks are made by users, they can be further labeled with the tag owl:sameAs. When multiple duplicates are found for one instance, it is then possible to determine which instance is the most complete and reliable. Therefore, it also helps to solve the problem of data incompleteness.

7 Conclusions and Future Work

This paper proposed a semantic wiki approach for managing ontology data on the Web. It discussed the desirable functions a semantic wiki should have to manage the data. A framework were presented to illustrate the design of such a semantic wiki system. These functions were then further explained and demon-strated in regard to the implementation of this framework. In particular, funda-mental functions of our implemented prototype system were discussed in detail to demonstrate the usability of the semantic wiki approach to web ontology data creation and management.

As an ongoing project, the future work includes the further improvement of the system interaction by using more advanced web techniques and the tighter and better integration with other function modules, for example, the function for data quality control. How to utilize the managed ontology data to provide certain services through web applications or agents will also be studied.

References

1. Berners-Lee, T., Hendler, J., Lassila, O.: The semantic web. Scientific Ameri-can 284(5), 34–43 (2001)
2. Hendler, J.: Agents and the semantic web. Intelligent Systems 16, 30–37 (2001)
3. Ding, L., Finin, T., Joshi, A., Pan, R., Cost, R.S., Peng, Y., Reddivari, P., Doshi, V., Sachs, J.: Swoogle: a search and metadata engine for the semantic web. In: Proceedings of the thirteenth ACM international conference on Information and knowledge management, pp. 652–659. ACM Press, New York (2004)
4. Baader, F., Calvanese, D., McGuinness, D., Nardi, D., Patel-Schneider, P. (eds.): The description logic handbook: theory, implementation, and applications. Cam-bridge University Press, New York (2002)
5. Gennari, J.H., Musen, M.A., Fergerson, R.W., Grosso, W.E., Crubezy, M., Eriks-son, H., Noy, N.F., Tu, S.W.: The evolution of protege: an environment for knowledge-based systems development. Int. J. Hum.-Comput. Stud. 58(1), 89–123 (2003)

6. Sure, Y., Erdmann, M., Angele, J., Staab, S., Wenke, R.S.D.: Ontoedit: Collaborative ontology development for the semantic web. In: Proceedings of the 1st International Semantic Web Conference, Sardinia, Italy (2002)
7. Kalyanpur, A., Parsia, B., Sirin, E., Cuenca-Grau, B., Hendler, J.: Swoop: A 'web' ontology editing browser. Journal of Web Semantics 4(2), 144–153 (2006)
8. Tazzoli, R., Castagna, P., Campanini, S.E.: Towards a semantic wiki wiki web. In: Proceedings of the 3rd International Semantic Web Conference (2004)
9. Auer, S.: Powl - a web based platform for collaborative semantic web development. In: Proceedings of the First Workshop Scripting for the Semantic Web (2005), http://www.semanticscripting.org/SFSW2005/papers/Auer-Powl.pdf
10. Auer, S., Dietzold, S., Riechert, T.: Ontowiki - a tool for social, semantic collaboration. In: Cruz, I., Decker, S., Allemang, D., Preist, C., Schwabe, D., Mika, P., Uschold, M., Aroyo, L. (eds.) ISWC 2006. LNCS, vol. 4273, pp. 736–749. Springer, Heidelberg (2006)
11. Völkel, M., Krötzsch, M., Vrandecic, D., Haller, H., Studer, R.: Semantic wikipedia. In: Proceedings of the 15th international conference on World Wide Web, pp. 585–594. ACM Press, New York, NY, USA (2006)
12. David Aumueller, S.A.: Towards a semantic wiki experience c desktop integration and interactivity in wiksar. In: Proceedings of the 1st Workshop on the Semantic Desktop in conjuction with the 4th International Semantic Web Conference (2005)
13. Fischer, J., Gantner, Z., Stritt, M., Rendle, S., Schmidt-Thieme, L.: Ideas and improvements for semantic wikis. In: Sure, Y., Domingue, J. (eds.) ESWC 2006. LNCS, vol. 4011, pp. 650–663. Springer, Heidelberg (2006)
14. Iorio, A.D., Presutti, V., Vitali, F.: Wikifactory: a web ontology-based application for creating domain-oriented wikis. In: Sure, Y., Domingue, J. (eds.) ESWC 2006. LNCS, vol. 4011, Springer, Heidelberg (2006)
15. McGuinness, D.L., Harmelen, F.v.: Owl web ontology language overview. w3c recommendation (2004), http://www.w3.org/tr/2004/rec-owl-features-20040210
16. Kleinberg, J.: Authoritative sources in a hyperlinked environment. In: Proceedings of 9th ACM-SIAM Symposium on Discrete Algorithms (1998)
17. Brin, S., Page, L.: The anatomy of a large-scale hypertextual web search engine. In: Proceedings of the seventh international conference on World Wide Web, pp. 107–117 (1998)
18. Prudhommeaux, E., Seaborne, A.: Sparql query language for rdf (2007), http://www.w3.org/tr/rdf-sparql-query/
19. Wang, C., Lu, J., Zhang, G.: Integration of ontology data through learning instance matching. In: Proceedings of the 2006 IEEE/WIC/ACM International Conference on Web Intelligence, pp. 536–539 (2006)
20. Wang, C., Lu, J., Zhang, G.: A constrained clustering approach to duplicate detection among relational data. In: Proceedings of the 11th Pacific-Asia Conference on Knowledge Discovery and Data Mining, pp. 308–319 (2007)

Web Service Composition: A Reality Check

Jianguo Lu[1], Yijun Yu[2], Debashis Roy[1], and Deepa Saha[1]

[1] School of Computer Science, University of Windsor
{jlu, roy17, sahag@cs.uwindsor.ca
[2] Computing Department, The Open University
y.yu@open.ac.uk

Abstract. Automated web service composition is one of the major promises of service-oriented architecture, where services can be discovered and composed dynamically and automatically. To investigate the methods for composite web service construction, we conducted an experiment on creating useful composite web services from real existing web services where semantic annotations are not available. The empirical study reveals the difficulties and research challenges in the discovery, invocation, and composition of web services. The automation of web service composition requires the inputs from both services providers and service consumers. Service providers need to develop high quality services in a disciplined and collaborative way, and service consumers need to be equipped with tools providing helps such as service discovery and matching.

Keywords: Web service discovery, web service composition, empirical study.

1 Introduction

Web service is designed to be reused and composed with other web services, manually or automatically. The ultimate goal of service oriented architecture is the automated discovery and automated composition of web services. There have been substantial researches on service discovery [2] [17] and composition [4] [16], based on various formal methods and AI technologies [3] [8] [9] [11] [12]. In the mean while, on the web there are already tens of thousands useful web services that are accessible to the public [5] [21]. However, most of the research and the resulting prototypes target on imaginary web services, usually with semantic annotation, instead of real ones. As a result, there are few tools available that assist the creation of real composite web services from existing publicly available one.

To identify the research challenges in the whole process of web service discovery, invocation, and composition, we conducted an experiment involving 23 graduate students. They are requested to create novel and useful web services out of existing ones on the web, and report their experience on web service discovery, invocation, and composition. The purpose of the experiment is to identify and evaluate the existing methods and tools that can be used in real web service discovery and composition, and identify the difficulties and research problems in real web service composition.

The study shows that the construction of composite web service is a difficult task that requires creativity. The most difficult part in developing new web services is the

B. Benatallah et al. (Eds.): WISE 2007, LNCS 4831, pp. 523–532, 2007.

discovery of the pertinent web services to achieve the goal. Due to various restrictions in existing web services, currently it is almost impossible to have automated web service composition. What is even worse is that manual composition is much more difficult than writing a conventional program because of the ad hoc nature of existing web services which lacks disciplined development and maintenance.

2 The Experiment

23 graduate students are formed into 11 groups. Many of them have excellent programming skills and IT industry experience. Each group is required to produce one or more novel and useful composite service from existing ones. Table 1 lists the composite services they generated. Please notice that although only 12 composite services are generated, many compositions consist of several atomic services. Before a successful composition scenario and atomic service can be decided, students have to investigate many existing services. Hence a large number of web services are tackled.

Before the experiment, students read extensively on both practical and theoretical aspects of web service and semantic web, and are exposed to a variety of web service discovery and composition methods.

In the process the students record the difficulties they encounter in service discovery, invocation, and composition.

Table 1. List of Composite Web Services

	Composite service(s)	Component web services
1	Truck driver route planning and notification	Mappoint, across communications, fast weather
2	Medicare map	Medicare, google map
3	Price comparison and conversion	Amazon, Barnsnoble, currency converter
4	Trip map	Yahoo trip, google map
5	e-commerce product rating	Amazon, google, msn search, currency converter
6	Price comparison	Ebay, amazon, currency converter
7	Geo tag	Flicker, google map, weather
8	Nearest airport	Airport, distance, Zipcode
9	Encrypted email	Encryption, email
10	Composite calendar	Calendar, reminder, call
11	Airport weather	Airport, google map, weather
12	Phone weather	Cell phone text message, weather

3 Web Service Discovery

The first step in web service composition is to locate the pertinent web services that can be used in our composition task. There has been substantial research in web service discovery and the UDDI [2] [18] [22]. In practice, students prefer two approaches to discovering web services manually, i.e., from web service portals, or from generic search engines such as google.

Most students used web service portals such as XMethods.com to locate web services, although they are familiar with researches in web service discovery. None of the students tried to use web service discovery tools or prototypes reported in literature. The reasons may be that most web service discovery researches focus on the semantic annotation of web services and semantic and capability matching of web services. However, existing web services on the web usually are not equipped with semantic descriptions or capability specifications.

Although the number of web services in each portal is limited (a few hundreds in general), most of those web services are valid or active ones. Two groups of students used programmable web APIs [13] instead of WSDLs, from which they generated more complex and interesting applications. Strictly speaking, some APIs from ProgrammableWeb are not the traditional web services. For example, they may not provide wsdl descriptions.

Here are the methods students use to find web services. Many groups use a combination of different methods. For example, they may search for google first to have a rough idea whether there are certain type of services, then go to a service portal to locate the ones that are active.

Table 2. Places to find web services

Portals to find web services	Number of groups used the portal
Google.com	3
Xmethods.com	5
ProgrammableWeb[13]	3
WebserviceList.com	2
WebserviceX.com	4
strikeIron.com	1
Trynt.com	1
Wsindex.com	1
remoteMethods.com	1

Given the fact that none of the web service portals contains large number of web services, and yet they are still the most popular ways to discover pertinent services, there is a need to build a larger web service repository where the quality of web service is put in the first priority.

4 Web Service Invocation

Once a web service is located, we need to call the web service. To automate the composition process, first web services have to be invoked automatically.

In theory, given the WSDL description of a web service, there is no problem to invoke it automatically. WSDL is designed to enable automated invocation. Indeed there are various tools supporting this in various programming languages. For the example of Java programming, Apache Axis can generate the corresponding support Java classes to call the web services with minimal programming effort.

During the experiment, there are a few factors that make the automated invocation impossible, including registration key and soap head change. In fact, many times students found that even the manual invocation is difficult due to lack of documentation and quick evolution of web services.

4.1 Difficulties for Automated Invocation

4.1.1 Registration Key

In the experiment we find that the current web services are not easy to invoke even when the WSDLs are available and the services are active. The main reason is that most web services require a registration key that has to be obtained manually. Almost all the web services that are investigated in our experiment need manual registration. Those service providers require users fill in registration forms manually, so that we can receive the registration keys by email. Then the key has to be provided as a parameter in the operation each time the web service is called. In addition to manual registration, some services even require payments. This requirement from service providers practically prohibits any automated web service invocation.

4.1.2 Tools May Fail

Except the registration key issue, most web services can be invoked by using tools such as Apache Axis. Basically, given the WSDL as input, Axis can generate a set of java classes corresponding to the Schemas in the WSDL description, and other classes supporting the remote call. Then the invocation of web service is simply to initiate some classes that are generated by the tool.

However, sometimes web services require something that can't be generated by the tools.

Example 1:   *Geoplaces* web service requires the authentication ID to be sent in SOAP header instead of in a parameter of the operation. But WSDL2Java in Axis does not generate the corresponding java code filling in the SOAP head properly. As a result we had to manually create the SOAP header element and attach it to the stub object before calling any function.

The format of the SOAP request for the *GetPlaceDetails* operation in *Geoplaces* is given below. The italic part has to be manually injected.

4.2 Difficulties for Manual Invocation

All the students found that invoking web services is not an easy task even when it is done manually. The task is much more difficult than writing a traditional program for a few reasons.

4.2.1 Lack of Documentation

Web services are designed as self-explanatory components in order to increase their interoperability. However, the information conveyed in WSDL is not adequate for a programmer to invoke the service.

```
POST /services/PlaceLookup.asmx HTTP/1.1
Host: codebump.com
Content-Type: text/xml; charset=utf-8
Content-Length: length
SOAPAction: "http://skats.net/services/GetPlaceDetails"

<?xml version="1.0" encoding="utf-8"?>
<soap:Envelope xmlns:xsi="http://www.w3.org/2001/XMLSchema-
    instance" xmlns:xsd="http://www.w3.org/2001/XMLSchema"
    xmlns:soapenc="http://schemas.xmlsoap.org/soap/encoding/"
    xmlns:tns="http://skats.net/services/"
    xmlns:types="http://skats.net/services/encodedTypes"
    xmlns:soap="http://schemas.xmlsoap.org/soap/envelope/">
  <soap:Header>
    <tns:AuthenticationHeader>
      <SessionID xsi:type="xsd:string">string</SessionID>
    </tns:AuthenticationHeader>
  </soap:Header>
  <soap:Body soap:encodingStyle=
    "http://schemas.xmlsoap.org/soap/encoding/">
    <tns:GetPlaceDetails>
      <place xsi:type="xsd:string">string</place>
      <state xsi:type="xsd:string">string</state>
    </tns:GetPlaceDetails>
  </soap:Body>
</soap:Envelope>
```

The most difficult part in web service invocation is to determine the parameters of the operations. In conventional programming languages, there are always explanations for operations and parameters in the operation. Sometimes there are even sample codes to illustrate the usages of the classes. In addition, with the class structure and the relationship with other classes, it is easy to derive a conceptual model of the problem at the hand.

For WSDL description, in general there are no comments in the document element to explain the parameters and operations. In addition, it is hard to find the web page containing use instructions that corresponds to the WSDL.

Web services usually are coarse grained and standalone, with the overall structure of the software hidden behind. Although most WSDLs are automatically generated from conventional programs, it seems that the comments in programs are not carried over to WSDL. For example, when we run java2wsdl in Apache Axis, the comments in the Java program are filtered out.

4.2.2 Versions of WSDLs
Web services are notorious for their fast evolution. For example, eBay service has a fundamental release every two weeks. The constant change makes even experienced programmers difficult to invoke the service. Some students have to contact the service providers to figure out the correct parameters to send out, because existing documents are for the older versions. With different versions of WSDLs scattered around on the web, web service invocation is like exploring a labyrinth. The situation is exacerbated with the scarcity of documentation for web services and the connection between them.

Since conventional software has mature version control and upgrading system, web services need to have a proper management system as well.

5   Web Service Composition

In the experiment we found the difficulties of service composition can be classified into the following categories: schema mapping between data types of web services, large data problem, data quality and optimization, and volatility of web services.

5.1 Schema Mapping

In web service composition literature, it is common to assume that we can use one service's output as another's input. In our experiment, the most common problem reported is the disparity of schemas between web services to be composed. Web services are developed by different organizations that use different conceptual models and vocabulary. Inevitably the resulting schemas in web services are different even if they are meant to be similar or the same.

Example 2:   We have two services *getAirport* and *getAirportCoordinate* to be combined: *getAirport* is of the type
$$state \rightarrow airport(code, city, country, name)$$
and *getAirportCoordinate* is of the type
$$airportCode \rightarrow (latitude, longitude).$$
In this case, we need to map the output data
$$<airport> <code/>...<name/></airport>$$
from one service to the input data *<airportCode/>* of another service.

This kind of schema mapping is not easy to be automated [6]. There are substantial researches on XML Schema mapping, with the aim to identify the correspondences between the elements of schemas, so that the data can be integrated. Schema mapping is also actively studied in migrating legacy systems. In web service composition, every web service is like a legacy system: it is developed by a third party that does not have adequate documentation. Web service composition is similar to legacy system integration, where we need to build the mappings between the parameters of the services. In our experiment, it is almost never possible to directly use one service's output as another's input without any change of the data.

5.2 Large Data and Quality of Data

Several web services in our experiment return huge number of data, which stalls the composition program, and makes the composite service inefficient and practically not possible to run.

Example 3:   Searching airports by country will result very large data, which makes the composite web service practically impossible to run. Hence we have to select an alternative, i.e., obtaining the airports in a state instead of a country.

Another common problem is the low quality of the data returned. Quite often, we need to write some code to process the data before it can be passed to another service.

Example 4:   Although the return of the Airport web service is an XML document, it is not well formed. As a result, it could not be validated using any XML schema. The problem with the XML document was that, the GMT Offset tag was missing in several airport elements. On top of that, it returns each airport information twice.

5.3 Optimization of Composite Service

Some web service composition seems perfect logically, and the data transferred is not big. Still, the composite service is extremely inefficient. The main reason is the involvement of remote invocation. While there are plenty of work on programming code optimization and database query optimization, there is little investigation on the optimization for composite web services.

Example 5:   Given two services
  book(Keywords?, Price, Currency),
  and
  exchange(FromCurrency?, ToCurrency?, FromAmount?, ToAmount).

The first returns the price and currency when given as input the keywords of the books. The second service accepts the amount the currencies to be converted, and gives as output the equivalent amount in another currency. It is easy to generate a composite service that gives the price in local currency:
  localBook(Keywords?, LocalCurrency?, LocalPrice)
  :- book(Keywords?, Price, Currency),
      exchange(Currency?, LocalCurrency?, Price?, LocalPrice).
Straightforward implementation of this composite web service will need to invoke exchange services as many times as the number of returns of the book search. For each *Price* for a book, the exchange service is called to get the corresponding *LocalPrice*.

However, one call to the exchange service is enough. To optimize this composite service, we need to reformulate the composite service as below, so that the calls to web services can be minimized:

  localBook(Keywords?, LocalCurrency?, LocalPrice)
  :- book(Keywords?, Price, Currency),
  *exchange(Currency?, LocalCurrency?, 1, X), LocalPrice=X*Price.*

5.4 Sporadic and Inactive Web Service

Almost all web services, including commercial ones, are not constantly available during our experiment period that lasted two months. Some times they could be off line for a few days. For example, *Medicare* and *YahooTravel* services experienced two days blackout.

Although it is common for a web site to be off line for a while for maintenance or network disruption reasons, the only affected place is the web site itself. Volatile web services will ripple its unreliability throughout all the applications that use them.

To engineer a robust composite web service, it is paramount to have backup web services. Almost all the groups built backup service just to make sure that the composite web service would work properly on the demonstration day. For functions that are served by several services, such as text messenger, we keep all of them. For functions that are provided by only one vendor, such as airport information, students replicate the service on our own machine by downloading part of the data and deploy our own service, so that it can be used as a contingent plan.

Sporadic and inactive web services impose a serious problem for composite web services, as there are abundant inactive web services on the web. Before trying out the web services, there is no way to tell whether they can be used. Table 3 lists the ratio between active web services from two groups. Please notice that most web services are from service portal. If the WSDLs are collected from UDDI or google, the activeness ratio would be much lower.

Table 3. Ration of active web services

	WSDLs checked	Active web service	Active Web Service Ratio
Group 1	11	5	0.45
Group 2	14	9	0.64

6 Conclusions

This is the first empirical study that we are aware of on web service composition based on publicly available real web services. Since this study is based on real web services available, the semantic approach to web service annotation, discovery, and composition is not covered. There have been substantial empirical studies on the state of the art of web services [5] [22], but not the compositions. We programmed using more than 100 web services, and constructed 12 composite services. Some composite services contain several operations.

The study shows that web service composition is a creative activity, whose automation is a daunting task that requires the efforts from service provider as well as service consumers.

Service providers need to develop and maintain high quality web services in a disciplined and collaborative way. In particular, providers need to

1) *Apply software engineering principles*: Most web services are developed in an ad hoc manner, disregarding all the software engineering principles. Even tools developed for web service generation ignored the importance of documentation. For example, Apache Axis removed comments in Java classes when generating WSDLs from those classes.

2) *Develop services collaboratively*: Collaborative web services need to be developed collaboratively. For example, if an XML Schema is already developed or used in other web services, it should be reused, instead of being reinvented every time a similar schema is needed. This way, schema heterogeneity can be minimized. Just as writing conventional programs,

developers should reuse existing components instead of re-develop similar classes and functions every time you need it. To achieve this goal, it is paramount to build and maintain a repository for web services and schemas to enhance the reuse.

Partially due to the inadequacy of WSDLs observed above, some students in our experiment preferred programmable web APIs to create mashups [13]. Similar to web service composition, when programmable web APIs are selected, it is relatively easy to compose them, as the process has little difference from the conventional programming. The difficulty is how to recommend appropriate services, and how to support end-users to create composite services with minimal programming effort [19].

Acknowledgements. We would like to thank all the students who also participated in the experiment, and the anonymous reviewers for their helpful comments. This work is supported by NSERC.

References

1. Agarwal, V., Dasgupta, K., Karnik, N., Kumar, A., Kundu, A., Mittal, S., Srivastava, B.: A service creation environment based on end to end composition of Web services. In: Proceedings of the 14th international Conference on World Wide Web, pp. 128–137 (2005)
2. Benatallah, B., Hacid, M., Leger, A., Rey, C., Toumani, F.: On automating Web services discovery. The VLDB Journal 14(1), 84–96 (2005)
3. Bultan, T., Su, J., Fu, X.: Analyzing Conversations of Web Services. IEEE Internet Computing 10(1), 18–25 (2006)
4. Dustdar, S., Schreiner, W.: A survey on web services composition. Int. J. Web and Grid Services 1, 1–30 (2005)
5. Fan, J., Kambhampati, S.: A snapshot of public web services. SIGMOD Rec. 34(1), 24–32 (2005)
6. Lu, J., Wang, J., Wang, S.: XML Schema Matching, IJSEKE, International Journal of Software Engineering and Knowledge Engineering (in press)
7. Lu, J., Yu, Y., Mylopoulos, J.: A Lightweight Approach to Semantic Web Service Synthesis. In: ICDE Workshop on Challenges in Web Information Retrieval and Integration, Tokyo (2005)
8. Matskin, M., Rao, J.: Value-Added Web Services Composition Using Automatic Program Synthesis. In: Revised Papers From the international Workshop on Web Services, E-Business, and the Semantic Web, pp. 213–224 (2002)
9. McIlraith, S.A., Son, T.C.: Adapting golog for composition of semantic web services. In: KR 2002. Proc. of the 8th Int. Conf. on Principles and Knowledge Representation and Reasoning, Toulouse, France (2002)
10. Medjahed, B., Bouguettaya, A., Elmagarmid, A.K.: Composing web services on the semantic web. The VLDB Journal 12, 333–351 (2003)
11. Milanovic, N., Malek, M.: Current solutions for web service composition. Internet Computing 8 (2004)
12. Mao, Z.M., Brewer, E.A., Katz, R.H.: Fault-tolerant, scalable, wide-area internet service composition. Technical Report UCB//CSD-01-1129, University of California, Berkeley, USA (2001)

13. ProgrammableWeb, http://www.programmableweb.com
14. Rahm, E., Bernstein, P.A.: A survey of approaches to automatic schema matching. VLDB J. 10(4), 334–350 (2001)
15. Ponnekanti, S.R., Fox, A.: SWORD: A developer toolkit for web service composition. In: Proc. of the 11th Int. WWW Conf. (2002)
16. Rao, J., Su, X.: A survey of automated web service composition methods. In: Cardoso, J., Sheth, A.P. (eds.) SWSWPC 2004. LNCS, vol. 3387, Springer, Heidelberg (2005)
17. Sirin, E., Parsia, B., Hendler, J.: Composition-driven filtering and selection of semantic web services. In: AAAI Spring Symposium on Semantic Web Services (2004)
18. UDDI, http://www.uddi.org/
19. Wong, J., Hong, J.I.: Making mashups with marmite: towards end-user programming for the web. In: CHI 2007. Proceedings of the SIGCHI Conference on Human Factors in Computing Systems, San Jose, California, USA, April 28 - May 03, 2007, pp. 1435–1444. ACM Press, New York, NY (2007)
20. Wu, D., Parsia, B., Sirin, E., Hendler, J.A., Nau, D.S.: Automating DAML-S web services composition using SHOP2. In: ISWC2003. Proc. of the 2nd Int. Semantic Web Conf, Sanibel Island, FL, USA (2003)
21. Yu, Y., Lu, J., Fernandez-Ramil, J., Yuan, P.: Comparing Web Services with Other Software Components. In: ICWS. International Conference on Web Services, International Conference on Web Services (2007)
22. Zhang, L., Chao, T., Chang, H., Chung, J.: XML-Based Advanced UDDI Search Mechanism for B2B Integration. Electronic Commerce Research 3(1-2), 25–42 (2003)
23. Zhang, R., Arpinar, I.B., Aleman-Meza, B.: Automatic composition of semantic web services. In: Proc. of the 2003 Int. Conf. on Web Services (2003)

MyQoS: A Profit Oriented Framework for Exploiting Customer Behavior in Online e-Commerce Environments

Ahmed Ataullah

David R. Cheriton School of Computer Science
University of Waterloo, Canada
aataulla@uwaterloo.ca

Abstract. Work related to improving the performance of web based e-commerce servers has largely focused on providing better response times and higher throughput. However the most important performance metric for businesses is profitability. Thus it is imperative that online businesses identify valuable user sessions and ensure their completion in transient and persistent overload conditions. This paper introduces MyQoS, an extensible and easy to deploy framework for identifying valuable user sessions in e-commerce settings. MyQoS is specifically geared towards maximizing profitability through continuous monitoring of user behavior in online retail scenarios. It provides online retailers with the ability to offer differentiated levels of service based on the individual and collective usage pattern of customers. Our tests with the TPC-W retail setting demonstrate that MyQoS can provide the benefits of existing approaches and enable service providers to implement profit maximizing techniques that have so far been ignored in e-commerce applications.

1 Introduction

The rapid growth in the popularity of the World Wide Web has made server behavior in high load conditions an active area of research. Phenomena such as the slashdot effect have underscored the importance of admission control measures and stress testing well beyond the expected load of Internet servers. Even in transient overloads lasting only a few minutes, businesses can lose substantial revenue if valuable customers become frustrated and leave. Consequently, it is of utmost importance to identify valuable customers in unpredictable load conditions and ensure that they accomplish their session objectives.

MyQoS is a framework through which online retailers can design customer valuation strategies for their unique requirements and offer better response times and throughput to valuable sessions. It acts as middleware between application servers and database servers in the 3-tier architecture and prioritizes requests to the database server based on the value of the session from which they originate. Session valuation in MyQoS is done by comparing known profitable user behavior against individual queries issued by sessions and retrieved results. We

B. Benatallah et al. (Eds.): WISE 2007, LNCS 4831, pp. 533–542, 2007.

present results of our test implementation and demonstrate that the additional layer of MyQoS can increase the profitability of an e-commerce system in high load conditions with a negligible performance penalty. Finally we conclude our discussion with a brief overview of the wide array of session valuation strategies that can be employed using MyQoS and highlight the avenues of future research that they open in computer science and online consumer behavioral analysis.

2　Related Work

Session Based Admission Control (SBAC) as introduced by Cherkasova and Phaal [1] was one of the earliest techniques which attempted to assign a value to a session. The main contribution of SBAC is the recognition that user interactions with web services are usually task oriented and its proposal that requests for new sessions be gracefully rejected if their completion can not be guaranteed. SBAC did not aim to directly increase profitability but it provided a mechanism by which older sessions, presumably those close to completion, were considered to be of higher value due to the resources already invested in serving their requests.

The work of Menascé et al. [2] is also worth mentioning as it was the first to introduce the notion of online customer behavior modeling. The main contribution of their work was the representation of user sessions as state transition graphs, where each state represents either a web page or a stage of session progression (such as searching, account creation/management or checkout). They associated a probability value for transitioning from one state to another, and demonstrated that in retail environments, customers with varying propensities to purchase can be grouped into different customer behavior modeling graphs (CBMGs). Frequent buyers are expected to have a higher combined probability of quickly transitioning into the final stages of the session and the unconvinced users may leave the online store (exit the graph) from any state. CBMG represents a significant improvement over SBAC in terms of valuation, as it allows user behavior, such as frequent buyer, regular visitor, and information seeker to be classified as session types and provides a framework for predicting the value of a particular session.

More recently Totok and Karamcheti proposed an extension to the CBMG model, by introducing the notion of *Reward Driven Request Prioritization* (RDRP) [3], where a numeric value associated with each session can be used to gauge its importance. The model adopted for assigning reward values to sessions is based on request types. The objective of their session prioritizing scheme is to maximize the expected reward from all sessions. For example, if service providers associate a reward of 1 with the CheckOutSuccess request type and 0 for all other request types, then RDRP would give preference to requests from sessions that are probabilistically more likely to end in the CheckOutSucces state. Similarly, businesses could associate a value with each request type in the CBMG to get richer models for profit maximization, such as giving higher priority to customers that have added multiple items to their shopping cart.

Although RDRP provides an innovative use of session inspection, it lacks the ability to accurately valuate individual sessions in e-commerce scenarios. For example, as pointed out by Menascé et al. [4] it would make business sense to give priority to customers who have more profitable items in their shopping carts, rather than those that intend to purchase low margin items. The RDRP approach to profit maximization defines reward over request types and does not deal with such business level details. Unfortunately, the suggestion of shopping cart valuation also holds little merit, as not all e-commerce situations have a cart oriented framework. Furthermore, information collected from recent surveys [5] suggests that the majority of customer shopping carts are abandoned before checkout is ever completed.

The problem of maximizing 'profit' and at the same time developing a framework that can accommodate a broad range of applications has proven to be very difficult. The main idea in existing techniques is to define a notion of profit and use widely accepted business level knowledge to prioritize sessions in an attempt to maximize it. Unfortunately, most techniques are situation specific and difficult to implement. Our work shares its goal with others in the technical area of electronic commerce but we refrain from proposing our own notion of profit and offer no suggestions on how to maximize it. Instead we present a system that provides a detailed view of user sessions, using which businesses can easily develop their own session valuation schemes for profit maximization. The novelty of MyQoS lies in the fact that it allows businesses to exploit (for QoS purposes) a much wider array of correlations between customer behavior and profitability, than existing web-request analytic approaches.

3   The MyQoS Framework

MyQoS is a controller that acts as middleware (proxy) in the 3-tier architecture between the application server and the database server. The goal of MyQoS is to manage the primary bottleneck resource, which in a large number of e-commerce scenarios, is the database server. We adopted a middleware approach in order to develop a solution which would require minimal changes to existing systems and yet be powerful enough to implement a wide array of QoS policies. The use of middleware in the 3-tier architecture is itself not novel [3,6]. Elnikety et al. [6], demonstrated using the GateKeeper proxy, that a transparent middleware utilizing estimates of overload limits and request response rates can serve as an admission controller and guarantee stable performance in severe overload conditions. Although our approach to QoS shares the same motivation of being easy to implement, it has the objective of profit maximization based on exploitable knowledge about the users of the system and is therefore not transparent.

Our reason for prioritizing requests for only the database server is based upon the general consensus in the published literature [6,7,8] that the resource most likely to become the bottleneck in e-commerce applications is the database server. Work done by Amza et al. [7] proves this point by showing that even if the entire database is memory resident (workload is not disk bound), queries

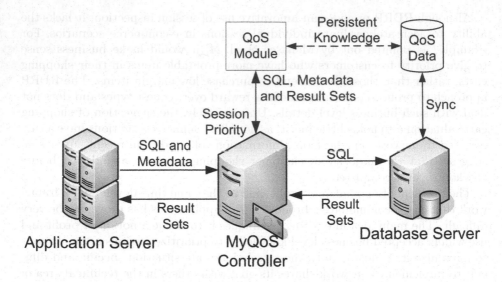

Fig. 1. Overview of the MyQoS Framework

requiring expensive joins and complex computations will cause the database server CPU to become the bottleneck before the web or application servers reach capacity. We acknowledge that under provisioned web and application servers can be the cause of the system wide bottleneck and describe in Section 5 how MyQoS can be extended to provide QoS hints to these layers. However, the implementation of MyQoS described in this paper considers the database server to be the primary bottleneck in the e-commerce system.

3.1 Implementing MyQoS

Implementing MyQoS requires no changes to the database server or the web server. The only modification required for implementing MyQoS is that of associating sessions maintained by the application server with queries received by the MyQoS proxy (Figure 1). Database interfaces for e-commerce applications are traditionally designed to accept well formed SQL statements and return the resulting data set. However, the MyQoS proxy requires explicit knowledge of each session's existence so that the each received query can be associated with the session in which it originated. Consequently the database interface used by the e-commerce application must be modified to not only issue SQL queries to the database but also attach metadata identifying the session itself.

There exist several techniques to accomplish this task. Simply appending metadata as XML after the query text is one option. Modifying the database connectivity packages used by applications (such as JDBC) to append session related metadata transparently is a more elegant and perhaps the least application invasive approach. We anticipate that most well designed applications handle database connectivity uniformly through the application, therefore changes to accommodate additional metadata along side the SQL will be a one-time cost and relatively insignificant when weighed against the benefits of this approach.

3.2   The MyQoS Controller

For our discussion we assume that the session related metadata discussed above will always contain a `session identifier`, uniquely identifying the web session issuing the query. In practice all trappable information at the web/application server levels such as the user's browser, operating system and IP address can be passed to MyQoS for analysis and request prioritization. Adding this additional metadata can subsequently help us analyze usage patterns in greater detail to make better QoS decisions.

The MyQoS controller maintains a table of identifiers for sessions that are in progress and an associated integer value representing their priority. All incoming queries are queued and then forwarded to the database server in order of the priority of the session where they originated. The incoming requests (SQL and metadata) are also processed by a QoS module, which attempts to re-calculate the priority value of each session based on the newly received request. By default all sessions start with a priority of zero and their priority can only be modified by the QoS module. Symmetrically, the results of each query forwarded to the database server are returned to the session and also processed by the QoS module. The QoS module in conjunction with the controller, essentially becomes a "Gatekeeper" which, based on predefined QoS rules analyzes queries and their results to determine respective priorities of each session.

3.3   QoS Modules

The sole purpose of a QoS module is to analyze session related metadata, queries and returned datasets to determine a priority for user sessions. For a query and result oriented framework such as ours, QoS modules with a wide array of objectives can be easily built by exploiting some basic characteristics common to most e-commerce applications. Firstly, we note that e-commerce sessions for a particular retail setting are very similar in nature. There are a fixed number of tasks that users can perform such as searching, purchasing, authenticating and checking order status. Consequently we assume that there are a constant number of types (templates) of queries that are issued by application level sessions. A very simple string analysis of these queries can often reveal valuable information about a session. For example consider a typical query generated by a search request for an online book store:

```
SELECT BNAME, BQTY, BPRICE FROM BOOK NATURAL JOIN AUTHOR
WHERE BCAT = 'Music' AND BAUTHOR LIKE 'Helen Drew'
LIMIT 75,100
```

A glance at the query reveals that the user is searching for music related books written by a specific author and that results requested are between the 75th and the 100th records (perhaps implying that the user has previously browsed 3 pages of the result each with 25 records). Even if the query itself does not reveal much, examining query results can often reveal the direction or intent of the user. For example if the majority of the search results of a particular query are books of a certain category then we can confidently ascertain the general category of products in which the user has exhibited interest.

We remind the reader that the primary goal of MyQoS is to increase profitability of the e-business by giving priority to valuable sessions as identified by their querying behavior. By continuously monitoring (or mining) the queries being issued and the resulting datasets, we can gain insight into a session allowing us to make statistically viable judgments about a user's intent. We can then base prioritization decisions on specific information such as the availability of products, the depth of user search in the session and even pricing/profitability data of the products for which the user has searched. We also have the ability to dynamically track session progress and if user authentication requests (as queries) are also detected, we can also adjust session priority according to the users' past behavior.

Dynamically tracking individual sessions allows businesses to more objectively differentiate valuable customers and prioritize their requests appropriately. The situation is similar to a physical retail store wherein sales associates often pay special attention to customers looking for expensive products and judge the customer's inclination to purchase within a few interactions. Unfortunately as is the case in the physical retail environment, determining whether or not a customer is valuable is not always easy.

There are several well known statistical correlations in e-commerce environments which can be naturally exploited using MyQoS. For example, the fact that returning customers are more likely to make a purchase than new visitors is widely accepted in the industry. MyQoS can potentially examine authentication queries and their results (looking for successful authentication requests) to boost priority to individual sessions. IP and location based heuristics can also be implemented using MyQoS for geo-targeting, wherein higher priority can be given to sessions originating from specific geographic locations. Similarly, studies [9] have shown that user sessions related to certain categories of products such as clothing last much longer when compared to other categories and are less likely to end up in successful purchases. By using MyQoS in such scenarios, customers browsing through more profitable items (for example electronics) can be given priority over those browsing through low margin items such as apparel. There are also unique situations such as the launch of a new product which can generate significant amounts of traffic. Usually the product is quickly sold out, but traffic levels may sustain for a prolonged period, affecting the response times afforded to other customers. MyQoS can be employed in such situations to detect and reduce priority for those sessions that are only searching for the particular sold out product.

In short, with MyQoS e-businesses have the freedom to dynamically enforce session level request prioritization based on any inferable information from the session related metadata (IP-address, web browser and operating system), individual query strings and the datasets returned to each session. There is a broad range of information that can be extracted from each of these three elements and immediately exploited for better QoS for more profitable customers. For brevity we have offered only a few examples of the QoS schemes that can be built using MyQoS.

4 Performance Results

It must be emphasized that performance and profitability are very different notions when examined from a business perspective. To the best of our knowledge there does not exist an e-commerce benchmark for profitability. Researchers usually accept the logical conclusion that less request-timeouts will lead to more sessions successfully checking out which will then lead to higher overall profitability. Our goal was similar, to demonstrate that if given sufficient business level knowledge to identify valuable customers, then MyQoS can detect and offer better response times to such customers, consequently maximizing profit (minimizing loss) in cases where all customer requests can not be accommodated by the e-commerce application.

TPC-W is a widely accepted benchmark for online e-commerce applications which attempts to model the workload of an online book store. Unfortunately TPC-W is strictly a performance oriented benchmark and it does not specify any business level correlations between the types of user sessions and the value of those sessions. Consequently we were forced to introduce our own business level knowledge for QoS purposes. We claim that studies conducted [5,9], which attempt to isolate online purchasing behavior between different categories of products (such as apparel and electronics) can lead to some basic rules regarding profitability. For our hypothetical business case of a TPC-W bookstore we assumed that customers browsing through books categorized as 'computers' are most likely to purchase them and that the customers browsing through books in the 'cooking' category are least likely to make a purchase. Consequently we deigned our QoS module to give the least priority to users searching for cooking related books and give the highest priority to sessions browsing through computer related books. To get comparable results we were also forced to adjust the TPC-W workload from being random to such that 10% of sessions always searched for computer related books and 10% always searched for cooking related books, while the remaining had no inclination towards any particular category of books. In our case the choice of categories of profitable products and user distribution among them was largely arbitrary. However we believe that detecting user search behavior and making prioritization decisions based on that behavior represents a very realistic business oriented QoS policy and serves as an ideal test for MyQoS.

We used a relatively small TPC-W database (72,000 customers and 10,000 items) on a PostgreSQL (8.2.3) server with the browsing mix as our workload. The total database size in our case was deliberately kept small at 138 MB so

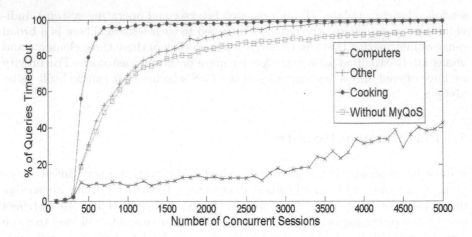

Fig. 2. Even in extreme overload conditions and with the behavior detection delay factored in, sessions of high priority customers (those searching for computer related books) timed out at a significantly lower rate when using MyQoS. Without MyQoS all customers are treated equally causing them to time out at similar rates (an average of the timeout rates of the three classes of customers is shown in this case).

that the database server (Centrino Duo 1.83GHz with 1GB RAM) was able to accommodate it entirely in memory. The reason for doing so was to ensure that the database server was not under provisioned for its load relative to the server running MyQoS. It is important to note that because e-commerce queries generated by web requests are short in nature (rarely retrieving a large number of records), increasing the size of the database does not increase the workload for MyQoS. However, increasing the number of concurrent sessions directly impacts the number of requests MyQoS is expected to analyze. By keeping the size of the database small we were able to test our framework with a very large number of simulated user sessions with varying behavior. A relatively under provisioned system (Pentium 4 1.6GHz with 512MB RAM) was used to run a simple Java based implementation of the controller, our above described QoS module and a session oriented query load generator. To ensure that the QoS module itself does not become the bottleneck resource in the system, our implementation gave the highest priority to acting as a proxy between the database and the simulated application server, and then performing analysis of intermediate data for QoS.

Figure 2 shows the test results summarizing the effect on response times offered to the different types of sessions. MyQoS was able to identify all sessions related to cooking and computer books, by continuously examining the issued queries and the returned datasets. As was expected, requests from high priority sessions timed out at a significantly lower rate when compared to other types of sessions. The data presented in Figure 2 for each session type includes results for the queries issued before a session's category was determined. Consequently, it is worth noting that even with the delay in detecting session types factored in, MyQoS can generally perform well by sacrificing response times for lower priority sessions and providing better response times to those with high priority.

5   Future Work

We believe that there are several aspects of e-commerce that can benefit from additional research. In our work we separated the issues of prioritization and session value identification through QoS modules as user programmed heuristics for profit maximization. While prioritization is a well understood concept, there is significant work that still needs to be done to determine which QoS strategies can be used for prioritization in various e-commerce situations. More specifically, we need to answer how can we use existing business level knowledge regarding identification, acquisition and retention of valuable customers and the patterns which they exhibit, for QoS purposes in online situations such as retailing, online auctioning and online stock trading.

Another area of future work is highlighted from the fact that the web and application servers can also benefit from knowing the value of a particular user session. Although our work strictly assumes the database server to be the bottleneck, it is possible that under provisioned application or web servers become the system wide bottleneck resource. A possible improvement which we foresee is cooperation between MyQoS and other layers of the 3-tier architecture. MyQoS controller could pass session priority to the application and database servers to allow enforcement of complex prioritization schemes. Furthermore, each of the players in the 3-tier architecture could also suggest priority values for sessions to each other and even disregard prioritization suggestions locally depending on the situation. Our work also assumes that e-commerce queries are short in nature and we ignore the issue of determining the precise opportunity cost of a query. MyQoS, having no statistics about the e-commerce database nor a query optimizer, lacks a method for accurately determining the cost of each query. If the assumption about the e-commerce only workload is taken away and we include lock contentious queries with the e-commerce workload, then we may need to extend MyQoS perhaps as global profitability controller embedded in the database server.

6   Conclusion

In this paper we presented MyQoS, a framework for QoS in e-commerce environments based on identification of valuable sessions through fine grained session inspection. Instead of proposing our own notion of profitability we proposed a system which is general enough to enforce a broad range of profit maximization heuristics for a large number of e-commerce situations. We believe that MyQoS represents a substantial improvement over existing techniques of prioritization and session valuation for two main reasons. Foremost, it provides a significantly greater fine grained view of user sessions than existing techniques. Consequently richer models of QoS can be built by identifying valuable sessions based on their behavior. Secondly, it separates request prioritization from session value identification which allows service providers to quickly deploy new QoS strategies based on continuously changing nature of their target market. To maintain focus in our work we addressed only how MyQoS can be used for session valuation and prioritization in online retail environments. However, we believe that our contribution is equally valuable for other e-commerce situations, such as stock trading and online auctions, where accurate valuation of user sessions is important.

Acknowledgments

The author would like to acknowledge the technical discussions, comments and the feedback of Ken Salem, Tim Brecht and Alexander Totok.

References

1. Cherkasova, L., Phaal, P.: Session based admission control: a mechanism for improving the performance of an overloaded web server. HPL-98-119, HP Labs Technical Reports (1998)
2. Menascé, D.A., Almeida, V.A.F., Fonseca, R., Mendes, M.A.: A methodology for workload characterization of e-commerce sites. In: EC 1999. Proceedings of the 1st ACM conference on Electronic commerce, pp. 119–128. ACM Press, New York, NY, USA (1999)
3. Totok, A., Karamcheti, V.: Improving Performance of Internet Services through Reward-Driven Request Prioritization. In: IWQoS. 14th IEEE International Workshop on Quality of Service, pp. 60–71 (2006)
4. Menascé, D.A., Fonseca, R., Almeida, V.A.F., Mendes, M.A.: Resource management policies for e-commerce servers. ACM SIGMETRICS Performance Evaluation Review 27(4), 27–35 (2000)
5. Shim, S., Eastlick, M.A., Lotz, S.L., Warrington, P.: An online prepurchase intentions model: The role of intetion to search. Journal of Retailing 77(3), 397–416 (2001)
6. Elnikety, S., Nahum, E., Tracey, J., Zwaenepoel, W.: A method for transparent admission control and request scheduling in e-commerce web sites. In: WWW 2004. Proceedings of the 13th international conference on World Wide Web, pp. 276–286. ACM Press, New York, NY, USA (2004)
7. Amza, C., Chanda, A., Cox, A.L., Elnikety, S., Gil, R., Rajamani, K., Zwaenepoel, W.: Specification and implementation of dynamic web site benchmarks. In: WWC-5. Fifth Annual IEEE International Workshop on Workload Characterization (2002)
8. Zhang, Q., Riska, A., Riedel, E., Smirni, E.: Bottlenecks and their performance implications in E-commerce systems. In: Chi, C.-H., van Steen, M., Wills, C. (eds.) WCW 2004. LNCS, vol. 3293, pp. 273–282. Springer, Heidelberg (2004)
9. Bhatnagar, A., Ghose, S.: Online information search termination patterns across product categories and consumer demographics. Journal of Retailing 80(3), 221–228 (2004)

Task Assignment on Parallel QoS Systems[*]

Luis Fernando Orleans[1,2], Carlo Emmanoel de Oliveira[2], and Pedro Furtado[1]

[1] Departamento de Engenharia Informática, Universidade de Coimbra,
Portugal
[2] Núcleo de Computação Eletrônica, UFRJ,
Brazil
{lorleans, pnf}@dei.uc.pt, carlo@nce.ufrj.br

Abstract. Request Processing Systems should exhibit predictable behavior by guaranteeing Quality-of-Service parameters. One of the important requirements of predictability is that requests should have maximum acceptable response time thresholds, denoted as deadlines in this paper. In order to provide QoS, the system should try to guarantee these deadlines. This way, it is desirable to use a load-balancing algorithm that tries to both maximize the throughput and minimize the missed deadlines. Assuming that all requests durations are known *a priori*, this paper shows, with the help of a simulator, how the control over the big tasks plays a crucial role in the pursuit of the objective. Finally, this paper presents a derived form of the traditional Least-Work-Remaining algorithm, named Task Assignment by Isolating Big Tasks, that proved to be a better alternative in such scenarios.

Keywords: QoS, parallel systems, load-balancing, deadlines.

1 Introduction

Requests Processing Systems should exhibit predictable behaviour by guaranteeing Quality-or-Service parameters. One of the important requirements of predictability is that requests (tasks) should have maximum acceptable response time threshold, which we denote as deadlines in this paper. In order to provide QoS with a Parallel Request Processing System (P-RPS), the P-RPS should try to guarantee deadlines. Such policy affects the way the system should be designed, because all task durations must be under the deadline, and also how tasks are distributed among the servers. This way, it is desirable to use a load-balancing algorithm that tries to both maximize the throughput and minimize the missed deadlines. Assuming that all tasks durations are known *a priori*, this paper presents the performance of well-known load-balancing algorithms when tasks with deadlines are submitted and shows, with the help of a simulator, how the control over the big tasks plays a crucial role in the pursuit of the objective. Finally, this paper presents a derived form of the traditional Least-Work-Remaining (LWR) algorithm that is supposed to be a better alternative when tasks have deadlines.

---

[*] This work was supported by the Adapt-DB project – FCT POSC/EIA/55105/2004.

B. Benatallah et al. (Eds.): WISE 2007, LNCS 4831, pp. 543–552, 2007.
© Springer-Verlag Berlin Heidelberg 2007

Both traditional LWR and Round-Robin (RR) provide no control over the concurrent executors (CE) number on the system, which means that every request that arrives is immediately put into execution, no matter how many requests are already running on the servers. This policy is known as *best-effort*, i.e., for all requests that arrive, the system will do its best to handle them, though providing no response time guarantees. Using a simple linear analysis, if a request would take 1 second to run alone, in a server that already has 100 requests running it will take 100 seconds to complete. In a system where tasks have deadlines, this behaviour is unacceptable, as the number of missed-deadlines highly depends on the CE number.

To reduce the number of missed-deadlines in P-RPS, one can add new servers to the cluster and upgrades on all servers, improving the computing capacity of the system as a whole. These are modifications of the installed capacity. They do not happen instantaneously, may lead to expensive over-capacity options to deal with peaks 0.5% of the time and on themselves do not provide the answer to meeting the deadlines with heavy traffic. No matter what the installed capacity is, the problem occurs whenever the rate of requests arrivals is sufficiently large (traffic peaks) to mean that each node will have to process several requests simultaneously.

A logical solution to reduce the number of missed-deadlines is to adapt the load-balancing algorithms, making them to limit the CE number of the system. This way, there will be only an adaptable number of concurrent requests interfering on each others execution. The requests that arrive while the system is saturated, i.e., the maximum CE value has been reached, give rise to a system busy notification. The actual CE depends on the deadline tightness (how tight is the deadline). In addition to the CE control, an algorithm could go even further: since all task durations are known, the algorithm could avoid that small tasks (with short durations) get stuck waiting for big tasks to complete. In other words, there can also be a control over the number of big tasks getting executed on each server.

This paper describes a model where requests may be accepted or rejected by the system at the moment of their arrival, depending on the current load. This mechanism of acceptance or rejection is called *admission control mechanism*. Our objective is to design approaches that are completely external to the controlled system. Since a request is accepted, it has to be fully executed before its deadline. Otherwise, its execution will be cancelled (or killed). This way, long tasks will not keep interfering with the execution of smaller ones. It is supposed that all requests are not *preemptible*. This way, once a task is assigned to a server, it will run on the same server until its completion. Finally, all tasks are independent, as their executions do not block each other.

This work is about load-balancing in QoS systems, so the best algorithm will be the one which combines good throughput and low number of killed tasks.

2 Related Work

There are plenty of works about load-balancing and QoS, most of them leading to already accepted and consolidated conclusions. Although these are almost exhausted themes, their combination seems to be an area where there are very few research results.

2.1 Load-Balancing

Many load-balancing algorithms have been studied, where most of them trying to maximize throughput or minimize the mean response time of requests. Reference [1] proposes an algorithm called TAGS, which is supposed to be the best choice when task sizes are unknown and they all follow a heavy-tailed distribution. This is not the case for the scenario analyzed in this paper, because task sizes must be below a deadline threshold. It is also shown in [1] that, when task sizes are not heavy-tailed, Least Work Remaining has a higher throughput then TAGS. In fact, [2] and [3] claim that Least-Work Remaining is optimal when task sizes are exponential and unknown.

The algorithm SITA E [4] has the best performance when task sizes are known and are heavy-tailed. In cases where task sizes are not heavy-tailed, Least-Work-Remaining presents a better throughput.

2.2 Quality of Service (QoS)

In real distributed systems, task sizes are heavy-tailed. This means that a very small portion of all tasks are responsible for half of the load [5]. In models with deadlines, like the one analyzed in this paper, occurs a similar distribution. Most of all tasks are very small ones. And there is a small number of big tasks, but these all are under the deadline limit, i.e., they're maximum duration is the same as the deadline's.

Reference [6] presents a model where the number of concurrent requests within the system is restricted. When this number is reached, the subsequent requests are enqueued. But this model has no concerns about deadlines or rejection of requests. It also does not show a way to load-balance the arriving tasks, since it is a single-server architecture.

Quality-of-Service was also studied for Web Servers. In [15] the authors propose session-based Admission Control (SBAC), noting that longer sessions may result in purchases and therefore should not be discriminated in overloaded conditions. They propose self-tunable admission control based on hybrid or predictive strategies. [14] uses a rather complex analytical model to perform admission control. There are also approaches proposing some kind of service differentiation: [12] proposes architecture for Web servers with differentiated services; [13] defines two priority classes for requests, with corresponding queues and admission control over those queues. [17] proposes an approach for Web Servers to adapt automatically to changing workload characteristics and [16] proposes a strategy that improves the service to requests using statistical characterization of those requests and services. Comparing to our own work, all the works referenced above except [16] are either intrusive, require extensive modifications to systems, or use analytical models that may not provide guarantees against real system. The work in [16] does not consider deadlines and does not adapt constantly as ours does, instead it fixes the maximum throughput capacity when it runs the test runs.

3 Load Balancing Alternatives and the TAIBT Algorithm

Load balancing is a fundamental basic building block for construction of scalable systems with multiple processing elements. There are several proposed load-balancing algorithms, but on of the most common in practice is also one of the simplest ones - Round-Robin (RR). This algorithm produces a kind of "static load-balancing" functionality, as the tasks are distributed round-the-table with no further considerations. We consider not only RR as also Least-Work-Remaining (LWR), as this algorithm is supposed to be the best choice when task sizes are not heavy-tailed. Our proposal is for an adapted version of LWR called Task Assignment by Isolating Big Tasks (TAIBT).

We are concerned with guaranteeing specified acceptable response time limits, which we denote as deadlines. The number of concurrent executions is a crucial variable when deadlines are involved, because as we increase the number of concurrent executions we have a larger probability of missing the deadlines. For this reason, in our work all the three algorithms have a control over the maximum number of concurrent executors (CE) within the system, which varies from 1 to 60. The TAIBT also includes another parameter: the minimum duration of requests that are considered "big". In this paper, tasks with duration greater than 1 second are classified as big tasks. This way, only very fast jobs have some kind of privilege, since they do not have to pass through another admission control mechanism that big tasks do.

In the following we describe each load balancing algorithm we compare considering deadlines and rejection:

Round-Robin (RR)

```
for each task that arrives:
            if SUM(tasks) <= EC

              next_server := server_list ->first

              send (task, next_server)

              server_list -> move_to_end(next_server)

            else

              NOT_ADMITTED(task)
```

Least-Work-Remaining (LWR)

```
  for each task that arrives:
            if SUM(tasks) <= EC

              next_server := server_list ->least_utilized

              send (task, next_server)

            else

              NOT_ADMITTED(task)
```

Task Assignment by Isolating Big Tasks (TAIBT)

```
  for each task that arrives:
            if SUM(tasks) > EC

              NOT_ADMITTED(task)

            if task is a big task
```

```
    if exists a server (x) with no big task
        next_server := server (x)
    else
        NOT_ADMITTED(task)
else
    next_server := server_list ->least_utilized
    send (task, next_server)
```

The TAIBT algorithm is an adaptation of Least-Work-Remaining to deal efficiently with deadlines and big tasks. In fact, they both have the same performance for small-sized tasks. The main difference is the control over big tasks included in TAIBT. If a big task is submitted to the system, it will only get executed if there is a server running no other big task. Otherwise, the task will be rejected. This algorithm clearly tries to reduce the number of killed tasks, by admitting only those that can be fully executed.

4 Experimental Analysis

Due to the large number of parameters involved, it becomes necessary to formally describe the simulation model that was used in this work. The simulator, constructed in University of Coimbra, has the following parameters :

- Number of servers.
- Processing capacity of each server.
- Tasks arrival rate.
- Mean task size.
- Number of maximum tasks being processed concurrently.
- Maximum amount of time a task can execute (deadline).
- Load-balancing algorithm.
- Minimum size of "big tasks".

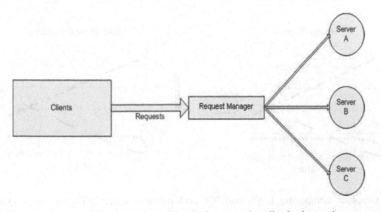

Fig. 1. Simulated architecture: three identical servers handle the incoming requests

In this paper, a system with 3 identical servers was simulated. The deadline time was set to 10 seconds and the duration of each request follows an exponential distribution with mean 2 (we generated tasks with durations between 0.1 and 10 seconds). In addition, all tasks have the same priority. The concurrency model is linear, which means that a task will take twice longer if it shares the server with another task, it will take three times longer in case of two other tasks and so forth.

The simulator implements the three algorithms: Least-Work-Remaining (LWR), Round-Robin (RR) and TAIBT. We will compare the performance of these and all results are presented as an arithmetic mean of 3 rounds of simulation. To eliminate the transient phase, the data obtained in the first 10 minutes of the simulation was discarded. Only the results obtained in the next 30 minutes were considered.

4.1 Simulation Results

This section contains the results obtained with the simulator. Figure 2 shows how the system performs when the chosen algorithms are LWR and RR, as they were described. As can be noticed, the performances is very similar for all alternatives, due to the low variability of task sizes, provided by the exponential distribution. Since the system is under high load, with 10 tasks arriving per second, the number of killed tasks increases as the number of the CE increases. Similarly, the number of rejected tasks decreases for higher CE values. In both algorithms the throughput is low: not even 20% of the submitted tasks get fully executed.

On the other hand, the TAIBT algorithm (figures 3a and 3b) tries to prevent tasks from being cancelled. Its control over the number of big tasks within each server prevents the small tasks from getting their execution indefinitely delayed. This control not only reduces the number of kills, but it also increases the throughput and diminishes the rejection. One important thing to be noticed here in figure 3a is that the maximum throughput is reached with only 9 concurrent executors and, no matter how much this number increases, the throughput remains the same. This occurs because only the small, fast-executable tasks pass through the admission control mechanism and to serve them only 6 CE are necessary (since the other 3 CE are reserved to the big tasks). So, even a high arrival rate like 10 tasks per second can be handled with only 9 CE with practically no loss.

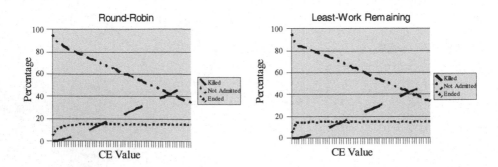

Fig. 2. Graphics comparing LWR and RR performances when submitted to a system with admission control

If the arrival rate is raised to 20 tasks per second (tasks durations keep following the exponential distribution with mean value 2) as shown in figure 3b, we can notice that the maximum throughput is reached with CE between 15 and 31, when the number of cancelled tasks begins to increase, which means that the number of small tasks in each server is so high that they are interfering on each others execution and making to each other miss their deadlines.

But how do these algorithms behave if tasks durations are not exponential? In fact, [6] claims that tasks durations in a distributed/parallel system are not exponential, but they follow a heavy-tailed distribution, such as a Pareto distribution. This means that a minuscule fraction of the very largest jobs comprise half of the total system load. This distribution is not applied in the scenario being studied in this paper, because we limit the task duration to a value below 10 seconds (variability is limited). But what would happen if the number of small tasks is much greater than the number of big tasks? Figures 4 and 5 show how the algorithms would perform in such a situation, where tasks durations follow a bounded-Pareto [4] distribution, with the exponent of the power law $\alpha = 0.6$, the smallest possible observation, $k = 0.1$ and the largest possible observation, $p = 8$.

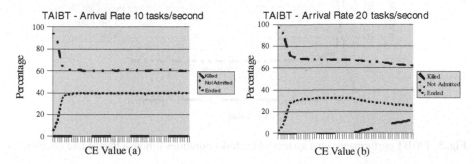

Fig. 1. TAIBT performance in a scenario where tasks durations are exponential. In (a) the arrival rate is 10 tasks per second, which is unable to reduce the performance of the system, even with high CE values. In (b) the arrival rate is 20 tasks per second and the performance starts to degrade with CE values higher than 31.

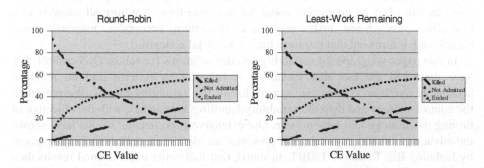

Fig. 4. Graphics comparing LWR and RR performances when submitted to the same system with admission control and tasks durations follow a Pareto-like distribution, although they still have deadlines

As can be noticed, the performance of LWR and RR are still close. Since the number of small tasks is much, much greater than the number of big tasks, the throughput is better than the exponential case. Actually, in no moment the throughput decreased, even at high CE values. But even in this scenario, with a great number of small tasks, the number of cancelled tasks is high, with almost 25%of the submitted tasks being killed.

The TAIBT performance is better in this last, more realistic scenario, too. The TAIBT algorithm keeps the number of killed tasks at a very low rate (0.1%) and reaches its optimal throughput (75,37%) with 18 CE . The low number of killed tasks clearly identify the reasons that leads TAIBT to have a better performance than LWR and RR: its control of the number of big tasks on each server.

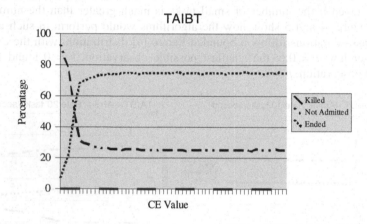

Fig. 5. TAIBT performance on a system where tasks durations follow a Pareto-like distribution

5   Conclusions and Future Work

Distributing the load on parallel systems following QoS adaptability principles is a new, challenging task. Most of the existing algorithms try to equally load-balance the requests, but offer no guarantees about the response time, handling all requests in a best-effort way. Since tasks may have maximum acceptable response times (deadlines), it turns out that response time is a crucial concern.

In this paper we proposed variability-solving solutions for robust QoS control. We have discussed the fact that variability is inherent to systems and no practical acceptable overhead control can account for all the variability factors. We presented the control strategy and then the variability-handling approaches, with the objective of turning the strategy into a robust one. Our extensive experimental results have shown the advantage of our proposals. We presented an algorithm, called Task Assignment by Isolating Big Tasks (or TAIBT, in short), that had better experimental results than traditional load-balancing techniques, such as Round-Robin and Least-Work-Remaining. This algorithm uses an admission control mechanism to accept or reject tasks at the moment of their arrival, limiting the number of concurrent executors within the system as a whole. The algorithm also uses another admission control layer to restrict the number of requests with long durations – in fact, there can be only one big task per server at the same time.

In our experimental results, the TAIBT algorithm was the be the best choice when requests duration follow not only an exponential distribution, but also when they follow a Pareto-like distribution. In all cases, the number of requests that break their deadlines was very small, while that number was even greater than the throughput in some cases when considering other algorithms.

As future works, we expect to study an efficient way to determine the duration of requests. This could be added to an TAIBT implementation in order to identify and classify the tasks according with their sizes. . We would also like to generalize the "a single large task per server" condition of TAIBT to a more adaptable limit on that number.

References

1. Harchol-Balter, M.: Task assignment with unknown duration. Journal of the ACM 49 (2002)
2. Nelson, R., Philips, T.: An approximation to the response time for shortest queue routing. Performance Evaluation Review, 181–189 (1989)
3. Nelson, R., Philips, T.: An approximation for the mean response time for shortest queue routing with general interarrival and service times. Performance Evaluation, 123–139 (1993)
4. Harchol-Balter, M., Crovella, M., Murta, C.: On choosing a task assignment policy for a distributed server system. Journal of Parallel and Distributed Computing 59(2), 204–228 (1999)
5. Harchol-Balter, M., Downey, A.: Exploiting process lifetime distributions for dynamic load-balancing. ACM Transactions on Computer Systems (1997)
6. Schroeder, B., Harchol-Balter, M.: Achieving class-based QoS for transactional workloads. ICDE 60 (2006)
7. Crovella, M., Bestavros, A.: Self-similarity in World Wide Web traffic: Evidence and possible causes IEEE/ACM Transactions on Networking, 835–836 (1997)
8. Knightly, E., Shroff, N.: Admission Control for Statistical QoS: Theory and Practice. IEEE Network 13(2), 20–29 (1999)
9. Barker, K., Chernikov, A., Chrisochoides, N., Pingali, K.: A Load Balancing Framework for Adaptive and Asynchronous Applications. IEEE Transactions on Parallel and Distributed Systems 15(2) (2004)
10. Serra, A., Gaïti, D., Barroso, G., Boudy, J.: Assuring QoS Differentiation and Load-Balancing on Web Servers Clusters. In: IEEE Conference on Control Applications, pp. 885–890 (2005)
11. Cardellini, V.C., Colajanni, M., Yu, P.S.: The State of the Art in Locally Distributed Web-Server Systems. ACM Computing Surveys 34, 263–311 (2002)
12. Bhatti, N., Friedrich, R.: Web server support for tiered services. IEEE Network 13(5), 64–71 (1999)
13. Bhoj, R., Singhal: Web2K: Bringing QoS to Web servers. Tech. Rep. HPL-2000-61, HP Labs (May 2000)

14. Chen, X., Mohapatra, P., Chen, H.: An admission control scheme for predictable server response time forWeb accesses. In: Proceedings of the 10th World Wide Web Conference, Hong Kong (May 2001)
15. Cherkasova, Phaal: Session-based admission control: A mechanism for peak load management of commercial Web sites. IEEE Req. on Computers 51(6) (2002)
16. Elnikety, S., Nahum, E., Tracey, J., Zwaenepoel, W.: A Method for Transparent Admission Control and Request Scheduling in E-Commerce Web Sites. In: WWW2004. The Thirteenth International World Wide Web Conference, New York City, NY, USA (May 2004)
17. Pradhan, P., Tewari, R., Sahu, S., Chandra, A., Shenoy, P.: An observation-based approach towards self managing Web servers. In: International Workshop on Quality of Service, Miami Beach, FL (May 2002)

Autonomic Admission Control for Congested Request Processing Systems

Pedro Furtado

University of Coimbra, CISUC
pnf@dei.uc.pt

Abstract. Consider a Transaction Server or any other kind of Request Processing System under congested workloads. The most probable outcome under peak congestion is for the system to degrade ungracefully and most sessions to receive timeouts after waiting for some piece of processing for a long period of time. Wasteful computation occurs as the system is unable to service in-time. Setting a session limit is the most common approach to minimize the problem, but while this approach is useful, it is neither autonomic nor dynamic. How do we determine a good value for the limit? Can we also have a strategy that automatically adapts to changes (e.g. the performance of backend database accesses can vary widely as it grows, as hardware is upgraded or as administrative tasks tune and modify settings)? In this paper we propose external adaptive control. We use a simulator environment to design and compare the strategies and also test against a transactional system.

Keywords: qos-broker, Transactional Systems, QoS, Admission Control.

1 Introduction

One traditional way for software systems to handle requests is to try to optimize the service to each incoming request, accommodating all of them as fast as possible. This could be described as a best-effort approach. However, while trying to guarantee the best to all requests, systems may well end up providing very bad service to each request when they are in congested states. The first thing that needs to be acknowledged to prevent this problem is that individual resources have finite capacities and therefore cannot provide guarantees to an arbitrary amount of requests, so they must not admit too large a workload (the only way to serve more requests than the admissible quota would be to load-balance among multiple resources). This can be accommodated by an Admission Control (AC) strategy. Considering a resource, manually setting a limit for the number of simultaneous sessions is the typical, simple AC strategy. But what is a good session limit? If this limit is set too high, many requests will take too much time; if it is set too low, the system will be under-utilized. A good session limit is dynamic and depends heavily on the system and workload being executed at each time. There are a few previous works proposing some form of autonomic control. Strategies may use analytic models to try to model execution conditions, feedback control over the miss rate or some way to find a presumably maximum throughput session limit. Analytic models have limited capacity because they represent a very crude representation for a system; feedback control on itself,

B. Benatallah et al. (Eds.): WISE 2007, LNCS 4831, pp. 553–562, 2007.

although useful, has limitations in this context due to the required length of the control cycle, as the consequence of a modification must be observed before the next control cycle begins. This was our motivation to investigate and propose prediction-based autonomic approaches that rely on empiric observations of runs instead of models and adjusts the session limit dynamically. The problem being posed is an optimization one, and while the objective function for many of the approaches is to achieve a maximum throughput, we reason that the objective should be implicit or explicit time constraints on the time taken to service a request. We assume that contracts may be provided specifying optional limit value for the response time of transactions (this may be a constant for all transactions or vary among transactions). This point is very relevant and practical, as it makes sense to specify for instance that users accessing a site should not have to wait more than 10 seconds for an action to return to them.

Our main contribution in this paper is to propose the use observed runtime information for autonomic admission control, to make the approach robust to request runtime variability and to analyze it.

The paper is structured as follows: section 2 presents related work. Section 3 presents a high-level overview of the QoS-Broker for controlling the execution environment and Contracts. Then section 4 uses the simple multiprocessing linear model to reason about admission control and deadline issues and section 5 proposes the alternative strategies for Admission Control. Section 6 discusses experimental simulation results, their achievements and comparison with non-controlled behavior. Section 7 concludes the paper.

2 Related Work

This work is related to Quality-of-Service approaches, which typically include not only admission control but also other strategies such as Service Differentiation, Resource Reservation and Load Balancing. The works in [7, 8, 9] focus QoS for transactions, in this case in the context of real-time databases specially aimed at applications with strict time-requirements such as military ones [1, 8]. Those works propose strategies to meet deadlines for individual transactions in a in-memory operations context because execution times in a memory-only system are much more predictable and controllable than in more complex contexts. In [11, 14] the authors propose the use of feedback control to adapt the admission procedure to guarantee deadlines in real time systems, using control theory strategies such as PID.

Admission control has also been studied in the Web Servers realm. In [6] the authors propose session-based Admission Control (SBAC), noting that longer sessions may be the ones resulting in purchases and therefore should not be discriminated in overloaded web server conditions. They also propose self-tunable admission control based on hybrid or predictive strategies. In [5] the authors use a rather complex analytical model to perform admission control. There are also several approaches proposing some kind of service differentiation, including [2] and [3]. In this work the authors define two priority classes for requests, with corresponding queues and admission control over those queues. In [12] the authors propose an approach for Web Servers to adapt automatically to changing workload characteristics and [10] proposes a strategy that improves the service to requests using statistical characterization of requests and services. Measurement-based Admission control is considered in [6], where the system maintains online estimates of request service times based on recent executions, a server capacity as the maximum load level that produces the highest throughput, and either a FIFO or SJF queue to keep the requests until they can be served. Most of these related works have not provided answers to the problem of determining the most appropriate execution parameters for requests, simply because they did not consider acceptable response time thresholds (deadlines) together with measures on the actual response times. We consider two main objectives: maximizing the throughput and providing the correct context for transactions to end within the deadlines. We provide dynamic strategies to use maximum throughput, deadline satisfaction and a balance between both and we solve a small-transaction blocking concern that raises with the dynamic approaches.

3 Basic Model for Multi-transaction Systems

It is easy to see that the *best-effort* approach fails in congested situations. While trying to offer service to all requests simultaneously, the system tends to trash most requests in congested states. This can be seen using a very simple linear model in what concerns multiple simultaneous executions (we will elaborate more on the actual effect and improved model for simultaneous executions later on):

Multi Transaction Execution Linear Model (MTE-LM): The model intends to reason about an approximate expectation on the time transactions would take to execute in a multi-transaction scenario, outside factors not considered. We consider a *standalone duration* d_{ij} being a statistic measure of the time transaction T_i would take to execute a run r_j if it were alone in the system. For simplicity we denote d_{ij} as d from now on. The execution in a multi-transaction environment is modeled as follows: transaction T_i ends run r_j as soon as it consumes the duration d. This duration is consumed at a rate that is dependent on the number of transactions executing simultaneously. If n transactions are executing simultaneously in the time interval Δt, T_i consumes at most $\Delta t / n$ of d in Δt. If we apply this model to a transaction running alone, $n=1$ and d is consumed precisely in time d, as expected. The model is linear because the duration is proportional to the number of simultaneous transactions (e.g. it predicts a duration proportional to $n \times d$ when n transactions with the same duration d are executed simultaneously). Given the MTE-LM model and a contract specifying that transactions should not take more than D to execute (limit threshold or deadline), under the *best*-effort approach a transaction with standalone duration d will for certain miss the deadline if there are in (time-weighted) average at least D/d transactions in the system (e.g. if the standalone duration is 1 second and the transaction should not take more than 10 seconds, it should for sure not run with more than 9 other simultaneous transactions, and a smaller limit on this value can be imposed to provide added guarantees). Under this logic, deadlines are missed when there is congestion, with bad results to the user or application that has to wait a long time to get an answer or a system busy timeout.

Setting a session or transaction limit is both an intuitive and an effective strategy to overcome the limitation of *best-effort*. The issue then is how to determine the appropriate value of n_i in an uncertain context as the one we are targeting. The value chosen would typically be either too conservative (allowing a lower number of simultaneous executions than it could be under the desired contract(s)) or wrongly chosen (transactions may take too long under congested conditions). Moreover, the system dynamism means that the setting for this value degrades as conditions change (e.g. database tuning and configuration, load, workload, hardware, system load).

4 Adaptable Strategies for Autonomic Admission Control

We consider a system with a FIFO queue. Queue wait times are limited by a fixed amount (we assumed transactions may seat in the queue for a fraction of their deadline, which we fixed at 50%). Then the system controls the number of transactions executing simultaneously using one of the following alternatives:

EC_{TPM} **Strategy:** consider a congested system with request-limit admission control in-place. If we want to maximize the fraction of requests that actually gets executed, it makes sense to optimize the throughput (TPMs). From now on we denote by "Execution Cardinality", or EC, the number of requests that is executing simultaneously at any given moment in time and we want to set EC_{max}. Therefore, we want to find the EC that maximizes TPMs, which we denote as $EC_{max} = EC_{TPM}$. Offline tests are yield which increase the number of simultaneous requests and TPMs are measured for each case, obtaining pairs <EC, TPM>. The EC_{TPM} value is chosen accordingly. In our experiments we made offline tests with the TPCC benchmark [15] and Figure 2 shows results for the test run using with 100 warehouses on a Pentium 4 with 5 disks, 1GB of memory and PostgreSQL version 8.1 and the default TPC-C transaction mix ($EC_{TPM}=21$).

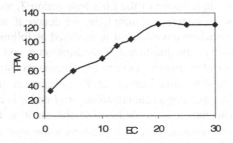

Fig. 1. Test Run for Determining $EC_{max} = EC_{TPM}$ for TPC-C workload

EC_{TPM+RT} **Strategy:** EC_{TPM} fixes EC_{max} and does not take into consideration the response times of requests. EC_{TPM+RT} is an evolution of EC_{TPM} to adapt automatically as the system executes and the workload changes. It is based on the maximum desirable response time parameter for transactions – the deadline. The EC_{TPM+RT} strategy can start with $EC_{max}=EC_{TPM}$, but it keeps track of request response times and compares those to the deadlines. Requests have contracts with the form $C(T, l\%, d\%, d)$, where T is the request, d is the deadline and $d\%$ is a fraction of d, which is the target runtime for request T. The value $l\%$ is a (lower) fraction of d, used to signal that the request is running sufficiently fast so that EC can be raised (we used default values of 30% and 90% for $l\%$ and $d\%$). An example of a contract would be $C(PAYMENT, 30\%, 90\%, 5000ms)$: the payment transaction should take less than 5 seconds and the strategy considers 4.5 seconds to set EC. If the transaction runs in 1.5 seconds or less, EC can be raised, whereas if it takes more than 4.5 seconds EC must be lowered.

The controller considers every request that missed its deadline and predicts the value of EC that would be necessary to avoid missing the deadline. If request T_i took time t_i to execute and its contract was $C(T_i, d_i\%, d_i)$, then:

> $EC_{max}=EC_{TPM}$;
> For each recently executed request T_i, compute EC_i
>   = time-weighted average of execution cardinality while the request was executing
> For each T_i which missed the deadline d_i
>   (miss value: $m_i = t_i - d_i\% \times d_i$):
>   compute desired EC to avoid missing the deadline:
> $EC_{inew} = EC_i \times (d_i\% \times d_i) / t_i$;
> Determine a new EC_{max} value to adjust to =
> $heuristic(EC_{1new},..., EC_{mnew})$;

The heuristic function decides new EC values based on proposals for requests that missed their deadlines. Conservative and average-value proposals are:

> $Heuristic_{conservative}(EC_{1new},..., EC_{mnew})=min(EC_{1new},..., EC_{mnew})$
> $Heuristic_{avg-value}(EC_{1new},..., EC_{mnew})=avg(EC_{1new},..., EC_{mnew})$

The conservative proposal is better at minimizing missed deadlines, but it may constrain admission too much (we opted for the optimistic alternative).

This algorithm presented above uses a "linear" model to adjust EC according to response time and desired deadlines $[(d_i\% \times d_i) / t_i]$. One improvement is to replace the $(d_i\% \times d_i) / t_i$ factor by a more accurate conversion factor taken from the $<EC_i,TPM_i>$ pairs obtained in the EC_{TPM} determination, as these pairs describe how performance (TPMs or RT) evolves with EC in average. The idea is that the average response time of requests with EC_i is EC_i/TPM_i and T_i took t_i for that EC_i value. Therefore, for T_i to take $(d_i\% \times d_i)$, we look for the EC_j value from the pairs $<EC_j,TPM_j>$ such that $EC_j/TPM_j = (d_i\% \times d_i) / t_i \times EC_i/TPM_i$.

The EC_{TPM+RT} must also be able to increase EC_{max} when the response time is good. This is the reason for the $l\%$ parameter in the contract, which determines that if EC_{max} requests execute simultaneously and the response time of all those requests is below $l\%$, the controller can increase EC_{max} to the value that would result in response times of $(d_i\% \times d_i)$.

EC_{RS} Strategy: in workloads with heterogeneous transactions, heavy transactions (transactions taking much more time to execute than the remaining ones) are frequently responsible for blocking fast ones when the execution cardinality is restricted. We observed this with exponential-like workloads and to prevent this kind of blocking behavior we took too additional steps: transactions with excessive outlier durations (larger than $\mu+\sigma$, μ and σ being the mean and standard deviation of response times respectively) count as having a duration $\mu+\sigma$ for the algorithm (this avoids excessive restriction of the EC parameter), and we made the system reserve admission slots to small transactions as well. For that the system needs information on the mean and variance of workload durations and of specific transactions. Each time a transaction executes, the system collects the runtime and maintains the up-to-date statistics based on those values. In the following we denote by E(X) the expected or average execution time and STDEV(X) as the standard deviation of execution time.

We define the parameters maxEECs (maximum extra execution cardinality) and maxD (maximum distance), where w denotes workload:

$$maxEECs = floor(\#EC_{TPM} \times \lambda), \text{ where } 0 <= \lambda <= 1 \text{ is a configurable parameter (default=8)};$$

$$maxD = (E(X)_w - \beta \times STDEV(X)_w), 0 <= \beta \text{ is a configurable parameter (default=1)};$$

Given a transaction T with duration smaller than E(X) and no free execution slot, the new strategy EC_{RS} determines the number of extra ECs that can be used by a transaction as,

$$d_T = E(X)_w - E(X)_T$$

$$nrExtraECs = ceil[(min\{d_T, maxD_w\} / maxD_w)^n \times maxEECs] \qquad (1)$$
$$0 <= n < 1;$$

If the number of simultaneous executions at the moment is smaller than *nrExtraECs*, transaction T is admitted. The parameter n in Equation (1) provides a simple way for configuring access to the extra slots differently (if n is smaller, lower distances will have more slots available than if n is larger).

Besides this modification, we introduced a reservation of a small configurable fraction of EC_{max} for large transactions (with default value floor(($\#EC_{TPM} \times 0.25$), as otherwise the previous modification could now block large running transactions instead of short ones.

5   Experimental Results

Our experimental results are organized as follows: we first compare the alternatives against the TPCC benchmark with a specific setting to show the advantage of the proposed strategies when deadlines are present. Then we engage in sensitivity analysis using an event-driven simulator in which we experiment with varied inter-arrival and duration distributions to test the approaches.

5.1   Running EC_{RS} with a Transaction-Processing System

Figure 3 shows our results when running the strategies with the TPCC benchmark [15]. We measured T_{noMiss} (the throughput considering only transactions that did not miss their deadline) for different deadline specifications. Offline test runs with increasing EC values were used to determine the EC providing maximum throughput, which was EC=21 for this particular system. We compare the results when no limit is imposed on the execution cardinality T_{noMiss}(no control) (*Best Effort* case) to the results with EC_{TPM} and with EC_{RS}. If no deadlines are specified, the maximum throughput is achieved whether we use EC_{TPM} or EC_{RS} (EC_{RS} maintains the execution cardinality at the maximum throughput point when there are no deadlines), while the "no control" alternative exhibits a smaller throughput (saturation and excessive overhead of running many simultaneous threads). As we specify deadlines for transactions the T_{noMiss} throughput decreases significantly, but EC_{RS} maintains the best throughput among the alternatives. For larger deadlines (25 seconds in the Figure), EC_{TPM} is able to achieve a comparable throughput to EC_{RS}.

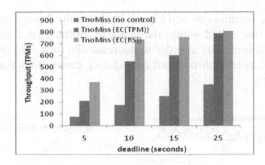

Fig. 2. Throughput with Deadlines

5.2 Sensitivity Analysis

For these experiments we used our event-driven simulator. The simulator uses a transaction inter-arrival distribution and a transaction standalone duration distribution together with an <EC,TPM> map (the throughput versus number of simultaneous executions) to simulate execution and obtain throughput and response time results while applying admission control approaches. We used the <EC,TPM> map we obtained experimentally for TPC-C (Figure 2).

For the next experiments we have set transaction deadlines to 5 seconds and varied either transaction duration (100 ms to 1.5 seconds) or arrival rates (between 10 ms and 500 ms), fixing the other parameter (the duration at 1 second, the inter-arrival time at 100ms). In every case these were always mean values for each of a distribution (Uniform, Exponential, Gaussian, Poisson). Figure 4a shows the mean and standard deviations of the distributions tested (considering a mean of 1 second). The normal (4b) and exponential (4c) are good representatives of the two classes of workload characteristics we wanted to test – homogeneous workload (small standard deviation) and highly varying workload (large standard deviation).

DISTR.	avg	stdev
Random	1006	588
Poisson	1013	92
Exponential	1038	976
Gaussian	1000	33
Uniform	1024	578

(a) Mean and standard deviations (msecs)   (b) Pdf for Gaussian   (c) Pdf for Exponential

Fig. 3. Characteristics of distributions used in the experiments

Each simulation run submitted 10000 transactions according to the arrival rate distribution and statistics. The results were consistent over the above distributions and we have chosen a subset of the results to show how we analyzed the efficiency and robustness of the strategies. The results we show in Figure 5 concern expected arrival rate of 100 ms (uniformly distributed) and varying duration rates from 100 ms (uncongested) to 1.4 seconds (congested) (Gaussian distribution).

Figure 5a shows the miss rate and Figure 5c shows the throughput considering only those transactions that ended within the deadline. Admission control is expected to restrict admission (Figure 5b) in order to decrease the miss rate (Figure 5a) and this way maintain an acceptable throughput considering transactions that ended within the deadline (Figure 5c).

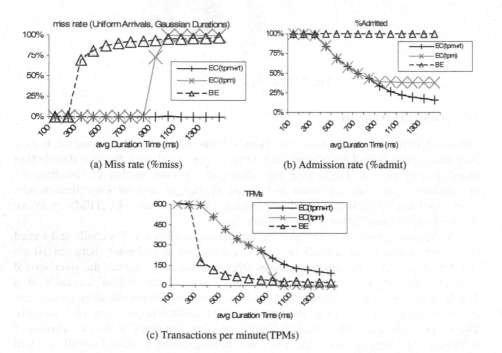

(a) Miss rate (%miss)

(b) Admission rate (%admit)

(c) Transactions per minute(TPMs)

Fig. 4. Performance measures for Uniform arrivals and Gaussian durations

It is clear from the figure that the Best Effort strategy BE quickly fails in congested state (large miss rate in Figure 5a), as it accepts a large number of transactions (Figure 5b). The two EC-limiting strategies (EC_{TPM} and EC_{TPM+RT}) restrict admissions (Figure 5b) and as a result maintain almost zero miss rates (Figure 5a). However, the figure also shows that above a mean of 0.9 seconds, EC_{TPM} is not sufficiently restrictive and consequently the miss rate becomes practically 100%, while EC_{TPM+RT} simply continues restricting admission below EC_{TPM} to keep an acceptable miss rate.

These results show that EC_{TPM+RT} is able to cope with congestion much better than EC_{TPM} when considering the Uniform arrivals with Gaussian distribution and a low standard deviation. This is a fairly homogeneous workload. We also tested several other combinations and distribution statistics and show next the results for an heterogeneous context. This time the duration distribution was changed from Gaussian to Exponential, with large variance. Figure 6 shows the results (the figures show only the first values concerning BE because it fails completely beyond 500 ms durations).

These results differ from the previous ones in some important aspects. Miss rates for EC strategies are a bit higher even in uncongested states, especially for EC_{TPM}, because there are some very large transactions in the large variance exponential distribution that miss the deadlines often. EC_{TPM+RT} has a better miss rate than EC_{TPM}, but at a high cost of admitting much fewer transactions (6b). The most important observation is that EC_{TPM} seems to outperform the supposedly more adaptable EC_{TPM+RT} strategy in relevant metrics. By looking at the nature of transactions that ended within the deadline we observed that many more small transactions would end well with EC_{TPM} than with EC_{TPM+RT}. We concluded that EC_{TPM+RT} would restrict EC to a point that would block fast transactions behind executing slow ones. To solve this limitation of the dynamic EC_{TPM+RT} approach we applied the modified approach EC_{RS}, also shown in the results. We tested this strategy with 10 extra execution slots, $\beta=0.5$ and n=0.75 and applied it to the previous Exponential distribution experiment and the results also show that EC_{RS} was the best strategy on most accounts. The improvement is mostly due to the fact that slow transactions are no longer blocking fast ones Although EC_{TPM} also exhibits acceptable performance results, Figure 6(a) shows that the miss rates can be as high as 25 to 50%, which are quite high. EC_{RS} avoids this problem and still delivers very good performance.

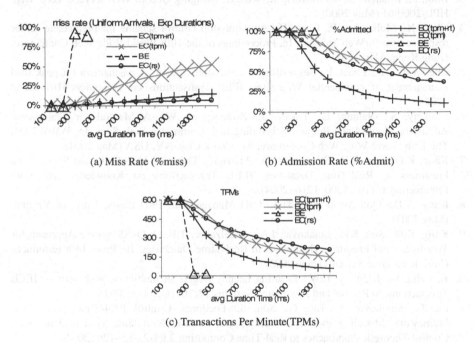

(a) Miss Rate (%miss)

(b) Admission Rate (%Admit)

(c) Transactions Per Minute(TPMs)

Fig. 5. Performance Uniform arrivals and Exponential durations

6 Conclusions

In this paper we proposed and analyzed external autonomic adaptable strategies for admission control in request processing systems that can be readily deployed in any context. We have described a TPM-optimizing strategy and the design of a more adaptable alternative. Besides applying it to a transactional benchmark context, we tested the design for efficiency and robustness (handling varied workload distributions efficiently) using appropriate metrics. Different arrival and duration distributions were generated and tested using an event-driven simulator. With this approach we were able to determine the weakness of the adaptable strategy and proposed a robust alternative.

Acknowledgments. This research was partially funded by FCT project POSC/EIA/55105/2004.

References

1. Abbott, R., Garcia-Molina, H.: Scheduling Real-Time Transactions: A Performance Evaluation. ACM Trans. Database System 17, 513–560 (1992)
2. Bhatti, N., Friedrich, R.: Web server support for tiered services. IEEE Network 13(5), 64–71 (1999)
3. Bhoj, P., Rmanathan, S., Singhal, S.: Web2K: Bringing QoS to Web servers. Tech. Rep. HPL-2000-61 (May 2000)
4. Chen, X., Mohapatra, P., Chen, H.: An admission control scheme for predictable server response time forWeb accesses. In: Proceedings of the 10th World Wide Web Conference, Hong Kong (May 2001)
5. Cherkasova, L., Phaal, P.: Session-based admission control: A mechanism for peak load management of commercial Web sites. IEEE Transactions on Computers 51(6) (June 2002)
6. Elnikety, S., Nahum, E., Tracey, J., Zwaenepoel, W.: A Method for Transparent Admission Control and Request Scheduling in E-Commerce Web Sites. In: WWW2004. The 13th World Wide Web Conference, New York City, NY, USA (May 2004)
7. Kang, K.D., Son, S.H., Stankovic, J.A.: Managing Deadline Miss Ratio and Sensor Data Freshness in Real-Time Databases. IEEE Transactions on Knowledge and Data Engineering 16(10), 1200–1216 (2004)
8. Kang, K.D.: QoS-Aware Real-Time Data Management. PhD thesis, Univ. of Virginia (May 2003)
9. Kang, K.D., Son, S.H., Stankovic, J.A., Abdelzaher, T.F.: A QoSSensitive Approach for Timeliness and Freshness Guarantees in Real-Time Databases. In: Proc. 14th Euromicro Conf. Real-Time Systems (June 2002)
10. Kanodia, V., Knightly, E.W.: Ensuring latency targets in multiclass Web servers. IEEE Transactions on Parallel and Distributed Systems 13(10) (October 2002)
11. Lu, C., Stankovic, J., Tao, G., Son, S.: Feedback Control Real-Time Scheduling: Framework, Modeling and Algorithms. special issue of Real-Time Systems Journal on Control-Theoretic Approaches to Real-Time Computing 23(1/2), 85–126 (2002)
12. Pradhan, P., Tewari, R., Sahu, S., Chandra, A., Shenoy, P.: An observation-based approach towards self managing Web servers. In: International Workshop on Quality of Service, Miami Beach, FL (May 2002)
13. Schroeder, B., Harchol-Balter, M., Iyengar, A., Nahum, E., Wierman, A.: How to determine a good multi-programming level for external scheduling. In: ICDE 2006. 22nd International Conference on Data Engineering (2006)
14. Stankovic, J., Lu, C., Son, S., Tao, G.: The Case for Feedback Control Real-Time Scheduling, EuroMicro Conference on Real-Time Systems (June 1999)
15. TPCC: www.tpc.org

Towards Performance Efficiency in Safe XML Update

Dung Xuan Thi Le, and Eric Pardede

Department of Computer Science and Computer Engineering,
La Trobe University, Bundoora VIC 3083, Australia
{dxlle@students., E.Pardede@}latrobe.edu.au

Abstract. The amount of research related to XML Updates has increased in the last few years. The works range from proposals for the updating of languages to studies of the implications of XML updates. While one major implication of XML update operations is the validity of data update, the other is the performance of update operations. Sometimes the validity and the performance are mutually exclusive. In this work, we present a real-life experimental result on the update which promotes validity, and affects the performance. At the end, we propose novel algorithms that we believe will be effective in reducing the cost of XML Updates without sacrificing the validity of updated data.

Keywords: XML Updates, XML Schema, Schema Validation, Update Performance.

1 Introduction

An XML Update was once perceived as an unnecessary operation in XML data management. XML documents were considered as static and even if an update was necessary, the users just replaced the whole document with a new one. However, with the increased usage of XML as a data format, there is a new attitude among XML communities towards XML updates. As for any traditional database, an XML document should be updateable and all issues that arise with update operations have emerged as research topics.

There is a common understanding that XML update is an expensive operation. However, since XML update has now become a core requirement and is not merely an option, XML communities still have to perform XML update during their data management. The question is how to find the cheapest way of performing the update.

Some earlier works propose primitive update methods where there is very little or no concern about the validity of updated XML documents. Some recent works have embedded constraint a checking mechanism in their update process, which we can consider as safe updates.

In this paper, we will first perform an experiment to determine how safe updates perform in comparison with the primitive updates. The updates themselves can be differentiated into processes such as insertion, deletion, replacement and rename. Next, using the result, we propose a novel algorithm to resolve the remaining issues in terms of update performance.

B. Benatallah et al. (Eds.): WISE 2007, LNCS 4831, pp. 563–572, 2007.
© Springer-Verlag Berlin Heidelberg 2007

2 XML Updates

Many works have investigated the importance of efficient XML storage, indexing and query processing [10,9]. However, it is widely known that update remains the main weakness in many XML databases [1].

It is very understandable that the XML communities expect the research initiative on XML update to come from the XML main standard body, the W3C. The first draft of the update requirements was released in February 2005, and the latest draft was released in June 2005 [19]. Even though we have seen some attempts from the standardization body to unify the XML update process, most existing works have put forward proposals based on the implementation environment.

2.1 XML Update Operations

We identify XML Update as a sequence of primitive operations that can be performed on a target. The target is a node or a collection of nodes/sub-tree in the XML data. The node content can be a scalar that contains a data value or a content that encodes structural information of the data.

The operations can be divided into three types. *Deletion* is the process of removing a target from the database. *Insertion* is the process of adding a target into the database. *Replacement* is the process of exchanging a target in the database with another one. [18] identifies two other operations namely *InsertBefore* and *Rename*. These operations actually have the same structure as the insertion operation and replacement operation, only with additional constraints or conditions. The definition of our XML Update is shown as follows.

```
XML Update = {[(Deletion (target))],
              [(Insertion (target))] |
              [(Replacement (target₁, target₂))]]}
Target = {(node(s) name), [node(s) value]}
```

2.2 XML Update: Existing Works

XML Update in XEnDB. The earlier papers that discussed XML Update focused on XML Update in established DBMS, such as XEnDB. The advantages of using this database for XML storage is the full database capability. One of the advantages is full support for update processes. Different manipulation techniques have been implemented by DBMS products and, once we store the XML in a DBMS, we should be able to utilize the update processes.

When people use XEnDB for XML storage, it is very likely that they will use the same language for updating both the simple data and the XML data. This is possible since, during the storing process, the XML data will be transformed into the underlying data format such as tables and classes. Some of the works that fall into this category are [5,16] for RDBMS and [20] for ORDBMS.

XML update using the basic DBMS language is straightforward and, therefore, is popular. However, there are two issues in this option. First, the DBMS language is standardized to cater for a specific data format. For example, SQL was developed to deal with tuples and tables. Even though we can use SQL, it does not exactly match the nature of XML query language. Second, we have to deal with the transformation process during the storage. If the mapping of XML into tuple or object is not correct, the update will not be optimal. Depending on the complexity of the XML data, the transformation can be expensive and some of the XML semantic might be diminished.

Some works propose methods to translate XML query languages (and update for that matter) into basic query languages such as SQL. These works seek to address the problem mentioned before, which is the different nature of the XML query language and the basic DBMS query languages.

This option covers the expressive power of the XML query languages. Many proposals [8, 3, 4] have been made regarding the adjusted query languages, some of which are implementation independent.

Despite the current research on adjusted DBMS query languages to accommodate XML features, these are mostly discussed only within the research community. In addition, none of them covers the entire update operations or considers the constraint preservation. In addition, the adjusted query languages are not standard languages, and thus, will be unlikely to gain support from the majority of the XML community.

XML Update in NXD. XML Update in NXD is still an open problem and provides no standard that can be used as a guideline. Many NXD products use their own proprietary update languages that will allow XML updating within the server [6]. Despite their update ability, the language is not a standard language that can be accepted and used by the whole XML community. It is observed that more recent products no longer use proprietary language. The conformance to standard XML query languages like XQuery has become a basic requirement.

Other products, such as eXist [13] and dbXML [2], use a special language called XUpdate [21]. The specification defines the syntax and semantics of the language. It is designed to be used independently of any kind of implementation. In the implementation, XUpdate is used with another language for retrieval purposes.

The other strategy that is followed by most NXD products is to retrieve the XML document and then update using XML API. After update, the document is returned to the database. This option is the most widely used option for XML update. However, the main reason behind this is the current limitation of XML query languages for update purposes. In addition, having a separate API for update is costly.

The last option for XML update in NXD is by embedding the update processes into XML language. This is the latest development and the interest is growing. The first work that started the idea was [18], which embedded simple update operations into XQuery and thus could be used for any NXD that supported this language. Since then, few works use [18] as the template for their proposals [12, 7].

Regardless of the environment, the earlier techniques raise concerns about the validity of the documents being updated. We do not know how the update operations can affect the semantic correctness of the updated XML documents.

2.3 Safe Update

Regardless of the environment, very few of the existing works on XML Updates seem to be concerned about the validity of the documents being updated. We do not know how the update operations can affect the semantic correctness of the updated XML documents. In our previous works [14, 15], we propose algorithms to perform constraint checking before an XML update.

Instead of direct update of XML documents, the constraint checking is performed, both at instance level and schema level.

To illustrate the difference between primitive and safe updates, we use the following sample XML document. It shows the structure of XML documents with root element *company*. Each document has various child elements that can be attributes such as *location* and *id*, leaf node or simple elements such as *address* and *city*, or complex elements such as *perm* and *contract*.

```
<company >
  <department location="ID001">
    <staffList>
      <perm id="ID000001">
        <address>23 Latham St, Ivanhoe</address>
        <name>
          <firstname>Alan</firstname>
            <lastname>Cameron</lastname>
        </name>
        <age>35</age>
        <city>Melbourne</city>
        <email>a.cameron@email.com</email>
        <supervisor>0000-AA</supervisor>
      </perm>
      <contract id="ID000100">
        <address>12 Stanley Ave</address>
        <supervisor>0000-AA</supervisor>
        <name>
          <firstname>Richard</firstname>
          <lastname>Mao</lastname>
        </name>
        <ages>55</ages>
        <city>Sydney</city>
        <status>Active</status>
      </contract>
    </staffList>
    <name>Marketing</name>
  </department>
</company>
```

In Figure 1, we show an example of updates using the abovementioned XML document. Assume that we want to insert a new department ID002 with name "Marketing". Another department in the database has already been named as the "Marketing" department. A content duplication of a node will obviously affect the storage use. However, this problem is minimal compared to the consequences of this action such as an referential integrity problem. The referential integrity problem will not occur during the insertion, but it will arise once there is a reference to the duplicated node. For example, if we want to insert a contract staff under "Marketing" department, the staff will be inserted into two separate departments. Using safe update, the instance and schema can be checked before the actual update takes place.

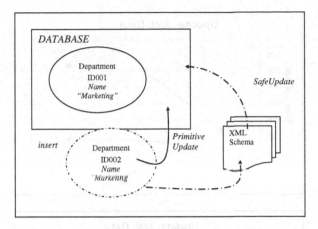

Fig. 1. Primitive Update and Safe Update Illustration

3   Safe Update and Primitive Update: Performance Experimentation

In this section, we discuss the preliminary results that we have obtained using our Safe update technique. The system is implemented on a machine that has Windows XP installed and hardware configuration of AMD64 (989) 3200+ with 1.0 GB of RAM disconnected from the network.

For this series of experiments, we adopted the schema used for the company XML document shown in the previous section. We used three data sets (compliant with this schema) of varying sizes: 15, 30 and 45 mega bytes. We measured, for each XML query, the performance of the two sets of results. The first result set is the query time without schema validation and the second result set is the query time with the schema validation.

The measurements are collected using the most efficient optimization tool of an off-the-shelf most reputable commercial database management system, and are averaged (out-layers are eliminated) over several attempts so that the possible interferences from the operating system and other system processes can be avoided.

The performance analysis, which deals with the primitive and safe updates, is depicted in Figure 2. These results are for two updating statements where, in each statement, the condition element is a simple data type where the element's given value for checking conflicts with its schema defined values. As expected, the performance shows that the primitive update performance outperforms the safe update performance by 50% to 80%. This is common and expected as we all know that checking a constraint would take longer. However, when this is applied to a non-conflict constraint, which results in the checks and updates being carried out one after another, a great deal of accessing of the database is required.

In our XML query, we test only the condition element associated with a value that conflicts with its schema defined value. The result shows that the system still toggles between the checking and updating, and normal accessing of the database seems to occur here. Of course, we also take into consideration the database storage syntax dependency and how the XML schema is pre-processed.

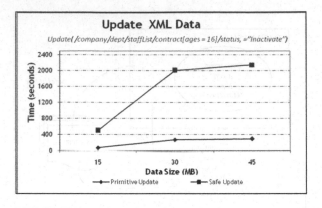

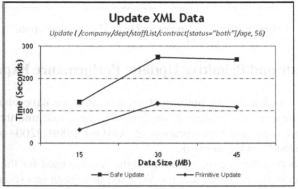

Fig. 2. Updating a conditional element

On the other hand, we also studied the performance of the complex XPath where the predicate can be another complex XPath. The result is presented in Figure 3 where we performed a deletion of parent node and checked whether the condition is the path formed by the children and the grand-children. This is interesting as we found that the condition is another path, which must be checked for validity before the structure of the deletion node and the dependent path are checked. Therefore, it actually does two-task checking for this type of XPath. The performance before checking the constraint for deletion of XML data in this test shows that it is about 80% better than the one with constraint checking activated. In addition to this, we also observed that for both update or deletion of data, the performance relies on the data size. The larger the data size, the more time is required.

On any off-the-shelf commercial DBMSs, this update statement is processed in two phases. First, the schema validation is done within the database engine itself. This normally activates the system function or trigger. Second, upon the completion of first phase, the XML document is updated *iff* the function/trigger is not fired.

The foregoing results motivated us to propose a new technique where we can avoid the checking constraints in the logical level to release the engine being held up for unnecessary jobs and thus, a new algorithm is proposed in the next section.

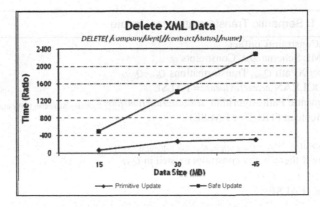

Fig. 3. Deleting Complex XPath

4 Integrated Semantic Transformation Concept Before Updating

There are many works addressed updating XML using XQuery [17]. On the other side, the use of XPath has not been carefully evaluated, even though XPath is the core of most XML query languages. In our proposed algorithm, we will use XPath not only for improved time performance but also to ensure consistency of the data.

Our updating/deleting concept has been extended to focus on the following main areas.

- **Constraint-conflict.** It occurs when a query has predicates in which the conditional test element value conflicts with the constraint element value.

    For example, using the XML document sample, the constraint of element of *16 < age <35* has been specified in the XML schema. Suppose a user given query requires to **delete** or **update** the contract employee record if the age is beyond 36. Thus, very often we see the query pattern such as *UPDATE company/department/staffList/contract [age > 16 or age<35];* or *UPDATE company/department/staffList/contract[age > 36].*

    According to the schema definition, if information has been validated correctly before it is inserted into the database, there should not be anyone who is beyond 36 existed in the database. This must be a confirmation of constraint-conflict and then, performing the constraint checking on a database engine is very costly since it holds up some job processes in database engine.

- **Path complement.** It applies when the XML query is specified using one or more path operator XP={*, .., //}.

    As a result, it may allow one or more predicates in which the conditional test elements are another series of XPaths or multiple paths within a whole XPath, *e.g UPDATE(company/department [//contract/status]/name, new-name).*

    The definition of path locations specified in the XML Schema has imposed that a department must have at least one contract employee; therefore, by adopting our previous work [11], we can simply transform the semantic to *UPDATE(company/department/name, new-name).* As we can see, the conditional predicate has been eliminated, and as a result, the processing time is improved with this method and schema constraints are more useful.

Algorithm 1. Semantic Transformation for Update

1: **Name:** IsConstraintConflict
2: **Given:** XML Schema $\mathcal{D}_{pi}$, Constraints c;
3:          User XPath Q_{xml}; Transformations Q_{xml}', $Q_{xml(c)}'$
4:          BOOLEAN $accessToUpdate$:= FALSE;
5:          Semantic Path Algorithm $semanticPathTransform$ () //Using Algorithm In [Le et.al., 2007]
6: **Return:** Boolean (TRUE or FALSE)
7: **Begin**
8:      Pre-processing $\mathcal{D}_p$ //Using SchemaAlgorithm In [Le et.al., 2007]
9:      //checking if there is any constraint existed in Q_{xml}
10:   **IF** $Q_{xml}(c) \notin \mathcal{D}_p(c)$ **THEN**
11:        Return FALSE
12:   **ELSE**
13:   **FOR EACH** c **in** $Q_{xml}(c)$ **Loop**
14:   //Check if XPath constraint is a single test element and its value is valid
15:       **IF** $Q_{xml}(c)$ is not Path structure && $Q_{xml}(c.val) \subseteq \mathcal{D}_p(c.val)$ **THEN**
16:           $Q_{xml}(c)' := Q_{xml}(c)$
17:           $Q_{xml}' = Q_{xm}' + Q_{xml(c)}'$;
18:           $accessToUpdate$:= TRUE
19:       **ELSE**
20:           $accessToUpdate$:= FALSE
21:       **END IF**
22:   //Check if XPath constraint is a single test element and its value is valid
23:       **IF** $Q_{xml}(c)$ is Path structure $Path(c)$ is true **THEN**
24:           $Q_{xml(c)}' = semanticPathTransform(Q_{xml(c)})$
25:           $Q_{xml}' = Q_{xm}' + Q_{xml(c)}'$;
26:           $accessToUpdate$:= TRUE
27:       **ELSE**
28:           $accessToUpdate$:= FALSE
29:       **END IF**
30:   **END LOOP**
31:   //The accessing of database for updating is relied on the status of $accessToUpdate$
32:   **IF** $accessToUpdate$ **THEN**
33:        Update database using Q_{xml}'
34:   **ELSE**
35:        Display constraint conflict message
36:   **END IF**
37: **End**

In [11], the authors have proposed a semantic transformation algorithm by pre-processing the XML schema and stored the schema information in a well managed, efficient data structure. The difference between the existing techniques [15], that had already implemented at the engine level, and this proposed technique, is that we are going to adopt the technique of [11] and integrate it with our work.

The benefit we obtain is that this technique has provided a schema validation which can be quickly processed at the conceptual level in order to avoid the overhead of processing query at the engine level. In addition, it provides an efficient methodology to check the XPath structure and transforming it to a more efficient one to minimize the processing time. The roadmap to our proposed algorithm is shown in **Algorithm 1.** This algorithm has 4 functionalities:

1. Pre-processing the schema using *SchemaAlgorithm* proposed in [11] at Line 8
2. Checking for the whole XML query structure, at Line 10 to 12, to eliminate the unnecessary further checking of predicates or constraint if the query structure does not satisfy the XML data tree structure
3. Checking for the validity of constraints specified in the XPath against the constraints specified in the XML schema. If the constraint is the single test element, then we check the value of the constraint. The return here is the transformation of the XPath which is used for updating. All are done in Line 15 to Line 20
4. Checking for the validity of the path structure if the constraint is the path implementation in the test element. If the path is valid, Line 23 to Line 29, we then call on the semantic transformation algorithm to transform the path to the more efficient and semantic equivalent one before we access the database. This is because, based on the test result, some XPaths are specified with many path operators { *,//,..} that in the end require a very high processing time.

Upon the completion of steps 2 and 3 we then access the database for updating only if the return status is required to update and the updating will use the new transformed XPath.

Due to limitation of space, the evaluation of this incorporated technique is not presented in this initial stage. Qualitatively, this algorithm will incur some initial cost for pre-processing and checking the initial schema. However, it is done once and for all for update operations that use the same schema. Therefore, no logical level checking is done in the database engine. It will significantly reduce the cost of safe update.

5 Conclusion and Future Works

In this paper, we have made a comparison between primitive update and safe update operations. From the experiment, we can see that a safe update using our previous algorithm is significantly higher in cost than primitive update. With a larger data set, the cost can increase rapidly. Therefore, we perform another safe algorithm, which checks the schema at the conceptual level and thus, no checking is done in the database engine. This will reduce the cost significantly, without jeopardizing the validity of updated XML data.

For future work, we are running more experiments using the proposed algorithm and propose to further refine the algorithm. The experiments will be done in various database environments such as XEnDB and NXD. It is expected that the results can be used for XML database selection given the update nature of the XML data.

References

1. Bourett, R.: XML Database Products (last updated March 2007), http://www.rpbourret.com/xml/XMLDatabaseProds.htm
2. dbXML Group: dbXML (2005), http://www.dbxml.com/product.html

3. DeHaan, D., Toman, D., Consens, M.P., Özsu, M.T.: A Comprehensive XQuery to SQL Translation using Dynamic Interval Encoding. In: SIGMOD2003, pp. 623–634. ACM Press, New York (2003)
4. Du, F., Amer-Yahia, S., Freire, J.: ShreX: Managing XML Documents in Relational Databases. In: VLDB2004, pp. 1297–1300. Morgan Kauffman, San Francisco (2004)
5. Florescu, D., Kossmann, D.: Storing and Querying XML Data using an RDMBS. IEEE Data Engineering Bulletin 22(3), 27–34 (1999)
6. Goldman, R., McHugh, J., Widom, J.: From Semistructured Data to XML: Migrating the Lore Data Model and Query Language. In: WebDB1999, INRIA, pp. 25–30
7. Kanc, B., Su, H., Rundensteiner, E.A.: Consistently updating XML documents using incremental constraint check queries. In: WIDM2002, pp. 1–8. ACM Press, New York (2002)
8. Khan, L., Rao, Y.: A Performance Evaluation of Storing XML Data in Relational Database Management Systems. In: WIDM2001, pp. 31–38. ACM Press, New York (2001)
9. Klettke, M., Meyer, H.: XML and Object-Relational Database Systems - Enhancing Structural Mappings Based on Statistics. In: Suciu, D., Vossen, G. (eds.) WebDB 2000. LNCS, vol. 1997, pp. 151–170. Springer, Heidelberg (2001)
10. Lahiri, T., Abiteboul, S., Widom, J.: Ozone: Integrating Structured and Semistructured Data. In: Connor, R.C.H., Mendelzon, A.O. (eds.) DBPL 1999. LNCS, vol. 1949, pp. 297–323. Springer, Heidelberg (2000)
11. Le, D., Bressan, S., Taniar, D., Rahayu, J.W.: Semantic XPath Query Transformation: Opportunities and Performance. In: DASFAA 2007, pp. 910–912. Springer, Heidelberg (2007)
12. Lehti, P.: Design and Implementation of a Data Manipulation Processor for an XML Query Language, Technische Universitat Darmstadt (2001)
13. Meier, W.M.: eXist Native XML Database. In: Chauduri, A.B., Rawais, A., Zicari, R. (eds.) XML Data Management: Native XML and XML-Enabled Database System, pp. 43–68. Addison Wesley, Reading (2003)
14. Pardede, E., Rahayu, J.W., Taniar, D.: Preserving Conceptual Constraints During XML Updates. Int'l J. Web Inf. Syst (IJWIS) 1(2), 65–82 (2005)
15. Pardede, E., Rahayu, J.W., Taniar, D.: XML Update Management in XML-Enabled Relational Database. J. of Comp. & Syst. Sciences (JCSS) (in press, 2007)
16. Shanmugasundaram, J., Tufte, K., Zhang, C., He, G., DeWitt, D.J., Naughton, J.F.: Relational Databases for Querying XML Documents: Limitations and Opportunities. In: VLDB1999, pp. 302–314. Morgan-Kauffman, San Francisco (1999)
17. Su, H., Rundensteiner, E., Mani, M.: Semantic Query Optimization for XQuery over XML Streams. In: VLDB 2005, pp. 277–282 (2005)
18. Tatarinov, I., Ives, Z.G., Halevy, A.Y., Weld, D.S.: Updating XML. In: SIGMOD 2001, pp. 413–424. ACM Press, New York (2001)
19. Chamberlin, D., Robie, J. (eds.): W3C. XQuery Update Facility Requirements, W3C working draft (June 3, 2005), http://www.w3.org/TR/xquery-update-requirements/
20. Widjaya, N.D., Taniar, D., Rahayu, J.W.: Transformation of XML Schema to Object-Relational Database. In: Taniar, D., Rahayu, J.W. (eds.) Web Information Systems, pp. 141–189. Idea Group Publishing, USA (2004)
21. XML DB: XUpdate – XML Update Language (2004), http://www.xmldb.org/xupdate/

From Crosscutting Concerns to Web Systems Models

Pedro Valderas[1], Vicente Pelechano[1], Gustavo Rossi[2], and Silvia Gordillo[2]

[1] Department of Information Systems and Computation
Technical University of Valencia, Spain
{pvalderas, pele}@dsic.upv.es
[2] LIFIA, Facultad de Informática, UNLP, La Plata
{gustavo, gordillo}@lifia.info.unlp.edu.ar

Abstract. In this paper we present a novel approach for dealing with crosscuting concerns in Web applications from requirements to design. Our approach allows to clearly decoupling requirements that belong to different concerns; these concerns are separately modeled and specified by using the task-based notation proposed by OOWS Web Engineering approach to specify requirements; we next show how we integrate task descriptions corresponding to different concerns to obtain a unified requirements model that is the source of a model-to-model and model-to-code generation process that allows us to obtain fully operative web application prototypes that are built from tasks descriptions.

1 Introduction

Even simple Web applications must deal with a myriad of concerns (functional and non-functional), each one of them encompassing multiple requirements. Some of these concerns crosscut and consequently the corresponding software artifacts may be tangled. This problem has been addressed in the field of Aspect-oriented software development (AOSD) [7,13] which encapsulate crosscutting concerns in separate modules, known as *aspects*, and composition mechanisms are later used to weave them back with other core modules, at loading time, compilation time, or run-time.

In our research we are interested however in the fact that crosscutting concerns are present well before the implementation, such as in requirements engineering as shown in [3]. Separating concerns from requirements allows modularizing those concerns that can not be easily specified as a single use case or task. Composition, on the other hand, apart from allowing the developers to picture the whole system, allows them to identify conflicting situations whenever a concern contributes negatively to others. Unfortunately so far mature Web engineering methods such as [5,14,8] do not offer primitives and composition mechanisms for advanced separation of concerns.

In [9] we presented an approach for identifying and composing navigational concerns in Web applications using concepts borrowed from aspect-oriented software. Using our approach we can detect cross-cuttings among concerns early in the development process and assess the impact of the crosscutting in the corresponding design models. Though our research was performed in the context of the OOHDM [12] design framework and uses UIDs [10] to describe application requirements, the ideas can be applied with other design methods as is the OOWS method. The OOWS

B. Benatallah et al. (Eds.): WISE 2007, LNCS 4831, pp. 573–582, 2007.

method introduces a notation to capture Web application requirements which is based in the task metaphor. Task descriptions are enriched with information about the interaction between the user and the system. It also includes a method to derive a OOWS navigational model automatically from these task based descriptions. In this way, the OOWS models can be further transformed into fully operative Web applications therefore allowing to completely bridging the gap between requirements elicitation and web application development.

In this paper we present a result of our research work combining techniques for separation of concerns in requirements engineering [16], with approaches for automatic derivation of Web design models from requirement models. The main contributions of the paper are the following:

- We present a novel approach to model requirements of Web applications as task-based representations of different concerns.
- We show how to obtain an integrated representation that can be later transformed into a unified design model.

Section 2 introduces the re-interpretation that we have done of the task-based method for Web application requirements specification in order to be used as technique for the description of concerns. Furthermore, an overview of the strategy proposed to automatically obtain Web application prototypes from task descriptions is also presented. Section 3 explains how the techniques for the integration of concerns are applied in the tasks-based method. Section 4 explains the different strategies that can be followed to support the integration of concerns in the generation of Web application prototypes. Finally, conclusions and further work are presented in Section 5.

2 Background: The Task Based Approach of OOWS

Section 2.1 introduces a technique for describing concerns by using the OOWS task-based notation for specifying requirements. Section 2.2 presents an overview of the strategy that allows us to obtain Web application prototypes from task descriptions.

2.1 Using Tasks for Describing Concerns

Following [1] we consider a concern to be a cohesive set of requirements which deal with the same application theme. Considering the interactive nature of Web applications, we model each concern by modeling the underlying tasks of the concern's requirements. We focus particularly on those functional concerns which involve interaction and navigation (called navigational concerns in [9]), because they define the skeleton of a Web information system.

Each concern is described by following three main steps: (1) Definition of a task taxonomy, (2) description of task performance and (3) specification of information requirements. Next, we introduce a brief description of each step. See [11] for more detailed information.

Step 1: Definition of a Task taxonomy. We define the different tasks that support the requirements of a concern in a task taxonomy. To do this, we consider a concern to be a task and then we perform a progressive refinement in order to obtain more specific tasks (see the upper left side of Figure 1). Furthermore, we can specify temporal relationships between subtasks for ordering them according to a specific logic. To represent these temporal relationships, we use those ones introduced by the CTT approach [15]. The upper left side of Figure 1 shows the task taxonomy which supports the requirements of the concern *Collection of CDs*. According to this taxonomy, in order to collect products (root task *Collect CD*) users must search them (subtask *Search CD*) and next (the temporal relationship []>> implies a sequence of tasks) user must add them to the shopping cart (subtask *Add CD*).

Step 2: Description of Tasks Performances. We describe how users must perform each task defined as a leave node in the task taxonomy. To do this, we propose a technique based on the description of the interaction that users require from the system to perform each task. This type of description is done by using UML activity diagrams. In each activity diagram we specify (see the lower left side of Figure 1): (1) the actions that the system must perform in order to correctly support the task (nodes depicted by dashed lines). And (2) a set of Interaction Points (IPs) (nodes depicted by solid lines) that represents the moments during a task where the system and the user interact to each other. In each of these interactions, either the system provide the user with both information and access to operations (IPs stereotyped with the keyword «output») or the user introduce information into the system (IPs stereotyped with the keyword «input»). The information and operations that are provided in each IP are related to a specific entity. Furthermore, for each Output IP, the number of information instances that it includes (cardinality) is depicted as a small circle in the top right side of the primitive.

Figure 1 shows (in the lower left side) the description of the task Search CD. The task Search CD starts with an Output IP where the system provides the user with a list (cardinality *) of music categories. From this list, the user can select a category. If the category has subcategories, the system provides again the user with a list of (sub) categories. If the selected category does not have subcategories the system informs about the CDs of the selected category by means of another Output IP. From this IP, the user has two possibilities to continue the task: (1) to select a CD, and then the system provides the user with a description of the selected CD or (2) to activate a system action which searches for the CDs of an artist. To do this search, the user must introduce the artist (the search criterion) by means of an Input IP. If the search returns only one CD, the system provides the user with its detailed description. Otherwise, the system provides the user with a set of CDs.

Step 3: Specification of Information Requirements. We describe the information that the system must store in order to correctly support the requirements associated to each concern (e.g. the information that the system must store to allow users to correctly collect CDs). To do this, we associate an information template to each entity identified in the task performance descriptions. These information templates are inspired by techniques such as CRC Cards [4]. In each template, we indicate an identifier, the entity, and a *specific data* section. In this section, we describe the information in detail by means of a list of specific features associated to the entity. We provide a name and a description for each feature. In addition, we use these templates to indicate the information exchanged in each IP. We indicate the IP/s (if there are any) where each feature is shown (Output) or requested (Input). To identify an IP, we use the following notation: *Output (Entity, Cardinality)* for Output IPs, and *Input (Entity, System Action)* for Inputs IPs.

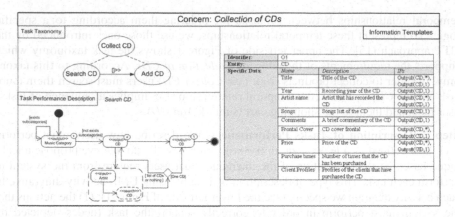

Fig. 1. Partial Description of the concern Collection of CDs

For instance, according to the template in Figure 1, the information that the system must store about a CD is (see the specific data section): the CD title, the artist's name, the front cover, and the price which are shown in the IPs Output(CD,1) and Output(CD,*); the recording date, some comments about the CD and the list of songs which are only shown in the IP Output(CD,1); and finally, the number of times that a CD has been bought and the profiles of the customer which usually purchase it are also stored. These two last features are not shown in any IP.

2.2 Obtaining Web Application Prototypes from Task Descriptions

In this section, we describe the OOWS strategy that allows us to automatically obtain Web application prototypes from task descriptions (see Figure 2).

According to Figure 2, a model-to-model transformation is applied first in order to derive a Web application conceptual model from the task-based requirements specification [11]. These transformations are defined by means of graph transformations. Graph transformations are rewriting rules that are made up of a Left Hand Side (LHS) and a Right Hand Side (RHS). They are applied as follow: When the LHS is found in a host graph, it is replaced by the RHS.

The obtained conceptual model is defined by means of the OOWS method [8]. The OOWS method proposes several models in order to describe the different aspects of a Web application: The system static structure and the system behaviour are described

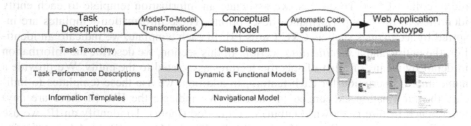

Fig. 2. The OOWS development process

in three models (*class diagram* and *dynamic*-and *functional* models) that are borrowed from an object-oriented software production method called OO-Method [2]. The navigational aspects of a Web application are described in a *navigational model*.

Next, a strategy of automatic code generation is applied to the conceptual model in order to obtain code. By following directives specified in design templates [8] a Web application prototype is automatically generated by the OOWS case tool [6].

3 Our Approach in a Nutshell

In this section, we explain how requirements of crosscutting concerns can be integrated. Three kinds of integration are proposed according to the elements used to perform it: (1) Integration at task taxonomy level, (2) Integration at task performance level and (3) Integration at task information level.

3.1 Integration at Task Taxonomy Level

We use the task taxonomy to perform the integration. To do this, we consider concerns to be tasks that can be connected to the task taxonomy of other concerns. For instance, let's consider the concern *Inspection of the Shopping Cart* which involves those requirements that allow users to inspect their shopping cart. We want to integrate the requirements involved in this concern with the requirements involved in the concern *Collect CDs*. In particular, we want that when users are collecting CDs they have always the possibility of inspecting the shopping cart. In order to do this integration, we connect the concern *Inspection of the Shopping Cart* (considering it to be a task) to the task Collect CD defined in the task taxonomy of the concern *Collect CDs*.

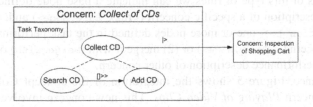

Fig. 3. Example of integration at task taxonomy level

Figure 3 shows this integration. As we can see in Figure 3, the temporal relationship that has been used to perform this integration is Suspend/Resume (▷). We use this relationship because its semantics (see step 1 explanation in Section 2.1) allows us to perform the integration in the way that we need: according to the relationship semantics, users can interrupt the task Collect CD at any time in order to perform the task-considered concern Inspection of the Shopping Cart. Thus, while users are collecting CDs they are always able to inspect the shopping cart. The implementation results of the integration presented in Figure 3 are shown in Figure 4. We can see how the Web pages that support the task Search CD (subtask of the task Collect CD) allow users to inspect the shopping cart.

Shopping Cart always available

Fig. 4. Example of concern integration implementation

3.2 Integration at Task Performance Level

In this case, we use the description of a task performance in order to do the integration. This type of integration is based on the one presented in [9]. This type of integration allows us to connect an activity diagram node defined in the task performance description of a specific concern with a node defined in the task performance description of another concern. To do this, we propose the use of a set of composition rules which has the form:

```
Compose <Task_Base> with <Task₁, ..., Taskₓ>
{
  <Task_Base.Node>
  [Merge | AddConnection] [to | with]
  < Taskᵢ.Node>
}
```

By means of this type of rules we can indicate a base node defined in a task performance description of a specific concern (`<Task_Base.Node>`) and: (1) connect it to (`AddConnection to`) one or more nodes defined in the task performance description of other concern (`< Taskᵢ.Node>`) or (2) merge it with (`Merge with`) one node defined in the task performance description of other concern.

For instance, Figure 5 shows the integration of the concept Collection of CDs with the concern *Playing of Video Clips*. This new concern involves those requirements that allow users to search the video clip of a song. Its description (task taxonomy and task performance description) is shown in the bottom left side of Figure 5. The task taxonomy of this concern is defined by an only one task. This task must be performed as follow: Users first access a list of video clips (IP Output(Video Clip,*)). From this list users can either select one an then access its description (IP Output (Video Clip,1)) or perform a search. The criterion of the search is introduced by the user throughout the Input IP. The results of this search are shown to users in a list.

The integration performed between these two concerns is the following: When users are collecting CDs and they access the list of CDs, users must have the possibility of search video clips. This integration is performed (see composition rule in the top left side of Figure 5) by connecting the IP Output(CD,*) defined in the description of the task Search CD (concern Collection of Products) with the search system action defined in the description of the task Play Video Clip (concern Playing of Video Clips). The solid arrow in Figure 5 graphically shows the connection that the composite rule is defining. Figure 6 shows the implementation results of the integration presented in Figure 5. As we can see in Figure 6, the page that provides users with a list of CDs (which support the concern Collection of CDs) also provides users with the possibility of search vide clips. If users use this search, results will be shown in the Web pages that support the concern Playing of Video Clips.

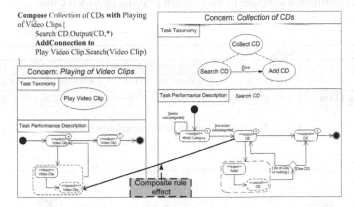

Fig. 5. Example of concern integration at task performance level

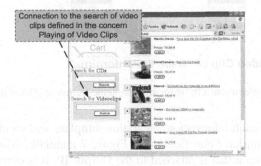

Fig. 6. Implementation result of the integration at task performance level

3.3 Integration at Task Information Level

We use this kind of integration when we want to extend some Output IP defined in a task performance description of a specific concern with information defined in information templates that belong to other concerns.

To do this, we use *information integration templates*. These templates are associated to an Output IP. They allow us to indicate that the IP must provide users with additional features that have been defined in an information template of other concern. In each information integration template (see Figure 7), we indicate an identifier, the IP that is extended, and a feature section. In this section, we describe the new features that are incorporated to the IP. For each feature, we indicate the name, the feature's entity and the identifier of the information template associate to the entity, and the concern in which the information template is defined. Figure 7 shows an information integration template that allows us to integrate the concern Collection of CDs with the concern Playing of Video Clips. According to this template, the information that is shown in the IP Output(CD,1), which is defined in the performance description of the task Search CD in the concern Collection of CDs, is extended with the feature *media file,* which is defined in an information template of the concern Playing of Video Clips.

Identifier:	Id1		
IP:	Output (CD,1)		
Features:	*Name*	*Entity and Template*	*Concern*
	Media file	Video Clip, Id1	Playing of Video Clips
Population Filter:	Self.artist.name="Melendi" and Video Clip. song="Calle la Pantomima"		

Fig. 7. Example of information integration template

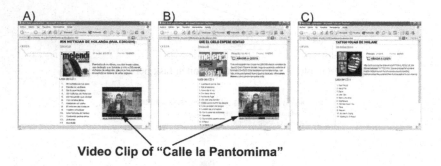

Video Clip of "Calle la Pantomima"

Fig. 8. Example of concern integration implementation at information task level

Furthermore, in each information integration template we can indicate a population filter (see the *Population Filter* field in the Figure 7 template). This filter allows us to restrict the information that is attached to the Output IP. It is defined by means of an OCL condition. The element *self* is used in this case to refer to the entity associated to the extended IP (CD in the example presented in Figure 7). The dot notation is used to access the different features defined for an entity. For instance, the information template in Figure 7 indicates that when users access the IP Output(CD,1), if the accessed CD is by the artist "Melendi", then the video clip of his song "Calle la Pantomima" must be shown. Figure 8 shows the implementation result of this integration. This figure shows three Web pages. Page A and Page B show information about two different CDs by the artist Melendi. In these pages we can see how the video clip "Calle la Pantomima" is shown. Page C shows information about a CD by The Rolling Stones. In this case the video clip is not shown.

4   Concern Integration in Prototyping Activities

In this section, we introduce how the integration of concerns defined in Section 3 can be taken into account in the process of generation of Web application prototypes (see Section 2.2). Two strategies can be followed to do this:

1. Perform the integration in the model-to-model transformation. We must extend the model-to-model transformation in order to interpret the concern integration defined in the task-based description and then derive the proper conceptual elements to support it. In this case, we obtain a unified conceptual model that supports the requirements of each concern already integrated. With this solution the generation of code from the conceptual model does not need to be changed.
2. Perform the integration in the automatic generation of code. In this case, we must: (1) Apply the model-to-model transformation for each concern. This provides us with a set of partial conceptual models that support the different concerns independently to each other. (2) Define the concern integration at the conceptual level. We must use a mechanism that allows us to define the same integration defined in the task descriptions but, in this case, in the conceptual model. Furthermore, we must define the proper transformation in order to automatically do this. (3) Finally, we must extend the strategy of automatic code generation in order to interpret the integration defined in the conceptual model and then generate the proper code. This solution allows us to perform the integration of concerns in later stages of the development process.

Due to the characteristics of the OOWS method, which does not provide mechanisms to define concern integration at the conceptual level, we have initially chosen the first strategy. In fact, we are currently extending the model-to-model transformation used in the prototype generation process in order to consider the integration of concerns presented in Section 3. However, we left as further work the development of the second strategy with a Web engineering method such as OOHDM, which provides us with mechanisms to define at conceptual level concern integration.

5   Concluding Remarks and Further Work

We have presented a novel approach for the representation and composition of Web application concerns at the requirements engineering stage. Functional concerns are modeled using tasks, and then integrated by applying task composition operators. The integrated model can then be mapped into an OOWS conceptual model and a prototype can be generated using existing transformation-based tools. We have shown with simple examples of archetypical Web applications how this process proceeds from requirements into design and prototype generation.

We are currently working on several research lines: first we are improving the concerns composition language to support more complex cross-cutting behaviors; besides we are exploring the generation of partial OOHDM models from concern models to experiment with composition at the conceptual and/or navigational levels.

References

1. Araújo, J., Whittle, J., Kim, D.: Modeling and Composing and Validating Scenario-Based Requirements with Aspects. In: Proceedings of the 12th International Requirements Engineering Conference, Kyoto, Japan (September 2004)
2. Pastor, O., Gómez, J., Insfran, E., Pelechano, V.: The OO-Method Approach for Information Systems Modelling: From Object-Oriented Conceptual Modeling to Automated Programming. Information Systems 26, 507–534 (2001)
3. Baniassad, E., Clements, P., Araújo, J., Moreira, A., Rashid, A., Tekinerdogan, B.: Discovering Early Aspects. IEEE Software 23(1), 61–70 (2006)
4. Wirfs-Brock, R., Wilkerson, B., Wiener, L.: Designing Object-Oriented Software. Prentice-Hall, Englewood Cliffs (1990)
5. Ceri, S., Fraternali, P., Bongio, A.: Web Modeling Language (WebML): A Modeling Language for Designing Web Sites. Computer Networks and ISDN Systems 33(1-6), 137–157 (2000)
6. Valverde, P., Valderas, P., Fons, J., Pastor, O.: A MDA-based Environment for Web Applications Development: From Conceptual Models to Code. In: 6th International Workshop on Web-Oriented Software Technologies (2007)
7. Filman, R., Elrad, T., Clarke, S., Aksit, M. (eds.): Aspect-Oriented Software Development. Addison-Wesley, Reading (2004)
8. Fons, J., Pelechano, V., Albert, M., Pastor, O.: Development of Web Applications from Web Enhanced Conceptual Schemas. In: Song, I.-Y., Liddle, S.W., Ling, T.-W., Scheuermann, P. (eds.) ER 2003. LNCS, vol. 2813, Springer, Heidelberg (2003)
9. Gordillo, S., Rossi, G., Moreira, A., Araujo, J., Urbieta, M., Vairetti, C.: Modeling and Composing Navigational Concerns in Web Applications. In: Proceedings of LA-Web 2006, IEEE Press, Los Alamitos (2006)
10. Güell, N., Schwabe, D., Vilain, P.: Modeling Interactions and Navigation in Web Applications. In: Laender, A.H.F., Liddle, S.W., Storey, V.C. (eds.) ER 2000. LNCS, vol. 1920, pp. 115–127. Springer, Heidelberg (2000)
11. Valderas, P., Pelechano, V., Pastor, O.: A Transformational Approach to Produce Web Application Prototypes from a Web Requirements Model. In: IJWET. International Journal on Web Engineering and Technology (2007)
12. Schwabe, D., Rossi, G.: An Object-Oriented Approach to Web-Based Application Design. Theory and Practice of Object Systems (TAPOS) 4, 207–225 (1998)
13. Kiczales, G., Lamping, J., Mendhekar, A., Maeda, C., Lopes, C., Loingtier, J., Irwin, J.: Aspect-Oriented Programming. In: Aksit, M., Matsuoka, S. (eds.) ECOOP 1997. LNCS, vol. 1241, Springer, Heidelberg (1997)
14. Koch, N., Kraus, A., Hennicker, R.: The Authoring Process of UML-based Web Engineering Approach. In: IWWOST 2001. Proceedings of the 1st International Workshop on Web-Oriented Software Construction, pp. 105–119 (June 2001)
15. Paterno, F., Mancini, C., Meniconi, S.: ConcurTaskTree: A diagrammatic notation for specifying task models. In: Interact 1997, pp. 362–369. Chapman&Hall, Australia (1997)
16. Moreira, A., Rashid, A., Araújo, J.: Multi-Dimensional Separation of Concerns in Requirements Engineering. In: RE 2005. Proceedings of the 13th IEEE International Requirements Engineering Conference, IEEE Computer Society, Los Alamitos (2005)

Generating Extensional Definitions of Concepts from Ostensive Definitions by Using Web

Shin-ya Sato[1], Kensuke Fukuda[2], Satoshi Kurihara[3],
Toshio Hirotsu[4], and Toshiharu Sugawara[5]

[1] NTT Network Innovation Labs, 180-8585 Tokyo, Japan
shin-ya.sato@acm.org
[2] National Institute of Informatics, 101-8430 Tokyo, Japan
kensuke@nii.ac.jp
[3] Osaka University, 567-0047 Osaka, Japan
kurihara@sanken.osaka-u.ac.jp
[4] Toyohashi University of Technology, 441-8580 Aichi, Japan
hirotsu@tut.ac.jp
[5] Waseda University, 169-8555 Tokyo, Japan
sugawara@isl.cs.waseda.ac.jp

Abstract. We present GEO (Generating an Extensional definition from an Ostensive definition), a method to exhaustively gather items falling under an ostensively defined concept from the Web. By utilizing structural information about HTML documents, GEO automatically and efficiently gathers thousands of items from Web pages taking only 2 or 3 items as input. GEO also yields high precision (0.99 at maximum, 0.97 in average over a set of inputs). We also introduce a new style of searching information, called Item Search, in which GEO plays an essential role. Item Search can look for items in a targeted category that are the best matches against a given query. Some examples of Item Search are presented as the proof-of-concept of the idea.

1 Introduction

The quality of the web searches was significantly improved by taking interrelationships among web pages into consideration as well as content of each page[1]. One such technique popularly used in current search engines is to associate the text of a link (anchor text) with the page the link points to. The idea of this technique is that the page most relevant to a query, "ajax" for example, must be the page most often referred to as "ajax". In this way, we are able to efficiently retrieve information about things or concepts from the Web.

This idea does not work, however, when the targeted concept cannot be well defined by query terms. This problem happens when the concept shares the same name with other concepts (lexical ambiguity), or the concept is difficult to describe. As an example, let us think of a case of searching for software whereby structural characteristics of networks are analyzed and visualized, such as Pajek[1]

---

[1] http://vlado.fmf.uni-lj.si/pub/networks/pajek/

B. Benatallah et al. (Eds.): WISE 2007, LNCS 4831, pp. 583–592, 2007.

and UCINET[2]. This need for information is not fulfilled by submitting the query "network analysis tools" to a search engine, which results in finding pages mostly about tools for analyzing communication networks (e.g., measuring traffic). One solution for this problem is to reformulate the query, which is not, however, an easy task for nonexperts.

We propose a new style of search called *Item Search* for the above-mentioned problem. Item Search provides a means for users to make a query like "find something like Pajek and UCINET." We believe that Item Search is quite useful to find something that users cannot clearly identify its appropriate keywords.

For implementing Item Search, the items that fall under the targeted concept must be exhaustively and efficiently enumerated without requesting a user to define the concept. In fact, in the query, the user defines the concept by giving only a few examples of items that belong to the concept. This is called an *ostensive definition*. Therefore, it is required for Item Search to generate extensional definitions[3] (exhaustively enumerating items) of concepts from ostensive definitions given by users. We call this approach *GEO* (Generating an Extensional definition from an Ostensive definition), in which structural data in HTML documents are analyzed to extract other items relevant to the given items.

In this paper, we introduce the idea of GEO in section 3 after reviewing related work in section 2. Its performance is measured in section 4. In section 5, Item Search is introduced with some examples. We have used Web pages mostly in Japanese for the experiments. Therefore, some terms or phrases in examples, such as item names, are not the original terms or phrases, but English translations of corresponding terms or phrases in Japanese.

2   Related Work

Finding items that fall under the same concept is equivalent to acquiring coordinate terms, which are the terms sharing the same hypernym. This task has been studied in a number of works, including a technique that uses statistics calculated from (co-occurrent) frequencies to measure relatedness among terms[2]. Another type of major approach is to analyze syntactic dependency relationships[4].

With development of the Web, it is getting the attention of researchers as a rich source of information in which coordinate terms can be found. A newly introduced technique for dealing with web pages is to utilize structural information about HTML documents to extract candidates of coordinate terms[6]. Still, statistics such as co-occurrent frequencies of terms are also commonly used to estimate relationships among the candidate terms[5,6]. Furthermore, machine

---

[2] http://www.analytictech.com/ucinet/ucinet.htm

[3] In general, there are two ways to explicitly define a concept: by an intensional definition and by an extensional definition. An intensional definition gives the meaning of the concept by giving all the properties (the necessary and sufficient conditions) required of what come under the concept. On the other hand, an extensional definition, which is the opposite approach to the internal definition, formulates its meaning by specifying every instance that falls under the definition of the concept.

learning techniques such as SVM are used in filtering out irrelevant terms in some studies[6].

Our approach, GEO, also makes use of structural information about HTML documents to extract coordinate terms. Our method, unlike [5,6], requires neither statistical information nor machine learning techniques to choose semantically coherent terms. The key idea of GEO is to focus on specific part in a list or table (e.g., the n-th column of a table) where terms belonging to that part are semantically coherent, whereas existing approaches try to extract coordinate terms from an entire list or table. With this approach, a relatively large number (several thousand) of coordinate terms can be obtained quite efficiently, as shown in section 4.2. This remarkable feature of GEO is indispensable in implementing Item Search.

3   GEO

3.1   Detailed Path Notation

To introduce the idea of GEO, we at first define the concept of *detailed path notation* of texts in HTML documents, using the table in Fig. 1 as an example. This is a variation of the path notation described in [6].

In the original path notation, each text (a sequence of characters between two tags) in an HTML document is identified by a sequence of nested HTML tags, which is called a *path*. Then, a set of texts associated with the same path can be considered as candidates of coordinate terms. For example, from the table in Fig. 1, the following texts sharing the same path are obtained:

⟨(TABLE,TR,TD) Pajek⟩
⟨(TABLE,TR,TD) free⟩
⟨(TABLE,TR,TD) UCINET⟩
⟨(TABLE,TR,TD) non-free⟩
⟨(TABLE,TR,TD) Jung⟩
⟨(TABLE,TR,TD) free⟩

As in this example, irrelevant terms ("free" and "non-free") are probably extracted together with targeted terms ("Pajek", "UCINET" and "Jung"[4]). Therefore, it is required in the previous works to verify whether obtained terms are appropriate as coordinate terms.

In our approach, each text is identified by a sequence of tags, but in a more detailed fashion so that a specific part in a table can be distinguished from other parts by comparing the sequence patterns. This sequence is generated in such a way that suffix numbers are sequentially assigned to sibling tags, i.e., the same name of tags that belong to the same parent tag. This is the *detailed path notation*. With this notation, a difference in the position of text in a table can be recognized as a difference in suffix numbers, as shown below in the case of the table in Fig. 1.

---

[4] http://jung.sourceforge.net/

name	availability
Pajek	free
UCINET	non-free
Jung	free

```
<table>
<tr><th>name</th><th>availability</th></tr>
<tr><td>Pajek</td><td>free</td></tr>
<tr><td>UCINET</td><td>non-free</td></tr>
<tr><td>Jung</td><td>free</td></tr>
</table>
```

Fig. 1. Rendered table and its HTML source code

$\langle(\text{TABLE}_1,\text{TR}_1,\text{TD}_1)\ \text{Pajek}\rangle$
$\langle(\text{TABLE}_1,\text{TR}_1,\text{TD}_2)\ \text{free}\rangle$
$\langle(\text{TABLE}_1,\text{TR}_2,\text{TD}_1)\ \text{UCINET}\rangle$
...

Furthermore, a wider area in a table, such as "the n-th column", can be represented as a path with wildcards. For example, the first column of the table in Fig. 1 can represented as the following pattern expression where * represents a wildcard that matches any number.

$(\text{TABLE}_1,\text{TR}_*,\text{TD}_1)$

If multiple items (texts) that belong to the targeted area are known before-hand ("Pajek" and "UCINET" for example), the above pattern can be obtained by analyzing the resemblance and difference among their paths:

$\begin{array}{l}\langle(\text{TABLE}_1,\text{TR}_1,\text{TD}_1)\ \text{Pajek}\rangle\\\langle(\text{TABLE}_1,\text{TR}_2,\text{TD}_1)\ \text{UCINET}\rangle\end{array} \rightarrow (\text{TABLE}_1,\text{TR}_*,\text{TD}_1)$

Then, other relevant items ("Jung" in this case) can be extracted from the table by matching each path against the generated expression.

3.2   Algorithm

In this way, based on a few examples of items, relevant items can be gathered from web pages. This gives an answer to our original problem of providing a means for users to make a query like "find something like Pajek and UCINET". To summarize the points described so far, we show the pseudo-code of GEO below:

1. I = example items given by a user;
2. $O_I = \phi$;
3. W = at most N web pages that contain all items in I gathered by using a search engine;
4. **foreach** w in W
5.   **foreach** i in I
6. P_i = paths of i extracted from w;
7.   **end**
8. E = pattern expressions generated by comparing elements in P_is;
9. M = items in w that match any pattern in E;
10.   Add elements of M to O_I;

11. **end**
12. **return** O_I;

There are multiple ways to generate patterns (i.e., to define E in line 8 of the above code). Here, we consider two of them below, where $p \sim x$ means the path p matches the pattern x.

$$E = \{x \mid {}^\exists p_1 \in P_i, {}^\exists p_2 \in P_j \ (i \neq j), \ p_1 \sim x, \ p_2 \sim x\} \qquad (1)$$

$$E = \{x \mid {}^\forall i, \ {}^\exists p \in P_i, \ p \sim x\} \qquad (2)$$

Definition (1) places a relatively looser condition that requires that a pattern must be matched by at least a pair of paths (any match). On the other hand, definition (2) places a strict condition that requires that a pattern should be matched by all the paths each of which is associated with each element of I (all match). Both (1) and (2) are equivalent to each other if $|I| = 2$.

4 Experiments

4.1 Finding Network Analysis Tools

We conducted experiments to validate our approach. The first experiment is to see if we can actually find network analysis and visualization tools by using GEO. In the experiment, "Pajek" and "UCINET" were given as the example, i.e., $I = \{$ "Pajek", "UCINET"$\}$. Some path patterns generated by the method and matching items are shown in Table 1.

The patterns of (i) and (ii) in Table 1 match a series of anchor texts in the first column of a table and in a list, respectively, which are hyperlinks pointing to (official) Web sites of each tool. This result demonstrates that our method can mine this kind of information that is well-maintained by people.

4.2 Gathering Names of Plants

We conducted a second experiment to measure the performance of GEO. For the experiment, we tried gathering names of plants. At first, we prepared a number

Table 1. Network analysis tools found by GEO

(i) TABLE$_1$ TR$_*$ TD$_1$ A$_1$
Agna Project, GraphViz, Grappa, JGraph - Professional Open Source Java Graph Visualization, JUNG - Java Universal Network/Graph Framework, Java for Social Networks, LEDA, NetStat(Plus), ORA (Organization Risk Analyzer), Pajek, SNA package for R, The KrackPlot home page, UCINET, VGJ, yFiles - product of yWorks
(ii) UL$_2$ LI$_*$ A$_1$ EM$_1$
FATCAT, GRADAP, GraphEd, GraphViz, InFlow, Javvin Network Packet Analyzer 4.0, KrackPlot, Moviemol, MultiNet, NEGOPY, NetVis, Noldus, Pajek, Tom Sawyer Software, UCINET, daVinci

Table 2. Some of Is used in the experiment and the values of $s(I)$

I	$s(I)$
carnation, tulip	$7.42 \cdot 10^5$
cyclamen, hydrangea	$8.06 \cdot 10^4$
hyacinth, Japanese enkianthus, lily of the valley	$4.68 \cdot 10^3$
acacia, calendula	$7.90 \cdot 10^2$

of item sets (each set consists of 2 or 3 names of plants) for input I. Then, we gathered items using GEO, and counted the number of obtained items $n(I)$, and calculated the ratio $r(I)$ of valid plant names with respect to the total number of obtained items. In this experiment, the parameter N was set to 100.

For an item set I, let $w(I)$ be the number of Web pages containing all items in I. This can be thought of as an index of generality of the concept ostensively defined by I; if $w(I)$ is large (small), the concept defined by I is probably general (specific). As a concrete example, let us consider the case where $I = \{$ "rose" $\}$. The value of $w(I)$ is expected to be very large, and I is too vague to define the concept of "plant". The implication of this example is that the performance of the proposed method would depend on the choice of I and the value of $w(I)$.

Although measuring the exact value of $w(I)$ is practically impossible, an approximation of this quantity can be obtained as the number of web search results $s(I)$ returned by a search engine. Some examples of I used in the experiment and values of $s(I)$ are listed in Table 2.

Relationships between $n(I)$ and $s(I)$ on a logarithmic scale are shown in Fig. 2. Circles indicate the case where $|I| = 2$ (this case is referred to as (a) hereafter). Filled and empty rectangles are plots for $|I| = 3$, and show cases where patterns are generated by any match (case (b)) and all match (case (c)), respectively.

As we expected, $n(I)$ tends to decay as $s(I)$ increases. Furthermore, $n(I)$ has a peak around $s(I) = 100$. That is, $n(I)$ also decreases as $s(I)$ decreases if

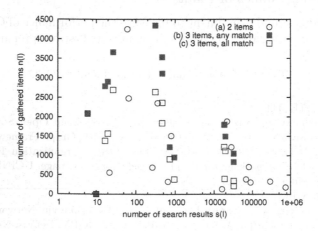

Fig. 2. Number of gathered items $n(I)$ vs number of search results $s(I)$ in logarithmic scale

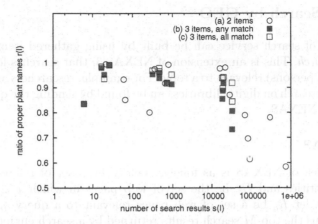

Fig. 3. Precision vs number of search result

$s(I) < 100$. Items are extracted from N pages selected by a search engine out of a total of $s(I)$ pages, so having a peak at around $s(I) = N$ is reasonable. Notably, that three cases ((a), (b), and (c)) share a similar distribution. This implies that $s(I)$ is a primary parameter to control the number of gathered items.

Another important aspect of performance is the precision, which is the ratio $r(I)$ of valid plant names to the total number of gathered items. To calculate $r(I)$, we went through all the obtained items and subjectively judged their validity. The results are shown in Fig. 3. This is a scatter plot of $r(I)$ against $s(I)$ on a logarithmic scale.

Like in the case of $n(I)$, $r(I)$ reaches a peak near $s(I) = 100$. In fact, the $s(I)$ of the peak $r(I)$ is less than 100. This finding becomes clearer by comparing the average of $r(I)$ taken over two intervals, $(10, 100)$ and $(10, 1000)$, of $s(I)$, as shown in Table 3. From this table, clearly (b) and (c) outperform (a). This is a reasonable result because, in general, three items are more informative than two items for defining a concept. A somewhat unexpected result is that (b) performs as well as (c). This implies that the key point of this approach is to find tables or lists that are of good quality. Once we find good tables of lists, the choice of the pattern generation method (any match or all match) matters little.

After having gathered a set of items O_I, obtaining more items by iterating the procedure is possible; that is to invoke a new gathering procedure where the new input I' is selected from the previous output O_I. We control quality (precision) of the new output by selecting elements of I' so that $s(I')$ is in the appropriate interval (i.e., $10 < s(I') < 100$).

Table 3. Average precisions over different intervals of $s(I)$

	(a)	(b)	(c)
(10, 1000)	0.923	0.951	0.961
(10, 100)	0.926	0.974	0.970

5 Item Search by GEO

A new type of search service can be built by using gathered items, which we call *Item Search*. This is an extension of NEXAS[3] that searches for real-world entities (e.g., persons) relevant to a topic. For example, researchers and librarians engaged in research on digital libraries can be found by sending the query "digital libraries" to NEXAS.

5.1 NEXAS

The basic idea of NEXAS is as follows. Let X be a set of references (proper names) of entities and D_x be the set of web pages in which $x \in X$ appears. Furthermore, let R_q be a set of web pages relevant to a query q, which is in concrete terms the top M search results returned by a search engine in response to the query q. Then, the relevance between x and q is estimated by a function $f(x, q)$ such as the Jaccard index:

$$f(x,q) = \frac{|D_x \cap R_q|}{|D_x \cup R_q|}.$$

NEXAS generates a list of entity names in descending order of this relevance score. Then, NEXAS extracts the first K names as entities having highest relevance to the query q.

In NEXAS, dictionaries are used to gather elements of X. We need to prepare and maintain a specific dictionary for each category of entities (people, organizations, and places), which is a considerably high cost.

5.2 Item Search

GEO solves the problem of dictionary maintenance by efficiently gathering names of entities belonging to a category on demand. Furthermore, the target of a search is not limited to named entities. For any ostensively defined concept, we can look for items falling under the concept that are relevant to a query. We call this style of search *Item Search*. In this section, we present some examples of Item Search.

The first example is to find plants relevant to queries "sweet-sour" and "insect deterrent", where X is defined as follows using $\{O_I\}$ obtained in the experiment described in section 4.2.

$$X = \bigcup_{I, 10 < s(I) < 100} O_I$$

Item Search yields satisfactory results as shown in Table 4. For the query "sweet-sour", names of sweet-sour fruits are obtained. For "insect deterrent", plants that deter insects by emitting a distinctive scent are mostly extracted.

The next example is more complicated (thus, more interesting) than the previous example in the sense that the relationship between query and items is not straightforward. In the previous example, items are extracted according to

Table 4. Plants found for "sweet-sour" and "insect deterrent"

"sweet-sour"	"insect deterrent"
passion fruit, raspberry, cassis, bergamot, lychee, cranberry, citrus, blueberry, muscat, grapefruit	eucalyptus, lemongrass, geranium, ilang-ilang, clary sage, bergamot, clove, peppermint, chamomile, St. John's wort

Table 5. Names of plants found for "Gogh"

name of plants	reason for the relevance
poppy	"Poppies" (1886)
durian	pseudonym of a blogger
sunflower	"Sunflowers" (1888)
planetree	"The Large Plane Trees" (1889)
white birch	literary journal
grape	"The Red Vineyard" (1890)
chestnut tree	"Blossoming Chestnut Branches" (1890)
julian	Julian Mitchell, a screenwriter of "Vincent and Theo"
oleander	"Majolica Jar with Branches of Oleander" (1888)
iris	"Irises" (1889)

their attributes or characteristics specified by queries. The following example is not the case. Targets of the search are also plants, but the query is "Gogh", a famous, Dutch Post-Impressionist artist.

The top 10 relevant names of plants generated in response to the query are listed in Table 5. In the second column of the table, phrases in quotation marks followed by a year in parentheses are titles of works by Gogh. Most of the plants are related to his work except "durian", "white birch", and "julian". "Durian" is the pseudonym of a blogger who writes many articles about the works of Gogh. "White birch" is the name of a literary journal that introduced Gogh and his works to Japan in early 1900s. "Julian" is the name of the screenwriter of the film "Vincent and Theo". To eliminate these incorrect answers, disambiguation techniques should be also used.

6 Conclusion

We have presented GEO, a method to exhaustively gather items falling under an ostensively defined concept by utilizing structural information about HTML documents. As an experiment, we tried gathering names of plants by using GEO. We successfully gathered thousands of plant names from 100 Web pages taking only 2 or 3 items as input. We confirmed that GEO yielded high precision: 0.99 at maximum, 0.97 in average over a set of inputs.

We have also introduced a new style of search called Item Search, whereby we can look for items in a targeted category that are the best matches against a given query. In Item Search, GEO plays an essential role for gathering large number of items to be searched.

References

1. Brin, S., Page, L.: The Anatomy of a Large-Scale Hypertextual Web Search Engine. In: Proceedings of the 7th International World Wide Web Conference (1998)
2. Church, K.W., Hanks, P.: Word association norms, mutual information, and Lexi cography. In: Proceedings of the 27th Annual Meeting of the Association for Computational Linguistics, pp. 76–83 (1989)
3. Harada, M., Sato, S., Kazama, K.: Finding authoritative people from the web. In: Proceedings of the 4th ACM/IEEE-CS joint conference on Digital libraries, pp. 306–313 (2004)
4. Lin, D.: Automatic retrieval and clustering of similar words. In: Proceedings of the 17th International Conference on Computational Linguistics, pp. 768–774 (1998)
5. Ohshima, H., Oyama, S., Tanaka, K.: Searching Coordinate Terms with Their Context from the Web. In: Aberer, K., Peng, Z., Rundensteiner, E.A., Zhang, Y., Li, X. (eds.) WISE 2006. LNCS, vol. 4255, pp. 40–47. Springer, Heidelberg (2006)
6. Shinzato, K., Torisawa, K.: A Simple WWW-based Method for Semantic Word Class Acquisition. In: Proceedings of International Conference on Recent Advances in Natural Language Processing, pp. 493–500 (2005)

Modeling Distributed Events in Data-Intensive Rich Internet Applications

Giovanni Toffetti Carughi[1,2], Sara Comai[1], Alessandro Bozzon[1], and Piero Fraternali[1]

[1] Politecnico di Milano, Milano, Italy
{giovanni.toffetti,sara.comai,alessandro.bozzon,
piero.fraternali}@polimi.it
[2] Faculty of Informatics, University of Lugano, Lugano, Swizerland
toffettg@lu.unisi.ch

Abstract. Rich Internet applications (RIAs) enable novel usage scenarios by overcoming the traditional paradigms of Web interaction. Conventional Web applications can be seen as reactive systems in which events are 1) produced by the user acting upon the browser HTML interface, and 2) processed by the server hosting the application state and logic. In RIAs, distribution of data and computation across client and server broadens the classes and features of the produced events as they can originate, be detected, notified, and processed in a variety of ways. In this work, we investigate how events can be explicitly described and coupled to the other concepts of a Web modeling language in order to specify collaborative Rich Internet applications.

1 Introduction and Motivation

Rich Internet Applications (RIAs) are fostering the growth of a new generation of Web applications providing reactive user interfaces and bidirectional client-server communication. Figure 1 depicts the general architecture of a RIA: the system is composed of a (possibly replicated) Web server and a set of *user applications* (implemented as JavaScript, Flash scripts, or applets) running in browsers. The latter are downloaded from the server and executed following the *code on demand* paradigm of code mobility [4].

In a traditional Web application the server holds the complete application state and receives all update invocations from the users: at each request the server responds with the rendering of the latest application state. In a RIA instead, the application state is scattered among different client-running components, user interactions are intercepted on the client, and relevant occurrences have to be explicitly signalled to interested applications either for state synchronization or to trigger a possible reaction. Client applications have to be notified of events occurring *outside* their execution environment, either triggered by other clients or by the server: examples of events that can trigger a reaction on a user client application include users' interactions on the interface, server internal or temporal events, and Web service invocations.

B. Benatallah et al. (Eds.): WISE 2007, LNCS 4831, pp. 593–602, 2007.

Our Contribution. As Rich Internet Application adoption is experiencing constant growth, a multitude of programming frameworks have been proposed to ease their development (e.g., [1,2]). While frameworks increase developer productivity, they fail in providing instruments that abstract from implementation details and that represent at a higher level the final application; this becomes necessary when tackling the complexity of large, data-intensive applications. Web engineering methodologies offer sound solutions to model traditional Web applications at a conceptual level and to automatically generate their code from the conceptual specification, but they generally lack the concepts to address Rich Internet Application development. In this paper we propose a model and a methodology to support the design of *event-driven* RIAs. Our contribution is an extension of the conceptual model of a Web engineering methodology for the high-level specification of distributed events and of new RIA-specific interaction paradigms and application functionalities. In addition, we report on our prototype implementation of the proposed model extensions in the framework of an existing commercial CASE tool (WebRatio).

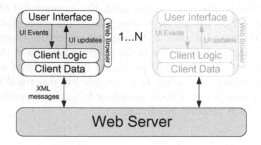

Fig. 1. A Rich Internet Application architecture

Case Study. In order to better discuss relevant aspects of the proposed approach, we introduce as case study a collaborative on-line project management application. The application aim is to let users communicate and organize projects and tasks. Application users impersonate different roles: project managers and project participants. A manager can create projects and tasks, assign tasks to participants, and define precedence relationships among the tasks of a project. Participants execute assigned tasks, possibly exchanging messages with their contacts.

Although the proposed approach applies to most Web engineering methodologies, we will illustrate it using the WebML notation [5]. In WebML content to be published is modeled using Entity-Relationship (E-R) diagrams. Figure 2 shows the data model for the case study application: the *user* entity represents application users, a self-relationship connects each user with his contact list. A user can participate in one or more *projects*, while each project can be directed by a unique manager. Users are assigned *tasks*: each task belongs to a project and can have precedence constraints w.r.t. other tasks. *Messages* can be exchanged between users, having a sender and a set of recipients.

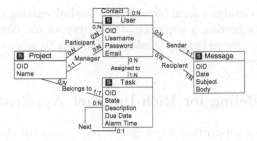

Fig. 2. Data Model of the project management application

Upon the same data model, it is possible to define different hypertexts (also called *site views*), targeted to different user roles. A site view is a graph of *pages*, possibly hierarchically organized into sub-pages. Pages comprise *content units*, i.e., components for content publishing: the content displayed by a unit comes from an *entity* of the data model, and can be determined by means of a *selector*, which is a logical condition filtering the entity instances to be published. Units are connected to each other through *links*, which carry parameters and allow the user to navigate the hypertext. WebML also allows specifying *operations* implementing arbitrary business logic (e.g., to manipulate entity instances).

In [3] we extended the modeling primitives of WebML to support the design of RIAs considering data and computation distribution among server and clients: pages, entities, logical conditions, operations, etc. can be marked as *client* or *server*: in the former case the client will compute them, in the latter they are computed at the server.

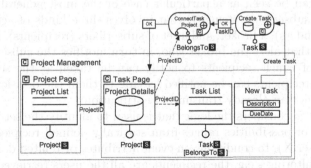

Fig. 3. Example of hypertext model for the project management application

Figure 3 depicts a fragment of the hypertext model for the case study. It represents a client (marked with an uppercase C) page *Project Management* including two sub-pages. *Project Page* contains a list of projects (*Project List* index unit), showing all instances of the entity *Project* on the server. When a project from the list is selected, the outgoing link carries the project identifier to (*Project Details*) showing the details of the selected instance. A list of project tasks is displayed, through the *Task List* index unit.

Task Page also contains a form (*New Task*): the link exiting the unit represents form submission, triggering a sequence of two operations: *CreateTask* creates a new instance of the (server) entity *Task*, and *BelongsTo* associates such instance to the selected project.

2   Event Modeling for Rich Internet Applications

The design of RIAs supporting event notifications implies the consideration of different aspects, like the identification of notification recipients, or the persistence of the communication. In this section we first introduce the possible behaviors that we want to model (represented by semantic dimensions in Table 1); then, we show the primitives for modeling distributed events in a RIA.

Table 1. Semantic dimensions in RIA event notification

Dimension Name	Possible Values
Filter location	Sender, Broker, Recipient
Filter logic	Static, Rule-based, Query-based
Communication Persistence	Persistent (asynchronous), Transient (synchronous)

Event Filtering: Not all users need to be notified of all event occurrences; in a generic distributed event system the process of defining the set of recipients is generally indicated as *event filtering* [9] and two dimensions can influence it: *where* it takes place and *how*.

Filtering location defines where the set of recipient is identified. The RIA architecture can be seen as a particular case of the most general architecture for publish / subscribe systems [7] that involves three kinds of actors: a set of publishers (senders), a broker, and a set of subscribers (recipients). Events occur at publishers that alert the broker who, in turn, notifies the subscribers. Thus, the decision of which recipients to notify can be taken at the sender, at the broker, or all recipients can be notified leaving to them the decision of whether to discard or not an already transmitted notification.

Filtering logic considers the logic that is used to identify the set of recipients: the spectrum of possibilities ranges from statically defined recipients (e.g., the recipient is User X), to conditions on event attributes or related domain entities (*query-based* filtering - e.g. the recipients are all the users retrieved from entity *User* and participating to Project Y), to interpreted run-time defined rules (*rule-based* filtering, e.g., using a rule engine to detect composite event patterns).

Communication persistence: Depending on the application, some events may need to be communicated in a persistent way to prevent their loss, others only need transient communication. We refer to this semantic dimension as *communication persistence*. RIAs behave like generic distributed systems, where message communication can be [12]: (i)*Persistent*, when the message that has been submitted for transmission is stored by the communication system as long as it takes to deliver it to the recipient. It is therefore not necessary for the sending application to continue execution after submitting the message. Likewise, the receiving application need not be executing while the message is submitted. Alternatively, a message is (ii) *Transient* when it is stored by the communication system only as long as the sending and receiving applications are executing. Therefore the recipient application receives a message only if it is executing, otherwise the message is lost.

2.1 Proposed Extensions

In this section we show how to extend the WebML model to take into account the previously introduced requirements for event management in RIAs. The extensions apply to the data and hypertext model (common to most Web modelling approaches) and therefore can be taken as a basis also for other Web Engineering methodologies.

The event model. One issue to be considered when designing event-driven solutions is the choice of relevant occurrences for the system which are, in general, application-specific. Considering our case study application, each action upon a task instance can be considered a specific *event type*: for example an event associated to the assignment of a task to a specific project participant, or the completion of a task by a project participant. Each event type may expose specific parameters: for example, for the task assignment event its parameters can be the task to be assigned and the user to which it has to be assigned.

In Web Engineering methodologies, the data model plays a central role in the specification of the structure and behavior of data-intensive Web application. We therefore extended the existing Data model for a RIA application with an appropriate *Event model* (expressed in an E-R like notation), in order to (i) model the event types (and their parameters) requested by the application and (ii) represent and instantiate relations between event occurrences and domain model entities. With respect to the E-R model, we define an *Event Entity* to represent the common features of a set of events meaningful for the application; an *event instance* is the occurrence of a given Event Entity. *Event Entities* are characterized by a set of *event parameters*, representing properties useful for the application's purposes. *Event entities* can be organized in specialization hierarchies to inherit common parameters, and can have relationships with domain entities from the data model. Consider, for example the event types "Task

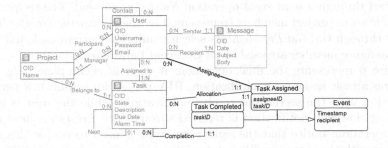

Fig. 4. The data model of Figure 2 extended with event types

assigned", and "Task completed" in our case study. They are represented in the event hierarchy in Figure 4, connected to the data model of Figure 2.

Event primitives in the hypertext model. In order to support event notifications, two basic primitives are introduced: the *send event* and the *receive event* primitives. The former allows one to send an event notification to a (set of) recipient(s); the latter is used to receive the notification and trigger the corresponding reaction. Each primitive must be associated with an *event entity*, as defined in the event model. The specified event entity provides both a mapping between the send and the corresponding receive event operation (i.e., operations defined on the same event type trigger each other) as well as their specific parameter set. Send and receive primitives can be combined with other operations to obtain patterns covering all the possible combinations of filtering and communication needs.

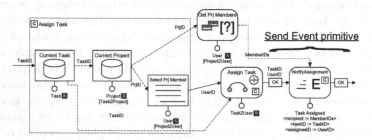

Fig. 5. The hypertext to assign a task and send a notification

Figure 5 shows a usage example of the send event primitive: it signals the assignment of a task to a project participant, triggered from a client page. The *Assign Task* RIA page shows the details of a selected task (*Current Task* data unit); the task belongs to a particular project, retrieved through the *Current Project* data unit. The current project allows to retrieve the list of users participating to that project (both in the *Project Members* index unit and in the *Get Prj Members* selector unit). The user can select a project member to whom the task should be assigned from the *Select Prj Member* index unit. This selection triggers the chain of operations represented outside the page: first, the selected project member is associated with the selected task (*Assign Task* operation); then this assignment is notified through a send event operation *NotifyAssignment*. The recipients are the whole set of project members (represented by the parameter *MemberIDs*) retrieved through the *Get Prj Members* selector unit; the current task and the selected assignee member are used to set the event parameters.

Figure 6 represents the dual operation: it models client page (*My Tasks*) receiving a task assignment notification. RIA pages can establish a persistent connection with the server and receive notifications while the user is on-line. The *Task Assigned* notification is received through the *RecAssignment* receive event operation: notice that the same event type is used to couple this receive primitive with the corresponding send primitive shown in the previous example. Then, the *Is it mine?* switch operation compares the assignee of the task with

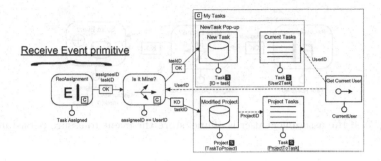

Fig. 6. A notification is received and the user interface is updated

the current user identifier: if they match, the task has been assigned to the current user and the *OK-link* is triggered, otherwise the *KO-link* is followed. In the former case the *New Task* data unit that will retrieve from the server the assigned task details to be shown in a pop-up window; in the same page, the *Current Tasks* index unit shows the updated task assignments list for the current user. Otherwise, the other part of the page is displayed: it shows the details of the project to which the novel task belongs (represented by the *Modified Project* data unit) with the updated list of its tasks (*Project Tasks* index unit). Notice that in this example filtering is performed by the recipients: the message is sent to all the members of the project and each client instance will check if the task has been assigned to the current user. Filter logic is based on a query, performed by the *Is it mine?* unit. Finally, communication persistence is transient (synchronous): once the event is sent, each on-line running application will immediately receive an event notification through the receive primitive.

In order to model persistent (asynchronous) communications, event data need to be explicitly stored in the event model, as we will see in the next example, where we consider the generation and reaction to events on the server, instead of on the user interfaces. Reactions to events received on the server are modeled by means of operation chains starting with a receive-event operation. They can trigger any server-side operation such as invoking Web services, sending emails, performing CRUD operations on the database, or signaling new event occurrences. In particular, it is possible to make the notification persistent using the database, to asynchronously signal the event when the intended recipient is back on-line; moreover, using conditional and query operations it is possible to specify *broker filtering* (Section 2).

Figure 7 depicts a server-side operation chain for the running example. The application requires that a notification be sent to a user when a task is assigned to her. When the user is off-line, she cannot receive notifications with server push as in the previous example: on the server, upon the reception of the notification of the Task Assigned event (represented by the *RecAssignment* receive event) the *User Online?* switch-operation triggers a create operation (*Store Event*) to persistently store the event occurrence if the assignee is not immediately reachable. When the assignee is back online, all the event occurrences stored on

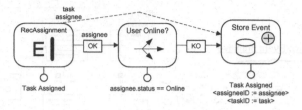

Fig. 7. If the assignment user is offline, store the event to assure persistance

the server are notified, activating the associated reactions on the client. The same primitives can be used to represent other classes of system events such as, for instance, temporal events, external database updates, or Web service invocations (for a complete reference see [13]).

3 Implementation

The implementation of the presented concepts builds on the WebRatio runtime architecture we developed for the work presented in [3]. The prototype automatically generates the code and the run-time framework for the WebML notation extended with distributed data and computation; it has been implemented using the OpenLaszlo [2] technology. We extended it with the needed components for distributed event notification: the resulting architecture is shown in Figure 8.

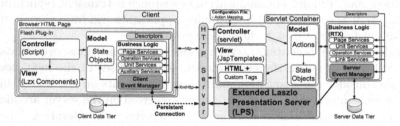

Fig. 8. The runtime architecture of our prototype implementation

OpenLaszlo natively provides a framework for enabling *persistent connections* and server to client communication. The implementation of the *receive event* and *send event* operations leveraged on such feature by extending both on the client and on the server side the existing architecture. On the client-side, an event handler (the client-side *Event Manager* in the architecture) is triggered whenever the persistent connection receives a message (i.e., an event notification) from the server. Message communication is performed by means of XML snippets, structured accordingly to the associated event type. The *Event Manager* is configured at compile time with a descriptor file, containing information about the event types managed by the client; it checks the received message for the carried event type and triggers the *receive event* operations associated with it, passing a reference to the received message. The *receive event* operation extracts relevant parameters from the message, sets the values for its outgoing parameters and calls the next operation in chain.

When a client-side *send event* operation is invoked, the client-side *Event Manager* builds an XML snippet reflecting the associated event type structure and whose content stems from the operation input parameters values upon invocation; then, it invokes a method of the persistent connection to send such message to the server. The server-side stub of the persistent connection is a servlet, provided by the Laszlo Presentation Server (LPS). We extended such servlet in order to bind the existing WebRatio run-time framework with the LPS server: it intercepts event reception on the server and invokes the server-side *Event Manager* component, also configured at compile time with a descriptor file, which, in turns, matches for the received event and triggers the appropriate server-side *receive event* operations.

4   Related Work

Although RIA interfaces aim at resembling traditional desktop applications, RIAs are generally composed of task-specific client-run pages deeply integrated with the architecture of a traditional Web application. Web Engineering approaches build on Web architectural assumptions to provide simple but expressive notations enabling complete specifications for code generation. Several methodologies have been proposed in literature like, for example, OOHDM [11] and WAE [6] but, to our knowledge, none of them addresses the design and development of RIAs considering them under the light of distributed event systems. This work, instead, considers system reaction to a collection of different events depending on *any* user interaction, Web service calls, temporal events, and data updates. Besides defining the generic concept of event in a Rich Internet Application, our approach also individuates the primitives and patterns that can be used to notify and trigger reactions upon external events. This is something that, to our knowledge, all Web engineering approaches lack, as reactions are only considered in the context of the typical HTTP request-response paradigm.

Both a survey and a taxonomy of distributed events systems are given in [9]: most of the approaches bear the concepts individuated in [10] concerning event detection and notification, or in [7] w.r.t. the publish-subscribe paradigm. With respect to these works, our proposal addresses events and notifications in a single *Web application* where on-line application users are the actors to be notified. In contrast, the above proposals aim at representing Internet-scale events with the traditional problems of wide distributed systems such as, for instance, clock synchronization and event ordering [8]. The system we are considering, instead, is both smaller in terms of number of nodes, and simpler in terms of topology: the server acts as a centralized broker where all events are ordered according to occurrence or notification time at the server. Nevertheless, it enables the specification and the automatic code generation of complex collaborative on-line applications accessible with a browser.

5   Conclusions and Future Work

RIAs provide the technological means to implement a new generation of on-line collaborative applications accessible from a Web browser. Notification-based communication of events (e.g., publish / subscribe, distributed event systems) and server-push technologies (e.g., based on AJAX) are both well-known and established: in this work we have presented the requirements and the primitives needed to *conceptually* model Rich Internet applications leveraging on them.

We plan to extend the event primitives to provide support for composite event recognition and explicit modeling of traditional Event-Condition-Action semantic dimensions such as event consumption and rule granularity, by means of a distributed event base across the server and the clients. Future work will also consider the integration of external rule engines or hierarchical distributed event systems to cater for scalability.

References

1. Google Gears, http://gears.google.com/
2. OpenLaszlo, http://www.openlaszlo.org
3. Bozzon, A., Comai, S., Fraternali, P., Toffetti Carughi, G.: Conceptual modeling and code generation for Rich Internet Applications. In: Wolber, D., Calder, N., Brooks, C., Ginige, A. (eds.) ICWE, pp. 353–360. ACM, New York (2006)
4. Carzaniga, A., Picco, G.P., Vigna, G.: Designing distributed applications with a mobile code paradigm. In: Proceedings of the 19th International Conference on Software Engineering, Boston, MA, USA (1997)
5. Ceri, S., Fraternali, P., Bongio, A., Brambilla, M., Comai, S., Matera, M.: Designing Data-Intensive Web Applications. Morgan Kaufmann Publishers Inc., San Francisco, CA, USA (2002)
6. Conallen, J.: Building Web applications with UML, 2nd edn. Addison-Wesley, Reading (2002)
7. Eugster, P.T., Felber, P.A., Guerraoui, R., Kermarrec, A.-M.: The many faces of publish/subscribe. ACM Comput. Surv. 35(2), 114–131 (2003)
8. Lamport, L.: Time, clocks, and the ordering of events in a distributed system. Commun. ACM 21(7), 558–565 (1978)
9. Meier, R., Cahill, V.: Taxonomy of Distributed Event-Based Programming Systems. The Computer Journal 48(5), 602–626 (2005)
10. Rosenblum, D.S., Wolf, A.L.: A design framework for Internet-scale event observation and notification. In: Jazayeri, M. (ed.) ESEC 1997 and ESEC-FSE 1997. LNCS, vol. 1301, pp. 344–360. Springer, Heidelberg (1997)
11. Schwabe, D., Rossi, G., Barbosa, S.D.J.: Systematic hypermedia application design with OOHDM. In: Hypertext, pp. 116–128. ACM, New York (1996)
12. Tanenbaum, A.S., Steen, M.V.: Distributed Systems: Principles and Paradigms. Prentice Hall PTR, Upper Saddle River, NJ, USA (2001)
13. Toffetti Carughi, G.: Conceptual modeling and code generation of data-intensive Rich Internet applications. PhD thesis, Politecnico di Milano, Italy (2007)

Privacy Inspection and Monitoring Framework for Automated Business Processes

Yin Hua Li, Hye-Young Paik, and Jun Chen

School of Computer Science and Engineering
University of New South Wales
Sydney, Australia
{yinhual,hpaik,jche340}@cse.unsw.edu.au

Abstract. More and more personal data is exposed to automatic and programmatic access, making it more difficult to safeguard the personal information from unauthorised access at every step. We introduce a privacy inspection and monitoring framework provides two functions: (i) the process designers can look over the privacy aspects of the automated business processes and resolve potential problems before deploying, and (ii) the obligation management technique ensures when certain actions performed on personal data, appropriate obligations are carried out.

1 Introduction

The recent emergence of Web services means that more and more business processes and their associated data are exposed to programmatic access, making it more difficult for human being to act as checkpoints that safeguard the personal information from unauthorised access at every step. This calls for a technical solution that will aid privacy officers by automatically (or semi-automatically) monitoring and detecting possible violation of privacy policies in the automated business processes.

One of the widely accepted standards in business process automation is Business Process Execution Language for Web Services (BPEL4WS, or BPEL for short) [1]. BPEL represents a programming language and its execution environment for creating composite Web services, in which interactions between autonomous entities (Web services) are defined and orchestrated. Our work focuses on the business processes written in BPEL. We summarise our contributions as follows:

- *Privacy inspection tool:* We have built a privacy inspection tool that allows the process designers to examine the privacy aspects of a BPEL process with respect to the organisational policies and resolve potential problems before deployment. This will remove the necessity for checking every activity for privacy conformance at runtime;
- *Privacy monitoring framework:* We have built an obligation management framework for a BPEL process which ensures that when certain action is performed on personal data, appropriate obligations are automatically carried out. We use Aspect-Oriented programming technique which is applied to an open source BPEL engine.

B. Benatallah et al. (Eds.): WISE 2007, LNCS 4831, pp. 603–612, 2007.

2   Privacy Model in BpelPrivacyMonitor

2.1   Privacy Data Model

Privacy Dictionaries. Different business processes collect different types of personal data for varying purposes and they involve different types of individuals or group who access the data (e.g., banking process vs. medical examination process). For this reason, we assume every organisation builds *privacy dictionaries*, in which the following information is defined in their own terms (e.g., using a language such as EPAL[1]):

- *Personal Data* (or data, for short) refers to information associated with an individual that can potentially identify the individual.
- *Data User* (or user, for short) represents a category of legal entities who can access personal data.
- *Data Usage Purpose* (or purpose, for short) describes the reason for which the data is collected and used.

Privacy Actions and Obligations. We define generic *Privacy Actions* (or actions for short) that are performed on data, namely: (i) collect – collecting the data for the first time, (ii) read – accessing the data as read-only, (iii) update – accessing the data as read-write, (iv) distribute – accessing the data to forward to partners and (v) delete – removing the data from the storage.

We also use the concept *Obligations* to dictate extra measures that must be taken when certain privacy actions occur. For example, when certain data is collected for the first time, the organisation may be obligated to log the details.

The term obligations are used in different context in security & privacy [2]. In BpelPrivacyMonitor, we observe the following measures as obligations: (i) sending a notification to the data subject (i.e., customer) about the usage of his/her data, (ii) logging the details of the usage of data, (iii) setting an alarm to automatically notify a possible over retention of data and (iv) regular reporting (e.g., monthly or yearly) to the data subject of the usage of her data. These are currently implemented in BpelPrivacyMonitor.

2.2   Privacy Policies

Based on the above abstraction, we model privacy policies as the organisation's *rights* and *obligations* on personal data. To be more precise, we define the following concepts[2]:

Definition 1 (Rights). *Let $\mathcal{U}, \mathcal{D}, \mathcal{P}$ be the privacy dictionary of users, data and purposes, respectively; and let $\mathcal{A} = \{collect, read, update, distribute, delete\}$. A right over $d(\in \mathcal{D})$, is a tuple (u, a, p), where $u \in \mathcal{U}, a \in \mathcal{A}$ and $p \in \mathcal{P}$. It is read that u can perform a on d for the purpose of p. We note it as $R_d : (u, a, p)$.*

---

[1] EPAL, http://www.w3.org/Submission/EPAL
[2] This model is based on our previous work [7,5].

For example, $R_{(email)}$: (customer manager, read, validate transactions) can be read as "an entity who plays a customer manager role can read the data email for the purpose validate transactions".

Definition 2 (Obligations). *Let $\mathcal{A}_{ob}$ be obligation actions, which consists of {send notification, write log, alert retention, regular report}. An obligation is a tuple (a_{ob}, μ_{ob}), where $a_{ob} \in \mathcal{A}_{ob}$ and μ_{ob} is the time property associated with a_{ob}.*

Each obligation is associated with a set of properties. One of which is time[3]. For example, write log is associated with a time property whose default value is 'immediately'. regular report also has a time property that determine how often the report should be sent to the customer (e.g., last day of each month). alert retention uses its time property to send the reminder of possible over retention of personal data (e.g., 90 days).

Definition 3 (Privacy Policy). *A privacy policy consists of a set of rights and obligation pairs. We note it as $R_d : (u, a, p) \rightarrow \{o_1, ..., o_n\}$, where R_d is the right over data d, $o_i (1 \leq i \leq n)$ is an obligation. The reading of each policy is that when a right is obtained, a set of obligations must follow.*

The example of privacy policies is given in the next section.

2.3   A Running Example in BPEL

Let us assume a company called BestMortgage who offers finance products such as personal, car or home loans. The company automated their home loan application process via a BPEL process homeLoanService. The process consists of a number of partners, each carrying a portion of the complete process.

The building blocks of BPEL are activities. In particular, receive, pick, invoke and reply activities provide the means of communication among the participants. Figure 1 shows the coordination of the communication.

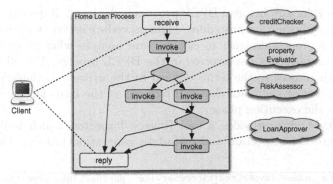

Fig. 1. homeLoanService and its partners

---

[3] To be generic, we associate a set of properties to an obligation. Currently, only a time property is implemented.

The process receives a home loan application from an applicant, then it invokes *CreditChecker* partner who will assess applicant's credit score. If the score is low, the application will be rejected. Otherwise, the process invokes *PropertyEvaluator* and *RiskAssessor* to obtain a property evaluation and risk evaluation, respectively. If the risk level is low, the application will be approved. Otherwise, the process invokes *LoanApprover* who will go through thorough assessment and decide whether to approve the application or not. Finally, the process replies the result to the applicant.

One thing to note is that during the communication, as shown in Fig. 2, the BPEL process sends/receives messages (referred to as variables). Some of the messages inherently contain personal data (such as income, home address, phone, etc.). The crux of the privacy issue in BestMortgage is to make sure that the personal data in the messages are exchanged in accordance with the privacy policies of the organisation. The company's privacy dictionaries and policies are as follows:

Privacy Dictionaries. The data dictionary contains {home postal, home phone, mobile phone, income}; the user dictionary includes {ours, credit checker, property evaluator, risk assessor, loan approver, public}; The purpose dictionary has {homeloan processing, risk assessment, property evaluation, send email, call customer}. The term ours refers to the organisation itself.

Privacy Policies. Here is a fragment of the privacy policies of the company which shows the company's rights on data and their associated obligations.

$R_{(home\ postal)}$:(ours, collect, homeloan processing) → {(alert retention, 180 days)}, $R_{(home\ postal)}$:(ours, distribute, risk assessment) → {(write log, immediately)}, $R_{(income)}$:(risk assessor, read, risk assessment) → {(write log, immediately), (regular report, last Friday of each month)}, ... *more policies* ...

3   Privacy Inspection Tool

In BpelPrivacyMonitor, before a BPEL process is deployed, it goes through what we call "privacy inspection procedure". The procedure serves two purposes: first, it enables the process designer to capture and analyse what types of personal data, user and purposes are involved in the BPEL process. The BPEL code is analysed with respect to the terms defined in the privacy dictionaries. Second, it produces a set of metadata that will be used as the basis for the monitoring procedure in the execution phase.

BpelPrivacyMonitor provides a tool Privacy Inspector, which is designed to semi-automate the inspection and metadata generation process. The tool has

```
<bpel:invoke  name="InvokeCreditScoreService" partnerLink="creditScoreLink"
    portType="ns1:creditScorePT" operation="checkCreditScore"
    inputVariable="CreditScoreRequest" outputVariable="CreditScoreReturn" />
```

Fig. 2. A BPEL activity with input/output messages (variables)

a "wizard-type" user interface, providing a step-by-step guide for the process designer.

3.1 Privacy Inspection Steps

Step 1: Associating Variables with Personal Data. The first step of the inspection involves examining the BPEL activities and their variables and marked the ones that carry personal data. At the end of this step, the identification of the activities for further inspection is done.

For instance, the invoke activity in Figure 2 takes creditScoreRequestMessage as input. For space reasons, we cannot present all message definitions in the BPEL code here. In our current implementation of the running example scenario, the message contains personal income and home address. Therefore, the process designer will mark the activity as being relevant to the inspection process and map its message part names to the data dictionary terms.

Step 2: Associating Activities with Actions. In the next step, the inspection tool presents, on the right hand side, the activities carried over from Step 1. The list of actions from the privacy dictionary is displayed on the left hand side. The developer will map each activity to appropriate actions. For example, the invoke activity in Figure 2 performs read and distribute actions on the data contains in its message. Figure 3(a) shows a partial screenshot of the inspector tool in Step 2 of the inspection process.

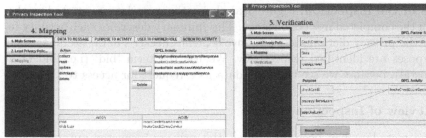

(a) Inspector Tool - Step 2: Associating Activities with Actions

(b) Review of the inspection results: resolving conflicts

Fig. 3. Privacy Inspector Tool

Step 3: Associating Activities with Purposes. In this step, the tool presents the purposes defined in the privacy dictionary. Each action identified in Step 2 is now mapped with its purpose by the process designer. For example, read action may be performed for the homeloan processing purpose and distribute is there for the credit check purpose.

Step 4: Associating PartnerRoles with Users. Finally, the users of the personal data are identified. The partnerRoles listed in the BPEL definition are potential receivers of the personal data. The process designer will map each

```
<InspectionResults name="homeLoanService">
 <activity name="invokeCreditChecker">
  <bid over="home postal">
   <user>ours</user>
   <action>collect</action>
   <purpose>homeloan processing</purpose>
  </bid>
  <bid over="income">
   <user>ours</user>
   <action>distribute</action>
   <purpose>check credit</purpose>
  </bid>
  ...
 </activity>
 <activity name="invokeRiskAssessor">
  ...
 </activity>
</InspectionResults>
```

```
<PrivacyTrails name="homeLoanService">
 <activity name="ReceiveHomeLoanRequest">
  :
 </activity>
 <activity name="invokeCreditChecker">
  <trail>
   <right over="home_postal">
    <user>ours</user>
    <action>collect</action>
    <purpose>homeloan processing</purpose>
   </right>
   <obligations>
    <alert_retention activate="180_days"/>
   </obligations>
  </trail>
  <trail>
   <right over="income">
    <user>ours</user>
    <action>distribute</action>
    <purpose>check credit</purpose>
   </right>
   <obligations>
    <write_log activate="immediately"/>
    :
   </obligations>
  </trail>
```

(a) The results of inspection for invokeCred- (b) The privacy trails generated for home-
itChecker activity   LoanService

Fig. 4. Inspection Results / Privacy Trails

partnerRole to the user dictionary. For example, the partnerLink in Figure 2 is linked with a parterRole who can be mapped to the user credit checker.

Results of the inspection. Figure 4(a) shows the metadata generated as the results of the completed inspection steps. The summary describes each activity in terms of the privacy dictionaries: e.g., invokeCreditChecker's data, users, actions and purposes are identified respectively. We use the term 'bid' (i.e., bid over home postal), because we see this metadata as a request for access to the data.

3.2   Review of Inspection Results

After the metadata has been generated, the privacy inspector automatically compares the bidding content with the rights stated in the privacy policies of the organisation. When the tool finds any conflicts between the content of metadata and the rights in the privacy policies, it highlights the problem which has to be rectified before the BPEL process can be deployed.

As a simple example, Figure 4(a) shows that invokeCreditChecker is bidding for data income to perform distribute action for the purpose of check credit. Let us assume that the combination does not appear in the privacy policies. Then the tool will raise an issue by displaying red coloured lines (e.g., Figure 3(b)). We assume that such problem should be resolved by either changing the privacy policies to reflect the reality of the business process or modifying the implementation of the business process to conform to the policies.

4   Privacy Monitoring Framework

Once the BPEL process has passed the inspection, it is deployed to the execution environment. The main responsibility of BpelPrivacyMonitor during the execution of the deployed process is to monitor its usage of personal data and implement the obligations. For this purpose, we first generate a set of privacy trails which will be attached to the BPEL process for monitoring.

4.1   Generating Privacy Trails for a Business Process

The process of generating privacy trails for a BPEL process is straightforward. The results obtained from the inspection tool are used as the basis. Simply put, each bid over data d in InspectionResults becomes right over d; and the action, purpose and user are appropriately annotated for the right. Then, for each right, obligations are attached. Which obligation should be attached to a right can be determined from the privacy policies.

Figure 4(b) shows the privacy trails generated for the homeLoanService process. It shows, for instance, that the invokeCreditChecker activity will collect home postal data for the purpose of processing homeloan, but it must activate alert retention obligation after 180 days.

4.2   Implementing Obligations

Without referring to the specific implementation platform and architecture we explain how obligations are meant to do.

In the obligation management framework, activities in the BPEL engine execution queue are monitored. As soon as an activity listed in the privacy trails appears, the obligation manager is activated. The manager is passed in enough information to carry out appropriate operations (i.e., the details of the activity and associated obligations).

When the obligation is set to be carried out 'immediately', the obligation manager will run the appropriate class that implements each obligation (e.g., writing log). When the obligation has a time property and if it is greater than 'immediately', the obligation manager registers it as a new timer event with the Time Manager. This sets an alarm with the Time Manager who will trigger the obligation manager when the timer is expired. This will in turn trigger the obligation manager to run the appropriate obligation implementation class. We currently support four types of obligations, however, more obligations can be added as needed. Each obligation implementation class does the following:

Write log: The implementation class of this obligation will write the details of the current action to the logs. Although it is possible to associate activation time to this obligation, it is more likely that 'write log' will be carried out immediately. The log includes the <right> section (i.e., data, action, purpose and user) of the privacy trails for the activity and a time-stamp.

Send notification: This obligation is implemented to notify the owner of the data of the actions performed on their data. This is likely to be used when the organisation is disclosing the data to a third party (e.g., credit checker in

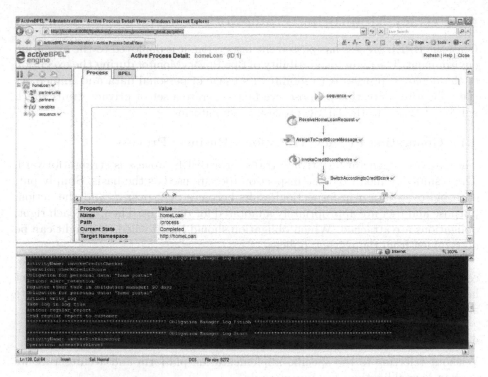

Fig. 5. Screenshot of ActiveBPEL engine and write_log obligations

homeloanService). Our current implementation uses email for notification. Similar to logging, the email contains the <right> section of the privacy trails for the activity and a time-stamp.

Alert retention: This obligation is designed to alert the system administrator of possible over retention of the personal data collected by a business process. For example, assume that an activity a performs 'collect' action and has (alert retention, 180 days) as an obligation. When activity a occurs, the obligation management framework registers the activity and its obligation alert_retention with the Time Manager. After 180 days from the event, the Time Manager will trigger the obligation manager to run the alert_retention class which sends a warning message to the system administrator.

Regular report: This obligation implements dispatching of regular reports to the data subject about the usage. The time property associated indicates how often the reports are sent. Each report contains the summary of the usage over the given period (e.g., last day of each year).

Implementation. For the execution environment, we use ActiveBPEL[4], a well-known open source BPEL engine. To implement our obligation framework, we extend the engine to run in an Aspect-Oriented environment.

---

[4] www.activebpel.org

Figure 5 shows the homeloanService process deployed into ActiveBPEL engine which is already extended to run the obligation implementation classes. The console terminal at the bottom shows the results of a few write_log obligations.

5 Related Work and Conclusion

We broadly define two related areas: monitoring BPEL processes and privacy enforcement.

Monitoring BPEL Processes. To our best knowledge, there is no work specif ically designed to monitor BPEL process for privacy. However, there are some work done in monitoring the execution of BPEL for generic non-functional properties. For example, Baresi et. al. [4] proposed a proxy-based solution for monitoring QoS agreements set between the service client and service provider. A monitoring rule consists of monitoring location, monitoring parameters and monitoring expressions. In Mahbub et. al. [8] presented a framework for monitoring requirements of a composite service. Event calculus are used to describe the requirements that need to be monitored. Requirements can be behavioural properties of a composite service or assumptions about the process and all of its underlying services. It is noted that none of these work directly consider privacy as their primary issue. It will require some ad-hoc privacy data modelling and significant extension for these systems to cater for privacy. An interesting work is proposed by Bettini et. al. [6], which we believe can be complimentary to our future work. The work introduces a way to manage obligations by associating data users with a level of trustworthiness. In case a user fails to fulfil an obligation, a penalty is applied (e.g., decreasing of trustworthiness). We believe, for example, logs generated by BpelPrivacyMonitor can be used to automatically detect the violation of obligations and apply the penalties.

Privacy Policy Enforcement. Enforcing privacy policies on the business applications has been largely approached from traditional access control view point. For example, Agrawal et. al. have introduced the concept of Hippocratic databases in [3]. The core idea is to exploit an SQL rewriting technique to incorporate policy evaluation into a database query to ensure that data is only disclosed to appropriate people. The implementation assumes relational database systems and it requires analysis of each SQL query for re-writing. Another database-centric approach is presented in [9], in which the author focuses on the management and enforcement of privacy obligations in a closed enterprise environment. The approach does not take purpose of access, the core concept of privacy, into consideration.

The drawback of these approaches is that they focus on the database level (relational databases in particular). We believe that a higher-level view (i.e., business process and orchestration) consideration is required to cater for the privacy concerns that may span across different systems and different types of data repositories.

Schunter et. al [10] has proposed the Privacy Injector system which leverages the Aspect-Oriented programming paradigm to implement privacy enforcement.

The system associates privacy metadata to each piece of personal data that is collected. All data-manipulation operations should work within the boundary of what the metadata dictates. All usage and disclosure of particular personal data are checked and the necessary actions are performed. One thing to note is the system performance aspect. Checking every action against the privacy metadata will undoubtfully be inefficient. In our approach, we move some of the checking to the design phase through the inspection process to avoid checking every action at runtime. Also, our approach proposes a generic obligation management whereas the privacy injector system mainly focuses on management of data disclosure.

To conclude, we introduced BpelPrivacyMonitor, a privacy inspection and monitoring framework for automated business processes. In particular, we considered processes implemented in BPEL. BpelPrivacyMonitor provides two functions: (i) the process designers can look over the privacy aspects of an automated business process and resolve potential problems before deployment, and (ii) the obligation management technique ensures when certain actions performed on personal data, appropriate obligations are carried out. We are currently working on how to exploit the logs generated by our framework. The aim is to assist the privacy audit process by using the knowledge obtained from the analysis of the process logs with respect to privacy.

References

1. BPEL 1.1, www.ibm.com/developerworks/library/specification/ws-bpel
2. DAML services: Security and privacy,
   www.daml.org/services/owl-s/security.html
3. Agrawal, R., Kiernan, J., Srikant, R., Xu, Y.: Hippocratic databases. In: VLDB, pp. 143–154 (2002)
4. Baresi, L., Guinea, S.: Towards dynamic monitoring of ws-bpel processes. In: Benatallah, B., Casati, F., Traverso, P. (eds.) ICSOC 2005. LNCS, vol. 3826, pp. 269–282. Springer, Heidelberg (2005)
5. Benbernou, S., Meziane, H., Li, Y., Hacid, M-S.: Privacy-agreement for web services. In: SCC (to appear, 2007)
6. Bettini, C., Jajodia, S., Wang, X., Wijesekera, D.: Obligation monitoring in policy management. In: POLICY 2002. 3rd International Workshop on Policies for Distributed Systems and Networks, IEEE Computer Society, Los Alamitos (2002)
7. Li, Y., Paik, H-Y., Benatallah, B., Benbernou, S.: Formal consistency verification between bpel process and privacy policy. In: Privacy Security and Trust, pp. 212–223 (2006)
8. Mahbub, K., Spanoudakis, G.: A framework for requirents monitoring of service based systems. In: ICSOC, pp. 84–93 (2004)
9. Mont, M.C.: Towards scalable management of privacy obligations in enterprises. In: Fischer-Hübner, S., Furnell, S., Lambrinoudakis, C. (eds.) TrustBus 2006. LNCS, vol. 4083, pp. 1–10. Springer, Heidelberg (2006)
10. Schunter, M., Berghe, C.V.: Privacy injector: Automated privacy enforcement through aspects. In: Workshop on Privacy Enhancing Technologies, pp. 99–117 (2006)

Digging the Wild Web:
An Interactive Tool for Web Data Consolidation

Max Goebel, Viktor Zigo, and Michal Ceresna

Database and Artificial Intelligence Group
Technische Universitat Wien
A-1040 Vienna, Austria
{goebel,zigo,ceresna}@dbai.tuwien.ac.at

Abstract. With a rapidly growing pool of applications and services available, the World Wide Web has become the single largest source of information. Yet the lack of structure and semantic annotation of individual data items within the Web poses a severe restriction on its usefulness to the end user. Web mashups and other service-oriented software try to remedy this dilemma in a pragmatic way by consolidating related data from different sources, yet their creation requires expertise on the user-side. We present a system that allows novice users to create data consolidation applications through interaction. Isolated tasks are performed as Web services. A simple and flexible Web service workflow language is defined that composes the isolated tasks into a workflow which models the user's interactions. As all this is done encapsulated away from the user, the user effort is kept minimal.

1 Introduction

With a rapidly growing pool of applications and services available, the World Wide Web has become the single largest source of information. On the other hand, the lack of structure and semantic annotation of individual data items within the Web pose a severe restriction on its usefulness to the end user. Both corporate and personal interest has recently turned to Web data consolidation via Web services, where different sources are composed into new content with added value based on situational needs [5,13]. Such data mash-ups are promoted through the rapid development of new tools and protocols (see e.g. [1]), yet they still require a certain degree of expertise in API programming.

In this paper we present a system targeted at Web data consolidation for real-world commercial use case scenarios[1]. Such scenarios typically are complex in nature and thus need to be expressed by a multitude of different Web services. Composition and orchestration of such services is achieved via business process management, where the execution of individual subtasks is controlled as workflows by business logic. There exist a number of workflow description frameworks and languages that do exactly that [3,4,10,11]. For our work we defined a workflow description language based on *JavaScript* that is modeled specifically to our task of interactive data consolidation from

---

[1] A download for research evaluation purposes is available at http://www.dbai.tuwien.ac.at/staff/ceresna/wrap/workflow-demo/

B. Benatallah et al. (Eds.): WISE 2007, LNCS 4831, pp. 613–622, 2007.

the Web. Its advantage lies in its light-weight implementation lending itself particularly well to the Web domain.

The rest of this paper is organized as follows. Section 2 gives links to some related work and outlines our contributions. In Section 3 we identify core tasks for Web data consolidation and show how these can be both modeled as workflows and orchestrated interactively. A technical definition of our workflow language is given in Section 4. Finally, Section 5 looks at action libraries that complete the core workflow language with functions for interactive Web scenarios and thus adding record-and-replay capabilities.

2   Related Work

A key difference from most existing solutions is our inside-browser (or browser-oriented) approach to Web interaction recording (see [14]). Such approach allows building very robust and expressive interactions. Similar implementations can be found in the following projects: Chickenfoot [6], TestGen4Web [2], iMacros [9].

TestGen4Web is a Firefox extension that automates Web navigation sequences. It allows recording of user performed navigation and to replay them later, with the goal to automate regression testing of Web portals. Similar to TestGen4Web is iMacros, a partially commercial product which offers general Web automation. Additionally, the project employs offset-based techniques that enable navigation in Flash and Java Applets. The solution is closely tied to Windows and Internet Explorer, although a Firefox extension has been released recently.

The Piggy bank project focuses on the conversion of legacy Web information to semantic structures such as RDF, and their storage for later use. The project is similar to the activities around the Haystack semantic browsing [12] and Thresher [8]. Piggy bank, however, requires additional installation of a quite heavy Java machinery on the client's computer. Moreover, it offers only very basic data extraction techniques and lacks any navigation aspect.

Technologically speaking, the project most closely related to our work is Chickenfoot [6]. Chickenfoot is a programming environment running inside the Firefox browser. It allows users to develop and run scripts (in JavaScript syntax with additional function libraries) that automate Web operations, and to modify content and behavior of Web pages. From the application point of view, Chickenfoot is better suited for small scenarios (e.g. short, linear processes). More influential is its introduction of pattern matching techniques which simplify the location of information or document elements on Web pages. The project uses keyword patterns and text constraint patterns. Our work differs from Chickenfoot in that the workflow language we present herein is capable of handling much more complex scenarios. This is due to the addition of complete parallel process synchronisation handling, giving our language the full expressive power of standard workflow languages. A common benchmark for evaluation workflow laguages is its comparison to the widely accepted collection of workflow patterns [4]. Our languages ranks very high as it implements all 21 control-flow patterns (almost all natively).

3   Web Services as Workflow Tasks

The main contribution of this paper is the specification of a light-weight workflow model capable of describing complex Web service flows (e.g. nested, cyclic) as often required by real-world scenarios. Whereas in related work such tasks are described in isolation, using Web wrappers or query interface learners alone, we introduce a model that can incorporate a range of different tools to describe more complex tasks.

3.1   Workflow Orchestration

We identify three core enterprise tasks that fully qualify to describe typical commercial scenarios (see below) in a flexible and user-oriented way: *Navigation, Collection-based Analysis* (clustering), and *Document-based Analysis* (extraction). Each task can be interpreted as a separate Web service. We present a workflow model that handles the orchestration of such services from a user perspective, requiring a minimal amount of user expertise: user navigations (crawling, form filling, query submissions) are recorded automatically and thus encapsulated from the user by the *Navigation* module; data items to be collected are marked interactively in the user's browser via mouse clicks. Easy-to-interpret extraction rules are infered from the labeled examples in the *Analysis* modules. of other. extraction A workflow task can be anything ranging from a primitive action (e.g. 'click here', 'load page') to more complicated actions (e.g. filling and submiting forms, interacting with dynamic widgets, classifying pages, executing wrappers, or invoking external services) to handling of arbitrary scenarios using custom scripts.

Navigation. Navigation is encapsulated directly in the workflow language and therefore requires no external Web service.

Collection-level Analysis. This module takes as input the documents (or document fragments) retrieved by the Web interaction module. It usually consists of a large number of HTML documents taken from the same domain. The module reorganizes its input as a set of *clusters*. One possible method is to group the input documents into clusters according to the similarity of their DOM tree structure. Although, quite naturally, the structure of the input documents may be assumed to be similar across the input set, frequent exceptions require this additional clustering step. Clustering ensures better extraction results over large, heterogeneous Web document collections.

Document-level Analysis. This module also takes as input a set of documents, usually obtained from clustering or Web interaction directly. Our system uses a learning algorithm that is based on the tree structure of HTML documents. We learn a combination of filters expressing tests on node properties and node path characteristics. In [7], we have demonstrated the effectiveness of such a learning algorithm on a number of well-known extraction tasks. Alternatively, other approaches may be used interchangibly.

3.2   Interaction Record-and-Replay

In our system we differentiate between two phases: an interactive workflow generation phase and a fully-automatic production phase. The level of automation in the latter

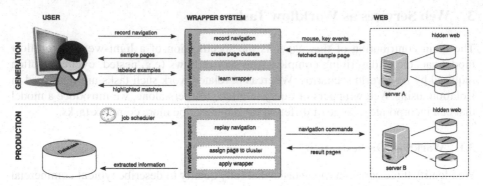

Fig. 1. Semi-supervised workflow design phase (top), and a fully-automated extraction phase (bottom)

is understood so that the workflow sequence—once defined—is applied without any further user involvement. As a typical scenario one can imagine a scheduled process manager invoking the workflow sequence every so often to capture changes in the underlying source data. Figure 1 depicts an abstraction of the wrapper learning system in terms of the user's involvement in specifying the wrapper.

The interactive phase of the system starts with the user navigating through the Web (using her browser) and recording her navigation. Such navigation may contain both simple crawling tasks and more complex query navigations. Query navigations are user interactions where the user is prompted to specify parameters for further navigation (e.g. log-in fields, forms, etc.). The wrapper here acts as a mediator between the user and the relevant web servers accessing the hidden Web.

In the production phase, the workflow sequence is simply executed either by the user or some process manager. For the latter, frequent executions are possible using automated task scheduling (e.g. weekly). The workflow herefore gets decomposed into its individual subtasks, which are then applied sequentially, one after the other. The resulting data fields returned by the extraction part of the workflow are stored in a separate database.

4   Workflows

The concept of a workflow as used herein arised from close partnership with the LUMBERJACZK project [15]. LumberJaczk is a light-weight solution to interactive Web information acquisition. Its main objective is to provide an engine and development environment for full-fledged applications operating on top of existing Web applications (so-called *on-top-of-Web applications*). The system promotes a very pragmatic approach: it focuses on expressiveness, functionality, and agility. Our workflow model therefore is more a programming language extension than a visual system[2]. This approach has proven particularly suitable in the challenging domain of Web information

---

[2] Various visual modelling front-ends, usually qualitatively more restrictive, will be provided additionally.

acquisition. The original specification of the workflow language is language agnostic (see [15]). As our system operates in the Web domain, however, we chose a *JavaScript* implementation for practical reasons.

4.1 Workflow Structure

The basic building block of a workflow is a *task*, which is defined by means of associated script code. Scripts are executed sequentially and are the atomic units of the modeling language. Using scripts at the *programming in the small* level brings significant expressiveness and completeness as it enables a task to convey flexible algorithmic behavior. In contrast to the above are tasks and *transitions*, which operate at the *programming in the large* level. They model the complementary aspects of complex processes using asynchronous transitions, concurrent execution, and stackless task chaining. These aspects are apparently present in Web interactions, whose nature is mainly asynchronous, non-deterministic, and unreliable.

Sequences of tasks constitute so-called *flows*. Each flow is bound to an *execution context*. Execution contexts represent a shared workspace for all tasks within a flow. They mediate access to relevant resources, functions, and data (e.g. Web browser windows, documents, variables). Moreover, they acts as a scope for calling the functions available in all modules, i.e. extraction, clustering, Web interaction, and other pluggable custom action libraries (see example below). The functions can be called directly from a task script.

```
function myTask1() {
  load('http://google.com')
  name=classify(window)
  return task('myTask2')
}
```

4.2 Transition Primitives

A task can decide what transition to follow next by passing the transition as a script return value. As this is done in run-time, the graph structure of the workflow is dynamic, i.e. not statically wired in the system's construction phase.

A workflow transition can be expressed using *transition primitives*. The following set of primitives is available in our language:

- task (direct transition), delay (deferred transition);
- join/joinM (a simple, conjunctive, and generalized conjunctive synchronization transitions);
- condition (an expression conditioned transition), timeout (a time guarded transition);
- choice (an implicit choice transition)

By composing transition primitives we can build control-flow macros. For example, the macro

```
timeout(join(bookCar,bookFlight,'pay'),10000,'fail')
```

waits 10 seconds for tasks 'bookCar' and 'bookFlight' to complete. It turns out that the above transition primitives suffice to express all workflow patterns identified in [4]. Three common patterns are described next using the transition primitives defined above.

Example 1. Synchronization Pattern (AND-join)

```
jobs=spawn('bookCar','bookFlight')
join(jobs,'pay')
```

or using subflows (nested workflows)

```
subflow('bookCar')
subflow('bookFlight')
return 'pay'
```

Example 2. Deferred choice pattern

```
service1=loadNew('http://service1/')
service2=loadNew('http://service2/')
userCancelation=interact("Abort service")
return choice(
      join(service1,'process'),
      join(service2,'process'),
      join(userCancelation,'canceled'), )
```

Example 3. Discriminator Pattern

```
return joinM(service1, service2, userCancelation, 1, 'done')
```

Execution Control. The execution control is responsible for synchronization of separate workflow tasks. All common process controls are implemented:

- spawn A task can spawn parallel flows of execution using the keyword to make more efficient use of resources or reduce response time. Each new flow runs in its own execution context, cloned from the context of the spawning task. Concurrent flows can, of course, be synchronized and joined using transitions. Parallel iteration through menu items or parallel search queries on different portals can be implemented using spawn.
- subflow This construct allows modularization and decomposition of workflows. A complex workflow fragment can be considered as a black-box, and treated as a single task, which completes only on completion of the associated sub-workflow.
- eventual It allows to schedule tasks to be done 'eventually', i.e. once there are no more scheduled tasks.

As the Web resources we are accessing are relatively unreliable and change frequently, there is a need for robust error handling. Erroneous and unexpected situations can be handled either inside tasks (at Javascript level), on workflow level by the specially devised tasks onError, onAllError, or with the interruption keywords cancel, break, or resume.

5   Action Libraries for Web Interaction

We previously defined a powerful workflow language based on a scripting language. We now enrich this language to allow for native integration of common user actions as found during typical interactive Web sessions. They let us model non-trivial interactions to access information hidden behind the walls of dynamic, stateful technologies (commonly dubbed as Web 2.0, e.g. rich client logic, AJAX), submittable forms, and restricted areas. With the Web interaction module being 'inside' the Web browser, it interacts with the fully rendered and live Web pages. Several techniques and actions are implemented that can simulate the interactions as if they were performed by a human user. We can generalize them into four use case patterns:

1. **Crawling** - uninformed link following trying to reach all pages on the web site, for example searching for all email addresses;
2. **Navigation** - traveling through a web site's interaction path known in advance, for example logging into a bank account;
3. **Form mapping/filling** - filling and submitting Web forms, e.g. searching via a search engine;
4. **In-page interaction** - driving dynamic Web pages (e.g. AJAX), e.g. expanding dynamic menus or clicking page components that change page content without issuing a new document load.

The actions—orchestrated by the workflow framework—can capture complex Web navigations and model arbitrary user interactions. Technically, all actions are available as methods that can be used by workflow tasks, i.e. called by the task scripts. They are executed in the context of the current task, and thereby can directly interact with the relevant Web pages. An execution context is associated with one working browser window, and thus also has access to a live document. Next, we present collections of the most frequently used methods available in the Web interaction module.

- `load(url, [postParams]),back(),forward()`
- `loadNew(url, [postParams]),close(),`

The first set of functions adds standard depth-first crawling capabilities. The functions `loadNew` and `close` serve for spawning parallel branches in a navigation sequence (usually handled by concurrent paths in the workflow). In the implementation this is achieved with opening or closing a browser tab or window.

- `click([target]),type(keys, [target])`
- `set(target, values)`
- `formfill({})`

The first two functions serve for sending mouse clicks and simulating key sequences. Most often they are used for dealing with dynamic page components such as menus or AJAX handled buttons. The function `set` assigns values to form input widgets, e.g. textfield, combo box, radio button, etc.

The convenience function `formFill` automatically fills form fields using various heuristics. Currently we locate the input fields by matching the input widget candidates

Fig. 2. Example scenario for monitoring flights on the flight search platform *Expedia*. Shown are two navigation jobs (*A* and *B*), and user-defined learning examples (framed in light-gray and labeled positive and negative).

to a tuple of relevant features (label text, id, tag name). Values are assigned via the set method.

The last group of functions deals with the integration of the extraction and clustering (classify) modules.

- find(*patternName*, [*context*]),
- extract(*patternName*, [*context*])
- classify([*document*])

Given a pattern name as the first parameter the find function returns the first matched DOM node. The extract function extracts information and stores it. The function classify is used for finding and returning the document cluster that best describes its input document.

In addition, these functions may take an optional parameter context that specifies an input sub-tree to be used for evaluation. If the parameter is omitted, the current document is used. In Algorithm 1 the extra parameter enables us to set the previous extraction results as context for the next extraction tasks for the find and extract functions (in contrast to classify which uses the entire document).

Web Interaction Workflow Example. Although the workflow is capable of handling very complex scenarios, we use a significantly simpler extraction scenario for demonstration purposes. Assume it is our goal to extract flight data from the travel portal *Expedia*[3]. As is typical for travel portals, the user is first prompted to specify the departure and destination of the trip together with the date of travel (Figure 2). When submitting our information, the portal crawls a set of airline databases, queries each for the travel details provided, and returns all matching offers. Note that the data to be extracted is only to be found on the dynamically generated result page, making the form querying step an integral part of the extraction task.

Algorithm 1 shows how the Expedia scenario is composed from the workflow and Web interaction primitives. A new workflow process is started with with two user-given

---

[3] http://www.expedia.co.uk

Algorithm 1 Workflow for the example scenario in Figure 2

```
 1:  function start() {                    19:  function handleResults() {
 2:    p1 = loadNew(`www.expedia.co.uk')   20:    //decide how to handle loaded page
 3:    p2 = loadNew(`www.expedia.co.uk')   21:    var cluster = classify()
 4:    return joinFirst(p1,p2,`sendQuery') 22:    switch(cluster) {
 5:  }                                      23:      case `normalPage':
 6:                                         24:        //find all flight entries
 7:  function sendQuery(page) {             25:        var eitems = findAll(`flights')
 8:    window = page                        26:        spawnEach(eltems,`handleDetails')
 9:    formfill({                           27:      case `noflightPage':
10:      `Departing from:' : `London',     28:        //ignore, i.e. terminate
11:      `Going to:' : `San Francisco',    29:    }
12:      `Depart:' : `params.DepartureDate', 30:  }
13:      `Return:' : `params.returnDate'   31:
14:    })                                   32:  function handleDetails(item) {
15:    var ebutton = find(`submitButton')  33:    extract(`price',item)
16:    click(ebutton)                       34:    //follow to flight detail page
17:    return join(window,`handleResults') 35:    var elink = find(`detailLink',item)
18:  }                                      36:    click(elink)
                                            37:    return join(window,`wrapDetail')
                                            38:  }
                                            39:
                                            40:  function wrapDetail() {
                                            41:    //extract flight details
                                            42:    extract(`detailPage')
                                            43:  }
```

---

parameters: `departureDate` and `returnDate`. The initial task, called 'start', opens two separate windows for different Expedia portals. The workflow continues with the portal that responds first, ignoring the slower one (through discriminating join transition *joinFirst*). For the second task, `sendQuery`, the search form gets filled out and, using the `find` function, the extraction module locates the search button. A subsequent mouse click sent to the search button then triggers a transition to the next task `handleResult`. Again, the next task needs to be synchronized with the page being loaded (`join` tansition). Several types of search result pages can occur. First, the standard results page depicted in Figure 2 may be loaded. Alternatively, a second type of Web page saying no flight connections are available may be loaded. The clustering module gets called to classify the loaded page into either category (`classify`).

In case of the standard results page the extraction module locates all flight entries and proceeds to the extraction of their details, each on a separate detail page. In order to speed up the process, the details are handled in parallel, each in a new task flow (`handleDetails`) spawned by the `spawnEach` command. On each detail page, additional flight data is extracted and stored by the `wrapDetails` task. The workflow completes when there is nothing more to be done.

6 Conclusion

We presented an integrated solution for complex, real-world data consolidation scenarios from Web documents. In a nutshell, our system models user behaviour as a flowgraph of simple extraction-related subtasks. We have introduced a workflow language that allows to describe extraction scenarios as orchestrated compositions of individiual sub-modules. The proposed language compares to state-of-the-art workflow languages in terms of expressiveness, but it is implemented natively in Java Script. This allows for an easy extension through customized action libraries for the Web domain. We introduced a rich set of such action libraries to model complex user behaviour on stateful Web sites and demonstrated their power on a running example.

References

1. http://programmableweb.com/
2. Testgen4web, http://developer.spikesource.com/wiki/index.php
3. Aalst, W., Hee, K.: Workflow Management: Models, Methods, and Systems. MIT press, Cambridge, MA (2002)
4. Aalst, W., Hofstede, A., Kiepuszewski, B., Barros, A.P.: Workflow patterns. Distributed and Parallel Databases 14(1), 5–51 (2003)
5. Ankolekar, A., Krotzsch, M., Tran, T., Vrandecic, D.: The two cultures: mashing up web 2.0 and the semantic web. In: WWW 2007. Proceedings of the 16th international conference on World Wide Web (2007)
6. Bolin, M., Webber, M., Rha, P., Wilson, T., Miller, R.C.: Automation and customization of rendered Web pages. In: UIST 2005. Proceedings of the 18th annual ACM symposium on User interface software and technology, ACM Press, New York, NY, USA
7. Carme, J., Ceresna, M., Goebel, M.: Web wrapper specification using compound filter learning. In: Proceedings of the IADIS International Conference WWW/Internet 2006 (2006)
8. Hogue, A., Karger, D.: Thresher: automating the unwrapping of semantic content from the World Wide Web. In: WWW '05: Proceedings of the 14th international conference on World Wide Web, pp. 86–95. ACM Press, New York, NY, USA (2005)
9. iOpus. imacro (2006), http://iopus.com
10. Leymann, F., Roller, D., Schmidt, M.-T.: Web services and business process management. IBM Systems Journal 41 2, 198–211 (2002)
11. Papazoglou, M.P.: The world of e-business: Web-services, workflows, and business transactions. In: Bussler, C.J., McIlraith, S.A., Orlowska, M.E., Pernici, B., Yang, J. (eds.) CAiSE 2002 and WES 2002. LNCS, vol. 2512, Springer, Heidelberg (2002)
12. Quan, D., Karger, D.R.: How to make a semantic web browser
13. Sabbouh, M., Higginson, J., Semy, S., Gagne, D.: Web mashup scripting language. In: Poster Proceedings of the 16th International World Wide Web Conference, pp. 1305–1306. ACM Press, Banff, Alberta (2007)
14. Wu, I.-C., Su, J.-Y., Chen, L.-B.: A web data extraction description language and its implementation. In: COMPSAC, pp. 293–298. IEEE Computer Society, Los Alamitos (2005)
15. Zigo, V.: A Light-weight Processing Model for Interactive Web Information Acquisition. PhD thesis, Database and Artificial Intelligence Group, Institute of Information Systems, Vienna University of Technology, Austria (2007)

A Web-Based Learning Information System - AHKME

Hugo Rego, Tiago Moreira, and Francisco José Garcia

University Of Salamanca, Plaza de la Merced s/n
37008 Salamanca, Spain
hugo.rego06@gmail.com, thm@mail.pt, fgarcia@usal.es

Abstract. Our system's main aim is to provide a system with adaptive and knowledge management abilities for students and teachers using the IMS specifications to represent information through metadata, granting semantics and meaning to all contents in the system. The system's tools along with metadata are used to satisfy requirements like reusability, interoperability and multipurpose. The system provides tools to define learning methods with adaptive characteristics, and tools to create courses allowing users with different roles, promoting several types of collaborative and group learning. It includes tools to retrieve, import and evaluate learning objects based on metadata, allowing students to use quality educational contents fitting their characteristics, and teachers may use quality educational contents to structure their courses. In this paper we will present the metadata management and quality evaluation components of the system since they play an important role in order to get the best results in the teaching/learning process.

Keywords: Web-Based Learning Information Systems, e-Learning, IMS Specifications, Learning Object, Knowledge Management, Metadata.

1 Introduction

In learning environments, information has to be perceived and processed into knowledge but one problem that emerged was its representation. Thus, standardization was indispensable to provide knowledge semantic representation through ontologies where concepts are clearly and unambiguously identified, providing a set of semantic relations allowing meaning representation by linking concepts together, the characterization of learning environments and structuring of pedagogical contents [16][6][22].

Here we present Adaptive Hypermedia Knowledge Management E-learning (AKHME) System that supports both knowledge representation and knowledge management based on metadata described by the IMS specifications in order to reach our objectives of multipurpose, independence of the learning domain, reusability and interoperability of resources and courses. In this system teachers have can create didactic materials and to evaluate, import and retrieve quality educational resources, and students can acquire knowledge through quality learning objects (LO), as well as through the most appropriate learning technique based on their characteristics, the available learning activities, the instructional design, their learning style and LO characteristics.

B. Benatallah et al. (Eds.): WISE 2007, LNCS 4831, pp. 623–632, 2007.

The goals of AHKME and main contributions are: LO management and quality evaluation, where we tried to introduce some intelligence through the usage of intelligent agents; Usage of the IMS specifications to standardize all the resources of the platform; And the interaction between all subsystems through feedback allowing the platform to adapt to students and teachers characteristics and to new contexts.

In this paper we will start to present an analysis of e-learning current approaches and a comparative analysis of standards and specifications in order to find the one to develop our system. Then we will give an overview and context the system and we will focus on tools that provide the management and quality evaluation of LOs through metadata which is the main goal of this article. Finally we'll present how it can be integrated with other systems, take conclusions and future work.

2 Current Approaches

Nowadays, there are several solutions to support e-learning, where most of them are content-centred neglecting some important educational issues. Before we started to develop our platform we have done an analysis, regarding the support of important features, on reference commercial and freeware/open-source current approaches to e-learning platforms/systems, like Blackboard, WebCT, IntraLearn, Angel, Atutor, Moodle, Sakai and DotLRN, shown in Table 1, in order to identify strong points and weaknesses, so we could try to use them in the development of our system [8][11][4].

Table 1. Analysis of e-learning systems

Tools/Features	Platforms							
	Comercial				Open Source			
	BB	WCT	IL	AG	AT	MD	SK	.LRN
Technical Aspects								
Interoperability/integration	✓	✓	✓	✓	✓	✓	✓	✓
Standards and specs. compliance	(1) (2) (3)	(6) (1)	(1) (2) (3) (4) (5)	(1) (6)	(1) (2)	(1)	(6)	(6)
Extensibility	x	x	x	x	✓	✓	✓	✓
Adaptation and Personalization								
Interface custom. and personalization	✓	✓	✓	✓	x	✓	✓	✓
Choose interface language	✓	✓	✓	✓	✓	✓	x	✓
Students previous knowledge	x	x	x	x	x	x	x	x
Courses and resources adaptability	x	x	x	x	x	x	x	x
Administrative								
Student manage. / monitor. tools	✓	✓	✓	✓	✓	✓	✓	✓
Database access mechanisms	x	x	✓	✓	✓	✓	✓	✓
Produce reports	✓	x	✓	✓	✓	✓	✓	✓
Admin. workflows quality & functio.	✓	✓	✓	✓	✓	✓	✓	✓
Tracking users	✓	✓	✓	✓	✓	✓	x	x
Resources Management								
Content authoring and editing	✓	✓	✓	✓	✓	✓	✓	✓
LOs and other types of content mng.	x	✓	x	x	x	x	x	x
Templates to aid on content creation	x	✓	✓	✓	✓	✓	✓	✓
LO search and indexation	x	x	x	x	✓	x	x	x

Table 1. (*continued*)

File upload/download mechanisms	✓	✓	✓	✓	✓	✓	✓	✓
Evaluation of quality of resources	x	x	x	x	x	x	x	x
Learning objects sharing/reuse	x	x	x	x	✓	x	x	x
Communication								
Forum	✓	✓	✓	✓	✓	✓	✓	✓
Chat	✓	✓	✓	✓	✓	✓	✓	x
Whiteboard	✓	✓	x	✓	✓	x	x	x
Email	✓	✓	✓	✓	✓	✓	✓	✓
Audio and video streaming	x	x	x	✓	x	x	x	x
Evaluation								
Self assessments	✓	✓	✓	✓	✓	✓	✓	✓
Tests	✓	✓	✓	✓	✓	✓	✓	✓
Inquiries	✓	✓	✓	x	x	✓	x	x
Costs	H	H	H	H	N	N	N	N
Documentation	✓	✓	✓	✓	✓	✓	✓	✓

SCORM-(1);IMS-(2);AICC-(3);LRN-(4);Section 508-(5);Some IMS Specifications-(6);High–H;None–N; BB – BlackBoard; WCT-WebCT; IL-IntraLearn; AG-Angel; AT-ATutor; MD-Moodle; SK – Sakai;

Analysing Table 1 we found that the majority of the platforms have good administrative and communication tools, the compliance with standards like SCORM, AICC and some IMS specifications, they have high implementation level (education, enterprise, medicine, etc) and good documentation. Although we noticed some problems regarding LO management, sharing, reusability and quality evaluation, adaptation of resources to the students' characteristics among others since they don't include any kind of adaptation engine or feature. The commercial platforms have more difficulty integrating with other systems, stating they support a standard/specification but they could not import files that came from other systems and the costs.

All these weaknesses are traduced in problems in terms of interoperability, reusability and quality of learning resources, learning domain independence and extensibility of the platforms, what meets some of our goals already presented before.

So, in order to solve these issues and from the analysis between commercial and open-source/freeware systems, we have decided to develop a modular open source system focused on issues like adaptation, LO metadata management and evaluation.

3   Standards and Specifications Comparative Analysis

One of the biggest difficulties of e-learning systems/platforms is in structuring content and information using nowadays pedagogical models so they can reach a wider range of educational systems and obtain a greater teaching quality, that is why there has been the development of several standards and specifications, like *Sharable Content Object Reference Model* (SCORM) [1], a project from *Advanced Distributed Learning* (ADL) and the IMS specifications developed by the IMS consortium [13], to make everything cross systems providing common knowledge, helping to achieve more stable systems, reducing development and maintenance time, allowing backward compatilibity and validation, increasing search engine success, among many other advantages [20].

Having detected the main problems of current e-learning approaches, we've started to analyse several aspects of several standards and specifications to choose the one(s) that would best fit our needs, like presented in Table 2.

Table 2. Standards and specifications comparative analysis

Features		IMS	AICC	SCORM	Dublin Core
Metadata		✓		✓	✓
Learner Profile		✓			
Content Packaging		✓	✓	✓	
Q&T Interoperability		✓			
DR Interoperability		✓			✓
Content structure		✓	✓	✓	
Content Communication			✓	✓	
Learning Design		✓			
Simple Sequencing		✓		✓	
Accessibility		✓			
Bindings	XML	✓		✓	✓
	RDF	✓			✓
Implementation handbooks		✓		✓	✓
Learner registration		✓			

We have analyzed the IMS Specifications, AICC[3], SCORM and Dublin Core [9], regarding the support of the issues shown on Table 3 and we've chosen the IMS specifications, since they allow most of the aspects we've analyzed and that we considered important to reach our goals.

4   AHKME Description

AHKME is an e-learning system that is divided in four different subsystems: Learning Object Manager and Learning Design subsystem, Knowledge Management subsystem, Adaptive subsystem and Visualization and Presentation subsystem. These subsystems were structured taking into account a line of reasoning, where first we have the LOs creation and management processes, which are followed by the course creation process through the learning design (LD). In parallel with these two processes the Knowledge Management subsystem evaluates the quality of the available LOs and courses. Then they pass through an adaptive process based on the students' characteristics to be presented to them, as we can see in Figure 1.

All subsystems use XML as standard for file storage and knowledge representation. This standard allows the interchange of contents between different applications and platforms, facilitating the publishing of contents [7].

We will now focus on the components of the system that provide the management and evaluation of LOs objects through metadata.

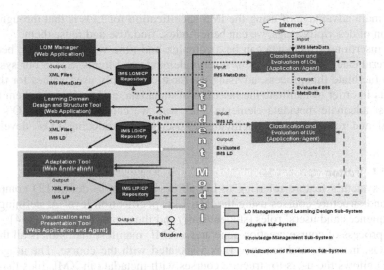

Fig. 1. AHKME's structure

4.1 LOM and Learning Design Subsystem

The Learning Object Management and Learning Design subsystem is mostly used by teachers, providing several features to develop, search, retrieve, import and analyze resources and also create courses. We will now describe the tools and features of this subsystem and how they are related with the IMS specifications.

4.1.1 LO Manager

The Learning Objects Manager tool, allows teachers to define/create/edit metadata to describe LOs. It uses the IMS *Learning Resource Metadata* (IMSLRM) specification [5] that is based on the IEEE LOM standard that allows the management/representation of knowledge through LOs [12].

This tool gives the possibility to create general metadata that can be associated to any LO and has an information packaging feature that allows the creation of packages with XML manifests, LOs and metadata and stores them in a MySQL database, enabling their management, so they can easily be transported and reused in other systems, going towards reusability and interoperability, using the IMS CP specification [19].

All files and packages pass through a validation process with schemas to check if they're in conformance with the IMS specifications, and all the communication between tools and databases is done through XML Document Object Model (DOM).

The LOs are in constant evaluation by the knowledge management (KM) subsystem that communicates with the LO Manager. After the LOs' evaluation, it may be needed to change the LO cataloguing or the way that a LO is related with other LOs, to get better LOs' associations, in order to obtain courses in a easier way taking into account the content models that were more efficient. This process is part of the adaptive process which is not explained on this article. So, this tool allows these changes to be reflected until the creation of the content package, taking into account the user's wishes, granting a higher level of flexibility.

The main advantage of using the IMS specification for LOs is that through the association of descriptive tags, we can better index, find, use and reuse them.

The insertion of metadata can be a complex and time-wasting process, because it has several categories and in them several elements and items. So, in our system we tried to facilitate this process, advising the most commonly used values for the LOs' elements in order to facilitate the insertion of metadata, and to describe them through the most adequate metadata elements. This way we can optimize the LO's search, retrieval and reusability and facilitate the user's task reducing LOs' development time.

4.1.2 LD Editor

The subsystem's feature referring to LD allows teachers to define LD components, create and structure courses using the IMS LD specification level A, defining activities, sequences and users' roles (student/staff) and the courses' metadata [14].

The process of course creation generates a XML manifest that gathers all the XML files, LOs, metadata and resource files associated with the course. The usage of the IMS LD allows the users to structure courses with metadata in XML files that can be reused in the construction of other courses making easier the portability of learning information to interact with *Learning Management Systems* (LMS).

This tool also provides the creation of packages with the courses integrating them in a data repository, to reach a more efficient management and communicates with the KM subsystem to evaluate the courses' quality. After the evaluation this tool allows the restructuring of the courses allowing the user to interact on the LD process.

4.2 Knowledge Management Subsystem

Knowledge management and e-learning are two concepts that are strictly related, as e-learning needs an adequate management of educational resources to promote quality learning, to allow students to develop in an active and efficient way.

The quality of the learning resources is becoming an aspect with great importance on e-learning environments due to the massive development of resources where features like reusability are discarded. Nowadays, there are already some criteria and aspects to consider to evaluate the quality of a learning resource. Vargo, *et. al* states that a systematic evaluation of LOs must become a valued practice if the promise of ubiquitous, high quality Web-based education is to become a reality [21].

Regarding this scenario we've decide to create a subsystem that evaluates the LOs quality through metadata to reach the best learning/teaching process.

4.2.1 LO Evaluation

To archive an LOs' optimal evaluation, it is necessary to consider quality criteria, for this reason the weighted criteria presented in Table 3 were proposed [17] where the final evaluation value is the sum of all the classifications of each category multiplied by their weight and has the following rating scale:0=not present;1=Very low; 2=Low;3=Medium;4=High;5=Very High. To use these criteria we have made a match between the IMSLRM educational category elements and the categories described on Table 3. For now we have just considered the educational category because it has most of the LOs' technical and educational aspects we found important to evaluate.

Table 3. Evaluation criteria categories and matching with the IMSLRM educational category

Eval. criteria categories	Weight	IMSLRM Ed. elements	Description
Psychopedagogical	30%	intended end user role; typical age range; difficulty	Criteria that can evaluate, for example, if the LO has the capacity to motivate the student for learning;
Didactic-curricular	30%	learning-resource type; context; typical learning time; description	Criteria to evaluate if the LO helps to archive the unit of learning objectives, etc;
Technical-aesthetic	20%	semantic density; language	Criteria to evaluate the legibility of the LO, the colors used, etc;
Functional	20%	interactivity type; inter activity level	Criteria to evaluate LOs accessibility among other aspects to guarantee that it doesn't obstruct the learning process;

With these criteria, we're developing two different tools to evaluate LOs' quality. One tool allows teachers and experts to analyze, change and evaluate LOs through a Web application and after the individual evaluation, all persons involved gather in a sort of online forum to reach to the LO final evaluation [17].The other tool is an intelligent agent that automatically evaluates LOs that acts when some kind of interaction is made on the LOs in order to readjust its quality evaluation. For example, if students have difficulties in using a LO, the quality evaluation will be recalculate in order to reflect them. Thus, the agent starts to import the LO to evaluate and others already evaluated, then applies data mining techniques (decision trees) to its educational characteristics defined in the IMSLRM specification to calculus its final evaluation.

With these two tool LOs are constantly being availed of their quality, playing an important role in the reusability of the LOs for different contexts. Meanwhile we are testing these tools in order to verify their reliability.

4.3 Visualization and Presentation Subsystem

This subsystem presents the educational contents to the students taking into account the adaptive meta-model generated for each student working as the system's front-end, defining the integration with LMSs. The objective is to give LMSs opportunity to benefit from this LD Back-Office system, as well as to give a Front-End to AHKME.

Our objective is to benefit from all the LMSs' strong points already mentioned on our analysis adding the tools we have developed by merging/integrating systems.

The integration can be done in several different ways depending on the integration tools the LMSs provide. For example if you have an open source system it can be directly integrated or if you have Blackboard it can be done through building blocks.

5 Application Scenarios

AHKME tool's application scenario could be both in educational and training contexts for instance in classes' laboratories. It is important to make a comparative analysis of AHKME tool's application versus similar tools, both a qualitative analysis like the level of feature support, and also a quantitative analysis like the level of user satisfaction in a application testing process; For instance analysing the AHKME LOM

Tool versus similar tools, the user may catalogue the LO with metadata, with and without the help of the automation metadata process, where he packages LO metadata through AHKME LOM feature and searches for a specific LO (with best quality classification in a specific application context, expressed in the metadata) with and without the use of the LOM search engine, also using evaluation tools to evaluate the LO's quality.

Regarding the qualitative level we have analyzed some key features of metadata tools confronting AHKME's LOM tool with some other similar LO metadata tools (LOM Editor [15], ADL SCORM [2], Reggie [18] and EUN [10]). To make this analysis we have defined a set of tasks mapping it to the study goals, like described on Table 4, and tested if the different tools supported them.

Table 4. Comparative analysis between AHKME LOM tool and similar tools

Task	Goal	LOM Editor	ADL SCORM	Reggie	AHKME LOM	EUN
Creation of new metadata files	LD Independence	✓	✓	✓	✓	✓
Modification of data in metadata files	LD Independence	✓			✓	
Support any metadata standard & specif.	LD Independence		✓	✓		
Modification of structure of metadata files	LD Independence				✓	
Validation in terms of data values	Interoperability		✓		✓	✓
Validation of structure of metadata	Interoperability				✓	
Support of the XML	Interoperability	✓	✓		✓	
Packaging of LOs metadata	Interoperability				✓	
Evaluation of LOs metadata	Resour. Quality				✓	
LO Search and Indexation	Reusab./ Res. Quality				✓	
Allow metadata document management	Reusability				✓	

The analysed tools can provide functionalities for meeting specific requirements like XML validation and support and creation of metadata files, lacking some important points like: Educational orientation, by not providing a list of available educational metadata; Require knowledge on XML; Lack of learning environment characterization; Don't provide resource's management.

Thus, AHKME LOM distinguishes itself from the others by introducing an abstraction level from the technical aspects of XML and is more focused on the users' needs, facilitating the LO's metadata annotation through a metadata automation process and the LO search and retrieval, for the user to reuse the LO in another scenarios. Because of AHKME's LO quality evaluation, the user may choose the best LOs that best fit his educational scenario and can introduce them in the design of a course, in its activities.

6   Conclusions

In this article we've presented how the platform AHKME uses metadata annotation for learning resource management and evaluation.

The IMS specifications, which use the combination of metadata and XML potentialities, are excellent to represent knowledge, dividing information in several meaningful chunks (LOs) providing their description through metadata and storage in XML files, therefore permitting their cataloguing, localization, indexation, reusability and interoperability, through the creation of information packages. These specifications grant the capacity to design learning units that simultaneously allow users with different roles promoting several types of both collaborative and group learning.

Through knowledge management we have a continuous evaluation of contents, granting quality to all the resources in the platform for teachers and students to use.

AHKME's main contributions are: the LO management and quality evaluation; the usage of the IMS specifications to standardize all the resources it tries to reach interoperability and compatibility of its learning components, and the interaction of all subsystems through the feedback between them allowing the platform to adapt to the students and teachers characteristics and to new contexts, using knowledge representation and management to grant success to the teaching/learning process.

Thus, it's very important to have the resources well catalogued, available and with quality to create quality courses, but quality courses don't just depend on quality resources, but also in the design of activities to reach determined learning objectives.

Being a multipurpose platform it can be applied to several kinds of matters, students, and learning strategies, in both training and educational environments being able to be fully integrated with other systems.

In terms of future work, we will include the IMS LD specification level B in the learning design tool, to include properties and generic conditions. In the adaptive subsystem we will add some functionality according to the IMS Question and Test Interoperability and Enterprise specification. In the KM subsystem we will add the feature of course quality evaluation through the development of some tools.

Acknowledgments. This work has been partly financed by Ministry of Education and Science, by the FEDER KEOPS project (TSI2005-00960) and the Junta de Castilla y León Project (SA056A07).

References

1. ADL: Sharable Content Object Reference Model (SCORM)® 2004 3rd Edition - Overview Version 1.0. Advanced Distributed Learning (2006)
2. ADL SCORM Metadata Generator (2005), http://www.adlnet.org
3. AICC, Aviation Industry CBT Committee (2005), http://www.aicc.org
4. ANGEL (2005), http://www.angellearning.com/
5. Barker, P., Campbell, L.M., Roberts, A., Smythe, C.: IMS Meta-data Best Practice Guide for IEEE 1484.12.1-2002 Standard for Learning Object Metadata - Version 1.3 Final Specification. IMS Global Learning Consortium, Inc. (2006)
6. Berners-Lee, T., Hendler, J., Lassila, O.: The Semantic Web. Scientific American 284(5), 34–43 (2001)
7. Bray, T., Paoli, J., Sperberg-MacQueen, C.M.: Extensible Markup Languaje (XML) 1.0, 3rd Edition, W3C Recommendation (2004)
8. Colace, F., De Santo, M., Vento, M.: Evaluating On-line Learning Platforms: a Case Study. In: HICSS 2003. Hawaii International Conference on System Science (2002)

9. Dublin Core Metadata Initiative (2005), http://dublincore.org
10. EUN (2005), http://www.en.eun.org/menu/resources/set-metaedit.html
11. Graf, S., List, B.: An Evaluation of Open Source E-Learning Platforms Stressing Adaptation Issues. In: ICALT 2005. The 5th IEEE International Conference on Advanced Learning Technologies (2005)
12. IEEE LTSC Working Group 12: Draft Standard for Learning Object Metadata. Institute of Electrical and Electronics Engineers, Inc. (2002)
13. IMS Specifications, IMS Global Learning Consortium, Inc. (2004), http://www.imsglobal.org/specifications.cfm
14. Koper, R., Olivier, B., Anderson, T.: IMS Learning Design Information Model - Version 1.0 Final Specification. IMS Global Learning Consortium, Inc. (2003)
15. LOM Editor, (2005), http://www.kom.e-technik.tu-darmstadt.de/ abed/lomeditor
16. Mendes, M.E.S., Sacks, L.: Dynamic Knowledge Representation for e-Learning Applications. In: Proceedings of the 2001 BISC International Workshop on Fuzzy Logic and the Internet, FLINT'2001, Memorandum No. UCB/ERL M01/28, University of California Berkeley, USA, pp. 176–181 (August 2001)
17. Morales, E., García, F.J., Moreira, T., Rego, H., Berlanga, A.: Units of Learning Quality Evaluation. In: González, J.R.H., de Mesa, J.A.G., de Miguel, R.V., Borda, R.M. (eds.) SPDECE 2004. Design (Guadalajara, Spain) CEUR Workshop Proceedings, 117th edn., pp. 1613–1673 (2004), http://ceur-ws.org/Vol-117 ISSN 1613-0073
18. Reggie Metadata Editor (2005), http://metadata.net/dstc
19. Smythe, C., Jackl, A.: IMS Content Packaging Information Model – Version 1.1.4 Final Specification. IMS Global Learning Consortium, Inc. (2004)
20. Totkov, G., Krusteva, C., Baltadzhiev, N.: About the Standardization and the Interoperability of E-Learning Resources. In: CompSysTech 2004. International Conference on Computer Systems and Technologies (2004)
21. Vargo, J., Nesbit, J.C., Belfer, K., Archambault, A.: Learning object evaluation: computer-mediated collaboration and inter-rater reliability. International Journal of Computers and Applications 25(3) (2003)
22. Wiley, D.: Connecting learning objects to instructional design theory. In: Wiley, D. (ed.) The Instructional Use of Learning Objects (May 10, 2003)

A Recommender System with Interest-Drifting

Shanle Ma[1], Xue Li[1], Yi Ding[1], and Maria E. Orlowska[1,2]

[1] School of Information Technology and Electrical Engineering,
University of Queensland, QLD 4072, Australia
[2] Polish-Japanese Institute of Information Technology, Faculty of IT,
Ul. Koszykowa 86, 02-008 Warsaw, Poland
{shanle, xueli,ding,maria}@itee.uq.edu.au

Abstract. Collaborative filtering and content-based recommendation methods are two major approaches used in recommender systems. These two methods have some drawbacks in dealing with situations such as sparse data and cold start problems. Recently, combined methods were proposed to overcome these problems. However, a highly effective recommender system may still face a new challenge on interest drift. In this case, customer interests may change over time. For example, more recent users' ratings on items may reflect more on users' current interests than those of long time ago. Unfortunately, current available combination approaches do not consider this important factor and training data sets are regarded as static and time-insensitive. In this paper, we present a novel hybrid recommender system to overcome the interest drift problem by embedding the time-sensitive functions into the recommendation process. The users' interests changing behaviours are considered with time function. Our experiments demonstrate a better performance than that of the collaborative filtering approaches considering interests drift and those of the combined approaches without considering interests drift.

1   Introduction

Recommender systems were introduced to help users to overcome information overload problem and have been widely used in e-commerce websites. Most prevalent recommender systems apply either content-based filtering or collaborative filtering approaches to predictions. Content-based filtering approach analyzes the similarity between items based on their contents and recommends the similar items based on users' previous preferences. On the other hand, collaborative filtering approach computes the similarity between users according to their historical ratings to items. Collaborative filtering systems recommend user with the items that were rated highly in the past based on a ranking approach of finding nearest neighbours [1].

Also, both content-based filtering and collaborative filtering approaches have shortcomings and perform badly in some situations. For example, for content-based approach, it needs to analyze profiles of items and then compute the similarity between items' profiles that were purchased or preferred by users as well as the profiles of items that are not rated or purchased. But profiles are hard

B. Benatallah et al. (Eds.): WISE 2007, LNCS 4831, pp. 633–642, 2007.

to extract from some data types such as audio/video products. Another problem inherent in content-based filtering approach is about the limitation of the scope of possible interests. The interesting items that system can recommend are only those, for which users have preferred or purchased. This is regarded as a "cold-start" problem [2]. Furthermore, the recommender system will not be able to achieve a global optimal situation since the system's ability is limited within the scope of only the interesting items that have ever been seen so far. On the other hand, collaborative filtering approach was introduced to overcome the limitations existing in content-based recommender systems. However, collaborative filtering approach still has its own disadvantages, typically the sparsity problem [3]. With a large number of items and a large number of users, an individual user can only express opinions on proportionally a very small amount of items. So the user-item matrix becomes very sparse. In this situation, the prediction accuracy will decrease along with the increase of the sparsity.

Above these mentioned problems, a more serious problem is about the interest drifting: uses' preferences may change with time. How can the recommender system reflect the changes of users' interests in the recommendation? In general people's interests will always change. User may dislike the items which he or she rated highly in the past. For example, a young couple had their first baby two years ago. And they were interested in items such as prenatal education, baby-related safe feeder, recipes, foods, and so on. When their baby became two years old, this couple may be interested in something else, not things for new-born babies. In personalized recommendations, if system were not aware of changes, it may result in low prediction accuracy. Therefore we need to build the system with ability to deal with dynamics. If ratings of items are all time-stamped, one hypothesis is that the item rated more recently by a user should have a larger impact on the prediction of user's preference than an item that was rated long time ago. Recent ratings should be able to reflect user's current interests more accurately.

Because both recommendation methods have their weaknesses and strengths, combined approaches have been proposed. The first hybrid recommender system Fab [4] was developed in the middle of 1990s. The combination approaches can be classified into three categories according to their ways of combinations: Combining Separate Recommenders, Combining Features, and Unification [1].

The method of Combining Separate Recommenders is to initially implement content-based and collaborative filtering recommenders separately and then the two results are combined by various of methods. For example, the P-Tong [5] system implements collaborative filtering and content-based filtering separately and then a linear combination method is used.

Feature combination method considers adding one feature of filtering methods to another. There are two main approaches: one is to put content-based features into collaborative filtering recommender system and another is to put collaborative filtering features into content-based recommender system. In [6], they used content-based filtering to predict the scores which unrated by the users, then produce the full rated pseudo matrix for the classic collaborative filtering.

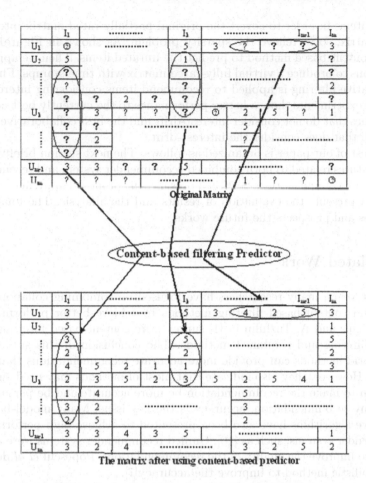

Fig. 1. Integrated Approach for Recommender systems

Unification method uses content-based and collaborative filtering in a single recommendation models. For example, probabilistic method [7] have been presented to unified content-based and collaborative filtering.

Unfortunately user's interest drift problem has not been considered in existing combined approaches.

In [8,9], authors consider time influence in classic collaborative filtering. The drawbacks remaining in the collaborative filtering are still there or even worse. For example, many literatures indicate that classic collaborative filtering suffers from sparsity problem. Since the recent ratings are a small part of the whole collected ratings, applying different weights on ratings might aggravate sparsity. It makes prediction harder and the accuracy decreases in some cases.

In this paper, we propose a novel combination approach for recommendations with interest-drifting. We consider recent ratings may reflect user's present interests more accurately. We propose using content-based method to predict unrated items in order to covert the original partially-rated matrix into a fully-rated matrix, so to relieve the sparsity problem. As shown in Figure1, firstly, we use content-based method to predict the unrated items. Then we apply these predictions to produce a virtual fully-rated matrix with time-stamps. Finally the collaborative filtering is applied to recommend items considering interest drifting. Our experiments show a better performance to the currently best combined approaches that do not consider interest-drift and the best collaborative filtering approach that does consider the interest-drift.

The rest of our paper is organized as follows. The next section briefly reviews the literatures related to both combined recommender system and recency-based concept in recommendation. Section 3 presents the proposed approach in details. Section 4 presents the evaluation of results and the analysis. The final section concludes and prospects the future works.

2   Related Work

In recent years, many researchers have focused on combining collaborative filtering and content-based filtering methods to achieve better performance. G. Adomavicius and A. Tuzhilin in [1] survey pure content-based filtering, collaborative filtering and combined methods. The conclusion of this survey shows that hybrid methods can provide more accurate recommendations than a pure method. However, they also indicate that most of methods may need significant extension to make the recommendation be more accurate and be personalized.

For any recommendation, accuracy is the key issue. Many model-based collaborative algorithms have also been presented to achieve high performance in recommender systems, such as [10]. Unified recommender systems were also considered to improve the quality of the recommendations. Popescul *et al* developed a probabilistic method to improve the accuracy[7].

However, to our knowledge, few papers have focused on the temporal feature of ratings in collaborative filtering process. In [11], the authors proposed that a movie's production year, which reflects the situational environment in which the movie is filmed, might significantly affect target users' future preferences. Kazunari Sugiyama et al explored a type of time-based collaborative filtering with detailed analysis of user's browsing history in one day [12]. However, the recency of ratings has not been studied so far. In fact, with the fast growth of e-commerce, collaborative filtering applications have been widely used for a long time. For example, in [13], the authors described a TiVo television-show collaborative recommendation system which had started four years ago. It has currently accumulated approximately 100 million user ratings, some of which are very old. Since the value of these very old ratings is questionable, we should seek to develop an algorithm that will decay the influence of these.

For sparsity problem, dimensionality reduction techniques are widely used [14]. In [3], the associative retrieval techniques are used to alleviate the sparsity problem. In hybrid approaches, M. Pazzani in [14] considered computing the similarity between two users based on demographic filtering. In [6], content-based filtering is used to produce predictions in order to convert sparse user-item matrix into a fully-rated matrix, so to solve the sparsity problem. Content-based filtering approaches would have no cold-start problem since they just compute the similarity between user profiles and item profiles. In [2], content-based and collaborative filtering methods are both used under a single probabilistic framework. The proposed algorithm was checked against navie Bayes classifiers on the cold-start problem.

The key issues still remain on how to effectively and efficiently deal with the user interest-drifting. The most current recommender systems do not provide satisfactory answers for this problem.

3   Proposed Approach

The main idea of our proposed approach is to improve the accuracy of the recommendations by integrating content-based and collaborative filtering with interest-drifting. We use time stamps to trace the users' interest changing. As we know, user-item rating matrix is normally very sparse and only a small number of items have been rated by users. Initially, content-based filtering will be implemented to predict the unrated items in the item-user rating matrix. Then the predicted ratings by the content-based filtering will be put into the original user-item matrix to create a virtual fully-rated matrix. The virtual fully-rated matrix includes the real rates by users and the rates predicted by content-based filtering approach. In the real item-user rating matrix, each rate is associated with a time stamp. Thus we can trace uses' interest changing over time. However, since the rates predicted by the content-based filtering method have no time stamps, we use average weights for their time stamps. When an active user enters the system, the prediction will be produced based on classic collaborative filtering using the virtual fully-rated matrix with the time-sensitive function.

We first present the content-based filtering predictor. Then we describe the pure collaborative with interest-drifting. Finally we propose our novel integrated approach.

3.1   Content-Based Predictor

We treat user ratings range 0-5 as one of six class labels. We used SVM method to learn a user's profile from labelled documents. Then use user's profile to predict the unrated items. Then we convert sparsity matrix to a virtual fully-rated matrix [6,15].

3.2   Interest-Drifting Collaborative Filtering

Collaborative filtering problem can be defined as follows: Given a database D as a tuple$\langle U_i, I_j, R_{ij}, T_{ij} \rangle$, where U_i identifies the i-th user of the system, I_j identifies the j-th items of the system, R_{ij} represents the i-th user's rate on the j-th item and T_{ij} represents producing time of the rate, find a list of k recommended items for each user U.

The user-based collaborative filtering can be summarized in the following two steps:

step1: compute the similarity between users and find a most mind-liked user to the active user.

step2: predict preference.

In the step1, the classic user-based collaborative filtering similarity function can be defined as follow. There are two main functions: cosine similarity and person correlation coefficient.

Cosine similarity: each user is treated as a vector in m dimensional space and the similarity between two users is measured by computing the cosine of angle between vectors, defines as follow:

$$Sim(U_a, U_b) = cos(\boldsymbol{a}, \boldsymbol{b}) = \frac{\boldsymbol{a} \cdot \boldsymbol{b}}{|\boldsymbol{a}||\boldsymbol{b}|} = \frac{\sum_{i \in I_{ab}} R_{ia} R_{ib}}{\sqrt{\sum_{i \in I_{ab}} R_{ia}^2} \sqrt{\sum_{i \in I_{ab}} R_{ib}^2}}, \tag{1}$$

Where U_a denotes the a-th user, U_b denotes the b-th user. R_{ij} denotes the i-th user rated the j-th item. I_{ab} denotes the set of all items co-rated by both user a and user b.

Pearson correlation coefficient: the similarity between two users is measured by computing the Pearson correlation. It defines as follow:

$$sim(U_a, U_b) = \frac{\sum_{i \in I_{ab}} (R_{ia} - \overline{R_a})(R_{ib} - \overline{R_b})}{\sqrt{\sum_{i \in I_{ab}} (R_{ia} - \overline{R_a})^2} \sqrt{\sum_{i \in I_{ab}} (R_{ib} - \overline{R_b})^2}} \tag{2}$$

Where $\overline{R_a}$ denotes the average rating of the a-th user rating itmes.

In the step2, the prediction of the preference for a given item can be computed by using the sum of the ratings of the user to items weighted by the similarity between different users as:

$$(a) P_{a,i} = \frac{\sum_{a \in \hat{U}} R_{a,i} \cdot sim(U_a, U_b)}{\sum_{a \in \hat{U}} sim(U_a, U_b)} \tag{3}$$

or

$$(b) P_{a,i} = \overline{R_a} + \frac{\sum_{a \in \hat{U}} (R_{a,i} - \overline{R_a}) \cdot sim(U_a, U_b)}{\sum_{a \in \hat{U}} |sim(U_a, U_b)|} \tag{4}$$

where U_a identifies the a-th user,
U_b identifies the nearest neighbors of the b-th user,
$P_{a,i}$ represents the a-th user's opinion on the i-th item
$\hat{U}$ denotes the set of N users that the most similar to user u who rated item i
$\overline{r_a}$ is average rate of user a rated items

3.3   Time Function

As we addressed, the users' interest changing is related to time. So we define the time function according to [8] as following:

$$f(t) = e^{-\lambda \cdot t} \tag{5}$$

Here, firstly, we define a half-life parameter T_0 as:

$$f(T_0) = (1/2)f(0) \tag{6}$$

That is to say, the weight reduces by 1=2 in T_0 days. Then we define the decay rate λ as:

$$\lambda = \frac{1}{T_0} \tag{7}$$

We can observe that the value of the time function is in the range (0, 1), and it reduces with time. The more recent the data, the higher the value of the time function is.

3.4 Recommendations with Interest-Drifting

In our proposed integrated approach for recommendations with interest-drifting, we first complement content-based predictor to create a virtual fully-rated matrix. Then we compute the similarity between two users by pearson correlation coefficient and use Equation (3) to produce the prediction with time weighting. We used Harmonic Mean Weighting [6]. It can be defined as follows.

$$hw_{a,i} = hm_{a,i} + sg_{a,i} \tag{8}$$

Here, $hm_{a,i}$ can be defined as following:

$$hm_{a,i} = \frac{2m_a \cdot m_i}{m_a + m_i} \tag{9}$$

$$m_j = \begin{cases} \frac{n_i}{50} : if\, n_i < 50 \\ 1 : otherwise \end{cases} \tag{10}$$

Where n_i is the number of items rated by user a.

Meanwhile $sg_{a,i}$ is the significance weighting factor and can be defined as follows:

$$sg_{a,i} = \begin{cases} \frac{n}{50} : if\, n \leq 50 \\ 1 : otherwise \end{cases} \tag{11}$$

Here n is the number of items rated by users.

The prediction for the active user with the interest-drifting is produced as follows:

$$P_{a,i} = \frac{\sum_{a \in \hat{U}} hw_{a,i} \cdot F_{a,i} \cdot sim(I_a, I_b) \cdot f(t_{a,i})}{\sum_{a \in \hat{U}} hw_{a,i} \cdot sim(I_a, I_b) \cdot f(t_{a,i})} \tag{12}$$

Here $F_{a,i}$ is the rate in virtual full rated matrix.

4   Experiments and Results

In order to demonstrate the effectiveness and the efficiency of our proposed approach, a few experiments are conducted. We compare our results with the approach of combined recommendation that did not consider interest-drifting [6,15] and the approach that uses collaborative filtering with interest-drifting [8]. We re-implemented these methods and compare these methods with our proposed method. The results are evaluated based on MAE and ROC-4 metrics.

4.1   Dataset

EachMovie dataset [1] is used to evaluate our proposed algorithms. The Each-Movie dataset recorded 72,916 users who rated 1,628 movies during 18 months period since 1996. Each vote is accompanied by a time stamp. We used a web crawler to have collected content of each movie from Internet Movie Database(IMDb)[2] followed the hyperlinks provided by EachMovie dataset. The content of a movie includes movie title, director, genre, plot summary, cast, award and user comment.

4.2   Metric

The measures for evaluating the performance of recommender systems can be partitioned into two main classes. One is statistical accuracy metrics and the other is the decision support accuracy metrics.

Statistical accuracy metrics evaluate the accuracy of recommendation by comparing predictions with users'ratings. Mean Absolute Error(MAE) is a widely applied metric for this category of metrics. It can be defined as the average absolute difference of recommendations from the real users' ratings. In Our experiments we computed the MAE on the test set for each user, and then averaged over the set of test of users.

Decision support accuracy metrics evaluate how effective recommendations helping uses select high-quality products. In our experiments, we use Receiver Operating Characteristic(ROC) sensitivity to scale decision support accuracy.

4.3   Experiment Setup

We used a subset of the ratings data from the EachMovie dataset. There were 7,650 users randomly selected and 1,438 movies rated in this subset. Ten percent of users were used in the test - all of them have rated at least forty movies. From each user in the test set, ratings for 25 % of items were withheld evaluation. We used different algorithms to predict the withheld items. The quality of the various prediction algorithms were measured by comparing the predicted ratings with the withheld ratings.

4.4   Experiment Results

The results of our experiments are show in Table1. It can be seen that from the Table 1, the performance of our approach for recommender systems with interest-drifting is better than other approaches on both MAE and ROC-4 metrics. For MAE metric, our approach performs 4.8% better than collaborative filtering with interest-drifting and 1.7% better than hybrid approach without interest-drifting.

For the ROC-4 metrics, our integrated approach with interest-drifting performs 5.5% better than collaborative filtering with interest-drifting and 1.4%

---

[1] www.research.compaq.com/SRC/eachmovie
[2] www.imdb.com

Table 1. Experiment Results

algorithms	MAE	ROC-4
User-based CF with interest-drifting	0.9714	0.668
Hybrid approach without interest-drifting	0.939	0.684
integrated approach with interest-drifting	0.923	0.698

better than hybrid approach without interest-drifting. These statistics shows that our integrated approach with interest-drifting can constantly perform better and provide users with higher quality recommendations.

5   Conclusions

In this paper, we present an integrated recommendation approach with interest-drifting. Although both content-based and collaborative filtering approaches have their own advantages, they still perform badly in many situations. Among those identified performance problems, sparsity, cold start, and scalability are the most widely addressed problems in literature. We use content-based predictor to relieve sparsity and cold start problems. Meanwhile, we consider users' interests changes over time, and the problem is regarded as interest-drifting. The scores rated more recently should reflect user's current interests. In our proposed algorithms, we give the recent ratings higher than that rated long time ago. Empirical evaluations of our approach show a better performance than that of a collaborative filtering approach with the interest-drifting and the combined approach without considering the interest-drifting.

Our further work will consider the periodical times and reoccurring interest changes that may appear in the recommender systems. The time-sensitive recommendations regarding both the individual user's interest-drifting as well as the item value-drifting will also be considered together.

Acknowledgement

This work is partially founded by the Australian ARC Large Grant DP0558879.

References

1. Adomavicius, M.G., Tuzhilin, M.A.: Toward the next generation of recommender systems: A survey of the state-of-the-art and possible extensions. IEEE Transactions on Knowledge and Data Engineering 17, 734–749 (2005)
2. Schein, A.I., Popescul, A., Ungar, L.H., Pennock, D.M.: Methods and metrics for cold-start recommendations. In: SIGIR 2002. Proceedings of the 25th annual international ACM SIGIR conference on Research and development in information retrieval, pp. 253–260. ACM Press, New York, NY, USA (2002)

3. Huang, Z., Chen, H., Zeng, D.: Applying associative retrieval techniques to alleviate the sparsity problem in collaborative filtering. ACM Trans. Inf. Syst. 22, 116–142 (2004)
4. Balabanovi, M., Shoham, Y.: Fab: content-based, collaborative recommendation. Commun. ACM 40, 66–72 (1997)
5. Claypool, M., Gokhale, A., Miranda, T., Murnikov, P., Netes, D., Sartin, M.: Combining content-based and collaborative filters in an online newspaper (1999)
6. Melville, P., Mooney, R.J., Nagarajan, R.: Content-boosted collaborative filtering for improved recommendations. In: Eighteenth national conference on Artificial intelligence, pp. 187–192. American Association for Artificial Intelligence, Menlo Park, CA, USA (2002)
7. Popescul, A., Ungar, L., Pennock, D., Lawrence, S.: Probabilistic models for unified collaborative and content-based recommendation in sparse-data environments. In: 17th Conference on Uncertainty in Artificial Intelligence, pp. 437–444. Seattle, Washington (2001)
8. Ding, Y., Li, X.: Time weight collaborative filtering. In: CIKM 2005. Proceedings of the 14th ACM international conference on Information and knowledge management, pp. 485–492. ACM Press, New York, NY, USA (2005)
9. Li, X., Barajas, J.M., Ding, Y.: Collaborative filtering on streaming data with interest-drifting. International Journal of Intelligent Data Analysis 11(1), 75–87 (2007)
10. Hofmann, T.: Collaborative filtering via gaussian probabilistic latent semantic analysis. In: SIGIR 2003. Proceedings of the 26th annual international ACM SIGIR conference on Research and development in informaion retrieval, pp. 259–266. ACM Press, New York, NY, USA (2003)
11. Tang, T.Y., Winoto, P., Chan, K.C.C.: On the temporal analysis for improved hybrid recommendations. In: WI 2003. Proceedings of the IEEE/WIC International Conference on Web Intelligence, p. 214. IEEE Computer Society, Washington, DC, USA (2003)
12. Sugiyama, K., Hatano, K., Yoshikawa, M.: Adaptive web search based on user profile constructed without any effort from users. In: WWW 2004. Proceedings of the 13th international conference on World Wide Web, pp. 675–684. ACM Press, New York, NY, USA (2004)
13. Ali, K., Stam, W.V.: TiVo: Making Show Recommendations Using a Distributed Collaborative Filtering Architecture. In: KDD 2004. Proceedings of The Tenth ACM SIGKDD Conference, Seattle, WA (2004)
14. Sarwar, B., Karypis, G., Konstan, J., Riedl, J.: Application of dimensionality reduction in recommender systems–a case study (2000)
15. Rojsattarat, E., Soonthornphisaj, N.: Hybrid Recommendation: Combining Content-Based Prediction and Collaborative Filtering. In: Liu, J., Cheung, Y.-m., Yin, H. (eds.) IDEAL 2003. LNCS, vol. 2690, pp. 337–344. Springer, Heidelberg (2003)

Improving Revisitation Browsers Capability by Using a Dynamic Bookmarks Personal Toolbar

José A. Gámez, Juan L. Mateo, and José M. Puerta

Computing Systems Department
Intelligent Systems and Data Mining Group – $i^3\mathcal{A}$
University of Castilla-La Mancha
Albacete, 02071, Spain

Abstract. In this paper we present a new approach to add intelligence to Internet browsers user interface. Our contribution is based on improving browsers revisitation capabilities by learning a model from user's navigation behaviour, that later is used to predict a set of bookmarks likely to be used next. These set of bookmarks must be a list of moderate size (≤ 10) because our goal is to show them in the browser bookmarks personal toolbar. We think that dealing with this part of the user interface is beneficial for revisitation because it is always visible and on the contrary to history or bookmarks list (tree) the user can access the desired web page by using a single mouse click. In this work we focus on performing the comparison of several (computationally) simple classifiers in order to identify a good candidate to be used as user navigation model. From the experiments carried out we identify that a combination of Naive Bayes with OneR could be a good choice.

1 Introduction

The World Wide Web (web) hypertext system is a large, distributed repository of information and resources that in recent years has become a wonderful tool for working, studying and just having fun. However, web's accelerated growing has turned it into a vast amount of information poorly structured, from which retrieving relevant information is often a quite laborious and complex task even for experienced users. Because of this, once a web page has been found and judged as interesting by a given user, it is easier for her to memorize its address in order to revisit it, than having to look for it again in the future (according to [1] 60% of the pages a person sees are revisits). Current browsers incorporate standard tools for web pages revisitation. Apart from *back/forward* buttons we can distinguish:

- *History.* This is a list automatically constructed by the browser that stores the pages seen by the user in the same order they were visited. The main problem with *history* is that the user must go to the corresponding menu, open the (usually long) history list and visually scan it in order to look for the desired web page. According to [1] *history* represents less than 1% of all navigational acts (*back* button represents 30%).

B. Benatallah et al. (Eds.): WISE 2007, LNCS 4831, pp. 643–652, 2007.

– *Bookmarks.* On the contrary of history, bookmarks represents a *personal web information space* in which the user adds web pages by marking them explicitly at the time they are being visited. In addition, bookmarks can be organised by using tree-shaped structures in order to access them more efficiently. However its use only represents around 3% of all navigational acts [1]. Takano and Winograd [2] identify user management overhead as the main reason of bookmarks infrequent usage.

In the literature we can find different works that try to improve browser's revisitation capabilities. Thus, in [3] the authors carry out a static analysis about the bookmarks file of 50 users in order to discover common ways of ordering/structuring them, with the aim of providing some recommendations about how to improve the organisation, visualisation and representation of bookmarks. A different approach is followed in [2] and [4] where bookmarks are collected automatically by recording every visited web page, and the emphasis is put in how the bookmarks are organised: recency list [4] and a three layer structure [2] obtained by learning (clustering plus page-rank) user browsing preferences. Notice that our approach is different of the one of developing user interface agents to assist user browsing or search, where usually web page content and links are the basis to guide users during their navigation [5,6,7], and web pages are usually modified by inserting new links or annotating/reordering the existing ones.

Our approach to improve browser revisitation capabilities lies on the philosophy of [2] and [4] in the sense that we propose to collect web pages automatically and to learn from the user browsing behaviour which pages should be shown as bookmarks. However, our approach largely differs from these works because we do not modify browser bookmarks file and/or history but only the bookmarks personal toolbar content. Thus, as only a few bookmarks can be shown in that toolbar, instead of working about how to structure the bookmarks, our approach focuses in forecasting (by using classifiers) what are the more probable pages to be visited next.

In the next sections we detail our proposal and all its components. Section 2 describes the general idea of our proposal. In Section 3 we identify the data we need for our study and how it is collected in order to perform our experiments. Section 4 describes the data mining algorithms used to create the users navigation models encoded as classifiers. In Section 5 we describe the experiments carried out in order to test our approach. Finally we present our conclusions and future research lines in Section 6.

2   Data Mining Based Dynamic Personal Toolbar

As a part of the bookmarks structure some browsers as Mozilla Firefox [8] and Konqueror [9] contain a bookmarks personal toolbar (BPTB) (figure 1). The main advantage of a BPTB is that because all the bookmarks it contains are visible at any moment, there is no need for searching and so only one mouse click is required from the user to get the desired web page. On the other hand, its main disadvantages are: (1) they are manually collected, so the user must interrupt

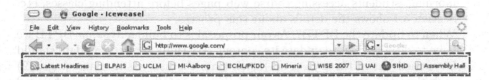

Fig. 1. Top fraction of Mozilla Firefox browser including (dashed box) the bookmarks personal toolbar

navigation to add a new one; (2) if we want to maintain all the bookmarks contained in the BPTB visible, then only a few bookmarks can be stored in it, typically from 5 to 10. Because of this, users tend to store in the BPTB a few web pages with a high revisitation degree.

In our opinion the advantages of BPTB make it a very attractive tool to improve browsers revisitation capability, and what we propose in this paper is how to overcome its disadvantages. Thus, we propose to fight them by automatically collecting the web pages to be included in the BPTB, and transforming the BPTB in a dynamic store whose content is automatically modified according to browsing sequence. To do this, our goal is to develop a software add-on for web browsers able to learn automatically from the web pages every user used to see and when. So the information we use for this task is made by the sequence of pages visited and the time in which they were visited. In previous work, authors only considered what web pages are visited, but we propose to enrich this information by including temporal information because it could be very useful in order to discover navigation patters. This add-on should provide a new toolbar with a list of a few links to web pages, besides, this list must be updated periodically, that is, the output is a dynamic bookmarks list.

The following are the main points to be considered for the development of our proposal:

- How web pages are automatically collected? and, what web pages should be considered as candidate pages to be included in the BPTB?. Section 3.1 presents our answers to these questions.
- How to build a model to encode the user navigation behaviour?. We propose to use data mining techniques in order to analyse user's navigation data and to discover a predictive model able of answering queries about the user navigation behaviour. Concretely we propose to learn a classifier being next page to be visit the class variable. Section 4 describes the proposed classifiers.
- What information should be used in order to build the user navigation model?. Because this information is collected at the same time the user is navigating we think that simple (easy to obtain and small size) data should be collected. Thus, we discard to use web page content, instead we collect the previous visited page and temporal data (hour, day of the week, etc.). We think that temporal information can be quite useful to predict user navigation behaviour, because users tend to associate some navigation acts with time and date, e.g., accessing to news papers during the first hours in the

morning, students accessing course web pages before attending classes, etc. Section 3.1 shows the details about this process.

- When to update the BPTB?. As our proposal is to build a dynamic BPTB we have to decide when to update its content. We think that this update should be carried out in three cases: (1) each time the user visits a new page; (2) each time the user model is modified (incrementally or from scratch); and (3) at least one time per hour because of using temporal information as predictive attributes.
- How to validate the proposal?. To validate our approach we have carried out an experimental study over a set of users (Section 5).

Then our intention is to build a model able to encode users navigation behaviour, and to do this we think we can base that model on temporal references and last visited page as predictive information. The question is to use as less information as possible in order to control model complexity but with the maximum descriptive power. The problem of researching in this topic is that there is no information or standard databases about user interaction with Internet.

At the same time, it is reasonable as well that the processes needed to get and use the model should not be too complex and time demanding, because in that way we will cause the opposite effect we want.

3 Collecting Data

In this section we address the problem of collecting data for our proposal. As aforementioned when collecting data we must record the address of the web page being visited (i.e. the *candidate* bookmark) and also its time-stamp, i.e. the information to be used later in order to construct the user navigation behaviour model. Perhaps the first choice to collect data could be the from browser history list, but the data it contains is not enough for our goal because only contains a list of web pages and its last access date.

Because of this and to the fact that we need to gather individual data for each user in order to learn a personalised model for it, here we have chosen an approach that is transparent for users and allows us to obtain the information required to carry out the study. Our solution is to attach to the browser a small program (an *extension* using Mozilla notation) which automatically saves the current information we need about user navigation to a local file.

3.1 Extension

As development platform we have chosen Mozilla Firefox because it is an open source web browser, so we can get access to all its facilities easily. Thus, we have developed a Mozilla Firefox extension that records user interaction in background. Although our code is quite simple and we have not needed to use the whole framework, we think it is interesting to begin our development with this type of software component because in the future we plan to use it as the starting point to develop the whole application.

The extension listen all events related with the history object in the browser, and when a new entry is going to be added or modified then it saves the current date, the previous page visited and the new one. In addition we also save the name of the server for each URL, because we will use this address as bookmark instead of the concrete web page address in some cases. The reason to do this is twofold:

- Semantic: by bookmarks we understand pages likely to be revisited, so we aim to avoid considering as bookmarks web pages rarely visited or that even disappear after a short period of time (e.g. newspapers concrete news).
- Modeling: by considering web pages that rarely or never are revisited we only can expect to learn noisy models about the user behaviour, i.e. models without any predictive capability.

Thus, under this consideration, we only include in the BPTB those (concrete) web pages likely to be revisited or those servers (portals) whose (possibly dynamic) content is also usually revisited.

4 Modeling the User Navigation Behaviour

In this section we deal with the problem of building a model of user navigation behaviour. As noted earlier we propose to use data mining techniques to build such a model, and concretely we put our emphasis in learning a *predictive* model [10] because our main goal here is to *predict* those web pages (or servers) likely to be revisited next, given some user navigation acts and background knowledge provided by the model. Concretely, as we aim to predict as output a (or some) web page(s) from a finite (though large) number of web pages (or servers) previously identified as bookmarks, we set the problem as a *classification* one, and so our goal is to learn as user model a *classifier*.

A *classifier* [11] is a function:

$$f : A_1 \times A_2 \times \cdots \times A_n \longrightarrow C \qquad (1)$$

where $A_1, \ldots, A_n$ are predictive attributes and C is the class variable. In this work, as a first study, our goal is to make a comparison over several classifiers based on different approaches. Nonetheless, as the ultimate goal of this comparison is to determine a proper classifier to carry out the target task, and so it must be integrated in the browser add on, then we only consider low memory and CPU-time demanding classifiers, both during learning and inference.

The classifiers tested here are: Naive Bayes (NB) [11], and also we consider a simpler version which only uses temporal information, that is, it ignores LAST_SERVER attribute, we call it NB_woLS; IB1 [12]; One Rule (OneR) [13]; C4.5 [14]; and a combination between NB and OneR (NB_OR), we propose to obtain the k best class labels by using a mixture of OneR and NB_woLS. Thus, we give as output the set of labels obtained by combining the (single) output of OneR with the first $k - 1$ pages returned by NB_woLS.

The dataset is obtained by our extension as described in Section 5.1, but by the moment is enough to know that the text file is transformed in a bidimensional matrix where each row corresponds to one of the entries registered by our extension and where columns represents the following variables extracted from such entry:

- CLASS. This is a nominal variable taken as values the web pages and servers previously identified as candidate bookmarks.
- LAST_SERVER. This is also a nominal variable that identifies the server from which we jump to the current web page contained in CLASS.
  DAY. This is an integer value from the date field and taking values in {1..31}.
- WEEKDAY. This is a nominal value also obtained from date and taking values in {Mon, Tue, Wen, Thu, Fri, Sat, Sun}.
- HOUR. This is a real value also obtained from date, where minutes have been transformed into a decimal range.

5   Model Evaluation

In this section we evaluate our proposal. To do this we first present the datasets used in the study, then the way in which the learnt classifiers are validated and, finally, we analyse the results.

5.1   Datasets

We have collected data from six different users by installing our extension in their Mozilla-Firefox browsers. The six users belong to our department and the information gathered comes from the computers they use in their offices. We have collected data in a two months time window, concretely November and December of 2006 for all users.

Table 1 shows some data about the datasets collected for the six users. The first two columns refer to the number of different web pages visited by each user and the number of instances/entries reported by our extension, that is, the number of visits. The third column is computed from them and show us the percentage of visits that actually are revisitations. As we can see there are different ranges among the users both in number of visited pages and navigational acts, but in all the cases more than the 75% of visits are in fact revisitations. This figure reinforces the utility of the task identified as target in this paper.

As we can see in table 1 the number of URLs is quite large and as discussed in Section 3.1 we do not think that all of them must be considered as *bookmarks*. In fact, when averaging over the six users we get that 50% of the web pages receives only one or two visits. In our opinion pages with low number of visits cannot be considered as candidate bookmarks and its consideration will introduce a considerable amount of noise in the process, leading to low quality models (classifiers).

Our approach to deal with this problem has been to apply a threshold, i.e. a minimal number of visits to a page in order to consider it a bookmark, that is, a good candidate for revisitation. However, those pages that do not pass the threshold are not deleted, but reduced to their server. Thus, we can have that the page www.mnewspaper.com/sports has many visits, while URLs www.mnewspaper.com/sport/news1.html, www.mnewspaper.com/weather/today.html, and www.mnewspaper.com/economy/currency.html, only have one visit. Then we consider the first one (sports) as a bookmark and integrate the remaining ones in their server (www.mnewspaper.com). For those URLs that are reduced to their servers, we replace such information in the dataset. In our study we have experimented with three different thresholds: T=10, T=20 and T=30. In the case of some URLs that do not reach the threshold but there are no more pages with the same server, then we do not make any change, and the original URLs are considered as class labels. Table 1 shows under columns T10, T20 and T30 the new number of class labels (URLs + servers) considered when applying such thresholds. The most important reduction is carried out when applying the first threshold (T10), being the subsequent ones less noticeable (specially in the number of servers).

Table 1. Description for the dataset uses in our experiments

	original			T10			T20			T30		
	class states	inst.	%rev.	Urls	Servers	Tot.	Urls	Servers	Tot.	Urls	Servers	Tot.
usr1	666	5605	88	179	97	276	141	97	238	134	97	231
usr2	742	3809	80	159	97	256	126	97	223	110	97	207
usr3	530	2240	76	70	50	120	41	50	91	38	51	89
usr4	942	15681	94	223	75	298	156	77	233	127	77	204
usr5	529	5015	89	167	64	231	105	66	171	85	66	151
usr6	111	676	83	42	17	59	37	17	54	35	17	52

5.2 Model Validation

In order to evaluate the goodness of the selected classifier over the target problem and taking into account a real scenario in which the sequence of the entries in the dataset is important because it is a temporal ordering[1], we have selected as validation technique a time-split cross-validation [15].

Given that the data is chronologically ordered the validation scheme is carried out in five steps (folds): In the initial step we use the first half of the instances in the dataset to train our model and we use the next 10% (i.e., instances betwen 50% and 60% in the dataset) to test it; The second step considers the first 60% of the dataset as training and the next 10% as test; and the next steps use respectively the 70%,80% and 90% of the dataset as training and the the subsequent 10% as test set. Then, the accuracy is computed as the average over

---

[1] Notice that it is not of interest trying to predict the web page the user visited yesterday given the behaviour observed today.

Table 2. Results averaged over the set of users

		k=1	k=3	k=5	k=7	k=10
T10	NB	0,19	0,30	0,34	0,38	0,40
	NB_OR	**0,32**	0,38	**0,40**	**0,43**	**0,45**
	NB_woLS	0,12	0,21	0,26	0,29	0,32
	IB1	0,20	0,20	0,20	0,20	0,21
	OneR	0,31	0,31	0,31	0,31	0,31
	C4.5	0,28	**0,39**	**0,40**	0,41	0,42
T20	NB	0,25	0,37	0,40	0,43	0,46
	NB_OR	**0,39**	0,46	**0,48**	**0,50**	**0,53**
	NB_woLS	0,15	0,25	0,29	0,33	0,37
	IB1	0,26	0,27	0,27	0,27	0,28
	OneR	**0,39**	0,40	0,40	0,40	0,40
	C4.5	0,35	**0,47**	**0,48**	0,49	0,50
T30	NB	0,29	0,39	0,42	0,45	0,48
	NB_OR	**0,44**	0,49	0,51	**0,54**	**0,56**
	NB_woLS	0,18	0,26	0,32	0,35	0,39
	IB1	0,29	0,30	0,31	0,31	0,32
	OneR	**0,44**	0,44	0,44	0,44	0,47
	C4.5	0,39	**0,51**	**0,52**	0,53	0,55

the five test sets. This validation scheme (based on [15]) respect the principles of cross-validation but following a temporal sequence similar to the expected operation mode in a real environment.

All the classifiers are tested over the six users and evaluated according to their classification accuracy. As our intention is not to predict exactly the next page to visit[2], but the k more probable pages, we consider the classifier provides a correct answer if the (actual) next page to be visited is included in the set of k pages it returns. In our case k is obviously related with the number of bookmarks we can expect to be able to allocate in the BPTB, so we have experimented with different values: 1, 3, 5, 7 and 10.

5.3 Results Analysis

Table 2 shows the results obtained for the different configurations ⟨classifier, threshold, k⟩ but averaged over the six users. From these results we can extract the following observations:

– Increasing the number of retrieved class labels (k=1,3,5,7,10) always increases the accuracy of the classifier. Greater improvements are obtained when passing from 1 to 5, than when passing from 5 to 7 or 10. Given our goal, we think that 5, 7 and 10 are adequate (and moderate) list size values, and that as this value has effect on the visual appearance of the browser it should be a parameter customizable by the user.

---

[2] That is, some thing similar go *Google I'm felling lucky*.

- Increasing the threshold (10, 20 and 30) always improves the accuracy of the classifiers. As happens with k a greater difference is obtained when passing from 10 to 20 than when passing from 20 to 30. What is clear is that increasing this parameter gets a higher improvement in accuracy than increasing k. In this case we think that 20 or 30 are appropriate thresholding values.
- With respect to the level of accuracy it also depends on the users behaviour, but in general it is in interval [.45,.52] with threshold 10 ($k = 7$ or 10), and in interval [.53,.62] with threshold 30. We think these results are really good taking into account the goal of our approach.
- In general, from the averaged data we can observe that NB_OR is almost ever the best classifier, specially in the recommended values of k (5, 7, 10).

In order to reinforce this conclusion we have carried out a statistical analysis based on Wilcoxon signed rank test ($\alpha = 0.05$). Thus, the comparison of NB_OR with C4.5 over all the configurations ⟨threshold, k⟩ returns a p-value of 0.04, while comparisons of NB_OR with any other classifier always returns a p-value smaller than 1.7e-3. From these results we can conclude that the difference between NB_OR and the rest of tested classifiers is statistically significant.

6   Conclusions

In this work we have presented an approach to improve Internet browsers revisitation capabilities. Our approach is based on automatically catching data about user navigation behaviour and then data mining algorithms are used to learn a user navigation model encoded as a classifier. Later, the classifier is used to predict the set of bookmarks to be placed at the browser's bookmark toolbar based on current navigation scenario, which is composed by the web server the user come from and current time and date. The main contributions of our work are the use of temporal information as predictive attributes, the way in which data is collected by using an extension, and a study about the accuracy of different classification algorithms over six different users. From the results, the classifier that best fits our goal, both because its accuracy and computational simplicity is our proposal of using a mixture about Naive Bayes and OneR.

In the next future we plan to extend this works as follows: (1) enriching model taking into account as class labels (bookmarks) other URLs apart from final web pages and servers, for example, extracting page www.mnewspaper.com/sports/ from address www.mnewspaper.com/sport/today/news100s0.html instead of just the server; (2) combination of the identification of bookmarks collection with incremental learning of the classifier/model, instead of learning it from scratch as did in the current experiments; and (3) including all these development in the programmed software add on, having in this way a full Mozilla Firefox extension.

Acknowledgments

This work has been partially supported by Spanish Ministerio de Educación y Ciencia (project TIN2004-06204-C03-03); Junta de Comunidades de Castilla-La

Mancha (project PBI-05-022) and FEDER funds. Special thanks are given to all our work mates who have helped us providing the data used in the experiments.

References

1. Tauscher, L., Greenberg, S.: How people revisit web pages: empirical findings and implications for the design of history systems. Int. J. Human-Computer Studies 47, 97–138 (1997)
2. Takano, H., Winograd, T.: Dynamic bookmarks for the WWW. In: Proceedings of the ninth ACM conference on Hypertext and hypermedia, pp. 297–298 (1998)
3. Abrams, D., Baecker, R., Chignell, M.: Information archiving with bookmarks: personal Web space construction and organization. In: Proceedings of the SIGCHI conference on Human factors in computing systems, pp. 41–48 (1998)
4. Kaasten, S., Greenberg, S.: Integrating back, history and bookmarks in web browsers. In: CHI 2001 extended abstracts on Human factors in computing systems, pp. 379–380 (2001)
5. Armstrong, R., Freitag, D., Joachims, T., Mitchell, T.: Webwatcher: A learning apprentice for the world wide web. In: AAAI Spring Symposium on Information Gathering from Heterogeneous, Distributed Environments (1995)
6. Lieberman, H.: Letizia: An agent that assists web browsing. In: IJCAI-1995. Proceedings of the 14th International Joint Conference on Artificial Intelligence, pp. 924–929 (1995)
7. Bailey, C., El-Beltagy, S.R., Hall, W.: Link augmentation: A context-based approach to support adaptive hypermedia. In: Proceedings of Hypermedia: Openness, Structural Awareness, and Adaptivity, pp. 239–251 (2001)
8. http://www.mozilla.org
9. http://www.konqueror.org
10. Weiss, S., Indurkhya, N.: Predictive data mining: a practical guide. Morgan Kaufmann Publishers Inc., San Francisco, CA, USA (1998)
11. Duda, R.O., Hart, P.E., Stork, D.G.: Pattern Classification. Wiley-Interscience Publication, Chichester (2000)
12. Aha, D., Kibler, D.: Instance-based learning algorithms. Machine Learning 6, 37–66 (1991)
13. Holte, R.: Very simple classification rules perform well on most commonly used datasets. Machine Learning 11, 63–91 (1993)
14. Quinlan, R.: C4.5: Programs for Machine Learning. Machine Learning 16(3), 235–240 (1994)
15. Bekkerman, R., McCallum, A., Huang, G.: Automatic categorization of email into folders: Bechmark experiments on enron and sri corpora. Technical Report IR-418, Department of Computer Science. University of Massachusetts, Amherst. (2004)

Hierarchical Co-clustering for Web Queries and Selected URLs

Mehdi Hosseini and Hassan Abolhassani

Web Intelligence Research Laboratory,
Computer Engineering Department,
Sharif University of Technology, Tehran, Iran
me_hosseini@ce.sharif.edu, abolhassani@sharif.edu

Abstract. Recently query log mining is extensively used by web information systems. In this paper a new hierarchical co-clustering for queries and URLs of a search engine log is introduced. In this method, firstly we construct a bipartite graph for queries and visited URLs, and then to discover noiseless clusters, all queries and related URLs are projected in a reduced dimensional space by applying singular value decomposition. Finally, all queries and URLs are iteratively clustered for constructing hierarchical categorization. The method has been evaluated using a real world data set and shows promising results.

1 Introduction

The exponential growth of available information on the Internet has given rise to greater demands for efficient search engines. In order to satisfy the users needs, a new generation of search engines has appeared that tries to understand users queries. Such search engine has a repository of query log which can be used to facilitate understanding of users' needs [1]. Given millions of queries and selected URLs, how can the search engine judge if URLs are related or not? One popular method is to apply clustering to the search engine log.

In this paper a Hierarchical co-clustering method is introduced. In it, a bipartite graph for a query log is created where the left nodes are queries, right nodes show the selected URLs, and an undirected link connects a query to the visited URLs. Since the formations of the bipartite graph is highly sensitive to the noisy selections of users, Singular Value Decomposition (SVD) [2] has been used to reduce noise inside the graph. Finally to find main clusters of queries and URLs a new hierarchical co-clustering is introduced. Therfore our main contribution is: using information of similar queries for clustering URLs and concurrently using information of similar URLs for clustering queries. In what follows, we discuss on related works in section 2 and introduce our method in detail in section 3. Evaluation of the method is discussed in section 4. Finally section 5 presents conclusion.

B. Benatallah et al. (Eds.): WISE 2007, LNCS 4831, pp. 653–662, 2007.

2 Related Works

Query clustering methods can be divided in tree groups: The first assumes that if two queries share the same keywords, then they are similar. The more keywords the two queries share, the higher the similarity. The second exploits query logs kept by the search engine. Given a large set of user query logs, similar queries can be extracted out by similar clicked URLs, also similar URLs can be detected by common related queries. However [3] states that there is another work which uses a combination of the above two methods for this task.

When a user selects a limited number of URLs among a large number of results and such a selection is repeated by other users, one can infer that there should be a semantic similarity between the selected links. On the other hand when for some queries similar URLs are selected that means that those queries are similar even if they are not similar in lexical terms. A major advantage of this approach is the ability to cluster URLs as well as queries without any need to have costly processing of contents of the pages. Also the other important advantages of this method is discovery of semantic relations that can be used to find synonymy and polysemy of queries. For the first time in [4] a bipartite graph for synchronous clustering of URLs and queries were used where clustering is done in an iterative approach. In each iteration two queries having maximum similarity are put in a cluster. Then based on the similarity of queries two URLs having maximum similarity are put in a cluster. Such steps are iterated and finally each connected component is recognized as a agglomerative cluster. The similarity of queries are calculated by equation 1 where $N(q_k)$ represents the set of URLs selected for query q_k.

$$Sim(q_1, q_2) = \begin{cases} \frac{(N(q_1) \cap N(q_2))}{N(q_1) \cup N(q_2)} & if N(q_1) \cup N(q_2) > 0 \\ 0 & else \end{cases} \quad (1)$$

In [5] the similarity measure of [4] (After this we refer to [4] as Beeferman)is modified to account the number of times a URL has been selected. Equation 2 shows this similarity computation. In this method $L(q_1, q_2)$ is the set of links connecting q_1 and q_2 to the same documents, $L(q_1)$ and $L(q_2)$ are all the URLs connecting to q_1 and q_2, respectively, and $|L()|$ is the cardinally of $L()$.

$$Sim(q_1, q_2) = \begin{cases} \frac{|L(q_1, q_2)|}{|L(q_1) \cup L(q_2)|} & if L(q_1) \cup L(q_2) > 0 \\ 0 & else \end{cases} \quad (2)$$

Considering the fact that selection of links by users is not always deterministic and accurate [1] a connected component may contain queries that are related to different topics which harm the quality of clustering. FIGURE 1(a) shows a sample of noise affects on clustering results. Labels on the edges show the number of times a URL is selected for a query. As shown in this figure actually there are two query clusters which are connected together by a noisy selection. Using algorithm of Beeferman as shown in FIGURE 1(b), the noisy selection causes two different clusters merge together and similarity between them rise to

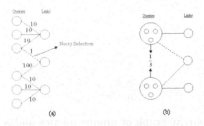

Fig. 1. (a) Query-Link bipartite graph with a noisy selection. (b) Query clustering by using algorithm in Beeferman.

$\frac{1}{3}$ by Equation 1. Using a method like Equation 2 increases the similarity value to $\frac{100+1}{100+1+30+30} = \frac{101}{161} \cong 0.63$. In what follows we introduce a new method to construct hierarchical categorization of queries and URLs concurrently.

3 Clustering Web Queries and Selected URLs Hierarchically

The http protocol allows commercial search engines the ability to record a great deal of information about their users such as the name and IP address of the machine which sent the request, the type of web browser running on the machine, the screen resolution of the machine, and so on. Here we are interested only in the sequence of characters comprising the query submitted by the user, and the URL selected by that user from among the choices presented by the search engine. Based on these type of information a bipartite graph between queries and URLs is constructed, where the vertices on one side correspond to unique queries, the vertices on the other side to unique URLs, and edges exist between a query and URL vertex when they co-occurred in a clickthrough record. After constructing graph, using a fast algorithm the connected components of the graph are detected, then for each high density connected component a query-URLs relation matrix is constructed, and finally an iterative clustering is applied to find hierarchical categorization of queries and URLs. Generally we divide the process of clustering in two steps, first is preprocessing and the second is hierarchical co-clustering. In the remaining of this section we explain each step in details.

3.1 Preprocessing

First from a query log all the records with a same query are combined to a single record. Therefore, for each unique queries between all sessions only one record could be existed which all selected URLs in different sessions have been collected in the record. Then a bipartite graph is constructed from the records that the nodes one side are unique URLs and the node in the other side unique queries. Figure 2 portrays a sample of such graph where the right nodes are selected URLs and the left nodes are unique queries. Some preprocessing processes could

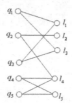

Fig. 2. A bipartite graph of unique queries and selected URLs

be used for raising quality of detected queries. For example in our experiment each query containing only a symbol character was eliminated, and all misspelled quires had been corrected. In addition for facility in next processes, each query is represented as a vector where element l_i^k shows the ratio of the number of times that i^{th} URL is selected for k^{th} query to the number of times which has selected for all queries. Also each URL is represented a vector of queries where element q_k^i shows the ratio of the number of selections of ith URL for k^{th} query to the total number of selection for this query. These definitions are shown in equations 3 and 4.

$$\vec{q_k} = (l_1^k, l_2^k, l_3^k, ..., l_m^k); l_i^k = \frac{|l_i^k|}{|l_i|} \quad i = 1, 2, .., m \tag{3}$$

$$\vec{l_i} = (q_1^i, q_2^i, q_3^i, ..., q_n^i); q_k^i = \frac{|q_k^i|}{|q_k|} \quad k = 1, 2, .., n \tag{4}$$

For showing up the weight of relation between queries and URLs we use average of each l_i^j and q_j^i, for example the amount of relation between i^{th} query and j^{th} URL is indicated by A_{ij} as equation 5. Therefore in bipartite graph corresponding to calculated weights, the edge between each query and URL could be labeled.

$$A_{ij} = \frac{l_i^j + q_j^i}{2} \tag{5}$$

To recognize connected components of a bipartite graph we can treat the graph as a forest where a tree represents a connected component. In such trees, nodes in alternate levels belong to queries or URLs. To build such a tree, we start from a query node and add all its linked nodes which are not visited in earlier period, in the other side of the graph, as its children in the next level. This process is repeated until all nodes of a connected component are added to the tree. Then we select another query node which not visited earlier and repeat the above routine to build another tree. Formally the algorithm for connected component recognitions has been shown in FIGURE 3. In each period, firstly a new (not traversed) query node is selected then all URLs and queries, there is a path from the first query to them, are selected and putted in a new connected component such as C_f. Since the algorithm is according to traverse breadth first, it is look like Breadth First Search (BFS) algorithm on a graph, so it can be easily shown

Q is the list of all queries, L the list of all links and C are queries and links related to the f-th connected component of the bipartite graph.

$$Q = \{\overrightarrow{q_1}, \overrightarrow{q_2}, ..., \overrightarrow{q_n}\}$$

$$L = \{\overrightarrow{l_1}, \overrightarrow{l_2}, ..., \overrightarrow{l_m}\}$$

$$C = \{Q, L\} \quad , \quad Q'_f \subseteq Q, L'_f \subseteq L$$

Also R is an empty set for queries and R is an empty set for links.

Steps of the algorithm:
1. $r=1$
2. Select an unmarked node q from Q put it in R and do the following steps. If there is not such a q stop the algorithm.
3. While R is not empty do:
    1. while R is not empty do the following steps
        a. Remove the first element of R let it be $\overrightarrow{q_{first}}$
        b. If $\overrightarrow{q_{first}}$ is not marked then:
            i. Put all of its elements (links) in R.
            ii. Mark $\overrightarrow{q_{first}}$ and put it in C.
    2. While R is not empty do the following steps
        a. Remove the first element from R let it be $\overrightarrow{l_{first}}$
        b. If $\overrightarrow{l_{first}}$ is not marked then:
            i. Put all its elements (queries) in R.
            ii. Mark $\overrightarrow{l_{first}}$ and put it in C.
4. Let $r=r+1$ and go to step 2.

Fig. 3. Connected Component recognition algorithm

the complexity of this algorithm is of $\theta(E + V)$ [6] which E is the number of edges and V is the number of nodes.

3.2 Hierarchical Co-clustering

After doing preprocessing, constructed bipartite graph has been used for clustering queries and URLs concurrently. For do this, firstly each connected component with low number of nodes are considered as noises, because based on our experiments if we have a massive query log, there are only a few connected components containing majority of queries and URLs. In other words, different clusters would be connected together by incorrect user selection and form a massive connected component. Also our experiments show nearly all related queries and all related URLs belong to same connected component. The other result of our experiments show if we have massive query log, clusters of different connected components do not have a same context, so it is not required to check context of different cluster for combination to get final clusters.

We use top down method for clustering hierarchically, so at first each connected component considered as a cluster in top level. For next level, a relation matrix is build for each connected component which the number of rows is equal to its query nodes and the number of columns equal to its URL nodes. In addition, each cell denotes the weight of corresponding query and URL. For each relation matrix, SVD is applied to project vectors of queries and URLs into reduced dimension space. In following SVD is described in detail:

Singular Value Decomposition (SVD): SVD is a method for dimensionality reduction. Using it, we obtain singular vectors that have most dependency of data. Such vectors correspond to largest singular values. According to SVD, matrix A can be decomposed as: $A = Q_{n \times r} \times S_{r \times r} \times L_{r \times m}^T$
Where:

- $r \leq \min(n, m)$
- Q is an $n \times r$ matrix where columns are Orthogonal singular vectors of $(AA^T)_{n \times n}$.
- S is a diagonal $r \times r$ matrix with singular values sorted in descending on the main diagonal line of the matrix. Those singular values belong to $(AA^T)_{n \times n}$ and $(A^T A)_{m \times m}$ matrixes.
- L is an $m \times r$ matrix where columns of it are Orthogonal singular vectors of $(A^T A)_{m \times m}$.

Eliminating weak singular values and corresponding singular vectors, we obtain a $Q'_{n \times r'}$ low rank of $Q_{n \times r}$ where $r' \leq r$. Since the number of dimensions in new space is r', each row of $Q'_{n \times r'} \times S_{r' \times r'}$ is a query vector in reduced dimension space, and each column of $S_{r' \times r'} \times L_{r' \times m}^T$ is a URL vector in reduced dimension space. We use Frobenius norm [7] to approximate best value for r'.

The K-mean method [8] is used to extract the clusters of queries and URLs in reduced dimensions space. We use K-Mean because of its simplicity and its appropriateness for text documents clustering. Sub clusters of next levels are

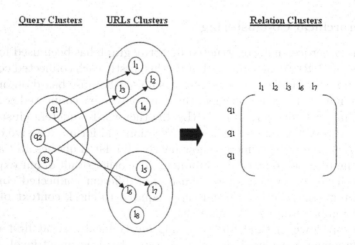

Fig. 4. Creating relation matrix to extract sub clusters

build applying similar processes, although there is a few differences for building relation matrixes. Creating relation matrix of each cluster of queries, all belonged queries are considered as the rows and all selected URLs for these queries are considered as the columns of the matrix. Figure 4 depicts creating relation matrix for building sub clusters of a query cluster, as can be seen selected URLs belong to different clusters. In addition the relation matrixes for URL clusters are build in similar manner, so each relation matrix is build by URLs of specific cluster and related queries. Using SVD each relation matrix is mapped into reduced dimension space and finally K-mean algorithm is applied.

4   Experimental Results

To evaluate our method we used two different query logs of queries, one is a collection from ACM KDDCUP 2005 which is extracted from MSN search engine query log (Available at: http://www.acm.org/kdd/kddcup). This collection contains 800,000 queries classified into 7 main categories and 67 target categories. In addition 800 queries of this collection are labeled by 3 different internet users. An illustration of the hierarchical structure for the target categories is shown in Figure 5. The second collection is from AOL search engine query log consist of ~26 millions web queries collected from ~650,000 users from 01 march to 31 may 2006.

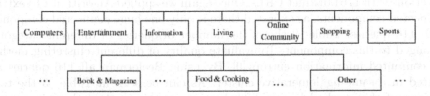

Fig. 5. Illustration of the hierarchy structure of 7 main categories and 67 target categories

4.1   Formation of Connected Components

As mentioned before the formation of connected components is highly sensitive to users selections. To illustrate it we extract randomly 5 sub collections of AOL data set with different number of queries. Table 1 depicts the numbers of queries in each collection. Algorithm 1(Figure 3) was applied on each collection to find its connected components. For each collection there was only one huge connected component and the others have low density. Figure 6 demonstrates the density of the hugest connected component in percentage for each collection. As can be seen the connected component of C5 with hugest amount of query is greater than the other hugest connected components of other collection. It could be inferred that if we have a large collection, there is only one huge connected component and the most of queries is belonged to it.

Table 1. 5 different query collections of AOL query log

Collection	C1	C2	C3	C4	C5
Numbers of Queries	20,000	40,000	60,000	80,000	100,000

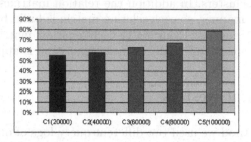

Fig. 6. The hugest connected components of different query collections

4.2 Evaluating Hierarchical Co-clustering

For evaluate our clustering method we used 800 labeled queries of ACM KDD-CUP 2005. This data set was distributed between the group of users (24 internet users) and they used Google to search them. Then all queries and selected URLs are putted into a collection which consists of 800 labeled queries and 2151 users selections with 1761 distinct URLs. Once again we applied Algorithm 1 to extract connected components, as expected there is only one huge connected component with 116 queries. Table 2 shows the number of queries of different categories belonged to this component. To evaluate quality of different clustering method we computed information entropy[9]. By using Beeferman all 116 queries are putted in a same agglomerative cluster. As can be seen in Table 3 the total entropy of this method is very high because in this method there is not any distinction between clusters in same level of hierarchy, and sometimes by effect of noisy user selection, it causes the relation between queries of different clusters is grater than the relation between queries in a same cluster. Also we applied our clustering method over this connected component. In this case the total levels of

Table 2. The number of queries of different categories belonging to the connected component

Category Name	Number of Related Queries in C1
Computer	3
Entertainment	32
Information	34
Living	23
Online Community	2
Shopping	19
Sport	3

Table 3. The total entropy of different query clustering methods

Method	Total entropy
Beeferman	0.832
Hierarchical co-Clustering	0.453

our hierarchical clustering is 3. The first level is the whole connected component, in second level there are 4 different main clusters and in third levels there are 9 different target clusters which is shown in Table 4.

Table 4. Different main categories and target categories in the connected component

Main Categories	Entertainment	Information	Living	Shopping
Target Categories	Music, Movie, Game	Art, Science	Health, Family	Store, Auto

The information entropy for each cluster has been computed by Equation 6. Where $|C|$ is the number of clusters, $|C_r^i|$ is the number of queries of rth cluster which are labeled by i^{th} category and $|C_r|$ is the total number of queries in r^{th} cluster.

$$E(C_r) = -\frac{1}{\log(|C|)} \sum \frac{|C_r^i|}{|C_r|} \log(\frac{|C_r^i|}{|C_r|}) \tag{6}$$

$E(C_r)$ is a real value between 0 and 1. If $E(C_r)$ is close to zero , it seems quality of rth cluster is high otherwise the quality is low. In addition we use equation 7 to find total entropy for all cluster of web site, where C_r is the number of query in section r and C is the total number of query.

$$E_r = \sum_{r=1}^{k} \frac{C_r}{C} E(C_r) \tag{7}$$

As can be seen in Table 3 the total entropy of our method is lower than the total entropy of the Beeferman method.

In Web search, an important application is to organize the large number of URLs in the search result, after the user issues a query, according to the potential categories of the results. Furthermore, in web search, many search engine companies are interested to help their users by recommending related queries toward attaining user demands. As seen, our method by clustering queries and URLs concurrently can be useful to organize these results.

5   Conclusion and Future Works

In this article we proposed a hierarchical co-clustering method for clustering queries and URLs which has promising results. In future, our aim is applying the algorithm in a much larger query log and evaluate our algorithm and make use our clustering in different applications.

References

1. Joachims, T., Granka, L., Pan, B., Hembrooke, H., Gay, G.: Accurately interpreting clickthrough data as implicit feedback. In: SIGIR 2005. Proceedings of the 28th annual international ACM SIGIR conference on Research and development in information retrieval, pp. 154–161. ACM Press, New York, NY, USA (2005)
2. Golub, G., Loan, C.V.: Matrix Computation, 2nd edn. Johns Hopkins, Baltimore (1989)
3. Wen, J.R., Zhang, H.: Query clustering in the web context (2003)
4. Beeferman, D., Berger, A.: Agglomerative clustering of a search engine query log. In: Knowledge Discovery and Data Mining, pp. 407–416 (2000)
5. Chan, W.S., Leung, W.T., Lee, D.L.: Clustering search engine query log containing noisy clickthroughs. In: SAINT, pp. 305–308. IEEE Computer Society, Los Alamitos, CA, USA (2004)
6. Cormen, T.H., Leiserson, C.E., Rivest, R.L., Stein, C.: Introduction to Algorithms, 2nd edn. The MIT Press, Redmond, Washington (2001)
7. Christopher, D., Manning, P.R., Schutze, H.: Introduction to Information Retrieval (2007)
8. Zhao, Y., Karypis, G.: Evaluation of hierarchical clustering algorithms for document datasets. In: CIKM, pp. 515–524. ACM, New York, NY, USA (2002)
9. Quinlan, R.R.: C4.5: programs for machine learning. Morgan Kaufmann Publishers Inc, San Francisco (1993)

Navigating Among Search Results: An Information Content Approach

Ramón Bilbao and M. Andrea Rodríguez

Department of Computer Science, University of Concepción
Center for Web Research, University of Chile
Edmundo Larenas 215, 4070409 Concepción, Chile
{rbilbao,andrea}@udec.cl

Abstract. Total or partial duplication of documents affects the effectiveness of the visualization of search results. In this paper we propose a navigation strategy that sorts a list of documents such that the first documents contain more information content decreasing considerably duplication. The strategy defines a content relation between documents based on their *equivalence* and *omission* and estimates the new information content obtained from visiting documents. In this paper, we describe the strategy and experimentally evaluate it. These results indicate the potential use of this strategy for the visualization of thematically related documents that are relevant to a query.

1 Introduction

The continuous growth in data volume on the World Wide Web makes it difficult for users to search and access information. The problem does not only involve the process of ranking documents that match some keywords, but also the visualization of the results to these searches. From one side, classical search engines give results as a large ranked list of documents, which may include total or partial duplication of documents. From another side, most users are not willing to review more than 3 pages of results [14], and they expect to obtain more information among the first documents in the list of results.

An important problem for the visualization of search results are duplicate documents [2], an issue that has gained the attention of several studies [7,17,3,9,12,4]. These studies have proposed strategies to compare documents that can be used by crawlers to detect duplication before indexing documents. Thus, they aim to avoid retrieving duplicate documents that would lower the number of valid responses provided to a user.

The work in this paper pursues a different approach to handling *partial* or *total duplication* in the visualization of search results. This approach analyzes the new information content obtained when visiting documents, an approach we have named Information Content Navigation (ICN). Indeed, we assume that although crawlers may eliminate some duplication, users still face the challenge of selecting documents from a large list of documents which may contain partial duplication.

B. Benatallah et al. (Eds.): WISE 2007, LNCS 4831, pp. 663–672, 2007.

Under this scenario, we define a visualization of results that could guide users to documents with new information. Our strategy considers the visualization of search results as an information navigation process, where we define a path (or sequence of documents) which provides users with more information in the first visited documents.

The approach taken by this work differs from alternative visualization strategies based on clustering or summaries [14]. Instead of highlighting the similarity between documents to form groups or clusters, we assume a set of documents to be thematically related and relevant to a user, and make the differences among them a determinant for estimating the amount of new information one could obtain by visiting them. In all of this, we argue that users are not interested in reading all documents with the same or partially the same information, but in obtaining all information relevant to a search as fast as possible. To define this strategy, we still compare documents, as clustering-based strategies do, but we concentrate not only on what is similar, but also on what is different among documents.

The organization of this papers is as follows. Section 2 summarizes related work concerning visualization of search results and methods for comparing the information content between documents. Section 3 describes, at an abstract level, the proposed strategy for navigating among search results. This description is independent of the method to compare documents, which is analyzed in Section 4. Some results of the implementation are discussed in Section 5, which are followed by conclusions in Section 6.

2   Related Work

In the area of information systems, the concept of information navigation has been associated with the visualization of search results. Besides classical ranked lists of documents, alternative navigation strategies use overviews of documents to guide users in the retrieval process [8]. These strategies can be classified into two important approaches: (1) category hierarchies and (2) clustering techniques. Category hierarchies assign metadata to documents based on their categories, which are then organized into a hierarchy. Examples of this approach are BoW [4], an on-line bibliographical repository based on a hierarchical concept index, the computer classification system of the ACM [1], which has developed a hierarchy of 1200 categories, and the popular search site Yahoo [15], which organizes documents in many categories.

There are several cluster-based solutions for helping users in searching and navigating documents. They attempt to display overview information derived from the metadata or the extraction of common features in a collection. Then, clusters group documents based on the similarity to one another. Examples of such strategies can be found in search engine VIVISIMO [11] and Kartoo [10]. An interesting approach that groups documents was implemented in THESUS [6], which exploits the connections between documents through links and a hierarchy of concepts to organize documents into thematically related subsets.

Detecting document duplication or performing automatic clustering of documents require comparing the information content of documents. The comparison of documents usually takes one of the following two approaches: shingle techniques and similarity measures. Shingle techniques, such as [9], take a set of contiguous terms or shingles of documents and compare the number of matchings among them. They rely on hash values to reduce the number of comparisons and allow us to obtain the overlapping segments of documents. Approaches to computing similarity measures between documents, on the other hand, use similarity computations to group potentially duplicate or similar documents. All pairs of documents are compared, that is, each document is compared to each other document. The similarity measure for comparing documents are usually adaptations of the vector model for information retrieval [2], where the query is here a document. A different and interesting work compares documents by using phrases in documents represented in bipartite graphs [16]. On the same lines, in [13] documents are also compared by matching phrases. The results of this last work indicate that the comparison by phrases has only sense for large documents; otherwise, the computational cost does not justify its use.

3   Information Content Navigation

This work proposes a strategy that suggests a sequence of documents that reduces the cost of obtaining the greatest amount of information contained in these documents. Here reducing cost is equivalent to accessing or visiting less documents, and information refers to data that decreases uncertainty. Thus, we aim to define a sequence of (a possible subset of) documents that avoids duplication, includes all information contained in documents, and gives an order in which we get the greatest amount of information as quick as possible. The basic assumption we have made is that documents are thematically related and relevant to a query, which in most cases mean that we are taking only the most relevant documents from the list of search results. We impose this assumption because we are not using the user query in defining the navigation among documents.

We start to describe our strategy by defining an information space that relates the information content of documents and that is represented as a graph. In this description, we first concentrate on defining the strategy at an abstract level, making it independent of the method we use to obtain the content relation between documents. Later, we will combine and explore alternative methods to construct the information space.

3.1   The Information Space

The information space characterizes the relation between the information content of documents in terms of two measures: *equivalence* and *omission*.

Definition 1 (Equivalence). *Let D_{d_i} and D_{d_j} be sets of distinguishing components of documents d_i and d_j, respectively, we define the equivalence between d_i and d_j, denoted by $E(d_i, d_j)$, as the number of components in D_{d_i} that match components in D_{d_j}.*

Definition 2 (Omission). *Let D_{d_i} and D_{d_j} be sets of distinguishing components of documents d_i and d_j, respectively, we define the difference of omission between d_i and d_j, denoted by $O(d_i, d_j)$, as the number of components in D_{d_j} that do not match components in D_{d_i}.*

In these definitions, distinguishing components can be words, sequence of words, or concepts that are part of a document. At this moment, we just refer to them as distinguishing components. Note that $E(d_i, d_j)$ is asymmetric, since one element in d_i can match two elements in d_j, counting one matching for d_1 and two for d_2.

The cardinality of the set of distinguishing components in a document can be derived by the equivalence of the document with itself. So the following property between content-relation measures holds:

$$E(d_i, d_i) = E(d_i, d_j) + O(d_j, d_i) \tag{1}$$

An important issue here is to know what these distinguishing components are and when they match. This will be addressed in Section 4, since the navigation algorithm among documents can be defined independently of the way the matching between distinguishing components is determined.

3.2 Navigation Algorithm

We explore three different alternatives for the navigation algorithm that where experimentally compared in section 3.3. To illustrate the different alternatives, consider a simple case of four documents (Figure 1), and the corresponding measure of equivalence between them. We then present three basic algorithms for navigating among documents in this graph.

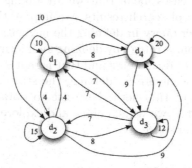

Fig. 1. Content-relation graph with four documents

– *Size-based strategy.* A very simple strategy sorts documents based on their size, that is, based on the number of their distinguishing components. To use this strategy all documents should be equally relevant, their information content is independent to each other, and all of them are retrieved. In the case of the example in Figure 1, the navigation sequence is d_4, d_2, d_3, and d_1.
– *Forward-based strategy.* A forward-based strategy takes the current position in the navigation (initially, the largest document or the one with the highest ranking) and selects the next document with respect to which the current document has the largest omission. Consider the example in Figure 1 and derive the values of omission between documents using equation 1. The forward-based strategy takes the largest document (in the example, document d_4) and from that, it selects the *non previously visited* document with respect to which the current document has the largest difference of omission. This strategy avoids retrieving documents with no new information. In this case, we start with document d_4 and obtain the following sequence of documents: d_4, d_2, d_1, and d_3.
– *History-based strategy:* This strategy uses the information about content relationships of all previously visited documents, and not only the current document, to define which document is the next in the navigation process. Like the forward-based strategy, history-based strategy avoids accessing documents with no new information. The history-based strategy estimates the amount of expected new information when accessing a non-previously visited document. To do that, it assumes that all distinguishing components have the same chance of being in the subset of distinguishing components that match components in another document. In particular, let d_i and d_j be documents, then $E(d_i, d_j) / E(d_i, d_i)$ estimates the *likelihood* that a distinguishing component in d_i matches a component in d_j.

To illustrate our derivation of the estimated new data in a next document of the navigation sequence, consider first a case with a non-previously visited document d_i and two already visited documents d_j and d_k. A possible scenario is that some of the distinguishing component in d_i match components in d_j and some components in d_k, where the distinguishing components in d_i that match components in d_j are not necessarily the same than the components in d_i that match components in d_k. Since we consider the same probability for distinguishing components in d_i to match any component in d_j and d_k, then the estimated number of new distinguishing components ($\mathcal{N}(d_i)$) that are obtained when visiting d_i is given by the simultaneous occurrence that these components in d_i do not match any component in both d_j and d_k.

$$\mathcal{N}(d_i) = E(d_i, d_i) \times \left(\frac{O(d_j, d_i)}{E(d_i, d_i)} \times \frac{O(d_k, d_i)}{E(d_i, d_i)} \right) = \frac{O(d_j, d_i) \times O(d_k, d_i)}{E(d_i, d_i)}$$

Note that if one of the previously visited documents (i.e., d_j or d_k) omits nothing with respect to d_i (i.e., $O(d_j, d_i) = 0$ or $O(d_k, d_i) = 0$), then $\mathcal{N}(d_i) = 0$.

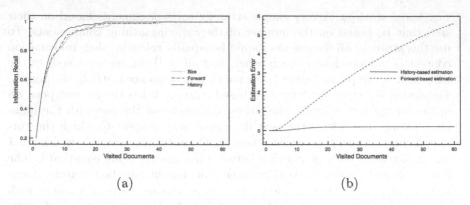

(a)   (b)

Fig. 2. Evaluation of navigation algorithms: (a) information recall obtained in a navigation guided by the size-based, forward-based, and history-based algorithm, (b) error of the estimated information recall by the history- and forward-based algorithms with respect to the real information recall

The general expression of the expected new information (new distinguishing components) in a non-visited document d_0, when m documents $\{d_1, \ldots, d_m\}$ have been previously visited is:

$$\mathcal{N}(d_0) = \frac{\prod_{0 < i \leq m} O(d_i, d_0)}{E(d_0, d_0)^{m-1}} \tag{2}$$

In the example of Figure 1, this strategy takes the largest document and continues accessing the non-visited documents in the following sequence: d_4, d_2, d_3 and d_1.

3.3   Experimental Evaluation of Navigation Algorithms

We performed a first evaluation to show the effectiveness of the algorithms for obtaining a sequence of documents that could obtain more information first. To evaluate this, we needed a set of documents where the common and different components of the documents were known. We obtained such data set by creating 60 documents as a random combination of 100 different paragraphs. Thus, we used paragraph as distinguishing components, since they represent a semantic unit of information.

Once we created documents, we could know how many paragraphs were in common and how many different. From these data, the information space as a graph was constructed and the algorithms were applied. We illustrate the results in a figure that compares the *information recall* obtained with each visited document using the different navigation algorithms (Figure 2a). We define the information recall as the ratio between the new information obtained when visiting a document and the total information in the collection of documents, measured as the total number of paragraphs in documents.

The results show that the history-based algorithm overcomes the forward- and size-based algorithms by retrieving the greatest amount of information first. Indeed, by visiting only 30% of the 60 documents, we could obtain over 98% of total information. The history-based algorithm does not only get more information first, but also indicates when we do not get new information, something that the size-based algorithm is unable to detect.

The quality of the estimation of new information by the history-based (Equation 2) and by the forward-based algorithm (the omission of the next documents respect to the previous document) are shown in figure 2b, where y-axis represents the error of the estimated with respect to the real information recall. We omit here the size-based algorithm, since it gives large errors with respect to the real information recall. In the following, we will refer as ICN to the strategy that uses a history-based algorithm to suggest the order in which documents should be visited in a navigation process.

4   Comparing Documents

The previous section shows a navigation strategy that works well if we are able to determine what is in common and what is different between documents. To do so, we adapt some of the classical models for comparing documents and experimentally evaluate them.

A simple strategy would be to consider documents as classical sets of words, but this does not characterize the relevance of a term within a document and has serial limitations since documents are compared in terms of purely syntactic matching between words. Another alternative is to derive sets of concepts within documents instead of set of words. In this case, concepts group several words that semantically represent the same or are semantically related. Then, we could compare common and different concepts between documents. After an implementation of this strategy we discarded it, since the computational effort to determine these concepts is not justified unless we work in predefined domains where it is possible to apply natural-language techniques and have well-defined training corpus. We finally analyzed two types of comparisons between documents that have good computational properties and consider more than individual terms: shingles and an adaptation of the vector model. Both strategies could be applied to documents or paragraphs, that is, we could try to compare whole documents or compare documents as the aggregated comparison of paragraphs. In all cases and before comparing documents, we eliminated stop-words and applied stemming [2].

4.1   Shingle-Based Model

The basic distinguishing components of documents are 4-shingles [9] which are compared syntactically using hash values. The matching function will return the number of common shingles between documents, as well as the number of shingles in each document. Since a document may contain duplicate shingles

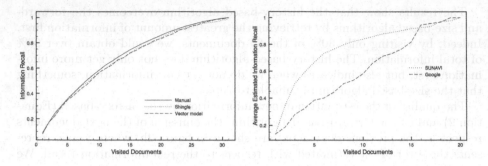

Fig. 3. Evaluations: (1) Comparing ICN with shingle- and vector-based comparisons of documents, and (2) Applying ICN over Google results: ratio between the information recall of the lists suggested by Google and by ICN

that match a single shingle in another document, then the equivalence measure is an asymmetric value.

4.2 Vector-Based Model

The basic distinguishing components of documents are terms or words, which are weighted by their occurrence within a document and within the whole collection of documents. Unlike taking the syntactic matching between terms in different documents d_i and d_j, we estimate $E(d_i, d_j)$ to be the number that results from using the similarity value of the vector model as a percentage of equivalence. Formally, let $S(d_i, d_j)$ be the similarity value derived from the vector model between documents d_i and d_j (a value between 0 and 1) and let T_i be the total number of different terms in document d_i, then we estimate the equivalence between terms in d_i with respect to terms in d_j by $E(d_i, d_j) = S(d_i, d_j) * T_i$. This method gives asymmetric values of equivalence.

4.3 Experimental Evaluation

We experimentally evaluated both strategies with the purpose of applying our navigation strategy. To do so, we use a set of 30 documents extracted from different Chilean news sources on the WWW. All these documents describe a particular event occurred in Chile during 2005 (the strongest storm in the last 20 years in the South of Chile). Paragraphs in each of these documents were manually analyzed and compared with paragraphs in each other document. In total, we have 235 paragraphs of different sizes. The comparison of paragraphs was not syntactic, but semantic, since we analyze the sense and meaning of sentences. As decision criteria, we consider than if over 80% of the information content of a paragraph was contained in another paragraph, we will consider the first paragraph to match the second one.

Using the manual determination of matching between documents and the comparison using shingles and vector model, we applied our proposed ICN and compared the *information recall* obtained in each case (Figure 3(a)). The information recall is here calculated as the estimated value obtained by equation 2 using the manually determined equivalence and omission between documents, and using the sequence of visited documents obtained with the algorithm and each of the different comparison methods (manual, shingles, and vector model). The results show no significant difference between shingles and vector model. We consider for the following implementation, however, a vector model for comparing documents, since it does not require additional implementation for most of the current information retrieval systems.

5 Implementation

We finally implemented our navigation strategy as a stand-alone interface, whose input is a collection of documents obtained by, for example, a search engine. The interface suggests to users the next document to visit, starting from the most relevant document given by the search engine or by the largest document. When users visited a document, the system suggests dynamically a next document.

We used this interface to visualize the first 20 results obtained from Google [5] in different searches, and compared the list given by Google with the sequence of documents suggested by ICN. We only used these first 20 documents, since we wanted to deal with thematically related documents, where it is useful to apply a strategy that reduces the duplication of results. Indeed, we found significant differences, but also duplication, in the documents given by Google. The graph in figure 3(b) presents the average estimated information recall obtained with the list suggested by ICN and Google. The results indicate that the first documents in the sequence proposed by our ICN provide more information content than the first documents in the list given by Google. Although at the end both lists of documents give the same information content, the sequence suggested by ICN allows users to retrieve most of the information content (over 80%) by visiting only half of the documents.

6 Conclusions

In this paper we have presented an algorithm that supports users in the navigation process across search results. It is based on the idea of minimizing duplication and collecting the greatest amount of information as quick as possible. The results of the experimental evaluation indicate that our algorithm could complement current search engines or information retrieval systems so that users could only need to visit few documents to obtain almost all the information contained in Web documents.

As future work, we expect to have a full implemented interface, with facilities for import sets of documents and the implementation of alternative methods for comparing documents.

Acknowledgment. This work has been funded by Nucleus Millenium Center for Web Research, Grant P04-067-F, Mideplan, Chile.

References

1. ACM (2006), http://www.acm.org/class
2. Baeza-Yates, R.A., Ribeiro-Neto, B.A.: Modern Information Retrieval. ACM Press, Addison-Wesley (1999)
3. Chowdhury, A., Frieder, O., Grossman, D.A., McCabe, M.C.: Collection statistics for fast duplicate document detection. ACM Trans. Inf. Syst. 20(2), 171–191 (2002)
4. Geffet, M., Feitelson, D.G.: Hierarchical indexing and document matching in bow. In: JCDL, pp. 259–267. ACM, New York, NY, USA (2001)
5. Google (2007), http://www.google.com
6. Halkidi, M., Nguyen, B., Varlamis, I., Vazirgiannis, M.: Thesus: Organizing web document collections based on link semantics. VLDB J. 12(4), 320–332 (2003)
7. Hammouda, K.M., Kamel, M.S.: Efficient phrase-based document indexing for web document clustering. IEEE Trans. Knowl. Data Eng. 16(10), 1279–1296 (2004)
8. Hearst, M.A.: Modern Information Retrieval, chapter User Interfaces and Visualization, pp. 257–324. ACM Press, New York, NY, USA (1999)
9. Pereira Jr., Á.R., Ziviani, N.: Syntactic similarity of web documents. In: LA-WEB, p. 194. IEEE Computer Society Press, Los Alamitos, CA, USA (2003)
10. Kartoo, S.A.: Kartoo Metaseach Engine (2006), http://www.kartoo.com
11. Koshman, S., Spink, A., Jansen, B.J.: Web searching on the vivisimo search engine. JASIST 57(14), 1875–1887 (2006)
12. Kou, H., Gardarin, G.: Similarity model and term association for document categorization. In: DEXA Workshops, pp. 256–260. IEEE Computer Society, Los Alamitos, CA, USA (2002)
13. Ouyang, B.: Delivering knowledge to nasa scientist and engineers: using phrase matching to determine document similarity. In: Guimarães, M. (ed.) ACM Southeast Regional Conference (1), pp. 384–385. ACM, New York, NY, USA (2005)
14. Roussinov, D., McQuaid, M.: Information navigation by clustering and summarizing query results. In: HICSS (2000)
15. Yahoo! (2007), http://www.yahoo.com
16. Zhang, Y., Ji, X., Chu, C.-H, Zha, H.: Correlating summarization of multi-source news with k-way graph bi-clustering. ACM SIGKDD Explorations Newsletter 6(2), 34–42 (2004)
17. Zhang, Y., Chu, C.-H., Ji, X., Zha, H.: Correlating summarization of multi-source news with k-way graph bi-clustering. SIGKDD Explorations 6(2), 34–42 (2004)

Author Index

Abolhassani, Hassan   653
Álvarez, Manuel   212
Amann, Bernd   87
Aschinger, Markus   273
Ataullah, Ahmed   533
Avagliano, Giuseppe   385

Babu, Suresh   461
Bahlor, Lisa   472
Bellas, Fernando   212
Bhattacharya, Kamal   484
Bilbao, Ramón   663
Borissov, Nikolay   335
Bouguettaya, Athman   38
Bozzon, Alessandro   593

Cacheda, Fidel   212
Calero, Coral   436
Carchiolo, Vincenza   50
Caro, Angélica   436
Caruso, Francesco   472
Ceresna, Michal   613
Chang, Junsheng   449
Chbeir, Richard   196
Chen, Jun   603
Cheng, Wenqing   310
Cobos, Carlos   298
Comai, Sara   593
Constantin, Camelia   87

Dai, Wenyuan   159
de Oliveira, Carlo Emmanoel   543
Di Martino, Sergio   385, 423
Díaz, Oscar   493
Ding, Yi   633
Domínguez, Eladio   184

Erradi, Abdelkarim   349

Fabregat, Ramon   298
Ferrucci, Filomena   385, 423
Filho, José Bringel   225
Fraternali, Piero   593
Fukuda, Kensuke   583
Furtado, Pedro   543, 553

Gámez, José A.   643
Garcia, Francisco José   623
Gensel, Jérôme   225
Gibbins, Nick   237
Goebel, Max   613
Gomez, Luis   298
Gordea, Sergiu   171
Gordillo, Silvia   573
Gravino, Carmine   423
Groppe, Sven   26
Grossniklaus, Michael   398
Gu, Jinguang   62
Gu, Zhimin   62
Guitart, Jordi   335

Hara, Takahiro   322
He, Hai   13
He, Yan   310
Herschel, Sven   99
Hirotsu, Toshio   583
Hoeller, Nils   26
Hosseini, Mehdi   653

Iwasaki, Hirotoshi   111

Jessenitschnig, Markus   273
Jin, Yu   62

Kadri, Reda   123
Kang, Byeong Ho   361
Kim, Yang Sok   361
Kukulenz, Dirk   26
Kurihara, Satoshi   583

Landberg, Anders H.   410
Lazovik, Alexander   373
Le, Dung Xuan Thi   563
Le Gloahec, Vincent   123
Li, Chuan   461
Li, Xue   633
Li, Yin Hua   603
Lin, Tao   461
Linnemann, Volker   26
Liu, Chengfei   1
Lloret, Jorge   184

Longheu, Alessandro   50
Lu, Jianguo   523
Lu, Jie   513
Ludwig, Heiko   373, 484

Ma, Jun   249
Ma, Shanle   633
Macías, Mario   335
Maheshwari, Piyush   349
Malgeri, Michele   50
Malik, Zaki   38
Mandreoli, Federica   285
Mangioni, Giuseppe   50
Mangkorntong, Piyanath   147
Martin, Hervé   225
Martoglia, Riccardo   285
Masutani, Osamu   111
Mateo, Juan L.   643
Mendes, Emilia   423
Mendoza, Martha   298
Meng, Weiyi   13
Micallef, Josephine   472
Molli, Pascal   503
Moreira, Tiago   623

Nakamura, Satoshi   135
Nakayama, Kotaro   322
Neumann, Dirk   335
Nicosia, Vincenzo   50
Niño, Miguel   298
Nishio, Shojiro   322
Norrie, Moira C.   398

Orleans, Luis Fernando   543
Orlowska, Maria E.   633

Paik, Hye-Young   603
Pan, Alberto   212
Paolino, Luca   385
Pardede, Eric   410, 563
Payne, Terry   237
Paz, Iñaki   493
Pelechano, Vicente   573
Penzo, Wilma   285
Pérez, Beatriz   184
Pérez, Sandy   493
Piattini, Mario   436
Puerta, José M.   643
Püschel, Tim   335

Rabhi, Fethi A.   147
Rahayu, J. Wenny   410

Raposo, Juan   212
Rego, Hugo   623
Rodríguez, Áurea   184
Rodríguez, M. Andrea   663
Röhm, Uwe   74
Rossi, Gustavo   573
Roy, Debashis   523
Rubio, Ángel L.   184

Saha, Deepa   523
Saleh, Ahmed   237
Sassatelli, Simona   285
Sato, Shin ya   583
Schewe, Klaus-Dieter   261
Schmidt, Sebastian   74
Sebillo, Monica   385
Setzer, Thomas   484
Shan, Ming-Chien   461
Shu, Liangcai   13
Stojanovic, Ljiljana   249
Stojanovic, Nenad   249
Sugawara, Toshiharu   583

Tanaka, Katsumi   135
Tang, Yangbin   449
Tekli, Joe   196
Thalheim, Bernhard   261
Tibermacine, Chouki   123
Toffetti Carughi, Giovanni   593
Torres, Jordi   335

Urso, Pascal   503

Valderas, Pedro   573
Viana, Windson   225
Villani, Giorgio   285
Villanova-Oliver, Marlène   225
Vitiello, Giuliana   385

Wang, Chao   513
Wang, Hai H.   237
Wang, Huaimin   449
Wang, Junhu   1
Weiss, Stéphane   503
Wu, Di   310

Xue, Gui-Rong   159

Yetongnon, Kokou   196
Yin, Gang   449
Yoshida, Taiga   135
Yu, Clement   13

Yu, Jeffrey Xu   1
Yu, Yijun   523
Yu, Yong   159

Zanker, Markus   171, 273
Zapata, María A.   184

Zeng, Xianyi   513
Zhang, Guangquan   513
Zhang, Xin   159
Zhao, Hongwu   62
Zigo, Viktor   613

Yu, Jeffrey Xu   1
Yu, Yinli   629
Yu, Yong   129

Zadiet, Marcus   173, 278
Zapata Maria A.   194

Zeng Xiaoyi   218
Zhang Guangnan   519
Zhang Xin   130
Zhao Hongwei   62
Zhao Yanfei   619

Lecture Notes in Computer Science

Sublibrary 3: Information Systems and Application, incl. Internet/Web and HCI

For information about Vols. 1–4480
please contact your bookseller or Springer

Vol. 4877: C. Thanos, F. Borri, L. Candela (Eds.), Digital Libraries: Research and Development. XII, 350 pages. 2007.

Vol. 4871: M. Cavazza, S. Donikian (Eds.), Virtual Storytelling. XIII, 219 pages. 2007.

Vol. 4857: J.M. Ware, G.E. Taylor (Eds.), Web and Wireless Geographical Information Systems. XI, 293 pages. 2007.

Vol. 4853: F. Fonseca, M.A. Rodríguez, S. Levashkin (Eds.), GeoSpatial Semantics. X, 289 pages. 2007.

Vol. 4836: H. Ichikawa, W.-D. Cho, I. Satoh, H.Y. Youn (Eds.), Ubiquitous Computing Systems. XIII, 307 pages. 2007.

Vol. 4832: M. Weske, M.-S. Hacid, C. Godart (Eds.), Web Information Systems Engineering – WISE 2007 Workshops. XV, 518 pages. 2007.

Vol. 4831: B. Benatallah, F. Casati, D. Georgakopoulos, C. Bartolini, W. Sadiq, C. Godart (Eds.), Web Information Systems Engineering – WISE 2007. XVI, 675 pages. 2007.

Vol. 4825: K. Aberer, K.-S. Choi, N. Noy, D. Allemang, K.-I. Lee, L. Nixon, J. Golbeck, P. Mika, D. Maynard, R. Mizoguchi, G. Schreiber, P. Cudré-Mauroux (Eds.), The Semantic Web. XXVII, 973 pages. 2007.

Vol. 4816: B. Falcidieno, M. Spagnuolo, Y. Avrithis, I. Kompatsiaris, P. Buitelaar (Eds.), Semantic Multimedia. XII, 306 pages. 2007.

Vol. 4813: I. Oakley, S. Brewster (Eds.), Haptic and Audio Interaction Design. XIV, 145 pages. 2007.

Vol. 4806: R. Meersman, Z. Tari, P. Herrero (Eds.), On the Move to Meaningful Internet Systems 2007: OTM 2007 Workshops, Part II. XXXIV, 611 pages. 2007.

Vol. 4805: R. Meersman, Z. Tari, P. Herrero (Eds.), On the Move to Meaningful Internet Systems 2007: OTM 2007 Workshops, Part I. XXXIV, 757 pages. 2007.

Vol. 4804: R. Meersman, Z. Tari (Eds.), On the Move to Meaningful Internet Systems 2007: CoopIS, DOA, ODBASE, GADA, and IS, Part II. XXIX, 683 pages. 2007.

Vol. 4803: R. Meersman, Z. Tari (Eds.), On the Move to Meaningful Internet Systems 2007: CoopIS, DOA, ODBASE, GADA, and IS, Part I. XXIX, 1173 pages. 2007.

Vol. 4802: J.-L. Hainaut, E.A. Rundensteiner, M. Kirchberg, M. Bertolotto, M. Brochhausen, Y.-P.P. Chen, S.S.-S. Cherfi, M. Doerr, H. Han, S. Hartmann, J. Parsons, G. Poels, C. Rolland, J. Trujillo, E. Yu, E. Zimányie (Eds.), Advances in Conceptual Modeling – Foundations and Applications. XIX, 420 pages. 2007.

Vol. 4801: C. Parent, K.-D. Schewe, V.C. Storey, B. Thalheim (Eds.), Conceptual Modeling - ER 2007. XVI, 616 pages. 2007.

Vol. 4797: M. Arenas, M.I. Schwartzbach (Eds.), Database Programming Languages. VIII, 261 pages. 2007.

Vol. 4796: M. Lew, N. Sebe, T.S. Huang, E.M. Bakker (Eds.), Human–Computer Interaction. X, 157 pages. 2007.

Vol. 4794: B. Schiele, A.K. Dey, H. Gellersen, B. de Ruyter, M. Tscheligi, R. Wichert, E. Aarts, A. Buchmann (Eds.), Ambient Intelligence. XV, 375 pages. 2007.

Vol. 4777: S. Bhalla (Ed.), Databases in Networked Information Systems. X, 329 pages. 2007.

Vol. 4761: R. Obermaisser, Y. Nah, P. Puschner, F.J. Rammig (Eds.), Software Technologies for Embedded and Ubiquitous Systems. XIV, 563 pages. 2007.

Vol. 4747: S. Džeroski, J. Struyf (Eds.), Knowledge Discovery in Inductive Databases. X, 301 pages. 2007.

Vol. 4744: Y. de Kort, W. IJsselsteijn, C. Midden, B. Eggen, B.J. Fogg (Eds.), Persuasive Technology. XIV, 316 pages. 2007.

Vol. 4740: L. Ma, M. Rauterberg, R. Nakatsu (Eds.), Entertainment Computing – ICEC 2007. XXX, 480 pages. 2007.

Vol. 4730: C. Peters, P. Clough, F.C. Gey, J. Karlgren, B. Magnini, D.W. Oard, M. de Rijke, M. Stempfhuber (Eds.), Evaluation of Multilingual and Multi-modal Information Retrieval. XXIV, 998 pages. 2007.

Vol. 4723: M. R. Berthold, J. Shawe-Taylor, N. Lavrač (Eds.), Advances in Intelligent Data Analysis VII. XIV, 380 pages. 2007.

Vol. 4721: W. Jonker, M. Petković (Eds.), Secure Data Management. X, 213 pages. 2007.

Vol. 4718: J. Hightower, B. Schiele, T. Strang (Eds.), Location- and Context-Awareness. X, 297 pages. 2007.

Vol. 4717: J. Krumm, G.D. Abowd, A. Seneviratne, T. Strang (Eds.), UbiComp 2007: Ubiquitous Computing. XIX, 520 pages. 2007.

Vol. 4715: J.M. Haake, S.F. Ochoa, A. Cechich (Eds.), Groupware: Design, Implementation, and Use. XIII, 355 pages. 2007.

Vol. 4714: G. Alonso, P. Dadam, M. Rosemann (Eds.), Business Process Management. XIII, 418 pages. 2007.

Vol. 4704: D. Barbosa, A. Bonifati, Z. Bellahsène, E. Hunt, R. Unland (Eds.), Database and XML Technologies. X, 141 pages. 2007.

Vol. 4690: Y. Ioannidis, B. Novikov, B. Rachev (Eds.), Advances in Databases and Information Systems. XIII, 377 pages. 2007.

Vol. 4675: L. Kovács, N. Fuhr, C. Meghini (Eds.), Research and Advanced Technology for Digital Libraries. XVII, 585 pages. 2007.

Vol. 4674: Y. Luo (Ed.), Cooperative Design, Visualization, and Engineering. XIII, 431 pages. 2007.

Vol. 4663: C. Baranauskas, P. Palanque, J. Abascal, S.D.J. Barbosa (Eds.), Human-Computer Interaction – INTERACT 2007, Part II. XXXIII, 735 pages. 2007.

Vol. 4662: C. Baranauskas, P. Palanque, J. Abascal, S.D.J. Barbosa (Eds.), Human-Computer Interaction – INTERACT 2007, Part I. XXXIII, 637 pages. 2007.

Vol. 4658: T. Enokido, L. Barolli, M. Takizawa (Eds.), Network-Based Information Systems. XIII, 544 pages. 2007.

Vol. 4656: M.A. Wimmer, J. Scholl, Å. Grönlund (Eds.), Electronic Government. XIV, 450 pages. 2007.

Vol. 4655: G. Psaila, R. Wagner (Eds.), E-Commerce and Web Technologies. VII, 229 pages. 2007.

Vol. 4654: I.-Y. Song, J. Eder, T.M. Nguyen (Eds.), Data Warehousing and Knowledge Discovery. XVI, 482 pages. 2007.

Vol. 4653: R. Wagner, N. Revell, G. Pernul (Eds.), Database and Expert Systems Applications. XXII, 907 pages. 2007.

Vol. 4636: G. Antoniou, U. Aßmann, C. Baroglio, S. Decker, N. Henze, P.-L. Patranjan, R. Tolksdorf (Eds.), Reasoning Web. IX, 345 pages. 2007.

Vol. 4611: J. Indulska, J. Ma, L.T. Yang, T. Ungerer, J. Cao (Eds.), Ubiquitous Intelligence and Computing. XXIII, 1257 pages. 2007.

Vol. 4607: L. Baresi, P. Fraternali, G.-J. Houben (Eds.), Web Engineering. XVI, 576 pages. 2007.

Vol. 4606: A. Pras, M. van Sinderen (Eds.), Dependable and Adaptable Networks and Services. XIV, 149 pages. 2007.

Vol. 4605: D. Papadias, D. Zhang, G. Kollios (Eds.), Advances in Spatial and Temporal Databases. X, 479 pages. 2007.

Vol. 4602: S. Barker, G.-J. Ahn (Eds.), Data and Applications Security XXI. X, 291 pages. 2007.

Vol. 4601: S. Spaccapietra, P. Atzeni, F. Fages, M.-S. Hacid, M. Kifer, J. Mylopoulos, B. Pernici, P. Shvaiko, J. Trujillo, I. Zaihrayeu (Eds.), Journal on Data Semantics IX. XV, 197 pages. 2007.

Vol. 4592: Z. Kedad, N. Lammari, E. Métais, F. Meziane, Y. Rezgui (Eds.), Natural Language Processing and Information Systems. XIV, 442 pages. 2007.

Vol. 4587: R. Cooper, J. Kennedy (Eds.), Data Management. XIII, 259 pages. 2007.

Vol. 4577: N. Sebe, Y. Liu, Y.-t. Zhuang, T.S. Huang (Eds.), Multimedia Content Analysis and Mining. XIII, 513 pages. 2007.

Vol. 4568: T. Ishida, S. R. Fussell, P. T. J. M. Vossen (Eds.), Intercultural Collaboration. XIII, 395 pages. 2007.

Vol. 4566: M.J. Dainoff (Ed.), Ergonomics and Health Aspects of Work with Computers. XVIII, 390 pages. 2007.

Vol. 4564: D. Schuler (Ed.), Online Communities and Social Computing. XVII, 520 pages. 2007.

Vol. 4563: R. Shumaker (Ed.), Virtual Reality. XXII, 762 pages. 2007.

Vol. 4561: V.G. Duffy (Ed.), Digital Human Modeling. XXIII, 1068 pages. 2007.

Vol. 4560: N. Aykin (Ed.), Usability and Internationalization, Part II. XVIII, 576 pages. 2007.

Vol. 4559: N. Aykin (Ed.), Usability and Internationalization, Part I. XVIII, 661 pages. 2007.

Vol. 4558: M.J. Smith, G. Salvendy (Eds.), Human Interface and the Management of Information, Part II. XXIII, 1162 pages. 2007.

Vol. 4557: M.J. Smith, G. Salvendy (Eds.), Human Interface and the Management of Information, Part I. XXII, 1030 pages. 2007.

Vol. 4541: T. Okadome, T. Yamazaki, M. Makhtari (Eds.), Pervasive Computing for Quality of Life Enhancement. IX, 248 pages. 2007.

Vol. 4537: K.C.-C. Chang, W. Wang, L. Chen, C.A. Ellis, C.-H. Hsu, A.C. Tsoi, H. Wang (Eds.), Advances in Web and Network Technologies, and Information Management. XXIII, 707 pages. 2007.

Vol. 4531: J. Indulska, K. Raymond (Eds.), Distributed Applications and Interoperable Systems. XI, 337 pages. 2007.

Vol. 4526: M. Malek, M. Reitenspieß, A. van Moorsel (Eds.), Service Availability. X, 155 pages. 2007.

Vol. 4524: M. Marchiori, J.Z. Pan, C.d.S. Marie (Eds.), Web Reasoning and Rule Systems. XI, 382 pages. 2007.

Vol. 4519: E. Franconi, M. Kifer, W. May (Eds.), The Semantic Web: Research and Applications. XVIII, 830 pages. 2007.

Vol. 4518: N. Fuhr, M. Lalmas, A. Trotman (Eds.), Comparative Evaluation of XML Information Retrieval Systems. XII, 554 pages. 2007.

Vol. 4508: M.-Y. Kao, X.-Y. Li (Eds.), Algorithmic Aspects in Information and Management. VIII, 428 pages. 2007.

Vol. 4506: D. Zeng, I. Gotham, K. Komatsu, C. Lynch, M. Thurmond, D. Madigan, B. Lober, J. Kvach, H. Chen (Eds.), Intelligence and Security Informatics: Biosurveillance. XI, 234 pages. 2007.

Vol. 4505: G. Dong, X. Lin, W. Wang, Y. Yang, J.X. Yu (Eds.), Advances in Data and Web Management. XXII, 896 pages. 2007.

Vol. 4504: J. Huang, R. Kowalczyk, Z. Maamar, D. Martin, I. Müller, S. Stoutenburg, K.P. Sycara (Eds.), Service-Oriented Computing: Agents, Semantics, and Engineering. X, 175 pages. 2007.

Vol. 4500: N.A. Streitz, A.D. Kameas, I. Mavrommati (Eds.), The Disappearing Computer. XVIII, 304 pages. 2007.

Vol. 4495: J. Krogstie, A. Opdahl, G. Sindre (Eds.), Advanced Information Systems Engineering. XVI, 606 pages. 2007.